HANDBOOK OF RESEARCH SYNTHESIS AND META-ANALYSIS
2ND EDITION

THE HANDBOOK OF RESEARCH SYNTHESIS AND META-ANALYSIS

2ND EDITION

Edited by

**HARRIS COOPER, LARRY V. HEDGES,
AND JEFFREY C. VALENTINE**

RUSSELL SAGE FOUNDATION   NEW YORK

THE RUSSELL SAGE FOUNDATION

Library of Congress Cataloging-in-Publication Data

The handbook of research synthesis and meta-analysis / Harris Cooper, Larry V. Hedges,
  Jeffrey C. Valentine, editors. — 2nd ed.
    p.   cm.
  Rev. ed. of: The handbook of research synthesis. c1994.
  Includes bibliographical references and index.
  ISBN 978-0-87154-163-5
  1. Research—Methodology—Handbooks, manuals, etc.   2. Information storage and
retrieval systems—Research—Handbooks, manuals, etc.   3. Research—Statistical methods—
Handbooks, manuals, etc.   I. Cooper, Harris M.   II. Hedges, Larry V.   III. Valentine, Jeff C.
IV. Handbook of research synthesis.
Q180.55.M4H35   2009
001.4'2—dc22   2008046477

RUSSELL SAGE FOUNDATION
112 East 64th Street, New York, New York 10065

10 9 8 7 6 5 4 3 2

To the authors of the *Handbook* chapters,
for their remarkable cooperation and commitment
to the advance of science and society

CONTENTS

FOREWORD

The past decade has been a time of enormous growth in the need for and use of research syntheses in the social and behavioral sciences. The need was given impetus by the increase in social research that began in the 1960s and continues unabated today. Over the past three decades, methods for the retrieval and analysis of research literatures have undergone enormous change. Literature reviewing used to be a narrative and subjective process. Today, research synthesis has a body of procedural and statistical techniques of its own. Because of the potential pitfalls that confront research synthesists, it is critical that the methods used to integrate the results of studies be rigorous and open to public inspection.

Like its predecessor, the second edition of the *Handbook of Research Synthesis and Meta-Analysis* is an attempt to bring together in a single volume state-of-the-art descriptions of every phase of the research synthesis process, from problem formulation to report writing. Readers will find that it is comprised of chapters intended for use by scholars very new to the research synthesis enterprise (for example, by helping them think about how to formulate a problem and how to construct a coding guide) as well as by experienced research synthesists looking for answers to highly technical questions (for example, options for addressing stochastically dependent effect sizes).

The content of this second edition of the *Handbook* is modeled on that of the very successful first edition. Determining this content required us to use multiple decision criteria. First, by relying on personal experience and knowledge of previous books in the area, we identified the central decision points in conducting a research synthesis. This led to the creation of several major divisions of the text as well as specific chapters—for example, vote-counting procedures, combining effect sizes, and correcting for sources of artifactual variation across studies. Next, we examined the issues involved within the major tasks and broke them down so that it would be manageable to cover the required information within a single chapter. Thus, within the topic of evaluating the literature, separate chapters are devoted to judging research quality and evaluating the reliability of coding decisions, among others. We supplemented the resulting list of chapters with others on special topics that cut across the various stages of research synthesis, but were important enough to merit focused attention. These topics included missing data, dependent effect sizes, and using research synthesis to develop theories. For this edition, twenty-nine chapters emerged with forty authors drawn from at least five distinct disciplines: information science, statistics, psychology, education, and social policy analysis. Nineteen of the chapter authors also served as peer reviewers for other chapters along with twenty-one scholars who took part in the review process only.

In the course of attempting to coordinate such an effort, we were faced with numerous editorial decisions. Perhaps the most critical involved how to handle overlap in coverage and content across chapters. Because we wanted the *Handbook* to be exhaustive, we defined topics for chapters narrowly, but anticipated that authors would likely define their domains more broadly. They did. The issue for us, then, became whether to edit out the redundancy or leave it in. We chose to leave it in for several reasons. First, there is considerable insight to be gained when the same issue is treated from multiple perspectives. Second, the repetition of topics helps readers understand some of the most pervasive problems that confront a research synthesist. Finally, the overlap allows chapters to stand on their own, as more complete discussions of each topic, thus increasing their individual usefulness for teaching.

Another important issue involved mathematical notation. Although some notation in meta-analysis has become relatively standard, few symbols are used universally. In addition, because each statistical chapter contains several unique mathematical concepts, we had no guarantee that the same symbols might not be used to represent different concepts in different chapters. To minimize these problems, we proposed that authors use the notational system that was developed for the first edition of the *Handbook*. This system is not perfect, however, and the reader may discover, when moving from one chapter to the next, that the same symbol has been used to represent different concepts or that different symbols are used for the same concept. However, each chapter introduces its notation, and we have worked to standardize the notation as much as possible in order to minimize any confusion.

In addition to the difficulties associated with mathematical notation, given the number of diverse disciplines contributing to the *Handbook*, we were faced with problems involving more general terminology. Again, we expected that authors would occasionally use different terms for the same concept and the same terms for different concepts. Further, some concepts would be new to readers in any particular discipline. Therefore, we asked authors to contribute to a glossary by defining key terms in their chapter, which we edited for style and overlap. The glossary for this second edition of the *Handbook* is significantly updated from the first edition's. Such instruments are always flawed and incomplete, but we hope that this one facilitates the reader's passage through the book.

Related to this notion, this edition of the *Handbook* uses the *Chicago Manual of Style*, required of all books published by the Russell Sage Foundation. Readers more familiar with the style used by the American Psychological Association will notice that first names appear in the references and the first time authors are mentioned in the text and the dates of publications are presented at the end of sentences rather than directly following the names of the authors.

Finally, a consideration that arises in any report of quantitative work concerns the conventions for deriving the numerical results given in the examples. Specifically, there are several possible ways that numerical results can be rounded for presentation. For example, when the final computations involve sums or quotients of intermediate results, the final numerical answer depends somewhat on whether and how the intermediate results are rounded. In the statistical examples, we have usually chosen to round intermediate computations, such as sums, to two places. Thus, hand computations using the intermediate values should yield results corresponding to our examples. If computations are carried out directly using a different rounding convention, the results might differ somewhat from those given. Readers who experience slight discrepancies between their computations and those presented in the text should not be alarmed.

The Russell Sage Foundation has been providing support for more than twenty years to efforts that lead directly to the production of this handbook. The Foundation, along with its president, Eric Wanner, initiated a program to improve and disseminate state-of-the-art methods in research synthesis. This is the fourth volume to emerge from that effort. The first, *The Future of Meta-Analysis*, was published by the Foundation in 1990. It contained the proceedings of a workshop convened by the Committee on National Statistics of the National Research Council. The second volume, *Meta-Analysis for Explanation: A Casebook*, presented a series of four applications of meta-analysis to highlight the potential of meta-analysis for generating and testing explanations of social phenomena as well as testing the effectiveness of social and medical programs. The third volume was the first edition of this handbook.

An effort as large as the *Handbook* cannot be undertaken without the help of many individuals. This volume is dedicated to the authors of the chapters. These scholars were ably assisted by reviewers, some of whom authored other chapters in the *Handbook* and some of whom kindly lent their expertise to the project. The chapter reviewers were:

Dolores Albarracin, Department of Psychology, University of Illinois at Urbana-Champaign

Pam M. Baxter, Cornell Institute for Social and Economic Research, Cornell University

Betsy Jane Becker, College of Education, Florida State University

Jesse A. Berlin, Johnson & Johnson Pharmaceutical Research and Development, Titusville, NJ

Michael Borenstein, Biostat, Englewood, New Jersey

Geoffrey D. Borman, School of Education, University of Wisconsin, Madison

Robert F. Boruch, Graduate School of Education, University of Pennsylvania

Brad J. Bushman, Institute for Social Research, University of Michigan, and VU University, Amsterdam, the Netherlands

Mike Clarke, U.K. Cochrane Centre, Oxford

Thomas D. Cook, Institute for Policy Research, Northwestern University

Alice H. Eagly, Department of Psychology, Northwestern University

Alan Gomersall, Centre for Evidence and Policy, King's College London

Lesley Grayson, Centre for Evidence and Policy, King's College London

Sander Greenland, Departments of Epidemiology and Statistics, University of California, Los Angeles

Adam R. Hafdahl, Department of Mathematics, Washington University in St. Louis

Rebecca Herman, AIR, Washington, D.C.

Julian P. T. Higgins, MRC Biostatistics Unit, Cambridge

Janet S. Hyde, Department of Psychology, University of Wisconsin

Sema Kalaian, College of Technology, Eastern Michigan University

Spyros Konstantopoulos, Lynch School of Education, Boston College

Julia Littell, Graduate School of Social Work and Social Research, Bryn Mawr College

Michael A. McDaniel, School of Business, Virginia Commonwealth University

David Myers, AIR, Washington, D.C.

Ingram Olkin, Department of Statistics, Stanford University

Andy Oxman, Preventive and International Health Care Unit, Norwegian Knowledge Centre for the Health Services

Anthony Petrosino, WestED, Boston, Massachusetts

Mark Petticrew, Department of Public Health and Policy, London School of Hygiene and Tropical Medicine

Therese D. Pigott, School of Education, Loyola University Chicago

Andrew C. Porter, Graduate School of Education, University of Pennsylvania

Stephen W. Raudenbush, Department of Sociology, University of Chicago

Jeffrey G. Reed, School of Business, Marian University

David Rindskopf, Department of Psychology, City University of New York

Hannah R. Rothstein, Zicklin School of Business, Baruch College

Paul R. Sackett, Department of Psychology, University of Minnesota

Julio Sánchez-Meca, Faculty of Psychology, University of Murcia, Spain

William R. Shadish, Department of Psychological Sciences, University of California, Merced

Jack L. Vevea, Department of Psychology, University of California, Merced

Anne Wade, Centre for the Study of Learning and Performance, Concordia University

Howard D. White, College of Information Science and Technology, Drexel University

David B. Wilson, Department of Administration of Justice, George Mason University

We also received personal assistance form a host of capable individuals. At Duke University, Melissa Pascoe and Sunny Powell managed the financial transactions for the *Handbook*. At Northwestern University, Patti Ferguson provided logistical support. At the University of Louisville, Aaron Banister helped assemble the glossary. At the Russell Sage Foundation, Suzanne Nichols oversaw the project. To the chapter authors, reviewers, and others who helped us with our work, we express our deepest thanks.

Harris Cooper
Larry Hedges
Jeff Valentine

ABOUT THE AUTHORS

Pam M. Baxter is a data archivist at the Cornell Institute for Social and Economic Research (CISER) of Cornell University.

Betsy Jane Becker is professor of educational psychology and learning systems, College of Education, Florida State University.

Jesse A. Berlin is a researcher with Johnson & Johnson Pharmaceutical Research and Development.

Michael Borenstein is director of Biostat in Englewood, New Jersey.

Geoffrey D. Borman is professor of educational leadership and policy analysis, educational psychology, and educational policy studies at the University of Wisconsin, Madison.

Brad J. Bushman is research professor at the Institute for Social Research, University of Michigan and visiting professor, VU University, Amsterdam, the Netherlands.

Mike Clarke is a researcher at the U.K. Cochrane Centre in Oxford, United Kingdom.

Thomas D. Cook is professor of sociology, psychology, education, and social policy, Joan and Serepta Harrison Chair in Ethics and Justice, and faculty fellow of the Institute for Policy Research at Northwestern University.

Harris Cooper is professor of psychology and neuroscience at Duke University.

David S. Cordray is professor of public policy and of psychology at Vanderbilt University.

Alice H. Eagly is James Padilla Chair of Arts and Sciences, professor of psychology, faculty fellow of the Institute for Policy Research, and department chair of psychology at Northwestern University.

Joseph L. Fleiss was professor of biostatistics at the Mailman School of Public Health, Columbia University.

Leon J. Gleser is professor of statistics at the University of Pittsburgh.

Joel B. Greenhouse is professor of statistics at Carnegie Mellon University.

Jeffrey A. Grigg is project assistant at the Wisconsin Center for Education Research at the University of Wisconsin, Madison.

C. Keith Haddock is associate professor of behavioral health at the University of Missouri, Kansas City.

Larry V. Hedges is Board of Trustees Professor of Statistics and Social Policy and faculty fellow at the Institute for Policy Research, Northwestern University.

Sally Hopewell is a research scientist at the U.K. Cochrane Centre.

Satish Iyengar is professor of statistics in the Department of Statistics at the University of Pittsburgh.

Spyros Konstantopoulos is assistant professor at the Lynch School of Education, Boston College.

Huy Le is assistant professor of psychology at the University of Central Florida.

Mark W. Lipsey is director of the Center for Evaluation Research and Methodology, and a senior research associate at the Vanderbilt Institute for Public Policy Studies, Vanderbilt University.

Georg E. Matt is professor of psychology at San Diego State University.

Paul Morphy is a predoctoral fellow at Vanderbilt University.

In-Sue Oh is a doctoral candidate at the Henry B. Tippie College of Business, University of Iowa.

Ingram Olkin is professor of statistics and of education and CHP/PCOR Fellow at Stanford University.

Robert G. Orwin is a senior study director with Westat, Inc.

Therese D. Pigott is associate professor at the School of Education, Loyola University Chicago.

Stephen W. Raudenbush is Lewis-Sebring Professor in the Department of Sociology and the College Chair of the Committee on Education at the University of Chicago.

Jeffrey G. Reed is professor of management and dean of the School of Business at Marian University.

Hannah R. Rothstein is professor of management at the Zicklin School of Business, Baruch College, The City University of New York.

Frank L. Schmidt is Gary Fethke Professor of Management and Organizations at the Henry B. Tippie College of Business, University of Iowa.

William R. Shadish is professor and Founding Faculty at the University of California, Merced.

Alex J. Sutton is a reader in Medical Statistics in the Department of Health Sciences at the University of Leicester.

Jeffrey C. Valentine is assistant professor in the Department of Educational and Counseling Psychology at the College of Education and Human Development, University of Louisville.

Jack L. Vevea is associate professor of psychology at the University of California, Merced.

Morgan C. Wang is professor of statistics at the University of Central Florida.

Howard D. White is professor emeritus at the College of Information Science and Technology, Drexel University.

David B. Wilson is associate professor in the Department of Administration of Justice, George Mason University.

Wendy Wood is James B. Duke Professor of Psychology and Neuroscience and professor of marketing at Duke University.

PART

I

INTRODUCTION

1

RESEARCH SYNTHESIS AS A SCIENTIFIC PROCESS

HARRIS COOPER
Duke University

LARRY V. HEDGES
Northwestern University

CONTENTS

It is necessary, while formulating the problems of which in our advance we are to find the solutions, to call into council the views of those of our predecessors who have declared an opinion on the subject, in order that we may profit by whatever is sound in their suggestions and avoid their errors.

Aristotle, *De Anima*, Book 1, chapter 2

1.1 INTRODUCTION

The moment we are introduced to science we are told it is a cooperative, cumulative enterprise. Like the artisans who construct a building from blueprints, bricks, and mortar, scientists contribute to a common edifice, called knowledge. Theorists provide our blueprints and researchers collect the data that are our bricks.

To extend the analogy further yet, we might say that research synthesists are the bricklayers and hodcarriers of the science guild. It is their job to stack the bricks according to plan and apply the mortar that makes the whole thing stick.

Anyone who has attempted a research synthesis is entitled to a wry smile as the analogy continues. They know that several sets of theory-blueprints often exist, describing structures that vary in form and function, with few a priori criteria for selecting between them. They also know that our data-bricks are not all six-sided and right-angled. They come in a baffling array of sizes and shapes. Making them fit, securing them with mortar, and seeing whether the resulting construction looks anything like the blueprint is a challenge worthy of the most dedicated, inspired artisan.

1.1.1 Replication and Research Synthesis

Scientific literatures are cluttered with repeated studies of the same phenomena. Multiple studies on the same problem or hypothesis arise because investigators are unaware of what others are doing, because they are skeptical about the results of past investigations, or because they wish to extend (that is, generalize or search for influences on) previous findings. Experience has shown that even when considerable effort is made to achieve strict replication, results across studies are rarely identical at any high level of precision, even in the physical sciences (see Hedges 1987). No two bricks are exactly alike.

How should scientists proceed when results differ? First, it is clear how they should *not* proceed: they should not pretend that there is no problem, or decide that just one study (perhaps the most recent one, or the one they conducted, or a study chosen via some other equally arbitrary

criterion) produced the correct finding. If results that are expected to be very similar show variability, the scientific instinct should be to account for the variability by further systematic work.

1.2 RESEARCH SYNTHESIS IN CONTEXT

1.2.1 A Definition of the Literature Review

The American Psychological Association's *PsycINFO* reference database defines a literature review as "the process of conducting surveys of previously published material" (http://gateway.uk.ovid.com). Common to all definitions of literature reviews is the notion that they are "not based primarily on new facts and findings, but on publications containing such primary information, whereby the latter is digested, sifted, classified, simplified, and synthesized" (Manten 1973, 75).

Table 1.1 presents a taxonomy of literature reviews that captures six distinctions used by literature review authors to describe their own work (Cooper 1988, 2003). The taxonomy was developed and can be applied to literature reviews appearing throughout a broad range of both the behavioral and physical sciences. The six features and their subordinate categories permit a rich level of distinction among works of synthesis.

The first distinction among literature reviews concerns the focus of the review, or, the material that is of central interest to the reviewer. Most literature reviews center on one or more of four areas: the findings of individual primary studies; the methods used to carry out research; theories meant to explain the same or related phenomena; and the practices, programs, or treatments being used in an applied context.

The second characteristic of a literature review is its goals. Goals concern what the preparers of the review hope to accomplish. The most frequent goal for a review is to *integrate* past literature that is believed to relate to a common topic. Integration includes: formulating general statements that characterize multiple specific instances (of research, methods, theories, or practices); resolving conflict between contradictory results, ideas, or statements

Table 1.1 A Taxonomy of Literature Reviews

Characteristic	Categories
Focus	research findings research methods theories practices or applications

Goal	integration
	   generalization
	   conflict resolution
	   linguistic bridge-building
	criticism
	identification of central issues

Perspective	neutral representation espousal of position
Coverage	exhaustive exhaustive with selective citation representative central or pivotal
Organization	historical conceptual methodological
Audience	specialized scholars general scholars practitioners or policy makers general public

SOURCE: Cooper 1988. Reprinted with permission from Transaction Publishers.

of fact by proposing a new conception that accounts for the inconsistency; and bridging the gap between concepts or theories by creating a new, common linguistic framework.

Another goal for literature reviews can be to critically analyze the existing literature. Unlike a review that seeks to integrate existing work, a review that involves a critical assessment does not necessarily summate conclusions or compare the covered works one to another. Instead, it holds each work up against a criterion and finds it more or less acceptable. Most often, the criterion will include issues related to the methodological quality if empirical studies, the logical rigor, completeness or breadth of explanation if theories are involved, or comparison with the ideal treatment, when practices, policies or applications are involved.

A third goal that often motivates literature reviews is to identify issues central to a field. These issues may include questions that have given rise to past work, questions that

should stimulate future work, and methodological problems or problems in logic and conceptualization that have impeded progress within a topic area or field.

Of course, reviews more often than not have multiple goals. So, for example, it is rare to see the integration or critical examination of existing work without also seeing the identification of central issues for future endeavors.

A third characteristic that distinguishes among literature reviews, perspective, relates to whether the reviewers have an initial point of view that might influence the discussion of the literature. The endpoints on the continuum of perspective might be called neutral representation and espousal of a position. In the former, reviewers attempt to present all arguments or evidence for and against various interpretations of the problem. The presentation is meant to be as similar as possible to those that would be provided by the originators of the arguments or evidence. At the opposite extreme of perspective, the viewpoints of reviewers play an active role in how material is presented. The reviewers accumulate and synthesize the literature in the service of demonstrating the value of the particular point of view that they espouse. The reviewers muster arguments and evidence so that it presents their contentions in the most convincing manner.

Of course, reviewers attempting to achieve complete neutrality are likely doomed to failure. Further, reviewers who attempt to present all sides of an argument do not preclude themselves from ultimately taking a strong position based on the cumulative evidence. Similarly, reviewers can be thoughtful and fair while presenting conflicting evidence or opinions and still advocate for a particular interpretation.

The next characteristic, coverage, concerns the extent to which reviewers find and include relevant works in their paper. It is possible to distinguish at least four types of coverage. The first type, exhaustive coverage, suggests that the reviewers hope to be comprehensive in the presentation of the relevant work. An effort is made to include the entire literature and to base conclusions and discussions on this comprehensive information base. The second type of coverage also bases conclusions on entire literatures, but only a selection of works is actually described in the literature review. The authors choose a purposive sample of works to cite but claim that the inferences drawn are based on a more extensive literature. Third, some reviewers will present works that are broadly representative of many other works in a field. They hope to describe just a few exemplars that are descriptive of numerous other works. The reviewers discuss the characteristics that make the chosen works paradigmatic of the larger group. In the final coverage strategy, reviewers concentrate on works that were highly original when they appeared and influenced the development of future efforts in the topic area. This may include materials that initiated a line of investigation or thinking, changed how questions were framed, introduced new methods, engendered important debate, or performed a heuristic function for other scholars.

A fifth characteristic of literature reviews concerns how a paper is *organized*. Reviews may be arranged historically, so that topics are introduced in the chronological order in which they appeared in the literature, conceptually, so that works relating to the same abstract ideas appear together, or methodologically, so that works employing similar methods are grouped together.

Finally, the intended audiences of reviews can vary. Reviews can be written for groups of specialized researchers, general researchers, policy makers, practitioners, or the general public. As reviewers move from addressing specialized researchers to addressing the general public, they employ less technical jargon and detail, while often paying greater attention to the implications of the work being covered.

1.2.2. A Definition of Research Synthesis

The terms *research synthesis*, *research review*, and *systematic review* are often used interchangeably in the social science literature, though they sometimes connote subtly different meanings. Regrettably, there is no consensus about whether these differences are really meaningful. Therefore, we will use the term research synthesis most frequently throughout this book. The reason for this choice is simple. In addition to its use in the context of research synthesis, the term *research review* is used as well to describe the activities of evaluating the quality of research. For example, a journal editor will obtain research reviews when deciding whether to publish a manuscript. Because research syntheses often include this type of evaluative review of research, using the term *research synthesis* avoids confusion. The term *systematic review* is less often used in the context of research evaluation, though the confusion is still there, and also it is a term less familiar to social scientists than is research synthesis.

A research synthesis can be defined as the conjunction of a particular set of literature review characteristics. Most definitional about research syntheses are their primary focus and goal: research syntheses attempt to integrate empirical research for the purpose of creating generalizations. Implicit in this definition is the notion that seeking generalizations also involves seeking the limits of generalizations. Also, research syntheses almost always pay attention to relevant theories, critically analyze the research they cover, try to resolve conflicts in the literature, and attempt to identify central issues for future research. According to Derek Price, research syntheses are intended to "replace those papers that have been lost from sight behind the research front" (1965, 513). Research synthesis is one of a broad array of integrative activities that scientists engage in; as demonstrated in the quote leading this chapter, its intellectual heritage can be traced back at least as far as Aristotle.

Using the taxonomy described, we can make further specifications concerning the type of research syntheses that are the focus of this book. With regard to perspective, readers will note that much of the material is meant to help synthesists produce statements about evidence that are neutral in perspective, that is, less likely to be affected by bias or by their own subjective outlooks. The material on searching the literature for evidence is meant to help synthesists uncover all the evidence, not simply positive studies (that might be overrepresented in published research) or evidence that is easy for them to find (that might be overly sympathetic to their point of view or disciplinary perspective). The material on the reliability of extracting information from research reports and how methodological variations in research should be handled is meant to increase the interjudge reliability of these processes. The methods proposed for the statistical integration of findings is meant to ensure the same rules about data analysis are consistently applied to all conclusions about what is and is not significant, both statistically and practically. Finally, the material on explicit and exhaustive reporting of the methods used in syntheses is meant to assist users in evaluating if or where subjectivity may have crept into the process and to replicate findings for themselves if they choose to do so.

Finally, the term *meta-analysis* often is used as a synonym for research synthesis. However, in this volume, it will be used in its more precise and original meaning—to describe the quantitative procedures that a research synthesist may use to statistically combine the results of studies. Gene Glass coined the term to refer to "the statistical analysis of a large collection of analysis results from individual studies for the purpose of integrating the findings" (1976, 3). The authors of the *Handbook* reserve the term to refer specifically to statistical analysis in research synthesis and not to the entire enterprise of research synthesis. Not all research syntheses are appropriate for meta-analysis.

1.3. A BRIEF HISTORY OF RESEARCH SYNTHESIS AS A SCIENTIFIC ENTERPRISE

1.3.1 Early Developments

In 1971, Kenneth Feldman demonstrated remarkable prescience: "Systematically reviewing and integrating . . . the literature of a field may be considered a type of research in its own right—one using a characteristic set of research techniques and methods" (86). He described four steps in the synthesis process: sampling topics and studies, developing a scheme for indexing and coding material, integrating the studies, and writing the report.

The same year, Richard Light and Paul Smith (1971) presented what they called a cluster approach to research synthesis, meant to redress some of the deficiencies in the existing strategies for integration. They argued that if treated properly the variation in outcomes among related studies could be a valuable source of information—rather than merely a source of consternation as it appeared to be when treated with traditional cumulating methods.

Three years later, Thomas Taveggia struck a complementary theme:

> A methodological principle overlooked by [synthesists] . . . is that research results are probabilistic. What this principle suggests is that, in and of themselves, the findings of any single research are meaningless—they may have occurred simply by chance. It also follows that, if a large enough number of researches has been done on a particular topic, chance alone dictates that studies will exist that report inconsistent and contradictory findings! Thus, what appears to be contradictory may simply be the positive and negative details of a distribution of findings. (1974, 397–98)

Taveggia went on to describe six common problems in research synthesis: selecting research; retrieving, indexing, and coding information from studies; analyzing the comparability of findings; accumulating comparable findings; analyzing distributions of results, and; reporting of results.

The development of meta-analytic techniques extends back further in time but their routine use by research synthesists is also relatively recent. Where Glass gave us the term meta-analysis in 1976, Ingram Olkin pointed out that ways to estimate effect sizes have existed since the turn of

the century (1990). For example, Pearson took the average of estimates from five separate samples of the correlation between inoculation for enteric fever and mortality (1904). He used this average to better estimate the typical effect of inoculation and to compare this effect with that of inoculation for other diseases. Early work on the methodology for combination of estimates across studies includes papers in the physical sciences by Raymond Birge (1932) and in statistics by William G. Cochran (1937), Frank Yates and Cochran (1938), and Cochran (1954).

Methods for combining probabilities across studies also have a long history, dating at least from procedures suggested in Leonard Tippett's *The Method of Statistics* (1931) and Ronald Fisher's *Statistical Methods for Research Workers* (1932). The most frequently used probability-combining method was described by Frederick Mosteller and Robert Bush in the first edition of the *Handbook of Social Psychology* more than fifty years ago (1954).

Still, the use of quantitative synthesis techniques in the social sciences was rare before the 1970s. Late in that decade, several applications of meta-analytic techniques captured the imagination of behavioral scientists: in clinical psychology, Mary Smith and Gene Glass's meta-analysis of psychotherapy research (1977); in industrial-organizational psychology, Frank Schmidt and John Hunter's validity generalization of employment tests; in social psychology, Robert Rosenthal and Donald Rubin's integration of interpersonal expectancy effect research (1977), and, in education, Glass and Smith's synthesis of the literature on class size and achievement (1978).

1.3.2 Research Synthesis Comes of Age

Two papers that appeared in the *Review of Educational Research* in the early 1980s brought the meta-analytic and synthesis-as-research perspectives together. The first proposed six synthesis tasks "analogous to those performed during primary research" (Jackson 1980, 441). Gregg Jackson portrayed meta-analysis as an aid to the task of analyzing primary studies but emphasized its limitations as well as its strengths. Also noteworthy about his paper was his use of a sample of thirty-six articles from prestigious social science periodicals to examine the methods used in integrative empirical syntheses. He reported, for example, that only one of the thirty-six syntheses reported the indexes or retrieval systems used to locate primary studies. He concluded that "relatively little thought has been given to the methods for doing integrative reviews. Such reviews are critical to science and social policy making and yet most are done far less rigorously than is currently possible" (459).

Harris Cooper drew the analogy between research synthesis and primary research to its logical conclusion (1982). He presented a five stage model of synthesis as a research project. For each stage, he codified the research question asked, its primary function in the synthesis, and the procedural differences that might cause variation in synthesis conclusions. In addition, Cooper applied the notion of threats-to-inferential-validity—which Donald Campbell and Julian Stanley introduced for evaluating the utility of primary research designs—to the conduct of research synthesis (1966; also see Shadish, Cook, and Campbell 2002) . Cooper identified ten threats to validity specifically associated with synthesis procedures that might undermine the trustworthiness of a research synthesis' findings. He also suggested that other threats might exist and that the validity of any particular synthesis could be threatened by consistent deficiencies in the set of studies that formed its database. Table 1.2 presents a recent revision of this schema (Cooper 2007) that proposes a six stage model that separates the original analysis and interpretation stage into two distinct stages.

The first half of the 1980s also witnessed the appearance of four books primarily devoted to meta-analytic methods. In 1981, Gene Glass, Barry McGaw, and Mary Smith presented meta-analysis as a new application of analysis of variance and multiple regression procedures, with effect sizes treated as the dependent variable. In 1982, John Hunter, Frank Schmidt, and Greg Jackson introduced meta-analytic procedures that focused on comparing the observed variation in study outcomes to that expected by chance (the statistical realization of Taveggia's point) and correcting observed effect size estimates and their variance for known sources of bias (for example, sampling error, range restrictions, unreliability of measurements). In 1984, Rosenthal presented a compendium of meta-analytic methods covering, among other topics, the combining of significance levels, effect size estimation, and the analysis of variation in effect sizes. Rosenthal's procedures for testing moderators of variation in effect sizes were not based on traditional inferential statistics, but on a new set of techniques involving assumptions tailored specifically for the analysis of study outcomes. Finally, in 1985, with the publication of *Statistical Procedures for Meta-Analysis*, Hedges and Olkin helped to elevate the quantitative synthesis of research to

an independent specialty within the statistical sciences. This book, summarizing and expanding nearly a decade of programmatic developments by the authors, not only covered the widest array of meta-analytic procedures but also presented rigorous statistical proofs establishing their legitimacy.

Another text that appeared in 1984 also helped elevate research synthesis to a more rigorous level. Richard Light and David Pillemer focused on the use of research syntheses to help decision-making in the social policy domain. Their approach placed special emphasis on the importance of meshing both numbers and narrative for the effective interpretation and communication of synthesis results.

Several other books appeared on the topic of meta-analysis during the 1980s and early 1990s. Some of these treated the topic generally (for the most recent editions, see, for example, Cooper 2009; Hunter and Schmidt 2004; Lipsey and Wilson 2001; Wolf 1986), some treated it from the perspective of particular research design conceptualizations (for example, Mullen 1989), some were tied to particular software packages (for example, Johnson 1993), and some looked to the future of research synthesis as a scientific endeavor (for example, Wachter and Straf 1990; Cook et al. 1992). The first edition of the *Handbook of Research Synthesis* appeared in 1994.

Readers interested in a popular history of meta-analysis can consult Morton Hunt's 1997 *How Science Takes Stock: The Story of Meta-Analysis*. In 2006, Mark Petticrew and Helen Roberts provided a brief "unsystematic history of systematic reviews" that includes discussion of early attempts to appraise the quality of studies and synthesize results. In 1990, Olkin did the same for meta-analysis.

Literally thousands of research syntheses have been published since the first edition of the *Handbook* in 1994. Figure 1.1 presents some evidence of the increasing impact of research syntheses on knowledge in the sciences and social sciences. The figure is based on entries in the Science Citation Index Expanded and the Social Sciences Citation Index, according to the Web of Science reference database (2006). It charts the growth in the number of citations to documents including the terms *research synthesis*, *systematic review*, *research review*, and *meta-analysis* in their title or abstract during the years 1995 to 2005 and this number of citations divided by the total number of documents in the two databases. The figure indicates that both the total number of citations and the number of citations per document in the databases has risen every year without exception. Clearly, the role that research syntheses play in our knowledge claims is large and growing larger.

Table 1.2 Research Synthesis Conceptualized as a Research Process

Stage	Stage Characteristics		
	Research Question	Primary Function	Procedural Variation
Define the Problem	What research evidence will be relevant to the problem or hypothesis of interest in the synthesis?	Define the variables and relationships of interest so that relevant and irrelevant studies can be distinguished	Variation in the conceptual breadth and detail of definitions might lead to differences in the research operations deemed relevant and/or tested as moderating influences
Collect the Research Evidence	What procedures should be used to find relevant research?	Identify sources (e.g., reference databases, journals) and terms used to search for relevant research and extract information from reports	Variation in searched sources and extraction procedures might lead to systematic differences in the retrieved research and what is known about each study
Evaluate the Correspondence between Methods and Implementation of Studies and the Desired Synthesis Inferences	What retrieved research should be included or excluded from the synthesis based on the suitability of the methods for studying the synthesis question or problems in research implementation?	Identify and apply criteria to separate correspondent from incommensurate research results	Variation in criteria for decisions about study inclusion might lead to systematic differences in which studies remain in the synthesis
Analyze (Integrate) the Evidence from Individual Studies	What procedures should be used to summarize and integrate the research results?	Identify and apply procedures for combining results across studies and testing for differences in results between studies	Variation in procedures used to analyze results of individual studies (narrative, vote count, averaged effect sizes) can lead to differences in cumulative results
Interpret the Cumulative Evidence	What conclusions can be drawn about the cumulative state of the research evidence?	Summarize the cumulative research evidence with regard to its strength, generality, and limitations	Variation in criteria for labeling results as important and attention to details of studies might lead to differences in interpretation of findings
Present the Synthesis Methods and Results	What information should be included in the report of the synthesis?	Identify and apply editorial guidelines and judgment to determine the aspects of methods and results readers of the synthesis report need to know	Variation in reporting might lead readers to place more or less trust in synthesis outcomes and influences others ability to replication results

SOURCE: Cooper 2007.

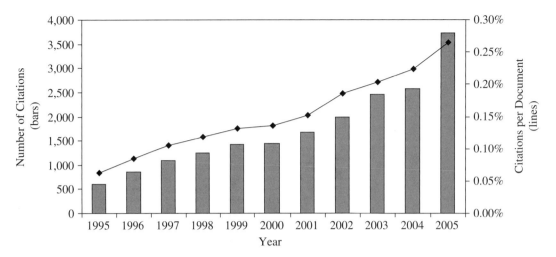

Figure 1.1 Citations to Articles, per Document

SOURCE: Authors' compilation.

NOTE. Based on entries in the Science Citation Index Expanded and the Social Sciences Citation Index, according to the Web of Science reference database (retrieved June 28, 2006). Bars chart the growth in the number of citations to documents including the terms *research synthesis*, *systematic review*, *research review* or *meta-analysis* in their title or abstract during the years following the publication of the first edition of the *Handbook of Research Synthesis*. Lines chart this number of citations divided by the total number of documents in the two databases.

In the past two decades the use of research synthesis has spread from psychology and education through many disciplines, especially the medical sciences and social policy analysis. Indeed, the development of scientific methods for research synthesis has its own largely independent history in the medical sciences (see Chalmers, Hedges, and Cooper 2002). A most notable event in medicine was the establishment of the U.K. Cochrane Centre in 1992. The Centre was meant to facilitate the creation of an international network to prepare and maintain systematic reviews of the effects of interventions across the spectrum of health care practices. At the end of 1993, an international network of individuals, called the Cochrane Collaboration (http://www.cochrane.org/index.htm), emerged from this initiative (Chalmers 1993; Bero and Rennie 1995). By 2006, the Cochrane Collaboration was an internationally renowned initiative with 11,000 people contributing to its work, in more than ninety countries. It is now the leading producer of research syntheses in health care and is considered by many to be the gold standard for determining the effectiveness of different health care interventions. Its library of systematic reviews numbers in the thousands. In 2000, an initiative called the Campbell Collaboration

(http://www.campbellcollaboration.org/) was begun with similar objectives for the domain of social policy analysis, focusing initially on policies concerning education, social welfare, and crime and justice.

Because of the efforts of scholars who chose to apply their skills to how research syntheses might be improved, syntheses written since the 1980s have been held to standards far more demanding than those applied to their predecessors. The process of elevating the rigor of syntheses is certain to continue into the twenty-first century.

1.3.3 Rationale for the *Handbook*

The Handbook of Research Synthesis and Meta-Analysis is meant to be the definitive volume for behavioral and social scientists intent on applying the synthesis craft. It distills the products of thirty years of developments in how research integrations should be conducted so as to minimize the chances of conclusions that are not truly reflective of the cumulated evidence. Research synthesis in the 1960s was at best an art, at worst a form of yellow journalism. Today, the summarization and integration of studies is viewed as a research process in its own right, is held to the standards of a scientific endeavor, and entails the application of data gathering and analyses techniques developed for its unique purpose.

Numerous excellent texts on research synthesis exist. None, however, are as comprehensive as this volume. Some texts focus on statistical methods. These often emphasize different aspects of statistical integration (for example, combining probabilities, regression-analog models, estimating population effects from sampled effects with known biases) and often approach research accumulation from different perspectives. Although these texts are complete within their domains, no single sourcebook describes and integrates all the meta-analytic approaches.

This volume incorporates quantitative statistical techniques from all the synthesis traditions. It brings the leading authorities on the various meta-analytic perspectives together in a single volume. As such, it is an explicit statement by its authors that all the statistical approaches share a common assumptive base. The common base is not only statistical but also philosophical. Philosophically, all the approaches rest on the presupposition that research syntheses need to be held to the same standards of rigor, systematicity, and transparency as the research on which they are based. The second and later users of data must be held as accountable for the validity of their methods as were the first.

Several problems arising in the course of conducting a quantitative synthesis have not received adequate treatment in any existing text. These include nonindependence of data sets, synthesis of multivariate data sets, and sensitivity analysis, to name just a few. Every research synthesist faces these problems, and strategies have been developed for dealing with them. Some of their solutions are published in widely scattered journals, others are passed on to colleagues through informal contacts. They have never received complete treatment within the same text. The *Handbook* brings these topics together for the first time.

Further, texts focusing on the statistical aspects of integration tend to give only passing consideration to other activities of research synthesis. These activities include: the unique characteristics of problem formulation in research synthesis, methods of literature search, coding and evaluation of research reports, and the meaningful interpretation and effective communication of synthesis results. The existing texts that focus on these aspects of research synthesis tend not to be comprehensive in their coverage of statistical issues. Fully half of the chapters in this volume deal with issues that are not statistical, evidencing the authors' collective belief that high quality syntheses require considerably more than just the application of quantitative procedures.

Finally, the handbook is meant for those who carry out research syntheses. Discussions of theory and proof are kept to a minimum in favor of descriptions of the practical mechanics needed to apply well the synthesis craft. The chapters present multiple approaches to problem solving and discuss the strengths and weaknesses of each approach. Readers with a comfortable background in analysis of variance and multiple regression and who have access to a research library should find the chapters accessible. The *Handbook* authors want to supply working synthesists with the needed expertise to interpret their blueprints, to wield their mortar hoe and trowel.

1.4 STAGES OF RESEARCH SYNTHESIS

The description of the stages of research synthesis presented in table 1.2 provides the conceptual organization of the handbook. In this section, we raise the principal issues associated with each stage. This allows us to briefly introduce the content of each of the chapters that follow.

1.4.1 Problem Formulation

Formulating a problem in research synthesis is constrained by one major factor: primary research on a topic must exist before a synthesis can be conducted. How much research? If the research question is important, it would be interesting to know how much research there is on the problem, even if the answer was none at all. The methods of meta-analysis can be applied to literatures containing as few as two hypothesis tests. Under certain circumstances—for instance, researchers synthesizing a pair of replicate studies from their own lab—the use of meta-analysis in this fashion might be legitimate. Yet, most scientists would argue that the benefits of such a synthesis would be limited (and its chances for publication even more limited).

A more general answer to the "how much research" question is that it varies depending on a number of characteristics of the problem. All else being equal, conceptually broad topics would seem to profit from a synthesis only after the accumulation of a more varied and larger number of studies than would narrowly defined topics. Similarly, literatures that contain diverse types of operations also would seem to require a relatively large number of studies before relatively firm conclusions could be drawn from a synthesis. Ultimately, the arbiter of whether a synthesis is needed will not be numerical standards, but the fresh insights a synthesis can bring to a field. Indeed, although a meta-analysis cannot be performed without data, many social scientists see value in empty syntheses that point to important gaps in our knowledge. When done properly, empty syntheses should have proceeded through the stages of research synthesis, including careful problem formulation and a thorough literature search.

Once enough literature on a problem has collected, the challenge—and promise—of research synthesis becomes evident. The problems that constrain primary researchers—small and homogeneous samples, limited time and money for turning constructs of interest into operations—are less severe for synthesists. They can capitalize on the diversity in methods that has occurred naturally across primary studies. The heterogeneity of methods across studies may permit tests of theoretical hypotheses concerning the moderators and mediators of relations that have never been tested in any single primary study. Conclusions about the population and ecological validity of relations uncovered in primary research may also receive more thorough tests in syntheses.

Part II of the handbook focuses on issues in problem formulation. In chapter 2, Harris Cooper discusses in detail the issues mentioned above. In chapter 3, Larry Hedges looks at the implications of different problem definitions for how study results will be statistically modeled. The major issues involve the populations of people and measurements that are the target of a synthesis' inferences; how broadly the key constructs are defined, especially in terms of whether fixed or random effect models are envisioned; and how choices among models influence the precision of estimates and the statistical power of meta-analytic tests.

1.4.2. Literature Search

The literature search is the stage of research synthesis that is most different from primary research. Still, culling through the literature for relevant studies is not unlike gathering a sample of primary data. The target of a literature search that is part of a synthesis attempting exhaustive coverage would be all the research conducted on the topic of interest.

Cooper found that about half of the reviewers he surveyed claimed that they conducted their search with the intention of identifying all or most of the relevant literature (1987). About three-quarters said they did not stop searching until they felt they had accomplished their goal.

About one-third said they stopped their search when they felt their understanding and conclusions about the topic would not be affected by additional material. About one-sixth said they stopped searching when retrieval became unacceptably difficult.

In contrast to the relatively well-defined sampling frames available to primary researchers, literature searchers confront the fact that any single source of primary reports will lead them to only a fraction of the relevant studies, and a biased fraction at that. For example, among the most egalitarian sources of literature are the reference databases, such as PsycINFO, ERIC, and Medline. Still, these broad, non-evaluative systems exclude much of the unpublished literature. Conversely, the least equitable literature searching technique involves accessing one's close colleagues and other researchers with an active interest in the topic area. In spite of the obvious biases, there is no better source of unpublished and recent works. Further complicating the sampling frame problem for synthesist is the fact that the relative utility and biases associated with any single source will vary as a function of characteristics of the research problem, including for example how long the topic has been the focus of study and whether the topic is interdisciplinary in nature.

These problems imply that research synthesists must carefully consider multiple channels for accessing literature and how the channels they choose complement one another. The three chapters in part III are devoted to helping the synthesist consider and carry out this unique task. In chapter 4, Howard White presents an overview of searching issues from the viewpoint of an information scientist. Both chapter 5, by Jeffrey Reed and Pam Baxter, and chapter 6, by Hannah Rothstein and Sally Hopewell, present in detail the practical considerations of how to scour our brickyards and quarries.

1.4.3 Data Evaluation

Once the synthesists have gathered the relevant literature, they must extract from each document those pieces of information that will help answer the questions that impel research in the field. The problems faced during data coding provide a strong test of the synthesist's competence, thoughtfulness, and ingenuity. The solutions found for these problems will have considerable influence on the contribution of the synthesis. Part IV contains four chapters on data coding and evaluation.

The aspect of coding studies that engenders the most debate involves how synthesists should represent differences in the design and implementation of primary studies. What is meant by quality in evaluating research methods? Should studies be weighted differently if they differ in quality? Should studies be excluded if they contain too many flaws? How does one rate the quality of studies described in incomplete research reports? In chapter 7, Jeff Valentine examines the alternative approaches available to synthesists for representing primary research methodology. Beyond quality judgments, synthesists make decisions about the classes of variables that are of potential interest to them. These can relate to variables that predict outcomes, potential mediators of effects, and the differences in how outcomes are conceptualized (and, therefore, measured). If a synthesist chooses not to code a particular feature of studies, then it cannot be considered in the analysis of results. General guidelines for what information should be extracted from primary research reports are difficult to develop, beyond those that are very abstract. Instead, direction will come from the issues that arise in the particular literature, coupled with the synthesist's personal insights into the topic. Still, there are commonalities that emerge in how these decisions are forged. Mark Lipsey, in chapter 8 ("Identifying Variables and Analysis Opportunities"), and David Wilson, in chapter 9 ("Systematic Coding"), present complementing templates for what generally should be included in coding frames.

Once decisions on what to code have been made, synthesists needs to consider how to carry out the coding of the literature and how to assess the trustworthiness with which the coding frame is implemented. There are numerous indexes of coder reliability available, each with different strengths and weaknesses. Chapter 10, by Robert Orwin and Jack Vevea, describes strategies for reducing the amount of error that enters a synthesis during the coding of the literature's features. Orwin and Vevea's description of reliability assessment focuses on three major approaches: measuring interrater agreement; obtaining confidence ratings from coders; and conducting sensitivity analysis, or using multiple coding strategies to determine whether a synthesis' conclusions are robust across different coding decision rules.

1.4.4 Data Analysis

As our brief history of research synthesis revealed, the analysis of accumulated research outcomes has been a separate area of specialization within statistics. Three decades ago, the actual mechanics of integrating research

usually involved intuitive processes taking place inside the heads of the synthesists. Meta-analysis made these processes public and based them on explicit, shared, statistical assumptions (however well these assumptions were met). We would not accept as valid a primary researcher's conclusion if it were substantiated solely by the statement "I looked at the treatment and control scores and I think the treated group did better." We would demand some sort of statistical test (for example, a t-test) to back up the claim. Likewise, we no longer accept "I examined the study outcomes and I think the treatment is effective" as sufficient warrant for the conclusion of a research synthesis.

Part V contains chapters that cover the components of synthesis dealing with combining study results. Brad Bushman and Morgan Wang, in chapter 11, describe procedures based on viewing the outcomes of studies dichotomously, as either supporting or refuting a hypothesis. The next two chapters cover methods for estimating the magnitude of an effect, commonly called the effect size. Cohen defined an effect size as "the *degree* to which the phenomenon is present in the population, or the degree to which the null hypothesis is false" (1988, 9–10). In chapter 12, Michael Borenstein describes the parametric measures of effect size. In chapter 13, Joseph Fleiss and Jesse Berlin cover the measures of effect size for categorical data, paying special attention to the estimate of effect based on rates and ratios.

To most research synthesists, the search for influences on study results is the most exciting and rewarding part of the process. The four chapters of part VI all deal with techniques for analyzing whether and why there are differences in the outcomes of studies. As an analog to analysis of variance or multiple regression procedures, effect sizes can be viewed as dependent (or criterion) variables and the features of study designs as independent (or predictor) variables. Because effect-size estimates do not all have the same sampling uncertainty, however, they cannot be combined using traditional inferential statistics. In chapter 14, William Shadish and Keith Haddock describe procedures for averaging effects and generating confidence intervals around these population estimates. In chapter 15, Spyros Konstantopoulos and Larry Hedges discuss analysis strategies when fixed-effects models are used. In chapter 16, Stephen Raudenbush describes the strategies appropriate for random-effects models.

Further, effect-size estimates may be affected by factors that attenuate their magnitudes. These may include, for example, a lack of reliability in the measurement instruments or restrictions in the range of values in the subject sample. Attenuating biases may be estimated and corrected using the procedures that Frank Schmidt, Huy Le, and In-Sue Oh describe in chapter 17.

Part VII includes three chapters that focus on some special issues in the statistical treatment of research outcomes. In chapter 18, Larry Hedges discusses combining special types of studies, in particular the increasing use of cluster- or group-randomized trials in social research and the implication of this for the definition of effect sizes, their estimation, and particularly the computation of standard errors of the estimates. Chapter 19 tackles the vexing problem of what to do when the same study (or laboratory, or researcher, if one wishes to push the point) provides more than one effect size to a meta-analysis. Because estimates from the same study may not be independent sources of information about the underlying population value, their contributions to any overall estimates must be adjusted accordingly. In chapter 19, Leon Gleser and Ingram Olkin describe both simple and more complex strategies for dealing with dependencies among effect sizes.

Sometimes, the focus of a research synthesis will not be a simple bivariate relation but a model involving numerous variables and their interrelations. The model could require analysis of more than one simple relation and/or complex partial or second-order relations. In chapter 20, Betsy Becker looks at what models in meta-analysis are, why they should be used, and how they affect meta-analytic data analysis.

1.4.5 Interpretation of Results

Estimating and averaging effect sizes and searching for moderators of their variability is how the interpretation of cumulative study results begins. However, it must be followed by other procedures that help the synthesists properly interpret their outcomes. Proper interpretation of the results of a research synthesis requires careful use of declarative statements regarding claims about the evidence, specification of what results warrants each claim, and any appropriate qualifications to claims that need to be made. In part VIII, three important issues in data interpretation are examined.

Even the most trustworthy coding frame cannot solve the problem of missing data. Missing data will arise in every research synthesis. Reporting of study features and results, as now practiced, leaves much to be desired. Although the emergence of research synthesis has led to

many salutary changes, most primary researchers still do not think about a research report in terms of the needs of the next user of the data. Also, problems with missing data would still occur even if the needs of the synthesist were the paramount consideration in constructing a report. There is simply no way primary researchers can anticipate the subtle nuances in their methods and results that might be identified as critical to piecing together the literature five, ten, or more years later. In chapter 21, Theresa Piggott describes how lost information can bear on the results and interpretation of a research synthesis. She shows how the impact of missing data can be assessed by paying attention to some of its discoverable characteristics. She also describes both simple and complex strategies for handling missing data in research synthesis. Joel Greenhouse and Satish Iyengar, in chapter 22, describe sensitivity analysis in detail, including several diagnostic procedures to assess the sensitivity and robustness of the conclusions of a meta-analysis. Alex Sutton, in chapter 23, examines publication bias and describes procedures for identifying whether data censoring has had an impact on the meta-analytic result. He also details some imputation procedures that can help synthesists estimate what summary effect size estimates might have been in the absence of the selective inclusion of studies.

The chapters in part IX bring us full circle, returning us to the nature of the problem that led to undertaking the research synthesis in the first place. Collectively, these chapters attempt to answer the question as to what can be learned from research synthesis. In chapter 24, Wendy Wood and Alice Eagly examine how theories can be tested in research synthesis. In chapter 25, David Cordray and Paul Morphy look at research synthesis for policy analysis. These chapters are meant to help synthesists think about the substantive questions for which the readers of their syntheses will be seeking answers.

1.4.6 Public Presentation

Presenting the background, methods, results, and meaning of a research synthesis' findings provide the final challenges to the synthesists' skill and intellect. In chapter 26, the first of part X, Geoffrey Borman and Jeffrey Grigg address the visual presentation of research synthesis results. Charts, graphs, and tables, they explain, should be used to summarize the numbers in a meta-analysis, along with a careful intertwining of narrative explication to contextualize the summaries. Proper use of interpretive devices can make a complex set of results easily accessible to the intelligent reader. Likewise, inadequate use of these aides can render incomprehensible even the simplest conclusions. Chapter 27 describes the reporting format for research synthesis. As with the coding frame, there is no simple reporting scheme that fits all syntheses. However, certain commonalities do exist and Mike Clarke discusses them. Not too surprisingly, the organization that emerges bears considerable resemblance to that of a primary research report although, also obviously, the content differs dramatically.

1.5 CHAPTERS IN PERSPECTIVE

In part XI, the final two chapters of the handbook are meant to bring the detailed drawings of the preceding chapters onto a single blueprint. In chapter 28, Georg Matt and Thomas Cook present an expanded analysis of threats to the validity of research syntheses, provide an overall appraisal of how inferences from syntheses may be restricted or faulty, and bring together many of the concerns expressed throughout the book.

Finally, chapter 29 also attempts to draw this material together. In it, Harris Cooper and Larry Hedges look at some of the potentials and limitations of research synthesis. Special attention is paid to possible future developments in synthesis methodology, the feasibility and expense associated with conducting a sound research synthesis, and a broad-based definition of what makes a research synthesis good or bad.

No secret will be revealed by stating our conclusion in advance. If procedures for the synthesis of research are held to standards of objectivity, systematicity, and rigor, then our knowledge edifice will be made of bricks and mortar. If not, it will be a house of cards.

1.6 REFERENCES

Bero, Lisa, and Drummond Rennie. 1995. "The Cochrane Collaboration: Preparing, Maintaining, and Disseminating Systematic Reviews of the Effects of Health Care." *Journal of the American Medical Association* 274(24): 1935–8.

Birge, Raymond T. 1932. "The Calculation of Error by the Method of Least Squares." *Physical Review* 40(2): 207–27.

Campbell, Donald T., and Julian C. Stanley. 1966. *Experimental and Quasi-Experimental Designs for Research.* Chicago: Rand McNally.

Chalmers, Iain. 1993. "The Cochrane Collaboration: Preparing, Maintaining and Disseminating Systematic Reviews of the Effects of Health Care." *Annals of the New York Academy of Sciences* 703: 156–63.

Chalmers, Iain, Larry V. Hedges, and Harris Cooper. 2002. "A Brief History of Research Synthesis." *Evaluation and the Health Professions* 25(1): 12–37.

Cochran, William G. 1937. "Problems Arising in the Analysis of a Series of Similar Experiments." *Journal of the Royal Statistical Society* 4(Supplement): 102–18.

———. 1954. "The Combination of Estimates from Different Experiments." *Biometrics* 10(1, March): 101–29.

Cohen, Jacob. 1988. *Statistical Power Analysis for the Behavioral Sciences.* Hillsdale, NJ: Lawrence Erlbaum.

Cook, Thomas D., Harris M. Cooper, David S. Cordray, Heidi Hartmann, Larry V. Hedges, Richard J. Light, Thomas Louis, and Frederick Mosteller. 1992. *Meta-Analysis for Explanation: A Casebook.* New York: Russell Sage Foundation.

Cooper, Harris M. 1982. "Scientific Guidelines for Conducting Integrative Research Reviews." *Review of Educational Research* 52(2, Summer): 291–302.

———. 1987. "Literature Searching Strategies of Integrative Research Reviewers: A First Survey." *Knowledge: Creation, Diffusion, Utilization* 8: 372–83.

———. 1988. "Organizing Knowledge Synthesis: A Taxonomy of Literature Reviews." *Knowledge in Society* 1: 104–26.

———. 2009. *Research Synthesis and Meta-Analysis: A Step-by-Step Approach.* Thousand Oaks, Calif.: Sage Publications.

———. 2003. "Editorial." *Psychological Bulletin* 129(1): 3–9.

———. 2007. *Evaluating and Interpreting Research Syntheses in Adult Learning and Literacy.* Boston, Mass.: National College Transition Network, New England Literacy Resource Center/World Education, Inc.

Feldman, Kenneth A. 1971. "Using the Work of Others: Some Observations on Reviewing and Integrating." *Sociology of Education* 4(1, Winter): 86–102.

Fisher, Ronald A. 1932. *Statistical Methods for Research Workers*, 4th ed. London: Oliver & Boyd.

Glass, Gene V. 1976. "Primary, Secondary, and Meta-Analysis." *Educational Researcher* 5(10): 3–8.

Glass, Gene V., and Mary L. Smith. 1978. *Meta-Analysis of Research on the Relationship of Class Size and Achievement.* San Francisco: Far West Laboratory for Educational Research and Development.

Glass, Gene V., Barry McGaw, and Mary L. Smith. 1981. *Meta-Analysis in Social Research.* Beverly Hills, Calif.: Sage Publications.

Hedges, Larry V. 1987. "How Hard Is Hard Science, How Soft is Soft Science?" *American Psychologist* 42(5): 443–55.

Hedges, Larry V., and Ingram Olkin. 1985. *Statistical Methods for Meta-Analysis*. Orlando, Fl.: Academic Press.

Hunt, Morton. 1997. *How Science Takes Stock: The Story of Meta-Analysis*. New York: Russell Sage Foundation.

Hunter, John E., and Frank L. Schmidt. 2004. *Methods of Meta-Analysis: Correcting Error and Bias in Research Findings*, 2nd ed. Thousand Oaks, Calif.: Sage Publications.

Hunter, John E., Frank L. Schmidt, and Greg B. Jackson. 1982. *Meta-Analysis: Cumulating Research Findings Across Studies*. Beverly Hills, Calif.: Sage Publications.

Jackson, Gregg B. 1980. "Methods for Integrative Reviews." *Review of Educational Research* 50: 438–60.

Johnson, Blair T. 1993. *DSTAT: Software for the Meta-Analytic Review of Research*. Hillsdale, N.J.: Lawrence Erlbaum.

Light, Richard J., and David B. Pillemer. 1984. *Summing Up: The Science of Reviewing Research*. Cambridge, Mass.: Harvard University.

Light, Richard J., and Paul V. Smith. 1971. "Accumulating Evidence: Procedures for Resolving Contradictions Among Research Studies." *Harvard Educational Review* 41(4): 429–71.

Lipsey, Mark W., and David B. Wilson. 2001. *Practical Meta-Analysis*. Thousand Oaks, Calif.: Sage Publications.

Manten, A. A. 1973. "Scientific Literature Review." *Scholarly Publishing* 5: 75–89.

Mosteller, Frederick, and Robert Bush. 1954. "Selected Quantitative Techniques." In *Handbook of Social Psychology*, vol. 1, *Theory and Method*, edited by Gardner Lindzey. Cambridge, Mass.: Addison-Wesley.

Mullen, Brian. 1989. *Advanced BASIC Meta-Analysis*. Hillsdale, N.J.: Lawrence Erlbaum.

Olkin, Ingram. 1990. "History and Goals." In *The Future of Meta-Analysis*, edited by Kenneth W. Wachter and Miron L. Straf. New York: Russell Sage Foundation.

Pearson, Karl. 1904. "Report on Certain Enteric Fever Innoculation Statistics." *British Medical Journal* 3: 1243–6.

Petticrew, Mark, and Helen Roberts. 2006. *Systematic Reviews in the Social Sciences: A Practical Guide*. Oxford: Blackwell Publishing.

Price, Derek J. de Solla. 1965. "Networks of Scientific Papers." *Science* 149(3683): 56–64.

Rosenthal, Robert. 1984. *Meta-Analytic Procedures for Social Research*. Beverly Hills, Calif.: Sage Publications.

Rosenthal, Robert, and Donald B. Rubin. 1978. "Interpersonal Expectancy Effects: The First 345 Studies." *The Behavioral and Brain Sciences* 3: 377–86.

Schmidt, Frank L., and John E. Hunter. 1977. "Development of a General Solution to the Problem of Validity Generalization." *Journal of Applied Psychology* 62: 529–40.

Shadish, William. R., Thomas D. Cook, and Donald T. Campbell. 2002. *Experimental and Quasi-Experimental Designs for Generating Causal Inference*. Boston, Mass.: Houghton Mifflin.

Smith, Mary L., and Gene V. Glass. 1977. "Meta-Analysis of Psychotherapy Outcome Studies." *American Psychologist* 32(9): 752–60.

Taveggia, Thomas C. 1974. "Resolving Research Controversy Through Empirical Cumulation: Toward Reliable Sociological Knowledge." *Sociological Methods & Research* 2(4): 395–407.

Tippett, Leonard Henry Caleb. 1931. *The Methods of Statistics*. London: Williams & Norgate.

Wachter, Kenneth W., and Miron L. Straf, eds. 1990. *The Future of Meta-Analysis*. New York: Russell Sage Foundation.

Wolf, Frederic M. 1986. *Meta-Analysis: Quantitative Methods for Research Synthesis*. Beverly Hills, Calif.: Sage Publications.

Yates, Frank, and William G. Cochran. 1938. "The Analysis of Groups of Experiments." *Journal of Agricultural Science* 28: 556–80.

PART
II

FORMULATING A PROBLEM

2

HYPOTHESES AND PROBLEMS IN RESEARCH SYNTHESIS

HARRIS COOPER
Duke University

CONTENTS

2.1 INTRODUCTION

Texts on research methods often list sources of research ideas (see, for example, Cherulnik 2001). Sometimes ideas for research come from personal experiences, sometimes from pressing social issues. Sometimes a researcher wishes to test a theory meant to help understand the roots of human behavior. Yet other times researchers find topics by reading scholarly research.

Still, sources of ideas for research are so plentiful that the universe of possibilities seems limitless. Perhaps we must be satisfied with Karl Mannheim's (1936) suggestion that the ideas researchers pursue are rooted in their social and material relations, from what the researcher deems important (Coser 1971). Then at least, we have a course of action for studying their determinants.

I will not begin my exploration of problems and hypotheses in research synthesis by examining the determinants of choice. Nor will I describe the existential bounds of the researcher. Instead, I start with a simple statement, that the problems and hypotheses in research syntheses generally are drawn from those that already have had data collected on them in primary research. Then, my task becomes much more manageable. In this chapter I examine the characteristics of research problems and hypotheses that make them empirically testable and are similar and different for primary research and research synthesis.

That syntheses for the most part are in fact tied to only those problems that have previously generated data does not mean that research synthesis is not a creative exercise. The process of gathering and integrating research often requires the creation of explanatory models to help make sense of related studies that produced incommensurate data. These schemes can be novel, having never appeared in previous theorizing or research. The cumulative results of studies are much more complex than the results of any single study. Discovering why two studies that appear to be direct replications of one another have produced conflicting results is a deliberative challenge for any research synthesist.

2.2 DEFINING VARIABLES

In its most basic form, a research problem includes a clear statement of what variables are to be related to one another and an implication of how the relationship can be tested empirically (Kerlinger and Lee 1999). The rationale for a research problem can be that some practical consideration suggests any discovered relation might be important. Or, the problem can contain a prediction about a particular link between the variables—based on theory or previous observation. This type of prediction is called a hypothesis.

Either rationale can be used for undertaking primary research or research synthesis. For example, we might want to test the hypothesis that teachers' expectations concerning how a student's intelligence will change over the course of a school year will cause the students' intelligence to change in the expected direction. This hypothesis might be based on a well-specified model of how interpersonal communications change as a function of the teachers' beliefs. Another example might be that we are interested in how students' perceptions of their instructors in high school and college correlate with achievement in the course. Here, we may know the problem is important, but may have no precise theory that leads to a hypothesis about whether and how perceptions of the instructor are related to achievement in the course. This research problem is more exploratory in nature.

2.2.1 Conceptual and Operational Definitions

The variables involved in social science research must be defined in two ways. First, each variable must be given a conceptual or theoretical definition. This describes qualities of the variable that are independent of time and space but which can be used to distinguish events that are and are not relevant to the concept (Shoemaker, Tankard, and Lasorsa 2004). For instance, a conceptual definition of intelligence might be a capacity for learning, reasoning, understanding, and similar forms of mental activity (Random House 2001, 990).

Conceptual definitions can differ in breadth, or in the number of events to which they refer. Thus, if we define *achievement* as "something accomplished esp. by superior ability, special effort, great courage, etc." (Random House 2001, 15), the concept is broader than if we confine the domain of achievement to academic tasks, or activities related to school performance in cognitive domains. The broader definition of achievement would include goals reached in physical, economic, and social spheres, as well as academic ones. So, if we are interested in the relationship between student ratings of instructors and achievement writ large we would include ratings taken in gym, health, and drama classes. If we are interested in academic achievement only, these types of classes would fall outside our conceptual definition. The need for well-specified conceptual definitions is no different in research synthesis than it is in primary research. Both primary researchers and research synthesists must clearly specify their conceptual definitions.

To relate concepts to concrete events, the variables in empirical research must also be operationally defined. An operational definition is a description of the characteristics of observable events that are used to determine whether the event represents an occurrence of the conceptual variable. So, for example, a person might be deemed physically fit (a conceptual variable) if they can run a quarter-mile in a specified time (the operational definition of the conceptual variable). Put differently, a concept is operationally defined by the procedures used to produce and measure it. Again, both primary researchers and research synthesists must specify the operations included in their conceptual definitions.

An operational definition of the concept *intelligence* might first focus on scores from standardized tests meant to measure reasoning ability. This definition might be specified further to include only tests that result in Intelligence Quotient (IQ) scores, for example, the Stanford-Binet and the Wechsler tests. Or, the operational definition might be broadened to include the Scholastic Aptitude Test (SAT), the Graduate Record Exam (GRE), or the Miller Analogies Test (MAT). This last group might be seen as broadening the definition of intelligence because these tests are relatively more influenced by the test-takers' knowledge than their native ability to reason and understand. Operations can also include the ingredients of a treatment. For example, an experimenter might devise an intervention for teachers meant to raise their expectations for students. The intervention might include information on how malleable intelligence might be and

techniques for recognizing when students have experienced a change in mental acuity.

2.2.1.1 Distinctions Between Operational Definitions in Primary Research and Research Synthesis

The first critical distinction between operational definitions in primary research and research synthesis is that primary researchers cannot start data collection until the variables of interest have been given a precise operational definition, an empirical reality. Otherwise, the researcher does not know what data to collect. A primary researcher studying intelligence must pick the measure or measures of intelligence they wish to study, ones that fit nicely into their conceptual definition, before the study begins.

In contrast, research synthesists need not be quite so conceptually or operationally precise, at least not initially. The literature search for a research synthesis can begin with only a broad, and sometimes fuzzy, conceptual definition and a few known operations that measure the construct. The search for studies then might lead the synthesists not only to studies that define the construct in the manners they specified, but also to research in which the same construct was studied with different operational definitions. Then, the concept and associated operations can grow broader or narrower—and hopefully more precise—as the synthesists grow more familiar with how the construct has been defined and measured in the extant research. Synthesists have the comparative luxury of being able to evaluate the conceptual relevance of different operations as they grow more familiar with the literature. They can even modify the conceptual definition as they encounter related alternative concepts and operations in the literature. This can be both a fascinating and anxiety-arousing task. Does intelligence include what is measured by the SAT? Should achievement include performance in gym class?

I do not want to give the impression that research synthesis permits fuzzy thinking. Typically, research synthesists begin with a clear idea of the concepts of interest and with some a priori specification of related empirical realizations. However, during a literature search it is not unusual to come across definitions that raise issues about concept boundaries they may not have considered or operations that they did not know existed but seem relevant to the construct being studied. Some of these new considerations may lead to a tweaking of the conceptual definition. In sum, primary researchers need to know exactly what events will constitute the domain to be sampled before beginning data collection. Research synthesists may discover unanticipated elements of the domain along the way.

Another distinction between the two types of inquiry is that primary studies will typically involve only one, and sometimes a few, operational definitions of the same construct. In contrast, research syntheses usually include many empirical realizations. For example, primary researchers may pick a single measure of intelligence, if only because the time and economic constraints of administering multiple measures of intelligence will be prohibitive. A few measures of academic achievement might be available to primary researchers, if they can access archived data (for example, grade point averages, achievement test scores). On the other hand, research synthesists can uncover a wide array of operationalizations of the same construct within the same research area.

2.2.2 The Fit Between Concepts and Operations

The variety of operationalizations that a research synthesist may uncover in the literature can be both a curse and a blessing. The curse concerns the fit between concepts and operations.

2.2.2.1 Broadening and Narrowing Concepts Synthesists may begin a literature search with broad conceptual definitions. However, they may discover that the operations used in previous relevant research have been confined to a narrower conceptualization. For instance, if we are gathering and integrating studies about how student perceptions of instructors in high school and college correlate with course achievement we might discover that all, or nearly all, past research has measured only liking the instructor, and not the instructor's perceived expertise in the subject matter or ability to communicate. Then, it might not be appropriate for us to label the conceptual variable as student perceptions of instructors. When such a circumstance arises, we need to narrow our conceptual definition to correspond better with the existing operations, such as liking the instructor. Otherwise, our conclusions might appear to apply more generally, to more operations, than warranted by the evidence.

The opposite problem can also confront synthesists—that is, they start with narrow concepts but then find measures in the literature that could support broader definitions. For example, this might occur if we find many studies of the effects of teachers' expectations on SAT scores when we initially intended to confine our operational definition of intelligence to IQ scores. We would then face the choice of either broadening the allowable measures of intelligence, and therefore perhaps the

conceptual definition, or excluding many studies that others might deem relevant.

Thus, it is not unusual for a dialogue to take place between research synthesists and the research literature. The dialogue can result in an iterative redefining of the conceptual variables that are the focus of the synthesis as well as of the included empirical realizations. As the literature search proceeds, it is extremely important that synthesists take care to reevaluate the correspondence between the breadth of their concepts and the variation in ideas and operations that primary researchers have used to define them.

Before leaving the issue of concept-to-operation-fit, it is important to make one more point. The dialogue between synthesists and research literatures is one that must proceed on the basis of an attempt to provide precise conceptual definitions that lead to clear linkages between concepts and operations in a manner that will allow meaningful interpretation of results by the audience of the synthesis. What this means is that synthesists must never let the results of the primary research dictate what operations will be used to define a concept. For example, it would be inappropriate for us to decide to operationally define intelligence as cognitive abilities measured by IQ tests and leave out results involving the Miller Analogies Test because the IQ tests revealed results consistent with our hypothesis and the MAT did not. The relevance of the MAT must be based on how well its contents correspond to our conceptual definition, and that alone. If this test of intelligence produces results different from other tests, then the reason for the discrepancy should be explored as part of the research synthesis, not used as a rationale for the exclusion of MAT studies.

2.2.2.2 Multiple Operations and Concept to Operation Fit Multiple operations also create opportunities for research synthesists. Eugene Webb and his colleagues presented strong arguments for the value of multiple operations (2000). They define multiple operationism as the use of many measures that share a conceptual definition but that have different patterns of irrelevant components. Multiple operationism has positive consequences because

once a proposition has been confirmed by two or more independent measurement processes, the uncertainty of its interpretation is greatly reduced. . . . If a proposition can survive the onslaught of a series of imperfect measures, with all their irrelevant error, confidence should be placed in it. Of course, this confidence is increased by minimizing error in each instrument and by a reasonable belief in the different and divergent effects of the sources of error. (Webb et al. 2000, 35)

So, the existence of a variety of operations in research literatures offers the potential benefit of stronger inferences if it allows the synthesists to rule out irrelevant sources of influence. However, multiple operations do not ensure concept-to-operation correspondence if all or most of the operations lack a minimal correspondence to the concept. For example, studies of teacher expectation effects may manipulate expectations in a variety of ways—by providing teachers with information on students in the form of tests scores, by reports from psychologists—and may measure intelligence in a variety of ways. This means that patterns of irrelevant variation in operations are different for different manipulations of expectations and measures of intelligence. Teachers might find tests scores more credible than psychologist's assessments. If results are consistent regardless of the operations used then the finding that expectations influence intelligence has passed a test of robustness. However, if the results differ depending on operations then qualifications to the finding are in order. For example, if we find that teacher expectations influence SAT scores but not IQ scores we might postulate that perhaps expectations affect students' knowledge acquisition but not mental acuity. This inference is less strong because it requires us to develop a post hoc explanation for our findings.

2.2.2.3 The Use of Operations Meant to Represent Other Concepts Literature searches of reference databases typically begin with keywords that represent the conceptual definitions of the variables of interests. Thus we are much more likely to begin a search by crossing the keyword teacher expectations with intelligence than we are by crossing Test of Intellectual Growth Potential (a made-up test that might have been used to manipulate teacher expectations) with Stanford-Binet and Wechsler. It is the abstractness of the keywords that allows unexpected operationalizations to get caught in the search net.

By extending the keywords even further, searchers may uncover research that has been cast in conceptual frameworks different from their own but which include manipulations or measures relevant to the concepts they have in mind. For instance, several concepts similar to interpersonal expectancy effects appear in the research literature—one is behavior confirmation—but come from different disciplinary traditions. Even though these concepts are labeled differently the operations used in the

studies they generate may be very similar, and certainly relevant to one another. When relevant operations associated with different constructs are identified, they should be included in the synthesis. In fact, similar operations generated by different disciplinary traditions often do not share other features of research design—for example, they may draw participants from different populations—and therefore can be used to demonstrate the robustness of results across methodological irrelevancies. Synthesists can improve their chances of finding broadly relevant studies by searching in reference databases for multiple disciplines and using the database thesauri to identify related, broader, and narrower terms.

2.2.3 Effects of Multiple Operations on Synthesis Outcomes

Multiple operations do more than introduce the potential for more robust inferences about relationships between conceptual variables. They are an important source of variability in the conclusions of different syntheses meant to address the same topic. A variety of operations can affect synthesis outcomes in two ways.

First, the operational definitions covered in two research syntheses involving the same conceptual variables can be different from one another. Thus, two syntheses claiming to integrate the research on the effects of teacher expectations on intelligence can differ in the way intelligence is operationally defined. One might include the SAT and GRE along with IQ tests and the other only IQ tests. Each may contain some operations excluded by the other, or one definition may completely contain the other.

Second, multiple operations affect outcomes of syntheses by leading to variation in the way study operations are treated after the relevant literature has been identified. Some synthesists pay careful attention to study operations. They identify meticulously the operational distinctions among retrieved studies and test for whether results are robust across these variations. Other synthesists pay less attention to these details.

2.2.4 Variable Definitions and the Literature Search

I have mentioned one way in which the choice of keywords in a reference database search can influence the studies that are uncovered. I pointed out that entering databases with abstract terms as keywords will permit the serendipitous discovery of aligned areas of research, more so than will entering with operational descriptions. Now I want to extend this recommendation further: synthesists should begin the literature search with the broadest conceptual definition in mind. While reading study abstracts, they should use the tag *potentially relevant* as liberally as possible. At later stages, it is okay to exclude particular operations on the basis of their lack of correspondence with the precise conceptual definition. However, in the early stages, when conceptual and operational boundaries may still be a bit fuzzy, the synthesist should err on the overly inclusive side, just as primary researchers collect some data that might not later be used in analysis.

This strategy initially creates more work for synthesists, but it has several long-term benefits. First, it will keep within easy reach operations that on first consideration may be seen as marginal but later jump the boundary from irrelevant to relevant. Reconstituting the search because relevant operations have been missed consumes far more resources than first putting them in the In bin but excluding them later.

Second, a good conceptual definition speaks not only to which operations are considered relevant but also to which are irrelevant. By beginning a search with broad terms, the synthesists are forced to struggle with operations that are at the margins of their conceptual definitions. Ultimately, this results in conceptual definitions with more precise boundaries. For example, if we begin a search for studies of teacher expectation effects with the keyword intelligence rather than IQ, we may decide to include research using the MAT, thus broadening the intelligence construct beyond IQ, but exclude the SAT and GRE, because they are too influenced by knowledge, rather than mental acuity. But, by explicitly stating these tests were excluded, readers gets a better idea of where the boundary of the definition lies, and can argue otherwise if they choose. Had we started the search with IQ, we might never have known that others considered the SAT and GRE tests of intelligence in teacher expectation research.

Third, a broad conceptual search allows the synthesis to be carried out with greater operational detail. For example, searching for and including IQ tests only permits us to examine variations in operations such as whether the test was group versus individually administered and timed versus untimed. Searching for, and ultimately including, a larger range of operations permits the synthesists to examine broader conceptual issues when they cluster findings according to operations and look for variation in results. Do teacher expectations produce different effects on IQ tests and the SAT? If so, are there plausible explanations for this? What does it mean for the

definition of intelligence? Often, these analyses produce the most interesting results in a research synthesis.

2.3 TYPES OF PROBLEMS AND HYPOTHESES

Once researchers have defined their concepts and operations, they next must define the problem or hypothesis of interest. Does the problem relate simply to the prevalence or level of a phenomenon in a population? Does it suggest that the variables of interest relate as a simple association or as a causal connection? Does the problem or hypothesis refer to a process that operates within an individual unit[1]—how it changes over time—or to general tendencies between groups of units?

The distinctions embodied in these questions have critical implications for the choice of appropriate research designs. I assume the reader is familiar with research designs and their implications for drawing inferences. Still, a brief discussion of the critical features of research problems as they relate to research design is needed in order to understand the role of design variations in research synthesis.

2.3.1 Three Questions About Research Problems and Hypothesis

Researchers need to answer three questions about their research problems or hypotheses to be able to determine the appropriateness of different research designs for addressing them (for a fuller treatment of these questions, see Cooper 2006):

- Should the results of the research be expressed in numbers or narrative?

- Is the problem or hypothesis seeking to uncover a description of an event, an association between events, or an explanation of an event?

- Does the problem or hypothesis seek to understand how a process unfolds within an individual unit over time, or what is associated with or explains variation between units or groups of units?

Because we are focusing on quantitative research synthesis, I will assume the first question was answered numbers and provide further comment on the latter two questions only. However, the gathering and integration of narrative, or qualitative, research is an important area of methodology and scholarship. The reader is referred to works by George Noblit and Dwight Hare (1988) and Barbara Paterson and her colleagues (2001) for detailed examinations of approaches to synthesizing qualitative research.

2.3.1.1 Descriptions, Associations, and Explanations First, a research problem might largely be descriptive. Such problems typically ask what is happening. As in survey research, the desired description could focus on a few specific characteristics of the event or events of interest and broadly sample those few aspects across multiple event occurrences (Fowler 2002). Thus, we might ask a sample of undergraduates, perhaps drawn randomly across campuses and in different subject areas, how much they like their instructors. From this, we might draw an inference about how well-liked instructors generally are on college campuses.

A second type of problem might be what events happen together. Here, researchers ask whether characteristics of events or phenomena co-occur with one another. A correlation coefficient might be calculated to measure the degree of association. For example, our survey of undergraduates' liking for their instructors might also ask the students how well they did in their classes. Then, the two variables could be correlated to answer the question of whether students' liking of instructors are associated with their class grades.

The third research problem seeks an explanation for an event. In this case, a study is conducted to isolate and draw a direct productive link between one event (the cause) and another (the effect). In our example, we might ask whether increasing the liking students have for their instructors causes them to get higher grades in class.

Three classes of research designs are used most often to help make these causal inferences. I call the first modeling research. It takes simple correlational research a step further by examining co-occurrence in a multivariate framework. For example, if we wish to know whether liking an instructor causes students to get higher grades, we might construct a multiple regression equation or structural equation model that attempts to provide an exhaustive description of a network of relational linkages (Kline 1998). This model would attempt to account for, or rule out, all other co-occurring phenomena that might explain away the relationship of interest. Likely, the model will be incomplete or imperfectly specified, so any casual inferences from modeling research will be tentative, at best. Studies that compare groups of participants to one another—such as men and women or different ethnic groups—and control for other co-occurring phenomena are probably best categorized as modeling studies.

The second class involves quasi-experimental research (Shadish, Cook, and Campbell 2002). Here, the researchers (or some other external agent) control the introduction of an event, often called an intervention in applied research or a manipulation in basic research, but they do not control precisely who may be exposed to it. Instead, the researchers use some statistical control in an attempt to equate the groups receiving and not receiving the intervention. It is difficult to tell how successful the attempt at equating groups has been. For example, we might be able to train college instructors to teach classes using either a friendly or a distant instructional approach. However, we might not be able to assign students to classes so we might match students in each class on prior grades and ignore students who did not have a good match.

Finally, in experimental research both the introduction of the event (for example, instructional approach) and who is exposed to it (the students in each class) are controlled by the researcher (or other external agent). The researcher uses a random procedure to assign students to conditions, leaving the assignment to chance (Boruch 1997). In our example, because the random assignment procedure minimizes average existing differences between classes with friendly and distant instructors we can be most confident that any differences between the classes in their final grades are caused by the instructional approach of the teacher rather than other potential explanations. Of course, there are numerous other aspects of the design that must be attended to for this strong inference to be made—for example, standardizing testing conditions across instructors and ensuring that instructors are unaware of the experimental hypothesis. For our purposes, however, focusing on the characteristics of researcher control over the conditions of the experiment and assignment of participants to conditions captures what is needed to proceed with the discussion.

2.3.1.2 Examining Change Within or Across Units Some research problems make reference to changes that occur within a unit over time. Other research problems relate to the average differences in and variability of a characteristic between groups of units. This latter problem is best addressed using the designs just discussed. The former problem—the problem of change within a unit—would best be studied using the various forms of time series designs, a research design in which units are tested at different times, typically equal intervals, during the course of a study. For example, we might ask a student in a class how much he or she liked one of his or her instructors after each weekly class meeting. We might also collect the student's grade on that week's homework assignment. A concomitant time series design would then reveal the association between change in liking and homework grades.

If the research hypothesis seeks to test for a causal relationship between liking and grades, it would call for charting a student's change in grades, perhaps on homework assignments or in-class quizzes, before and after an experimental manipulation changed the way the instructor behaved, perhaps from being distant to being friendly. Or, the change from distant to friendly instruction might involve having the instructor add a different friendly teaching technique with each class session, for example, leaving time for questions during session 2, adding revealing personal information during session 3, adding telling jokes during session 4. Then, we would chart how each additional friendly teaching technique affected a particular student's grades on each session's homework assignment.

Of course, for this type of design to allow strong causal inferences other experimental controls would be needed— for example, the random withdrawal and reintroduction of teaching techniques. For our purposes however, the key point is that it is possible that how each individual student's grades change as a function of the different teaching techniques could look very different from the moving group average of class-level homework grades. For example, the best description of how instructor friendliness affected grades based on how individual students react might be to, say, each student's performance improved precipitously after the introduction of a specific friendly teaching technique that seemed to appeal to that student while other techniques produced little or no change. However, because different students might have found different approaches appealing, averaged across the students in the class, the friendliness manipulations might be causing gradual improvements in the average homework grade. At the class level of analysis the best description of the results might be that gradual introduction of additional friendly teaching techniques led to a gradual improvement in the classes' average homework grades. So, the group-averaged data can be used to explain improvement in grades only at the group level; it is not descriptive of the process happening at the individual level, nor visa versa.

It is important to note that the designation of whether an individual or group effect is of interest depends on the research question being asked. So, the group performance in this example is also a description of how one classroom might perform. Thus, the key to proper inference is that the unit of analysis at which the data from the study is being analyzed corresponds to the unit of interest in the problem or hypothesis motivating the primary research or research synthesis. If not, an erroneous conclusion can be drawn.

2.3.2 Problems and Hypotheses in Research Synthesis

2.3.2.1 Synthesis of Descriptive Research When a description of the quantitative frequency or level of an event is the focus of a research synthesis, it is possible to integrate evidence across studies by conducting what Robert Rosenthal called aggregate analysis (1991). Floyd Fowler suggests that two types of descriptive questions are most appropriately answered using quantitative approaches (2002). The first involves estimating the frequency of an event's occurrence. For example, we might want to know how many college students receive A grades in their classes. The second descriptive question involves collecting information about attitudes, opinions, or perceptions. For example, our synthesis concerning liking of instructors might be used to answer how well on average college students like their instructors.

Gathering and integrating research on the question involving frequency of grades would lead us to collect from each relevant study the number of students receiving A grades and the total number of students in each study. The question regarding opinions would lead us to collect the average response of students on a question about liking their instructor.

The primary objective of the studies this data might be collected from might not have been to answer our descriptive questions. This evidence could have been reported as part of a study examining the two-variable association between liking of the instructor and grades. However, it is not unusual for such studies to report, along with the correlation of interest, the distribution of grades in the classes under study and the average liking of the instructors. This might be done because the primary researchers wanted to examine the data for restrictions in range (Was their sufficient variation in grades and liking to permit a strong test of their association?) or to describe their sample for purposes of assessing the generality of their findings (How were the students in this study doing in school compared to the more general population of interest?).

The difficulty of aggregating descriptive statistics across studies depends on whether frequencies or attitudes, opinions, and perceptions are at issue. With regard to frequencies, as long as the event of interest is defined similarly across studies, it is simple to collect the frequencies, add them together and, typically, to report them as a proportion of all events. But there are still two complications.

First, by using the proviso "defined similarly" I am making the assumption that, for example, it is the act of getting a grade of A that interests us, not that getting an A means the same thing in every class used in every study. If instructors use different grading schemes, getting an A might mean very different things in different classes and different studies. Of course, if different studies report different frequencies of A grades, we can explore whether other features of studies (for example, the year in which the study was conducted) covary with the grading curve used in their participating classrooms.

Second, it might be that the aggregation of frequencies across studies is undertaken to estimate the frequency of an event in a population. So, our motivating question might be what the frequency is that the grade of A is assigned in American college classes. The value of our aggregate estimate from studies will then depend on how well the studies represented American college classes. It would be rare for classes chosen because they are convenient to be fortuitously representative of the nation as a whole. But, it might be possible to apply some sampling weights that would improve our approximation of a population value.

Aggregating attitudes, opinions, and perceptions across studies is even more challenging. This is because social scientists often use different scales to measure these variables, even when the variables have the same conceptual definition. Suppose we wanted to aggregate across studies the reported levels of liking of the instructor. It would not be unusual to find that some studies simply asked, "Do you like your instructor, yes or no?" while others asked, "How much do you like your instructor?" but used different scales to measure liking. Some might have used ten-point scales and others five-point or seven-point scales. Scales with the same numeric gradations might also differ in the anchors they used for the liking dimension. For example, some might have anchored the positive end of the dimension with "a lot," others with "very much," and still others with "my best instructor ever." In this case, even if all the highest numerical ratings are identical, the meaning of the response is different.

Clearly, we cannot simply average these measures. One solution would be to aggregate results only across studies using the identical scales and than we report results separately for each scale type. Another solution would be to report simply the percentage of respondents using one side of the scale or the other. For example, our cumulative statement might be "Across all studies, 85 percent of students reported positive liking for their instructor." This percentage is simply the number of students whose ratings are on the positive side of the scale, no matter what the scale, divided by the total number of raters. The number can be derived from the raw rating frequencies, if they are given, or from the mean and variance of the ratings (for an example, see Cooper et al. 2003).

It is rare to see research syntheses in social science that seek to aggregate evidence across studies to answer descriptive questions quantitatively. When this occurs, however, it is critical that the synthesists pay careful attention to whether the measures being aggregated are commensurate. Combining incommensurate measures will result in gibberish.

2.3.2.2 Synthesis of Group-Level Associations and Causal Relationships
The accumulation and integration of group-level research for purposes of testing associations and causal relationships has been by far the most frequent objective of research syntheses in social science. In primary research, a researcher selects the most appropriate research design for investigating the problem or hypothesis. It is rare for a single study to implement more than one design to answer the same research question, although there may be multiple operationalizations of the constructs of interest. More typical would be instances in which primary researchers intended to carry out an optimal design for investigating their problem but, due to unforeseen circumstances, ended up implementing a less-than-optimal design. For example, if we are interested in asking, "Does increasing the liking students have for their instructors cause students to do better in class?" we might begin by intending to carry out an experiment in which high school students are randomly assigned to friendly and distant instructors (who are also randomly assigned to teaching style). However, as the semester proceeds, some students move out of the school district, others change sections because of conflicts with other courses, and others change sections because they do not like the instructor. By the end of the semester, it is clear that the students remaining in the sections were not equivalent, on average, when the study began. To rescue the effort, we might use post hoc matching or statistical controls to

re-approximate equivalent groups. Thus, the experiment has become a quasi-experiment. If students in the two conditions at the end of the study are so different on pre-tests that procedures to produce post hoc equivalence seem inappropriate (for example, with small groups of students sampled from different tails of the class distributions), we might simply correlate the students' ratings of liking (originally meant to be used as a manipulation check in the experiment) with grades. So, a study searching for a causal mechanism has become a study of association. The legitimate inferences allowed by the study began as strong causal inference, degraded to weak causal inference, and ended as simple association.

In contrast, research synthesists are likely to come across a wide variety of research designs that relate the concepts of interest to one another. A search for studies using the term *instructor likeability* and examining them for measures of achievement should identify studies that resulted in simple correlations, multiple regressions, perhaps a few structural equation models, some quasi-experiments, a few experiments, and maybe even some time series involving particular students or class averages.

What should the synthesists do with this variety of research designs? Certainly, our treatment of them depends first and foremost on the nature of the research problem or hypothesis. If our problem is, "Is there an association between the liking students have for their instructors and grades in the course?" then it seems that any and all of the research designs address the issue. The regressions, structural equation models, quasi-experiments, and experiments ask a more specific question about association—these seek a causal connection or one with some other explanations ruled out—but they are tests of an association nonetheless. Thus, it seems that we would be well advised to include all the research designs in our synthesis of evidence on association.

The issue is more complex when our research question deals with causality: Does increasing the liking students have for their instructors cause higher student grades? Here, the different designs produce evidence with different capabilities for drawing strong inferences about our problem. Again, correlational evidence addresses a necessary but not sufficient condition for causal inference. As such, if this were the only research design found in the literature, it would be appropriate for us to assert that the question remained untested. When an association is found, multiple regressions statistically control for some alternative explanations for the relationship, but probably not all. Structural equation models relate to the plausibility of causal networks but do not address causality in the generative sense, that is, whether manipulating one variable will produce a change in the other variable. Well-conducted quasi-experiments may permit weak causal inferences, made stronger through multiple and varied replications. Experiments using random assignment of participants to groups permit the strongest inferences about causality.

How should synthesists interested in problems of causality treat the various designs? At one extreme, they can discard all studies but those using true experimental designs. This approach applies the logic that these are the only studies that directly test the question of interest. All other designs either address association only or do not permit strong inferences. The other extreme would be to include all the research evidence while carefully qualifying inferences as the evidence for causality moves farther from the ideal. A less extreme approach would be to include some but perhaps not all designs while again carefully qualifying inferences.

There are arguments for each of these approaches. The first obligation of synthesists is to clearly state the approach they have employed and the rationale for it. In research areas where strong experimental designs are relatively easy to conduct and plentiful—for example research on the impact of drug treatments on the social adjustment of children with attention disorders (Ottenbacher and Cooper 1984)—I have been persuaded that including only designs that permit relatively strong causal inferences was an appropriate approach to the evidence.

In other instances, experiments may be difficult to conduct and rare—for example the impact of homework on achievement (Cooper et al. 2006). Here, the synthesists are forced to make a choice between two philosophically different approaches to evidence. If the synthesists believe that discretion is the better part of valor, then they might opt for including only the few experimental studies or stating simply that no credible evidence on the causal relationship exists. Alternatively, if they believe that any evidence is better than no evidence at all then they might proceed to summarize the less-than-optimal studies, with the appropriate precautions, of course.

Generally speaking, when experimental evidence on causal questions is lacking or sparse I think a more inclusive approach is called for, assuming that the synthesists pay careful and continuous attention to the impact of research design on the conclusions that they draw. In fact, the inclusive approach can provide some interesting benefits to inferences. Returning to our example of instructor liking, we might find a small set of studies in which the likeability of the instructor has been manipulated and students assigned randomly to conditions. However, in order to accomplish the manipulation, these studies might have been conducted in courses that involved a series of guest lecturers, used to manipulate instructor's likeability. Grades were operationalized as scores on homework assignments turned in after each lecture. These studies might have demonstrated that more likeable guest lecturers produced higher student grades on homework. Thus, in order to carry out the manipulation it was necessary to study likeability in short-term instructor-student interactions. Could it be that over time—more similar to the real-world relationship that develops between instructors and students—likeability becomes less important and perceived competence becomes more important? Could it be that the effect of likeability is short-lived—it appears on homework grades immediately after a class but does not affect how hard a student studies for exams and therefore has much less, if any, effect on test scores and final grades?

These issues, related to construct and external validity, go unaddressed if only the experimental evidence is permitted into our synthesis of instructor likeability and homework grades. Instead, we might use the nonexperimental evidence to help us gain tentative, first approximations about how the likeability of instructors plays out over time and with broader constructions of achievement. The quasi-experiments found in the literature might use end-of-term class grades as outcome measures. The structural equation models might use large, nationally-representative samples of students and relate ratings of liking of instructors in general to broad measures of achievement, such as cumulative GPAs and SAT scores.

By employing these results to form a web of evidence, we can come to more or less confident interpretations of the experimental findings. If the non-experimental evidence reveals relationships consistent with the experiments, we can be more comfortable in suggesting that the experimental results generalize beyond the specific operations used in the experiments. If the different types of evidence are inconsistent, it should be viewed as a caution to generalization.

In sum then, it is critical in both primary research and research synthesis that the type of relationship between the variables of interested be clearly specified. This specification dictates whether primary researchers have used the appropriate research designs. Designs appropriate to gather data on one type of problem may or may not provide information relevant for investigating another type of problem. Typically, in primary research only a single research design can be employed in each study. However, in research synthesis, a variety of research designs relating the variables of interest are likely to be found. When the relation of interest concerns an association between variables, designs that seek to rule out alternative explanations or establish causal links are still relevant. When the problem of interest concerns establishing a causal link between variables, designs that seek associations or seek to rule out a specified set of alternative explanations, but not all alternative explanations, do not match the question at hand. Still, if the constraints of conducting experiments place limits on these studies' ability to establish construct and external validity, a synthesis of the nonexperimental work can give a tentative, first approximation of the robustness, and limits, of inferences from the experimental work.

2.3.2.3 Synthesis of Studies of Change in Units Across Time The study of how individual units change over time, and what causes this change, has a long history in the social sciences. Single-case research has developed its own array of research designs and data analysis strategies (for an exposition of single-case research designs, see Barlow and Hersen 1985; for a discussion of data analysis issues, Thomas Kratochwill and Joel Levin 1992). The issues I have discussed regarding descriptive, associational, and causal inferences and how they relate to different research designs play themselves out in the single-case arena in a manner similar to that in group-level research. Different time series designs are associated with different types of research problems and hypotheses. For example, the question of how a student's liking for the instructor changes over the course of a semester would be most appropriately described using a simple time series design. The question of whether a student's liking for the instructor is associated with the student's grades on homework as the semester progresses would be studied using a concomitant time series design. The question of whether a change in a student's liking for the instructor causes a change in the student's homework grade would be studied using any of a number of interrupted time series designs. Much like variations in group designs, variations in these interrupted time series would result in stronger or weaker inferences about the causal relationship of interest.

The mechanics of quantitative synthesis of single-case designs requires its own unique toolbox (for examples of some approaches, see Busk and Serlin 1992; Faith, Alison, and Gorman 1997; Scruggs, Mastropieri, and Casto 1987; Van den Noortgate and Onghena 2003; White et al. 1989). However, the logic of synthesizing single-case research is identical to that of synthesizing group-level research. It would not be unusual for synthesists to uncover a variety of time series designs relating the variables of interest to one another. If we are interested in whether a change in a student's liking of the instructor caused a change in the student's course grade we may very well find in the literature some concomitant time series, some interrupted time series with a single liking intervention, and perhaps some designs in which liking is enhanced and then lessened as the semester progresses. Appropriate inferences drawn from each of these designs correspond more or less closely with our causal question. The synthesists may choose to focus on the most correspondent designs, or to include designs that are less correspondent but that do provide some information. The decision about which of these courses of action are most appropriate again should be influenced by the number and nature of studies available. And, of course, regardless of the decision rule adopted concerning the inclusion of research designs, synthesists are obligated to carefully delimit their inferences based on the strengths and weaknesses of the included designs.

2.3.3 Problems and Hypothesis Involving Interactions

At their broadest level, the problems and hypotheses that motivate most research syntheses involve main effects, relations between two variables. Does liking of the instructor cause higher class grades? Do teachers' expectations influence intelligence test scores? This is due to the need to establish such fundamental relationships before investigating the effects of third variables on them.

Of course, the preponderance in practice of main effect questions does not mean that a focus on an interaction cannot or should not motivate a research synthesis. For example, it may be the case that a bivariate relationship is so well established that an interaction hypothesis has become the focus of attention in a field. The fact that liking the instructor causes higher grades in high school might be a finding little in dispute, perhaps because a previous research synthesis has shown it to be so. Now, the issue is whether the causal impact is equally strong across classes dealing with language or mathematics. So, we undertake a new synthesis to answer the interaction question of whether the effect of liking on grades is equally strong across different subject areas.

Also, undertaking a synthesis to investigate the existence of a main effect relationship should in no way diminish the attention paid to interactions within the same project. Indeed, in my earlier example the fact that the previous research synthesis on liking of the instructor causing higher grades in high school did not look at the moderating influence of subject matter might be considered a shortcoming of that work. Typically, when main effect relationships are found to be moderated by third variables, these findings are given inferential priority and are viewed as a step toward understanding the processes involved in the causal chain of events.

Examining interactions in research syntheses presents some unique opportunities and unique problems. To fully understand both, I must introduce yet another critical distinction in the types of evidence that can be explored when research is gathered and integrated.

2.4 STUDY-GENERATED AND SYNTHESIS-GENERATED EVIDENCE

Research synthesis can contain two sources of evidence about the research problem or hypothesis. The first type is called study-generated evidence. Study-generated evidence is present when a single study contains results that directly test the relation being considered. Research syntheses also contain evidence that does not come from individual studies but rather from the variations in procedures across studies. Such evidence, called synthesis-generated, is present when the results of studies using different procedures to test the same hypothesis are compared to one another.

Any research problem or hypothesis—a description, simple association, or causal link—can be examined through either study-generated or synthesis-generated evidence. However, only study-generated evidence based on experimental research allows us to make statements concerning causality.

2.4.1 Study-Generated Evidence, Two-Variable Relationships

Suppose we were interested in gathering and integrating the research on whether teacher expectations influence IQ. We conduct a literature search and discover twenty studies that randomly assigned students to one of two conditions. Teachers for one group of students were given information indicating they could expect unusual intellectual growth during the school year from the students in their class. The other teachers were given no out-of-the-ordinary information about any students. These twenty studies each provide study-generated evidence regarding a simple two-variable relationship, or main effect. The cumulative results of these studies comparing the end-of-year IQ scores of students in the two groups of classes could then be interpreted as supporting or not supporting the hypothesis that teacher expectations influence IQ.

2.4.2 Study-Generated Evidence, Three-Variable Interactions

Next, assume we are interested as well in whether teacher expectations manipulated through the presentation of a bogus Test of Intellectual Growth Potential have more of an impact on student IQs than the same information manipulated by telling teachers simply that students' past teachers predicted unusual intellectual growth based on subjective impressions. We discover ten of the twenty studies used both types of expectation manipulation and randomly assigned teachers to each. These studies provide study-generated evidence regarding an interaction. Here, if we found that test scores produced a stronger expectation-IQ link than past teachers' impressions, we could conclude that the mode of expectation induction caused the difference. Of course, the same logic would apply to attempts to study higher order interactions—involving more than one interacting variable—although these are still rare in research syntheses.

2.4.2.1 Integrating Interaction Results Across Studies Often, the integration of interaction results in research synthesis is not as simple as combining significance levels or calculating and averaging effect sizes from each study. Figure 2.1 illustrates the problem by presenting the results of two hypothetical studies comparing the effects of manipulated (high) teacher expectations on student IQ that also examined whether the effect was moderated by the number of students in the class. In study 1, the change in student IQ was compared across a set of classrooms that ranged in size from ten to twenty-eight. In study 2 the range of class sizes was from ten to twenty. Study 1 might have produced a significant main effect and a significant interaction involving class size but study 2 might have reported a significant main effect only.

If we uncovered these two studies we might be tempted to conclude that they produced inconsistent results. However, a closer examination of the two figures illustrates why this might not be an appropriate interpretation. The

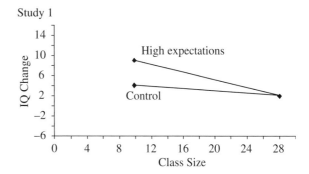

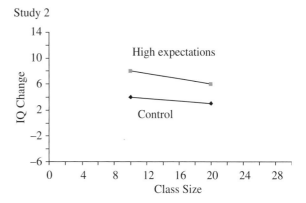

Figure 2.1 Effects of Teacher Expectations on IQ

SOURCE: Author's compilation.

results of study 2 probably would have closely approximated those of study 1 had the class sizes in study 2 represented the same range of values as those in study 1. Note that the slopes for the high expectations groups in study 1 and study 2 are nearly identical, as are those for the control groups.

This example demonstrates that synthesists should not assume that strengths of interaction uncovered by different studies necessarily imply inconsistent results. Synthesists need to examine the differing ranges of values of the variables employed in different studies, be they measured or manipulated. If possible, they should chart results taking the different levels into account, just as I did in figure 2.1. In this manner, one of the benefits of research synthesis is realized. Although one study might conclude that the effect of expectations dissipates as class size grows larger and a second study might not, the research synthesists can discover that the two results are in fact perfectly commensurate.

This benefit of gathering and integrating research also highlights the importance of primary researchers presenting detailed information concerning the levels of variables used in their studies. Research synthesists cannot conduct an across-study analysis similar to this example without the needed information. If the primary researchers in study 1 and study 2 neglected to specify their range of class sizes, perhaps because they simply said they compared smaller classes with larger classes, the commensurability of the results would have been impossible to demonstrate.

Variations in ranges of values for variables also can produce discrepancies in results involving two-variable, or main effect, relationships. This is the circumstance under which the problem is least likely to be recognized and is most difficult to remedy when it is discovered.

2.4.3 Synthesis-Generated Evidence, Two-Variable Relationships

Earlier I provided an example of a two-variable relationship studied with synthesis-generated evidence when I suggested we might relate the average ratings of liking students gave their instructors to the year in which the study was conducted. We might do this to obtain an indication of whether liking of instructors was increasing or decreasing over time. Here, we have taken two descriptive statistics from the study reports and related them to one another. This is synthesis-generated evidence for a two-variable relationship, or main effect.

It should be clear that only associations can be studied in this way, since we have not randomly assigned studies to liking conditions. It should also be clear that looking for such relationships between study-level characteristics opens up the possibility of examining a multitude of problems and hypotheses that might never have been the focus of individual primary studies. For example, even if no previous study has examined whether liking for instructors has gone up or down over the years, we can look at this relation in our synthesis.

The use of synthesis-generated evidence to study main effect relationships is rarely the primary focus of research synthesis, although there is no pragmatic reason why this should be the case. Study characteristics—that is, evidence at the synthesis level—most often come into play as variables that influence the magnitude of two-variable relationships examined within studies. Conceptually, the difference is simply that in this case—the interaction case—one of the two study characteristics being examined already is some expression of a two-variable relationship.

2.4.4 Synthesis-Generated Evidence, Three-Variable Interactions

Let's return to our teacher expectations and IQ example. Suppose we find no studies that manipulated the mode of expectation induction but we do discover that ten of the studies that experimentally manipulated expectations did so using test scores and that ten used the impressions of previous teachers. When we compare the magnitude of the expectation effect on IQ between the two sets of studies, we discover that the link is stronger in studies using test manipulations. We could then infer that an association exists between mode-of-expectation-induction and IQ but we could not infer a causal relation between the two. This is synthesis-generated evidence for an interaction.

When groups of effect sizes are compared, regardless of whether they come from simple correlational analyses or controlled experiments using random assignment, the synthesists can only establish an association between a moderator variable—a characteristic of the studies—and the outcomes of studies. They cannot establish a causal connection. Synthesis-generated evidence is restricted to making claims only about associations and not about causal relationships because it is the ability to employ random assignment of participants that allows primary researchers to assume third variables are represented equally in the experimental conditions.

The possibility of unequal representation of third variables across study characteristics cannot be eliminated because we did not randomly assign experiments to modes of expectation manipulation. For example, it might be that the set of studies using tests to manipulate IQ were also conducted in schools with smaller class sizes. Now, we do not know whether it was the use of tests or the size of the classes that caused the difference in the relationship between teacher expectations and IQ. The synthesists cannot discern which characteristic of the studies, or perhaps some unknown other variable related to both, produced the stronger link. Thus, when study characteristics are found associated with study outcomes the synthesists should report the finding as just that, an association, regardless of whether the included studies tested the causal effects of a manipulated variable or estimated the size of an association.

Synthesis-generated evidence cannot legitimately be used to rule out as possible true causes other variables that are confounded with the study characteristic of interest. Thus, when synthesis-generated evidence reveals a relationship that would be of special interest if it were causal, the report should include a recommendation that future research examine this factor using a more systematically controlled design, so that its causal impact can be appraised. In our example, we would call for a primary study to experimentally manipulate both the mode of expectation induction and the size of classes.

2.4.5 High and Low Inference Codes

The examples of study characteristics I have used this far might all be thought of as low inference codes. Variables such as class size or the type of IQ measure require the coders only to locate the needed information in the research report and transfer it to the database. In some circumstances, synthesists might want to make more inferential judgments about study operations. These high inference codes typically involve attempting to infer how a contextual aspect of the studies might have been interpreted by participants.

For example, I used the mode-of-manipulation as a study-level variable that might have been examined as a potential moderator of the link between teacher expectations and change in student IQ. I used bogus tests and previous teachers' impressions as the two modes of manipulation. Most manipulations of these types could be classified into the two categories with relatively little inference on the part of the coders. However, suppose that we found in the literature a dozen different types of expectation manipulations, some involving long tests taken in class, others short tests taken after school, some the impressions of past teachers, and others the clinical judgments of child psychologists. It might still be relatively simple to categorize these manipulations at an operational level but the use of the multiple categories for testing meaningful hypotheses at the synthesis-level becomes more problematic. Now, we might reason that the underlying conceptual difference between study operations that interests us is the credibility of the manipulation. So, in order to ask the question, "Does the magnitude of the teacher expectation-student IQ link vary with the credibility of the expectation manipulation?" we might want to examine each study report and score the manipulations on this dimension. This high inference code than becomes a study characteristic we relate to study outcomes. Here, we are dimensionalizing the manipulations by making an inference about how much credibility the participating teachers might have placed in the information given to them.

High inference codes create a special set of problems. First, careful attention must be paid to the reliability of inference judgments. So, it is important to show that these judgments are being made consistently both across and within the inferers making the judgments. Also, inferers are being asked to play the role of a research participant—in our example the teachers being given expectation information—and the validity of role-playing methodologies has been the source of much controversy (Greenberg and Folger 1988). However, Norman Miller, Ju-Young Lee, and Michael Carlson empirically demonstrated that high inference codes can lead to valid judgments (they did this by comparing inferences to manipulation checks) and can add a new dimension to synthesists' ability to interpret literatures and resolve controversies (1991). If high inference information can be validly extracted from articles and the benefit of doing so is clear, than this can be an important technique for exploring problems and hypotheses in research synthesis.

2.4.6 Value of Synthesis-Generated Evidence

In sum then, it is critical that synthesists keep in mind the distinction between study-generated and synthesis-generated evidence. Only evidence coming from experimental manipulations within a single study can support assertions concerning causality. However, I do not want my attention to the fact that synthesis-generated evidence cannot support causal inferences to suggest that this source of evidence should be ignored. As our examples suggest, the use of synthesis-generated evidence allows us to test relations which may have never been examined by primary researchers. In fact, it is typically the case that synthesists can examine many more potential moderators of study outcomes than have appeared as interacting variables in primary research. Often, these study-level variables are of theoretical importance. For example, it is easy to see how our mode-of-expectation-induction variable relates to the credibility of the information given to teachers and how class size relates to the teachers' opportunities to communicate their expectations to students. By searching across studies for variations in the operationalizations of theoretically relevant construct, synthesists can produce the first evidence on these potentially critical moderating variables. Even though this evidence is equivocal, it is a major benefit of research synthesis and a source of potential hypotheses (and motivation) for future primary research.

2.5 CONCLUSION

A good way to summarize what I have tried to convey in this chapter is to recast its major points by framing them in the context of how they impact the validity of the conclusions drawn in research syntheses. The most central decisions made during problem and hypothesis formulation concern the fit between the concepts used in variable definitions and the operational definitions found in the literature and the correspondence between the inferences about relationships permitted by the evidence in the studies at hand and those drawn by the synthesists.

First, research synthesists need to clearly and carefully define the conceptual variables of interest. Unlike primary researchers, synthesists need not complete this task before the search of the literature begins. Some flexibility permits them to discover operations unknown to them before the literature search began. It also permits them to expand or contract their conceptual definitions so that, in the end, the conceptual definitions appropriately encompass operations present in the literature. If relatively broad concepts accompany claims about the generality of findings not warranted by the operationalizations in hand then an inferential error may be made. Likewise, relatively narrow concepts accompanied by marginally related operations also may lead to inferential errors. Research synthesists need to expand or contract conceptual definitions and/or retain or eliminate operations so that the highest correspondence between them has been achieved.

When some discretion exists regarding whether to expand or contract definitions, the first decision rule should be that the results will be meaningful to the target audience. The second decision rule is pragmatic: Will expanding the definitions make the synthesis so large and unwieldy that the effort will outstrip resources? If both of these concerns are not negligible, sometimes synthesists will opt to use narrow definitions with only a few operations in order to define their concepts so as to ensure consensus about how their concepts are related to observable events. However, there is reason to favor using broader constructs with multiple realizations, as numerous rival interpretations for the findings may be tested and at least tentatively ruled out if the multiple operations produce similar results. Also, narrow concepts provide little information about the generality or robustness of the results. Therefore, the greater the conceptual breadth of the definitions, the greater is its *potential* to produce conclusions that are more general than when narrow definitions are used.

The word potential was highlighted because if synthesists only cursorily detail study operations, their conclusions may mask important moderators of results. An erroneous conclusion—that research results indicate negligible differences in outcomes across studies—can occur if different results across studies are masked in the use of very broad categories.

So, holding constant the needs of the audience, and assuming adequate resources are available to the synthesists, the availability of ample studies to support a fit between either narrow or broad conceptual definitions suggests it is most desirable for syntheses to employ the broadest possible conceptual definition. To test this possibility, they should begin their literature search with a few central operations but remain open to the possibility that other relevant operations will be discovered in the literature. When operations of questionable relevance are encountered, the synthesist should err toward making overly inclusive decisions, at least in the early stages of their project. However, to complement this conceptual broadness, synthesists should be thorough in their attention to the distinctions in study characteristics. Any suggestion that a difference in study results is associated with a distinction in study characteristics should be tested using moderator analyses.

The second set of validity issues introduced during problem and hypothesis formulation concerns the nature of the relationship between the variables under study. Only study-generated evidence can be used to support causal relationships. When the relationship at issue is a causal one, synthesists must take care to categorize research designs according to their strength of causal inference. They might then exclude all but the most correspondent study-generated evidence (that is to say, experiments using random assignment). Or, they might separately examine less correspondent study-generated evidence and use this to obtain first approximations of the construct and external validity of a finding.

Also, synthesis-generated evidence cannot be used to make causal claims. Thus, synthesists should be careful to distinguish study-generated evidence from synthesis-generated evidence. Of course, this does not mean that synthesis-generated evidence is of little or no use. In fact, synthesis-generated evidence is new evidence, often exploring problems and hypotheses unexplored in primary research. As such, it is the first, among many, unique contributions that research synthesis makes to the advance of knowledge.

2.6 NOTES

1. I use the term *unit* here to refer to individual people, groups, classrooms, or whatever entity is the focus of study.

2.7 REFERENCES

Barlow, David H., and Michel Hersen. 1985. *Single Case Research: Design and Methods.* Columbus, Oh.: Merrill.

Boruch, Robert F. 1997. *Randomized Experiments for Planning and Evaluation.* Thousand Oaks, Calif.: Sage Publications.

Busk, Patricia L., and Ronald C. Serlin. 1992. "Meta-Analysis for Single-Case Research." In *Single Case Research Design and Analysis: New Directions for Psychology and Education*, edited by Thomas R. Kratochwill and Joel R. Levin. Hillsdale, N.J.: Lawrence Erlbaum.

Cherulnik, Paul D. 2001. *Methods for Behavioral Research: A Systematic Approach.* Thousand Oaks, Calif.: Sage Publications.

Cooper, Harris. 2006. "Research Questions and Research Designs." In *Handbook of Research in Educational Psychology*, 2nd ed., edited by Patricia A. Alexander, Philip H. Winne, and Gary Phye. Mahwah, N.J.: Lawrence Erlbaum.

Cooper, Harris, Jeffrey C. Valentine, Kelly Charlton, and April Barnett. 2003. "The Effects of Modified School Calendars on Student Achievement and School Community Attitudes: A Research Synthesis." *Review of Educational Research* 73: 1–52.

Coser, Lewis. 1971. *Masters of Sociological Thought.* New York: Harcourt Brace.

Faith, Myles S., David B. Allison, and Bernard S. Gorman. 1997. "Meta-Analysis of Single-Case Research." In *Design and analysis of single-case research*, edited by Ronald D. Franklin, David B. Allison, and Bernard S. Gorman. Mahwah, N.J.: Lawrence Erlbaum.

Fowler, Jr., Floyd J. 2002. *Survey Research Methods*, 3rd ed. Thousand Oaks, Calif.: Sage Publications.

Greenberg, Jerald, and Robert Folger. 1988. *Controversial Issues in Social Research Methods.* New York: Springer-Verlag.

Kerlinger, Fred N., and Howard B. Lee. 1999. *Foundations of Behavioral Research*, 4th ed. New York: Holt, Rinehart & Winston.

Kline, Rex B. 1998. *Principles and Practices of Structural Equation Modeling.* New York: Guilford Press.

Kratochwill, Thomas R., and Joel R. Levin. 1992. *Single Case Research Design and Analysis: New Directions for Psychology and Education.* Hillsdale, N.J.: Lawrence Erlbaum.

Mannheim, Karl. 1936. *Ideology and Utopia.* London: Routledge & Kegan Paul.

Miller, Norman, Ju-Young Lee, and Michael Carlson. 1991. "The Validity of Inferential Judgments When Used in Theory-Testing Meta-Analysis." *Personality and Social Psychology Bulletin* 17: 335–43.

Noblit, George W., and R. Dwight Hare. 1988. *Meta-Ethnography: Synthesizing Qualitative Studies.* Newbury Park, Calif.: Sage Publications.

Ottenbacher, Kenneth, and Harris Cooper. 1984. "The Effect of Class Placement on the Social Adjustment of Mentally Retarded Children." *Journal of Research and Development in Education* 17(2): 1–14.

Paterson, Barbara L., Sally E. Thorne, Connie Canam, and Carol Jillings. 2001. *Meta-Study of Qualitative Health Research.* Thousand Oaks, Calif.: Sage Publications.

Random House. 2001. *The Random House Dictionary of the English Language, the Unabridged Edition.* New York: Random House.

Rosenthal, Robert. 1991. *Meta-Analytic Procedures for Social Research.* Newbury Park, Calif.: Sage Publications.

Scruggs, Thomas E., Margo A. Mastropieri, and Glendon Casto. 1987. "The Quantitative Synthesis of Single-Subject Research: Methodology and Validation." *Remedial and Special Education* 8: 24–33.

Shadish, William. R., Thomas D. Cook, and Donald T. Campbell. 2002. *Experimental and Quasi-Experimental Designs for Generalized Causal Inference.* Boston, Mass.: Houghton Mifflin.

Shoemaker, Pamela J., James William Tankard Jr., and Dominic L. Lasorsa. 2004. *How to Build Social Science Theories.* Thousand, Oaks, Calif.: Sage Publications.

Van den Noortgate, Wim, and Patrick Onghena. 2003. "Combining Single-Case Experimental Data Using Hierarchical Linear Models." *School Psychology Quarterly* 18: 325–46.

Webb, Eugene J., Donald T. Campbell, Richard D. Schwartz, and Lee Sechrest. 2000. *Unobtrusive Measures.* Thousand Oaks, Calif.: Sage Publications.

White, David M., Frank R. Rusch, Alan E. Kazdin, and Donald P. Hartmann. 1989. "Application of Meta-Analysis in Individual-Subject Research." *Behavioral Assessment* 11:2 81–96.

3

STATISTICAL CONSIDERATIONS

LARRY V. HEDGES
Northestern University

CONTENTS

3.1 INTRODUCTION

Research synthesis is an empirical process. As with any empirical research, statistical considerations have an influence at many points in the process. Some of these, such as how to estimate a particular effect parameter or establish its sampling uncertainty, are narrowly matters of statistical practice. They are considered in detail in subsequent chapters of this handbook. Other issues are more conceptual and might best be considered statistical considerations that impinge on general matters of research strategy or interpretation. This chapter addresses selected issues related to interpretation.

3.2 PROBLEM FORMULATION

The formulation of the research synthesis problem has important implications for the statistical methods that may be appropriate and for the interpretation of results. Careful consideration of the questions to be addressed in the synthesis will also have implications for data collection, data evaluation, and presentation of results. In this section I discuss two broad considerations in problem formulation: universe to which generalizations are referred and the number and source of hypotheses addressed.

3.2.1 Universe of Generalization

A central aspect of statistical inference is making a generalization from an observed sample to a larger population or universe of generalization. Statistical methods are chosen to facilitate valid inferences to the universe of generalization. There has been a great deal of confusion about the choice between fixed- and random-effects statistical methods in meta-analysis. Although it is not always conceived in terms of inference models, the universe of generalization is the central conceptual feature that determines the difference between these methods.

The choice between fixed- and random-effects procedures has sometimes been framed as entirely a question of homogeneity of the effect-size parameters. That is if all of the studies estimate a common effect-size parameter, then fixed-effects analyses are appropriate. However if there is evidence of heterogeneity among the population effects estimated by the various studies, or heterogeneity that remains after conditioning on covariates, then random-effects procedures should be used. Although fixed- and random-effects analyses give similar answers when there is in fact a common population effect size, the underlying inference models remain distinct.

I argue that the most important issue in determining statistical procedure should be the nature of the inference desired, in particular the universe to which one wishes to generalize. If the analyst wishes to make inferences only about the effect-size parameters in the set of studies that are observed (or to a set of studies identical to the observed studies except for uncertainty associated with the sampling of subjects into those studies), this is what I will call a *conditional* inference. One might say that conditional inferences about the observed effect sizes are intended to be robust to the consequences of sampling error associated with sampling of subjects (from the same populations) into studies. Strictly, conditional inferences apply to this collection of studies and say nothing about other studies that may be done later, could have been done earlier, or may have already been done but are not included among the observed studies. Fixed-effects statistical procedures can be appropriate for making conditional inferences.

In contrast, the analyst may wish to make a different kind of inference, one that embodies an explicit generalization beyond the observed studies. In this case the studies that are observed are not the only studies that might be of interest. Indeed one might say that the studies observed are of interest only because they reveal something about a putative population of studies that is the real object of inference. If the analyst wishes to make inferences about the parameters of a population of studies that is larger than the set of observed studies and that may not be strictly identical to them, I call this an unconditional inference. Random-effects (or mixed-effects) analysis procedures are designed to facilitate unconditional inferences.

3.2.1.1 Conditional Inference Model In the conditional inference model, the universe to which generalizations are made consists of ensembles of studies identical to those in the study sample except for the particular people (or primary sampling units) that appear in the studies. Thus the studies in the universe differ from those in the study sample only as a result of sampling of people into the groups of the studies. The only source of sampling error or uncertainty is therefore the variation resulting from the sampling of people into studies.

In a strict sense, the universe is structured—it is a collection of identical ensembles of studies, each study in an ensemble corresponding to a particular study in each of the other ensembles. Each of the corresponding studies would have exactly the same effect-size parameter (population effect size). In fact, part of the definition of identical (in the requirement that corresponding studies in different ensembles of this universe be identical) is that they have the same effect-size parameter.

The model is called conditional because it can be conceived as a model that holds fixed, or conditions on, the characteristics of studies that might be related to the effect-size parameter. The conditional model in research synthesis is in the same spirit as the usual regression model and fixed-effects analysis of variance in primary research. In the case of regression, the fixed effect refers to the fact that the values of the predictor variables are taken to be fixed (not randomly sampled). The only source of variation that enters into the uncertainty of estimates of the regression coefficients or tests of hypotheses is that due to the sampling of individuals (in particular, their outcome variable scores) with a given ensemble of predictor variable scores. Uncertainty due to the sampling of predictor variable scores themselves is not taken into account. To put it another way, the regression model is conditional on the particular ensemble of values of the predictor variables in the sample.

The situation is similar in the fixed-effects analysis of variance. Here the term *fixed effects* refers to the fact that the treatment levels in the experiment are considered fixed, and the only source of uncertainty in the tests for treatment effects is a consequence of sampling of a group of individuals (or their outcome scores) within a given ensemble of treatment levels. Uncertainty attributable to the sampling of treatment levels from a collection of possible treatment levels is not taken into account.

Inference to Other Cases. In conditional models, (including regression and analysis of variance (ANOVA) inferences, are, in the strictest sense, limited to cases in which the ensemble of values of the predictor variables are represented in the sample. Of course, conditional models are widely used in primary research and the generalizations supported typically are not constrained to predictor values in the sample. For example, generalizations about treatment effects in fixed-effects ANOVA are usually not constrained to apply only to the precise levels of treatment found in the experiment, but are viewed as applying to similar treatments as well even if they were not explicitly part of the experiment. How are such inferences justified? Typically they are justified on the basis of an a priori (extra-empirical) decision that other levels (other treatments or ensembles of predictor values) are sufficiently like those in the sample that their behavior will be identical. There are two variations of the

argument. One is that a level not in the sample is sufficiently like a level or levels in the sample that it is essentially judged identical to them (for example, this instance of behavior modification treatment is essentially the same as the treatment labeled behavior modification in the sample, or this treatment of twenty weeks is essentially identical to the treatment of twenty weeks in the sample). The other variation of the argument is that a level that is not in the sample lies between values in the sample on some implicit or explicit dimension and thus it is safe to interpolate between the results obtained for levels in the sample. For example, suppose we have the predictor values ten and twenty in the sample, then a new case with predictor value fifteen might reasonably have an outcome half way between that for the two sampled values, or a new treatment might be judged between two others in intensity and therefore its outcome might reasonably be assumed to be between that of the other two. This interpolation between realized values is sometimes formalized as a modeling assumption (for example, in linear regression, where the assumption of a linear equation justifies the interpolation or even extrapolation to other data provided they are sufficiently similar to fit the same linear model). In either case the generalization to levels not present in the sample requires an assumption that the levels are similar to those in the sample—one not justified by a formal sampling argument.

Inference to studies not identical to these in the sample can be justified in meta-analysis by the same intellectual devices used to justify the corresponding inferences in primary research. Specifically, inferences may be justified if the studies are judged a priori to be similar enough to those in the study sample. It is important to recognize, however, that the inference process has two distinct parts. One part is the generalization from the study sample to a universe of identical studies, which is supported by a sampling theory rationale. The second part is the generalization from the universe of studies that are identical to the sample to a universe of similar enough (but not identical) studies. This second part is not strictly supported by a sampling argument but by an extra-statistical one.

3.2.1.2 Unconditional Inference Model In the unconditional model, the study sample is presumed to be literally a sample from a hypothetical collection (or population) of studies. The universe to which generalizations are made consists of a population of studies from which the study sample is drawn. Studies in this universe differ from those in the study sample along two dimensions. First the studies differ from one another in study characteristics and in effect-size parameter. The generalization is not, as it was in the fixed-effects case, to a universe consisting of ensembles of studies with corresponding members of the ensembles having identical characteristics and effect-size parameters. Instead the studies in the study sample (and their effect-size parameters) differ from those in the universe by as much as might be expected as a consequence of drawing a sample from a population. Second, in addition to differences in study characteristics and effect-size parameters, the studies in the study sample also differ from those in the universe as a consequence of the sampling of people into the groups of the study. This results in variation of observed effect sizes about their respective effect-size parameters.

I can conceive of the two dimensions indicated above as introducing two sources of variability into the observed (sample) effect sizes in the universe. One is due to variation in observed (or potentially observable) study effect sizes about their effect-size parameters. This source of variability is a result of the sampling of people into studies and is the only source of variability conceived as random in the conditional model. The other is one due to variation in effect-size parameters across studies.

This model is called the unconditional model because, unlike the conditional model, it does not condition (or hold fixed) characteristics of studies that might be related to the effect-size parameter. The random-effects model in research synthesis is in the same spirit as the correlation model or the random-effects analysis of variance in primary research. In the correlation model both the values of predictor variable and those of the dependent variable are considered to be sampled from a population—in this case one with a joint distribution. In the random-effects analysis of variance, the levels of the treatment factor are sampled from a universe of possible treatment levels (and consequently, the corresponding treatment effect parameters are sampled from a universe of treatment effect parameters). There are two sources of uncertainty in estimates and tests in random-effects analyses. One attributable to sampling of the treatment effect parameters themselves and the other to the sampling of individuals (in particular, outcome scores) into each treatment.

Inference to Other Cases. In the unconditional model, inferences are *not* limited to cases with predictor variables represented in the sample. Instead the inferences, for example, about the mean or variance of an effect-size parameter, apply to the universe of studies from which the study sample was obtained. In effect, the warrant for generalization to other studies is via a classical sampling argument. Because the universe contains studies that differ in their characteristics, and those differences find their way into the study sample by the process of random selection, generalizations to the universe pertain to studies that are not identical to those in the study sample.

By using a sampling model of generalization, the random-effects model seems to avoid subjective difficulties that plagued the fixed-effects model in generalizations to studies not identical to the study sample. That is, we do not have to ask, "How similar is similar enough?" Instead we substitute another question: "Is this new study part of the universe from which the study sample was obtained?" If study samples were obtained from well defined sampling frames via overtly specified sampling schemes, this might be an easy question to answer. This however is virtually never the case in meta-analysis (and is unusual in other applications of random-effects models). The universe is usually rather ambiguously specified and consequently the ambiguity in generalization based on random-effects models is that it is difficult to know precisely what the universe is. In contrast, the universe is clear in fixed-effects models, but the ambiguity arises in deciding if a new study might be similar enough to the studies already contained in the study sample.

The random-effects model does provide the technical means to address an important problem that is not handled in the fixed-effects model, namely the additional uncertainty introduced by the inference to studies that are not identical (except for the sample of people involved) to those in the study sample. Inference to (nonsampled) studies in the fixed-effects model occurs outside of the technical framework and hence any uncertainty it contributes cannot be evaluated by technical means within the model. In contrast, the random-effects model does incorporate between-study variation into the sampling uncertainty used to compute tests and estimates.

Although the random-effects model has the advantage of incorporating inferences to a universe of studies exhibiting variation in their characteristics, the definition of the universe may be ambiguous. A tautological universe definition could be derived by using the sample of studies to define the universe as a universe from which the study sample is representative. Such a population definition remains ambiguous; moreover, it may not be the universe definition desired for the use of the information produced by the synthesis. For example, if the study sample includes many studies of short duration, high intensity treatments, but the likely practical applications usually involve low-intensity, long duration treatments, the universe defined implicitly by the study sample may not be the universe most relevant to applications.

One potential solution to this problem might be to explicitly define a structured universe in terms of study characteristics, and to consider the study sample as a stratified sample from this universe. Estimates of parameters describing this universe could be obtained by weighting each stratum appropriately. For example, if one-half of the studies in the universe are long-duration studies, but only one-third of the study sample have this characteristic, the results of each long-duration study must be weighted twice as much as the short-duration studies.

3.2.1.3 Fixed Versus Random Effects The choice between fixed (conditional) or random (unconditional) modeling strategies arises in many settings in statistics and has caused lengthy debates because it involves subtleties of how to formulate questions in scientific research and what data are relevant to answering questions. For example, the debate between R. A. Fisher and Yates versus Pearson on whether to condition on the marginal frequencies in the analysis of 2×2 tables is about precisely this issue (see Camilli 1990), as is the debate concerning whether word stimuli should be treated as fixed or random effects in psycho-linguistics (see Clark 1972).

Those who advocated the fixed-effects position argue that the basis for scientific inference should be only the studies that have actually been conducted and are observed in the study sample (for example, Peto 1987). Statistical methods should be employed only to determine the chance consequences of sampling of people into these (the observed) studies. Thus they would emphasize estimation and hypothesis testing for (or conditional on) this collection of studies. If we must generalize to other studies, they would argue that this is best done by subjective or extra-statistical procedures.

Those who advocate the random-effects perspective argue that the particular studies we observe are, to some extent, an accident of chance. The important inference question is not what is true about these studies, but what is true about studies like these that could have been done. They would emphasize the generalization to other studies or other situations that could have been studied and that these generalizations should be handled by formal statistical methods. In many situations where research is used to inform public policy by providing information about the likely effects of treatments in situations that have not been explicitly studied, this argument seems persuasive.

3.2.1.4 Using Empirical Heterogeneity Although most statisticians would argue that the choice of analysis procedure should be driven by the inference model, some researchers make that choice based on the outcome of a statistical test of heterogeneity of effect sizes. Larry Hedges and Jack Vevea, who studied the properties of such tests for both conditional and unconditional inferences, called this a conditionally random-effects analysis (1998). They found that the type I error rate of conditionally random-effects analyses were in between those of fixed- and random-effects tests, being slightly inferior to fixed-effects tests for conditional inferences (but better than random-effects tests) and slightly inferior to random-effects tests for unconditional inferences (but better than fixed-effects tests). Whatever the technical performance of conditionally random-effects tests, they have the disadvantage that they permit the user to avoid a clear choice of inference population, or worse to allow the data to (implicitly) make that determination depending on the outcome of a statistical test of heterogeneity.

3.2.2 Theoretical Versus Operational Parameters

Another fundamental issue in problem formulation concerns the nature of the effect-size parameter to be estimated. The issue can best be described in terms of population parameters (though each parameter has a corresponding sample estimate). Consider an actual study in which the effect-size parameter represents the true or population relationship between variables measured in the study. This effect size-parameter may be systematically affected by artifactual sources of bias such as restriction of range or measurement error in the dependent variable. Corresponding to this actual study, one can imagine a hypothetical study in which the biases due to artifacts were controlled or eliminated. The effect size-parameter in this hypothetical study would differ from that of the actual study because the biases from artifacts of design would not be present. A key distinction is between a theoretical effect size (reflecting a relationship between variables in a hypothetical study) and an operational effect size (the parameter that describes the population relationship between variables in an actual study) (see chapter 18, this volume). Theoretical effect sizes are often conceived as those corrected for bias or for some aspect of experimental procedure (such as a restrictive sampling plan or use of an unreliable outcome measure) that can systematically influence effect size. Operational effect-size parameters, in contrast, are often conceived as affected by whatever bias or aspects of procedure that happen to be present in a particular study.

Perhaps the most prominent example of a theoretical effect size is the population correlation coefficient corrected for attenuation due to measurement error and restriction of range. One can also conceive of this corrected coefficient as the population correlation between true scores in a population where neither variable is subject to restriction of range. The operational effect size is the correlation parameter between observed scores in the population in which variables have restricted ranges. Because the relation between the attenuated (operational) correlation and disattenuated (theoretical) correlation is known, it is possible to convert operational effect sizes into theoretical effect sizes.

Most research syntheses make use of operational effect size. However there are two reasons why theoretical effect sizes are sometimes used. One is to enhance the comparability and hence combinability of estimates from studies whose operational effect sizes would otherwise be influenced quite substantially (and differently) by biases or incidental features of study design or procedure. This has sometimes been characterized as "putting all of the effect sizes on the same metric" (Glass, McGaw, and Smith 1981, 116). For example, in research on personal selection, virtually all studies involve restriction of range which attenuates correlations (see Hunter and Schmidt 2004). Moreover, the amount of restriction of range typically varies substantially across studies. Hence correction for restriction of range ensures that each study provides an estimate of the same kind of correlation—the correlation in a population having an unrestricted distribution of test scores. Because restriction of range and many other consequences of design are incidental features of the studies, disattenuation to remove their effects is sometimes called *artifact correction.*

A more controversial reason that theoretical effect sizes are sometimes used is because the theoretical effect sizes are considered more scientifically relevant. For example, to estimate the benefit of scientific personnel selection using cognitive tests versus selection on an effectively random basis, we would need to compare the performance of applicants with the full range of test scores (those selected at random) with that of applicants selected via the test—applicants who would have a restricted range of test scores. Although a study of the validity of a selection test would compute the correlation between test score and job performance based on the restricted sample, the correlation that reflects the effectiveness of the test in predicting job performance is the correlation that would have been obtained with the full range of test scores—a

theoretical correlation. Another example might be the estimation of the standardized mean difference of a treatment intended for a general population of people, but which has typically been investigated with studies using more restricted groups of people. Since the scores in the individual studies have a smaller standard deviation than the general population, the effect sizes will be artifactually large—that is if the treatment produced the same change in raw score units, dividing by the smaller standard deviations in the study sample would make the standardized difference look artifactually large. Hence a theoretical effect size might be chosen—the effect size that would have been obtained if the outcome scores in each study had the same variation as the general population. This would lead to corrections for sampling variability. Corrections of this sort are discussed in chapter 17.

3.2.3 Number and Source of Hypotheses

Richard Light and David Pillemer distinguished between two types of questions that might be asked in a research synthesis (1984). One kind of question concerns a hypothesis that is specified precisely in advance, for example, whether on average the treatment works. The other type of question is specified only vaguely (for example, "under what conditions does the treatment work best"). This distinction in problem specification is similar to that between planned and post hoc comparisons in the analysis of variance which is familiar to many researchers. Although either kind of question is legitimate in research synthesis, the calculation of levels of statistical significance may be affected by whether the question was defined in advance or discovered post hoc by examination of the data.

Although it is useful to distinguish the two cases, sharp distinctions are not possible in practice. Some of the research literature is surely known to the reviewer prior to the synthesis, thus putatively a priori hypotheses are likely to have been influenced by the data. Conversely, hypotheses derived during exploration of the data may have been conceived earlier and proposed for explicit testing because the examination of the data suggested that they might be fruitful given the data. Perhaps the greatest ambiguity arises when a very large number of hypotheses are proposed a priori for testing. In this case it is difficult to distinguish between hypotheses selected by searching the data informally, then proposing a hypothesis a posteriori and simply proposing all possible hypotheses a priori. For this reason it may be sensible to treat large numbers of hypotheses as if they were post hoc. Despite the difficulty in drawing a sharp distinction, we still believe that the conceptual distinction between cases in which a few hypotheses are specified in advance and those in which there are many hypotheses (or hypotheses not necessarily specified in advance) is useful.

The primary reason for insisting on this distinction is statistical. When testing hypotheses specified in advance, it is appropriate to consider that test in isolation from other tests that might have been carried out. This is often called the use of a testwise error rate in the theory of multiple comparisons, meaning that the appropriate definition of the significance level of the test is the proportion of the time this test, considered in isolation, would yield a type I error.

In contrast, when testing a hypothesis derived after exploring the data, it may not be appropriate to consider the test in isolation from other tests that might have been done. For example, by choosing to test the most promising of a set of study characteristics (that is, the one that appears to be most strongly related to effect size), the reviewer has implicitly used information from tests that could have been done on other study characteristics. More formally, the sampling distribution of the largest relationship is not the same as that of one relationship selected a priori. In cases where the hypothesis is picked after exploring the data, special post hoc test procedures that take account of the other hypotheses that could have been tested are appropriate (see chapter 15; Hedges and Olkin 1983, 1985; more generally Miller 1981). This is often called the use of an experimentwise error rate because the appropriate definition of the statistical significance level of the test is the proportion of the time the group of tests would lead to selecting a test that made a type I error.

Post hoc test procedures are frequently much less powerful than their a priori counterparts for detecting a particular relationship that is of interest. On the other hand, the post hoc procedures can do something that a priori procedures cannot: detect relationships not suspected in advance. The important point is that there is a trade-off between the ability to find relationships that are not suspected and the sensitivity to detect those that are thought to be likely.

3.3 DATA COLLECTION

Data collection in research synthesis is largely a sampling activity which raises all of the concerns attendant on any other sampling activity. Specifically the sampling procedure must be designed so as to yield studies that are representative of the intended universe of studies. Ideally the sampling is carried out in a way that reveals aspects of the sampling (like dependence of units in the sample, see chapter 19, this volume), or selection effects (like publication bias, see chapter 23, this volume) that might influence the analytic methods chosen to draw inferences from the sample.

3.3.1 Representativeness

Given that a universe has been chosen, so that we know the kind of studies about which the synthesis is to inform us, a fundamental problem is ensuring that the study sample is selected in such a way that it supports inferences to that universe. Part of the problem is to ensure that search criteria are consistent with the universe definition. Given that search criteria are consistent with the universe definition, an exhaustive sample of studies that meet the criteria is often taken to be a representative sample of studies of the universe. However, there are sometimes reasons to view this proposition skeptically.

The concept of representativeness has always been a somewhat ambiguous idea (see Kruskal and Mosteller 1979a, 1979b, 1979c, 1980) but the concept is useful in helping to illuminate potential problems in drawing inferences from samples.

One reason is that some types of studies in the intended universe may not have been conducted. The act of defining a universe of studies that could be conducted does not even guarantee that it will be nonempty. We hope that the sample of studies will inform us about a universe of studies exhibiting variation in their characteristics, but studies with a full range of characteristics may not have been conducted. The types of studies that have been conducted therefore limits the possibility of generalizations, whatever the search procedures. For example, the universe of studies might be planned to include studies with both behavioral observations and other outcome measures. But if no studies have been conducted using behavioral observations, no possible sample of studies can, strictly speaking, support generalizations to a universe including studies with behavioral observations. The situation need not be as simple or obvious as suggested in this example. Often the limitations on the types of studies conducted arise at the level of the joint frequency of two or more characteristics, such as categories of treatment and outcome. In such situations the limitations of the studies available are evident only in the scarcity of studies with certain joint characteristics and not in the marginal frequency of any type of study.

A second reason that exhaustiveness of sampling may not yield a representative sample of the universe is that, although studies may have been conducted, they may not reported in the forums accessible to the reviewer. This is the problem of missing data, as discussed in chapter 21 of this volume. Selective reporting (at least in forums accessible to synthesists) can occur in many ways: entire studies may be missing or only certain results from those studies may be missing, and missingness may, or may not be correlated with study results. The principal point here is that exhaustive sampling of data bases rendered nonrepresentative by publication or reporting bias does not yield a representative sample of the universe intended.

3.3.2 Dependence

Several types of dependence may arise in the sampling of effect size estimates. The simplest form of dependence arises when several effect size estimates are computed from measures from identical or partially overlapping groups of subjects. Methods for handling such dependence are discussed in chapter 19. While this form of dependence most often arises when several effect size estimates are computed from data reported in the same study, it can also arise when several different studies report data on the same sample of subjects. Failure to recognize this form of dependence and to use appropriate analytic strategies to cope with it can result in inaccurate estimates of effects and their standard errors.

A second type of dependence occurs when studies with similar or identical characteristics exhibit less variability in their effect-size parameters than does the entire sample of studies. Such an intraclass correlation of study effects leads to misspecification of random-effects models and hence to erroneous characterizations of between-study variation in effects. This form of dependence would also suggest misspecification in fixed-effects models if the study characteristics involved were not part of the formal explanatory model for between-study variation in effects.

3.3.3 Publication Bias

A particularly pernicious form of missing data occurs when the probability that a result is reported depends on the result obtained (for example, on whether it is sta-

tistically significant). Such missing data would be called nonignorable in the Little-Rubin framework and can lead to bias in the results of a meta-analysis. The resulting biases are typically called publication bias, although the more general reporting bias is a more accurate term. Methods for detecting and adjusting for publication bias are discussed at length in chapter 23 of this volume.

3.4 DATA ANALYSIS

Much of this handbook deals with issues of data analysis in research synthesis. It is appropriate here to discuss three issues of broad application to all types of statistical analyses in research synthesis.

3.4.1 Heterogeneity

Heterogeneity of effects in meta-analysis introduces a variety of interpretational problems that do not exist or are simpler if effects are homogeneous. For example, if effects are homogeneous, then fixed- and random-effects analyses essentially coincide and problems of generalization are simplified. Homogeneity also simplifies interpretation by making the synthesis basically an exercise in simple triangulation with the findings from different studies behaving as simple replications.

Unfortunately, heterogeneity is a rather frequent finding in research syntheses. This compels the synthesist to find ways of representing and interpreting the heterogeneity and dealing with it in statistical analyses. Fortunately, there has been considerable progress in procedures for representing heterogeneity in interpretable ways, such as theoretically sound measures of the proportion of variance in observed effect sizes due to heterogeneity, the I^2 measure (see chapter 14, this volume). Similarly, there has been great progress in both theory and software for random and mixed effects analyses that explicitly include heterogeneity in analyses (see chapter 16, this volume).

Some major issues arising in connection with heterogeneity are not easily resolvable. For example, heterogeneity is sometimes a function of the choice of effect size index. If effect sizes of a set of studies can be expressed in more than one metric, they are sometimes more consistent when expressed in one metric than another. Which then (if either) is the appropriate way to characterize the heterogeneity of effects? This question has theoretical implications for statistical analyses (for example, how necessary is homogeneity for combinability, see Cochran

1954; Radhakrishna 1965). However it is more than a theoretical issue since there is empirical evidence that meta-analyses using some indexes find more heterogeneity than others (for example, meta-analyses using odds ratios are often more consistent across studies than risk differences; see Engels et al. 2000).

3.4.2 Unity of Statistical Methods

Much of the literature on statistical methodology for research synthesis is conceived as statistical methods for the analysis of a particular effect size index. Thus, much of the literature on meta-analysis provides methods for combining estimates of odds ratios, or correlation coefficients, or standardized mean differences. Even though the methods for a particular effect size index might be similar to those for another index they are presented in the literature as essentially different methods. There is however a set of underlying statistical theory (see Cochran 1954; Hedges 1983) that provides a common theoretical justification for analyses of the effect size measures in common use—for example, the standardized mean difference, the correlation coefficient, the log odds ratio, the difference in proportions, and the like.

Essentially all commonly used statistical methods for effect size analyses rely on two facts. The first is that the effect size estimate or a suitable transformation is normally distributed in large samples with a mean of approximately the effect-size parameter. The second is that the standard error of the effect size estimate is a continuous function of the within-study sample sizes, the effect size, and possibly other parameters that can be estimated consistently from within-study data. Statistical methods for different effect size indexes appear to differ primarily because the formulas for the effect size indexes and their standard errors differ.

I am mindful of the variety of indexes of effect size that have been found useful. Chapter 12 is a detailed treatment of the variety of effect size indexes that can be applied to studies with continuous outcome measures and gives formulas for their standard errors. Chapter 13 provides a similar treatment of the variety of different effect size measures that are used for studies with categorical outcome variables. However, in this handbook we have chosen to stress the conceptual unity of statistical methods for different indexes of effect size by describing most methods in terms of a generic effect size statistic T, its corresponding effect-size parameter θ, and

a generic standard error $S(T)$. This permits statistical methods to be applied to a collection of any type effect size estimates by substituting the correct formulas for the individual estimates and their standard errors. This procedure not only provides a compact presentation of methods for existing indexes of effect size, but provides the basis for generalization to new indexes that are yet to be used.

3.4.3 Large Sample Approximations

Virtually all of the statistical methods described in this handbook and used in the analysis of effect sizes in research synthesis are based on so-called large-sample approximations. This means that, unlike some simple statistical methods such as the t-test for the differences between means or the F-test in analyses of the general linear model, the sampling theory invoked to construct hypothesis tests or confidence intervals is not exactly true in very small samples. The use of large sample statistical theory is not limited to meta-analysis; in fact, it is used much more frequently in applied statistics than is exact (or small sample) theory, typically because the exact theory is too difficult to develop. For example, Pearson's chi-square test for simple interactions and log linear procedures for more complex analysis in contingency tables are large sample procedures, as are most multivariate test procedures, procedures using LISREL models, item response models, or even Fisher's z-transform of the correlation coefficient.

That large sample procedures are widely used does not imply that they are always without problems in any particular setting. Indeed, one of the major questions that must be addressed in any application of large sample theory is whether the large sample approximation is accurate enough in samples of the size available to justify its use. In meta-analysis, large sample theory is primarily used to obtain the sampling distribution of the sample effect size estimates. The statistical properties of combined estimates or tests depends on the accuracy of the (approximations) to the sampling distributions of these individual effect size estimates. Fortunately quite a bit is known about the accuracy of these approximations to the distributions of effect size estimates.

In the cases of the standardized mean difference, the large sample theory is quite accurate for sample sizes as small as five to ten per group (see Hedges 1981, 1982; Hedges and Olkin 1983). In the cases of the correlation coefficient, the large sample theory is notoriously inaccurate in samples of less than a few hundred, particularly if the population correlation is large in magnitude. However, the large sample theory for the Fisher z-transformed correlation is typically quite accurate when the sample size is twenty or more. For this reason I usually suggest that analyses involving correlation coefficients as the effect size index be performed using the Fisher z-transforms of the correlations.

The situation with effect size indexes for experiments with discrete outcomes is more difficult to characterize. The large sample theory for differences in proportions and for odds ratios usually seems to be reasonably accurate when sample sizes are moderate (for example, greater than fifty) as long as the proportions involved are not too near zero or one. If the proportions are particularly near zero or one, larger sample sizes may be needed to insure comparable accuracy.

A final technical point about the notion of large sample theory in meta-analysis concerns the dual meaning of the term. The total sample size N may be thought of as the sum of the sample sizes across k studies included in the synthesis. In studies comparing a treatment group with sample size n_i^E in the ith study and a control group with sample size n_i^C in the ith study, the total sample size is

$$N = \Sigma(n_i^E + n_i^C).$$

The formal statistical theory underlying most meta-analytic methods is based on large sample approximations that hold when N is large in such a way that *all* of n_1^E, n_1^C, n_2^E, n_2^C, $\cdots$, n_k^E, n_k^C are also large.

Formally, large sample theory describes the behavior of the limiting distribution as $N \rightarrow \infty$ in such a way that n^E/N and n^C/N are fixed as N increases. Much of this theory is not true when $N \rightarrow \infty$ by letting k increase and keeping the within-study sample sizes n^E and n^C small (see Neyman and Scott 1948). In most practical situations, this distinction is not important because the within-study sample sizes are large enough to support the assumption that all n^E and n^C are large. In unusual cases where all of the sample sizes are very small, real caution is required.

3.5 CONCLUSION

Statistical thinking is important in every stage of research synthesis, as it is in primary research. Statistical issues are part of the problem definition, play a key role in data collection, and are obviously important in data analysis. The careful consideration of statistical issues throughout a synthesis can help assure its validity.

3.6 REFERENCES

Camilli, Greg. 1990. "The Test of Homogeneity for 2×2 Contingency Tables: A Review and some Personal Opinions on the Controversy." *Psychological Bulletin* 108: 135–45.

Clark, Herbert H. 1973. "The Language-as-Fixed Effect Fallacy: A Critique of Language Statistics in Psychological Research." *Journal of Verbal and Learning Behavior* 12(4): 335–59.

Cochran, William G. 1954. "The Combination of Estimates from Different Experiments." *Biometrics* 10(1): 101–29.

Engels, Eric A., Christopher C. Schmid, Norma Terrin, Ingram Olkin, and Joseph Lau. 2000. "Heterogeneity and Statistical Significance in Meta-Analysis: An Empirical Study of 125 Meta-Analyses." *Statistics in Medicine* 19(13): 1707–728.

Glass, Gene V., Barry McGaw, and Mary L. Smith. 1981. *Meta-Analysis in Social Research.* Beverly Hills, Calif.: Sage Publications.

Hedges, Larry V. 1983. "Combining Independent Estimators in Research Synthesis." *British Journal of Mathematical and Statistical Psychology* 36: 123–31.

Hedges, Larry V. 1981. "Distribution Theory for Glass's Estimator of Effect Size and Related Estimators." *Journal of Educational Statistics* 6(2): 107–28.

Hedges, Larry V. 1982. "Estimating Effect Size from a Series of Independent Experiments." *Psychological Bulletin* 92: 490–99.

Hedges, Larry V., and Ingram Olkin. 1983. "Clustering Estimates of Effect Magnitude." *Psychological Bulletin* 93: 563–73.

———. 1985. *Statistical Methods for Meta-Analysis.* New York: Academic Press.

Hedges, Larry V. and Jack L. Vevea. 1998. "Fixed and Random Effects Models in Meta-Analysis." *Psychological Methods* 3(4): 486–504.

Hunter, John E. and Frank L. Schmidt. 2004. *Methods of Meta-Analysis.* Thousand Oaks, Calif.: Sage Publications.

Kruskal, William, and Fred Mosteller. 1979a. "Representative Sampling I: Nonscientific Literature." *International Statistical Review* 47(1): 13–24.

———. 1979b. "Representative Sampling, II: Scientific Literature, Excluding Statistics." *International Statistical Review* 47(2): 111–27.

————. 1979c. "Representative Sampling III: The Current Statistical Literature." *International Statistical Review* 47(3): 245–65.

————. 1980. "Representative Sampling IV: The History of the Concept in Statistics, 1895–1939." *International Statistical Review* 48(2): 169–95.

Light, Richard J., and David Pillemar. 1984. *Summing Up.* Cambridge, Mass.: Harvard University Press.

Miller, Rupert D. 1981. *Simultaneous Statistical Inferences.* New York: Springer Verlag.

Neymann, Jerzy, and Elizabeth L. Scott. 1948. "Consistent Estimates Based on Partially Consistent Observations." *Econometrika* 16(1): 1–32.

Peto, Richard. 1987. "Why Do We Need Systematic Overviews of Randomized Traits with Discussion." *Statistics in Medicine* 6(3): 233–44.

Radhakrishna, S. 1965. "Combination of the Results from Several 2×2 Contingency Tables." *Biometrics* 21(1): 86–94.

PART
III

SEARCHING THE LITERATURE

4

SCIENTIFIC COMMUNICATION AND LITERATURE RETRIEVAL

HOWARD D. WHITE
Drexel University

CONTENTS

4.1 COMMUNING WITH THE LITERATURE

Following analysts such as Russell Ackoff and his colleagues (1976), it is convenient to divide scientific communication into four modes: informal oral, informal written, formal oral, and formal written. The first two, exemplified by telephone conversations and mail among colleagues, are relatively free-form and private. They are undoubtedly important, particularly as ways of sharing news and of receiving preliminary feedback on professional work (Menzel 1968; Garvey and Griffith 1968; Griffith and Miller 1970). Only through the formal modes, however, can scientists achieve their true goal, which is recognition for claims of new knowledge in a cumulative enterprise. The cost to them is the effort needed to prepare claims for public delivery, especially to critical peers. Of all the modes, formal *written* communication is the most premeditated (even more so than a formal oral presentation, such as a briefing or lecture). Its typical products—papers, articles, monographs, and reports—constitute the primary literatures by which scientists lay open their work to permanent public scrutiny in hopes that their claims to new knowledge will be validated and esteemed. This holds whether the literatures are distributed in printed or digital form. Moreover, the act of reading is no less a form of scientific communication than talking or writing; it might be called *communing with the literature*.

The research synthesis emerges at a late stage in formal written communication, after scores or even hundreds of primary writings have appeared. Among the synthesist's tasks, three stand out: discovery and retrieval of primary works, critically evaluating their claims in light of theory, and synthesizing their essentials so as to conserve the reader's time. All three tasks require the judgment of one trained in the research tradition (the theory or methodology) under study. As Patrick Wilson put it, "The surveyor must be, or be prepared to become, a specialist in the subject matter being surveyed" (1977, 17). But the payoff, he argued, can be rich: "The striking thing about the process of evaluation of a body of work is that, while the intent is not to increase knowledge by the conducting of independent inquiries, the result may be the increase of knowledge, by the drawing of conclusions not made in the literature reviewed but supported by the part of it judged valid. The process of analysis and synthesis can produce new knowledge in just this sense, that the attempt to put down what can be said to be known, on the basis of a given collection of documents, may result in the establishment of things not claimed explicitly in any of the documents surveyed" (11). Greg Myers held that such activity also serves the well-known rhetorical end of persuasion: "The writer of a review shapes the literature of a field into a story in order to enlist the support of readers to continue that story" (1991, 45).

4.2 THE REVIEWER'S PROGRESS

As it happens, not everyone has been glad to do the shaping. In 1994, when this chapter first appeared, the introductory section was subtitled "The Reviewer's Burden" to highlight the oft-perceived difficulty of getting scientists to write literature reviews as part of their communicative duties. The Committee on Scientific and Technical Communication quoted one reviewer as saying, "Digesting the material so that it could be presented on some conceptual basis was plain torture; I spent over 200 hours on that job. I wonder if 200 people spent even one hour reading it" (1969, 181). A decade later, Charles Bernier and Neil Yerkey observed: "Not enough reviews are written, because of the time required to write them and because of the trauma sometimes experienced during the writing of an excellent critical review" (1979, 48–49). Still later, Eugene Garfield noted a further impediment—the view among some scientists "that a review article—even one that is highly cited—is a lesser achievement than a piece of original research" (1989, 113).

My summary in the original chapter was that "the typical review involves tasks that many scientists find irksome—an ambitious literature search, obtaining of documents, extensive reading, reconciliation of conflicting claims, preparation of citations—and the bulk of effort is centered on the work of others, as if one had to write chapter 2 of one's dissertation all over again" (1994, 42). Even so, it was necessary to add: "Against that backdrop, the present movement in research synthesis is an intriguing development. The new research synthesists, far from avoiding the literature search, encourage consideration of *all* empirical studies on a subject—not only the published but the unpublished ones—so as to capture in their syntheses the full range of reported statistical effects (Rosenthal 1984; Green and Hall 1984)" (42).

The new research synthesists were of course the meta-analysts, and their tradition is now a well-established alternative to the older narrative tradition (Hunt 1997; Petticrew 2001). Rather than shunning the rigors of the narrative review, meta-analysts seek to make it more rigorous, with at least the level of statistical sophistication found in primary research. Where narrative reviewers have been mute on how they found the studies under consideration, the research synthesists call for making one's sources and search strategies explicit. Where narrative reviewers accept or reject studies impressionistically, research synthesists want firm editorial criteria. Where narrative reviewers are inconsistent in deciding which aspects of studies to discuss, the research synthesists require consistent coding of attributes across studies. Where narrative reviewers use ad hoc judgments as to the meaning of statistical findings, the meta-analysts insist on formal comparative and summary techniques. Where reviewers in the past may have been able to get by on their methodological and statistical skills without extra reading, they now confront whole new meta-analytic textbooks (Wang and Bushman 1999; Lipsey and Wilson 2001; Hunter and Schmidt 2004). Meta-analysis has even been extended to qualitative studies (Jones 2004; Dixon-Woods et al. 2007) and nonstatistical integration of findings (Bland, Meurer, and Maldonado 1995). Moreover, while the meta-analytic method is not without critics (Pawson 2006), the years since the early 1990s have seen remarkable improvements for its advocates on several fronts, and it is only fair to sketch them in a new introduction.

4.3 THE WEB AND DOCUMENTARY EVIDENCE

The Internet and the Web have spectacularly advanced science in its major enterprise, the communication of warranted findings (Ziman 1968). They have sped up interpersonal messaging (e-mail, file transfer, listservs), expanded the resources for publication (e-journals, preprint archives, data repositories, pdf files, informational websites), created limitless library space, and, most important, brought swift retrieval to vast stores of documents. The full reaches of the Web and its specialized databases are accessible day or night through search engines, notably Google. In many cases, one can now pass quickly from bibliographic listings to the full texts they represent, closing a centuries-old gulf. Retrievals, moreover, are routinely based on all content-bearing words and phrases in documents and not merely on global descriptions like titles and subject headings.

The new technologies for literature retrieval have made it easier for fact-hungry professionals to retrieve something they especially value: documentary evidence for claims, preferably statistical evidence from controlled experimental trials (Mulrow and Cook 1998; Pettiti 2000; Egger, Smith, and Altman 2001). Evidence-based practice

has become a movement, especially in medicine and healthcare but also in quantitative psychology, education, social work, and public policy studies. Naturally, this movement has converged and flourished with the somewhat older research synthesis movement (Cooper 2000, appendix), as represented by this handbook. The meta-analysts use special statistical techniques to test evidence from multiple experimental studies and then synthesize and publish the results in relatively short articles. Members of the evidence-based community are their ideal consumers (compare Doig and Simpson 2003). The two movements exemplify symbiotic scientific communication in the Internet era.

4.4 THE COCHRANE COLLABORATION

Both movements, however, were gathering strength before the Internet took off and would exist even if it had not appeared. In their area of scientific communication, a striking organizational improvement also antedates most of the Internet's technological advances. It is the Cochrane Collaboration, headquartered in the United Kingdom but international, which began in 1993. Cooper (2000) gives the background:

> In 1979, Archie Cochrane, a British epidemiologist, noted that a serious criticism of his field was that it had not organized critical summaries of relevant randomized controlled trials (RCTs). In 1987, Cochrane found an example in health care of the kind of synthesis he was looking for. He called this systematic review of care during pregnancy and childbirth "a real milestone in the history of randomized trials and in the evaluation of care," and suggested that other specialties should copy the methods (Cochrane 1989). In the same year, the scientific quality of many published syntheses in medicine was shown to leave much to be desired (Mulrow 1987).

The Cochrane Collaboration was developed in response to the call for systematic, up-to-date syntheses of RCTs of healthcare practices. Funds were provided by the United Kingdom's National Health Service to establish the first Cochrane Center. When the center opened at Oxford in 1992, those involved expressed the hope that there would be a collaborative international response to Cochrane's agenda. This idea was outlined at a meeting organized six months later by the New York Academy of Sciences. In October 1993, at what was to become the first in a series of annual Cochrane Colloquia, seventy-seven people from eleven countries co-founded the Cochrane Collaboration.

A central idea of the founders is that, in matters of health, the ability to find studies involving RCTs is too important to be left to the older, relatively haphazard methods. To produce what they call *systematic reviews*—another term for the integrative review or research synthesis—they have built a sociotechnical infrastructure for assembling relevant materials. Criteria for selecting studies for meta-analysis, and for reporting results are now explicit. Teams of specialists known as review groups are named to monitor literatures and to contribute their choices to several databases (the Cochrane Library) that are in effect large-scale instruments of quality control in meta-analytic work. The Library includes databases of systematic reviews already done, of methodology reviews, of abstracts of reviews of effects, and of controlled trials available for meta-analysis.

The 265-page *Cochrane Handbook for Systematic Reviews of Interventions*, which is available online, prescribes a set of procedures for retrieving and analyzing items from these databases (Higgins and Green 2005). Websites such as that of the University of York's Centre for Reviews and Dissemination back up its precepts on literature retrieval. The Centre offers detailed guidance on how to cover many bases in searching, including the hand searching of journals for items that computerized search strategies may miss (2005).

A typical set of injunctions for retrieval of medical literature in the specialty of back and neck pain was prepared by the Cochrane Back Group (van Tulder et al. 2003). They recommend:

1. A computer-aided search of the MEDLINE and EMBASE databases since their beginning. The highly sensitive search strategies for retrieval of reports of controlled trials should be run in conjunction with a specific search for spinal disorders and the intervention at issue.

2. A search of the Cochrane Central Register of Controlled Trials (CENTRAL) that is included in the latest issue of the Cochrane Library.

3. Screening references given in relevant systematic reviews and identified RCTs.

4. Personal communication with content experts in the field. It is left to the discretion of the synthesists to identify who the experts are on a specific topic.

5. Citation tracking of the identified RCTs (use of Science Citation Index to search forward in time which subsequent articles have cited the identified RCTs)

should be considered. The value of using citation tracking has not yet been established, but it may be especially useful to identify additional studies on topics that are poorly indexed in MEDLINE and EMBASE.

There was a time when the way to protect a synthesist's literature search from criticism was simply not to reveal what had been done. For example, Gregg Jackson (1980) surveyed thirty-six research syntheses in which only one stated the indexes (such as *Psychological Abstracts*) used in the search and only three mentioned drawing upon the bibliographies of previous review articles. "The failure of almost all integrative review articles to give information indicating the thoroughness of the search for appropriate primary sources," Jackson wrote, "does suggest that neither the reviewers nor their editors attach a great deal of importance to such thoroughness" (444). The different standards of today are seen even in the abstracts of systematic reviews, many of which mention the literature search underlying them. The protocol-based method section of an abstract by Rob Smeets and his colleagues illustrates:

> *Method.* Systematic literature search in PUBMED, MEDLINE, EMBASE and PsycINFO until December 2004 to identify observational studies regarding deconditioning signs and high quality RCTs regarding the effectiveness of cardiovascular and/or muscle strengthening exercises. Internal validity of the RCTs was assessed by using a checklist of nine methodology criteria in accordance with the Cochrane Collaboration. (2006, 673)

This is a good example of scientific communication interacting with literature retrieval to produce change (but see also Branwell and Williams 1997).

The rigorous Cochrane approach upgrades reviewing in general. The various stakeholders in medical and healthcare research, including funders, are likely to want Cochrane standards applied wherever possible—Gresham's law in reverse. The reviewers themselves seem to accept the higher degree of structure that has been imposed, as demonstrated over the last decade or so by leaping productivity (Lee, Bausell, and Berman 2001). Figure 4.1 presents data from Thomson Scientific's Web of Science—specifically, the combined Science Citation Index (SCI) and Social Sciences Citation Index (SSCI) from 1981 through 2005—that show the number of articles per year with "meta-analy*," "integrative review," or "systematic review" in their bibliographic descriptions. They now total more than 23,000, the great majority of them in

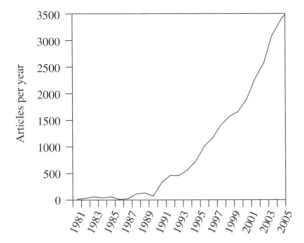

Figure 4.1 Growth of Meta-Analytic Literature in SCI and SSCI

<small>SOURCE: Author's compilation.</small>

journals of medicine and allied fields, though many non-medical journals are represented as well.

4.5 THE CAMPBELL COLLABORATION

Medicine and its allied fields rest on the idea of beneficial intervention in crucial situations. Given that crucial here can mean life and death, it is easy to see why health professionals want their decisions to reflect the best evidence available and how the Cochrane Collaboration emerged as a response. But professionals outside medicine and healthcare also seek to intervene beneficially in people's lives, and their interventions may have far-reaching social and behavioral consequences even if they are not matters of life and death. The upshot is that, in direct imitation of the Cochrane, the international Campbell Collaboration (C2) was formed in 1999 to facilitate systematic reviews of intervention effects in areas familiar to many of the readers of this handbook. Cooper again:

> The inaugural meeting of the Campbell Collaboration was held in Philadelphia, Pennsylvania, on February 24 and 25, 2000. Patterned after the Cochrane Collaboration, and championed by many of the same people, The Campbell Collaboration aims to bring the same quality of systematic evidence to issues of public policy as the Cochrane does to health care. It seeks to help policy makers, practitioners, consumers, and the general public make informed decisions

by preparing, maintaining, and promoting access to systematic reviews of studies on the effects of public policies, social interventions, and educational practices. (2000)

The Campbell Collaboration was named after the American psychologist and methodologist, Donald Campbell, who drew attention to the need for society to assess more rigorously the effects of social and educational experiments. These experiments take place in education, delinquency and criminal justice, mental health, welfare, housing, and employment, among other areas.

Like its predecessor, the Campbell Collaboration offers specialized databases to meta-analysts—the C2 Social, Psychological, Education, and Criminological Trials Registry, and the C2 Reviews of Interventions and Policy Evaluations—and has organized what it calls *coordinating groups* to supervise the preparation of reviews in specialties of these topical areas.

Also like its predecessor, C2 has a high presence on the Web. For example, the Nordic Campbell Center (2006) is strong on explanatory Web pages, such as this definition of a systematic review:

> Based on one question, a review aims at the most thorough, objective, and scientifically based survey possible, collected from all available research literature in the field. The reviews collocate the results of several studies, which are singled out and evaluated from specific systematic criteria. In this way, the reviews give a thorough aperçu of a specific question posed and answered by a variety of studies. In some cases, statistical methods (meta analysis) is employed to analyse and summarize the results of the comprised studies. (2006)

This accords with my 1994 account, which viewed systematic reviews as a way of coping with information overload:

> What had been merely too many things to read became a population of studies that could be treated like the respondents in survey research. Just as pollsters put questions to people, the new research synthesists could 'interview' existing studies (Weick 1970) and systematically record their attributes. Moreover, the process could become a team effort, with division and specialization of labor (for example, a librarian retrieves abstracts of studies, a project director chooses the studies to be read in full and creates the coding sheet, graduate students do the attribute coding). (42–43)

But whereas I pictured lone teams improvising, the Campbell Collaboration (2006) has now put forward crisply standardized procedures, as in this table of contents

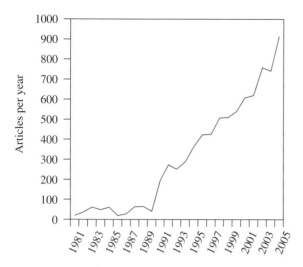

Figure 4.2 Growth of Meta-Analytic Literature in SSCI
<small>SOURCE:</small> Author's compilation.

describing what the methods section of a C2 protocol should contain:

- criteria for inclusion and exclusion of studies in the review
- search strategy for identification of relevant studies
- description of methods used in the component studies
- criteria for determination of independent findings
- details of study coding categories
- statistical procedures and conventions
- treatment of qualitative research

Despite occasional resistance and demurrals (such as Wallace et al. 2006), the new prescriptiveness seems to have helped reviewers. Figure 4.1 combined data from both SSCI and the much larger SCI. For readers whose specialties are better defined by the Campbell Collaboration than the Cochrane, figure 4.2 breaks out the almost 8,000 articles that were covered by SSCI and plots their growth from 1981 to 2005. Again, the impression is of great vigor in the years since the first edition of this handbook and since the founding of C2.

For the same SSCI data, journals producing more than 25 articles during 1981–2005 are rank-ordered in table 4.1. Psychology and psychiatry journals predominate.

SSCI selectively covers a fair number of medical journals, and some of those are seen as well, but topics extend beyond medicine and psychology. This is even plainer in the hundreds of journals not shown.

Although the problems of literature retrieval are never entirely solved, the Cochrane and Campbell materials make it likelier they will be addressed in research synthesis from now on. It is not that those materials contain anything new about literature searching; librarians, information scientists, and individual methodologists have said it all before. The first edition of this handbook said it and so have later textbooks (see, for example, Fink 2005; Petticrew and Roberts 2006). The difference is that power relations have changed. Whereas the earlier writers could advise but not prescribe, their views are gradually being adopted by scientists who, jointly, do have prescriptive power and who also have a real constituency, the evidence-based practitioners, to back them up. Because scientists communicate not only results but norms, this new consensus strikes me as a significant advance.

4.6 RECALL AND PRECISION

In the vocabulary of information science, for which literature retrieval is a main concern, research synthesists are unusually interested in high recall of documents. Recall is a measure used in evaluating literature searches: it expresses (as a percentage) the ratio of relevant documents retrieved to all those in a collection that should be retrieved. The latter value is, of course, a fiction: if we could identify all existing relevant documents in order to count them, we could retrieve them all, and so recall would always be 100 percent. Difficulties of operationalization aside, recall is a useful fiction in analyzing what users want from document retrieval systems.

Another major desideratum in retrieval systems is high precision, where precision expresses (as a percentage) the ratio of documents retrieved and judged relevant to all those actually retrieved. Precision measures how many irrelevant documents—false positives—one must go through to find the true positives or hits.

Precision and recall tend to vary inversely. If one seeks high recall—complete or comprehensive searches—one must be prepared to sift through many irrelevant documents, thereby degrading precision. Conversely, retrievals can be made highly precise, so as to cut down on false positives, but at the cost of missing many relevant documents scattered through the literature (false negatives), thereby degrading recall.

Table 4.1 SSCI Journals with More than Twenty-Five Articles Relevant to Meta-Analysis, 1981–2005

195	Psychological Bulletin	40	Perceptual and Motor Skills
179	Journal of Applied Psychology	39	Journal of Clinical Psychiatry
136	Schizophrenia Research Association	38	JAMA—Journal of the American Medical Association
84	Journal of Advanced Nursing	38	Journal of Vocational Behavior
82	Personnel Psychology	38	Nursing Research
76	Journal of Consulting and Clinical Psychology	35	International Journal of Psychology
68	Review of Educational Research	35	Psychological Medicine
67	American Journal of Psychiatry	34	American Psychologist
66	Addiction	34	Psychological Methods
64	Clinical Psychology Review	34	Psychology and Aging
58	British Journal of Psychiatry	32	Online Journal of Knowledge Synthesis For Nursing
55	British Medical Journal	31	Academy of Management Journal
54	Educational and Psychological Measurement	31	Health Psychology
51	Journal of Personality and Social Psychology	31	Medical Care
49	Clinical Psychology—Science and Practice	31	Sex Roles
49	Journal of Organizational Behavior	30	Personality and Social Psychology Bulletin
45	Acta Psychiatrica Scandinavica	29	Human Relations
44	Evaluation & the Health Professions	29	Journal of Psychosomatic Research
44	Journal of Management	28	American Journal of Public Health
44	Patient Education and Counseling	28	Australian and New Zealand Journal of Psychiatry
42	Psychological Reports	28	Journal of Parapsychology
42	Social Science & Medicine	28	Journal of Research in Science Teaching
42	Value in Health	28	Personality and Individual Differences
41	American Journal of Preventive Medicine	27	Child Development
41	Journal of Affective Disorders	27	Gerontologist
41	Journal of the American Geriatrics Society	27	Psychosomatic Medicine
40	Journal of Applied Social Psychology	26	Accident Analysis and Prevention
40	Journal of Occupational and Organizational Psychology	26	Journal of Business Research

SOURCE: Author's compilation.

Most literature searchers actually want high-precision retrievals, preferring relatively little bibliographic output to scan and relatively few items to read at the end of the judgment process. This is one manifestation of least effort, an economizing behavior often seen in information seekers (Mann 1993, 91–101). The research synthesists are distinctive in wanting (or at least accepting the need for) high recall. As Jackson put it, because "there is no way of ascertaining whether the set of located studies is representative of the full set of *existing* studies on the topic, the best protection against an unrepresentative set is to locate as many of the existing studies as is possible" (1978, 14). Bert Green and Judith Hall called such attempts "mundane and often tedious" but "of the utmost importance" (1984, 46). Some research synthesists doubt whether comprehensive searches are worth the effort (for example, Laird 1990), but even they seem to have

more rigorous standards for uncovering studies than librarians and information specialists typically encounter. The only other group likely to be as driven by a need for exhaustiveness are PhD students in the early stages of their dissertations.

The point is not to track down every paper that is somehow related to the topic. Research synthesists who reject this idea are quite sensible. The point is to avoid missing a useful paper that lies outside one's regular purview, thereby ensuring that one's habitual channels of communication will not bias the results of studies obtained by the search. In professional matters, most researchers find it hard to believe that their habitual channels, such as subscriptions to certain journals and conversations with certain colleagues, can fail to keep them fully informed. But the history of science and scholarship is full of examples of mutually relevant specialties that were unaware of each other for years because their members construed their own literatures too narrowly and failed to ask what other researchers—perhaps with different technical vocabularies (Grupp and Heider 1975)—were doing in a similar vein.

More to the point, it is likely that one has missed writings in one's own specialty. The experience of the research team reported in Joel Greenhouse and his colleagues shows that even the sources immediately at hand may produce an unexpectedly large yield (1990). The further lesson to be drawn is that, just as known sources can yield unknown items, there may also be whole sources of which one is unaware. A case can be made that research synthesists primarily concerned with substantive and methodological issues should get professional advice in literature retrieval, just as they would in statistics or computing if a problem exceeded their expertise (Wade et al. 2006).

4.7 IMPROVING THE YIELD

The most obvious way to improve recall is for researchers to avail themselves more of bibliographic databases (often called abstracting and indexing or "A and I" services, or reference databases). Harris Cooper described fifteen ways in which fifty-seven authors of empirical research reviews actually conducted literature searches (1985, 1987). Most preferred to trace the references in review papers, books, and nonreview papers already in hand (presumably many were from their own files) and to ask colleagues for recommendations. These are classic least-effort strategies to minimize work and maximize payoff in literature retrievals. For those purposes, they are unexceptionable; that is why academics and other intelligent people use them so often. Nevertheless, they are also almost guaranteed to miss relevant writings, particularly if literatures are large and heterogeneous. Moreover, because they are unsystematic, they cannot be replicated.

How many modes of searching are there? In an independent account from library and information science, Patrick Wilson identified five major ones (1992). Harris Cooper illustrated all of them. In Mann (1993), a Library of Congress reference librarian names eight modes as an efficient way to teach people how to do literature retrieval, but his eight are simply variants on Cooper's fifteen. They, too, illustrate Wilson's five categories, even though Thomas Mann wrote independently of either him or Cooper. Together, the three accounts illustrate Wilson's 1977

claim that writings in combination may yield new knowledge not found in any of them separately.

To show these experts converging, table 4.2 places Cooper's and Mann's modes verbatim under Wilson's headings. Discussing computer searching, Mann combines citation databases with other kinds, and I have put them with his example of citation searches in printed sources; otherwise, his text is unaltered. Even among those who use these techniques, however, it is an open question as to whether all likely sources are being consulted; researchers are often unaware of the full range of bibliographic tools available. Nowadays, the Cochrane and Campbell databases would be primary places to look, but they were not available when Cooper was doing his study, and even now they may not figure in some researchers' plans.

The retrieval modes in table 4.2 pretty well exhaust the possibilities. All ought to be considered—and most tried—by someone striving for high recall of documents (Beahler et al. 2000; Conn et al. 2003; Schlosser et al. 2005). The different modes can be used concurrently, thus saving time. Research synthesists disinclined to go beyond their favorite few can still extend their range by delegating searches to information specialists, including librarians. This would particularly hold for less favored strategies, such as searching reference and citation databases, browsing library collections, and discovering topical bibliographies.

The other obvious way to improve recall is to do what are sometimes called forward searches in the citation indexes of Thomson Scientific, now available in the Web of Science and on Dialog. The footnote chasing entries in table 4.2 reflect citation searches that go backward in time; one moves from a known publication to the earlier items it cites. In the contrasting kind of search in Dialog or the Web of Science, one moves forward in time by looking up a known publication and finding the later items that cite it. Although the citation indexes will not infallibly yield new items for one's synthesis, it is fatuous to ignore them—roughly equivalent to not keeping abreast of appropriate statistical techniques.

The result of accepting higher retrieval standards is that, rather than veiling one's strategies, one can state them candidly in the final report as documentation (Rothstein, Turner, and Lavenberg 2004; Campbell Education Coordinating Group 2005; Wade et al. 2006). Among other benefits, such statements help prevent unnecessary duplication of effort by later searchers, because they can replicate the original searcher's results. Whether these

Table 4.2 Five Major Modes of Searching

Footnote Chasing
Cooper 1985
References in review papers written by others
References in books by others
References in nonreview papers from journals you
subscribe to
References in nonreview papers you browsed through
at the library
Topical bibliographies compiled by others
Mann 1993
Searches through published bibliographies (including
sets of footnotes in relevant subject documents)
Related Records searches

Consultation
Cooper 1985
Communication with people who typically share
information with you
Informal conversations at conferences or with students
Formal requests of scholars you knew were active in
the field (e.g., solicitation letters)
Comments from readers/reviewers of past work
General requests to government agencies
Mann 1993
Searches through people sources (whether by verbal
contact, E-mail, electronic bulletin board, letters, etc.)

Searches in Subject Indexes
Cooper 1985
Computer search of abstract data bases (e.g., *ERIC,
Psychological Abstracts*)
Manual search of abstract data bases
Mann 1993
Controlled-vocabulary searches in manual or printed
sources
Key word searches in manual or printed sources
Computer searches—which can be done by subject
heading, classification number, key word. . .

Browsing
Cooper 1985
Browsing through library shelves
Mann 1993
Systematic browsing

Citation Searches
Cooper 1985
Manual search of a citation index
Computer search of a citation index (e.g., *SSCI*)
Mann 1993
Citation searches in printed sources
Computer searches by citation

SOURCE: Adapted from Cooper (1985), Wilson (1992), Mann
(1993). The boldfaced headings are from Wilson.

statements appear in the main text or as an endnote or
appendix, Bates argued that they should include

- notes on domain—all the sources used to identify the
  studies on which a research synthesis is based (including sources that failed to yield items although
  they initially seemed plausible)

- notes on scope—subject headings or other search
  terms used in the various sources; geographic, temporal, and language constraints

- notes on selection principles—editorial criteria used
  in accepting or rejecting studies to be meta-analyzed.
  (1992a)

Because the thoroughness of the search depends on
knowledge of the major modes for retrieving studies, a
more detailed look at each of Wilson's categories follows.
Cooper's and Mann's examples appear in italics as they
are woven into the discussion. The first two modes, footnote chasing and consultation, are understandably attractive to most scholars, but may be affected by personal
biases more than the other three, which involve searching
relatively impersonal bibliographies or collections; and
so the discussion is somewhat weighted in favor of the
latter. It is probably best to assume that all five are needed,
though different topics may require them in different
proportions.

4.8 FOOTNOTE CHASING

This is the adroit use of other authors' footnotes, or, more
broadly, their references to the prior literature on a topic.
Trisha Greenhalgh and Richard Peacock discussed the
use of snowballing—following up the references of documents retrieved in an ongoing literature search to accumulate still more documents (2005). The reason research
synthesists like footnote chasing is that it allows them to
find usable primary studies almost immediately. Moreover, the footnotes of a substantive work do not come as
unevaluated listings (like those in an anonymous bibliography), but as choices by an author whose critical judgment one can assess in the work itself. They are thus more
like scholarly intelligence than raw data, especially if one
admires the author who provides them.

Footnote chasing is obviously a two-stage process: to
follow up on someone else's references, the work in
which they occur must be in hand. Some reference-
bearing works will be already known; others must be discovered. Generally, there will be a strong correlation

between familiarity of works and their physical distance from the researcher; known items will tend to be nearby (for example, in one's office), and unknown items, farther away (for example, in a local or nonlocal library collection or undiscovered on the Web).

The first chasing many researchers do is simply to assemble the publications they already know or can readily discover. Indeed, from the standpoint of efficiency, the best way to begin a search is, first, to pick studies from one's own shelves and files and, second, to follow up leads to earlier work from their reference sections. Cooper's example was *references in nonreview papers from journals you subscribe to* (1985).

According to Green and Hall, meta-analysts doing a literature search should page through the volumes of the best journals for a topic year by year—presumably not only those to which they personally subscribe, but also those in the library (1984, 46). An alerting service like Thomson Scientific's *Current Contents* can be used as well. Despite its apparent cost in time, this hand searching may actually prove rather effective, given the small and focused set of titles in any particular table of contents. Items found in this fashion may lead as a bonus to earlier studies, through what Cooper called *references in nonreview papers you browsed through at the library* (1985).

Existing syntheses are probably the type of works most likely to be known to a researcher. If not, the first goal in a literature search should be to discover previous *review papers or books by others* on a topic, because they are likely to be both substantively important in their own right and a rich source of references to earlier studies. There is a burgeoning literature on the retrieval of systematic reviews (see, for example, Harrison 1997; Boynton et al. 1998; White et al. 2001; Montori et al. 2004; InterTASC Information Specialists' Sub-Group 2006).

Another goal should be to discover *topical bibliographies compiled by others.* The entries in free-standing bibliographies are not footnotes, and they may be harder than footnotes to judge for relevance (especially if abstracts are lacking). Nevertheless, if a bibliography exists, tracking down its preassembled references may greatly diminish the need for further retrievals (compare Wilson 1992; Bates 1992a). Mann has a nice chapter on bibliographies published in book form (1993, 45–56).

A way of searching online for reviews and bibliographies will be given later. A good mindset at the start of a search assumes the existence of at least one that is unknown. It is also important to ascertain the nonexistence of such publications (if that is the case), given that one's

strategy of presentation will depend on whether one is the first to survey the literature of a topic.

Mann's *Related Records searches* are a sophisticated form of footnote chasing that became practical only with computerization. They make it possible to find articles that cite identical references. When viewing the *full record of a single article* in the Web of Science, one will see the number of references it makes to earlier writings, the number of times it has been cited by later articles, and a button Find Related Records. Clicking the button will retrieve all articles that refer to at least one of the same earlier writings. They are ranked high to low by the number of references they share with the seed article. (Information scientists call this their bibliographic coupling strength.) According to the help screen,

> the assumption behind Related Records searching is that articles that cite the same works have a subject relationship, regardless of whether their titles, abstracts, or keywords contain the same terms. The more cited references two articles share, the closer this subject relationship is.

The shared cited works may themselves be retrievable through the Web of Science interface.

Footnote chasing is an inviting method—so inviting that the danger lies in stopping with it, on the pretext that any other method will quickly produce diminishing returns. If one is serious about finding primary studies to consider, one will not assume without checking that they do not exist or that diminishing returns begin outside one's office door. The substantive reason for concern is that one may fail to capture the full range of studies—and reported effects—that exist. Just as one's own footnotes reflect personal biases, so does one's collection of books and journals. The authors whose footnotes one pursues will tend to cite works compatible with biases of their own. In journals—the key publications in science—this tendency produces citation networks that link some journals tightly and others loosely or not at all; the result is specialties and disciplines that fail to communicate despite interests in common (for an example, see Rice 1990). Thus footnote chasing may simply reinforce the homogeneity of one's findings, and other methods are required to learn whether unknown but relevant writings exist. The next strategy is only a partial solution, as we shall see.

4.9 CONSULTATION

Most researchers trust people over bibliographies for answers on what is worth reading. Wilson's (1992) *con-*

sultation—finding usable studies by talking and writing to others rather than by literature searching—is illustrated in table 4.2. Cooper gives examples such as *informal conversations at conferences or with students*, and Mann mentions ways of getting in touch. Actually, one is still searching bibliographies; they are simply inside people's heads. Everyone, including the most learned, does this; there is no more practical or fruitful way of proceeding. The only danger lies in reliance on a personal network to the exclusion of alternate sources—a danger similar to overreliance on a personal library. Bibliographies, however dull, will help one search a literature more thoroughly than colleagues, however responsive.

A Google search on *meta-analysis* and *experts* (one may add *consult* or *contact*) reveals that many papers now add consultation of experts to their search strategies—not as something desirable but as something the authors successfully did. (Related search terms would no doubt add more examples.) Some papers mention finding unpublished work through this means, but there is no ready summary of what expert advice led to. A brief article in support of consultation claimed that, on the basis of a survey of experts for a review of primary medical care, forty of the references selected for review were identified by domain experts (McManus et al. 1998). "Of the 102 unique eligible references," they wrote, "50 (49%) were identified by one of the electronic databases, 40 (39%) by people working in the field, and 31 (30%) by hand searches" (1562). The percentages exceed 100 because some references were identified by more than one means. "Twenty four references," they added, "would have been missed entirely without the input of people working in the field." A librarian complained that their article, billed as a review, was rather a case study that knocks down a straw man no librarian has ever erected (Due 1999). R. J. McManus and colleagues countered (following Due's letter) that what is self-evident to librarians is not necessarily so to clinicians. This may be true; one does hear of innocents who think that database or Web searches retrieve everything that exists on a topic.

The quality of response that researchers can get by consulting depends, of course, on how well connected they are, and also perhaps on their energy in seeking new ties. Regarding the importance of personal connections, a noted author on scientific communication, the late Belver Griffith, told his students, "If you have to search the literature before undertaking research, you are not the person to do the research." He was being only slightly ironic. In his view, you may read to get to a research front, but you cannot stay there by waiting for new publications to appear; you should be in personal communication with the creators of the literature and other key informants before your attempt at new research—or synthesis—begins. Some of these *people who typically share information with you* may be local—that is, colleagues in your actual workplace with whom face-to-face conversations are possible every day. Others may be geographically dispersed but linked through telephone calls, e-mail, attendance at conferences and seminars, and so on, in your "invisible college" (Cronin 1982; Price 1986).

Diana Crane described invisible colleges as social circles whose members intercommunicate because they share an intense interest in a set of research problems (1972). Although the exact membership of an invisible college may be hard to define, a nucleus of productive and communicative insiders can usually be identified. If one is not actually a member of this elite, one can still try to elicit their advice on the literature by making *formal requests of scholars active in the field* (for example, solicitation letters). Scholarly advice in the form of *comments from readers/reviewers of past work* seems less likely unless one is already well-published. The *American Faculty Directory,* the *Research Centers Directory,* and handbooks of professional organizations are useful when one is trying to find addresses or telephone numbers of persons to contact. Web searches are often surprisingly productive, although beset by the problem of homonymic personal names.

While consultation with people may well bring published writings to light, its unique strength lies in revealing unpublished writings (compare Royle and Milne 2003). To appreciate this point, one must be clear on the difference between published and unpublished. Writings such as doctoral dissertations and Educational Resources Information Center (ERIC) reports are often called unpublished. Actually, any document for which copies will be supplied to any requestor has been published in the legal sense (Strong 1990), and this would include reports, conference proceedings, and dissertations. It would also include some machine-readable data files. The confusion occurs because people associate publication with a process of editorial quality control; they write *unpublished* when they should write *not independently edited* or *unrefereed*. Documents such as reports, dissertations, and data files are published, in effect, by their authors (or other relatively uncritical groups), but they are published nonetheless. Conference proceedings, over which editorial quality control may vary, have a similar status.

Preprints—manuscripts sent to requestors after being accepted by an editor—are simply not published yet.

Many printed bibliographies and bibliographic databases cover only published documents. Yet research synthesists place unusual stress on including effects from *un*published studies in their syntheses, to counteract a potential bias among editors and referees for publishing only effects that are statistically significant (see chapter 6 and chapter 23, this volume; for broader views, see Chalmers et al. 1990; McAuley et al. 2000; Rothstein, Sutton, and Borenstein 2005). Documents of limited distribution—technical reports, dissertations, and other grey literature—may be available through standard bibliographic services on the Web (such as *Dissertation Abstracts International* and *National Technical Information Service*). To get documents or data that are truly unpublished, one must receive them as a member of an invisible college (for example, papers in draft from distant colleagues) or conduct extensive solicitation campaigns (for example, *general requests to government agencies*). There is no bibliographic control over unreleased writings in file drawers except the human memories one hopes to tap (see chapter 6, this volume).

Consultation complements footnote chasing, in that it may lead to studies not referenced because they have never been published; and, like footnote chasing, it produces bibliographic advice that is selective rather than uncritical. But also like footnote chasing, it may introduce bias into one's search. The problem again is too much homogeneity in the studies that consultants recommend. As Cooper put it, "A synthesist gathering research solely by contacting hubs of a traditional invisible college will probably find studies that are more uniformly supportive of the beliefs held by these central researchers than are studies gathered from all sources. Also, because the participants in an invisible college use one another as a reference group, it is likely that the kinds of operations and measurements employed in the members' research will be more homogeneous than those employed by all researchers who might be interested in a given topic" (1998, 47). These observations should be truer still of groups in daily contact, such as colleagues in an academic department or teachers and their students.

The countermeasure is to seek heterogeneity. The national conferences of research associations, for example, are trade fairs for people with new ideas. Their bibliographic counterparts—big, open, and diversified—are the national bibliographies, the disciplinary abstracting and indexing services, the online library catalogs, the classi-fied library stacks. What many researchers would regard as the weakness of these instruments—that they are so editorially inclusive—may be for research synthesists their greatest strength. We turn now to the ways in which one can use large-scale, critically neutral bibliographies to explore existing subject literatures, as a complement to the more selective approaches of footnote chasing and consultation.

4.10 SEARCHES WITH SUBJECT INDEXING

Patrick Wilson called subject searching "the strategy of approaching the materials we want indirectly by using catalogs, bibliographies, indexes: works that are primarily collections of bibliographical descriptions with more or less complete representations of content" (1992, 156). Seemingly straightforward, retrieval of unknown publications by subject has in fact provided design problems to library and information science for over a century. The main problem, of course, is to effect a match between the searcher's expressed interest and the documentary description.

Only a few practical pointers can by given here. The synthesist wanting to improve recall of publications through a computer search of abstract databases (for example, ERIC, *Psychological Abstracts*) needs to understand the different ways in which topical literatures may be broken out from larger documentary stocks. One gets what one asks for, and so it helps to know that different partitions will result from different ways of asking. What follows is intended to produce greater fluency in the technical vocabulary of document retrieval and to provide a basis for the next chapter on computerized searching (see chapter 5, this volume; see also Cooper 1998; Grayson and Gomersall 2003; Thames Valley Literature Search Standards Group 2006).

Different kinds of indexing bind groups of publications into literatures. These are authors' natural-language terms, indexers' controlled-vocabulary terms, names of journals (or monographic series), and authors' citations. While all are usable separately in retrievals from printed bibliographic tools, it is important to know that they can also be combined, to some degree, in online searching in attempts to improve recall or precision or both. If the research synthesist is working with a professional searcher, all varieties of terms should be discussed in the strategy-planning interview. Moreover, the synthesist would do well to try out various search statements personally, so as to learn how different expressions perform. Mann, the librarian, brings out a distinction not in Cooper between keyword searches and controlled-vocabulary searches.

4.10.1 Natural-Language Terms

As authors write, they manifest their topics with terms such as *aphasia*, *teacher burnout*, or *criterion-referenced education* in titles, abstracts, and full texts. Because these terms are not assigned by indexers from controlled vocabularies but emerge naturally from authors' vocabularies, librarians call them natural language or keywords. (*Keywords* is a slippery term. By it, librarians usually mean all substantive words in an author's text. Authors, in contrast, may mean only the five or six phrases with which they index their own writings at the request of editors. Retrieval systems designers often use it to mean all nonstoplisted terms in a database, *including* the controlled vocabulary that indexers add. It can also mean the search terms that spontaneously occur to online searchers. More on this later.) Retrieval systems such as Dialog are capable of finding any natural-language terms of interest in any or all fields of the record. Thus, insofar as all documents with "teacher burnout" in their titles or abstracts constitute a literature, that entire literature can be retrieved. The Web also now makes it possible to retrieve documents on the basis of term occurrence in authors' full texts.

Google has so accustomed people to searching with natural language off the top of their heads that they may not realize there is any other way. Other ways include controlled-vocabulary and forward-citation retrievals. The Google search algorithm, though clever, cannot yet distinguish among different senses of the same natural-language query, nor can it expand on that query to pick up different wordings of the same concept, such as "gentrification" versus "renewal" versus "revitalization," or "older people" versus "senior citizens" versus "old-age pensioners." The first failure explains why Google retrievals are often both huge and noisy; the second, why Google retrievals may miss so much potentially valuable material. The first is a failure in precision; the second, a failure in recall. Long before Google appeared, a famous study by David Blair and M. E. Maron showed that lawyers who trusted computerized natural-language retrieval to bring them all documents relevant to a major litigation got only about one-fifth of those that were relevant (1985). The lawyers thought they were getting three-fourths or more. Google taps the largest collection of documents in history—a wonderful thing—but it has not solved the problem that the Blair-Maron study addresses.

Research synthesists have a special retrieval need in that they generally want to find empirical writings with measured effects and acceptable research designs (compare Cooper and Ribble 1989). This is a precision problem within the larger problem of achieving high recall. That is, given a particular topic, one does not want to retrieve literally all writings on it—mainly those with a certain empirical content. Therefore, the strategy should be to specify the topic as broadly as possible (with both natural-language and controlled-vocabulary terms), but then to qualify the search by adding terms designed to match very specific natural language in abstracts, such as *findings*, *results*, *ANOVA*, *random-*, *control-*, *t test*, *F test*, and *correlat-*. Professional searchers can build up large groupings of such terms (called hedges), with the effect that if any one of them appears in an abstract, the document is retrieved (Bates 1992b). The *Cochrane Handbook* devotes an appendix to hedges, and the Cochrane review groups add specific ones (Higgins and Green 2005).

This search strategy assumes that abstracts of empirical studies state methods and results, which is not always true. It would thus be most useful in partitioning a large subject literature into empirical and nonempirical components. Of course, it would also exclude empirical studies that lacked the chosen signals in their abstracts, and so would have to be used carefully. Nevertheless, it confers a potentially valuable power.

4.10.2 Controlled Vocabulary

These are the terms added to the bibliographic record by the employees of abstracting and indexing services or large research libraries—broadly speaking, indexers. The major reason for controlled vocabulary in bibliographic files is that authors' natural language in titles, abstracts, and full texts may scatter related writings rather than bringing them together. Controlled vocabulary is supposed to counteract scattering by re-expressing the content of documents with standard headings, thereby creating literatures on behalf of searchers who would otherwise have to guess how authors might express a subject. (Blair and Maron's lawyers failed on just this count, and they are not alone.)

Controlled vocabulary consists of such things as classification codes and subject headings for books, and descriptors for articles and reports. Classification codes reflect a hierarchically arranged subject scheme, such as that of the Library of Congress. With occasional exceptions, they are assigned one to a book, so that each book will have a single position in collections arranged for browsing. On the other hand, catalogers may assign more than one heading per book from the *Library of Congress Subject Headings.* Usually, however, they assign no more than three, because tables of contents and back-of-the-book indexes are presumed as complements.

Subject headings are generally the most specific terms (or compounds of terms, such as *Reference services—Automation—Bibliographies*) that match the scope of an entire work. In contrast, descriptors, taken from a list called a thesaurus, name salient concepts in writings rather than characterizing the work as a whole. Unlike compound subject headings, which are combined by indexers before any search occurs, descriptors are combined by the searcher at the time of a computerized search. Because the articles and reports they describe often lack the internal indexes seen in books, they are applied more liberally than subject headings, eight or ten being common.

Descriptors are important because they are used to index the scientific journal and report literatures that synthesists typically want, and a nodding familiarity with such tools as the *Thesaurus of ERIC Descriptors,* the *Medical Subject Headings,* or the *Thesaurus of Psychological Index Terms* will be helpful in talking with librarians and information specialists when searches are to be delegated. Thesauri, created by committees of subject experts, enable one to define research interests in standardized language. They contain definitions of terms (scope notes), cues from nonpreferred terms to preferred equivalents (for *criterion-referenced education* use *competency-based education*), and links between preferred terms (*aversion therapy* is related to *behavior modification*). Their one drawback is that, because the standardization of vocabulary takes time, they are always a bit behind authors' expressions in natural language. *Evidence-based medicine*, for example, was not a MeSH (Medical Subject Headings) term until 1997 (Harrison 1997), though it appeared in journals at least five years earlier. Therefore, anyone doing a search should be able to combine both descriptors (or other controlled vocabulary) and natural language to achieve the desired fullness of recall.

The social sciences present searchers with graver retrieval problems than medicine. According to Lesley Grayson and Alan Gomersall, these include "a more diverse literature; the greater variety and variability of secondary bibliographical tools; the increasing availability of material on the internet; and a less precise terminology"

(2003, 2). Abstracts may be missing or inadequate. Some literatures may lack vocabulary control altogether. If thesauri exist, they may be inconsistent in the descriptors and other indexing features they offer, and indexers may apply them inconsistently. (For example, some indexers might apply, and others omit, terms indicating research methods.) An upshot of these failings is that searchers may find it harder to distinguish analyzable studies in the social sciences than in medicine.

Supplements to descriptors include *identifiers* (specialized terms not included in the thesaurus) and *document types* (which partition literatures by genres, such as article or book review, rather than content). Being able to qualify an online search by document type allows research synthesists to break out one of their favorite forms, past reviews of research, from a subject literature. For example, Marcia Bates united descriptors and document types in this statement in an ERIC search on Dialog (1992b):

Select Mainstreaming AND (Deafness OR Hearing Impairments OR Partial Hearing) AND (Literature Reviews OR State of the Art Reviews)

One could add *Annotated Bibliographies* to the expression if that document type was also wanted. In the Web of Science, document types such as reviews can be retrieved by checkbox. As noted, relevant review papers and bibliographies are among the first things one should seek when preparing a new synthesis.

Raya Fidel stated that some online searchers routinely avoid descriptor lookups in thesauri in favor of natural language (1991). That was before the Web and Google; her observation would be more widely true today. Few nonprofessional searchers ever learn to distinguish between natural language and controlled vocabulary; everything is just keywords, a blurring that system designers perpetuate. For example, in the Cochrane Database of Systematic Reviews, the box for entering search terms is labeled Keywords, and so is the field in bibliographic records where reviews are characterized by subject. The latter is actually filled for the most part with MeSH descriptors, which are definitely controlled vocabulary. But one can also retrieve Cochrane records with almost any keyword from natural language. The word *steadily*, obviously not a MeSH descriptor, retrieves sixty as I write, for example. Contemporary retrieval systems reflect the principle of least effort for the greatest number; they cater to searchers who do not know about controlled vocabulary and probably would not use thesauri even if they did. But that does not mean that most searchers are *good* searchers where both high recall and high precision are concerned. To retrieve optimally for a research synthesis, searchers should be able to exploit all the resources available to them.

4.10.3 Journal Names

When editors accept contributions to journals, or to monographic series in the case of books, they make the journal or series name a part of the writing's standard bibliographic record. Writings so tagged form literatures of a kind. Abstracting and indexing services usually publish lists of journals they cover, and many of the journal titles may be read as if they were broad subject headings.

To insiders, however, names such as *American Sociological Review* or *Psychological Bulletin* connote not only a subject matter but also a level of authority. The latter is a function of editorial quality control, and it can be used to rank journals in prestige, which implicitly extends to the articles appearing in their pages. Earlier, a manual search through "the best" journals was mentioned—probably for items thought to be suitably refereed before publication. Actually, if one wants to confine a subject search to certain journals, the computer offers an alternative. Journal and series names may be entered into search statements just like other index terms. Thus one may perform a standard subject or citation search but restrict the final set to items appearing in specific journals or series. For example:

> Select Aphasia AND JN = Journal of Verbal Learning and Verbal Behavior

One might still have a fair number of abstracts to browse through, but all except those from the desired journal would be winnowed out.

4.11 BROWSING

Browsing, too, is a form of bibliographic searching. Browsing through library shelves is not one of the majority strategies in table 4.2. The referees for a draft of the present chapter claimed that no one browses in libraries any more—only on the Web. But wherever one browses, the principle is the same:

- Library shelves: face book spines classed by librarians and then look for items of interest.
- Journals: subscribe to articles bundled by editors and then look for items of interest.

- Web: use search terms to form sets of documents and then look for items of interest.

Library classification codes, journal names, and terms entered into search engines are all ways of assembling documents on grounds of their presumed similarity. People who browse examine these collections in hopes they can recognize valuable items, as opposed to seeking known items by name. In effect, they let collections search them, rather than the other way around. Of course, this kind of openness has always been a minority trait; browsing is not simply one more strategy that research synthesists in general will adopt. Even delegating it to someone such as a librarian or information specialist is problematical, because success depends on recognizing useful books or other writings that were not foreknown, and people differ in what they recognize.

Patrick Wilson called browsing book stacks "a good strategy when one can expect to find a relatively high concentration of things one is looking for in a particular section of a collection; it is not so good if the items being sought are likely to be spread out thinly in a large collection" (1992, 156). The Library of Congress (or Dewey) classification codes assigned to books are supposed to bring subject-related titles together so that, for example, knowledge of one book will lead to knowledge of others like it. Unfortunately, in the system as implemented—diverse classifiers facing the topical complexity of publications over decades—it is common for dissimilar items to be brought together and similar items to be scattered. But Google's algorithms make these mistakes, too, and so do indexers of journal articles; one must be able to tolerate low-precision retrievals at best.

Still, serendipitous finds do take place. It is also true that research synthesists and their support staffs do more than gather data-laden articles for systematic reviews. They need background knowledge of many kinds, and even old-fashioned library browsing may further its pursuit. For example, several textbooks on meta-analysis can be found if one traces the class number of, say, Kenneth Wachter and Miron Straf's book (1990) to the H62 section ("Social sciences (general). Study and teaching. Research.") in large libraries using the Library of Congress classification scheme.

4.12 CITATION SEARCHES

The last type of strategy identified in table 4.2 is, in Cooper's language, *the manual or computer search of a citation index* (for example, SSCI).

Authors' citations make substantive or methodological links between studies explicit, and networks of cited and citing publications may thus constitute literatures. In citation indexes, as noted, earlier writings become indexing terms by which to trace the later journal articles that cite them. Whereas the earlier writings may be of any sort—books, articles, reports, government documents, conference papers, dissertations, films, and so on—the later writings retrieved will be drawn solely from the journals (or other serials) covered by Thomson Scientific; no non-journal items, such as books, will be retrieved. Nevertheless, forward citation searching, still underutilized, has much to recommend it, because it tends to produce hits different from those produced by retrievals with natural language or controlled vocabulary. In other words, it yields writings related to one's topic that have relatively little overlap with those found by other means (Pao and Worthen 1989).

The reasons for lack of overlap are, first, that authors inadvertently hide the relevance of their work to other studies by using different natural language, which has a scattering effect, and, second, that indexers often fail to remedy this when they apply controlled vocabulary. Luckily, the problems of both kinds of terminology are partly corrected by citation linkages, which are vocabulary-independent (as noted in the earlier discussion of Related Records). They are also authors' linkages rather than indexers', and so presumably reflect greater subject expertise. Last, because Thomson's citation databases are multidisciplinary, they may cut across standard disciplinary lines. (In traditional footnote chasing, we saw, one retrieves many writings cited by one author, whereas in forward citation searching one starts with one writing and retrieves the many authors who cited it, which improves the chances for disciplinary diversity.) Hence, forward citation searching should be used with other retrieval procedures by anyone who wants to achieve high recall.

Several kinds of forward citation searching are possible in Scisearch, Social Scisearch, and Arts and Humanities Search on Dialog. In printed form, these are respectively *Science Citation Index*, *Social Sciences Citation Index*, and *Arts and Humanities Citation Index*. Thomson's own Web of Science makes the same three databases available for joint or separate searching. One needs only modest bibliographic data to express one's interests: retrievable literatures can be defined as all items that cite a known document, a known author, or a known organization. If need be, such data can be discovered by searching initially on natural language from titles and abstracts of citing articles. (Thomson's citation databases now have descriptors, but these are not governed by a publicly available thesaurus.) Title terms may also be intersected with names of cited documents or cited authors to improve precision.

Thomson is now testing a Web Citation Index for Web documents, but a similar tool, Google Scholar, has already won wide acceptance because of its ready availability. Given an entry term such as a subject description, an author's name, or the title of a document, Scholar will retrieve documents on the Web bearing that phrase as well as the documents that cite them. The documents retrieved are ranked high to low by the number of documents citing them, and the latter can themselves be retrieved through a clickable link. Because the Web is so capacious, retrievals are likely to include both journal articles and more informally published documents (such as dissertations or technical reports released in pdf) that have found their way onto the Web. This makes Scholar a wonderful device for extending one's reach in literature retrieval. Precisely because its use is so seductive, however, one should know that its lack of vocabulary control can make for numerous false drops.

Thomson's Web of Science databases also have problems of vocabulary control. For example, where authors are concerned, different persons can have the same name (homonyms), and the same person can have different names (allonyms, a coinage of White 2001). Homonyms affect precision, because works by authors or citees other than the one the searcher intended will be lumped together with it in the retrieval and must be disambiguated. The practice in these databases of using only initials for first and middle names worsens the problem. (For instance, *Lee AJ* unites what *Lee, Alan J.* and *Lee, Amos John* would separate.) Allonyms affect recall, because the searcher may be unable to guess all the different ways an author's or citee's name has been copied into the database. The same author usually appears in at least two ways (for example, *Small HG* and *Small H*). Derek J. de Solla Price's tricky name is cited in more than a dozen ways.

One other matter. Thomson Scientific is the former Institute for Scientific Information (ISI) in Philadelphia and continues ISI's policy of including—and excluding—journals in the citation indexes on the basis of their international citedness (Testa 2006). This imperfectly understood policy has often led to accusations of bias toward U.S. or North American or Anglo-American sources, since many thousands of the world's journals do not make the cut. It is true that the ISI/Thomson indexes have an English-language bias, because English is so dominant in the literatures of science and scholarship. Even so, leading journals in other languages are covered. Yet researchers routinely lament the absence of journals, in both English and other languages, they deem important. Given the economic constraints on ISI/Thomson and its subscribers, many journals must always be omitted. New tools such as Google Scholar, Scopus, and the Chinese Social Sciences Citation Index will be needed to extend researchers' capabilities for retrieval across relatively less-cited journals, languages, and genres.

4.13 FINAL OPERATIONS

The synthesist's ideal in gathering primary studies is to have the best possible pool from which to select those finally analyzed. In practice, most would admit that the selection of studies has a discretionary element. Research syntheses are constructs shaped by individual taste and by expectations about the readership. Different synthesists interested in the same topic may differ on the primary studies to be included. If a particular set of studies is challenged as incomplete, a stock defense would claim that it was not cost-effective to gather a different set. The degree to which the synthesist has actually tried the various avenues open will determine whether such a defense has merit.

4.13.1 Judging Relevance

The comprehensiveness of a search may be in doubt, but it is likely that judging the resulting documents for relevance will be constrained by the fact that research synthesis requires comparable, quantified findings. The need for studies with data sharply delineates what is relevant and what is not: the presence or absence of appropriate analytical tables gives an explicit test. Of course, if nonquantitative studies are included, relevance judgments may be more difficult. Presumably, the judges will be scientists or advanced students with subject expertise, high interest in the literature to be synthesized, and adequate time in which to make decisions. Further, they will be able to decide on the basis of abstracts or even full texts of documents if titles leave doubts. That limits the potential sources of variation in relevance judgments to factors such as how documents are to be used (Cuadra and Katter 1967), order of presentation of documents (Eisenberg and Barry 1988), subject knowledge of judges, and amount of

bibliographic information provided (Cooper and Ribble 1989). A major remaining cognitive variable is openness to information, as identified by David Davidson (1977). Although research synthesists as a group seem highly open to information, considerable variation still exists in their ranks: some are more open than others. This probably influences not only their relevance judgments, but their relative appetites for multimodal literature searches and their degree of tolerance for studies of less than exemplary quality. Some want to err on the side of inclusiveness, and others want stricter editorial standards. The tension between the two tendencies is felt in their own symposia, sometimes even within the same writer.

Where service to practitioners is concerned, a study by Ruth Lewis, Christine Urquhart, and Janet Rolinson reports that health professionals doubt librarians' ability to judge the relevance of meta-analyses for evidence-based practice (1998). However, other studies show librarians and information specialists gearing up for their new responsibilities on this front (Cumming and Conway 1998; Tsafrir and Grinberg 1998).

4.13.2 Document Delivery

The final step in literature retrieval is to obtain copies of items judged relevant. The online vendors cooperate with document suppliers so that hard copies of items may be ordered as part of an online search on the Web, often through one's academic library. Frequently the so-called grey literature, such as dissertations, technical reports, and government documents, can also be acquired on the Web or through specialized documentation centers. Interlibrary loan services are generally reliable for books, grey publications, and photocopies of articles when local collections fail. The foundation of international interlibrary loan in North America is the OCLC system, which has bibliographic records and library holdings data for more than 67 million titles, including a growing number of numeric data files in machine-readable form. In the United Kingdom, the British Library at Boston Spa serves an international clientele for interlibrary transactions. Increasingly the transfer of texts, software, and data files is being done between computers. Even so, organizations such as OCLC and the British Library are needed to make resources discoverable in the new electronic wilderness.

The key to availing oneself of these resources is to work closely with librarians and information specialists, whose role in scientific communication has received more emphasis here than is usual, in order to raise their visibility (Smith 1996; Petticrew 2001). Locating an expert in high recall may take effort; persons with the requisite motivation and knowledge are not at every reference desk. But such experts exist, and it is worth the effort to find them.

4.14 REFERENCES

Ackoff, Russell L., et al. 1976. *Designing a National Scientific and Technological Communication System.* Philadelphia: University of Pennsylvania Press.

Bates, Marcia J. 1992a. "Rigorous Systematic Bibliography." In *For Information Specialists; Interpretations of Reference and Bibliographic Work,* by Howard D. White, Marcia J. Bates, and Patrick Wilson. Norwood, N.J.: Ablex.

———. 1992b. "Tactics and Vocabularies in Online Searching." In *For Information Specialists; Interpretations of Reference and Bibliographic Work,* by Howard D. White, Marcia J. Bates, and Patrick Wilson. Norwood, N.J.: Ablex.

Beahler, C. C., J. J. Sundheim, and N. I. Trapp. 2000. "Information Retrieval in Systematic Reviews: Challenges in the Public Health Arena." *American Journal of Preventive Medicine* 18(4): 6–10.

Bernier, Charles L., and A. Neil Yerkey. 1979. *Cogent Communication; Overcoming Reading Overload.* Westport, Conn.: Greenwood Press.

Blair, David C., and M. E. Maron. 1985. "An Evaluation of Retrieval Effectiveness for a Full-Text Document Retrieval System." *Communications of the ACM* 28(3): 289–99.

Bland, Carole J., L. N. Meurer, and George Maldonado. 1995. "A Systematic Approach to Conducting a Non-Statistical Meta-Analysis of Research Literature." *Academic Medicine* 70(7): 642–53.

Boynton, Janette, Julie Glanville, David McDaid, and Carol Lefebvre. 1998. "Identifying Systematic Reviews in MEDLINE: Developing an Objective Approach to Search Strategy Design." *Journal of Information Science* 24(3): 137–54.

Branwell, V. H., and C. J. Williams. 1997. "Do Authors of Review Articles Use Systematic Methods to Identify, Assess and Synthesize Information?" *Annals of Oncology* 8(12): 1185–95.

Campbell Collaboration. 2006. "Guidelines." http://www.campbellcollaboration.org/guidelines.html.

Campbell Education Coordinating Group. 2005. "Information Retrieval Methods Group Systematic Review Checklist." http://www.campbellcollaboration.org/ECG/PolicyDocuments.asp.

Centre for Reviews and Dissemination, University of York. 2005. "Finding Studies for Systematic Reviews: A Checklist for Researchers." http://www.york.ac.uk/inst/crd/revs.htm

Chalmers, T. C., C. S. Frank, and D. Reitman. 1990. "Minimizing the Three Stages of Publication Bias." *Journal of the American Medical Association* 263(10): 1392–5.

Cochrane, Archie L. 1989. Foreword to *Effective Care in Pregnancy and Childbirth,* edited by I. Chalmers, M. Enkin, and M. Keirse. Oxford: Oxford University Press.

Committee on Scientific and Technical Communication. 1969. *Scientific and Technical Communication: A Pressing National Problem and Recommendations for Its Solution.* Washington, D.C.: National Academy of Sciences.

Conn, Vicki S., Sang-arun Isaramalai, Sabyasachi Rath, Peeranuch Jantarakupt, Rohini Wadhawan, and Yashodhara Dash. "Beyond MEDLINE for Literature Searches." *Journal of Nursing Scholarship* 35(2): 177–82.

Cooper, Harris M. 1985. "Literature Searching Strategies of Integrative Research Reviewers." *American Psychologist* 40(11): 1267–9.

———. 1987. "Literature-Searching Strategies of Integrative Research Reviewers: A First Survey." *Knowledge: Creation, Diffusion, Utilization* 8(2): 372–83.

———. 1998. *Synthesizing Research: A Guide for Literature Reviews,* 3rd ed. Thousand Oaks, Calif.: Sage Publications.

———. 2000. "Strengthening the Role of Research in Policy Decisions: The Campbell Collaboration and the Promise of Systematic Research Review." In *Making Research a Part of the Public Agenda,* MASC Report 104, edited by Mabel Rice and Joy Simpson. Lawrence, Kan.: Merrill Advanced Studies Center. http://www.merrill.ku.edu/publications/2000whitepaper/cooper.html.

Cooper, Harris M., and Ronald G. Ribble. 1989. "Influences on the Outcome of Literature Searches for Integrative Research Reviews." *Knowledge: Creation, Diffusion, Utilization* 10(3): 179–201.

Crane, Diana. 1972. *Invisible Colleges; Diffusion of Knowledge in Scientific Communities.* Chicago: University of Chicago Press.

Cronin, Blaise. 1982. "Invisible Colleges and Information Transfer; A Review and Commentary with Particular Reference to the Social Sciences." *Journal of Documentation* 38(3): 212–36.

Cuadra, Carlos A., and Robert V. Katter. 1967. "Opening the Black Box of Relevance." *Journal of Documentation* 23(4): 291–303.

Cumming, Lorna, and Lynn Conway. 1998. "Providing Comprehensive Information and Searches with Limited Resources." *Journal of Information Science* 24(3): 183–85.

Davidson, David. 1977. "The Effect of Individual Differences of Cognitive Style on Judgments of Document Relevance." *Journal of the American Society for Information Science* 28(5): 273–84.

Dixon-Woods, Mary, Andrew Booth, and Alex J. Sutton. 2007. "Synthesizing Qualitative Research: A Review of Published Reports." *Qualitative Research* 7(3): 375–422.

Doig, Gordon Stuart, and Fiona Simpson. 2003. "Efficient Literature Searching: A Core Skill for the Practice of Evidence-Based Medicine." *Intensive Care Medicine* 29(12): 2119–27.

Due, Stephen. 1999. "Study Only 'Proves' What Librarians Knew Anyway." *British Medical Journal* 319: 259.

Egger, Matthias, George Davey-Smith, and Douglas G. Altman, eds. 2001. *Systematic Reviews in Health Care: Meta-Analysis in Context.* London: BMJ Publshing.

Eisenberg, Michael, and Carol Barry. 1988. "Order Effects: A Study of the Possible Influence of Presentation Order on User Judgments of Document Relevance." *Journal of the American Society for Information Science* 39(5): 293–300.

Fink, Arlene. 2005. *Conducting Research Literature Reviews: From the Internet to Paper*, 2nd ed. Thousand Oaks, Calif.: Sage Publications.

Fidel, Raya. 1991. "Searchers' Selection of Search Keys: III. Searching Styles." *Journal of the American Society for Information Science* 42(7): 515–27.

Garfield, Eugene. 1989. "Reviewing Review Literature." In *Essays of an Information Scientist,* vol. 10. Philadelphia, Pa.: ISI Press.

Garvey, William D., and Belver C. Griffith. 1968. "Informal Channels of Communication in the Behavioral Sciences: Their Relevance in the Structure of Formal or Bibliographic Communication." In *The Foundations of Access to Knowledge; A Symposium*, edited by Edward B. Montgomery. Syracuse, N.Y.: Syracuse University Press.

Grayson, Lesley, and Alan Gomersall. 2003. "A Difficult Business: Finding the Evidence for Social Science Reviews." *ESRC* Working Paper 19. London: ESRC UK Centre for Evidence Based Policy and Practice. http://evidencenetwork.org/Documents/wp19.pdf.

Green, Bert F., and Judith A. Hall. 1984. "Quantitative Methods for Literature Reviews." *Annual Review of Psychology* 35: 37–53.

Greenhalgh, Trisha, and Richard Peacock. 2005. "Effectiveness and Efficiency of Search Methods in Systematic Reviews of Complex Evidence: Audit of Primary Sources." *British Medical Journal* 331(7524): 1064–65.

Greenhouse, Joel B., D. Fromm, S. Iyengar, M. A. Dew, A.L. Holland, and R. E. Kass. 1990. "The Making of a Meta-Analysis: A Quantitative Review of the Aphasia Treatment Literature." In *The Future of Meta-Analysis,* edited by Kenneth W. Wachter and Miron L. Straf. New York: Russell Sage Foundation.

Griffith, Belver C., and A. James Miller. 1970. "Networks of Informal Communication among Scientifically Productive Scientists." In *Communication among Scientists and Engineers,* edited by Carnot E. Nelson and Donald K. Pollock. Lexington, Mass.: Heath Lexington Books.

Grupp, Gunter, and Mary Heider. 1975. "Non-Overlapping Disciplinary Vocabularies. Communication from the Receiver's Point of View." In *Communication of Scientific Information,* edited by Stacey B. Day. Basel: S. Karger.

Harrison, Jane. 1997. "Designing a Search Strategy to Identify and Retrieve Articles on Evidence-Based Health Care Using MEDLINE." *Health Libraries Review* 14: 33–42.

Higgins, Julian P. T., and Sally Green, eds. 2005. *Cochrane Handbook for Systematic Reviews of Interventions*, vers. 4.2.5. Chichester: John Wiley & Sons. http://www.cochrane.org/resources/handbook/hbook.htm.

Hunt, Morton. 1997. *How Science Takes Stock: The Story of Meta-Analysis.* New York: Russell Sage Foundation.

Hunter, John E., and Frank L. Schmidt. 2004. *Methods of Meta-Analysis: Correcting Error and Bias in Research Findings*, 2nd ed. Thousand Oaks, Calif.: Sage Publications.

InterTASC Information Specialists' Sub-Group. 2006. "Search Filter Resource." http://www.york.ac.uk/inst/crd/intertasc/index.htm.

Jackson, Gregg B. 1978. *Methods for Reviewing and Integrating Research in the Social Sciences.* Springfield, Va.: National Technical Information Service.

———. 1980. "Methods for Integrative Reviews." *Review of Educational Research* 50(3): 438–60.

Jones, Myfanwy Lloyd. 2004. "Application of Systematic Methods to Qualitative Research: Practical Issues." *Journal of Advanced Nursing* 48(3): 271–78.

Laird, Nan M. 1990. "A Discussion of the Aphasia Study." In *The Future of Meta-Analysis,* edited by Kenneth W. Wachter and Miron L. Straf. New York: Russell Sage Foundation.

Lee, Wen-Lin, R. Barker Bausell, and Brian M. Berman. 2001. "The Growth of Health-Related Meta-Analyses Published from 1980 to 2000." *Evaluation and the Health Professions* 24(3): 327–35.

Lewis, Ruth A., Christine J. Urquhart, and Janet Rolinson. 1998. "Health Professionals' Attitudes towards Evidence-Based Medicine and the Role of the Information Professional in Exploitation of the Research Evidence." *Journal of Information Science* 24(5): 281–90.

Lipsey, Mark W., and David B. Wilson. 2001. *Practical Meta-Analysis.* Thousand Oaks, Calif.: Sage Publications.

Mann, Thomas. 1993. *Library Research Models: A Guide to Classification, Cataloging, and Computers.* New York: Oxford University Press.

McAuley, Laura, Ba Pham, Peter Tugwell, and David Moher. 2000. "Does the Inclusion of Grey Literature Influence Estimates of Intervention Effectiveness Reported in Meta-Analyses?" *Lancet* 356(9237): 1228–31.

McManus, R. J., S. Wilson, B. C. Delaney, D. A. Fitzmaurice, C. J. Hyde, R. S. Tobias, S. Jowett, and F. D. R. Hobbs. 1998. "Review of the Usefulness of Contacting Other Experts When Conducting a Literature Search for Systematic Reviews." *British Medical Journal* 317: 1562–3.

Menzel, Herbert. 1968. "Informal Communication in Science: Its Advantages and Its Formal Analogues." In *The Foundations of Access to Knowledge; A Symposium,* edited by Edward B. Montgomery. Syracuse, N.Y.: Syracuse University Press.

Montori, Victor M., Nancy L. Wilczynski, Douglas Morgan, R. Brian Haynes. 2004. "Optimal Search Strategies for Retrieving Systematic Reviews from Medline: Analytical Survey." *British Medical Journal.* http://bmj.bmjjournals.com/cgi/content/short/bmj.38336.804167.47v1.

Mulrow, Cynthia D. 1987. "The Medical Review Article: State of the Science." *Annals of Internal Medicine* 106(3): 485–88.

Mulrow, Cynthia D., and Deborah Cook. 1998. *Systematic Reviews: Synthesis of Best Evidence for Health Care Decisions.* Philadelphia, Pa.: American College of Physicians.

Myers, Greg. 1991. "Stories and Styles in Two Molecular Biology Review Articles." In *Textual Dynamics of the Professions; Historical and Contemporary Studies of Writing in Professional Communities,* edited by Charles Bazerman and James Paradis. Madison: University of Wisconsin Press.

Nordic Campbell Center, Danish National Institute of Social Research. 2006. "What Is a Systematic Review?" http://www.sfi.dk/sw28581.asp.

Pao, Miranda Lee, and Dennis B. Worthen. 1989. "Retrieval Effectiveness by Semantic and Citation Searching." *Journal of the American Society for Information Science* 40(4): 226–35.

Pawson, Ray. 2006. *Evidence-Based Policy: A Realist Perspective*. London: Sage Publications.

Petitti, Diana B. 2000. *Meta-Analysis, Decision Analysis and Cost-Effectiveness Analysis: Methods for Quantitative Synthesis in Medicine.* New York: Oxford University Press.

Petticrew, Mark. 2001. "Systematic Reviews from Astronomy to Zoology: Myths and Misconceptions." *British Medical Journal* 322(7278): 98–101. http://bmj.bmjjournals.com/cgi/content/full/322/7278/98.

Petticrew, Mark, and Helen Roberts. 2006. *Systematic Reviews in the Social Sciences: A Practical Guide.* Malden, Mass.: Blackwell.

Price, Derek J. de Solla. 1986. "Invisible Colleges and the Affluent Scientific Commuter." In *Little Science, Big Science . . . and Beyond.* New York: Columbia University Press.

Rice, Ronald E. 1990. "Hierarchies and Clusters among Communication and Library and Information Science Journals, 1977–1987." In *Scholarly Communication and Bibliometrics,* edited by Christine L. Borgman. Newbury Park, Calif.: Sage Publications.

Rosenthal, Robert. 1984. *Meta-Analytic Procedures for Social Research.* Beverly Hills, Calif.: Sage Publications.

Rothstein, Hannah R., Alexander J. Sutton, and Michael Borenstein, eds. 2005. *Publication Bias in Meta-Analysis: Prevention, Assessment and Adjustments.* New York: John Wiley & Sons.

Rothstein, Hannah R., Herbert M. Turner, and Julia G. Lavenberg. 2004. *The Campbell Collaboration Information Retrieval Policy Brief, Prepared for the Campbell Collaboration Steering Committee.* Philadelphia, Pa.: Campbell Collabortion. http://www.duke.edu/web/c2method/IRMG-PolicyBriefRevised.pdf.

Royle, Pamela, and Ruairidh Milne. 2003. "Literature Searching for Randomized Controlled Trials Used in Cochrane Reviews: Rapid versus Exhaustive Searches." *International Journal of Technology Assessment in Health Care* 19(4): 591–603.

Schlosser, Ralf W., Oliver Wendt, Katie Angemeier, and Manisha Shetty. 2005. "Searching for Evidence in Augmentative and Alternative Communication: Navigating a Scattered Literature." *Augmentative and Alternative Communication* 21(4): 233–55.

Smeets, Rob J. E. M., Derick Wade, Alita Hidding, Peter Van Leeuwen, Johan Vlaeyen, and Andre Knottneruset. 2006. "The Association of Physical Deconditioning and Chronic Low Back Pain: A Hypothesis-Oriented Systematic Review." *Disability and Rehabilitation* 28(11): 673–93.

Smith, Jack T. 1996. "Meta-Analysis: The Librarian as a Member of an Interdisciplinary Research Team." *Library Trends* 45(2): 265–79.

Strong, William S. 1990. *The Copyright Book: A Practical Guide,* 3rd ed. Cambridge, Mass.: MIT Press.

Thames Valley Literature Search Standards Group. 2006. *The Literature Search Process: Protocols for Researchers.* http://www.Library.nhs.uk/SpecialistLibraries/Download.aspx?resID=101405.

Testa, James. 2006. "The Thomson Scientific Journal Selection Process." *International Microbiology* 9(2): 135–38.

Tsafrir, J., and M. Grinberg. 1998. "Who Needs Evidence-Based Health Care?" *Bulletin of the Medical Library Association* 86(1): 40–45.

van Tulder, Maurits, Andrea Furlan, Claire Bombardier, Lex Bouter, and the Editorial Board of the Cochrane Collaboration Back Review Group. 2003. "Updated Method Guidelines for Systematic Reviews in the Cochrane Collaboration Back Review Group." *Spine* 28(12): 1290–99.

Wachter, Kenneth W., and Miron L. Straf, eds. 1990. *The Future of Meta-Analysis.* New York: Russell Sage Foundation.

Wade, Anne C., Herbert M. Turner, Hannah R. Rothstein, and Julia G. Lavenberg. 2006. "Information Retrieval and the Role of the Information Specialist in Producing High-Quality Systematic Reviews in the Social, Behavioural and Education Sciences." *Evidence & Policy: A Journal of Research, Debate and Practice* 2(1): 89–108. http://www.ingentaconnect.com/content/tpp/ep/2006/00000002/00000001/art00006.

Wallace, Alison, Karen Croucher, Mark Bevan, Karen Jackson, Lisa O'Malley, and Deborah Quilgars. 2006. "Evidence for Policy Making: Some Reflections on the Application of Systematic Reviews to Housing Research." *Housing Studies* 21(2): 297–314.

Wang, Morgan C., and Brad J. Bushman. 1999. *Integrating Results Through Meta-Analytic Review Using SAS Software.* Raleigh, N.C.: SAS Institute.

Weick, Karl E. 1970. "The Twigging of Overload." In *People and Information,* edited by Harold B. Pepinsky. New York: Pergamon.

White, Howard D. 1994. "Scientific Communication and Literature Retrieval." *Handbook of Research Synthesis,* edited by Harris Cooper and Larry V. Hedges. New York: Russell Sage Foundation.

———. 2001. "Authors as Citers over Time." *Journal of the American Society for Information Science* 52(2): 87–108.

White, V. J., J. M. Glanville, C. Lefebvre, and T. A. Sheldon. 2001. "A Statistical Approach to Designing Search Filters to Find Systematic Reviews: Objectivity Enhances Accuracy." *Journal of Information Science* 27(6): 357–70.

Wilson, Patrick. 1977. *Public Knowledge, Private Ignorance; Toward a Library and Information Policy.* Westport, Conn.: Greenwood Press.

———. 1992. Searching: Strategies and Evaluation. In *For Information Specialists; Interpretations of Reference and Bibliographic Work,* by Howard D. White, Marcia J. Bates, and Patrick Wilson. Norwood, N.J.: Ablex.

Ziman, John. 1968. *Public Knowledge: The Social Dimension of Science.* Cambridge: Cambridge University Press.

5

USING REFERENCE DATABASES

JEFFREY G. REED
Marian University

PAM M. BAXTER
Cornell University

CONTENTS

5.1 INTRODUCTION

The objective of this chapter is to introduce standard literature search tools and their use in research synthesis. Our focus is on research databases relevant to the social, behavioral, and medical sciences. We assume that the reader

- is acquainted with the key literature in his or her own discipline (such as, basic journal publications, handbooks and other reference tools) and the most important sources on the topic of synthesis;

- intends to conduct an exhaustive search of the literature—that is he or she desires to be thorough and identify a comprehensive list of almost all citations and documents that provide appropriate data relevant to the subject under investigation;

- is familiar with the structure and organization of research libraries and the basic finding tools associated with them such as catalogs of their holdings and periodical indexes. It is important to note that the mission of a research library goes beyond the basic materials and services designed to support course-related instruction. It maintains printed and electronic materials that provide information for independent study and exploration, staff dedicated to locating resources beyond those available in the local institution, and services to link the individual researcher with those resources;

- has or will familiarize him- or herself with other chapters in this volume, especially those on scientific communication (chapter 4) and grey literature (chapter 6);

- has or will call on the expertise of a librarian skilled in the art of bibliographic searching in his or her discipline.

The reader is also encouraged to refer to a source such as *The Oxford Guide to Library Research* (Mann 2005), or *Library Use: Handbook for Psychology* (Reed and Baxter 2003) for an overview of the basics of library use and search processes.

Why offer this chapter? In our experience, researchers may fail to identify important citations to documents relevant to their project, either because they select inappropriate search tools or because their search strategy and process are incomplete. Our goal is to help researchers make the best possible use of bibliographic search tools (in this case, reference databases), so that they can avoid this problem. When conducting a research synthesis on a topic, the goal is, or should be, completeness of coverage. Failure to identify an important piece of research (for example, an article, paper, report, or presentation) that contains data relevant to the topic may mean that the results of the synthesis could be biased and the conclusions reached or generalizations made possibly incorrect.

This chapter is focused on reference databases, also known as bibliographic databases (sources such as Psyc-INFO and MEDLINE), that contain citations to documents. The citations in the reference databases are surrogates that represent the documents themselves. Citations in some sources are very brief; they provide only bibliographic information that enables one to locate the document—for example. author, title, publication date. Other sources contain more extensive surrogate records, including the citation and bibliographic information as well as an abstract or summary of the source (for example, Psyc-INFO). Some sources, such as PsycARTICLES, may contain electronic versions of the documents themselves. The documents are those materials that contain the information we are seeking—the results of research on our topic. These documents may be in many forms—for example, books, journals, conference papers, or research reports. These documents may be available in hardcopy form or electronically.

We began our professional work with reference databases in the 1970s, and wrote the first edition of *Library*

Use: Handbook for Psychology, which was published in 1983. Since that time, changes have taken place in information storage and retrieval that some might characterize as a paradigm shift. In 1965, for example, there was no online PsycINFO that the researcher could access directly; there was only *Psychological Abstracts*, a monthly printed journal. It contained the citations and abstracts to journal articles in the field of psychology, and both subject and author printed indexes. Success in identifying relevant literature was a function of time, creativity, persistence, vocabulary, and indexer precision—how much time did the user have to search the monthly, annual and cumulated author and subject indexes; how accurate was the user in identifying terms that describe the topic that were also used by the indexers; how persistent was the user in manually searching different sections of the monthly source that might contain a relevant citation; how complete was the controlled vocabulary of the indexer's thesaurus of subject descriptors; and how precise were the indexers in using subject terms that accurately described the content of each document. Unfortunately, there were often shortfalls in one or more of these areas.

By the late 1970s, publishers were creating electronic databases of citations used as the basis for their published indexes and abstracts; a few even made searching of these databases available for a fee. But these electronic databases were searched remotely by experienced searchers who accessed reference databases through a vendor aggregator (a service provider who bought the database and sold access time to your library). So you took your request to a librarian who used a vendor such as Dialog to search available reference databases. By the early 1980s, some reference databases were available on a CD-ROM that a library acquired from a vendor such as Silver Platter. By the 1990s, electronic versions of many databases became available online, with libraries subscribing to vendor aggregators such as EBSCO, Ovid, or ProQuest.

As time passed, the information retrieval world continued to change. Some databases disappeared; new ones were created. An increasing number of databases became available electronically. Some vendors that provided access to databases merged or went out of business. Publishers changed the appearance of database output, some databases expanded coverage in their field, and the types of documents included in databases changed (for example, by adding citations to books) over time. Some publishers expanded the content of their database by adding fields to provide more information about each citation. Some databases added abstracts to the citations. Early electronic databases only allowed searching by authors or subject terms—those terms that the indexers selected and added to documents to represent their contents. If the indexer made a judgment error in assigning subject terms and did not use a relevant key term, an electronic search would not identify the relevant document. As search engines became more sophisticated and users demanded better access, searching of titles was added. Some databases enabled searching of abstracts or even the entire text of documents if they were included in the database. Free text searching, also known as natural language searching, became available. Although important terms that reflect the content of a document may appear in the title or abstract of a document, they may not be terms assigned as subjects to a document or terms included in the controlled vocabulary of a reference database.

Today, many libraries do not subscribe to some printed indexes or abstracts, choosing to rely entirely on electronic database access to services. Some services have ceased publication of printed versions—for example, the American Psychological Association announced in December 2006 that it would no longer publish its monthly *Psychological Abstracts*, but instead rely entirely on the electronic PsycINFO database to access citations to documents in the field of psychology. Indexes and abstracts (reference databases) have changed in appearance, format, content, and what they cover. The primary message is that access to reference databases and their content has changed and will continue to change.

Online user guides and tutorials for particular reference databases are increasingly common. They are becoming more user friendly and are updated more frequently than print information. They are often available, like a help system, from a selection within the reference database. Consult these tools to understand better the scope and mechanics of the reference databases you use.

Why is this important? Specific details have changed over the past forty years, but general approaches to searching reference databases have remained relatively constant. Rather than providing elaborate illustrations of services, the appearance of which the publisher or database programmer may change tomorrow, we will present generic examples. Even this handbook, unfortunately, will become a bit dated by the time it is printed. We focus this chapter on general principles and techniques related to the use of reference databases, and offer examples that we hope will stand the test of time.

5.2 DATABASE SELECTION

The function, content, and structure of bibliographic databases vary. A few are merely indexes, containing only the most basic bibliographic information on publications. More are abstracting services, in which bibliographic information is accompanied by nonevaluative summaries of content. An increasing number of databases either contain the entire text of a referenced item or link to the complete text of the item. For example, the indexes to proceedings and conference papers described in the appendix generally provide a bibliographic citation and a limited number of access points such as subject codes. *The International Bibliography of the Social Sciences*, which indexes almost 3,000 publications worldwide, contains bibliographic information about each item, a nonevaluative abstract, as well as indexing terms to aid searching (for example, to limit a search by language of publication, subject headings, publication year, or format). On the other end of the spectrum, the PsycINFO database supplies abstracts, many additional indexing options, the ability to search by cited references within indexed articles for many recent years and (depending on the database vendor) links to the local holdings records of an individual library or the full text of an article. Its companion database, PsycARTICLES, permits searching for terms within the complete text of the articles themselves—a fulltext search.

It is important to note, however, that the availability of enhanced search features (notably cited-reference and fulltext search ability and links from a bibliographic citation to the fulltext) is a reasonably recent enhancement to database production. In addition, availability of features varies within a database itself or between versions of a database distributed by different vendors. As a result, one must carefully examine user documentation (for example, a tutorial) that accompanies a database to determine precisely what search features are available and how comprehensively they are applied.

A more recent publishing phenomenon, open source electronic publishing, complicates the landscape of published literature and requires some explanation here. Spearheaded by scholarly societies, universities, and academic libraries, these journals can provide all the characteristics of traditional commercially produced products at a lower cost and with greater flexibility (Getz 2004). However, exploiting their contents can represent a challenge to researchers. Acceptance of open-access publications as items worthy of indexing in traditional reference databases tends to vary within the social sciences. For example, EconLIT, a literature indexing tool in economics, cites a large number of working paper series, an important communication format among economists. These series are frequently hosted by universities and other research institutions and freely available over the Internet.

A subset of this phenomenon is self-archiving of scholarly material by academic departments or even individual researchers. This can be as simple as making an individual paper, body of original data, or computing program accessible on a public web site. Much of this material may never be published in the sense that it is peer reviewed, published in a recognized scholarly publication, and therefore accessible via a reference database (for discussion of such materials, known as grey literature, see chapter 6, this volume).

Awareness of the scope of relevant databases helps the researcher to formulate a plan of action. Conversely, there are lacunae in database subject and publication format coverage among databases. As a general rule, selection of a bibliographic database is dictated by a combination of many factors related to your research question. These include the following:

- Disciplinary Scope. In what disciplines is research on this topic being conducted? Through what reference databases is the literature of the relevant fields accessed? For example, accuracy of eyewitness testimony in a court of law is a topic of interest to both psychologists studying memory and attorneys desiring to win their cases.

- Access. What reference databases are available to you at your institution? Where else can you access an important reference database?

- Date. What period does your topic cover? Is its intellectual foundation a relatively recent development in the discipline, or has it been the subject of research and debate over a period of decades—or even longer?

- Language and country. In many areas, significant research has been conducted outside of the USA and published in languages other than English. Although many are expanding their scope of coverage, some reference databases limit their coverage of journals published in non-Western languages or even of those not published in English. Is there research on your topic published in languages other than English or outside of the United States?

• Unpublished Work. This encompasses a varied and ill-defined body of formats, frequently referred to as *grey literature*. Is there relevant research in the form of conference papers, unpublished manuscripts, technical reports, or other forms?

5.2.1 Subject Coverage

Each bibliographic database is designed for a particular audience and has a defined scope of subject coverage. The list of selected databases provided in the appendix reveals some obvious subject strengths among reference databases regarding subject coverage. Some, such as Psyc-INFO and MEDLINE, cover core publications within their disciplines as well as titles from related and interdisciplinary areas of study. Researchers and publications in areas such as psychology and medicine cover such diverse topics; the reference databases offer a broad scope of subject coverage. Other databases, such as Social Work Abstracts, Ageline, and Criminal Justice Periodical Index, have a narrower subject focus. Reference volumes such as the *Gale Directory of Databases* can be helpful in identifying bibliographic databases of potential interest. In many areas of research, it is not only advantageous; it is essential to access multiple reference database sources to ensure comprehensive coverage of the published literature. This is especially important for topics that are interdisciplinary (such as stress, communication, conflict, leadership, group behavior, learning, public policy). As Howard White notes in an earlier chapter, it is important to avoid missing relevant material that lies outside one's normal purview.

5.2.2 Years of Coverage

Depending on the nature of a topic, the number of years covered by an indexing source may be important to the literature search process. Many bibliographic databases began production in the late 1960s and early 1970s, thus, it is common that their coverage of the published journals extends for more than thirty years. For example, the ERIC database initiated in 1966, covering education and related areas, contained the same information as the printed indexes produced from that database. Many databases, however, were preceded by printed indexes, which were published on a monthly or quarterly basis for many years and cumulated annually before electronic databases were available, or even possible.

A complete and retrospective literature search often entailed using both hardcopy index volumes and a bibliographic database. To some extent, and in some areas, this has changed over the last ten years. Some index publishers have enhanced their online databases by retrospectively adding the contents of their hardcopy predecessors, PsycINFO being an excellent example. The PsycINFO database originally began with a coverage date back to 1967, whereas the printed volumes indexed publications back to 1887. PsycINFO gradually added the contents of *Psychological Abstracts* so that its coverage is truly retrospective. In recent years, PsycINFO coverage has been extended to include *Psychological Abstracts* 1927–1966, *Psychological Bulletin* 1921–1926, *American Journal of Psychology* 1887–1966, *Psychological Index* English language journals 1894–1935, *Classic Books in Psychology of the 20th Century*, and *Harvard List of Books in Psychology,* 1840–1971 (APA Online 2006). In fact, some publishers have discontinued their printed products in favor of online versions due to publishing costs and declining subscriptions to the print. APA ceased publication of *Psychological Abstracts* in favor PsycINFO in 2006 (Benjamin and VandenBos 2006).

The National Library of Medicine (2000, 2004) discontinued publication of its *Cumulated Index Medicus* and monthly *Index Medicus* in favor of MEDLINE access in 2000 and 2004, respectively. Although the MEDLINE database provided comprehensive coverage of the journal literature back to 1966, a variety of earlier reference tools (for example, *Quarterly Cumulative Index to Current Literature* covering 1916 to 1926 and *Bibliographica Medica* from 1900 to 1902) extends the reach of medical literature indexing much earlier in time, before production of the National Library of Medicine's *Index Medicus* and MEDLINE (National Library of Medicine 2006).

An additional illustration is *Index of Economic Articles in Journals and Collected Volumes*, which began publication in 1924 with some coverage extending back as far as 1886, whereas EconLIT coverage extends back to 1969. The American Economic Association (AEA) discontinued *Index of Economic Articles* in favor of EconLIT in 1996. Examples of printed indexes that precede the coverage provided by reference databases are numerous. Information about them is often included in help or tutorial guides that accompany reference databases. Additional sources of such information include the expertise provided by library reference staff, discipline-specific guides to reference literature, or more general reference guides. One example of a discipline-specific volume that provides such guidance is *Sociology: A Guide to Reference and Information Sources* (Aby, Nalen, and Fielding 2005). The broader social sciences are covered by several current guides, including those by Nancy Herron (2002), and David Fisher, Sandra Price, and Terry Hanstock (2002). The inclusive *Guide to Reference Books* (Balay 1996) is now in its eleventh edition.

Many database producers, however, have limited their retrospective coverage, because the cost is high relative to potential demand for information about older research material. Still others have elected to provide limited retrospective coverage; for example, by adding citations, but not summaries or abstracts of older documents. Thus, you must know both the timeframe over which research on your topic extends and the coverage dates of the reference database you are using.

The limits of database coverage can tempt a researcher to restrict search strategies to particular reference databases based on publication dates—for example, 1965 to present. Unless the topic is one that has emerged within the defined time frame and there is no literature published outside that time, this is a poor practice. Date limiting a project can lead to the failure to include important research results in a research synthesis, and thus potentially significantly bias its results.

When considering the criteria for inclusion and coverage of a reference database, carefully note the dates when a particular journal began to be included and whether titles are afforded cover-to-cover or selective indexing. Just to illustrate this point, *Acta Psychologica* began publication in 1935, though PsycINFO indexes its contents beginning with 1949 (APA Online 2007). At present, *Acta Psychologica* contents are indexed in their entirety. However, it is difficult to determine in what index, whether printed or electronic, the early issues of *Acta Psychologica* are indexed or if PsycINFO began its cover-to-cover indexing of its contents in 1949 or at some later time. Such journal titles might be candidates for hand searching if they are highly relevant to the project at hand and completeness of their indexing cannot be verified.

5.2.3 Publication Format

Some bibliographic databases index only one type of publication; for example, dissertations, books, research reports, conference papers. Others provide access to a wide variety of formats, although the coverage of every format may not be consistent across all years of a particular database. Suffice it to say that some research formats

are not as well represented bibliographically as others. Articles from well-established, refereed journal titles are by far the most heavily represented format in reference databases. The empirical literature, however, is not restricted to one publication type, so finding research in other formats and procuring the documents is critical, although it can be challenging. Although this topic is covered extensively in the context of the grey literature in chapter 6 of this volume, a few generalizations warrant mention here.

Subject access to information on book-length studies is available within several comprehensive, international databases. WorldCat, a service of the Online Computer Library Center (OCLC), allows you to search the catalogs of many libraries throughout the world. It not only contains records for books, manuscripts, and nonprint media held by thousands of libraries, it has also recently incorporated information about websites, datasets, media in emerging formats, and so forth. These are important sources because books may contain empirical evidence relevant to the research synthesis. However, finding information about content within individual chapters remains difficult, a situation exacerbated by inconsistent coverage both among and within disciplines. For example, PsycINFO (and to a more limited extent, APA's PsycBOOKS database) affords detailed coverage of book chapters but is by no means exhaustive in its retrospective coverage of books or publishers. Other reference databases, such as *International Bibliography of the Social Sciences*, incorporate book chapter coverage. However, no one or two databases provide adequate coverage of this publication format.

Coverage of research and technical reports in the social sciences is uneven. Their unpublished nature dictates limited bibliographic control and distribution. Interim or final reports produced by principle investigators on grants may eventually be indexed in the National Technical Information Service (NTIS) database, an index to more than 2 million reports issued since the 1960s (NTIS 2006). Established report series are most frequently indexed in databases such as NTIS and EconLIT and to a lesser extent, PsycINFO and ERIC. The launch of APA's PsycEXTRA product in 2004 was intended, at least in part, to address this gap. The large numbers of report series generated by university departments, research institutes, and associations are difficult to identify, let alone index and locate. Their wider availability on Web pages ameliorates this situation to some extent, though much content remains in the "deep Web" beyond the reach of most standard Internet search engines.

Like research reports, papers presented at scholarly conferences have traditionally been difficult to identify and even harder for researchers to acquire. Information about conference proceedings and individual papers remains difficult to locate, despite improved bibliographic access and the posting of papers on individual or institutional websites. A few databases, notably *Proceedings First* and *Papers First*, index this material exclusively. However, representation of this research communication format remains limited in most databases.

The literature review in many dissertations is extensive, and the quality of the empirical research is often extremely high. Ignoring such documents in a research synthesis is ill-advised. Although some consider this a part of the grey literature, for many years access has been provided through *Dissertation Abstracts International (DAI)*. More recently access is available through the electronic ProQuest Dissertations & Theses. Dissertations and theses, however, are often not a part of the local library's collections, and must be purchased. Although it is desirable to obtain a dissertation through interlibrary loan (ILL), some libraries do not lend theses and dissertations. In Europe, part of the process in many universities requires that a doctoral dissertation be published for the granting of the degree, making access to these documents a bit easier through searching a national bibliography of published monographs. As a general rule, however, researchers requiring access to a large number of doctoral dissertations should be prepared for a lengthy wait to obtain them through interlibrary loan, significant costs to purchase them, or both.

Locating materials in grey literature formats (such as technical reports, conference papers, grant reports, and other documents) can be difficult. They are often not included in major reference databases. In chapter 6, Rothstein and Hopewell provide valuable guidance on searching the grey literature, and suggest approaches that supplement your use of reference databases.

5.2.4 Database Availability

Ideally, selection of the reference databases is dictated solely by the research inquiry; that is, the subject scope, years of coverage, and so forth. An additional consideration is the accessibility of files to the individual researcher and, by extension, the availability of the publications they index.

Most academic libraries and larger public libraries subscribe to a suite of bibliographic databases, their selection tailored to the needs of their users (faculty, staff, students, and general users). These arrangements allow users of that particular institution to search and print or save citations, abstracts, or even the complete text of a publication to use in their research. Researchers at large, well-funded research institutions serving a wide-range of disciplines are more likely to have access to both core (such as, PsycINFO, Biological Abstracts) and specialized databases (for example, EI Compendex, Derwent Biotechnology Abstracts, Social Work Abstracts). However, differential pricing makes many of the core databases affordable even for smaller institutions or those with less comprehensive program offerings. For independent scholars, or those unaffiliated with an academic institution, there are other options that permit individual subscriptions. In addition, some major databases, notably ERIC and MEDLINE, are produced on behalf of the federal government and can be used on the Internet at no charge.

The cost of sources such as the electronic Web of Science, basis for the printed *Science Citation Index* and *Social Science Citation Index*, however, has become so high that many institutions have ceased subscribing. The cumulative cost of reference databases to libraries may limit your access to more specialized tools. You may have to find an alternate way to access an important, relevant reference database, such as contracting with a commercial service to conduct a search on your behalf or initiating short-term access as an individual subscriber. Professional associations may provide access to a limited number of databases as a membership benefit or at a reduced cost. For example, the American Economic Association provides discounted limited access to EconLit, and American Psychological Association has subscription options for individuals, whether or not they are APA members. Your college or university library may be able to provide you, as a scholar researcher, an introduction to a nearby research library that has the resources you need; ask a librarian for assistance.

5.2.5 Vendor Differences

Many databases are available from more than one vendor. As mentioned earlier, ERIC and MEDLINE are examples of two that are freely available for Internet searching. In addition, these files are also accessible from commercial vendors (for example, EBSCO, Ovid). In the case of files accessible via more than one source, one would assume that the same search strategy implemented on the same database would yield the same body of citations.

However, this is not always the case, for reasons that are difficult to identify. Frequency of updates or dates of coverage are obvious sources of differences. Other less obvious differences may be search defaults ("false memory" with quotation marks might yield a larger set than the same terms without the quotation marks) or the differential use of stop words (words that are ignored regardless of how they are used in the search). Just as different Internet search engines will yield differing results to the same search, reference database search engines employing different search algorithms provided by different vendors may result in different search results. Your primary strategies to ensure completeness will be based on the methods described shortly, such as a careful search strategy, multiple searches, and multiple reference databases.

5.3 STRUCTURING AND CONDUCTING THE SEARCH

Think of a bibliographic search as a treasure hunt. In the case of research synthesis and meta-analysis, however, there is no one single treasure chest. It is as though many pieces of treasure were dumped overboard and the task is to retrieve as many pieces as possible from the ocean. Unfortunately, there are reefs hiding pieces, currents have pushed pieces about, and since it was strewn about many years ago, much of the treasure is covered by sand and plants. It will take a lot of effort to retrieve each piece of treasure. But because the pieces of treasure complete a puzzle, it is important to find every piece of the puzzle that you possibly can to unlock the secrets of the puzzle.

5.3.1 Search Strategy

Unlocking the secrets of the puzzle requires a plan. Defining a plan for finding relevant research evidence to include in the research synthesis is a key element of the process. This plan is the search strategy. It is a road map that defines how one will pursue the process of seeking information. Developing a solid plan will aid you in proceeding logically, prevent unnecessary retracing of steps already covered, and aid in the quest for completeness. What should be included in the search strategy?

- Research Question: What is the specific topic? How is it constrained?
- Descriptors: What terms are used to describe key aspects of this topic?

- Search Profile: How will the search be structured (that is, how will search terms be combined logically)?
- Sources: What bibliographic sources will be searched—reference (bibliographic) databases, web sources? What time period will be covered? How will electronic sources be supplemented? What sources will be consulted to access the grey or fugitive literature?
- Methods: How will the search be conducted? In what order will elements of the search be pursued?

It is not uncommon for the search strategy to evolve during the research process. Failure to formulate a well-considered plan for research can result in

- a search process that is far too broad and wastes an inordinate amount of time by requiring examination and rejection of many false positives retrieved;
- a search that is far too narrow with many (unknown) false negatives and results in an incomplete or biased synthesis;
- a research syntheses paper that is poorly focused, neglects important sources, or is too complex because it attempts to accomplish too much.

In the sections that follow, we describe the process of conducting the search and illustrate how each of these may be pursued.

5.3.2 Steps in the Search Process

5.3.2.1 Topic Definition The first step in conducting a search is to clearly and precisely define the topic. Here one determines the scope of what aspects of the subject are included or excluded, the subject population, the methodology employed, and so forth. Defining the topic requires knowledge of the subject gained through preliminary examining and evaluating available literature. It may start with a textbook chapter, a literature review article, or a conference paper. It should be followed by more reading and study of related literature. Many researchers would support it with a quick preliminary literature search. The goal of this first search is to begin to understand the scope of the topic, where the controversies lie, what research has been conducted, and how large the body of literature is. One should then state the topic in the form of a sentence or question.

In this chapter we will use a sample topic of eyewitness testimony. There are many aspects of this topic we could consider. In their research synthesis of studies on the validity of police lineups for the purpose of eyewitness identification of perpetrators, Gary Wells and Elizabeth Olson considered characteristics of the witness, the event that occurred, and the testimony that was given (2003). They also considered system variables such as lineups, pre-lineup instructions given to witnesses, the way in which the lineup is presented, and so forth. Asher Koriat, Morris Goldsmith, and Ainat Pansky, in their synthesis of studies on the accuracy of memory, considered factors such as misattributions, false recall, and the provision of misleading postevent information that can affect eyewitness testimony (2000). Siegfried Sporer and his colleagues focused on confidence and accuracy in eyewitness identification (1995). Elizabeth Loftus differentiated among perceiving of events, retaining information, retrieving information, and difference among individuals as they affect accuracy of testimony (1996). One could look at anxiety and stress, arousal, use of leading questions, or focus on children as eyewitnesses (Siegel and Loftus 1978; Johnson 1978; Loftus 1975; Dale, Loftus, and Rathbun 1978). For this chapter, we start by defining a fairly broad sample topic—memory accuracy of adult eyewitness testimony. We have limited our study of the testimony given by eyewitnesses to the accuracy of the report given by adults and a focus on the memories of the eyewitnesses.

5.3.2.2 Search Terms The second step is to identify terms descriptive of the topic. In a typical bibliographic search, the goal is to use the right set of terms in the right way to accurately describe the topic at the appropriate level of specificity. Classically, we strive for high precision in order to recall a maximum of *hits* (relevant sources), a minimum of *false positives* (sources identified but irrelevant), and a very small number of *false negatives* (relevant sources not identified) to maximize efficiency. In research synthesis, however, the paradigm is different. Because the goal is thoroughness (completeness), the primary need is to reduce the number of false negatives (relevant sources not identified), so the research synthesist must be willing to examine a larger number of false positives. To reduce false negatives, one must expand the search by including all relevant synonyms and appropriate related terms. One must at the same time be concerned about precision; as the net expands with more alternate search terms (for example, synonyms), the number of citations retrieved can be expected to increase. Each must be checked to determine its relevance, a process that both increases the costs of the synthesis and the potential for errors to occur when judging relevance.

What types of terms should be included? In the field of health care, when examining health-care interventions, *The Cochrane Handbook* suggests that one should include three types of concepts in an electronic reference database search—the health condition under investigation, the intervention being evaluated, and the types of studies to be included (Higgins and Green 2006). This advice may be adapted for interventions in other areas, such as the social sciences, social policy, and education. In each, we are often concerned with a problem to be addressed (psychological, behavioral, developmental, social, or other), an approach being considered to address the problem, and methodological approaches that may provide acceptable data for the research synthesis. Not all research syntheses, however, are focused on interventions. In chapter 2 of this volume, Harris Cooper discusses alternative research approaches that may provide an assessment of descriptive or other nonquantitative research. The research synthesist must understand the topic well enough to cover its important domains.

It is important to note that literature on many topics is scattered. When researchers in different disciplines conduct research on a topic, their frame of reference and the vocabularies of their disciplines differ. Articles on similar or related topics are published in different journals, cited in different indexes, and read by different audiences. When a new research area is emerging, as different researchers coin their unique terms to describe a phenomenon, lack of consistency of vocabulary is a problem. Yet the research in these different fields may be related.

Because of this problem of inconsistent vocabulary, it is important that the research synthesist carefully construct a set of relevant search terms that represents different approaches to the topic, and includes synonyms that may be used in early stages of the development of the area. An example of this may be seen in the 1970s where lines of related psychological research used the terms *chunking* versus *unitizing*. Claudia Cohen and Ebbe Ebbesen used the term *unitizing* to describe schemas provided to subjects to be used in organizing streams of information to be processed (1979). They found that providing different unitizing schemas resulted in different processing of information and subsequently different assessments of personality of persons observed. Robert Singer, Gene Korienek, and Susan Ridsdale used different learning strategies (one of which was chunking) to assess retention and retrieval of information as well as transfer of information to other tasks (1980). Both studies involved cognitive processing and organization of information and its impact on

Table 5.1 Subject Terms in PsycINFO

Search	Citation	Subject Terms
Unitizing	Cohen, Ebbesen (1979)	Classification (cognitive process) Experimental instructions Impression formation Memory Personality traits
Chunking	Singer, Korienek, Ridsdale (1980)	Fine motor skill learning Imagery Retention Strategies Transfer (learning) Verbal communication

SOURCE: Authors' compilation.

later judgments. It is true that the two groups studied different facets of the psychological phenomenon and pursued differing theoretical propositions. However, though these lines of research are related, the search terms attached to the two studies in the PsycINFO database were totally different, as noted in table 5.1.

The definition of the topic should

- reflect the scope as well as the limits of the research precisely,
- include all important concepts and terms,
- indicate relationship among concepts, and
- provide criteria for including and excluding materials.

Returning to our sample topic on eyewitness testimony, several primary concepts lie at its core: memory, accuracy, adult, eyewitness, and testimony. To expand the list of possible search terms and identify those most frequently used, based on preliminary reading and quick preliminary searches, we started with several documents relevant to our topic. The citations to the Loftus book and the articles noted earlier were found in PsycINFO.

We examined the subject terms attached to these documents in their source citations by the source indexers. We found a wide variety of subject terms, which are presented in table 5.2. It is interesting to note that in five documents, terms such as eyewitness, witness, testimony, and accuracy were not among those subjects assigned by the PsycINFO indexers to the documents. Thus, retrieval of these documents on eyewitness testimony would have been difficult using these subject terms.

Note that for the purpose of accuracy, we include both the ISSN (International Standard Serial Number) for the serial publication (journal) in which each of these articles was published, and the accession number assigned to the citation within the index (for example, PsycINFO, Wilson Web OmniFile Full Text Mega) in which the citation was identified. The ISSN is a unique eight-digit number assigned to a print or electronic serial publication to distinguish it from all other serial publications. An accession number is a unique identifier assigned to each citation within a reference database that differentiates one citation from all others. (Keeping track of this is especially important if you need to request the Interlibrary Loan (ILL) of a document, to avoid confusion and increase the probability of your receiving exactly what you request. Some ILL departments may request either the ISSN or the name of the reference database and citation's accession number.)

To expand the list of subject search terms, we consult thesauri used by sources likely to contain relevant citations to documents. Many reference databases, such as PsycINFO and MEDLINE, use a controlled vocabulary of subject terms that describe the topical contents of documents indexed in the database. This controlled vocabulary defines what terms will or will not be used as subject indexing terms within the reference database and how they are defined for use. A thesaurus is an alphabetical listing of the controlled vocabulary (subjects, or descriptors) used within the reference database. A thesaurus may use a hierarchical arrangement to identify broader (more general), narrower (more specific) and related subject terms.

Table 5.2 Citations to Documents in a Preliminary Search on Eyewitness Testimony

Document	Source	Subject Terms
Buckhout, Robert. 1974. "Eyewitness Testimony." *Scientific American* 231(6): 23–31. ISSN: 0036-8733. PsycINFO Accession # 1975-11599-001.	PsycINFO	Legal processes Social perception
Johnson, Craig L. 1978. "The Effects of Arousal, Sex of Witness and Scheduling of Interrogation on Eyewitness Testimony." *Dissertation Abstracts International* 38(9-B): 4427–4428. Doctoral Dissertation, Oklahoma State University. ISSN: 0419-4217. PsycINFO Accession # 1979-05850-001.	PsycINFO	Crime Emotional responses Human sex differences Observers Recall (learning)
Leinfelt, Fredrik H. 2004. "Descriptive Eyewitness Testimony: The Influence of Emotionality, Racial Identification, Question Style, and Selective Perception." *Criminal Justice Review* 29(2): 317–40. ISSN: 0734-0168. Wilson Web OmniFile Full Text Mega Accession # 200429706580003.	Wilson Web OmniFile	Identification Witnesses-psychology Eyewitness identification Race awareness Recollection (psychology) Witnesses
Loftus, Elizabeth F., and Guido Zanni. 1975. "Eyewitness Testimony: The Influence of the Wording of a Question." *Bulletin of the Psychonomic Society* 5(1): 86–88. ISSN: 0090-5054. PsycINFO Accession # 1975-21258-001.	PsycINFO	Grammar Legal processes Memory Sentence structure Suggestibility
Loftus, Elizabeth F., Diane Altman, and Robert Geballe. 1975. "Effects of Questioning Upon a Witness' Later Recollections." *Journal of Police Science & Administration* 3(2): 162–165. ISSN: 0090-9084. PsycINFO Accession # 1976-00668-001.	PsycINFO	Interviewing Memory Observers
Wells, Gary L. 1978. "Applied Eyewitness-Testimony Research: System Variables and Estimator Variables." *Journal of Personality and Social Psychology* 36(12): 1546–1557. ISSN: 0022-3514. PsycINFO Accession # 1980-09562-001.	PsycINFO	Legal processes Methodology

SOURCE: Authors' compilation.

NOTE: Examples identified on eyewitness testimony with associated subject terms assigned by source indexers.

We first examined the online thesaurus of PsycINFO for the terms *eyewitness*, *testimony*, and *memory accuracy*. This identified for us the date when each search term was added to the index vocabulary. Before the date it was included, a search term would not have been assigned to any citations to document and thus would not be useful in retrieving citations in that reference database source. This is especially important if searching is limited to subject field searching. In a scope note, the thesaurus defined the meaning of the term and how it is used. It also identified preferred and related terms. Of interest is the fact that *eyewitness* is not used in PsycINFO; instead the term used is *witness*. *Expert testimony* is used, but we cannot find the term *testimony*. *Memory accuracy* is not used, but a variety of related terms are available—for example, *memory*,

early memories, *short-term memory*, *visuospatial memory*, and so forth. Examples are provided in table 5.3.

As noted earlier, different sources often use different subject search terms. This can be seen in table 5.4, where we compare terms used in PsycINFO, MEDLINE, Criminal Justice Periodicals Index, and Lexis-Nexis Academic Universe. (Academic Universe provides worldwide, full-text access to major newspapers, magazines, broadcast transcripts, and business and legal publications. It is included here as an example of a general information resource, as compared to the specialized, research-oriented services used by those searching the research literature.) Note that PsycINFO prefers to use the age group field to limit the search as *adult (18 years and older)* rather than using the subject term *adult*. Both PsycINFO and MED-

Table 5.3 Subject Terms in PsycINFO Thesaurus

Initial Term Investigated	Preferred Term	Year Added	Related Term if any	Scope
Eyewitness	Witness	1985	—	Persons giving evidence in a court of law or observing traumatic events in a nonlegal context. Also used for analog studies of eyewitness identification performance, perception of witness credibility, and other studies of witness characteristics having legal implications.
Testimony	—	—	—	Term is NOT included in the PsycINFO Thesaurus
Legal testimony	Legal Testimony	1982	—	Evidence presented by a witness under oath or affirmation (as distinguished from evidence derived from other sources) either orally or written as a deposition or an affidavit.
Memory accuracy	—	—	Memory Early memories Short-term memory Visuospatial memory Autobiographical memory Explicit memory False memory Iconic memory Implicit memory etc.	Term is NOT included in the PsycINFO Thesaurus

SOURCE: Authors' compilation.

LINE have carefully developed thesauri of subject search terms, allowing us to learn how subject terms are used within the PsycINFO and MEDLINE indexes. Neither Criminal Justice Periodical Index or Lexis-Nexis, however, have such a source available to the user. When a reference database does not provide a thesaurus, it is especially important to use preliminary searches to identify relevant citations and ascertain from them what subject terms are being used by the database to index documents relevant to your topic.

5.3.2.3 Search Profile The third step is to provide a logical structure for the search. Having identified an initial set of search terms using a variety of methods noted earlier, we next would use boolean operators to link concepts. Boolean algebra was invented by George Boole (1848), refined by Claude Shannon (1940) for use in digital circuit design, and linked with Venn Diagrams developed by John

Venn (1880) to provide what is known today as boolean logic. The boolean AND operator is used to link concepts as an intersection—both concepts must be attached to retrieve a document. For example, with *eyewitness AND testimony* both terms must be present for a citation to be retrieved. AND is generally used to restrict and narrow a search. The boolean OR operator is used to link concepts as a union—either concept being found related to the document would retrieve the document. For example, *reliability OR accuracy* would retrieve any citation that included either of these terms. OR is generally used to expand and broaden a search. The boolean NOT is used to exclude a particular subject term. For example, *NOT children* would exclude from a search terms that included reference to children. The boolean NEAR may be used in some indexes to link terms that may not be adjacent, but must be near each other within the text to be retrieved. In some

Table 5.4 Subject Search Terms in Different Sources

Source	Memory Accuracy	Adult	Eyewitness	Testimony
PsycINFO (Use thesaurus)	Memory Narrower terms: 　Autobiographical memory 　Early memories 　Episodic memory 　Explicit memory 　False memory 　Implicit memory 　Memory decay 　Memory trace 　Reminiscence 　Repressed memory 　Short-term memory 　Verbal memory 　Visual memory 　Visuospatial memory Related terms: 　Cues 　Forgetting 　Hindsight bias 　Procedural knowledge 　Serial recall	Limit search by: Age Group: 　Adult (18 years 　& older) AND Population: Human	Witness	Expert testimony
MEDLINE (Use MESH – Medical Subject Headings)	—	—	"Eyewitness" not used "Witness" not used	"Testimony" not used "Expert testimony" used
Criminal Justice Periodical Index – ProQuest (No thesaurus available)	Memory	Adult	Eyewitness Witness	Testimony Report Evidence
Lexis-Nexis (No thesaurus available)	Memory Reliability Accuracy	—	Eyewitness Witness	—

SOURCE: Authors' compilation.

reference databases a wild card such as an asterisk (*) may be used to include multiple variants using the same truncated root or stem. For example, using *adult** would retrieve citations containing any of the following terms: adult, adults, adultery, adulterer. The risk when using wild cards is that irrelevant terms may inadvertently become a part of the search strategy, as in the preceding example.

In our example, we have identified five important concepts. These are *eyewitness*, *testimony*, *memory*, *accuracy*, and *adult*. These would be logically joined by an AND. Additionally, there are alternate terms related to each of these concepts. Because different disciplines, authors, and reference databases use differing terminology, we need to include all of these, as noted conceptually in figure 5.1. These sets of alternative terms would be linked with OR operators within our search. The portion in the middle of the figure, the intersection where all five conceptual areas overlap, represents the set of items that we would expect should be retrieved from our search. Our search may be incomplete when it fails to include one or

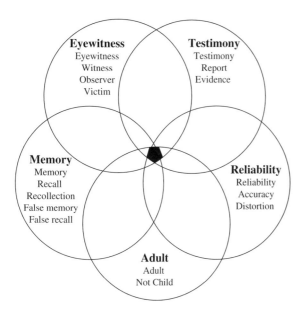

Figure 5.1 Venn Diagram Illustrating the Intersection of Five Concepts and Their Related Terms in Eyewitness Testimony Search Strategy

SOURCE: Authors' compilation.

more key terms in one of the conceptual areas, or when the citation itself includes a term or terms different than our search parameters because the authors or indexers are using a different vocabulary.

There may be an implied relationship among concepts in a search. A computer bibliographic search, however, will not generally deal with the relationship among concepts. For the most part, the current generation of bibliographic search tools available in 2008 is limited to boolean relationships among identified concepts. This will limit the precision of a search. For example, suppose the research topic had been "the impact of teacher expectations on academic performance of children in differing racial groups." The technology of the search software generally available today does not allow making a distinction between two possible topics: the influence of knowledge of academic performance scores on the expectations of teachers concerning their students' performance and the impact of teachers' expectations in influencing students' academic performance. We must examine each document identified to determine whether it is related to the former or the latter topic.

5.3.2.4 Main Searches The fourth step is to begin searching. The preliminary searches have begun to explore the field, led to some key documents, helped define the topic, explored which reference databases will be most useful, and identified important search terms. Now it is time to begin extensive searching to execute the search strategy developed. Each major discipline that has an interest in the topic has a reference database and that database must be searched to retrieve the relevant literature of the discipline.

Important terms that reflect the content of a document may appear in the title or abstract of a document, though they may not be terms assigned as subjects to a document or terms included in the controlled vocabulary of a reference database. In the age of the printed index and early years of electronic reference databases, one was limited to searching by subject search term. Searching the subject index to find terms the indexers used was limited by the controlled vocabulary of the index and the accuracy of indexers. Today the researcher has an ability with an electronic reference database to expand the search by looking beyond the subject or author index to scan the title or abstract (if your reference database includes abstracts), and in some cases even the document itself if it is available electronically, for relevant terms whether they are included in the controlled vocabulary (subject search terms) or not included (free-text or natural language search terms).

In a first pass of the main search phase, we conducted searches of four reference databases—PsycINFO, Criminal Justice Periodical Index, MEDLINE, and ProQuest Dissertations & Theses. Our first search used the strategy of *eyewitness testimony + memory accuracy*. These searches were disappointing. Only seven citations were identified in these databases. Because we knew that there are many other documents available, we also knew that these initial searches were too narrowly focused. We decided to expand our search strategy.

We knew that changing the search strategy would change the search results in a research database. Using the example of eyewitness testimony, we conducted several searches of the same database with slight modifications to the search structure in order to find additional citations to documents and to illustrate how changing search parameters can affect search results. This is illustrated in table 5.5, reporting the results of several different searches of the PsycINFO database. In each instance we conducted a free text search, included all dates of publication, and searched all text fields. In the six searches conducted, the number of citations retrieved ranged from

Table 5.5 Comparison of Eyewitness Testimony Search Results in PsycINFO

#	Search Terms	Field searched	Dates	Results
1	Eyewitness AND Testimony	All text fields	All	795 citations
2	Eyewitness AND Testimony AND Memory	All text fields	All	430 citations
3	Witness AND Testimony AND Memory	All text fields	All	244 citations
4	Eyewitness AND Testimony AND Adult	All text fields	All	142 citations
5	Eyewitness AND Testimony AND Memory accuracy	All text fields	All	8 citations
6	Witness AND Testimony AND Memory accuracy	Subject field	All	0 citations

SOURCE: Authors' compilation.

795 citations in the least restrictive search to zero citations in the most restrictive search.

By now, given the large number of citations identified, it should be clear that the sample search topic is far too broad and multifaceted to result in a focused synthesis. This would result in examining hundreds of documents, some that are not useful. The documents retrieved address multiple aspects of eyewitness testimony. For a credible research synthesis, the topic would probably need to be focused more narrowly. Earlier meta-analyses have focused on issues such as the relationship between confidence and accuracy (Sporer et al. 1995), eyewitness accuracy in police showups and lineups (Steblay, Dysart, and Lindsay 2003), magnitude of the misinformation effect in eyewitness memory (Payne, Toglia, and Anastasi 1994), modality of the lineup (Cutler et al. 1994), impact of high stress on eyewitness memory (Deffenbacher et al. 2004). We need to identify a unique aspect of the topic not previously investigated.

There is also a need to be aware of earlier research syntheses. As this handbook demonstrates, the field is growing rapidly and the number of people worldwide who have an interest in evidence based practice is increasing. At the same time that the number of published meta-analyses and research synthesis papers is increasing, collaborative international efforts to support this practice have been initiated. Along with these efforts, unique reference databases have been created by the Cochrane and Campbell Collaborations.

Cochrane and Campbell. Observing the rapid expansion of information in health care, the Cochrane Collaboration was established in 1993 to support improved health care decision making by providing empirical summaries of research. Funded by the British National Health Service, the U.K. Cochrane Centre fostered development of a multinational consortium. Guidelines for conducting and documenting research syntheses were developed. Team-based research syntheses on health care issues were coordinated. A database of clinical trials was developed as a repository of research studies that meet standards for inclusion in meta-analyses. A second database of research synthesis reports sponsored by the Cochrane Collaboration was developed. Today the Cochrane Collaboration is a worldwide collaborative research effort with offices in several countries. Information about the Cochrane Centre and Cochrane Library is available online (http://www.cochrane.org).

The Campbell Collaboration was established in 2000 to focus on education, criminal justice, and social welfare, and to provide summaries of research evidence on what works. The structure of the Campbell Collaboration is similar to Cochrane's, with multinational research synthesis teams and databases of resources (for more, see http://www.campbellcollaboration.org).

As part of the process, consider whether to consult the databases of either the Cochrane or Campbell Collaboration to find a research synthesis that may have been conducted or to identify studies that are relevant to your project.

Because the focus of this book is research synthesis and meta-analysis, it is expected that the researcher will focus on empirical research for inclusion in his or her meta-analysis. The types of control methods, data, and analysis that will be included in the research synthesis must be considered to define the standards for including and excluding research reports. Based on our experience, many citations identified and their associated documents do not report studies that provide empirical data or meet the criteria of the research synthesist for inclusion in a meta-analysis. Many citations retrieved in sample searches represent opinion pieces, comments, reviews of literature, case studies, or other nonempirical pieces, not

Table 5.6 Targeting a search of Eyewitness Testimony in PsycINFO

Search Parameters	Targeting	Results
Reference Database		
(Eyewitness OR Witness OR Observer OR Victim) AND (Testimony OR Report OR Evidence) AND (Memory OR Recall OR Recollection OR "False Memory" OR "False Recall") AND (Reliability OR Accuracy OR Distortion)	Age Group: Adult (18 years & older) AND Population: Human Methodology: Treatment Outcome / Clinical Trial	0 citations
(Eyewitness OR Witness OR Observer OR Victim) AND (Testimony OR Report OR Evidence) AND (Memory OR Recall OR Recollection OR "False Memory" OR "False Recall") AND (Reliability OR Accuracy OR Distortion)	Age Group: Adult (18 years & older) AND Population: Human Methodology: Quantitative Study	29 citations
(Eyewitness OR Witness OR Observer OR Victim) AND (Testimony OR Report OR Evidence) AND (Memory OR Recall OR Recollection OR "False Memory" OR "False Recall") AND (Reliability OR Accuracy OR Distortion)	Age Group: Adult (18 years & older) AND Population: Human Methodology: Empirical Study	153 citations

SOURCE: Authors' compilation.

reports of controlled research The search may need further refinement to identify the types of studies appropriate for a meta-analysis, a further refinement of the search profile discussed earlier in section 5.3.2.3. The Cochrane Collaboration views this as targeting or developing a highly sensitive search strategy.

The most extensive work on targeting bibliographic searches has been done in the field of health care. The appendix to the *Cochrane Handbook for Systematic Reviews of Interventions* (Higgins and Green 2006), provides a highly sensitive search strategy for finding randomized controlled trials in MEDLINE. There is differentiation between different vendors—SilverPlatter, Ovid, and PubMed—who provide access to MEDLINE. The Cochrane strategy suggests addition of certain key terms and phrases within the search to target it to relevant types of studies, such as *randomized controlled trials*, *controlled clinical trials*, *random allocation*, *double blind method*, and so forth. The terms selected have been validated in studies of the effect of retrieving known citations from the MEDLINE database using the prescribed approaches (Robinson and Dickersin 2002). This approach takes advantage of subject headings available and assigned to citations in MEDLINE. Julie Glanville and her colleagues have extended this work, validating that the "best discriminating term is 'Clinical Trial' (Publication Type)" when searching for randomized controlled trials in MEDLINE (2006). Giuseppe Biondi-Zoccai and his colleagues suggest a simple way to improve search accuracy,

by excluding from the search non-experimental, non-controlled reports such as literature reviews, commentaries, practice guidelines, meta-analyses, and editorials (2005).

When one expands beyond health care and MEDLINE, the process of refining the search to focus on sources that would be included in a meta-analysis is more difficult because methodology descriptors appear not to be consistently applied within reference databases. We used the sample topic on eyewitness testimony and focused the search within PsycINFO (the provider was EBSCO) by using the following limiters: age group = adult (over age eighteen), subject population = human, and methodology type. Table 5.6 reports the results of three searches specifying different methodologies—treatment outcome–clinical trial (no citations retrieved), quantitative study (twenty-nine citations retrieved), and empirical study (153 citations retrieved). Note that this research synthesis topic is not focused on clinical trials, and so this limiter produced no results. By not selecting methodologies such as qualitative study, or clinical case study, citations identified by PsycINFO indexers as using these methodological types are excluded from our search. One might find additional citations by using a key limiting term within the search itself instead of as a limiter.

Examining other research databases, the situation differs. ERIC (EBSCO vendor) and Social Work Abstracts (WebSpirs vendor) do not provide methodology fields as search limiters. To limit by methodology in the field of education or social work using these reference databases, one would need to specify methodology limiters within the set of search terms used. Lists such as those provided in *The Cochrane Handbook* (Higgins and Green 2006) provide a starting point for identifying terms that may be useful to qualify the search and increase its precision. The methodology that provides usable information for research synthesis may be described by terms such as: experiment, field experiment, or control group. A risk of methodology limits is that the term specified may not have been used by the author or subject indexer, resulting in a missed citation.

One must always be concerned with the trade-off between false positives (poor accuracy resulting in unnecessary work to examine and eliminate irrelevant research reports) and false negatives (the attempt to get better precision yielding more relevant citations). At present, given the lack of methodological clarity in many reference databases, gaining increased precision through the use of methodological limiters within the search is a challenge.

5.3.2.5 Document Evaluation The fifth step is to evaluate the various documents identified in the reference database searches. This will require time and effort. It may require examining many false positives if the search is overly broad. When evaluating search results, it is advised to begin by examining the citation itself (title and abstract) to determine if the cited document has some relevance. For example, does it meet the criteria for the synthesis? Is the methodology appropriate? Is the document's subject within the defined scope? If the study appears relevant, one needs to next examine the document itself. This may require an interlibrary loan if the local library does not have a document and it is not available electronically. In examining documents, watch for additional studies that were not identified in the reference database searches. Does the author cite an unpublished conference paper, an unpublished technical report, or the report of research conducted under a grant? Is there information about a body of work being done at a research institute in another part of the world? In both of these situations, one is advised to contact the author of the study. At this point, you may need to conduct hand searches (see chapters 4 and 6, this volume). When examining a chapter in a book, check to see if there is another chapter in the book that is relevant. We found additional research reports on eyewitness testimony in this way. The presence of any of these hints provides an alert that additional searches may be required of additional databases that have not yet been tapped. Participation in Cochrane or Campbell may provide access to an international network of scholars enabling you to collaborate with others to tap into that previously invisible research institute in another country.

Some researchers prefer to perform the examination of documents as they proceed—for example, conduct the search of a database and then evaluate the results before proceeding to the next search—rather than batching all of the this activity until after all of the searches have been conducted. This ongoing examination enables the researcher to continuously validate the quality of the search strategy itself. Identification of false negatives might suggest a need to modify the strategy.

5.3.2.6 Final Searches The sixth step, following examination of documents and possibly the analysis of data, is to fill in the gaps with supplemental focused searches. If the process of examining studies and conducting analyses has taken a significant amount of time, it may be important to revisit searches previously conducted to identify and recover recent documents. Here one can probably limit the search to the period of time since the search was last conducted, for example, the most recent year. This update is an important step, given that one probably does not want to publish a research synthesis and find that an important new study, or one with divergent results, has been overlooked.

5.3.3 Author Search

A useful complementary strategy is to conduct author searches in the selected reference databases. Identify several key authors who have published extensively on the topic. In the eyewitness testimony example, several authors are influential and prolific within the psychological literature—to illustrate, Elizabeth Loftus, Steven Penrod, and Gary Wells. Searching the author field of the reference database located additional citations. The drawback of this approach was the retrieval of many false positives. Each of these researchers has published in a number of areas, yielding many citations retrieved that were irrelevant to the topic at hand.

5.3.4 Citation Searching

Citation searching is an important complement to searching the literature by subject. It allows the researcher to avoid reliance on the subjectivity of indexers and avoids the inherent currency lag and biases of controlled vocabulary. Citation searching relies on three important assumptions: one or several key chronologically early sources exist on a topic, these key sources are well known within the field, and these key sources are cited by others conducting and publishing subsequent research on the topic.

A citation index can identify all articles, reports, or other materials in the body of literature it covers that have cited the key reference. To conduct a citation search, begin by identifying several important, seminal sources (articles or books) on the research topic (see, for example, Loftus 1975; Wells 1978). These works are highly likely to be cited by researchers working on problems related to eyewitness accuracy, and as such are good candidates for a citation search.

Three indexes of interest to social and behavioral scientists allow access to research literature by their cited references: Social Sciences Citation Index (SSCI), Science Citation Index (SCI), and PsycINFO. In the case of PsycINFO, inclusion of cited references in records and the ability to conduct searches on them is a reasonably recent development. Because cited references from source articles began only in 2001, the remaining two indexes, both published by ISI, will provide more extensive coverage. To conduct a citation search, one would specify an original document, for example, the Loftus article, within the search engine. The system would search its database and produce a list of other citations contained within its database that have made reference to your initial citation.

5.4 ISSUES AND PROBLEMS

5.4.1 Need for Searching Multiple Databases or Indexes

Each database or index is designed for a particular audience and has a particular mission, as noted in the section on databases. Sources that overlap in topic, publication formats covered, or source title coverage often have at least as many differences as they do similarities. On topics that are interdisciplinary (such as, stress, communication, conflict, leadership), it is essential to consult multiple databases. This is equally important when using databases that appear so comprehensive as to be duplicative. One may elect to consult both ERIC and the Wilson Education Index to ensure comprehensive coverage in an education project.

MEDLINE and EMBASE include extensive, international coverage of the life sciences and biomedical literature, indexing thousands of publications (Weill 2008). Although a comprehensive research synthesis demands exploiting the contents of both, Margaret Sampson and her colleagues suggested that researchers might make

measured trade-offs if using both databases proves impractical (2003). Failure to do so may result in an incomplete search and missed sources (numerous false negatives). Most databases provide a list of journals and sometimes serials indexed. Consulting such lists allows you to determine the overlap or independence of the indexing sources.

5.4.2 Need for Multiple Searches

In conducting an exhaustive search on a topic, multiple searches are generally needed. Searches also occur in phases.

As noted, searches in the first phase will be preliminary, exploratory searches to understand the topic, identify search terms, and confirm which reference databases are relevant for fuller searches.

The second phase will be much more extensive and thorough, and will require multiple searches of each relevant reference database. It will also require searching grey literature. Each search on the sample topic returned different results. For example, following examination of PsycINFO (described in table 5.5), we turned to the Criminal Justice Periodical Index provided by ProQuest. Several different searches were conducted using varying search configurations. As in PsycINFO, results were different for each search. Though there was some overlap between these searches, new searches yielded new citations that appeared potentially relevant. Results of six searches of Criminal Justice Periodical Index are noted in table 5.7. Because many journals indexed in this source are not included in PsycINFO, this set of searches yielded additional documents that should be considered for inclusion in our synthesis.

Subsequent searches of ProQuest Dissertations & Theses, Lexis-Nexis Academic Universe, and MEDLINE produced similar results. In consulting each, we conducted several searches, varying the terms included and the logic structure. The result is that in gradually expanded the fishing net, we gradually expanded the number of citations and documents of potential relevance to our research synthesis. Had we not done so, we would have failed to uncover relevant documents. It was interesting that a number of dissertations and theses have been focused on various aspects of eyewitness testimony. The first, a fairly broad search, identified 149 dissertations or theses that address some aspect of eyewitness testimony. These must be examined to ascertain which are relevant to the project.

Table 5.7 Eyewitness Testimony Search Results in Criminal Justice Periodical Index Pro Quest.

#	Search Terms	Field Searched	Dates	Results
1	Eyewitness AND Testimony	Citation Abstract	All	105 citations
2	Eyewitness AND Testimony AND Memory	Citation Abstract	All	13 citations
3	Witness AND Psychology	Citation Abstract	All	37 citations
4	Eyewitness OR Witness AND Testimony OR Report AND Memory AND Accuracy	Citation Abstract	All	6 citations
5	Eyewitness AND Reliability AND Memory	Citation Abstract	All	4 citations
6	Victim AND Testimony	Citation Abstract	All	205 citations

SOURCE: Authors' compilation.

Table 5.8 Comparable Eyewitness Testimony Searchers from Three Databases

Search Parameters	Results	Uniqueness
PsycINFO (Eyewitness OR Witness OR Observer OR Victim) AND (Testimony OR Report OR Evidence) AND (Memory OR Recall OR Recollection OR "False Memory" OR "False Recall") AND (Reliability OR Accuracy OR Distortion) AND Age Group: Adult (18 years & older) AND Population: Human	165 citations	Duplicated in Criminal Justice Periodical Index = 2 Duplicated in Dissertations & Theses = 13 Unduplicated = 150 (91%)
Criminal Justice Periodicals (Eyewitness OR Witness OR Observer OR Victim) AND (Testimony OR Report OR Evidence) AND (Memory OR Recall OR Recollection OR "False Memory" OR "False Recall" OR Reliability OR Accuracy OR Distortion)	196 citations	Duplicated in PsycINFO = 2 Duplicated in Dissertations & Theses = 0 Unduplicated = 194 (99%)
ProQuest Dissertations & Theses (Eyewitness OR Witness OR Observer OR Victim) AND (Testimony OR Report OR Evidence) AND (Memory OR Recall OR Recollection OR "False Memory" OR "False Recall") AND (Reliability OR Accuracy OR Distortion)	110 citations	Duplicated in PsycINFO = 13 Duplicated in Criminal Justice Periodical Index = 0 Unduplicated = 97 (88%)

SOURCE: Authors' compilation.

How much overlap is there across reference databases? The answer depends on the topic and field. Searches were conducted of three databases (PsycINFO, Criminal Justice Periodical Index, and ProQuest Dissertations & Theses) using the sample topic as articulated in figure 5.1. Each search was structured as close to the topic definition as possible, with slight variations resulting from the search configuration allowed by the database. Each database contained over one hundred citations to potentially relevant studies. The vast majority of studies (more than 88 percent) found in one reference database source were not found in the other two shown in table 5.8. Thus, for completeness, it is essential that the research synthesist investigating eyewitness testimony consult at least these three databases. Then recall that these three databases do not include conference papers, technical reports, and other formats, and thus additional searching is appropriate, if not critical.

5.4.3 Manual Searching Versus Electronic Databases

We are blessed with powerful search engines and extensive databases. However, these sources are relatively new.

Some sources, such as PsycINFO, have added older citations retroactively, however, not all have done so. As a result, you may need to conduct manual searches of the subject and author indexes reference databases that are only available in hardcopy form.

For example, if your topic predates the initiation of the electronic ERIC reference database in 1966, you may consult paper indexes such as Education Index to identify citations before 1966. Wilson Education Index began as an electronic index in 1983, and added abstracts in 1994, but its hardcopy predecessor was launched in 1929. Wilson has provided a new service, Education Index Retrospective, that covers indexing of educational literature from 1929 to 1983. In this case, one would supplement an ERIC search with a search of the Wilson products.

5.4.4 Limits of Vocabulary Terms

Before beginning a search, it is a good idea to check the controlled vocabulary of a database by referring to the published thesaurus for that database, if one is available. As noted earlier, the thesaurus will identify those terms used by indexers (in PsycINFO, they are called *subjects*) to describe the subject content of a source when entering citations into the database. In most databases, the record for a citation typically contains a listing of the subjects used to index that source, in addition to an abstract and other information. In many databases (such as PsycINFO), the subject terms are hyperlinked, enabling you to do a secondary search to find articles indexed using those subjects. The thesaurus can be helpful in identifying terms related to your topic that you had not thought of using.

There are limits to use of controlled vocabulary, however. In an emerging subfield, it may take several years for particular terms to emerge as recognized descriptors of the subject content, and to find their way into the controlled vocabulary of the thesaurus. For example, *Psychological Abstracts* (and PsycINFO) did not include the terms *intelligence* or *intelligence quotient* until 1967, and still does not include the term *IQ* in its controlled vocabulary. Because indexes rarely do retrospective revisions to subject terms in materials already indexed, you must use a variety of terms to describe the subject content of a new field. Human error also limits use of subject terms. Some indexers, for example, may not be familiar with an emerging field, or may have biases against certain terms, limiting reliance on the controlled vocabulary.

On the other hand, there are limits to free text searching, especially if one does not consult a thesaurus. The most significant problems are the occurrence of false negatives (failure to find relevant sources that exist) because important subject descriptors were not included in the search structure and identification of false positives.

The bottom line is that it makes sense to include both thesaurus subject terms, as well as relevant nonthesaurus terms in your search. This makes it even more critical that your early work and planning be thorough and grounded in reading and preliminary searching forays into the databases.

5.4.5 Underrepresented Research Literature Formats

As we discussed earlier in the context of database selection, some research publication formats are not as well represented in reference databases as others. Locating information about studies not published in the journal literature and in non-English languages—as well as obtaining such documents—can be challenging. It bears repeating that, after constructing or beginning a literature search strategy, locating such material takes persistence and a search strategy that casts a broad net for relevant literature.

5.4.5.1 Reports and Papers Some conferences publish proceedings that may be covered by a relevant indexing service (database). Some researchers may take the initiative to submit a conference paper to a service such as ERIC if their research is within the domain of a particular index. However, information on much of this material remains inaccessible as fugitive or grey literature in a discipline (Denda 2002).

5.4.5.2 Books and Book Chapters Information on book-length studies is not difficult to find, largely because of extensive, shared library cataloging databases used for many years by libraries. Primary among these is OCLC WorldCat, which contains records for books, manuscripts, and nonprint media held by thousands of libraries. More recently, it has incorporated information about Internet sites, datasets, media in emerging formats, and so forth.

Information on the individual chapters within books remains challenging, however, especially regarding edited volumes in which chapters are written by many individuals. PsycINFO is unique in that it includes citations to English-language books and chapters back to 1987, with selected earlier coverage. Once identified, most books published by either commercial or noncommercial publishers can be acquired more easily than unpublished material.

5.4.5.3 Conference Papers and Proceedings In addition to the format-specific databases, you may need to consult conference programs in a relevant field to identify important unpublished documents. Acquiring conference papers often requires time and persistence, especially when it comes to the complete text of a conference presentation (plus any presentation media) not subsequently published in a journal article, book chapter, or as part of conference proceedings book produced by a commercial publisher.

Although these materials may provide valuable information, as they are not published in the typical sense of the word, and may not have been through a review process, the results may be preliminary and quality of the research may be unknown. Thus, such research must be evaluated and used accordingly.

5.4.6 Inconsistent Coverage Over Time

From their inception in the 1960s, bibliographic databases suffered from uneven coverage and, to some extent, this continues as publication formats have evolved and research has become more interdisciplinary. For example, the PsycINFO bibliographic database provides researchers with retrospective coverage of the discipline's research literature, dates of coverage that span many decades, and a highly structured record format with a supporting controlled subject vocabulary. Given this, it does present many idiosyncrasies in coverage that bear consideration. The controlled vocabulary was uniformly applied only to citations indexed after 1967. Coverage of books and book chapters have been consistently added since 1987 but only those published in English. The ability to search using cited references enhances the usefulness of the database, but addition of this feature was relatively recent. Its coverage of open-source e-journals is inconsistent. The case of PsycINFO illustrates a basic caveat to keep in mind when using any bibliographic database: the researcher should examine coverage and usage notes when using a file.

5.4.7 Non-U.S. Publication

Many scholars rely heavily on sources published in the United States and rely on reference databases published in the United States. However, there is much research conducted and published outside the United States and in languages other than English, and studies published in English appear to be associated with larger effects (see chapter 6). In many disciplines, the researcher must find ways to identify documents outside of the limited scope of their language and nationality. Once a document has been identified that is written in German, or Italian, or Russian, or Chinese, you will have to examine it to determine whether it is relevant to your subject, and then extract the important details on methodology, results, and conclusions. This may require translation. At times, an available English language abstract may provide an initial clue whether the document is relevant. If a document abstract is not available in a language you read, you will need to find a way to ascertain whether the contents might be relevant.

Participation in a collaborative network such as the Cochrane or the Campbell may provide access to peers with related interests who may be willing to collaborate and assist in your task. They may be willing to assist in the search of non-U.S. databases. They may be willing to assist with translations of relevant non-English documents.

Consulting non-English or non-American documents is important in many disciplines and becomes especially important in exploring broad topics. For example, non-Western and non-U.S. literature becomes important in study of the effectiveness of health-care approaches in treatment of certain diseases. If studying a health care problem that may be treated with acupuncture, failure to consult sources that provide access to publications such as *Clinical Acupuncture & Oriental Medicine*, or *Journal of Traditional Chinese Medicine (Beijing)* may be a major error.

Non-U.S. literature is relevant in understanding many aspects of behavior in organizations—for instance, decision making, motivation, and negotiation. The field of organizational behavior was dominated for many years by researchers proposing theories and testing them in the United States. Geert Hofstede's studies of culture demonstrated that differences in national culture directly influence organizational culture (1987, 1993). When some theories developed in the United States have been tested in other countries and other cultures, conflicting data sometimes emerged calling into question the cross-cultural applicability of the theoretical propositions. Failure to include non-English and non-Western sources may lead to bias and questions of generalizability of the research synthesis. To explore the non-U.S. literature, you may need to consult a colleague in another country who has access to reference databases covering literature not locally available to you.

There may be limits, however. How much searching must one do? How much time must be spent to identify that last missing relevant document is a judgment call. You must do enough searching to make certain that you have found all of the important relevant sources. On the other hand, the task without end will never be completed and published. How many hours of additional searching are justified to find one more citation?

5.4.8 Adequate Search Documentation

In documenting your search, you must disclose enough information to enable the reader to replicate your work. The reader must also be able to identify the limits of your search, and ways in which she or he might build on your work. A researcher should make this information transparent to subsequent users of the research synthesis.

Unfortunately, too often researchers do not disclose enough information in publications to enable the reader to do either of these. As readers, we often do not know what sources (databases) were consulted or the search strategy used. Of concern is the possibility that a researcher does not keep records on what reference databases were searched on what dates with what search strategies. Additionally, poor record keeping of searches conducted may lead to repetition of searches, or holes in the search.

In section of 5.2.2 of its *Cochrane Handbook for Systematic Reviews of Interventions*, the Cochrane Collaboration provides guidance on record keeping and disclosure (Higgins and Green 2006). They suggest that for each search conducted, the researcher should record

- title of database searched,
- name of the host or vendor,
- date of the search,
- years covered by the search,
- complete search strategy with all search terms used,
- descriptive summary of the search strategy, and
- language restrictions.

In its handbook, Cochrane even provides examples of search strategies for electronic databases (Higgins and Green 2006). When arriving at the level of the individual citation and document, we find that it is sometimes helpful to record the unique reference database accession number of each citation identified, in addition to bibliographic information (author, title, date, and so on).

5.4.9 World Wide Web and Search Engines

Many people jump quickly to the information superhighway to find answers to questions. Some sources are available online. Most reference databases, however, are not available directly online. There is typically a subscription fee, paid by your local institution, for access to core reference databases. Conducting your search using an Internet search engine such as AltaVista, Google, Lycos, MSN, or Yahoo will not provide access to the contents of reference databases such as PsycINFO or Criminal Justice Periodical Index. Your search may provide other information addressed in other chapters.

A new service has appeared only recently, Google Scholar. Although it shows promise, its scope is limited to materials accessible on the Web at this writing; contents of proprietary databases such as PsycINFO are generally not accessed. It thus provides access to books that can be identified in publicly available library catalogs, journals that are published online (for example, publications of Springer), journal articles that are accessed through net available research databases such as ERIC and PubMED. Only time will tell whether additional content is added.

5.4.10 Deciding When to Stop Searching

Knowing when your task is complete is as difficult as defining the topic, structuring the search, and deciding what is truly relevant. We are unaware of any empirical guidelines for completeness in the social sciences. Mike Clarke, director of the UK Cochrane Center, describes this as one of the difficult balances that has to be achieved in doing a synthesis (Personal communication, November 26, 2006). It involves the trade-off between finishing the project in a timely way and being fully comprehensive and complete. One must be pragmatic. This is a judgment task the research synthesist must embrace. We assume that you have followed the procedures suggested in this chapter. We also assume that along the way you have used other methods such as hand searching and contacting key sources for grey literature. You might want to consider the following questions:

- Have you identified and searched all of the reference databases that are likely to contain significant numbers of relevant citations?

- Have you verified that there are no systematic biases of omission—for example, you have not excluded non-English language sources?

- Have you modified your search strategy by adjusting search terms as you identified and examined citations highly relevant to your topic?

- As you varied the parameters in searching any single reference database, how many new, unique, relevant citations did you identify in the last two searches?

- As you compare the results from searching complementary reference databases, how many new, unique, relevant citations were identified in the last two searches?

- If you conducted author searches on the three or four of the most prolific authors on your topic, how many new, unique, relevant citations were retrieved on the last search or two?

If your answers to these questions is that the rate of identification of new citations in your three most recent searches is less than 1 percent, you are probably finished searching, or very close. If the last ten hours of intense database searching have produced no new relevant citations you are probably finished. If, on the other hand, you are still finding significant numbers of new relevant citations with each new search, then the task is not complete.

5.5 APPENDIX: SELECTED DATABASES IN THE BEHAVIORAL AND SOCIAL SCIENCES

The databases listed represent a selection of bibliographic files indexing empirical research studies. Particulars such as years of coverage, update frequency, and specific features are subject to frequent change, so much of this information should be verified with the database producer or vendor or by consulting members of a library reference department. Because database vendors are subject to corporate mergers and producers complete or terminate contracts with vendors frequently, specific database vendors are not listed. To ensure accurate and up-to-date status of availability and specific features available, we recommend consulting the producer web sites provided below or library reference staff for this information. *Gale Directory of Databases* is a valuable tool for verifying changes in database scope, coverage, and availability.

Ageline

Years of coverage: 1978+, with selective earlier coverage

Producer: American Association of Retired Persons (AARP), Policy and Research, Washington, D.C. (http://www.aarp.org)

Updates: Bimonthly

Publication formats included: journal articles, books and book chapters, government reports, dissertations, policy and research reports, conference proceedings

Controlled vocabulary: *Thesaurus of Aging Terminology*, 8th ed. (2005) Available for download.

This file focuses broadly on the area of social gerontology; specifically the social, behavioral, and economic experiences of those over age fifty. It cites a limited amount of consumer-oriented material, although emphasis is research publications. Freely available for searching from the AARP website (http://www.aarp.org/research/ageline/ index.html).

Campbell Library

The goal of the Campbell Collaboration is to enable people to make well-informed decisions about the effects of interventions in education, criminal justice, and social welfare. The Campbell Collaboration (C2), established in 2000, is an international nonprofit organization, modeled after the Cochrane Collaboration. C2 prepares, maintains, and disseminates systematic reviews. The Campbell Collaboration Library consists of several databases including *C2 Reviews of Interventions*, and *Policy Evaluations (C2-RIPE)*, and *C2 Social, Psychological, Education*, and *Criminological Trials Registry (C2-SPECTR)*.

C2 Reviews of Interventions, and Policy Evaluations (C2-RIPE)

Years of coverage: Reviews since 2000

Producer: Campbell Collaboration

Updates: Ongoing

Publication formats included: Research syntheses

Availability: Online (http://www.campbellcollaboration .org/index.asp).

The C2-RIPE database contains approved Titles, Protocols, Reviews, and Abstracts for systematic reviews in education, crime and justice, and social welfare.

C2 Social, Psychological, Education, and Criminological Trials Registry (C2-SPECTR)

Years of coverage: Most research reports are post 1950, with the oldest published in 1927 (as of February 2007)

Producer: Campbell Collaboration

Updates: Ongoing

Publication formats included: Citation, availability information, and usually an abstract describing randomized or possibly randomized trials in education, social welfare and criminal justice

Availability: Online (http://www.campbellcollaboration.org/index.asp)

As of February 2007, the C2-SPECTR database contained nearly 12,000 entries on randomized and possibly randomized trials. C2-SPECTR was modeled after the Cochrane Central Register of Controlled Trials.

Cochrane Library

The goal of the library is to aid health-care decision making by providing high-quality, independent evidence of the effects of health-care treatments. It is supported by the Cochrane Collaboration, an international, nonprofit, independent organization established in 1993 that produces and disseminates systematic reviews of healthcare interventions. It also promotes the search for evidence in the form of clinical trials and other studies of interventions. The library consists of several databases including the *Cochrane Database of Systematic Reviews,* the *Cochrane Central Register of Controlled Trials (CENTRAL),* and the *Cochrane Database of Abstracts of Reviews of Effects (DARE).*

Cochrane Database of Systematic Reviews

Years of coverage: Systematic research reviews since 1993

Producer: Cochrane Collaboration

Updates: Ongoing

Publication formats included: Research syntheses

Availability: Online (http://www.cochrane.org/; http://www3.interscience.wiley.com/cgi-bin/mrwhome/106568753/HOME)

As of February 2007, the *Cochrane Database of Systematic Reviews* included 4655 reviews. Systematic reviews adhere to strict design methodology. They are comprehensive (including all known instances of trials on an intervention) to minimize the risk of bias in results. They assesses whether a particular health-care intervention works.

Cochrane Central Register of Controlled Trials (CENTRAL)

Years of coverage: Earliest identified study was in 1898 (as of February 2007)

Producer: Cochrane Collaboration

Updates: Ongoing

Publication formats included: Information on controlled trials of health-care interventions.

Availability: Online (http://www.cochrane.org; http://www3.interscience.wiley.com/cgi-bin/mrwhome/106568753/ HOME)

Description: As of February 2007, the CENTRAL database included information on 489,167 controlled trials. CENTRAL includes the details of published articles (not the articles themselves) taken from health-care bibliographic databases (notably MEDLINE and EMBASE), and other published and unpublished sources. It includes the results of hand searches of health care literature, and reports not contained in published indexes. It includes non-English language documents and reports produced outside the United States.

Cochrane Database of Abstracts of Reviews of Effects (DARE)

Years of coverage: Reviews since 1993

Producer: Cochrane Collaboration

Updates: Ongoing

Publication formats included: Reviews of research syntheses

Availability: Online (http://www.cochrane.org; http://www3.interscience.wiley.com/cgi-bin/mrwhome/106568753/HOME)

DARE contains abstracts of systematic reviews in the field of health care that have been assessed for quality. Each abstract includes a summary of the review together with a critical commentary about the overall quality. As of February 2007, DARE included 3,000 abstracts of systematic reviews.

CORK: A Database of Substance Abuse Information

Years of coverage: 1978+, with selected earlier coverage

Producer: Project CORK, Dartmouth University Medical School (http://www.projectcork.org)

Updates: Quarterly

Publication formats: Journal articles, books and book chapters, conference papers, reports generated from federal and state agencies as well as foundations and professional organizations

Controlled vocabulary: Thesaurus of subject terms is accessible as part of the database. Interdisciplinary coverage of addiction research in clinical medicine, law, social work, and the broader social sciences literature. Includes addictive behaviors and co-dependency research.

Criminal Justice Periodical Index

Years of coverage: 1974+

Producer: ProQuest, Ann Arbor, Mich. (http://www.proquest.com)

Updates: Weekly

Publication formats: Journal articles

Supplies citations and abstracts for more than 200 core titles in law enforcement and corrections practice, criminal justice, and forensics. Some indexed material is available in fulltext.

EconLit

Years of coverage: 1969+

Producer: American Economic Association (http://www.aeaweb.org)

Updates: Monthly

Publication formats included: Journal articles, books, book and book reviews, chapters in edited volumes, working papers, dissertations

Controlled vocabulary: American Economic Association Classification System, previously used in the *Journal of Economic Literature*. The subject descriptor and corresponding code list are linked from the EconLit site (http://www.econlit.org).

International coverage of literature of theoretical and applied topics in economics and related areas in mathematics and statistics, economic behavior, public policy and political science, geography and area studies.

EMBASE

Years of coverage: 1974+

Producer: Elsevier Science (http://www.elsevier.com)

Updates: Daily

Publication formats included: Journal articles

Controlled vocabulary: EMBASE thesaurus called EM-TREE annual

International coverage of the biomedical literature, with recognized strength in drug-related information and strong coverage of the international journal literature.

ERIC (Educational Resources Information Center)

Years of coverage: 1966+

Producer: Computer Sciences Corporation, under contract from the U.S. Department of Education, Institute of Education Sciences, Washington, DC (http://eric.ed.gov)

Updates: Weekly

Publication formats included: Journal articles, unpublished research reports, conference papers, curriculum and lesson plans, and books. Many unpublished reports may be downloaded from the web site at no cost.

Controlled vocabulary: ERIC Thesaurus, available online from the ERIC website

The primary reference database on all aspects of the educational process, from administration to classroom curriculum, preschool to adult and postdoctoral levels. Because its scope encompasses almost every aspect of the human learning process. The contents of this file are of interest to a broad range of social scientists. Freely available for searching from the ERIC site.

International Bibliography of the Social Sciences

Years of coverage: 1951+

Producer: London School of Economics and Political Science (http://www.lse.ac.uk)

Updates: Weekly

Formats: Journal articles, books, book chapters

Emphasis on the fields of anthropology, economics, politics, and sociology, with selected interdisciplinary coverage of related areas such as gender studies, human development, and communication studies. Noteworthy for coverage of journal literature published outside the English-speaking world, including abstracts in multiple languages as those abstracts are available.

MEDLINE

Years of coverage: 1966+ A version available from some vendors includes a retrospective component with citations back to 1950.

Producer: National Library of Medicine, Bethesda, Md.

Updates: Daily

Formats: Journal articles. An increasing number of cited articles can be downloaded.

Controlled vocabulary: Medical Subject Headings, revised annually and searchable online

Exhaustive in its coverage of biomedicine and related fields for example, life sciences, clinical practice, nursing. MEDLINE indexes approximately 4,800 journals worldwide. It also includes coverage of health care issues and policy, clinical psychology and psychiatry, and interdisciplinary areas such as gerontology. An increasing number of items indexed are available online as fulltext documents. Freely available for searching from the NLM's PubMed service (http://pubmed.gov).

National Criminal Justice Reference Service

Years of coverage: 1970+

Producer: Office of Justice Programs, U.S. Department of Justice (http://www.ncjrs.gov/library.html)

Updates: Quarterly

Formats: Journal articles and newsletters, government reports (federal, state, local), statistical compendia, books, published and unpublished research reports

Controlled vocabulary: NCJRS Thesaurus, searchable online as part of the database

Focus is on criminal justice and law enforcement literature, although there is additional coverage in the areas of forensics, problems related to criminal behavior; for example, drug addiction, social deviancy, juvenile and family behavior delinquency, victim and survivor responses to crime). Freely available for use from the producer's website.

National Technical Information Service

Years of coverage: 1969+, with selected earlier coverage

Producer: National Technical Information Service, U.S. Department of Commerce (http://www.ntis.gov)

Updates: Monthly

Format: Unpublished research reports, datasets

An important source of information on reports resulting from government-sponsored research. Although emphasis is on engineering, agriculture, energy, and the biological sciences, reports also cover personnel and management, public policy, and social attitudes and behavior. The version searchable from the NTIS web site covers 1990 to the present and many recent technical reports can be downloaded as fulltext documents.

Papers First and Proceedings First

Years of coverage: 1993+

Producer: OCLC Online Computer Library Center (http://www.oclc.org)

Updates: Semimonthly and semiweekly, respectively

Publication formats: Papers presented at professional conferences and symposia

These are complementary databases, information for both derived from the extensive conference proceedings holdings of the British Library Document Supply Centre. Papers First contains bibliographic citations to presented papers or their abstracts, whereas Proceedings First indexes a compilation of papers and includes a list of those presented at a given conference. The Directory of Published Proceedings database produced by InterDok (http://www.interdok.com), is not as extensive.

ProQuest Dissertations and Theses Database

Years of coverage: 1861+

Producer: ProQuest, Ann Arbor, MI (http://www.proquest.com)

Updates: Continuous

Publication formats: Doctoral dissertations, masters theses

PDTD is a multidisciplinary index of dissertations produced by from more than 1,000 American institutions and a growing number of international universities. As such, it is a unique access point to an important body of original research literature. The very earliest dissertations contain only basic bibliographic descriptions. Lengthy descriptive abstracts of dissertations are available from 1980 forward, with shorter abstracts of masters theses beginning in 1988. This product has been enhanced by addition of "page previews" of indexed documents, generally about twenty five pages in length. If available, the complete text of dissertations can be downloaded in PDF format. Otherwise, copies can be ordered in microformat or hardcopy from the database distributor.

PsycINFO

Years of coverage: pre-1900

Producer: American Psychological Association, Washington DC (http://www.apa.org)

Updates: Weekly

Formats: Journal articles, books, book chapters, dissertations, research reports

Controlled vocabulary: *Thesaurus of Psychological Index Terms*, 11th edition (2007)

Indexes more than 2,000 journal titles in both English and foreign languages, as well as selected English-language books and other formats. Coverage of publication formats has changed dramatically since the database began in 1967, as have search features; for example, cited reference searching began in 2002 but is not accessible in all years of coverage. Coverage of the discipline of psychology is very broad and highly interdisciplinary include publications in biology and medicine, sociology and social work, and applied areas such as human-computer interaction and workplace behavior.

APA launched a series of complementary databases in 2004. The purpose of PsycEXTRA is to index grey literature as represented in conference proceedings; reports produced by federal and state governments, nonprofit and research organizations, and professional associations; nonprint media; popular and trade publications. PsycBOOKS provides fulltext access to books published by APA as well as selected out-of-print books and classic texts in psychology originally published in the late-nineteenth and early twentieth centuries.

Science Citation Index

Years of coverage : 1900+

Producer : Thomson Corporation, Stamford, Conn. (http://www.isiknowledge.com).

Updates: Weekly

Formats: Journal articles

Social Sciences Citation Index

Years of Coverage: 1956+

Producer: Thomson Reuters, Philadelphia. (http://www.isiknowledge.com)

Updates: Weekly

Formats: Journal articles

Each of these products indexes the contents of thousands of journal publications and incorporates English-language abstracts as they appeared in the journals since the early 1990s. Although it's possible to search by keywords in titles and abstracts for indexed articles, the strength of these long-standing citation indexes remains the ability to search by cited reference; that is, using key references to discover other, more recent publications. Both SCI and SSCI are accessible via the Thomson/ISI Web of Science. Scopus, a competing product produced by Elsevier (http://www.scopus.com), began production in 2006 and promises expanded coverage of grey literature formats and increasing indexing of journal titles and trade publications.

Social Work Abstracts

Years of coverage: 1977+

Producer: National Association of Social Workers, Silver Spring, Md. (http://www.socialworkers.org)

Updates: Quarterly

Formats: Journal articles and dissertations

Despite significant overlap in subject scope with other files, Social Work Abstracts supplements other indexes with its coverage of substance abuse, gerontological, family and mental health literatures.

Sociological Abstracts

Years of coverage: 1952+

Producer: Cambridge Scientific Abstracts, Bethesda, Md. (http://www.csa.com)

Updates: Monthly

Formats: Journal articles, books, conference publications, dissertations

The primary indexing tool for journal literature (almost 2,000) titles in sociology as well as interdisciplinary topics such as criminal justice, social work, gender studies, population and gender studies.

Wilson Education Index

Years of coverage: 1929+

Producer: H. W. Wilson Company (http://www.hwwilson.com)

Updates: Daily

Formats: Journals, yearbooks, and monographic series

Wilson Education Index includes coverage of 475 core international English-language periodicals and other formats relevant to the field of education from 1983 to the present. It is supplemented by Education Index Retrospective covering the period of 1929 to 1983.

WorldCat

Years of coverage: comprehensive

Producer: Online College Library Center, Dublin, Oh. (http://www.oclc.org)

Updates: Daily

Formats: Books, manuscripts and special collections, dissertations, media

Consisting of records of holdings contributed by over 9,000 member libraries in the United States and abroad, WorldCat is noteworthy for its breadth of publication formats and years of coverage. Although helpful for locating books, report series, and non-print media, published articles and book chapters are seldom included.

5.6 REFERENCES

APA Online. 2006. PsycINFO Database Information. http://www.apa.org/psycinfo/products/psycinfo.html.

APA Online. 2007. PsycINFO Journal Coverage List. http://www.apa.org/psycinfo/about/covlist.html.

Aby, Stephen H., James Nalen, and Lori Fielding. 2005. *Sociology: A Guide to Reference and Information Resources*, 3rd ed. Westport, Conn.: Libraries Unlimited.

Balay, Robert, ed. 1996. *Guide to Reference Books*. Chicago: American Library Association.

Benjamin, Ludy T., and Gary R. VandenBos. 2006. "The Window on Psychology's Literature: A History of Psychological Abstracts. *American Psychologist* 61(9): 941–54.

Biondi-Zoccai, Giuseppe GL, Pierfrancesco Agostoni, Antonio Abbate, Luca Testa, and Francesco Burzotta. 2005. "A Simple Hint to Improve Robinson & Dickersin's Highly Sensitive PubMed Search Strategy for Controlled Clinical Trials." *International Journal of Epidemiology* 34(1): 224.

Boole, George. 1848. "The Calculus of Logic." *Cambridge and Dublin Mathematical Journal, III*: 183–98.

Cohen, Claudia E., and Ebbe B. Ebbesen. 1979. "Observational Goals and Schema Activation: A Theoretical Framework for Behavior Perception." *Journal of Experimental Social Psychology* 15(4): 305–29.

Cutler, Brian L., Garrett L. Berman, Steven Penrod, and Ronald P. Fisher. 1994. "Conceptual, Practical, and Empirical Issues Associated with Eyewitness Identification Test Media." In *Adult Eyewitness Testimony: Current Trends and Developments,* edited by David Frank Ross, Don J. Read, and Michael P. Toglia. New York: Cambridge University Press.

Dale, Philip S., Elizabeth F. Loftus, and Linda Rathbun. 1978. "The Influence of the Form of the Question on the Eyewitness Testimony of Preschool Children." *Journal of Psycholinguistic Research* 7(4): 269–77.

Deffenbacher, Kenneth A., Brian H. Bornstein, Steven D. Penrod, and E. Kiernan McGorty 2004. "A Meta-Analytic Review of the Effects of High Stress on Eyewitness Memory." *Law and Human Behavior* 28(6): 687–706.

Denda, Kayo. 2002. "*Fugitive Literature* in the Cross Hairs: An Examination of Bibliographic Control and Access." *Collection Management* 27(2): 75–86.

Fisher, David, Sandra Price, and Terry Hanstock, eds. 2002. *Information Sources in the Social Sciences*. Munich: K. G. Saur.

Getz, Malcolm. 2004. "Open-Access Scholarly Publishing in Economic Perspective." Working Paper 04-W14. Nashville, Tenn.: Vanderbilt University, Department of Economics.

Glanville, Julie M., Carol Lefebvre, Jeremy N. V. Miles, and Janette Cmosso-Stefinovic. 2006. "How to Identify Randomized Controlled Trials in MEDLINE: Ten Years On." *Journal of the Medical Library Association* 94(2): 130–36.

Herron, Nancy L., Ed. 2002. *The Social Sciences: A Cross-Disciplinary Guide to Selected Sources*, 3rd ed. Englewood, Colo: Libraries Unlimited.

Higgins, Julian, and Sally Green, eds. 2006. *Cochrane Handbook for Systematic Reviews of Interventions*, version 4.2.6. Chichester: John Wiley & Sons. http://www.cochrane.org/resources/handbook/hbook.htm.

Hofstede, Geert. 1987. "The Applicability of McGregor's Theories in South East Asia." *Journal of Management Development* 6(3): 9–18.

———. 1993. "Leadership: Understanding the Dynamics of Power and Influence in Organizations." *Academy of Management Executive* 7(1): 81–93.

Johnson, Craig L. 1978. "The Effects of Arousal, Sex of Witness and Scheduling of Interrogation on Eyewitness Testimony." *Dissertation Abstracts International* 38(9-B) (March): 4427–8.

Koriat, Asher, Morris Goldsmith, and Ainat Pansky. 2000. "Toward a Psychology of Memory." *Annual Review of Psychology* 51: 481–537.

Loftus, Elizabeth F. 1975. "Leading Questions and the Eyewitness Report." *Cognitive Psychology* 7(4): 560–72.

Loftus, Elizabeth F. 1979, 1996. *Eyewitness Testimony*. Cambridge, Mass.: Harvard University Press.

Mann, Thomas. 2005. *The Oxford Guide to Library Research*. Oxford: Oxford University Press.

National Library of Medicine. 2000. "2000 Cumulated Index Medicus: The End of an Era." *NLM Newsline*, 55(4). http://www.nlm.nih.gov/archive/20040423/pubs/nlmnews/octdec00/2000CIM.html

National Library of Medicine. 2004. "Index Medicus to Cease as Print Publication." NLS Technical Bulletin 338. http://www.nlm.nih.gov/pubs/techbull/mj04/mj04_im.html.

National Library of Medicine. 2006. Index Medicus Chronology. http://www.nlm.nih.gov/services/indexmedicus.html.

National Technical Information Service. 2006. *Government Research Center: NTIS Database*. Washington, D.C.: U. S. Department of Commerce, Technology Administration, National Technical Information Service. http://grc.ntis.gov/ntisdb.htm.

Payne, David G., Michael P. Toglia, and Jeffrey S. Anastasi. 1994. "Recognition Performance Level and the Magnitude of the Misinformation Effect in Eyewitness Memory." *Psychonomic Bulletin & Review* 1(3): 376–82.

Reed, Jeffrey G., and Pam M. Baxter. 2003. *Library Use: Handbook for Psychology*, 3rd ed. Washington, D.C.: American Psychological Association.

Robinson Karen A, and Kay Dickersin. 2002. "Development of a Highly Sensitive Search Strategy for the Retrieval of Reports of Controlled Trials Using Pub Med." *International Journal of Epidemiology* 31(1): 150–53.

Sampson, Margaret, Nicholas J. Barrowman, David Moher, Terry P. Klassen, et al. 2003. "Should Meta-analysts Search Embase in Addition to Medline?" *Journal of Epidemiology*, 56(October): 943–55.

Shannon, Claude Elwood. 1940. "A Symbolic Analysis of Relay and Switching Circuits." Master's thesis. Massachusetts Institute of Technology, Department of Electrical Engineering.

Siegel, Judith M., and Elizabeth F. Loftus. 1978. "Impact of Anxiety and Life Stress upon Eyewitness Testimony." *Bulletin of the Psychonomic Society* 12(6): 479–80.

Singer, Robert N., Gene G. Korienek, and Susan Risdale. 1980. "The Influence of Learning Strategies in the Acquisition, Retention, and Transfer of a Procedural Task." *Bulletin of the Psychonomic Society* 16(2): 97–100.

Sporer, Siegfried Ludwig, Steven Penrod, Don Read, and Brian Cutler. 1995. "Choosing, Confidence, and Accuracy: A Meta-Analysis of the Confidence-Accuracy Relation in Eyewitness Identification Studies." *Psychological Bulletin* 118(3): 315–27.

Steblay, Nancy, Jennifer Dysart, Solomon Fulero, and R. C. L. Lindsay. 2003. "Eyewitness Accuracy Rates in Police Showup and Lineup Presentations: A Meta-Analytic Comparison." *Law and Human Behavior* 27(5): 523–40.

Venn, John. 1880. "On the Diagrammatic and Dechanical Representation of Propositions and Reasonings." *Dublin Philosophical Magazine and Journal of Science* 9(59): 1–18.

Weill Medical College of Cornell University. (2008). *EMBASE*. http://library2.med.cornell.edu/Tutorials/RLS/ch4_BiomedDBClass/4.2.3_majorDB_embase.html#.

Wells, Gary L. 1978. "Applied Eyewitness Testimony Research: System Variables and Estimator Variables." *Journal of Personality & Social Psychology* 36:89–103.

Wells, Gary L, and Elizabeth A. Olson. 2003. "Eyewitness Testimony." *Annual Review of Psychology* 54: 277–95.

6

GREY LITERATURE

HANNAH R. ROTHSTEIN
Baruch College

SALLY HOPEWELL
U.K. Cochrane Centre

CONTENTS

6.1 INTRODUCTION

The objective of this chapter, and of the original in the 1994 edition of this handbook, on which this one is based, is to provide methods of retrieving hard to find literature and information on particular areas of research.[1] We begin by briefly outlining the many developments that have taken place in retrieval of hard-to-find research results over the past decade, including the formalization of this work as a distinct research specialization.

Although we continue to focus on how to retrieve hard to find literature, developments in research synthesis methodology allow us to place retrieval of this literature in the context of the entire research synthesis process. The key point is that there is a critical relationship between the reliability and validity of a research synthesis and the thoroughness of and lack of bias in the search for relevant studies.

Since 1994, changes in technology and in scholarship have created new challenges and new opportunities for anyone interested in reading or synthesizing scientific literature. As the Internet, the Web, electronic publishing, and computer-based searching have developed, grey literature has both grown and become more accessible to researchers. Grey literature comprises an increasing proportion of the information relevant to research synthesis; it is likely to be more current than traditional literature and channels for its distribution are improving. As a result, sifting through the grey literature, as well as locating it has become a major task of the information retrieval phase of a research synthesis.

The emergence of the idea of grey literature is not new and the concept is beautifully illustrated in a quotation by George Minde-Pouet in 1920, cited by Dagmar Schmidmaier: "No librarian who takes his job seriously can today deny that careful attention has also to be paid to the 'little literature' and the numerous publications not available in normal bookshops, if one hopes to avoid seriously damaging science by neglecting these" (1986). Acceptance of the term dates back to 1978. A seminar at the University of York in the UK was recognized as a milestone for grey literature as a primary information source and resulted in the creation of the System for Information on Grey Literature in Europe database (SIGLE), which is managed by the European Association for Grey Literature Exploitation (EAGLE). Unfortunately, this database has not been updated since 2004, and its future is uncertain.

The first scholarly society devoted to the study of grey literature, the Grey Literature Network Service (GreyNet), was founded in 1993. This group sponsors an annual international conference, produces a journal, and maintains a list of grey literature resources in the service of its mission to "facilitate dialog, research, and communication between persons and organizations in the field of grey literature. . . . [and] to identify and distribute information on and about grey literature in networked environments" (http://www.greynet.org). The work of this group has highlighted the fact that traditional published sources of scientific information can no longer be counted on as the primary repositories of scientific knowledge.

6.2 DEFINITION OF TERMS

Grey literature, gray literature, and fugitive literature are synonymous terms, grey literature the one in most common use as of this writing. The most widely accepted definition is the one agreed on at the Third International Conference on Grey Literature, held in Luxembourg in November 1997: "that which is produced on all levels of government, academics, business and industry in electronic and print formats not controlled by commercial publishers" (Auger 1998). The key aspect is that the literature's production is by a source for which publishing is not the primary activity. This definition and focus may have some relevance to research synthesists, but are not the primary reason we are interested in grey literature. Tim McKimmie and Joanna Szurmak have offered an alternate definition in proposing that any materials not identifiable through a traditional index or database be considered grey literature (2002). The key point here is that the literature is hard to retrieve. This second definition is closer to that of fugitive literature provided in the first edition and better reflects the purpose of this chapter. Broad views of grey literature from this angle include just about everything that is not published in a peer-reviewed academic journal, whether or not it is produced by those for whom publishing is the primary activity. According to

Irwin Weintraub, "scientific grey literature comprises newsletters, reports, working papers, theses, government documents, bulletins, fact sheets, conference proceedings and other publications distributed free, available by subscription or for sale" (2000). For present purposes, all documents except journal articles that appear in widely known, easily accessible electronic databases will be considered grey literature. Although not all forms of this literature are equally fugitive, all are in danger of being overlooked by research synthesists if they rely exclusively on searches of popular electronic databases to identify relevant studies. For example, for many purposes books and book chapters should be viewed as nearly grey literature, because though they are produced and distributed commercially by publishers and may be widely available, studies located in them are often difficult to identify through typical search procedures.

6.3 RETRIEVING GREY LITERATURE

6.3.1 The Impact of the Literature Search

The replicability and soundness of a research synthesis depend heavily on the degree to which the search for relevant studies is thorough, systematic, unbiased, transparent, and clearly documented. Lack of bias is the most essential feature; "if the sample of studies retrieved for review is biased, the validity of the results . . . no matter how systematic and thorough in other respects, is threatened" (Rothstein, Sutton, and Borenstein 2005, 2). Literature searching, or information retrieval, is therefore a critical part of the research synthesis process, albeit often underappreciated and underemphasized. Here we will describe various ways in which bias can enter a synthesis through an inadequate search, and then explain how a more comprehensive search can minimize some of those biases. In particular, we focus on including grey literature as an attempt to lessen publication bias in research syntheses, and we provide empirical examples from both the health and social sciences to illustrate our points.

The aim of a high quality literature search in a research synthesis is to generate as comprehensive a list as possible of both published and unpublished studies relevant to the questions posed in the synthesis (Higgins and Green 2008). Unless the searcher is both informed and vigilant, however, this aim can be subverted in various ways and bias can be introduced. Publication bias is the generally accepted term for what occurs when the research that appears in the published literature (or is otherwise readily

available) is systematically unrepresentative of the population of completed studies. This is not a hypothetical issue: evidence that publication bias has had an impact on research syntheses in many areas has been clearly demonstrated. Initially, the concern was that research was accepted for publication based on the statistical significance of the study's results, rather than on the basis of its quality (Song et al. 2000). Kay Dickersin provided a concise and compelling summary of the research showing that this type of publication bias exists in both the social and biomedical sciences, for experimental as well as observational studies (2005). This body of research includes direct evidence including editorial policies, results of surveys of investigators and follow up of studies registered at inception, all of which indicate that statistically significant studies are less likely to be published, and, if they are, published more slowly than significant ones, which is also known as time-lag bias (Hopewell and Clarke 2001). It also includes persuasive indirect evidence that publication bias exists, including the overrepresentation of significant results in published studies and negative correlations between sample size and effect size in several large sets of published data.

Readers should be aware that numerous potential information suppression mechanisms have the potential to bias how studies relevant to a research synthesis are identified and retrieved. One of these is language bias, because of the selective use by synthesists of mainstream U.S. and UK databases, selective searching for only English-language material, or excluding material in other languages (Egger et al. 1997; Jüni et al. 2002). Another is availability bias, or the selective inclusion of studies that are easily accessible to the researcher. A third is cost bias, the selective inclusion of studies that are available free or at low cost. A fourth is familiarity bias, the selective inclusion of studies only from one's own discipline. A fifth is duplication bias, that is, studies with statistically significant results are more likely to be published more than once (Tramer et al. 1997). Last is citation bias, whereby studies with statistically significant results are more likely to be cited by others and therefore easier to identify (Gøtzsche 1997; Ravnskov 1992).

In addition to the evidence that whole studies are selectively missing are demonstrations by An-Wen Chan and his colleagues (Chan et al. 2004; Chan and Altman 2005) that reporting of outcomes within studies is often incomplete and biased, due to the selective withholding of statistically nonsignificant results. Some are biases introduced by the authors of primary studies; others are attributable to editorial policies of journals or of databases, while the researchers producing the synthesis introduce others. All lead to the same consequence, however, namely that the literature located for a synthesis may not represent the population of completed studies. The biases therefore all present the same threat to the validity of a synthesis. Although it would be more accurate to refer to all by the single, broad term *dissemination bias*, as Fujian Song and his colleagues proposed (2000), we will follow common usage and employ the term publication bias, while noting that readers should keep in mind the larger set of suppression mechanisms that we have listed here.

Those carrying out research syntheses need to ensure that they conduct as comprehensive a search as possible to keep themselves from introducing bias into their work. It is important that they do not limit their searches to those studies published in the English language, because evidence suggests that those published in other languages may have different results (Egger et al. 1997; Jüni et al. 2002). Similarly, it is important that synthesis authors go beyond databases they are most familiar with, those that are most easily accessible, and those that are free or inexpensive. In expanding their search,, researchers should be aware that exclusive reliance on the references cited by key relevant studies is likely to introduce bias (Gøtzsche 1987).

The identification of unfamiliar databases can be made relatively easy by using sources such as the *Gale Directory of Databases*; getting access to them, however, may be much more difficult. Research synthesists should consider attempting to get access to databases located at institutions or universities other than their own. Another possible option is to commission, for a fee, searches of databases to which one cannot easily get access. For example, GESIS (a German organization) can be commissioned to search a range of German language and other European databases (http://www.gesis.org/en/index.htm).

Appropriate staff, including librarians or information specialists, are critical to conducting an effective and efficient search. As Rosenthal pointed out, library reference specialists are more likely to be knowledgeable about new sources of information and about changes in searching techniques than the researcher is (1994). In addition, working with a librarian is likely to increase the number of terms that can be used in a search, because librarians are likely to be aware of differences in terminology that may exist in different fields. Librarians are also helpful in acquiring documents that have been located during a search, as well as in costing, documenting, and managing

searches (for an example of the value added by a competent information specialist, see Wade et al. 2006). Sufficient budget and time are also essential components of a good search. The likely difficulty of searching for grey literature and the probable need for issue by issue, page by page search of key journals (hand searching) also need to be taken into account when planning the time to completion of the entire synthesis.

Despite the increased costs likely to be needed to locate and retrieve grey literature, however, there is little justification for conducting a synthesis that excludes it.

6.3.2 Information Search

6.3.2.1 Search Terms and Search Strategies Before beginning the actual search, synthesis authors must clarify the key concepts imbedded in their research question, and attempt to develop an exhaustive list of the terms that have been used to describe each concept in the research literature. A comprehensive search strategy will invariably involve searching for multiple terms that describe the independent variable of interest, and may also include others that describe the dependent variables or the target population in which the synthesists are interested, or both (see chapter 5 this volume for an example of how this is carried out).

It is critical to take into account the fact that the search terms as well as strategies are likely to vary across different sources of information. Authors are advised to consult both standard reference tools such as subject encyclopedias and subject thesauri (for example, *The Encyclopedia of Psychology*, or the *Dictionary of Education*) as well as specialized thesauri that have been created for specific databases—such as the *Thesaurus of Psychological Index Terms* for PsycINFO, the ERIC *Thesaurus,* or the U.S. National Library of Medicine *Medical Subject Headings (MeSH)* for searching MEDLINE—as well as key papers that come up in preliminary searches. It is both efficient and effective to work with an information specialist at this point in the search. Our recommendation is a preliminary search for each term in multiple databases, to include databases that contain grey literature. This is to ensure that the terms retrieve both unpublished and published relevant studies, and to generate synonyms to use during the main search. This work is particularly important for synthesists in the social sciences and education because of the extremely variable quality of keywording, indexing, and abstracting in these literatures (Grayson and Gomersall 2003; Wade et al. 2006). Prospective authors, no matter what their field of research, should be aware that this is an iterative process, for which they must allocate adequate time and staff. Trisha Greenhalgh and Richard Peacock reported that their electronic search, including developing, refining, and adapting search strategies to the different databases they used, took about two weeks of specialist librarian time, and that their hand search of journals took a month (2005).

There are several guides to literature searching for research syntheses in health care that may be of use to both synthesists and librarians interested in retrieving grey literature, including *The Cochrane Handbook for Systematic Reviews of Interventions* (Higgins and Green 2008). In social science and education, prospective synthesis authors may wish to look at the *Information Retrieval Methods Policy Brief* (Rothstein, Lavenberg, and Turner 2003) produced by the Campbell Collaboration (www.campbellcollaboration.org/MG/IRMGPolicyBrief-Revised.pdf). More general sources that might be helpful include Arlene Fink's *Conducting Research Literature Reviews* (2004) and Chris Armstrong and Andrew Large's *Humanities and Social Sciences* (2001).

6.3.2.2 Information Sources It is obvious that, for individual syntheses, the research question will determine which information sources should be searched. For searches related to simple and straightforward questions, most of the studies may be identified by a subset of possible sources. As the topic of the synthesis gets more complex, or becomes more likely to draw on a variety of research literatures, the importance of multiple methods of search increases dramatically. That said, we cannot imagine a thorough search that did not consider at least the following sources:

- electronic bibliographic databases
- hand searching key journals
- conference proceedings
- reference lists and citation searching
- contacting researchers
- research registers
- the Internet

6.3.2.2.1 Electronic Bibliographic Databases. Searching multiple large electronic bibliographic databases such as MEDLINE, EMBASE, ERIC, PsycINFO, and Sociological Abstracts is in most cases the primary way to identify studies for possible inclusion in a research synthesis.

It is important to note, however, that limiting a search solely to electronic databases has the potential to introduce bias into a synthesis. A recent systematic examination of studies compared the results of manual cover-to-cover searching of journals (hand searching) with those resulting from searching electronic databases such as MEDLINE and EMBASE to identify reports of randomized trials. Hand searching identified 92 to 100 percent of all reports of randomized trials found. The Cochrane Highly Sensitive Search Strategy, which is used to identify reports of randomized trials in MEDLINE, identified 80 percent of the total number of reports of trials found (Dickersin Scherer, and Lefebvre 1994). On the other hand, complex electronic searches found 65 percent and simple ones found only 42 percent. The results of electronic searches were better when the search was limited to English language journals (62 percent versus 39 percent). A major reason that the electronic searches failed to identify more of the reports was that those reports published as abstracts or in journal supplements were not routinely indexed by the electronic databases. Restricting the search to full reports raised the retrieval rate of complex searches to 82 percent. An additional reason for the poor performance of the electronic searches was that some reports were published before 1991, when the National Library of Medicine's system for indexing trial reports was not as well developed as it is today (Hopewell et al. 2002). Herbert Turner and his colleagues, who conducted research comparing the retrieval ability of hand searching with searches of electronic databases to identify randomized controlled trials in education, found similar results (2004). Their results, based on a search of twelve education and education related journals, indicated that electronic searches identified, on average, only one-third of the trials identified by hand searching.

Tricia Greenhalgh and Richard Peacock examined the yield of different search methods for their synthesis of the diffusion of health-care innovations, and found that approximately one-fourth of their studies were produced by electronic searches, with a yield of one useful study every forty minutes (2005). David Ogilvie and his colleagues found in their synthesis of the literature on promoting a shift from driving a car to walking that only four of the 69 relevant studies were in first line health databases (2005). About half of the remaining studies were found in a specialized transportation database, and the others by Internet searches and chance.

Although searching large, general databases may provide only a fraction of relevant studies, such searches are generally not difficult. Large databases typically have controlled languages (thesauri), the full range of boolean search operators (for example, AND, OR, NOT), the ability to combine searches, and a range of options for manipulating and downloading searches. Smaller and more specialized databases often contain material not indexed in their larger counterparts. They are often created and operated with limited resources, however, and thus more focused on getting the information in than on retrieving it. Most have rudimentary thesauri, if any, and a limited set of search tools. These limitations can make searching of small, specialized databases time-consuming and inefficient, but their potential yield can make them valuable. Diane Helmer and colleagues, for example, found that searches of specialized databases were the most effective strategy when producing potentially relevant studies for two syntheses of randomized trials of health-care interventions, but their definition of specialized databases included several large trial registries as well as dissertation abstracts (2001).

6.3.2.2.2 Hand Searching Key Journals. Targeted searching of relevant individual journals can overcome problems caused by poor indexing and incomplete database coverage, but is labor intensive and time consuming. Potentially relevant journals in every discipline are not indexed by database producers. Additionally, not all indexed articles are retrievable from databases. The proportion of studies relevant to a research synthesis that are likely to be found through hand searching only and time involved in finding these is hard to estimate. Greenhalgh and Peacock (2005) reported that for their synthesis, twenty-four of 495 useful sources were identified through hand searching, with a yield of one useful article for every nine hours of search time. Diane Helmer and her colleagues examined the search strategies of two research synthesis, and reported that hand searching identified a small, but unique, fraction of potentially useful studies (2001).

The Cochrane Collaboration has invested significant time and effort hand searching health-care journals to locate reports of randomized trials of health-care interventions (Dickersin et al. 2002). The results are included in the Cochrane Central Register of Controlled Trials (CENTRAL) and details of which journals have been hand searched are contained in their Master List of Journals being searched. (http://apps1.jhsph.edu/cochrane/masterlist.asp). The Campbell Collaboration's register (C2-SPECTR) includes reports of trials of social, educational, psychological or crime and justice interventions that have been located through hand searching; however this is a relatively small effort. Both Cochrane and Campbell

Collaborations have developed hand-search manuals and related materials to assist with this activity that are available for download (www.Cochrane.org and www.Campbellcollaboration.org).

6.3.2.2.3 Conference Proceedings. Conference proceedings are a very important source for identifying studies found in the grey literature for inclusion in research syntheses. These contain studies that are often not published as full journal articles and their abstracts are not generally included in electronic databases; thus relevant studies may be identified only by searching the proceedings of the conference. Some conference proceedings are available electronically from professional society websites, while others may be issued as compact discs, or only on paper. Many conferences do not even publish proceedings, and require a search of the conference program, and subsequent contact with the abstract authors in order to retrieve potentially relevant studies.

Considerable evidence shows that many studies reported at conferences do not get published in journals, and that this is often for reasons unrelated to the quality of the study. A recent research synthesis of assessments found that fewer than half (44.5 percent) of research studies presented as abstracts at scientific conferences were published in full. This figure was somewhat higher (63.1 percent) for abstracts that presented the results of randomized trials. Most studies that were eventually published appeared in full within three years of presentation at the meeting (Scherer, Langenberg, and von Elm 2005). For a large proportion of trials, then, the only available source of information about a trial and its results might be an abstract published in the proceedings of a conference. Consistent with these findings, a study of conference abstracts for research into the methods of health-care research, found that approximately half of the abstracts had not been or were unlikely ever to be published in full (Hopewell and Clarke 2001). The most common reason given by investigators as to why research submitted to scientific meetings had not been subsequently published in full was lack of time. Other reasons included the researchers' belief that a journal was unlikely to accept their study, or that the authors had decided the results were not important (Callaham et al. 1998; Weber et al. 1998; Donaldson and Cresswell 1996).

In the social sciences, Harris Cooper, Kristina DeNeve, and Kelly Charlton (1997) and Katherine McPadden and Hannah Rothstein (2006) reported a slightly more complex state of affairs, but one that is generally consistent with the picture from health care. Cooper and his colleagues surveyed thirty-three psychology researchers to follow up on studies that had received institutional review board (IRB) approval. About two-thirds of these completed studies did not result in published summaries, and half fell out of the process for reasons other than methodological quality. Cooper and his colleagues also found that research results that were statistically significant were more likely than nonsignificant findings to be submitted for presentation or publication (1997). In a selective sample of best papers (the top 10 percent as judged by the evaluators) published in proceedings from three years of Academy of Management conferences, McPadden and Rothstein found that 73 percent were eventually published in journals, with an average time to publication of twenty-two months from time of submission (2006). Nearly half of these papers (43 percent) included more or different data than was described in the conference proceedings. The most common changes were the type of analyses conducted and the number of outcomes reported. In a less selective sample of all papers presented at three years of annual conferences of the Society for Industrial and Organizational Psychology, they found that only 48 percent were eventually published, and that 60 percent of these papers contained data that was different than that contained in the conference paper. In neither McPadden and Rothstein sample was the statistical significance of findings the major reason given for nonsubmission to or nonpublication in a journal. Given the policies of these two conferences, however, it is, in our experience, unlikely that statistically nonsignificant findings would have been submitted or accepted for presentation.

Although conference proceedings and programs are an invaluable source of identifying grey studies, the research synthesist must be aware that the risk of duplication bias (with reports of the same study being found in different conference proceedings) may increase when he or she includes these information sources, and should take appropriate precautions to avoid this (see chapter 9, this volume).

6.3.2.2.4 Reference Lists and Citations. Searching reference lists of previously retrieved primary studies and research synthesis is nearly universally recommended as an effective method of identifying potentially relevant literature. This method, also called snowballing, calls for the reference sections of all retrieved articles to be examined, and for any previously unidentified possibly relevant material to be retrieved. This process continues until no new possibly relevant references are found.

Two studies of research synthesis provide some evidence of the effectiveness of this method. In Greenhalgh and Peacock's research synthesis of the diffusion of health-care innovations, searching the reference lists of previously identified references yielded the greatest number of useful studies (12 percent of the total useful material), and was also an efficient means of searching, with a yield of one useful paper for every fifteen minutes of searching (2005). Helmer and her colleagues found that scanning reference lists identified about a quarter of the potentially useful studies that were identified by means other than searching of electronic databases (2001).

Similarly to manually scanning the reference lists of relevant articles, researchers are now increasingly using electronic citation search engines as a method of identifying additional studies for inclusion in their research synthesis. Several examples of these include the Science Citation Index, for use in health care, the Social Sciences Citation Index and Social SciSearch, all of which are available through the Web of Knowledge (http://isiwebofknowledge.com/), and SCOPUS, which is available through Elsevier (www.scopus.com).

It is important, however, to note that scanning reference lists and citation searching should not be done in isolation and must only be done in combination with other search methods described in this chapter. Overreliance on this method of searching will introduce a real threat of citation bias (Gøtzsche 1987), because the more a research study is cited the easier it will be to identify and therefore include in the research synthesis.

6.3.2.2.5 Researchers. Research centers and experts are another important source of information that may not be available publicly. Contacting researchers who are active in a field, subject specialists and, both for-profit and nonprofit companies that conduct research—in health care, the pharmaceutical industry or medical device companies, in psychology or education, test publishers, consulting firms, government agencies, and so on—can be an important way to identifying ongoing or completed but unpublished studies. For example, Greenhalgh and Peacock reported success with using emails to colleagues for identification of studies in their synthesis (2005). Helmer and colleagues found that personal communication with other researchers yielded about 25 percent of the potentially useful studies that were identified by means other than searching of electronic databases (2001). Retrieving studies, once identified, is a separate task. Success in the latter endeavor may depend on the goodwill of the researchers, the time they have available to respond to these requests, and the time that has elapsed since the study was conducted as well as the policies of organizations with proprietary interests. Evidence suggests that requests for additional information about a study are not always successful (www.mrc-bsu.cam.ac.uk/firstcontact/pubs.html) (McDaniel, Rothstein, and Whetzel 2006).

6.3.2.2.6 Research Registers. Research registers are an important way of identifying ongoing research, some of which may never be formally published. There are many national registers such as the National Research Register in the United Kingdom, which contains ongoing health research of interest to, and funded by, the U.K. National Health Service (www.doh.gov.uk/research/nrr.htm). There are also numerous local and specialist registers, each of which is specific to its own area of interest. Identifying ongoing studies is important so that whenever a synthesis is updated, these studies can be assessed for possible inclusion. The identification of ongoing studies also helps the synthesist to assess the possibility of time-lag bias whereby studies with statistically significant results are more likely to be published sooner than those with nonsignificant results (Hopewell et al. 2001).

Some pharmaceutical companies are making public records of their ongoing and completed trials, either through the metaRegister of Controlled Trials (www.controlled-trials.com) or by providing registers of research on their own website. For example, an initiative by Glaxo Wellcome has led to the development of a register of phase II, III and IV studies on newly registered medicines. (http://ctr.glaxowellcome.co.uk). In 2004, the Canadian Institutes of Health Research hosted a meeting to develop a plan for international registration of all studies of health related interventions at their inception. That meeting led to issuance of the Ottawa Statement, a set of guiding principles for the development of research registers (Krleza-Jeric et al. 2005). Among the objectives of this movement are to identify and deter unnecessary duplication of research and publications, and to identify and eliminate reporting biases. The statement calls for results for all outcomes and analyses specified in the protocol as well as any adverse effects data to be publicly registered, whether or not the information is published elsewhere. The World Health Organization (WHO) is also working on standards for international trials registration. In August 2005, the WHO established the International Clinical Trials Registry Platform with the primary objective of ensuring that all clinical trials worldwide are uniquely identifiable through a system of public registers, and that a minimum set of results is made publicly available (www.who.int/ictrp). As these initiatives begin to influence governmental and pharmaceutical industry policies, the amount of grey health-care data available to synthesists from research registers should increase substantially. It seems unlikely at present that similar initiatives will take place in the social sciences or in other disciplines, though we would be happy to be wrong on this account.

6.3.2.2.7 Internet. The Internet is an increasingly important source of information about ongoing and completed research, particularly that which has not been formally published. Sources of grey literature that are accessed primarily by Internet searching include the websites of universities, major libraries, government agencies, nongovernmental organizations including nonprofit research organizations, think-tanks and foundations, professional societies, corporations and advocacy groups. Specific examples of some of these are listed in section 6.4.2.2.6. The Internet is often particularly helpful for identifying documents produced by local government authorities because these are increasingly published online.

Searching the Internet, can, however, be a major undertaking. According to the University of British Columbia Library (http:/toby.library.ubc.ca) almost half of the Internet is unreachable or invisible using standard search engines, and specific search tools are needed to access this content. The Internet searching skills of many synthesists are fairly rudimentary. Even when searchers are skilled, many of the general search engines do not allow sophisticated searching and may identify thousands of Internet sites to check, many of which are irrelevant or untrustworthy. Even when potentially relevant grey studies appear that are familiar to experts in a particular field, they often are subject to search conventions that are idiosyncratic to the site, or even to the specific collection in which they appear.

Despite the popularity of Internet searches among research synthesists, very few empirical studies have been conducted to establish whether searching the Internet is generally useful in identifying additional unpublished or ongoing studies. Gunther Eysenbach and colleagues conducted one of the few studies that examined this question (2001). These authors adapted search strategies from seven research syntheses in an attempt to find additional relevant trials. They searched 429 websites in twenty-one hours and located information about fourteen unpublished, ongoing, or recently published trials. A majority of these were considered relevant for inclusion in one of the research syntheses.

6.4 HOW TO SEARCH

A number of electronic resources, available and relatively easily accessible, may prove useful in searching for grey

Table 6.1 Databases, Websites, and Gateways

Source	Type of Material	Cost
General		
British Library's Inside Web	Includes references to papers presented at over 100,000 conference proceedings.	Subscription
Dissertation and Theses Database http://www.proquest.com/ products_pq/descriptions/pqdt .shtml/.	PQDT—Full text includes 2.4 million dissertation citations from around the world from 1861 to the present day together with full text dissertartion that are available for download in PDF format.	Subscription
Evidence Network www.evidencenetwork.org.	A service provided by the ESRC U.K. Centre for Evidence Based Policy and Practice. The Resources section of the website covers six major types of information resource: bibliographic and research databases; internet gateways; systematic review centers; EBPP centers; research centers; and library services.	Free access
Index to Theses http://www.theses.com/	Claims to have all U.K. theses back to 1716, as well as a subset of Irish theses.	Subscription
NTIS (National Technical Information Service) http://www.ntis.gov/.	Has a heavy technical emphasis, but is the largest central resource for U.S. government-funded scientific, technical, engineering, and business related information available. Contains over half a million reports in over 350 subject areas from over 200 federal agencies. There is limited free access, and affordable 24-hour access.	Subscription
System for Information on Grey Literature in Europe (SIGLE)	A project of EAGLE (European Association for Grey Literature Exploitation), SIGLE is a bibliographic database covering European grey literature in the natural sciences and technology, economics, social sciences, and humanities. It includes research and technical reports, preprints, working papers, conference papers, dissertations and government reports. All grey literature held by the British Library in the National Reports Collection can be retrieved from here. This database has not been updated since 2004; the current status of this database is uncertain, and it may cease to be available.	Free access
Healthcare, Biomedicine and Science		
BiomedCentral www.biomedcentral.com/.	Contains conference abstracts, papers and protocols of ongoing trials.	Free access
U.S. National Library of Medicine (NLM) Gateway http://gateway.nlm.nih.gov/gw/Cmd	Gateway for most U.S. National Library of Medicine holdings and includes meeting abstracts relevant to health care and health technology assessment.	Free access
Social Science, Economics and Education		
ESRC Society Today http://www.esrc.ac.uk/ ESRCInfoCentre/index.aspx.	Contains descriptions of projects funded by the U.K. Economic and Social Research Council and links to researchers' websites and publications.	Free access

SOURCE: Authors' compilation.

literature. They can be sorted into four main types. Grey literature databases include studies found only or primarily in the grey literature. General bibliographic databases contain studies published in the grey and conventional literature. Specific bibliographic databases contain studies published in the grey and conventional literature in a specific subject area. Ongoing trial registers offer information about planned and ongoing studies. Tables 6.1

Table 6.2 General Bibliographic Databases

Source	Type of Material	Cost
General		
Science Citation Index http://scientific.thomson.com/ products/sci/	Provides access to current and past bibliographic information, author abstracts, and cited references for a broad array of science and technical journals in more than 100 disciplines, and includes meeting abstracts that are not covered in MEDLINE or EMBASE.	Subscription
SCOPUS http://info.scopus.com/.	Claims to be the largest database of citations and abstracts with coverage of: Open Access journals, conference proceedings, trade publications and book series. Also provides access to many web sources such as author homepages, university websites and pre-print services. Currently has 28 million abstracts, and their references.	Subscription
Healthcare, Biomedicine and Science		
Database of Abstracts of Reviews of Effects (DARE) http://www.crd.york.ac.uk/crdweb/	Contains summaries of systematic reviews about the effects of interventions on a broad range of health and social care topics.	Free access
NHS Economic Evaluation Database (NHS EED) http://www.crd.york.ac.uk/crdweb/	Contains summaries of studies of the economic evaluation of healthcare interventions culled from a variety of electronic databases and paper-based resources.	Free access
Health Technology Assessment (HTA) Database http://www.crd .york.ac.uk/crdweb/	Contains records of the reviews produced by the members of the International Network of Health Technology Assessment, spanning 45 agencies in 22 countries.	Free access
Ongoing Reviews Database http://www.crd.york.ac.uk/crdweb/	A register of ongoing systematic reviews of health care effectiveness, primarily from the United Kingdom.	Free access
Social Science, Economics and Education		
Social Sciences Citation Index http://scientific.thomson.com/ products/ssci/	Provides access to current and past bibliographic information, author abstracts, and cited references for a large number of social science journals in over 50 disciplines. Includes conference abstracts relevant to the social sciences from 1956 to date.	Subscription
SOLIS (Social Sciences Literature Information System) http://www.cas.org/ONLINE/DBSS/ solisss.html	German language social science database including grey literature and books as well as journals – searchable in English. Access may be available via Infoconnex service at http://www.infoconnex.de/	Subscription
SSRN (Social Science Research Network) http://www.ssrn.com/index.html	Contains abstracts of over 100,000 working papers and forthcoming papers in accounting, economics, entrepreneurship, management, marketing, negotiation, and similar disciplines .Also has full text, downloadable text for about 100,000 documents in these fields.	Free access

SOURCE: Authors' compilation.

through 6.5 present annotated lists of important sources of grey literature in each of the four categories, as well as sources of grey literature that did not fit in any of the other categories. Those that are not of general interest primarily cover either health care, or one of the social sciences or education. Databases for some other fields in which research syntheses are commonly done are men-tioned when we are aware of them. These sources are correct at the time of press but may change.

Whichever sources are searched, it is important that the authors of the research synthesis are explicit and document in their synthesis exactly which sources have been searched (and when) and which strategies they have used so that the reader can assess the validity of the

Table 6.3 Subject- or Country-Specific Databases

Source	Type of Material	Cost
Healthcare, Biomedicine and Science		
Alcohol Concern http://www.alcoholconcern.org .uk/ servlets/wrapper/publications .jsp.	Holds the most comprehensive collection of books and journals on alcohol-related issues in the United Kingdom.	Free access
Biological Abstracts http://scientific.thomson.com/ products/ba/	Indexes articles from over 3,700 serials each year, indexes papers presented at meeting symposia and workshops in the biological sciences.	Subscription
CINAHL www.cinahl.com.	Includes nursing theses, government reports, guidelines, newsletters, and other research relevant to nursing and allied health.	Free access to some information
Computer Retrieval of Information on Scientific Projects (CRISP) http://crisp.cit.nih.gov/.	A searchable database of federally funded biomedical research projects, and is maintained by the U.S. National Institutes of Health. It includes projects funded by the National Institutes of Health, Substance Abuse and Mental Health Services, Health Resources and Services Administration, Food and Drug Administration, Centers for Disease Control and Prevention, Agency for Health Care Research and Quality, and Office of Assistant Secretary of Health.	Free access
DIMDI http://www.dimdi.de/dynamic/en/ index.html.	A gateway to 70 databases on medicine, drugs and toxicology sponsored by the German Institute of Medical Documentation and Information. Searchable in English and German.	Free access
DrugScope http://www.drugscope.org.uk/library/ librarysection/libraryhome.asp	Provides access to information and resources on drugs, drug misuse, and related issues.	Charge per individual article
Health Management Information Consortium Database (HMIC) http://www.ovid.com/site/catalog/ DataBase/99.jsp?top=2&mid= 3&bottom=7&subsection=10.	This database is updated bimonthly and contains essential information from two key institutions: the Library & Information Services of the Department of Health, England and the King's Fund Information & Library Service. It contains data from 1983 to the present.	Subscription
Social Science, Economics and Education		
British Library for Development Studies http://blds.ids.ac.uk/blds/	Calls itself 'Europe's largest collection on economic and social change in developing countries'	Free access
Cogprints http://cogprints.ecs.soton.ac.uk/.	Provides material related to cognition from psychology, biology, linguistics, and philosophy, and from computer, physical, social, and mathematical sciences. This site includes full-text articles, book chapters, technical reports, and conference papers.	Subscription, but provides free 30-day trial
Criminal Justice Abstracts http://www.csa.com/factsheets/cja-set-c.php?SID=c4fg6rpjg 3v5nmn09k2kml).	Contains comprehensive coverage of international journals, books, reports, dissertations and unpublished papers on criminology and related disciplines.	Subscription, but provides free 30-day trial
Education Line http://www.leeds.ac.uk/educol/.	Database of the full text of conference papers, working papers and electronic literature which supports educational research, policy and practice in the United Kingdom.	Free access

(Continued)

Table 6.3 (*Continued*)

Source	Type of Material	Cost
EconLIT http://www.econlit.org/	Provides a comprehensive index of thirty years worth of journal articles, books, book reviews, collective volume articles, working papers and dissertations in economics, from around the world.	Subscription
Education Resources Information Center (ERIC) http://www.eric.ed.gov/.	Contains nearly 1.2 million citations going back to 1966 and, access to more than 110,000 full-text materials at no charge. Includes report literature, dissertations, conference proceedings, research analyses, translations of research reports relevant to education and sociology.	Free access
International Transport Research Documentation (ITRD) http://www.itrd.org/.	A database of information on traffic, transport, road safety and accident studies, vehicle design and safety, and environmental effects of transport, construction of roads and bridges etc. Includes books, reports, research in progress, conference proceedings, and policy-related material.	Subscription
Labordoc http://www.ilo.org/public/english/support/lib/labordoc/index.htm	Produced by the International Labour Organisation, contains references and full text access to literature on work. Can be searched in English, French or Spanish.	Free access
National Criminal Justice Reference Service (NCJRS) http://www.ncjrs.gov/.	United States government funded resource policy, and program development worldwide. Includes reports, books and book chapters as well as journal articles.	Free access
Public Affairs Information Service (PAIS) http://www.csa.com/factsheets/pais-set-c.php	Contains references to more than 553,300 journal articles, books, government documents, statistical directories, research reports, conference reports, internet material from over 120 countries.	Subscription
Policy File http://www.policyfile.com.	Indexes research and publication abstracts addressing the complete range of U.S. public policy research.	Subscription
PsycINFO www.apa.org/psycinfo/.	Includes dissertations, book chapters, academic and government reports and journal articles in psychology, sociology and criminology, as well as other social and behavioral sciences.	Subscription
PSYNDEX http://www.zpid.de/retrieval/login.php	Abstract database of psychological literature, audiovisual media, and tests from German-speaking countries. Journal articles, books, chapters, reports and dissertations, are documented. Searchable in German and English.	Subscription
Research Papers in Economics (Repec) http://repec.org/.	International database of working papers, journal articles and software components.	Free access
Social Care Online http://www.scie-socialcareonline.org.uk/.	Includes U.K. government reports, research papers in social work and social care literature.	Free access
Social Services Abstracts & InfoNet http://www.csa.com/factsheets/ssa-set-c.php.	Includes dissertations in social work, social welfare, social policy and community development research from 1979 to date. This is a subscription-based compilation of content from U.K. databases Social Care Online, AgeINFO, Planex and Urbadoc, all of which include grey literature.	Subscription

Table 6.3 (*Continued*)

Source	Type of Material	Cost
Sociological Abstracts http://www.csa.com/factsheets/ socioabs-set-c.php.	Covers sociology as well as other social and behavioral sciences. It includes dissertations, conference proceedings and book chapters from 1952 to date.	Subscription
Other-Including Public Policy		
AgeInfo http://ageinfo.cpa.org.uk/scripts/ elsc-ai/hfclient.exe?A=elsc-ai&sk	Produced by the U.K. Centre for Policy on Ageing (free access, but there is also subscription-based version with better search facilities at http://www.cpa.org.uk/ ageinfo/ageinfo2.html.).	Free access
Australian Policy Online http://www.apo.org.au/.	This is a database/gateway of Australian research and policy literature.	Free access
ChildData http://www.childdata.org.uk/.	Produced by the U.K. National Children's Bureau.	Subscription
ELDIS http://www.eldis.org/.	This is a gateway to all kinds of full text grey literature on international development issues. It can be searched using simple Boolean logic, or sub-collections can be browsed.	Free access
IDOX Information Service database (formerly Planex) http://www.idoxplc.com/iii/ infoservices/idoxinfoservice/ index.htm.	Covers service areas of relevance to U.K. local government, especially in Scotland, including planning, transport, social care, education, leisure, environment, local authority management and soon.	Subscription
Directory Database of Research and Development Activities (READ) http://read.jst.go.jp/index_e.html	A database service due to promote cooperation among industry, academia, and government. A database service that provides information on research projects in Japan. Searchable in English and Japanese.	Free access
TRIS Online http://trisonline.bts.gov/search .cfm	Covers reports as well as journal papers in transportation and includes English language material from ITRD. It is mainly technical but also contains material relevant to, for example, transport-related health issues.	Free access
Urbadoc http://www.urbadoc.com/index .php?id=12&lang=en.	Similar to IDOX but focuses more on London and south of England.	Subscription

SOURCE: Authors' compilation.

conclusions of the synthesis based on the search that has been conducted.

6.5 GREY LITERATURE IN RESEARCH SYNTHESIS

6.5.1 Health-Care Research

Although one of the hallmarks of a research synthesis is the thorough search for all relevant studies, estimates vary regarding the actual use of grey literature, and the types of grey literature actually included. Mike Clarke and Teresa Clarke conducted a study of the references included in Cochrane protocols and reviews published in *The Cochrane Library* in 1999 (2000). They found that 91.7 percent of references to studies included in syntheses were to journal articles, 3.7 percent were conference proceedings that had not been published as journal articles, 1.8 percent were unpublished material (for example personal communication, in press documents and data on file), and 1.4 percent were books and book chapters.

Susan Mallett and her colleagues looked at the sources of grey literature included in the first 1,000 Cochrane systematic reviews published in issue 1, 2001 of *The Cochrane Library* (2000). Of these, 988 syntheses were

Table 6.4 Trial Registers

Source	Type of Material	Cost
Healthcare, Biomedicine and Science		
BiomedCentral www.biomedcentral.com/clinicaltrials/.	Publishes trials together with their International Standard Randomized Controlled Trial Number (ISRCTN) and publishes protocols of trials	Free access
ClincalTrials.gov http://clinicaltrials.gov	Covers U.S. trials of treatments for life-threatening diseases.	Free access
The Cochrane Central Register of Controlled Trials (CENTRAL) www.cochrane.org	Includes references to trial reports in conference abstracts, handsearched journals, and other resources.	Free access
Current Controlled Trials' metaRegister of Controlled Trials www.controlled-trials.com.	Contains records from multiple trial registers including the U.K. National Research Register, the Medical Editors' Trials Amnesty, the U.K. MRC Trials Register, and links to other ongoing trials registers.	Free access
Database of Promoting Health Effectiveness Reviews (DoPHER) http://eppi.ioe.ac.uk/cms/.	Produced by the U.K. EPPI-Centre and includes completed as well as unpublished randomized and nonrandomized trials in health promotion.	Free access
The Lancet (Protocol Reviews) www.thelancet.com	Contains protocols of randomized trials.	Free access
TrialsCentral www.trialscentral.org.	Contains over 200 U.S.-based registers of trials.	Free access
Social Science, Economics and Education		
The Campbell Collaboration Social, Psychological, Educational and Criminological Trials Register (C2-SPECTR and C-PROT) www.campbellcollaboration.org/	Includes references to trials reported in abstracts, handsearched journals, and ongoing trials.	Free access

SOURCE: Authors' compilation.

examined. All sources of information other than journal publications were counted as grey literature sources and were identified from the reference lists and details provided within the synthesis. The 988 syntheses together contained 9,723 included studies. Fifty-two percent (n = 513) included trials for which at least some of the information was obtained from the grey literature. The sources of grey information in these syntheses are listed in table 6.6. The majority (55 percent) was from unpublished information that supplemented published journal articles of trials. The high proportion of unpublished information in Cochrane reviews differs from other research syntheses in health care, where the major source of grey information was conference abstracts (Egger et al. 2003; McAuley et al. 2000). This may be because Cochrane reviewers are encouraged to search for and use unpublished information.

Within syntheses containing grey literature, studies containing some type of grey literature contributed a median of 28 percent of the included studies (interquartile range 13 to 50 percent). In addition to studies that were entirely grey, the grey literature also provided important supplemental information for a nontrivial proportion of published studies.

Nine percent of the trials (588/6,266) included in the 513 Cochrane reviews using grey literature were retrieved from grey literature sources only. Of these grey trials, about two-thirds were retrieved from conference abstracts, about 15 percent were from government or companies and about 5 percent were from theses or dissertations. The remaining grey trials were located through unpublished reference sources such as personal communication with trialists, in press documents, and data on file. An analysis by Laura McAuley and colleagues (2000) of 102 trials from grey literature sources in thirty-eight meta-analyses, found 62 percent were referenced by conference abstracts

Table 6.5 Other Electronic Sources

Source	Type of Material	Cost
General		
Association of College and Research Libraries http://www.ala.org/ala/acrl/acrlpubs/crlnews/ backissues2004/march04/graylit.htm.	Reprints an article that lists websites about grey literature and how to search for it, as well as free websites that contain grey literature (some full-text). Mostly scientific and technical literature, but also some resources for other fields.	Free access
CERUK (Current Educational Research in the UK) http://www.ceruk.ac.uk/ceruk/.	Database of current or ongoing research in education and related disciplines. It covers a wide range of studies including commissioned research and PhD theses, across all phases of education from early years to adults.	Free access
DEFF http://www.deff.dk/default.aspx?lang=english	Denmark's Electronic Research Library is a partnership between Denmark's research libraries co-financed by the Ministry of Science, Technology and Innovation, the Ministry of Culture and the Ministry of Education.	Subscription
GreyNet (Grey Literature Network Service) http://www.greynet.org/.	Website founded in 1993 and designed to facilitate dialog, research, and communication in the field of grey literature. Provides access to a moderated Listserv, a Distribution List, and the Grey Journal (TGJ).	Free access to some information
Health Technology Assessment- www.york.ac.uk/inst/crd/htadbase.htm	Contains details of nearly 5,000 completed HTA publications and around 800 ongoing INAHTA (International Network of Agencies for Health Technology Assessment) projects.	Free access
INTUTE: Health and Life Sciences http://www.intute.ac.uk/healthandlifesciences/	Internet resources in health and life sciences (Formerly known as BIOME).	Free access
INTUTE: Medicine http://www.intute.ac.uk/healthandlifesciences/ medicine/	Internet resources in medicine. (Formerly known as OMNI).	Free access
INTUTE: Social Sciences http://www.intute.ac.uk/socialsciences/	Internet resources in the social sciences.	Free access
NLH http://www.library.nhs.uk/Default.aspx	U.K. National Library for Health.	Free access to some information
NOD http://www.onderzoekinformatie.nl/en/oi/nod/.	Dutch Research Database contains information on scientific research, researchers and research institutes covering all scientific disciplines.	Free access
NY Academy of Medicine: Grey Literature Page http://www.nyam.org/library/pages/grey_ literature_report	Focuses on grey literature resources in medicine. It features a quarterly "Grey Literature Report," that lists many items available online.	Free access
School of Health and Related Research (ScHARR) www.shef.ac.uk/~scharr/ir/netting/.	Evidence-based resources in health care.	Free access
Trawling the net www.shef.ac.uk/~scharr/ir/trawling.html	Free databases of interest to healthcare researchers working in the United Kingdom.	Free access.
UK Centre for Reviews and Dissemination www.york.ac.uk/inst/crd/revs.htm	Provides a basic checklist for finding studies for systematic reviews.	Free access
US Government Information: Technical Reports http://www-libraries.colorado.edu/ps/gov/us/ techrep.htm.	Provides an annotated listing of U.S. government agencies and other U.S. government resources for technical reports.	Free access
Virtual Technical Reports Central http://www.lib.umd.edu/ENGIN/TechReports/ Virtual-TechReports.html.	Contains a large list of grey literature—producing institutions and technical reports, research reports, preprints, reprints, and e-prints.	Free access

SOURCE: Authors' compilation.

Table 6.6 Citations for Grey Literature in Cochrane Reviews

Source	Number of trial references*
Unpublished information	1,259 (55 percent)
Conference abstracts	805 (35 percent)
Government reports	78 (4 percent)
Company reports	66 (3 percent)
Theses and dissertations	63 (3 percent)
Total (1,446 trials)	2,271 (100 percent)

SOURCE: Authors' compilation.
* Trials can be referenced by more than one grey literature source. 4,820/6,266 trials were referenced only by published journal articles. Seventeen trials had missing citations.

and 17 percent were from unpublished sources, with similar proportions to the Mallett and colleagues (2000) study coming from government and company reports and theses. A study of 153 trials in sixty meta-analyses by Matthias Egger and his colleagues found that 45 percent of trials came from conference abstracts, 1 percent from book chapters, 3 percent from theses, and 37 percent from other grey literature sources including unpublished, company and government reports (2003). Overall, the results are similar across the studies: the majority of grey studies were retrieved from conference abstracts, though in all cases more than one-third of the studies were obtained from other sources.

6.5.2 Social Science and Educational Research

The use of grey literature in the social sciences and educational research may be even higher than in health-care research, although empirical evidence in this area is more limited. Based on survey data Diana Hicks presented in 1999, Lesley Grayson and Alan Gomersall suggested health-care and the social sciences differ greatly in the publication of research (2003). Social sciences research is published in a much broader set of peer-reviewed journals; additionally a larger proportion of social science research is exclusively published in books, governmental and organizational reports, and other sources that are not as widely available as peer-reviewed journals. In her international survey, Hicks identified books as the primary nonjournal source of information in the social sciences (1999).

In a synthesis of the health and psychological effects of unemployment, 72 percent of included studies were published and 28 percent were found in the grey literature (Loo 1985). In a research synthesis by Anthony Petrosino and colleagues of randomized experiments in crime reduction, approximately one-third of the 150 trials identified were reported in government documents, dissertations and technical reports (2000). Earlier, a synthesis of early delinquency experiments reported that about 25 percent of the included studies were unpublished (Boruch, McSweeney, and Soderstrom 1978).

In a study conducted by Hannah Rothstein, the ninety-five meta-analyses published in *Psychological Bulletin* between 1995 and 2005 were examined to see whether they included unpublished research, and if so, what types of research were included (2006). Of the ninety-five syntheses, sixty-nine clearly included unpublished data, twenty-three clearly did not, one apparently did, one apparently did not, and in one case the description was so unclear that it was impossible to tell. There did not appear to be any relationship between the publication date of the synthesis and whether or not it included unpublished studies. Complicating the matter was the fact that what was considered unpublished varied from study to study. All ninety-five meta-analyses included published journal articles, sixty-one included dissertations or theses, forty-two included books or book chapters, and twenty-six included conference papers. On the other hand, seven included government or technical reports, six incorporated raw data, four stated that they used data from monographs, three used information from in-press articles, and two each explicitly mentioned test manuals or national survey data. Unpublished data without a clear description of its source were included in thirty-three meta-analyses, some of which mentioned specific types of grey literature as well. The limited evidence suggests that in the social sciences, books, dissertations and theses and government/technical reports may be the most frequent sources of grey studies.

6.5.3 Impact of Grey Literature

We now examine the evidence showing that there may be systematic differences in estimates of the same effect depending upon whether the estimates are reported in published studies or in the grey literature.

6.5.3.1 Health Care An examination of health-care syntheses also shows systematic differences between the results of published studies and those found in the grey literature. An early example is a meta-analysis of preoperative parenteral nutrition for reducing complications from major surgery and fatalities (Detsky et al. 1987). In this study, the results of randomized trials published only as abstracts found the treatment to be more effective than trials published in full papers.

On the other hand, the exclusion of grey literature from meta-analyses in the health-care literature has been shown to produce both larger estimates of treatment effects and smaller effects. A recent synthesis by Sally Hopewell and her colleagues compared the effect of the inclusion and exclusion of grey literature on the results of a cohort of meta-analyses of randomized trials in health care (2006). Five met the criteria for the synthesis (Burdett, Stewart, and Tierney 2003; Egger et al. 2003; Fergusson et al. 2000; Hopewell 2004; McAuley et al. 2000). In all cases, there were more published trials included in the meta-analyses than grey trials—a median of 224, the interquartile range (IQR) was 108 to 365, versus one of 45, IQR 40 to 102. Additionally, the median number of participants included in published trials was also larger than those in the grey trials. One study also assessed the statistical significance of the trial results included in the meta-analyses (Egger et al. 2003). Here, the researchers found that published trials (30 percent) were more likely to have statistically significant results ($p < 0.05$) than grey trials (19 percent).

For all five included studies, the most common type of literature was abstracts (55 percent). Unpublished data were the second most common (30 percent). The definition of what constituted unpublished data varied slightly because Hopewell and her colleagues used the same definition the authors had used, which varied from study to study. However, it generally included data from trial registers, file-drawer data, and data from individual trialists. The third most common type was book chapters (9 percent). The remainder (6 percent) was a combination of unpublished reports, pharmaceutical company data, in-press publications, letters, and theses .

All five included studies found that published trials showed an overall greater treatment effect than grey trials (Burdett, Stewart, and Tierney 2003; Egger et al. 2003; Fergusson et al. 2000; Hopewell 2004; McAuley et al. 2000). This difference was statistically significant in one of the five studies (McAuley et al. 2000). Data could be combined formally for three of the five (Egger et al. 2003; Hopewell 2004; McAuley et al. 2000), which showed that, on average, published trials showed a 9 percent greater effect than grey trials. The study by Sarah Burdett and her colleagues (2003) included individual

patient data meta-analyses and showed that when published trials were included they produced a 4 percent greater treatment effect than trials as a whole did. Only one study assessed the type of grey literature and its impact on the overall results (McAuley et al. 2000). This study found that published trials showed an even greater treatment effect than grey trials if abstracts were excluded.

6.5.3.2 Social Sciences Much of the early work on the impact of grey literature comes from psychology. Gene Glass and his colleagues were the first to demonstrate that there was a difference in effect sizes between published and unpublished studies (1981). They analyzed eleven psychology meta-analyses published between 1976 and 1980, from areas including psychotherapy, the effects of television on antisocial behavior and sex bias in counseling. In each of the nine instances where a comparison could be made, the average effect from studies published in journals was larger than the corresponding average effect from theses and dissertations. On average, the findings reported in journals were 33 percent more favorable to the investigators' hypothesis than those reported in theses or dissertations. In one meta-analysis of sex bias in counseling and psychotherapy, the effect was of approximately equal magnitude, but in opposite directions in the published and the unpublished studies (Smith 1980). Mark Lipsey and David Wilson's oft-cited study of meta-analyses of psychological, educational and behavioral treatment research found that estimates of treatment effects (standardized differences in means between the treatment group and the control group) from published studies were larger than those from unpublished studies (1993). Across the ninety-two meta-analyses for which comparison by publication status was possible, the average treatment effect for published studies was .53 (expressed as a standardized mean difference effect size); for unpublished studies the average treatment effect was .39. Examinations of individual research syntheses show that in most cases in which effect sizes are broken down by publication status, differences between published and unpublished studies show smaller effects for unpublished studies. G. A. de Smidt and Kevin Gorey contrasted the results of an evaluation of social work interventions based on published studies with those based on doctoral dissertations and masters' theses only (1997). The meta-analysis based on published studies showed slightly more effective interventions than those in the dissertations and theses, specifically 78 percent to 73 percent. In a meta-analysis of sixty-two randomized trials of interventions designed to change behavioral intentions, Thomas Webb and Paschal Sheeran found that published studies reported larger effects for the average standardized difference in means between treatment and control groups than unpublished papers did: one-third of a standard deviation to one-fifth (2006). They also noted that studies reporting interventions not finding statistically significant differences in behavioral intentions were significantly less likely to be published (33 percent) than studies showing them (89 percent published). Bryce McLeod and John Weisz found similar results (2004). Their examination of 121 dissertations and 134 published studies in the general youth psychotherapy literature showed that the dissertations reported effects less than half the size of those in journal articles on the same topics. Their findings are particularly compelling because, overall, the dissertations were stronger methodologically, and there appeared to be no differences in the steps taken to ensure treatment integrity. Their conclusion was that they might have overestimated treatment effects because the majority of the meta-analyses included only published studies.

An exception to the general finding is a study by John Weisz, Carolyn McCarty, and Sylvia Valeri, which found that effects were consistent across published and non–peer-reviewed studies (2006). They noted however, that most youth depression psychotherapy literature is relatively recent, and they attribute their result to an increased focus by journals in the past few years on the quality of research procedures, and a decreased emphasis on statistically significant results.

In the crime and justice literature, Friedrich Losel and Martin Schmucker found that the mean effect size for published studies was greater than for unpublished studies, an odds ratio 1.62 versus one of 1.16 (2005). In a meta-analysis on child skills training to prevent antisocial behavior, Losel and Andreas Beelman reported no significant differences between the effects of published and unpublished studies, but neither the differences in effect sizes nor the power to detect a difference of a specific size were reported (2003).

In the education literature, the available evidence on this issue is very limited. A search of the ERIC database using the terms *education* and *meta-analysis* for 1999 through mid-July 2006, followed by examination of the abstracts of all resulting studies and of those full-text articles available electronically yielded only three that included grey or unpublished literature and reported effect sizes separately by information source. In a meta-analysis of the effects of classroom placement on the self concepts of children with learning disabilities, Batya

Elbaum found minimal effects, for both published (n = 20) and unpublished (n = 45) studies. The published studies, however, had a larger effect ($d = 0.05$) than the unpublished studies, which found no effect ($d = 0.00$) (2002). In a meta-analysis of the results of interventions to improve the reading skills of learning disabled students, Lee Swanson found no differences in average effects by publication status (1999. The average standardized difference in means between treatment and control groups for ninety-four independent samples from journal articles was $d = 0.51$, and the analogous effect for fourteen samples from dissertations and technical reports was $d = 0.52$. In a second meta-analysis, this time on interventions using single-subject designs, Swanson and Carole Sachse-Lee found an average treatment effect per study of $d = 1.42$ for journal articles while for dissertations and technical reports the analogous effect was slightly smaller, $d = 1.27$ (2000).

6.6 PROBLEMS WITH GREY LITERATURE

Some researchers have suggested that studies from the grey literature should not be included in research syntheses because their methodological quality is suspect. Many research synthesists who include only published journal articles seem to be concerned that other sources of data are lesser quality. John Weisz and his colleagues, for example, said this: "We included only *published* psychotherapy outcome studies, relying on the journal review process as one step of quality control" (1995, 452). Pamela Keel and Kelly Klump explained their study this way: "We restricted our review to published works. Benefits of this approach include the increased scientific rigor of published (particularly peer-reviewed) work compared with unpublished work and permanent availability of published work within the public domain for independent evaluation by others" (2003, 748). Because of this, they therefore confined the "search to studies published in peer-reviewed journals . . . to ensure that the studies met scientific standards and to facilitate access to the studies by people who want to examine the sources of our data" (366).

Other researchers believe that this view is unwarranted, arguing that publication in the academic peer-reviewed journal literature has as much, or more, to do with the incentive system for particular individuals than it does with the quality of the research. As Grayson and Gomersall pointed out, a report (not a journal article) is the primary publication medium for individuals working for such entities as independent research organizations, professional societies, and think-tanks (2003). They went on to say that "researchers working in these environments may produce work of high quality but are free from the intense pressure . . . . to publish in peer-reviewed journals" (7). We believe that incentives to publish among degree recipients influence the decision to submit a dissertation or thesis for publication at least as much as the quality of the work does. Furthermore, it has also been noted that though the traditional view of the peer review process is that it helps ensure the quality of published studies, peer review itself may be at times unreliable or biased (Jefferson et al. 2006). Grayson described some of the deficiencies of peer review in the biomedical literature; this work has possible implications for the social sciences as well (2002). Our view is that peer review cannot be taken as a guarantee that research is of good quality, nor can the lack of peer review be taken as an indicator that the quality is poor. The research synthesist must assess, de novo, the methodological quality of each study considered for possible inclusion in his or her synthesis.

A survey by Deborah Cook and Gordon Guyatt in 1993 showed that a substantial majority (78 percent) of meta-analysts and methodologists felt that unpublished material should definitely or probably be included in meta-analyses, but that journal editors were less enthusiastic about the inclusion of unpublished studies (only 47 percent felt that unpublished data should probably or definitely be included). Cook and Guyatt found that both editors' and methodologists' attitudes varied depending on the type of material, with attitudes less favorable toward including journal supplements, published abstracts, published symposia, book chapters, and dissertations and more favorable toward unpublished presentations or materials that had never been published or presented. This survey was recently updated by Jennifer Tetzlaff and her colleagues (2006). Their findings showed that journal editors still appear less favorable toward the grey (unpublished) literature than research synthesists and methodologists do, though these differences have diminished relative to what Cook reported. Further analyses by Tetzlaff indicated that variations between journal editors and research synthesists and methodologists are related in part to differences in research synthesis experience or previous familiarity with the grey literature.

Empirical evidence supports the contention that it is harder to assess the methodological quality of randomized trials of health-care interventions found in the grey literature than those reported in peer-reviewed journals (Egger et al. 2003; MacLean et al. 2003). In an analysis of sixty meta-analyses of health-care interventions, Egger and colleagues found that trials published in the grey literature were somewhat less likely to report adequate concealment of allocation than published trials, 33.8 percent versus 40.7 percent (2003). Trials published in the grey literature were also less likely to report blinding of outcome assessment compared to published trials, 45.1 percent versus 65.8 percent. In contrast, a study by Hopewell found no difference in the quality of reporting of allocation concealment or generation of the allocation sequence between published and grey trials; this information was often unclear in both groups (2004). Although quality and adequacy of reporting are not the same as quality and adequacy of the research, deficiencies in reporting make it difficult for a reviewer to assess a study's methodological quality.

Several studies have shown that many abstracts submitted to scientific conferences have inadequacies in the reporting aims, methods, data, results, and conclusions (Bhandari et al. 2002; McIntosh 1996; Panush et al. 1989; Pitkin and Branagan 1999). A study of trials reported as abstracts and presented at the American Society of Clinical Oncology conferences in 1992 and 2002 found that for most of the studies, it was not possible to determine key aspects of trial quality because of the limited information available in the abstract. Information was also limited on the number of participants initially randomized and the number actually included in the analysis (Hopewell and Clarke 2005).

6.7 CONCLUSIONS AND RECOMMENDATIONS

A number of key sources, then, are important to consider when carrying out a comprehensive search for a research synthesis. These include searching electronic bibliographic databases, hand searching key journals, searching conference proceedings, contacting researchers, searching research registers, and searching the Internet. These sources are likely to identify studies in both the published and grey literature. The optimal extent of the search depends very much on the subject matter and the resources available for the search. For example, those carrying out a research synthesis need to consider when it is appropriate to supplement electronic searches with a search by hand of key journals or journal supplements not indexed by electronic databases, and when it is to search by hand conference proceedings for a specific subject area. It is also critical to assess the quality of the studies identified from the search, regardless of whether they were found in the published or grey literature. Whatever is searched, it is important that the researcher is explicit and documents in the synthesis exactly which sources have been searched and which strategies have been used so that the reader can assess the validity of the conclusions of the synthesis based on the search conducted.

Ultimately, however, unless one knows the number of studies carried out for a particular intervention or other research question (be that in health care, social sciences, education or other fields), regardless of the final publication status, one will not know how many studies have been missed, despite the rigorous literature searches conducted. Thus, the biggest issue is how to collect, organize, and manage centralized repositories of grey literature. In the case of health care, one suggestion put forward many times is the prospective registration of randomized trials at their inception (Tonks 2002; Chalmers 2001; Horton and Smith 1999). This would enable research synthesists and others using the evidence from randomized trials to identify all trials conducted that are relevant to a particular health-care question. In addition, research-funding agencies (public, commercial, and charitable), will be able to make funding decisions in the light of information about trials that are already being conducted. This will, therefore, avoid unnecessary duplication of effort and help promote collaboration between individual organizations. Finally, if information about ongoing trials is publicly available, patients and clinicians will be better informed about trials in which they could participate, which will in turn also help increase trial recruitment rates. As far as we are aware, there are currently no plans related to the prospective registration of field trials conducted in the social sciences and in education. In theory, however, there is no reason why the same principles underpinning prospective registration in health care cannot be applied in the social sciences and in education, and similar benefits achieved. Research registries more broadly construed would also be of benefit for questions that do not involve interventions. In addition to listing ongoing studies, we can envision registries attached to clearinghouses that maintain archived copies of the reports from completed studies in different fields. This, however, would need collaboration among professional societies, universities, other research organizations, and most likely, governments. It is our hope that by the time this handbook is in its third edition, we will be able to report on the creation and growth of these structures.

6.8 ACKNOWLEDGMENTS

We would like to thank Anne Eisinga, Alan Gomersall, Anne Wade, and especially Lesley Grayson for their assistance with writing this chapter. We gratefully acknowledge the path Marylu Rosenthal set out for us in preparing the original handbook chapter, and hope that the current chapter will contribute as much as hers did to the improvement of research syntheses.

6.9 NOTES

1. In the near future, we may even have to amend the phrase from grey literature to grey material as written media may no longer be the only archival source of scientific information. Consider, for example, the potential for podcasts or video clips of scientific reports as sources of data relevant to a synthesis.

6.10 REFERENCES

Armstrong, Chris J., and J. Andrew Large, eds. 2001. *Manual of On-Line Search Strategies: Humanities and Social Sciences*, 3rd ed. Abingdon, U.K.: Gower.

Auger, Charles P. 1998. *Information Sources in Grey Literature. Guides to Information Sources,* 4th ed. London: Bowker Saur.

Bhandari, Mohit, P. J. Devereaux, Gordon H. Guyatt, Deborah J. Cook, Mark F. Swiontkowski, Sheila Sprague, and Emil H. Schemitsch. 2002. "An Observational Study of Orthopaedic Abstracts and Subsequent Full-Text Publications." *Journal of Bone and Joint Surgery, American* 84-A(4): 615–21.

Boruch, Robert F., A. J. McSweeney, and E. J. Soderstrom. 1978. "Randomized Field Experiments for Program Planning, Development, and Evaluation: An Illustrative Bibliography." *Evaluation Quarterly* 2(4): 655–95.

Burdett, Sarah, Lesley A.Stewart, and Jane F. Tierney. 2003. "Publication Bias and Meta-Analyses: A Practical Example." *International Journal of Technology Assessment in Health Care* 19(1): 129–34.

Callaham, Michael L., Robert L. Wears, Ellen J. Weber, Christopher Barton, and Gary Young. 1998. "Positive-Outcome Bias and Other Limitations in the Outcome of Research Abstracts Submitted to a Scientific Meeting." *Journal of the American Medical Association* 280(3): 254–57.

Chalmers, Iain. 2001. "Using Systematic Reviews and Registers of Ongoing Trials for Scientific and Ethical Trial Design, Monitoring, and Reporting." In *Systematic Reviews in Heath Care: Meta-Analysis in Context*, 2nd ed., edited by Matthias Egger, George Davey Smith and Douglas G. Altman. London: BMJ Books.

Chan, An-Wen, and Douglas G. Altman. 2005. "Identifying outcome reporting bias in randomised trials on PubMed: review of publications and survey of authors." *British Medical Journal* 330(7494): 753.

Chan, An-Wen, Asbjørn Hróbjartsson Mette T. Haahr, Peter C. Gøtzsche, and Douglas G. Altman. 2004. "Empirical Evidence for Selective Reporting of Outcomes in Randomized Trials: Comparison of Protocols to Published Articles." *Journal of the American Medical Association* 291(20): 2457–65.

Clarke, Mike, and Teresa Clarke. 2000. "A Study of the References Used in Cochrane Protocols And Reviews. Three Bibles, Three Dictionaries, and Nearly 25,000 Other Things." *International Journal of Technology Assessment in Health Care* 16(3): 907–9.

Cook, Deborah J., and Gordon H. Guyatt. 1993. "Should Unpublished Data Be Included in Meta-Analyses?" *Journal of the American Medical Association* 269(21): 2749–53.

Cooper, Harris, Kristina DeNeve, and Kelly Charlton. 1997. "Finding the Missing Science: The Fate of Studies Submitted for Review by a Human Subjects Committee." *Psychological Methods* 2(4): 447–52.

de Smidt, G.A., and Kevin M. Gorey. 1997. "Unpublished Social Work Research: Systematic Replication of a Recent Meta-Analysis of Published Intervention Effectiveness Research." *Social Work Research* 21(1): 58–62.

Detsky, Allan S., Jeffrey P. Baker, Keith O'Rourke, and Vivek Goel.1987. "Perioperative Parenteral Nutrition: A Meta-Analysis." *Annals of Internal Medicine* 107(2): 195–203.

Dickersin, Kay. 2005. "Publication Bias: Recognizing the Problem Understanding its Origins and Scope, and Preventing Harm." In *Publication Bias in Meta-Analysis: Prevention, Assessment and Adjustments*, edited by Hannah R. Rothstein, Alexander J. Sutton, and Michael Borenstein. Chichester: John Wiley & Sons.

Dickersin, Kay, Roberta Scherer, and Carol Lefebvre. 1994. "Identifying Relevant Studies for Systematic Reviews." *British Medical Journal* 309(6964): 1286–91.

Dickersin, Kay, Eric Manheimer, Susan Wieland, Karen A. Robinson, Carol Lefebvre, and Steven McDonald. 2002. "Development of the Cochrane Collaboration's CENTRAL Register of Controlled Clinical Trials." *Evaluation and the Health Professions* 25(1): 38–64.

Donaldson I. J, and Patricia A. Cresswell. 1996. "Dissemination of the Work of Public Health Medicine Trainees in Peer-Reviewed Publications: An Unfulfilled Potential." *Public Health*, 110(1): 61–63.

Egger, M, Peter Jüni, Christopher Bartlett, F. Holenstein, and Jonathan A.C. Sterne. 2003. "How Important are Comprehensive Literature Searches and the Assessment of Trial Quality in Systematic Reviews? Empirical Study." *Health Technology Assessment* 7(1): 1–76.

Egger, Matthias, Tanja Zellweger-Zähner, Martin Schneider, Christoph Junker, Christian Lengeler, and Gerd Antes. 1997. "Language Bias in Randomised Controlled Trials Published in English and German." *Lancet* 350: 326–29.

Elbaum, Batya. 2002. "The Self-Concept of Students with Learning Disabilities: A *Meta-Analysis* of Comparisons Across Different Placements." *Learning Disabilities: Research and Practice* 17(4): 216–26.

Eysenbach, Gunther, Jens Tuische, and Thomas L. Diepgen Diepgen. 2001. "Evaluation of the Usefulness of Internet Searches to Identify Unpublished Clinical Trials for Systematic Reviews." *Medical Informatics and the Internet in Medicine* 26(3): 203–18.

Fergusson, Dean, Andreas Laupacis, L. Rachid Salmi Finley A. McAlister, and Charlotte Huet. 2000. "What Should Be Included in Meta-Analyses? An Exploration of Methodological Issues Using the ISPOT Meta-Analyses." *International Journal of Technology Assessment in Health Care* 16(4): 1109–19.

Fink, Arlene. 2004. *Conducting Research Literature Reviews: From the Internet to Paper*, 2nd ed. Thousand Oaks, Calif.: Sage Publications.

Glass, Gene V, Barry McGaw, and Mary Lee Smith. 1981. *Meta-Analysis in Social Research*. Thousand Oaks, Calif.: Sage Publications.

Gøtzsche, Peter C. 1987. "Reference Bias in Reports of Drug Trials." *British Medical Journal* 295(6599): 654–56.

Grayson, Lesley. 2002. "Evidence-Based Policy and the Quality of Evidence: Rethinking Peer Review." ESRC Working Paper 7. London: ESRC UK Centre for Evidence Based Policy and Practice. http://www.evidencenetwork .org.

Grayson, Lesley, and Alan Gomersall. 2003. "*A Difficult Business: Finding the Evidence for Social Science Reviews.*" *ESRC* Working Paper 19. London: ESRC U.K. Centre for Evidence Based Policy and Practice.

Greenhalgh, Tricia, and Richard Peacock. 2005. "Effectiveness and Efficiency of Search Methods in Systematic Reviews of Complex Evidence: Audit of Primary Sources." *British Medical Journal* 331(7524): 1064–65.

Helmer, Diane, Isabelle Savoie Carolyn Green, and Arminee Kazanjian. 2001. "Evidence-Based Practice: Extending the

Search to Find Material for the Systematic Review." *Bulletin of the Medical Librarians Association* 89(4): 346–52.

Hicks, Diana. 1999. "The Difficulty of Achieving Full Coverage of International Social Science Literature and the Bibliometric Consequences." *Scientometrics* 44(2): 193–215.

Higgins, Julian T., and Sally Green. 2008. *Cochrane Handbook for Systematic Reviews of Interventions*, version 5.0.0. Chichester: John Wiley & Sons.

Hopewell, Sally. 2004. "Impact of Grey Literature on Systematic Reviews of Randomized Trials." PhD diss. Wolfson College, University of Oxford.

Hopewell, Sally, and Mike Clarke. 2001. "Methodologists and Their Methods. Do Methodologists Write Up Their Conference Presentations or Is It Just 15 Minutes of Fame?" *International Journal of Technology Assessment in Health Care* 17(4): 601–3.

———. 2005. "Abstracts presented at the American Society of Clinical Oncology conference: How Completely Are Trials Reported?" *Clinical Trials* 2(3): 265–68.

Hopewell, Sally, Mike Clarke, Carol Lefebvre, and Roberta Scherer. 2002. "Handsearching Versus Electronic Searching to Identify Reports of Randomized Trials." *Cochrane Database of Methodology Reviews* 2002(4): MR000001.

Hopewell, Sally, Mike Clarke, Lesley Stewart, and Jane Tierney. 2001. "Time to Publication for Results of Clinical Trials." *Cochrane Database of Methodology Reviews* 2001(3): MR000011.

Hopewell, Sally, Steve McDonald, Mike Clarke, and Matthias Egger. 2006. "Grey Literature in Meta-Analyses of Randomized Trials of Health Care Interventions." *Cochrane Database of Methodology Reviews* 2006(2): MR000010.

Horton, Richard, and Richard Smith. 1999. "Time to Register Randomised Trials." *Lancet* 354(9185): 1138–39.

Jefferson, Tom, Melanie Rudin, Suzanne Brodney Folse, and Frank Davidoff. 2006. "Editorial Peer Review for Improving the Quality of Reports of Biomedical Studies." *Cochrane Database of Methodology Reviews* 2006(1): MR000016.

Jüni, Peter, F. Holenstein, Jonathan AC Sterne, Christopher Bartlett, and Matthias Egger. 2002. "Direction and Impact of Language Bias in Meta-Analyses of Controlled Trials: Empirical Study." *International Journal of Epidemiology* 31(1): 115–23.

Keel, Pamela K, and Kelly L. Klump. 2003. "Are Eating Disorders Culture-Bound Syndromes? Implications for Conceptualizing their Etiology." *Psychological Bulletin* 129(5): 747–69.

Krleza-Jeric, Karmela, An-Wen Chan Kay Dickersin, Sim I, Jeremu Grimshaw, and Christian Gluud. 2005. "Principles for International Registration of Protocol Information and Results from Human Trials of Health-Related Interventions: Ottawa Statement." *British Medical Journal* 330(7497): 956–58.

Lipsey, Mark W., and David B. Wilson. 1993. "The Efficacy of Psychological, Educational, and Behavioral Treatment: Confirmation from Meta-Analysis." *American Psychologist* 48(12): 1181–1209.

Loo, John. 1985. "Medical and Psychological Effects of Unemployment: A 'Grey' Literature Search." *Health Libraries Review* 2(2): 55–62.

Losel, Friedrich, and Andreas Beelman. 2003. "Effects of Child Skills Training in Preventing Antisocial Behavior: A Systematic Review of Randomized Evaluations." *Annals of the American Academy of Political and Social Science* 587(1): 84–109.

Losel, Friedrich, and Martin Schmucker. 2005. "The Effectiveness of Treatment for Sexual Offenders: A Comprehensive Meta-Analysis." *Journal of Experimental Criminology* 1(1): 117–46.

MacLean, Catherine H., Sally C. Morton, Joshua J. Ofman, Elizabeth A.Roth, and Paul G. Shekelle. 2003. "How Useful Are Unpublished Data from the Food and Drug Administration in Meta-Analysis?" *Journal of Clinical Epidemiology* 56(1): 44–51.

Mallett, Susan, Sally Hopewell, and Mike Clarke. 2002. "The Use of Grey Literature in the First 1000 Cochrane Reviews." Presented at the Fourth Symposium on Systematic Reviews, "Pushing the Boundaries." Oxford (July 2–4, 2002).

McAuley, Laura, Ba' Pham, Peter Tugwell, and David Moher. 2000. "Does the Inclusion of Grey Literature Influence Estimates of Intervention Effectiveness Reported in Meta-Analyses?" *Lancet* 356(9237): 1228–31.

McDaniel, Michael A,, Hannah R. Rothstein, and Deborah Whetzel. 2006. "Publication Bias: A Case Study of Four Test Vendor Manuals." *Personnel Psychology* 59(4): 927–53.

McIntosh, Neil. 1996. "Abstract Information and Structure at Scientific Meetings." *Lancet* 347(9000): 544–45.

McKimmie, Tim, and Joanna Szurmak. 2002. "Beyond Grey Literature: How Grey Questions Can Drive Research." *Journal of Agricultural and Food Information* 4(2): 71–79.

McLeod, Bryce D., and John R. Weisz. 2004. "Using Dissertations to Examine Potential Bias in Child and Adolescent Clinical Trials." *Journal of Consulting and Clinical Psychology* 72(2): 235–51.

McPadden, Katherine, and Hannah R. Rothstein. 2006. "Finding the Missing Papers: The Fate of Best Paper Proceedings." Presented at AOM Conference, Academy of

Management Annual Meeting. Atlanta, Ga. (August 11–16, 2006).

Ogilvie, David, Richard Mitchell Nanette, Mutrie N, Mark Petticrew, and Stephen Platt. 2006. "Evaluating Health Effects of Transport Interventions: Methodologic Case Study." *American Journal of Preventive Medicine* 31(2): 118–26.

Panush, Richard S., Jeffrey C. Delafuente, Carolyn S. Connelly, N. Lawrence Edwards, Jonathan M. Greer, Sheldon Longley, and Fred Bennett. 1989. "Profile of a Meeting: How Abstracts Are Written and Reviewed." *Journal of Rheumatology* 16: 145–47.

Petrosino, Anthony, Robert F. Boruch, C. Rounding, Steven McDonald, and Iain Chalmers. 2000. "The Campbell Collaboration Social, Psychological, Educational and Criminological Trials Register C2-SPECTR." *Evaluation and Research in Education* 14(3–4): 206–18.

Pitkin, Roy M, and Mary Ann Branagan. 1998. "Can the Accuracy of Abstracts be Improved by Providing Specific Instructions? A Randomized Controlled Trial." *Journal of the American Medical Association* 280(3): 267–69.

Ravnskov, Uffe. 1992. "Cholesterol Lowering Trials in Coronary Heart Disease: Frequency of Citation and Outcome." *British Medical Journal* 305(6844): 15–19.

Rosenthal, Marylu C. 1994. "The fugitive literature." In *The Handbook of Research Synthesis*, edited by Harris Cooper and Larry V. Hedges. New York: Russell Sage Foundation.

Rothstein, Hannah R. 2006. "Use of Unpublished Data in Systematic Reviews in the *Psychological Bulletin* 1995–2005." Working paper.

Rothstein, Hannah R., Alexander J. Sutton, and Michael Borenstein. 2005. *Publication Bias in Meta-Analysis: Prevention, Assessment and Adjustments*. Chichester: John Wiley & Sons.

Rothstein, Hannah R., Herbert M. Turner, and Julia Lavenberg. 2004. *Information Retrieval Policy Brief*. Philadelphia, Pa.: International Campbell Collaboration.

Scherer, Roberta W., Patricia Langenberg, and Erik von Elm. 2005. "Full Publication of Results Initially Presented in Abstracts." *Cochrane Database of Methodology Reviews* 2005(2): MR000005.

Schmidmaier, Dagmar. 1986. "Ask No Questions and You'll Be Told No Lies: Or How We can Remove People's Fear of 'Grey Literature.'" *Librarian* 36(2): 98–112.

Smith, Mary Lee. 1980. "Sex Bias in Counseling and Psychotherapy." *Psychological Bulletin*, 87(2): 392–407.

Song, Fujian, Alison J. Eastwood, Simon Gilbody, Leila Duley, and Alexander J. Sutton. 2000. "Publication and Related Biases." *Health Technology Assessment* 4(10): 1–115.

Swanson, H. Lee, and Carole Sachse-Lee. 2000. "A Meta-Analysis of Single-Subject-Design Intervention Research for Students with LD." *Journal of Learning Disabilities* 33(2): 114–36.

Swanson, H. Lee. 1999. "Reading Research for Students with LD: A Meta-Analysis of Intervention Outcomes." *Journal of Learning Disabilities* 32(6): 504–32.

Tetzlaff, Jennifer, David Moher, Ba Pham, and Douglas Altman. 2006. "Survey of Views on Including Grey Literature in Systematic Reviews." Presented at the XIV Cochrane Colloquium. Dublin (October 23–26, 2006).

Tonks, Alison. 2002. "A Clinical Trials Register for Europe." *British Medical Journal* 325(7376): 1314–15.

Tramer, Martin R., D. John M. Reynolds, R. Andrew Moore, and Henry J. McQuay. 1997. "Impact of Covert Duplicate Publication on Meta-Analysis: A Case Study." *British Medical Journal* 315(7109): 635–40.

Turner, Herbert M, Robert Boruch, Julian Lavenberg, J. Schoeneberger, and Dorothy de Moya. 2004. "Electronic Registers of Trials." Presented at the Fourth Annual Campbell Collaboration Colloquium, "A First look at the Evidence." Washington, D.C. (February 18–20, 2004).

Wade, Anne, Herbert M. Turner, Hannah R. Rothstein, and Julia G. Lavenberg. 2006. "Information Retrieval and the Role of the Information Specialist in Producing High-Quality Systematic Reviews in the Social, Behavioral and Education Sciences." *Evidence and Policy: a Journal of Research, Debate and Practice* 2(1): 89–108.

Webb, Thomas L., and Paschal Sheeran. 2006. "Does Changing Behavioral Intentions Engender Behavior Change? A Meta-Analysis of the Experimental Evidence." *Psychological Bulletin* 132(2): 249–68.

Weber, Ellen J, Michael L. Callaham, Robert L. Wears, Christopher Barton, and Gary Young. 1998. "Unpublished Research from a Medical Specialty Meeting: Why Investigators Fail to Publish." *Journal of the American Medical Association* 280(3): 257–59.

Weintraub, Irwin. 2000. "The Role of Grey Literature in the Sciences." http://library.brooklyn.cuny.edu/access/greyliter.htm.

Weisz, John R., Carolyn A. McCarty, and Sylvia M. Valeri. 2006. "Effects of Psychotherapy for Depression in Children and Adolescents: a Meta-Analysis." *Psychological Bulletin* 132(1): 132–49.

Weisz, John R., Bahr Weiss, Susan S. Han, Douglas A. Granger, and Todd Morton. 1995. "Effects of Psychotherapy with Children and Adolescents Revisited: A Meta-Analysis of Treatment Outcome Studies." *Psychological Bulletin* 117(3): 450–68.

PART
IV

CODING THE LITERATURE

7

JUDGING THE QUALITY OF PRIMARY RESEARCH

JEFFREY C. VALENTINE
University of Louisville

CONTENTS

7.1 INTRODUCTION

Empirical studies vary in terms of the rigor with which they are conducted, and the quality of the studies comprising a research synthesis can have an impact on the validity of conclusions arising from that synthesis. However, the wide agreement on these points belies a fundamental tension between two very different approaches in research synthesis to addressing study quality and its impact on the validity of study conclusions. These approaches are the focus of this chapter. In the sections that follow I provide a definition for the term *study quality*, discuss two approaches for addressing study quality in a research synthesis, and conclude with specific advice for scholars undertaking a synthesis.

7.1.1 Defining Study Quality

One issue in applying the term *quality* to a study is that the word can have several distinct meanings. As William Shadish pointed out, asking a university administrator to define the quality of a piece of research might lead to the use of citation counts as an indicator (1989). Asking a school teacher the same question might lead one to examine whether the study resulted in knowledge that had clear implications for practice. As such, the answer to the question "What are the characteristics of a high quality study?" depends in part on why the judgment is being made. In this chapter, I use quality to refer to the fit between a study's goals and the study's design and implementation characteristics. As an example, if a scholar would like to know if the beliefs teachers hold about the academic abilities of their students (that is, teacher expectancies) can cause changes in the IQ of those students, then the best way to approach this question is through a well-implemented experiment in which some teachers are randomly assigned to experience change in their expectations about a student and other teachers are randomly

assigned to not experience such change. All else being equal, the closer a study on this question approaches the ideal of a well-implemented randomized experiment, the higher its quality.

7.1.2 Campbell's Validity Framework

In the social sciences, perhaps the most widely used system guiding thinking about the quality of primary studies is that articulated by Donald Campbell and his colleagues (Campbell and Stanley 1966; Cook and Campbell 1979; Shadish, Cook, and Campbell 2002). In the Campbell tradition, *validity* refers to the approximate truth of an inference (or claim) about a relationship (Cook and Campbell 1979; Shadish, Cook, and Campbell 2002). Although it is commonly stated that an inference is valid if the evidence on which it is based is sound and invalid if that evidence is flawed, the judgment about validity is neither dichotomous nor one-dimensional, features that have important implications for the manner in which validity is used to assess study quality. Instead of being dichotomous and unidimensional, different characteristics of study design and implementation lead to inferences that are more or less valid along one or more dimensions. Factors that might lead to an incorrect inference are termed threats to validity or plausible rival hypotheses. Threats to validity can be placed into four broad categories or dimensions: *internal validity*, *external validity*, *construct validity*, and *statistical conclusion validity*.

Internal validity refers to the validity of inferences about whether some intervention has caused an observed outcome. Threats to internal validity include any mechanism that might plausibly have caused the observed outcome even if the intervention had never occurred. External validity refers to how widely a causal claim can be generalized from the particular realizations in a study to other realizations of interest. That is, even if we can be relatively certain that a study is internally valid—meaning we can be relatively certain that observed changes in the outcome variable were caused by the intervention—we still don't know if the results of the study will apply, for example, to different people in different contexts using different treatment variations with outcomes measured in different ways. Construct validity refers to the extent to which the operational characteristics of interventions and outcome measures used in a study adequately represent the intended abstract categories. Researchers most often think of construct validity in terms of the outcome measures. It also refers, however, to the adequacy of other labels used in the study, such as the intervention and the labels applied to participants (for example, at-risk). A measure that purports to measure intelligence but actually measures academic achievement, for example, is mislabeled and hence presents a construct validity problem. Finally, statistical conclusion validity refers to the validity of statistical inferences regarding the strength of the relationship between the presumed cause and the presumed effect. Assume, for example, that a researcher manipulates teacher expectancies and, after conducting an analysis of variance, infers that if a teacher alters her academic expectancy about a student to be more positive it causes that student to achieve at higher levels. If the data that led to this inference did not meet the assumptions underlying the analysis of variance, then the inference may be compromised.

7.1.3 Relevance Versus Quality

From the outset, it is important to draw a distinction between relevance and quality. Relevance refers to study features that pertain to the decision about whether to include a given study in a research synthesis. These features may or may not relate to the fit between a study's goals and its methods used to achieve those goals. As an example, assume a team of scholars is interested in the impact of teacher expectancies on the academic achievement of low socioeconomic status (SES) students. If the researchers were to encounter a study that focuses on wealthy students, they would not include that study in their synthesis because its focus on wealthy students made it irrelevant to their goal. This would be true regardless of fit between the study's goals and the methods used to achieve those goals, and as such researchers should be careful not to frame it as a quality issue.

There is, however, a sense in which relevance and quality are intertwined. These issues will usually be encoun-

tered when considering the operational definitions that were employed in the studies that might be relevant to the research synthesis, and hence largely address issues of construct validity. As an example, David DuBois and his colleagues operationally defined mentoring, in part, as an activity that occurs with an older mentor and a single younger mentee, that was meant to focus on positive youth development in general, and to occur over a relatively long period (2002). This operational definition of mentoring was grounded in theoretical work in the area of youth development. During the literature search, the researchers found a study in which an adult was called a mentor; this adult met with children a few times in groups of two or three, and focused on helping them with their homework. Because of these characteristics, the activity the authors described—even though they used the term *mentoring*—did not fit the definition the synthesists used. This study is an example of how relevance and quality decisions sometimes cannot be separated. Ultimately, it was not included in the synthesis because it was deemed irrelevant. However, this decision could be labeled a quality judgment because the original study's authors used the term in a way that was not consistent with the common usage among youth development scholars, and hence presented a definitional (that is, construct validity) problem.

7.1.4 Role of Study Design

Before collecting data, research synthesists should decide what kinds of evidence will be included in the synthesis. The answer to this question is often complex, necessitating a thoughtful consideration of the trade-offs involved in each type of design the team is likely to encounter. For example, researchers interested in synthesizing the results of multiple political polls must decide whether to exclude all studies that did not use a random sampling technique, because in the absence of random sampling it is not clear whether the sample can be expected to be similar to the population (an external validity issue). Similarly, when random assignment is used to allocate participants to study groups in studies of the causal effects of an intervention, the groups are expected to be equivalent on all dimensions (an internal validity issue). As such, researchers synthesizing the effects of an intervention must decide whether to exclude all studies that did not use random assignment.

Empirical evidence also suggests that not using randomization when it would be appropriate to do so can lead to biased results. In medicine, for example, a body of evidence suggests that studies relying on nonrandom allocation methods are positively biased relative to random allocation. Thomas Chalmers and his colleagues, for example, found that studies using nonrandom allocation were about three times more likely to have at least one baseline variable that was not equally distributed across study groups; baseline variables also were about two times more likely to favor the treatment group (1983). These findings are highly suggestive of an assignment process that stacked the deck in favor of finding treatment effects. Taking a somewhat different approach, Graham Colditz, James Miller, and Frederick Mosteller found, through a synthesis of studies reported in medical journals comparing a new treatment to a standard treatment, that studies using a nonrandom sequential allocation scheme yielded treatment effects that were modestly larger (that is, favored the new treatment) than those using random assignment (1989). In a separate study, this time looking specifically at surgical interventions, they again found a small bias favoring nonrandomized relative to randomized trials (Miller, Colditz, and Mosteller 1989).

For synthesists working on research questions that would be appropriate for randomization, there is no easy blanket answer to the question: "Should we exclude all studies that do not employ randomization?" Relatively little empirical work has been done to investigate the specific selection mechanisms operating in different contexts. Further, the answer to this question probably depends heavily on the context of the research question and the types of research designs that have been used to address it. It is possible, for example, that in some contexts nonrandom selection mechanisms leading to an overestimate of the population parameter in one study might be balanced by different nonrandom selection mechanisms leading to an underestimate of the population parameter in another study (Glazerman, Levy, and Myers 2003; Lipsey and Wilson 1993). It is also possible that selection mechanisms will operate to relatively consistently over- or underestimate (bias) the population parameter (for example, Shadish and Ragsdale 1996). Specific recommendations for addressing this problem are presented in section 7.6 of this chapter.

7.1.5 Study Quality Versus Reporting Quality

A final consideration is the need to draw a distinction between study quality and reporting quality. In many studies, information vital to a research synthesist is not present. Study authors may fail to include important estimates of measurement reliability and validity, for example. It could also be that the authors of the study simply thought the information was irrelevant. Perhaps they chose not to present the information thinking that it revealed a weakness of the study. It is also possible that the information was intended for inclusion but eliminated by an editor due to space considerations. It is not always clear how a synthesist should proceed in these situations. Regardless, there is no escaping the reality that when synthesists attend to study quality issues they will find some of the information they want is missing and they will not know if this is because of poor study quality or poor reporting, or perhaps both. Requesting additional information from study authors should be the first approach to dealing with this problem.

7.2 OPTIONS FOR ADDRESSING STUDY QUALITY IN A RESEARCH SYNTHESIS

Generally, researchers undertaking a synthesis have used a blend of two strategies for addressing study quality: a priori strategies that involve setting rules for determining either the kinds of evidence that will be included or how different evidence will be weighted, and post hoc strategies that treat the impact of study quality as an empirical question.

The importance of how to address quality in research synthesis is indicated by the number of organizations that have become involved in setting standards for studies investigating causal relations. A thorough description of these is beyond the scope of this chapter, but three useful illustrations are the work of the Task Force on Community Preventative Services (Bris et al. 2000; Zara et al. 2000) as evident in the Guide to Community Preventative Services, the standards of evidence established by the What Works Clearinghouse (2006), and the quality assessment criteria established by the Social Care Institute for Excellence (Coren and Fisher 2006). I discuss each of these in turn.

7.2.1 Community Preventative Services

The Task Force on Community Preventative Services developed methods and standards to be used in evaluating and synthesizing research in community prevention (Bris et al. 2000; Zara et al. 2000). For studies that pass a design suitability screen, the Task Force set up a three-tiered grading system that characterizes the quality of the study execution; the categories are good, fair, and limited. Stud-

Table 7.1 Quality Assessment for Guide to Community Preventive Services

Quality Category	Example Items
Descriptions (study population and intervention)	Was the study population well described?
Sampling	Did the authors specify the sampling frame or universe of selection for the study population?
Measurement (exposure)	Were the exposure variables valid measures of the intervention under study?
Measurement (outcome)	Were the outcome and other independent (or predictor) variables reliable (consistent and reproducible) measures of the outcome of interest?
Data analysis	Did the authors conduct appropriate analysis by conducting statistical testing (where appropriate)?
Interpretation of Results: Participation	Did at least 80% of enrolled participants complete the study?
Interpretation of Results (comparability and bias)	Did the authors correct for controllable variables or institute study procedures to limit bias appropriately?
Interpretation of Results (confounders)	Describe all potential biases or unmeasured/contextual confounders described by the authors.
Other	Other important limitations of the study not identified elsewhere?
Study Design	Concurrent comparison groups and prospective measurement of exposure and outcome

SOURCE: Author's compilation; Zara et al. 2000.
NOTE: Adequacy of study design is assessed separately.

ies revealing zero to one limitation are characterized as good, two to four as fair, and five to nine as limited. Neither those that fail the design screen nor those judged as limited are used to develop recommendations, hence are not used in the synthesis that guides the development of recommendations. The nine dimensions relate to a variety of aspects of study design and implementation, and are presented in table 7.1.

7.2.2 What Works Clearinghouse

The What Works Clearinghouse (WWC) developed standards to be used in evaluating and synthesizing research in education (2006). To assess the quality of interventions, the WWC set up a three-tiered grading system which characterizes the quality of study execution. The categories are *meets evidence standards*, *meets evidence standards with reservations*, and *does not meet evidence standards*. To be included in a WWC review, studies must first use a research design appropriate for studying causal relations—that is, a randomized experiment, a quasi-experiment

with equating, a regression discontinuity design, or an appropriate single case design. They must not experience excessive loss of participants—that is, attrition, either overall or differential. They also must not confound teacher effects with intervention effects, must use equating procedures judged to have a reasonable chance of equating the groups, and must not experience an event that contaminates the comparison between study groups.

7.2.3 Social Care Institute for Excellence

The Social Care Institute for Excellence (SCIE) commissions systematic reviews with the aim of improving the quality of experiences by individuals who use social care. To address study quality, SCIE established a framework of quality dimensions, although synthesists may choose to use a different system. The dimensions addressed in SCIE's system are relevance of study particulars, appropriateness of study design, quality of reporting and analysis, and generalizability. Unlike the Task Force on Community Preventive Services and the WWC,

SCIE does not screen out evidence based on these scores. Rather, SCIE requires synthesists to weight study results by their quality scores.

7.3 SPECIFIC ELEMENTS OF STUDY QUALITY

The specific elements of study quality that will require the attention of the research synthesist will of course vary from topic to topic. However, regardless of the research question, Campbell's validity framework provides a convenient way of thinking about the dimensions of quality that ought to be considered. Many of the considerations outlined are represented in at least one of the three systems for evaluating study quality just identified.

7.3.1 Internal Validity

By definition, internal validity is only a concern in studies in which the goal is to investigate causal relations. Several different mechanisms are potentially plausible threats to validity in these studies. Section 7.1.4 already addressed the important role of random assignment to conditions. In addition, information about attrition (data loss and exclusions; see section 7.4.1) and crossover (participants change study conditions after random assignment) should be gathered from each study. In some cases, studies might report conducting an intention-to-treat analysis to address problems of attrition and crossover. Ideally, in an intent-to-treat analysis all participants are included in the analysis in their originally assigned conditions, regardless of attrition or cross-over. Unfortunately, there appears to be little consistency in the use of the term *intention-to-treat analysis* (Hollis and Campbell 1999), so synthesists need to carefully examine reports that use the term to determine what exactly was done.

In studies that do not use random assignment to conditions, synthesists should carefully attend to the baseline comparability of groups and to any procedures used by the researchers to attempt to equate groups. They should also be aware of increased risk for history, maturation, statistical regression, and testing threats to internal validity inherent in these designs (see Shadish, Cook, and Campbell 2002). Unfortunately, studies in the social sciences often report little information on these threats. Nonetheless, the synthesist should seriously consider how these threats might reveal themselves in the context of the research question, and should read and code studies with the intention of identifying these problems when they exist.

In the medical sciences, recent attention has been paid to the concealment of the allocation schedule from the individuals enrolling participants into trials as an integral part of protecting a random allocation scheme. Kenneth Schulz and his colleagues, for example, found that treatment effects were exaggerated by about one-third in studies with unclear or inadequate concealment relative to experiments in which the allocation schedule was protected (1995). This aspect of study quality has received virtually no attention in the social sciences, however.

7.3.2 Construct Validity

Construct validity pertains to the adequacy of the labels used in a study and as such, is relevant to virtually all studies. For outcome measures, synthesists should attend to indices of score reliability (see section 7.3.2.2) and validity (here, the extent to which a measure actually measures what it is supposed to), as well as the method used to assess the construct (for example, self-report vs. parent-report of grades). Synthesists should also attend to the labels applied to study participants. Some of these will be relatively straight-forward (for example, age and sex), but many will not. For example, if a team of synthesists is interested in investigating the effects of teacher expectancies on students labeled at-risk, they will likely find a wide range of methods that study authors have used to assign risk status; some of these methods may be more valid than others.

Finally, synthesists examining the effects of an intervention should attend to several aspects of the intervention. This might include, for example, an analysis of the extent to which the intervention as it was implemented matches the intervention's label. Synthesists might also determine whether intervention fidelity was monitored and if so, how faithfully the intervention was judged to have been implemented.

7.3.3 External Validity

A sampling framework in which units are randomly chosen from a carefully specified population provides the strongest basis for making inferences about the generalizability of study findings. Unfortunately, except in survey research, randomly selecting a sample from a population of potential participants is relatively rare across the social and medical sciences, and even when random sampling is used it rarely extends to study characteristics beyond the individuals or other units of interest. These characteristics mean that unless the synthesist is working in an area in which random sampling occurs frequently, study quality concerns related to external validity receive relatively less attention. However, synthesists should carefully consider and code elements of the study that might help readers make rational and empirically based decisions about the likely external validity of the findings of both individual studies and the research synthesis (see chapter 28, this volume).

7.3.4 Statistical Conclusion Validity

In a research synthesis, the results of individual studies are represented as effect sizes (see chapters 12 and 13, this volume). Effect sizes are not based on the probability values arising from the results of statistical tests in those individual studies, and as such research synthesists are usually only interested in some of the threats to statistical conclusion validity usually identified in the Campbell tradition. For example, both low statistical power and too many statistical tests can threaten the validity of statistical conclusions at the individual study level. However, these are usually not of primary interest to the research synthesist because they respectively affect type II and type I error rates but do not bias effect sizes themselves.

Other threats to statistical conclusion validity can affect the estimation of the magnitude of the study's effects or the stated precision with which the effects are estimated, or both. For example, violating the assumptions underlying the estimation of an effect size can lead to either an overestimate or an underestimate of the population effect. As such, synthesists should code studies for evidence pertaining to the assumptions underlying the effect size metric they are employing. Generally however, violations of the equinormality assumption (that is, populations that are normally distributed and have equal variances) do not have a great impact on effect size estimation, although the potential for problems becomes larger as sample size decreases (Fowler 1988). A violation of the assumption of independence can result in a sometimes dramatic overstatement of the precision of the effect size estimate, even when the degree of violation is relatively small. As such possible multilevel data structures—for example, students in classrooms and classrooms in school buildings—need to be investigated in every study (Hedges, 2007; Murray 1998).

Further, selective outcome reporting can lead to biased estimation of effect size (see also chapter 6, this volume). As an example, assume researchers measure IQ two ways but report only results for the one IQ measure that yielded statistically significant results. This will in all likelihood overestimate the magnitude of the effect of the intervention on the construct of IQ. When effects cannot be estimated for measured outcomes synthesists should make every effort to retrieve the information from study authors. Unfortunately, authors often do not exhaustively report the outcomes they measured, so to some extent this problem is likely to be invisible to synthesists (Williamson et al. 2005). Increasing use of prospective research registers should help synthesists diagnose the problem more effectively in the future (see chapter 6, this volume).

7.4 A PRIORI STRATEGIES

7.4.1 Excluding Studies

As evident in the quality assessment frameworks established by the Task Force on Community Preventive Medicine and the What Works Clearinghouse, one common approach to addressing study quality in a research synthesis is simply to exclude studies that have certain characteristics. Returning, as an example, to the random assignment of participants to conditions, synthesists who have relevant evidence or are operating under the assumption that a lack of random assignment will bias a body of evidence may choose to exclude all nonrandomized comparisons. Less extreme, other synthesists may exclude studies that rely on a within-groups comparison (for example, one group pretest-posttest designs) to estimate the treatment effect but include studies that use intact groups and attempt to minimize selection bias through matching or statistical controls. Other characteristics sometimes used to exclude studies are small sample size, high attrition rates, poor measurement characteristics, publication status (for example, include only published studies), and poor intervention implementations.

Ideally, researchers will attend to three concerns about their adopted rules. First, they need to adopt their rules before examining the data. Obviously, adopting a rule for excluding evidence after the data have been collected increases the likelihood (and perception) that the decision has been made to exclude studies that did not support the synthesists' existing beliefs and preferences. To the extent possible, therefore, synthesists should avoid making these types of decisions after the outcomes of studies are known. If the need for such a rule becomes apparent after the data have been collected, the synthesists should be clear about the timing of and rationale for the decision.

Further, they should consider conducting a sensitivity analysis to test the implications of the decision rule (see chapter 22, this volume).

Second, researchers should operationally define their exclusion rules. For example, it is not enough to state that studies with too much attrition were excluded because readers have no way of knowing how much attrition is too much, and as such cannot judge the adequacy of this rule for themselves. Further, absent an operational definition of too much attrition it becomes less likely that the individuals coding studies for the synthesis will consistently apply the same rule to all studies, leading to study ratings (and hence, inclusion decisions) that have less agreement than is desirable. So, for example, researchers wishing to exclude studies with too much attrition might operationally define too much as more than 30 percent loss from the original sample.

Finally, synthesists should provide empirical or theoretical evidence that their rules function to remove bias from the body of studies or otherwise improve the interpretability of the results of the synthesis. Regrettably, empirical evidence rarely exists in many research contexts and for most characteristics of study design and implementation. This creates a problem for synthesists wishing to employ a priori rules. In the absence of empirical evidence, the belief that the rules adopted serve to increase the validity of the findings has little basis.

If more attention were paid to the last two requirements (operationally defining the exclusion criteria and providing empirical evidence for them), fewer a priori exclusion criteria would be adopted. The process of operationally defining features of study design and implementation that will lead to exclusion highlights the arbitrary nature of such rules, and the lack of empirical evidence justifying them. To illustrate the difficulties in more detail, I discuss two dimensions on which evidence might be excluded: attrition and reliability.

7.4.1.1 Attrition Attrition occurs when some portion of participants who begin a study (for example, are randomly assigned to treatment conditions) are not available for follow-up measurements. A distinction can be made between overall attrition, which refers to the total attrition across study groups, and differential attrition, which occurs when the characteristics of the dropouts differ across study groups. (Note that researchers often use differences in dropout rates as a proxy for differences in participant characteristics.) Therefore, synthesists wishing to use attrition as a basis for exclusion must first decide whether they mean overall attrition, differential attrition,

or both. Further, the timing of attrition may matter. For example, assume that some loss of participants occurs after participants have been identified for the study but before they have been randomly assigned to conditions, because some identified participants decline to be part of the experiment. This loss has implications for external validity (results can be generalized only to participants like those who have agreed to participate), but not for internal validity (because only participants who consented were randomized). As such, if a synthesist is screening studies for attrition because of a concern about the effects of attrition on internal validity, that synthesist would need to be careful not to exclude studies that experienced attrition before assignment to conditions occurred. More generally, the timing of the participant loss is likely to be important in many research contexts.

Next, the synthesists have to decide how much attrition is too much—that is, the point at which nonzero attrition leads to exclusion. Setting a specific point is arbitrary and likely to lead to instances in which the decision rule produces unsatisfactory results. For example, suppose a synthesist sets 30 percent or greater overall attrition as the criterion for excluding studies. If one study lost 29 percent of its participants it would be included. If a second study lost 30 percent of its participants it would be excluded. However, most scholars would agree that these two studies experienced qualitatively indistinct degrees of overall attrition.

The essential problem facing synthesists who wish to use attrition as an exclusion criterion is that an informed judgment on where to place the cutoff requires a reasonable guess about the treatment outcome that would have been observed for the individuals who dropped out of the study had they remained in the study. This information is not knowable in most contexts. Therefore, in most contexts, little empirical evidence exists to help guide the decision about where to place the cutoff. For example, Jeffrey Valentine and Cathleen McHugh examined the impact of attrition in studies in education, mostly studies that took place in classrooms (2007). Although the findings were complex, these researchers found no evidence to justify a specific cutoff to guide exclusion decisions.

7.4.1.2 Reliability Like attrition, scholars will sometimes only include a study if the outcome measure is reliable, or generally consistent in measurement. This criterion reflects a fundamental misunderstanding about the nature of reliability. Reliability is a property of scores, not of measures, and therefore depends on sample and context (Thompson 2002). A measure yielding good reliability estimates with one sample might yield poor ones with a sample from a different population, or even with the same sample under different testing conditions. This misunderstanding is evident when researchers write that the measure has been shown to have acceptable reliability, provide a citation, and then fail to examine the reliability of the scores obtained. This error is further compounded when synthesists set a threshold for reliability, then rely on the citation to help them determine whether they should include the measure in their synthesis. Further, as with attrition, little empirical evidence supports a rule in which a precise threshold for acceptable reliability is established.

When synthesists use score reliability as an exclusion criterion, their concern might be about validity. In the context of measurement, validity is the extent to which a measure actually measures what it is supposed to. As an example, validity addresses the question of whether the test is a good measure of intelligence. Unfortunately, reliability and validity are not necessarily related. Although measurement reliability does set a limit on the size of the correlation that can be observed between two measures, it is possible for scores to be reliable on a measure that has no validity, or to have no reliability using a particular reliability metric for a measure that is perfectly valid. An example of the latter would be demonstrated by a 100-item questionnaire with no internal consistency but perfect validity if each item had a zero correlation with every other item but a correlation of $r = .10$ with the construct (Guilford 1954). As a somewhat less extreme example, suppose that students in a pharmacology class are given a test of content knowledge at the beginning of the semester and at the end of the semester. Assume that on the pretest the students know nothing about pharmacology (their responses are, in effect, random) but on the posttest they all receive scores in the passing range, that is, 70 to 100 percent correct. Test-retest reliability would suggest very little consistency. However, the test itself would seem to be measuring content taught during the course (indicating good validity). Even though these examples appear relatively implausible, they illustrate the point that reliability and validity are not necessarily related; therefore, a specific reliability coefficient should not be viewed as a proxy for validity.

7.4.2 Quality Scales

As an alternative to establishing various a priori exclusion rules, some scholars have used quality scales to determine what evidence gets included. Unfortunately, quality scales generally have little to recommend them, largely because they fail the two tests just outlined. That is, they generally lack adequate operational specificity and are based on criteria that have no empirical support. In a demonstration of these problems, Peter Jüni and his colleagues applied twenty-five quality scales to seventeen studies reporting on trials comparing the effects of low-molecular weight heparin (LMWH) to standard heparin the risk of developing blood clots after surgery (1999). The authors then performed twenty-five meta-analyses, examining in each case the relationship between study quality and the effect of LMWH (relative to standard heparin). They then examined the conclusions of the meta-analyses separately for high and low quality trials. For six of the scales, the high quality studies suggested no difference between LMWH and standard heparin, but the low quality studies suggested a significant positive effect for LMWH. For seven other quality scales, this pattern was reversed. That is, the high quality studies suggested a positive effect for LMWH, and the low quality studies suggested no difference. The remaining twelve scales showed no differences. Thus, Jüni and his colleagues suggested that the clinical conclusion about the efficacy of the two types of heparin depended on the quality scale used.

The dependence between the outcomes of the synthesis and the quality scale used reveals a critical problem with the scales. A team of scholars could have chosen quality scale A, used it to exclude low quality studies, and then concluded that LMWH is effective. Starting with the same research question, another team could have chosen scale B, used it to exclude low quality studies, then concluded that LMWH is no more effective than standard heparin. As such, the scales appear to have been at best useless and at worst misleading (Berlin and Rennie 1999).

Examination of the quality scales used makes it easy to understand why this result occurred. The twenty-five scales often focused on different dimensions or aspects of research design and implementation (for example, some focused on internal validity alone and others focused on all four classes of validity), and analyzed a number of dimensions (from three to thirty-four), including dimensions unrelated to validity in the traditional sense (for example, whether an institutional review board approved the study). Even when the same dimension was analyzed, the weights assigned to the dimension varied. For example, one scale allocated 4 percent of its total points to the presence of random assignment, and 12 percent of the points to whether the outcome assessor was unaware of the condition the participants were in (called masking). Another scale very nearly reversed the relative importance of these two dimensions, allocating 14 percent of the total points to randomization and 5 percent to masking. In sum, there is no empirical justification for these weighting schemes, and the result in the absence of that justification is a hopelessly confused set of arbitrary weights.

Further, the scales Jüni and his colleagues reviewed all rely on single scores to represent a study's quality (1999). Especially when scales focus on more than one aspect of validity, the single score approach results in a score that is summed across very different aspects of study design and implementation, many of which are not necessarily related to one another—for example, between the validity of outcome measures and the mechanism used to allocate units to groups. When scales combine disparate elements of study design into a single score, it is likely that important considerations of design are being obscured. A study with strong internal validity but weak external validity might have a score identical to a study with weak internal validity and strong external validity. If the quality of these studies is expressed as a single number, how would users of the synthesis know the difference between studies with such different characteristics?

To address these problems, Jeffrey Valentine and Harris Cooper created the Study Design and Implementation Assessment Device (Study DIAD) in conjunction with a panel that included experts in the social and medical sciences (2008). The Study DIAD is a multilevel and hierarchical instrument that results in a profile of scores rather that arriving at a single score to represent study quality. In addition, in recognition of the fact that important quality features may vary across research questions, the instrument was designed to allow for contextual variation in quality indicators. Further, it was written with explicit attention to operationally defining the quality indicators, something notably absent from most quality scales.

The Study DIAD is not a complete solution to the problems faced by individuals wishing to use a quality scale as part of the study screening process. It requires a great deal of both methodological and substantive expertise to implement, and relative to other scales is human resource intensive. In addition, all quality scales that I have seen, including the Study DIAD, pay no attention to the direction of bias associated with the quality indicator. To borrow an example from Sander Greenland and Keith O'Rourke (2001), it may be that due to the way selection mechanisms operate in a given area, the lack of random assignment of units results in an overestimate of the treatment effect, and both overall attrition and differential attrition result in an underestimate of the treatment effect. That is, different quality indicators might be associated with both different directions and degrees of bias. Thus, a study that does not randomly assign units but does experience overall attrition could experience two sources of bias that act in opposite directions. Unless a quality scale captures these directions, the resulting quality score will not properly rank studies in terms of their degree of bias (Greenland and O'Rourke 2001). Again, one problem is that in most areas of inquiry, we simply do not know enough to be able to predict the direction of bias associated with any given quality indicator. Another problem is that even if we had this information, quality scales as currently constructed do not capture the possibility that bias might cancel out across a set of potentially biasing design and implementation features.

7.4.3 A Priori Strategies and Missing Data

I mentioned earlier that synthesists will sometimes be unable to determine a study's status on a given quality indicator. The best case scenario for missing data occurs when the synthesists can make relatively safe assumptions about the quality dimension when that information is missing. For example, if a study does not mention that participants were randomly assigned to groups, it is often reasonable to assume that some process other than random assignment was used.

Unfortunately for synthesists using quality criteria as screens for determining whether include a particular piece of evidence, most decisions are not so straightforward. As such, synthesists will have to make difficult choices, often with little or no empirical evidence to guide their decisions. For example, assume a synthesist chooses to exclude nonrandomized experiments, as well as studies in which overall attrition was too high (arbitrarily defined as greater than 20 percent) or differential attrition was too high (arbitrarily defined as greater than a 10 percent difference). The synthesist might feel comfortable excluding studies that are silent about the process used to assign participants to groups, under the assumption that if the process was random the authors are highly likely to have mentioned it. Assumptions about attrition, however, are probably not as safe in most research contexts. Probably the best the synthesist can do is to be explicit about what rule was used and the rationale for using it, so that others can critically evaluate that decision.

7.4.4 Summary of A Priori Strategies

Earlier I referred to validity as the approximate truth of an inference, and noted that although we often refer to studies as high or low validity, in reality, validity is not a dichotomous characteristic of studies but rather a continuous characteristic of inferences. The continuous nature often makes using the concept in a dichotomous fashion difficult, such as deciding that some particular cutoff point on some criterion should lead to a study's exclusion. As a further complication, there is often little evidence or strong theory supporting either a specific cutoff point or even a general rule (for example, attrition biases a body of literature) for determining when a particular design or implementation problem warrants excluding a study from a synthesis, nor do most fields have a substantial evidence base on which to make predictions about the direction of bias for many quality indicators. These characteristics make the use of a priori strategies questionable in many areas of inquiry and for most criteria. Ironically, scholars who rely on a priori strategies may be excluding the evidence that might help them begin to establish the rules.

7.5 POST HOC ASSESSMENT

7.5.1 Quality As an Empirical Question

Because of the many problems associated with a priori approaches to addressing study quality, some scholars adopt a post hoc approach to examining the impact of design and implementation quality on study results. Typified by Mary Smith and Gene Glass's meta-analysis of the effects of psychotherapy, one method involves treating the impact of design and implementation features of study outcomes as an empirical question (1977; see also chapter 10, this volume). Using analogs to the analysis of variance and multiple regression, synthesists using this strategy attempt to determine statistically whether certain study characteristics influence study results. A team of synthesists might compare the results of studies using random assignment to those using nonrandom methods. If studies using random and nonrandom methods arrive at similar results, this might be taken as evidence that non random assignment did not result in consistent bias. On the other hand, if these assignment methods yield

different results might bias might be associated with nonrandom assignment.

The virtue of this approach is that it relies much less on arbitrary decision rules for determining what evidence ought to be counted. It has, however, has two potential drawbacks. First, like any other inferential test, tests of moderators related to quality in a research synthesis should be conducted under conditions of adequate statistical power (see Hedges and Pigott 2001, 2004). A failure to reject the null hypothesis is much more persuasive if the researchers can claim to have had good power to detect a meaningful difference. It is less persuasive if researchers were relatively unlikely to reject a false null hypothesis due to low power. As such, the viability of the empirical approach to the assessment of the impact of study quality depends greatly on having enough information to conduct fair (that is, well-powered) tests of the critical aspects of quality.

A second potential problem is that, in all likelihood, the quality characteristics were not randomly assigned to studies. This problem relates to what Cooper refers to as the distinction between study-generated evidence and synthesis-generated evidence (1998; see also chapter 2, this volume). Evidence generated from study-based comparisons between groups composed through random assignment of units permit causal inferences. Evidence generated from synthesis-based comparisons—that is, comparisons between studies with different design characteristics—was not constructed using random assignment and therefore yields evidence about association only, not about causality. Study features will likely be interrelated. For example, studies using random assignment may tend to be of shorter duration than studies using nonrandom methods to create groups. Further, it is generally not possible to rule out a spurious relationship between a quality indicator, the study's estimated effect size, and some unknown third variable. As a result, it is often difficult to tease apart the effects of the interrelated quality indicators and unknown third variables. Even when scholars have enough data for well-powered comparisons using techniques such as meta-regression, ambiguity about how to interpret the results will often be considerable.

7.5.2 Empirical Uses

If researchers can find or create a quality scale that avoids the problems noted, it might make sense to use quality scores in the empirical assessment of the impact of quality on study outcomes. For example, a quality scale focused on limited dimensions of internal validity might result in a score that could be used as a moderator in statistical analyses. If studies scoring well on that scale yielded different study outcomes than those not scoring well, it might be an indication that an internal validity problem was biasing study outcomes. Note, however, that as pointed out earlier, if the different quality indicators affect the direction of bias differently, a summary of those indicators is unlikely to be a valid indicator of quality unless the instrument can capture the complexity.

7.5.3 Post Hoc Strategies and Missing Data

Researchers who adopt a largely empirical approach to investigating the effect of study quality on study outcomes find themselves in a somewhat better position than a priori synthesists to address missing data. For example, they may be able to compare effect sizes for studies in which random assignment clearly occurred to those in which it clearly did not and to those in which it was unclear whether it did occur. Similar to their a priori counterparts, they could also adopt a rule governing what should be assumed if a quality indicator is missing. Then, as a sensitivity analysis, they could eliminate those studies from the analysis and examine the impact of the decision rule on the statistical conclusion. This approach provides a flexible method of dealing with the inevitable problem of missing data and is another point favoring the post hoc approach.

7.5.4 Summary of Post Hoc Assessment

The post hoc approach to addressing study quality in a research synthesis treats the impact of study quality as an empirical question. This strategy has a number of benefits. Most notably, it relies less on ad hoc, arbitrary, and unsupported decision rules. It also provides the synthesists with the ability to empirically test relationships that are plausibly related to study quality. However, when a literature does not have enough evidence to conduct a fair statistical test (that is, one with sufficient statistical power to detect an effect of some size deemed important) scholars will not be in a good position to evaluate the impact of quality on their conclusions. Further, even if the impact of quality can be investigated using credible statistical tests, the analyses will almost certainly be correlational, a fact limiting the strength of the inferences that can be drawn about the relation between study quality and study outcomes.

7.6 RECOMMENDATIONS

I have outlined two broad strategies for dealing with the near certainty that studies relevant to a particular synthesis will vary in terms of their research designs and implementation. Although most synthesists adopt a blend of the two strategies, some rely more on a priori strategies and others are more comfortable with post hoc. Neither approach is without problems. So, how should synthesists approach the issue of study quality? One approach is to use a few highly defensible criteria for excluding some studies. For example, a synthesist might be working in a field in which it seems clear that differential attrition positively biases estimates of effect. Ideally, the evidence for this assertion would be closely tied to the research context in which the synthesist is working. It would not be enough, for example, for a synthesist to assert that attrition results in a consistent bias, because that claim is neither specific nor tied to the synthesist's research question. A better assertion might be that research has shown that, in classroom studies of the type we are examining, differential attrition results in a consistent underestimate of the intervention effect, because high-need students are more likely to drop out of the comparison condition. Then, other characteristics of study quality that lack a solid empirical justification could be coded and addressed empirically. This approach allows synthesists to screen out studies that the empirical evidence suggests will bias the results of a synthesis yet allow them to empirically test the impact of other study features lacking a reasonable empirical basis.

For synthesists investigating causal relations, one important consideration that deserves special attention is whether or not to include studies in which units were not randomly assigned to conditions. Again, the relevant issue pertains to the effect that bias at the level of the individual study has on the estimation of effects in a collection of related studies. Because the nature of the relation between assignment mechanism and bias is likely to depend heavily on the research context, researchers should strongly consider collecting both randomized and nonrandomized experiments and empirically investigating both the selection mechanisms that appear to be operating in the nonrandomized experiments and the relationship between assignment mechanism and study outcomes. However, averaging effects across the assignment mechanisms is something that should be done very cautiously, if at all (Shadish and Myers 2001).

Of course, there are exceptions. If a team of synthesists has the luxury of a large number of randomized

Table 7.2 Relevance Screen

Coder initials: ___ ___ ___

Date: _____

Report Characteristics	
1. First author (Last, initials)	
2. Journal	
3. Volume	
4. Pages	
Inclusion Criteria	
5. Is this study an empirical investigation of the effects of teacher expectancies?	0. No 1. Yes
6. Is the outcome a measure of IQ?	0. No 1. Yes
7. Are the study participants in grades 1–5 at the start of the study?	0. No 1. Yes
8. Does the study involve a between-teacher comparison (i.e., some teachers hold an expectancy while others do not)?	0. No 1. Yes

SOURCE: Author's compilation.

experiments that appear to adequately represent the constructs of interest (for example, variations in participants and outcome measures), then a strong case can be made for excluding all but randomized comparisons. Further, as suggested earlier, the strong theoretical underpinnings of the preference for randomization in studies investigating causal relations could be viewed as a justification for excluding all studies that do not employ randomization. Even in the absence of empirical evidence, there is ample reason to be cautious, perhaps to the extent of excluding all but randomized comparisons.

7.6.1 Coding Study Dimensions

The specific aspects of study design and implementation related to study quality will vary as a function of the spe-

cific research question. Research syntheses that aim to address issues relating to uncovering causal relations, however, address many issues faced in all types of quantitative research. As such, what follows is an example set of codes designed to investigate the problem of whether teacher expectancies affect student IQ. Two aspects of study quality are examined: internal validity and the construct validity of the outcome measures. Table 7.2 provides an example of how studies can be screened for relevance. For included studies, table 7.3 shows how one might code issues related to research design for studies, and table 7.4 provides examples for coding the outcome measures.

The example set out, of course, represents only a fraction of the study characteristics one would actually code in a synthesis of this nature. For instance, synthesists also

Table 7.3 Coding for Internal Validity

9. Sampling strategy	1. Randomly sampled from a defined population 2. Stratified sampling from a defined population 3. Cluster sampling 4. Convenience sample 5. Can't tell
10. Group assignment mechanism	1. Random assignment 2. Haphazard assignment 3. Other nonrandom assignment 4. Can't tell
11. Assignment mechanism	1. Self-selected into groups 2. Selected into groups by others on a basis related to outcome (e.g., good readers placed in the expectancy group) 3. Selected into groups by others not known to be related to outcome (e.g., randomized experiment) 4. Can't tell
12. Equating variables	0. None 1. Prior IQ 2. Prior achievement 3. Other _____ 4. Can't tell
13. If equating was done, was it done using a statistical process (e.g., ANCOVA, weighting) or manually (e.g., hand matching)?	0. n/a, equating was not done 1. Statistically 2. Manual matching
14. Time 1 sample size, expectancy group (at random assignment)	___ ___ ___
15. Time 1 sample size, comparison group (at random assignment)	___ ___ ___
16. Time 2 sample size, expectancy group (for effect size estimation)	___ ___ ___
17. Time 2 sample size, comparison group (for effect size estimation)	___ ___ ___

SOURCE: Author's compilation.

Table 7.4 Coding Construct Validity

18. IQ measure used in study	1. Stanford-Binet 5 2. Wechsler (WISC) III 3. Woodcock-Johnson III 4. Other _____ 5. Can't tell _____
19. Score reliability for IQ measure (If a range is given, code the median value)	____ ___
20. Metric for score reliability	1. Internal consistency 2. Split-half 3. Test-retest 4. Can't tell 5. None given
21. Source of score reliability estimate	1. Current sample 2. Citation from another study 3. Can't tell 4. None given
22. Is the validity of the IQ measure mentioned?	0. No 1. Yes
23. If yes, was a specific validity estimate given?	0. No 1. Yes
24. If yes, what was the estimate?	___ ___ ___
25. If a validity estimate was given in the report, what was the source of the estimate?	1. Current sample 2. Citation from another study 3. Can't tell 4. Validity not mentioned in report
26. What was the nature of the validity estimate?	1. Concurrent validity 2. Convergent validity 3. Predictive validity 4. Other 5. Can't tell (simply asserted that the measure is valid) 6. n/a, validity was not mentioned in the report

SOURCE: Author's compilation.

might want to know as much as possible about how the expectancies were generated and the specific nature of the expectancy. These important substantive issues will vary from synthesis to synthesis. For example, some researchers may manipulate teacher expectancies through modifications of the students' school files, and others may do so by manipulating verbal feedback from the students' previous teachers. Further guidance can be found in many textbooks (see, for example, Cooper 1998; Lipsey and Wilson 2001) and chapters (for example, Shadish and Myers 2001, especially chapter 13), as well as in the policy briefs of the Campbell Collaboration (Shadish and Myers 2001), and the Open Learning materials of the Cochrane Collaboration (Alderson and Green 2002).

7.6.2 The Relevance Screen

In the example that follows, assume the synthesists have established that to be included studies must meet four criteria. A study must examine the impact of teacher expectancies on the IQ of students in grades one through five, and must do so by manipulating teacher expectancy conditions and comparing students for whom teachers were given varying expectancies. Any study not meeting these criteria would be screened out. So, for example, studies that investigate the impact of teacher expectancies on student grades, or that do so by examining naturally occurring variation in expectancies and how they covary with IQ, would be excluded. Table 7.2 presents a coding form that will help guide the relevance screening process. Note that if any one of questions five through eight is answered in the negative, the study will be excluded from further consideration.

7.6.3 Example Coding

Coding aspects of studies related to internal validity requires attention to both the study design and to how that study was carried out. Table 7.3 includes several design issues, such as how groups were formed and what variables were used to equate students, if applicable. Because this synthesis could also include nonrandomized studies, table 7.3 includes an item (#12) designed to capture procedures the researchers might have used to make the groups more comparable (though it would be good to know this information even in a synthesis limited to randomized comparisons). The table does not include a code designed to capture how the comparison group was treated, because the assumption is that all comparison

students would be receiving treatment as usual. If, however, this assumption were not accurate, the synthesists would have to add an item to the coding guide and recode all previous studies on that dimension.

Many categories of designs that might be present in other contexts are not in the list of possible designs. For example, within-group comparisons are not included, so categories for interrupted time series and one-group pretest-posttest designs do not appear on the coding form. Regression discontinuity designs also are not represented. An Other code, however, is included so that unexpected designs can be tracked.

Also of note in table 7.3 is that the coding sheet includes a mechanism for capturing sample sizes at the beginning and at the end of the study for both the expectancy and the comparison groups (questions 14 through 17). This information will allow the synthesists to compute both overall and differential attrition rates.

Finally, table 7.4 addresses aspects of coding that relate to issues of the construct validity of the outcome measures. Here, the name of the IQ measure and the score reliability estimate given in the study are coded, along with some aspects related to the validity of the IQ measure.

7.7 CONCLUSION

Although most synthesists use a combination of a priori and post hoc approaches and neither one works perfectly in every situation, it should be clear that I believe a priori rules are used too often in research synthesis. Absent empirical evidence that justifies the rules employed and their operational definitions, it seems unlikely that these rules serve to increase the validity of the inferences arising from a research synthesis. A better approach is to rely on these rules only when strong empirical justification exists to support them, and to attempt to investigate empirically the influence of varying dimensions of study quality on the inferences arising from those studies.

7.8 REFERENCES

Alderson, Phil, and Sally Green. 2002. "Cochrane Collaboration Open Learning Material for Reviewers." http://www.cochrane-net.org/openlearning.

Berlin, Jesse, and Drummond Rennie. 1999. "Measuring the Quality of Trials: The Quality of Quality Scales." *Journal of the American Medical Association* 282(11): 1053–55.

Bris, Peter A., Stephanie Zara, Marguerite Pappaioanou, Jonathan Fielding, Linda Wright-De Agüero, et al. 2000.

"Developing an Evidence-Based Guide to Community Preventive Services—Methods." *American Journal of Preventive Medicine* 18(1S): 35–43.

Campbell, Donald T. and Julian C. Stanley. 1966. *Experimental and Quasi-experimental Designs for Research.* Chicago: Rand McNally.

Chalmers, Thomas C., Paul Celano, Henry S. Sacks, and Harry Smith Jr. 1983. "Bias in Treatment Assignment in Controlled Clinical Trials." *New England Journal of Medicine* 309: 1358–61.

Colditz, Graham A., James N. Miller, and Frederick Mosteller. 1989. "How Study Design Affects Outcomes in Comparisons in Therapy, I: Medical." *Statistics in Medicine* 8(4): 441–54.

Cook, Thomas D., and Donald T. Campbell. 1979. *Quasi-Experimentation: Design and Analysis Issues for Field Settings.* Boston, Mass.: Houghton Mifflin.

Cooper, Harris. 1998. *Synthesizing Research: A Guide for Literature Reviews*, 3rd ed. Thousand Oaks, Calif.: Sage Publications.

Coren, Ester and Mike Fisher. 2006. "The Conduct of Systematic Research Reviews for SCIE Knowledge Reviews." London: Social Care Institute for Excellence. http://www.scie.org.uk/publications/researchresources/rr01.pdf.

DuBois, David L., Bruce E. Holloway, Jeffrey C. Valentine, and Harris Cooper. 2002. "Effectiveness of Mentoring Programs for Youth: A Meta-Analytic Review." *American Journal of Community Psychology* 30(2): 157–97.

Fowler, Robert L. 1988. "Estimating the Standardized Mean Difference in Intervention Studies." *Journal of Educational Statistics* 13(4): 337–50.

Glazerman, Steven, Dan M. Levy, and David Myers. 2003. "Nonexperimental Versus Experimental Estimates of Earnings Impacts." *Annals of the American Academy of Political and Social Science* 589(1): 63–93.

Greenland, Sander and Keith O'Rourke. 2001. "On the Bias Produced by Quality Scores in Meta-Analysis, and a Hierarchical View of Proposed Solutions." *Biostatistics* 2(4): 463–71.

Guilford, Joy P. 1954. *Psychometric Methods.* New York: McGraw-Hill.

Hedges, Larry V., and Therese D. Pigott. 2001. "The Power of Statistical Tests in Meta-Analysis." *Psychological Methods* 6(3): 203–17.

———. 2004. "The Power of Statistical Tests for Moderators in Meta-Analysis." *Psychological Methods* 9(4): 426–45.

Hedges, Larry V. 2007. "Correcting a Significance Test for Clustering." *Journal of Educational and Behavioral Statistics* 32(2): 151–79.

Hollis, Sally, and Fiona Campbell. 1999. "What Is Meant by Intention to Treat Analysis. Survey of Published Randomised Controlled Trials." *British Medical Journal* 319(7211): 670–74.

Jüni, Peter, Anne Witshci, Ralph Bloch, and Matthias Egger. 1999. "The Hazards of Scoring the Quality of Clinical Trials for Meta-Analysis." *Journal of the American Medical Association* 282(11): 1054–60.

Lipsey, Mark W., and David B. Wilson. 1993. "The Efficacy of Psychological, Educational, and Behavioral Treatment: Confirmation from Meta-Analysis." *American Psychologist* 48(12): 1181–1209.

Miller, James N., Graham A. Colditz, and Frederick Mosteller. 1989. "How Study Design Affects Outcomes in Comparisons of Therapy, II: Surgical." *Statistics in Medicine* 8(4): 455–66.

Murray, David M. 1998. "Design and Analysis of Group Randomized Trials." New York: Oxford University Press.

Schulz, Kenneth F., Iain Chalmers, Richard J. Hayes, and Douglas G. Altman. 1995. "Empirical Evidence of Bias: Dimensions of Methodological Quality Associated with Estimates of Treatment Effects in Controlled Trials." *The Journal of the American Medical Association* 273(5): 408–12.

Shadish, William R. 1989. "The Perception and Evaluation of Quality in Science." In *Psychology of Science: Contributions to Metascience*, edited by Barry Gholson, William R. Shadish Jr., Robert A. Niemeyer, and Arthur C. Houts. Cambridge: Cambridge University Press.

Shadish, William, and David Myers. 2001. "Campbell Collaboration Research Design Policy Brief." http://www.campbellcollaboration.org/MG/ResDesPolicyBrief.pdf.

Shadish, William R., and Kevin Ragsdale. 1996. "Random Versus Nonrandom Assignment in Psychotherapy Experiments: Do You Get the Same Answer?" *Journal of Consulting and Clinical Psychology* 64(6): 1290–1305.

Shadish, William R., Thomas D. Cook, and Donald T. Campbell. 2002. *Experimental and Quasi-Experimental Designs for Generalized Causal Inference.* Boston, Mass.: Houghton Mifflin.

Smith, Mary Lee, and Gene V. Glass. 1977. "Meta-Analysis of Psychotherapy Outcome Studies." *American Psychologist* 32(9): 752–60.

Thompson, Bruce, ed. 2002. *Score Reliability: Contemporary Thinking on Reliability Issues.* Newbury Park, Calif.: Sage Publications.

Valentine, Jeffrey C., and Harris Cooper. 2008. "A Systematic and Transparent Approach for Assessing the Methodological Quality of Intervention Effectiveness Research: The

Study Design and Implementation Assessment Device (Study DIAD)." *Psychological Methods* 13(2): 130–49.

Valentine, Jeffrey C., and Cathleen M. McHugh. 2007. "The Effects of Attrition on Baseline Group Comparability in Randomized Experiments in Education: A Meta-Analytic Review." *Psychological Methods* 12(3): 268–82.

What Works Clearinghouse. 2006. *What Works Clearinghouse Evidence Standards for Reviewing Studies*, revised September 2006. http://www.whatworksclearinghouse.org/reviewprocess/study_standards_final.pdf.

Williamson, P. R., Carrol Gamble, Douglas G. Altman, and Jane L. Hutton. 2005. "Outcome Selection Bias in Meta-Analysis." *Statistical Methods in Medical Research* 14(5): 515–24.

Zara, Stephanie, Linda Wright-De Agüero, Peter A. Bris, Benedict I. Truman, David P. Hopkins, Michael H. Hennessy, et al. 2000. "Data Collection Instrument and Procedure for Systematic Reviews in the Guide to Community Preventive Services." *American Journal of Preventive Medicine* 18(1S): 44–70.

8

IDENTIFYING INTERESTING VARIABLES AND ANALYSIS OPPORTUNITIES

MARK W. LIPSEY
Vanderbilt University

CONTENTS

8.1 INTRODUCTION

Research synthesis relies on information reported in a selection of studies on a topic of interest. The purpose of this chapter is to examine the types of variables that can be coded from those studies and to outline the kinds of relationships that can be examined in the analysis of the resulting data. It identifies a range of analysis opportunities and endeavors to stimulate the synthesist's thinking about what data might be used in a research synthesis and what sorts of questions might be addressed.

8.1.1 Potentially Interesting Variables

Research synthesis revolves around the effect sizes that summarize the main findings of each study of interest. As detailed elsewhere in this volume, effect sizes come in a variety of forms—correlation coefficients, standardized differences between means, odds ratios, and so forth—depending on the nature of the quantitative study results they represent. Study results embodied in effect sizes, therefore, constitute one major type of variable that is sure to be of interest.

Research studies, however, have other characteristics that may be of interest in addition to their results. For instance, they might be described in terms of the nature of the research designs and procedures used, the attributes of the subject samples, and various other features of the settings, personnel, activities, and circumstances involved. These characteristics taken together constitute the second major category of variables of potential concern to the research synthesist—study descriptors.

8.1.2 Study Results

Whatever the nature of the issues bearing on study results, one of the first challenges the synthesist faces is the likelihood that many of the studies of interest will report multiple results relevant to those issues. If we define a study to mean a set of observations taken on a subject sample on one or more occasions, there are three possible forms of multiple results that may occur. There may be different results for different measures, for different subsamples, and for different times of measurement. All of these variations have implications for conceptualizing, coding, and analyzing effect sizes.

8.1.2.1 Results on Multiple Measures Each study in a synthesis may report results on more than one measure. These results may represent different constructs, that is, different things being measured such as academic achievement, attendance, and attitudes toward school. They may also represent multiple measures of the same construct, for instance achievement measured both by a standardized achievement test and grade point average. A study of predictors of juvenile delinquency thus might report the correlations of one measure each of age, gender, school achievement, and family structure with a central delinquency measure. Each such correlation can be coded as a separate effect size. Similarly, a study of gender differences in aggression might compare males and females on physical aggression measured two different ways, verbal aggression measured three ways, and the aggressive content of fantasies measured one way, yielding six possible effect sizes. Moreover, the various studies eligible for a research synthesis may differ among themselves in the type, number, and mix of measures.

The synthesist must decide what categories of constructs and measures to define and what effect sizes to code in each. The basic options are threefold. One approach is to code effect sizes on all measures for all constructs in relatively undifferentiated form. This strategy would yield, in essence, a single global category of study results. For example, the outcome measures used in research on the effectiveness of psychotherapy show little commonality from study to study. A synthesist might, for some purposes, treat all these outcomes as the same—that is, as instances of results relevant to some general construct of personal functioning. An average over all the resulting effect sizes addresses the global question of whether psychotherapy has generally positive effects on the mix of measures typical to psychotherapy research. This is the approach used in Mary Smith and Gene Glass's classic meta-analysis (1977).

At the other extreme, effect sizes may be coded only when they relate to a particular measure of interest to the synthesist—that is, a specific operationalization of the construct of interest or, perhaps, a few such specific operationalizations. Mean effect sizes would then be computed and reported separately for each such measure. In a meta-analysis of the effects of transcranial magnetic stimulation on depression, for instance, Jose Luis Martin and his colleagues examined effect sizes only for outcomes measured on the Beck Depression Inventory and, separately, those measured on the Hamilton Rating Scale for Depression (2003).

An intermediate strategy is to define one or more sets of constructs or measurement categories that include different operationalizations but distinguish different content domains of interest. Study results that fell within those categories would then be coded but other results would be ignored. In a synthesis of the effects of exercise for older adults, for example, Yael Netz and her colleagues distinguished four physical fitness outcome constructs (cardiovascular, strength, flexibility, and functional capacity) and ten psychological outcome constructs (anxiety, depression, energy, life satisfaction, self-efficacy, and so forth), while allowing different ways of measuring the respective construct within each category (2005). These constructs were considered to represent distinctly different kinds of outcomes and the corresponding effect sizes were kept separate in the coding and analysis.

Whether global or highly specific construct categories are defined for coding effect sizes, the synthesist is likely to find some studies that report results on multiple measures within a given construct category. In such cases, one of several approaches can be taken. The synthesist can simply code and analyze all the effect sizes contributed to a category, including multiple ones from the same study. This permits some studies to provide more effect sizes than others and thus to be overrepresented in the synthesis results. It also introduces statistical dependencies among the effect sizes because some of them are based on the same subject samples. These are issues that will have to be dealt with in the analysis. Alternatively, criteria can be developed for selecting the single most appropriate measure within each construct category and ignoring the remainder (despite the related loss of data). Or multiple effect sizes within a construct category can be coded and then averaged (perhaps using a weighted average) to yield a single mean or median effect size within that category for each study (for further discussion of these issues, see chapter 19, this volume).

The most important consideration is the purpose of the synthesis. For some purposes, the study results of interest may involve only one construct, such as when synthesis of research evaluating educational innovations focuses on achievement test results. For others, the synthesist may be interested in a broader range of study results and welcome diverse measures and constructs. Having a clear view of the purposes of a research synthesis and adopting criteria for selecting and categorizing study results consistent with those purposes are fundamental requirements of good synthesis. These steps define the effect sizes the synthesist will be able to analyze and thus shape the findings of the synthesis.

8.1.2.2 Results for Subsamples In addition to overall results, many studies within a synthesis may report breakdowns for one or more subject subsamples. For example,

a validation study of a personnel selection test might report test-criterion correlations for different industries, or a study of the effects of drug counseling for teenagers might report results separately for males and females. Coding effect sizes for each subsample permits a more discriminating analysis of the relationship between subject characteristics and study results than is possible from overall study results alone.

Two considerations bear on the potential utility of effect sizes from subsamples. First, the variable distinguishing subsamples must represent a dimension of potential interest to the synthesist. Effect sizes computed separately for male and female subsamples, for instance, will be of concern only if there are theoretical, practical, or exploratory reasons to examine whether gender moderates the magnitude of effects. Second, a given subject breakdown must be reported widely enough to yield a sufficient number of effect sizes to permit meaningful analysis. If most studies do not report results separately for males and females, the utility in coding effect sizes for those subsamples is limited.

As with effect sizes from multiple measures within a single study, effect sizes for multiple subsamples within a study pose issues of statistical dependency if the subsamples are not mutually exclusive—that is, subsamples that share subjects. Although the female subsample in a breakdown by gender would not share subjects with the male subsample, it would almost certainly share subjects with any subsample resulting from a breakdown by age. In addition, however, any feature shared in common by subsamples within a study—for example, being studied by the same investigator—can introduce statistical dependencies into the effect sizes computed for those subsamples. Analysis of effect sizes based on subsamples, therefore, must be treated like any analysis that uses more than one effect size from each study and thus possibly violates assumptions of statistical independence (see chapter 19, this volume).

8.1.2.3 Results Measured at Different Times Some studies may report findings on the same variables measured at different times. A study of the effects of a smoking cessation program, for example, might report the outcome immediately after treatment and at six-month and one-year follow-ups. Such results might also be reported both for the overall subject sample and for various subsamples. Such time-series information potentially permits an interesting analysis of temporal patterns in outcome—decay curves for the persistence of treatment effects, for example.

As with effect sizes for subsamples, the synthesist must determine whether follow-up results are reported widely enough to yield enough data to be worth analyzing. Even in favorable circumstances, only a portion of the studies in a synthesis is likely to provide measures at different time points. A related problem is that the intervals between times of measurement may vary widely from study to study, making these results difficult to summarize across studies. If this occurs, one approach is for the synthesist to establish broad categories for follow-up periods and code each result in the category it most closely fits. Another approach is to code the time of measurement for each result as an interval from some common reference point—for instance, the termination of treatment. The synthesist can then analyze the functional relationship between the effect size and the amount of time that has passed since the completion of treatment.

Because the subjects at time 1 will be much the same (except for attrition) as those at follow-up time 2, the effect sizes for these different occasions will not be statistically independent. As with multiple measures or overlapping subsamples, these effect sizes cannot be included together in an analysis that assumes independent data points unless special adjustments are made.

8.1.2.4 The Array of Study Results   As the previous discussion indicates, the quantitative results from a study selected for synthesis may be reported for multiple constructs, for multiple measures of each construct, for the total subject sample, for various subsamples, and for multiple times of measurement. To effectively represent key study results and support interesting analyses, the synthesist must establish conceptual categories for each dimension. These categories will group those effect sizes judged to be substantially similar for the purposes of the synthesis, differentiate those believed to be importantly different, and ignore those judged irrelevant or uninteresting. Once the categories are developed, it should be possible to classify each effect size that can be coded from any study according to the type of construct, subsample, and time of measurement it represents. When enough studies yield comparable effect sizes, that set of results can be separately analyzed. Some studies will contribute only a single effect size (one measure, one sample, one time) to the synthesis. Others will report more differentiated results and may yield a number of useful effect sizes.

It follows that different studies may provide effect sizes for different analyses on different categories of effect sizes. Any analysis based on effect sizes from only a subset of the studies under investigation raises a question of the generalizability of the findings to the entire set because of the distinctive features that may characterize that subset. For instance, in a synthesis of the correlates of effective job performance, the subset of studies that break out results by age groups might be mostly those conducted in large corporations. What is learned from them about age differences might not apply to employees of small businesses. Studies that report differentiated results, nonetheless, offer the synthesist the opportunity to conduct more probing analyses of the issues of interest. Including such differentiation in a research synthesis can add rich and useful detail to its findings.

8.1.3 Study Descriptors

Whatever the set of effect sizes under investigation, it is usually of interest to also examine matters related to the characteristics of the studies that yield those results. To accomplish this, the synthesist must identify and code information from each study about the particulars of its subjects, methods, treatments, and the like. Research synthesists have shown considerable diversity and ingenuity in the study characteristics they have coded and the coding schemes they have used (for a specific discussion of the development, application, and validation of coding schemes, see chapter 9, this volume). Described here are the general types of study descriptors that the synthesist might consider in anticipation of conducting certain analyses on the resulting data.

To provide a rudimentary conceptual framework for study descriptors, we first assume that study results are determined conjointly by the nature of the substantive phenomenon under investigation and the nature of the methods used to study it. The variant of the phenomenon selected for study and the particular methods applied, in turn, may be influenced by the characteristics of the researcher and the research context. These latter will be labeled extrinsic factors because they are extrinsic to the substantive and methodological factors assumed to directly shape study results. In addition, the actual results and characteristics of a study must be fully and accurately reported to be validly coded in a synthesis. Various aspects of study reporting thus constitute a second type of extrinsic factor because they, too, may be associated with the study results represented in a synthesis though they are not assumed to directly shape those results.

This scheme, though somewhat crude, serves to identify four general categories of study descriptors that may interest the synthesist: substantive features of the matter the studies investigate, particulars of the methods and procedures used to conduct those investigations, characteristics of the researcher and the research context and attributes of the way the study is reported and the coding done by the synthesist. The most important of these, of course, are the features substantively pertinent to characterizing the phenomenon under investigation. In this category are such matters as the nature of the treatment provided; the characteristics of the subjects used; the cultural, geographical, or temporal setting; and those other influences that might moderate the events or relationships under study. It is these variables that permit the synthesist to identify differences among studies in the substantive characteristics of the situations they investigate that may account for differences in their results. From such information, the synthesist may be able to determine that one treatment variation is more effective than another, that a relationship holds for certain types of subjects or circumstances but not for others, or that there is a developmental sequence that yields different results at different time periods. Findings such as these, of course, result in better understanding of the nature of the phenomenon under study and are the objective of most syntheses.

Those study characteristics not related to the substantive aspects of the phenomenon involve various possible sources of distortion, bias, or artifact in study results as they are operationalized in the original research or in the coding for the synthesis. The most important are methodological or procedural aspects of the way in which the studies were conducted. These include variations in the designs, research procedures, quality of measures, and forms of data analysis that might yield different study results even if every study were investigating the same phenomenon. A synthesist may be interested in examining the influence of method variables for two reasons. First, analysis of the relationships between method choices and study results provides useful information about which aspects of research procedures make the most difference and, hence, should be most carefully selected by researchers. Second, method differences confound substantive comparisons among study results and may thus need to be included in the analysis as statistical control variables to help disentangle the influence of methods on those results.

Factors extrinsic to both the substantive phenomenon and the research methods include characteristics of the researcher (such as gender and disciplinary affiliation), the research circumstances (such as the nature of study sponsorship), and reporting (such as the form of publication). Such factors would not typically be expected to directly shape study results but nonetheless may be related to something that does and thereby correlate with coded effect sizes. Whether a researcher is a psychologist or a sociologist, for instance, should not itself directly determine study results. Disciplinary affiliation may, however, influence methodological practices or the selection of variants of the phenomenon to study, which, in turn, could affect study results.

Another type of extrinsic factor involves those aspects of the reporting of study methods and results that might yield different values in the synthesist's coding even if all studies were investigating the same phenomenon with the same methods. For example, studies may not report important details of the procedures, measures, treatments, or results. This requires coders to engage in guesswork or record missing values on items that, with better reporting, could be coded accurately. Inadequacies in the information reported in a study, therefore, may influence the effect sizes coded or the representation of substantive and methodological features closely related to effect sizes in ways that distort the relationships of interest to the synthesis.

A synthesist who desires a full description of study characteristics will want to make some effort to identify and code all those factors thought to be related, or possibly related, to the study results of interest. From a practical perspective, the decision about what specific study characteristics to code will have to reconcile two competing considerations. First, synthesis provides a service by documenting in detail the nature of the body of research bearing on a given issue. This consideration motivates a coding of a broad range of study characteristics for descriptive purposes. On the other hand, many codable features of studies have limited value for anything but descriptive purposes—they may not be widely reported in the literature or may show little variation from study to study. Finally, of course, some study characteristics will simply not be germane (though it may be difficult to specify in advance what will prove relevant and what will not). Documenting study characteristics that are poorly reported, lacking in variation, or irrelevant to present purposes surely has some value, as the next section will argue. However, given how time-consuming and expensive coding in research synthesis is, the synthesist inevitably must find some balance between coding broadly for descriptive purposes and coding narrowly around the specific target issues of the particular synthesis.

8.2 ANALYSIS OPPORTUNITIES

A researcher embarks on a synthesis to answer certain questions on the basis of statistical analysis of data developed by systematically coding the results and characteristics of the studies under consideration. There are, of course, many technical issues involved in such statistical analysis; these are discussed elsewhere in this volume. For present purposes it is more important to obtain an overview of the various types of analysis and the kinds of insights they might yield. With such an overview in mind, the synthesist should be in a better position to know what sorts of questions might be answered by synthesis and what data must be coded to address them.

Four generic forms of analysis are outlined here. The first, descriptive analysis, uses the coded variables to provide an overall picture of the nature of the research studies included in the synthesis. The other three forms examine relationships among coded variables. As described previously, variables coded in a synthesis can be divided into those that describe study results (effect sizes) and those that describe study characteristics (study descriptors). Three general possibilities for analysis of relationships emerge from this scheme: among study descriptors, between study descriptors and effect sizes, and among study effect sizes.

8.2.1 Descriptive Analysis

By its nature, research synthesis involves the collection of information describing key results and various important attributes of the studies under investigation. Descriptive analysis of these data can provide a valuable picture of the nature of a research literature. It can help identify issues that have already been studied adequately and gaps in the literature that need additional study. It can also highlight common methodological practices and provide a basis for assessing the areas in which improvement is warranted. The descriptive information provided by a synthesis deals with either study results or study descriptors. Each of these is considered in turn in the following sections.

8.2.1.1 Descriptive Information About Study Results Descriptive information about study results is almost universally reported in research synthesis. The central information, of course, is the distribution of effect sizes across studies. As noted earlier, there may be numerous categories of effect sizes; correspondingly, there may be numerous effect size distributions to describe and compare (see chapter 12 and chapter 13, this volume).

A useful illustration of descriptive analysis of effect sizes occurs in a synthesis of sixty-six studies that reported correlations representing the stability of vocational interests across the lifespan (Low et al. 2005). Douglas Low and his colleagues identified and reported descriptive statistics for distributions of effect sizes for eight different age categories ranging from twelve to forty. They also reported effect size means for different birth cohorts ranging from those born before 1930 to those born in the 1980s. These descriptive statistics allow the reader to assess the uniformity of the correlations across age and historical period. Furthermore, the associated numbers of effect sizes and confidence intervals also reported allow the depth and statistical precision of the available evidence for each mean to be appraised.

Description of effect size distributions may be more than descriptive in the merely statistical sense—inferential tests also may be reported regarding such matters as the homogeneity of the effect sizes and the confidence interval for their mean value (see chapter 14 this volume). Reporting other information closely related to effect sizes may also be appropriate. Such supplemental information might include the number of effect sizes per study, the basis for computation or estimation of effect sizes, whether effects were statistically significant, what statistical tests were used in the original studies, and coder confidence in the effect size computation.

8.2.1.2 Study Descriptors Research synthesists all too often report relatively little information about studies other than their results. It can be quite instructive, however, to provide breakdowns of the coded variables that describe substantive study characteristics, study methods, and extrinsic factors such as publication source. This disclosure accomplishes two things. First, it informs readers about the specific nature of the research that has been chosen for the synthesis so they may judge its comprehensiveness and biases. For example, knowing the proportions of unpublished studies, or studies conducted before a certain date, or studies with a particular type of design might be quite relevant to interpreting the findings of the synthesis. Also, synthesists frequently discover that studies do not adequately report some information that is important to the synthesis. It is informative for readers to know the extent of missing data on various coded variables.

Second, summary statistics for study characteristics provide a general overview of the nature of the research available on the topic addressed. This overview allows readers to ascertain whether researchers have tended to use restricted designs, measures, samples, or treatments and what the gaps in coverage are. Careful description of the research literature establishes the basis for critique of research practices in a field and helps identify characteristics desirable for future research.

Swaisgood and Shepherdson used this descriptive approach to examine the methodological quality of studies of the effects of environmental enrichment on stereotypic behavior among zoo animals (2005). They provided breakdowns of the characteristics of the studies with regard to sample sizes, observation times, research designs, the specificity of the independent variables, the selection of subject species, and the statistical analyses. This information was then used to assess the adequacy of the available studies and to identify needed improvements in methodology and reporting.

It is evident that much can be learned from careful description of study results and characteristics in research synthesis. Indeed, it can be argued that providing a broad description and appraisal of the nature and quality of the body of research under examination is fundamental to all other analyses that the synthesist might wish to conduct. Proper interpretation of those analyses depends on a clear understanding of the character and limitations of the primary research on which they are based. It follows that conducting and reporting a full descriptive analysis should be routine in research synthesis.

8.2.2 Relationships Among Study Descriptors

The various descriptors that characterize the studies in a synthesis may also have interesting interrelationships among themselves. It is quite unlikely that study characteristics will be randomly and independently distributed over the studies in a given research literature. More likely, they will fall into patterns in which certain characteristics tend to occur together. Many methodological and extrinsic features of studies may be of this sort. We might find, for example, that published studies are more likely to be federally funded and have authors with academic affiliations. Substantive study characteristics may also cluster in interesting ways, such as when certain types of treatments are more frequently applied to certain types of subjects. Analysis of across-study intercorrelations of descriptors that reveals these clusters has not often been explored in research synthesis. Nonetheless, such analysis should be useful both for data reduction—that is, creation of composite moderator variables—and to more fully describe the nature of research practices in the area of inquiry.

Table 8.1 Correlation of Selected Study Characteristics with Method of Subject Assignment (Nonrandom Versus Random) in a Synthesis of Cognitive-Behavioral Therapy for Offenders

Variables	Across-Studies Correlation (N = 58)
Academic author (no vs. yes)	.47*
Grant funded (no vs. yes)	.28*
Publication type (unpublished vs. published)	.26*
Practice versus demonstration program	.36*
Total sample size	−.26*
Manualized program (no versus yes)	.03
Brand name program (no versus yes)	.11
Implementation monitored (no versus yes)	.27*
Mean age of sample	−.36*
Sample percent male	.08
Sample percent minority	.20

SOURCE: Author's calculations.
* $p < .05$

Another kind of analysis of relationships among study descriptors focuses on key study features to determine if they are functions of other, temporally or logically prior features. For example, a synthesist might examine whether a particular methodological characteristic, perhaps sample size, is predictable from characteristics of the researcher or the nature of the research setting. Similarly, the synthesist could examine the relationship between the type of subjects in a study and the treatment applied to those subjects.

Analysis of relationships among study descriptors can be illustrated using data from Nana Landenberger and Mark Lipsey's meta-analysis of the effects of cognitive-behavioral therapy for offenders (2005). One important methodological characteristic in this research is whether a study employed random assignment of subjects to experimental conditions. Table 8.1 shows the across-study correlations that proved interesting between method of subject assignment and other variables describing the studies.

The relationships in table 8.1 show that studies using random assignment were more likely to have authors with an academic affiliation, be grant funded, formally published, and study a demonstration program rather than a real-world practice program. Randomized studies also tended to have smaller and younger samples and to monitor the treatment implementation. On the other hand, they were no more likely to use a manualized or brand name cognitive-behavioral program or to involve male or minority samples.

A distinctive aspect of analysis of the interrelations of study descriptors is that it does not involve the effect sizes that are central to most other forms of analysis in research synthesis. An important implication of this feature is that the sampling unit for such analyses is the study itself, not the subject within the study. That is, features such as the type of design chosen, nature of treatment applied, publication outlet, and other similar characteristics describe the study, not the subjects, and thus are not influenced by subject-level sampling error. This simplifies much of the statistical analysis when investigating these types of relationships because attention need not be paid to the varying number of subjects represented in different studies.

8.2.3 Relationships Between Effect Sizes and Study Descriptors

The effect sizes that represent study results often show more variability within their respective categories than would be expected on the basis of sampling error alone (see chapters 15 and 16, this volume). A natural and quite common question for the analyst is whether any study descriptors are associated with the magnitude of the effect, that is, whether they are moderators of effect size. It is almost always appropriate for a synthesis to conduct a moderator analysis to identify at least some of the circumstances under which effect sizes are large and small. Moderator analysis is essentially correlational, examining the covariation of selected study descriptors and effect size, though it can be conducted in various statistical formats. In such analysis, the study effect sizes become the dependent variables and the study descriptors become the independent or predictor variables.

The three broad categories of study descriptors identified earlier—those characterizing substantive aspects of the phenomenon under investigation, those describing study methods and procedures, and those dealing with extrinsic matters of research circumstances, reporting, coding, and the like—are all potentially related to study effect sizes. The nature and interpretation of relationships involving descriptors from these three categories, however, is quite different.

8.2.3.1 Extrinsic Variables Extrinsic variables, as defined earlier, are not generally assumed to directly shape actual study results even though they may differentiate studies that yield different effect sizes. In some instances, they may be marker variables that are associated, in turn, with research practices that do exercise direct influence on study results. It is commonly found in synthesis, for example, that published studies have larger effect sizes than unpublished studies (Rothstein, Sutton, and Borenstein 2005). Publication of a study, of course, does not itself inflate the effect sizes, but may reflect the selection criteria and reporting proclivities of the authors, peer reviewers, and editors who decide if and how a study will be published. Analysis of the relationship of extrinsic study variables to effect size, therefore, may reveal interesting aspects of research and publication practices in a field, but is limited in its ability to reveal why those practices are associated with different effect sizes.

A more dramatic example is a case that appears in the synthesis literature in which the gender of the authors was a significant predictor of the results of studies of sex differences in conformity. Alice Eagly and Linda Carli conducted a research synthesis of this literature and found that having a higher percentage of male authors was associated with findings of greater female conformity (1981). One might speculate that this reflects some bias exercised by male researchers, but it is also possible that their research methods were not the same as those of female researchers. Indeed, Betsy Becker demonstrated that male authorship was confounded with characteristics of the outcome measures and the number of confederates in the study. Both of these method variables, in turn, were correlated with study effect size (1986).

8.2.3.2 Method Variables The category of moderator variables that has to do with research methods and procedures is particularly important and too often underrepresented in synthesis. Experience has shown that variation in study effect sizes is often associated with methodological variation among the studies (Lipsey 2003). One reason these relationships are interesting is that much can be learned from synthesis about the connection between researchers' choice of methods and the results those methods yield. Such knowledge provides a basis for examining research methods to discover which aspects introduce the greatest bias or distortion into study results.

Investigation of method variables in their meta-analysis of the association of the NRG1 gene with schizophrenia, for example, allowed Marcus Munafo and his colleagues to determine that there was no difference in the findings of studies that used a family-based design and those that used a case control design (2006). Similarly, Rebecca Guilbault and her colleagues discovered that the results of studies of hindsight bias depended in part on whether respondents were given general information questions or questions about real world events and whether they were asked to provide an objective or subjective estimate of probability (2004). On measurement issues, Lösel and Schmucker (2005) found that self-report measures showed larger treatment effects on recidivism for sex offenders than measures drawn from criminal records. Findings such as these provide invaluable methodological information to researchers seeking to design valid research in their respective areas of study.

Another reason to be concerned about the relationship between methodological features of studies and effect sizes is that method differences may be correlated with substantive differences. In such confounded situations, a synthesist must be careful not to attribute effect size differences to substantive factors when those factors are also related to methodological factors. For example, in a meta-analysis of interventions for juvenile offenders, Mark Lipsey found that such substantive characteristics as the gender mix of the sample and the amount of treatment provided were not only related to the recidivism effect sizes but were also significantly related to the type of research design (randomized or nonrandomized) used in the study (2003). Such confounding raises a question about whether the differential effects associated with gender and amount of treatment reflect real differences in treatment outcomes or only spurious effects stemming from the correlated methodological differences.

8.2.3.3 Substantive Variables Once disentangled from method variables, substantive variables are usually of most interest to the synthesist. Determining that effect size is associated with type of subjects, treatments, or settings often has considerable theoretical or practical importance. A case in point is the meta-analysis of the effects of peer-assisted learning on the achievement of elementary school students that Cynthia Rohrbeck and her colleagues conducted (2003). Extensive moderator analysis revealed that the largest effect sizes were associated with study samples of younger, urban, low-income, and minority students. Moreover, effect sizes were larger for interventions that used rewards based on the efforts of all team members, individualized measures of student accomplishments, and provided students with more autonomy. These findings point research and practice toward program configurations that should produce the greatest gains for the participating students.

Where there are multiple classes of effect size variables as well as of substantive moderator variables, quite a range of analyses of interrelations is possible. For instance, investigation can be made of the different correlates of effects on different outcome constructs. Kenneth Deffenbacher, Brian Bornstein, and Steven Penrod used this approach in a synthesis of the effects of prior exposure to mugshots on the ability of eyewitnesses to identify offenders in police lineups (2006). They found that mugshot commitment, that is, identification of someone in a mugshot prior to the lineup, was a substantial moderator of both the effect on the proportion of correct responses and on the proportion of false positives. Another possibility for differentiated analysis is to explore effect size relationships for different subsamples. This option was illustrated in Sandra Wilson, Mark Lipsey, and Haluk Soydan's comparative analysis of the effects of juvenile delinquency programs separately for racial minority and majority offenders (2003).

The opportunity to conduct systematic analysis of the relationships between study descriptors and study results is one of the most attractive aspects of research synthesis. Certain limitations must be kept in mind, however. First, the synthesis can examine only those variables that can be coded from primary studies in essentially complete and accurate form. Second, this aspect of synthesis is a form of observational study, not experimental study. Many of the moderator variables that show relationships with effect size may also be related to each other. Confounded relationships of that sort are inherently ambiguous with regard to which of the variables are the more influential ones or, indeed, whether the influence stems from other unmeasured and unknown variables correlated with those being examined. As noted, this situation is particularly problematic when method variables are confounded with substantive ones. Even without confounded characteristics, the observational nature of synthesis tempers the confidence with which the analyst can describe the causal influence of study characteristics on their findings (for more discussion of the distinction between study-generated and synthesis-generated evidence, see chapter 2 this volume).

8.2.4 Relationships Among Study Results

Study results in research synthesis are reported most often either as an overall mean effect size or as mean effect sizes for various categories of results. When multiple classes of effect sizes are available, however, there is potential for informative analysis of the patterns of covariation among the effect sizes themselves. This can be especially fruitful when different results from the same study represent different constructs. In that situation, a synthesist can examine whether the magnitude of effects on one construct is associated, across studies, with the magnitude of effects on another construct. Some studies of educational interventions, for instance, may measure effects on both achievement and student attitude toward instruction. Under such circumstances, a synthesist might wish to ask whether the intervention effects on achievement covary with the effects on attitude. That is, if an intervention increased the achievement scores for a sample of students, did it also improve their attitude scores? The across-studies correlation, though potentially interesting, does not necessarily imply a within-study relationship. Thus, achievement and attitude might covary across studies, but not among the students in the samples within the studies.

Another example comes from the analysis Stacy Najaka, Denise Gottfredson, and David Wilson reported on the relationship between the effects of intervention on youth risk factors and the problem behaviors with which they are presumed to be associated (2001). Problem behaviors—delinquency, alcohol and drug use, dropout and truancy, aggressive and rebellious behavior—were the major outcome variables of interest in their meta-analysis of school-based prevention programs. Many studies, however, also reported intervention effects on academic performance, school bonding (liking school, educational aspirations), and social competency skills. These characteristics are known to be predictive risk factors for problem behaviors. The authors thus reasoned that treatment induced reductions on these factors should be associated with reductions in problem behaviors—that is, changes in risk should mediate changes in problem behavior. In some of the studies in their synthesis, outcomes were measured on one of these risk factors and also on problem behaviors. This made it possible to ask whether the effect sizes for the risk factor were correlated with the effect sizes for problem behaviors. That is, if intervention produced an increase in academic performance, school bonding, or social skills, was there a corresponding decrease in problem behavior?

Najaka and her colleagues found that the across-study correlations between the effect sizes for the risk factors and those for problem behaviors were in the expected direction (2001). In particular, those correlations showed that increased school bonding was strongly associated with decreased problem behaviors ($r = .86$), with academic performance associated to a lesser extent ($r = .33$) and social competency skills even smaller ($r = .12$). Moreover, those relationships were relatively unchanged when various potentially relevant between-study differences were introduced as control variables in multiple regression analyses. By examining the across-study covariation of effects on different outcome constructs, therefore, we obtain some insight into the patterning of the change brought about by intervention and the plausibility that some changes mediate others (Shadish 1996).

An even more ambitious form of synthesis may be used to construct a correlation (or covariance) matrix that represents the interrelations among many different variables in a research literature. Each study in such a synthesis contributes one or more effect sizes representing the correlation between two of the variables of interest. The resulting synthetic correlation matrix can then be used to test multivariate path or structural equation models. An example of such an analysis is a synthesis on the relationship between workers' perceptions of the organizational climate in their workplace and such work outcomes as employee satisfaction, psychological well-being, motivation, and performance (Parker et al. 2003). Christopher Parker and colleagues found 121 independent samples in ninety-four studies that reported the correlations among at least some of the variables relevant to their question. Those correlational effect sizes were synthesized into a single correlation matrix and corrected for measurement unreliability. Their final matrix contained eleven correlations, six relating to employee perceptions of the organizational climate and five relating to work outcomes, with each synthesized from between two and forty-seven study estimates. This matrix, then, became the basis for a path analysis testing their model. The final model showed not only relatively strong relationships between organizational climate variables and worker outcomes, but also that the relationships between climate and employee motivation and performance were fully mediated by employees' work attitudes.

A potential problem in analyzing multiple categories of study results is that all categories of effects will not be reported by all studies. Thus, each synthesized effect size examining a relationship will most likely draw on a different subset of studies creating uncertainty about their generality and comparability (for more on this and related issues having to do with analyzing synthesized correlation matrices, see chapter 20, this volume; Becker 2000).

8.3 CONCLUSION

Research synthesis can be thought of as a form of survey research in which the subjects interviewed are not people but research reports. The synthesist prepares a questionnaire of items of interest, collects a sample of research reports, interacts with those reports to determine the appropriate coding on each item, and analyzes the resulting data. The kinds of questions that can be addressed by synthesis of a given research literature are thus determined by the variables that the synthesist is able to code and the kinds of analyses that are possible given the nature of the resulting data.

This chapter has examined the broad types of variables potentially available from research studies, the kinds of questions that might be addressed using those variables, and the general forms of analysis that investigate those questions. Its goal was to provide an overview that might help the prospective research synthesist select appropriate variables for coding and plan a probing and interesting analysis of the resulting database. Contemporary research synthesis often neglects important variables and analysis opportunities. As a consequence, we learn less from such work than we might. Even worse, what we learn may be erroneous if confounds and alternative analysis models have been insufficiently probed. Current practice has only begun to explore the potential breadth and depth of knowledge that research synthesis might yield.

8.4 REFERENCES

Becker, Betsy J. 1986. "Influence Again: An Examination of Reviews and Studies of Gender Differences in Social Influence. In *The Psychology of Gender: Advances Through Meta-Analysis*, edited by Janet S. Hyde and Marcia C. Linn. Baltimore, Md.: Johns Hopkins University Press.

Becker, Betsy. J. 2000. "Multivariate Meta-Analysis." In *Handbook of Applied Multivariate Statistics and Mathematical Modeling* edited by Howard E. A. Tinsley and Steven D. Brown. New York: Academic Press.

Deffenbacher, Kenneth A., Brian H. Bernstein, and Steven D. Penrod. 2006. "Mugshot Exposure Effects: Retroactive Interference, Mugshot Commitment, Source Confusion, and Unconscious Transference." *Law and Human Behavior* 30(3): 287–307.

Eagly, Alice H., and Linda L. Carli. 1981. "Sex of Researchers and Sex-Typed Communications as Determinants of Sex Differences in Influenceability: A Meta-Analysis of Social Influence Studies." *Psychological Bulletin* 90(1): 1–20.

Guilbault, Rebecca L., Fred B. Bryant, Jennifer H. Brockway, and Emil J. Posavac. 2004. "A Meta-Analysis of Research on Hindsight Bias." *Basic and Applied Social Psychology* 26(2–3): 103–17.

Landenberger, Nana A., and Mark W. Lipsey. 2005. "The Positive Effects of Cognitive-Behavioral Programs for Offenders: A Meta-Analysis of Factors Associated with Effective Treatment." *Journal of Experimental Criminology* 1(4): 451–76.

Lipsey, Mark W. 2003. "Those Confounded Moderators in Meta-Analysis: Good, Bad, and Ugly." *Annals of the American Academy of Political and Social Science* 587(May): 69–81.

Lösel, Friedrich, and Martin Schmucker. 2005. "The Effectiveness of Treatment for Sexual Offenders: A Comprehensive Meta-Analysis." *Journal of Experimental Criminology* 1: 117–46.

Low, K. S. Douglas, Mijung Yoon, Brent W. Roberts, and James Rounds. 2005. "The Stability of Vocational Interests from Early Adolescence to Middle Adulthood: A Quantitative Review of Longitudinal Studies." *Psychological Bulletin* 131(5): 713–37.

Martin, Jose Luis R., Manuel J. Barbanoj, Thomas E. Schlaepfer, Elinor Thompson, Victor Perez, and Jaime Kulisevsky. 2003. "Repetitive Transcranial Magnetic Stimulation for the Treatment of Depression: Systematic Review and Meta-analysis." *British Journal of Psychiatry* 182(6): 480–91.

Munafo, Marcus. R., Dawn L. Thiselton, Taane G. Clark, and Jonathan Flint. 2006. "Association of the NRG1 Gene and Schizophrenia: A Meta-Analysis." *Molecular Psychiatry* 11(6): 539–46.

Najaka, Stacy S., Denise C. Gottfredson, and David B. Wilson. 2001. "A Meta-Analytic Inquiry into the Relationship Between Selected Risk Factors and Problem Behavior." *Prevention Science* 2(4): 257–71.

Netz, Yael, Meng-Jia Wu, Betsy J. Becker, and Gershon Tenenbaum. 2005. "Physical Activity and Psychological Well-Being in Advanced Age: A Meta-Analysis of Intervention Studies." *Psychology and Aging* 20(2): 272–84.

Parker, Christopher, Boris N. Baltes, Scott A. Young, Joseph W. Huff, Robert A. Altmann, Heather A. Lacost, and Joanne E. Roberts. 2003. "Relationships Between Psychological Climate Perceptions and Work Outcomes: A Meta-Analytic Review." *Journal of Organizational Behavior* 24(4): 389–416.

Rohrbeck, Cynthia. A., Marika D. Ginsburg-Block, John W. Fantuzzo, and Traci R. Miller. 2003. "Peer-Assisted Learning Interventions with Elementary School Students: A Meta-Analytic Review." *Journal of Educational Psychology* 95(2): 240–57.

Rothstein, Hannah R., Alexander J. Sutton, and Michael Borenstein, eds. 2005. *Publication Bias in Meta-Analysis: Prevention, Assessment and Adjustments.* New York: John Wiley & Sons.

Shadish, William R. 1996. "Meta-Analysis and the Exploration of Causal Mediating Processes: A Primer of Examples, Methods, and Issues." *Psychological Methods* 1(1): 47–65.

Smith, Mary L., and Gene V. Glass. 1977. "Meta-Analysis of Psychotherapy Outcome Studies." *American Psychologist* 32(9): 752–60.

Swaisgood, Ronald R., and David J. Shepherdson. 2005. "Scientific Approaches to Enrichment and Stereotypes in Zoo Animals: What's Been Done and Where Should We Go Next?" *Zoo Biology* 24(6): 499–518.

Wilson, Sandra J., Mark W. Lipsey, and Haluk Soydan. 2003. "Are Mainstream Programs for Juvenile Delinquency Less Effective with Minority Youth than Majority Youth? A Meta-Analysis of Outcomes Research." *Research on Social Work Practice* 13(1): 3–26.

9

SYSTEMATIC CODING

DAVID B. WILSON
George Mason University

CONTENTS

9.1 INTRODUCTION

A research synthesist has collected a set of studies that address a similar research question and wishes to code the studies to create a dataset suitable for meta-analysis. This task is analogous to interviewing, but a study rather than a person is interviewed. Interrogating might be a more apt description, though it is usually the coder who loses sleep because of the study and not the other way around. The goal of this chapter is to provide synthesists with practical advice on designing and developing a coding protocol suitable to this task. A coding protocol is both the coding forms, whether paper or computerized, and the coding manual providing instructions on how to apply coding form items to studies.

A well-designed coding protocol will describe the characteristics of the studies included in the research synthesis and will capture the pertinent findings in a fashion suitable for comparison and synthesis using meta-analysis. The descriptive aspect typically focuses on ways in which the studies may differ from one another. For example, the research context may vary across studies. In addition, studies may use different operationalizations of one or more of the critical constructs. Capturing variations in settings, participants, methodology, experimental manipulations, and dependent measures is an important objective of the coding protocol, not only for careful description but also for use in the analysis to explain variation in study findings (see chapters 7 and 8, this volume). More fundamentally, the coding protocol serves as the foundation for the transparency and replicability of the synthesis.

9.2 IMPORTANCE OF TRANSPARENCY AND REPLICABILITY

A bedrock of the scientific method is replication. Research synthesis exploits this feature by combining results across a collection of studies that are either pure replications of one another or conceptual replications (that is, studies examining the same basic empirical relationship, albeit with different operationalizations of key constructs and variation in method). A detailed methods section that is part of a research write-up allows others to critically assess the procedures used (*transparency*) and to conduct an identical or similar study to establish the veracity of the original findings (*replicability*). Research synthesis should in turn have these features, allowing others to duplicate the synthesis if they question the findings or wish to augment it with additional studies. This should not be interpreted to imply that research synthesis is merely the technical application of a set of procedures. The numerous decisions made in the design and conduct of a research synthesis require thoughtful scholarship. A high quality research synthesis, therefore, explicates these decisions clearly enough to allow others to fully understand the basis for the findings and to be able to replicate the synthesis.

The coding protocol synthesists develop is a critical aspect of this transparency and replicability. Beyond detailing how studies will be characterized, it documents the procedures used to extract information from studies and gives guidance on how to translate the narrative information of a research report into a structured and (typically) quantitative form.

Coding studies is challenging and requires subjective decisions. A decision that may on the surface seem straightforward, such as whether a study used random assignment to experimental conditions, often requires judgments based on incomplete information in the written reports. The goal of a good coding protocol is to provide guidance to coders on how to handle these ambiguities. Appropriately skilled and adequately trained coders should be able to independently code a study and produce essentially the same coded data. Transparency and replicability are absent if only the author of the coding protocol can divine its nuances. In the absence of explicit coding criteria, a researcher who claims to know a good study when he or she sees one, or to know a good measure of leadership style by looking at the items, or to determine whether an intervention was implemented well by reading the report may be a good scholar, but is not engaged in scientific research synthesis. Because it is impossible to know what criteria were invoked in making these judgments, the decisions lack both transparency and replicability. The expertise of an experienced scholar should be reflected in the coding protocol (the specific coding items and decision rules for how to apply those items to specific studies) and not in opaque coding that others cannot scrutinize.

9.2 CODING ELIGIBILITY CRITERIA

The studies a synthesist identifies through the search and retrieval process should be coded against explicit eligibility criteria, though this is not technically part of the coding protocol. Assessing each retrieved study against the eligibility criteria and recording this information on a screening form or into an eligibility database is an important record of the nature of studies excluded from a research synthesis. With this information, the synthesist can report on the number of studies excluded and the reasons for the exclusions. These data not only serve an internal audit function, but are also useful when a synthesist is responding to a colleague about why a particular study was not included in the synthesis.

The eligibility criteria should flow naturally from the research question or purpose of the synthesis (see chapter 2, this volume). Your research question may be narrow or broad, and this will be reflected in the eligibility criteria. For example, a research synthesis might focus narrowly on social psychological laboratory studies of the relationship between alcohol and aggressive behavior that use the competitive reaction time paradigm or on the

effectiveness of a specific drug relative to a placebo for a clearly specified patient population. At the other end of the continuum would be a research syntheses focused on broad research questions such as the effect of class size on achievement or juvenile delinquency interventions on criminal behavior. In each case, the eligibility criteria should clearly elucidate the boundaries for the synthesis, indicating the characteristics studies must have to be included and those that would exclude them.

9.3.1 Study Features to Be Explicitly Defined

Eligibility criteria should address several issues, including the defining feature of the studies of interest, the eligible research designs, any sample restrictions, required statistical data, geographical and linguistic restrictions, and any time frame restriction. I briefly elaborate on each of these issues.

9.3.1.1 Defining Features Research questions typically address a specific empirical relationship or empirical finding. This relationship or finding is the defining feature of the studies included in the synthesis. Eligibility criteria should clearly elaborate on the nature of the empirical finding that defines the parameters for study inclusion. For example, a meta-analysis of the effectiveness of an intervention, such as cognitive-behavioral therapy for childhood anxiety, needs to clearly delineate the characteristics that an intervention must have to be considered cognitive-behavioral. Similarly, a meta-analysis of group comparisons, such as the relative math achievement of boys and girls, needs to define the parameters of the groups being compared and the dependent variables of interest. Earlier examples focused on an empirical relationship, such as the effect of an intervention or social policy on one or more outcomes or the correlation between an employment test and job performance. In some cases, a single statistical parameter may be of interest, such as the prevalence rate of alcohol or illegal drugs in homicide victims, though meta-analyses of this type are less common.

Specifying the defining features of the eligible studies involves specifying both the nature of an independent variable, such as an intervention or employment test, and one or more dependent variables. The level of detail needed to adequately define these will vary across syntheses. The more focused your synthesis, the easier it will be to clearly specify the construct boundaries. Broad syntheses will require more information to establish the guidelines for determining which studies to include and which to exclude. In some research areas, it is easier to specify the essential features of the independent and dependent variables as separate criteria.

9.3.1.2 Eligible Designs and Required Methods What is the basic research design that the studies have in common? If multiple design types are allowable, what are they? A research synthesis of intervention studies may be restricted to experimental (randomized) designs or may include quasi-experimental comparison group designs. Similarly, a synthesis of correlational studies may be restricted to cross-sectional designs, or only longitudinal designs, or it may include both types. Some research domains have a plethora of designs examining the same empirical relationship. The eligibility criteria need to specify which of these designs will be included and which will not, and justify both inclusion and exclusion.

It is easy to argue that only the most rigorous research designs should be included in a synthesis. There are trade-offs, however, between rigor and inclusiveness and these trade-offs need to be evaluated within the context of the research question and the broader goals of the synthesis. For example, external validity may be reduced by including only well-controlled experimental evaluations of a social policy intervention, as the level of control needed to mount such a study may change the nature of the intervention and the context within which the intervention occurs. In this situation, quasi-experimental studies may provide valuable information about the likely effects of the social policy in more natural settings.

9.3.1.3 Key Sample Features This criterion addresses the critical features of the sample or study participants. For example, are only studies involving humans eligible? In conducting a research synthesis of the effects of alcohol on aggressive behavior, Mark Lipsey and his colleagues found many studies involving fish—intoxicated fighting fish to be more precise (1997). Most syntheses of social science studies will not need to specify that they are restricted to studies of humans, but may well need to specify which age groups are eligible and other characteristics of the sample, such as diagnostic status (high on depression, low achievement in math, involved in drug use, adjudicated delinquent) and gender. A decision rule may need to be developed for handling studies with samples that include a small number of participants outside the target sample. For example, studies of juvenile delinquency programs are not always clearly restricted to youths age twelve to eighteen. Some studies include a small number of nineteen- and twenty-year-olds. The sample criterion should specify how these ambiguous

cases are to be handled. The researchers could decide to exclude these studies, to include them only if the results for youth age twelve to eighteen are presented separately, or to include them if the proportion of nineteen- and twenty-year-olds does not exceed some threshold value.

9.3.1.4 Required Statistical Data A frustrating aspect of conducting a meta-analysis is missing data. It is common to find an otherwise eligible study that does not provide the statistical data needed for computing an effect size. With some work it is often possible to reconfigure the data provided to compute an effect size, or at least a credible estimate. However, there are times when the necessary statistical data are simply not provided.

Eligibility criteria must indicate how to handle studies with inadequate statistical data. There are several possibilities: excluding the study from the synthesis, including it (either in the main analysis or as a sensitivity analysis; see chapter 22, this volume) with some imputed effect size value, or including it in the descriptive aspects of the synthesis but excluding it from the effect size analyses. The advantages and disadvantages of different options is beyond the scope of this chapter but missing data are dealt with elsewhere in this volume (see chapter 21).

9.3.1.5 Geographical and Linguistic Restrictions It is important to specify any geographical or linguistic restrictions for the synthesis. For example, will only English language publications be considered? Does the locale for the study matter? Juvenile delinquency is a universal problem: there are juvenile delinquents in every country and region of the world. However, the meaning of delinquency and the societal response to it are culturally embedded. As such, a research synthesis of juvenile delinquency programs must consider the relevance of studies conducted outside the synthesist's cultural context. Any geographical or linguistic restriction should be based on theoretical and substantive issues. Many non-English journals provide an English abstract, enabling the identification of potentially relevant studies in these journals by those who are monolinguistic. Free online translation programs may also be useful, at least for assessing eligibility (such as www.freetranslation.com).

9.3.1.6 Time Frame It is important that a synthesist specify the timeframe for eligible studies. Furthermore, the basis for this timeframe should be theoretical rather than arbitrary. Issues to consider are whether studies before a particular date generalize to the current context and whether constructs with similar names have shifted in meaning sufficiently over time to reduce comparability. Similarly, social problems may evolve in ways that reduce the meaningfulness of older research to the present. In some research domains, the decision regarding a timeframe restriction is easy: research on the topic began at a clearly defined historical point and only studies from that point forward are worth considering. In other research domains, a more theory-based argument may need to be put forth to justify a somewhat arbitrary cut point.

An important issue to consider is whether the year of publication or year the data were collected is the critical issue. For some research domains, the two can be quite different. Unfortunately, it is not always possible to determine when data were collected, necessitating a decision rule for these cases.

I have often seen published meta-analyses that restrict the timeframe to studies published after another meta-analysis or widely cited synthesis. The logic seems to be to establish what the newer literature "says" on the topic. This makes sense only if the research in this area changed in some meaningful way following the publication of the synthesis. The goal should be to exclude only studies based on some meaningful substantive or methodological basis, not an arbitrary one such as the date of an earlier synthesis.

Note that publication type has been excluded as an eligibility dimension in need of specification. A quality research synthesis will include all studies meeting the explicit eligibility criteria independent of publication status. For a complete discussion of the publication selection bias issue, see chapters 6 and 23, this volume.

9.3.2 Refining Eligibility Criteria

It can be difficult to anticipate all essential dimensions that must be clearly specified in the eligibility criteria. I have had the experience of evaluating a study for inclusion and found that it clearly does not fit with the literature being examined yet it meets the eligibility criteria. This generally arises if the criteria are not specific enough in defining the essential features of the studies. In these cases, the criteria can be modified and reapplied to studies already evaluated. It is important, however, that any refinement not be motivated by a desire to include or exclude a particular study based on its findings. It is for this reason that the Cochrane Collaboration, an organization devoted to the production of systematic reviews in health care (see www.cochrane.org for more information), requires that eligibility criteria be specified beforehand and not modified once study selection has begun.

9.4 DEVELOPING A CODING PROTOCOL

Developing a good coding protocol is much like developing a good survey questionnaire: it requires a clear delineation of what you wish to measure and a willingness to revise and modify initial drafts. Before beginning to write a coding protocol, the synthesist should list the constructs or study characteristics that are important to measure, much as one would when constructing a questionnaire. Many of these constructs may require multiple items. An overly general item is difficult to code reliably, as the generality allows each coder to interpret the item differently. For example, asking coders to evaluate the internal validity of a study on a five-point scale is a more difficult coding task than asking a series of specific questions related to internal validity, such as the nature of assignment to conditions, extent of differential attrition, the use of statistical adjustments to improve baseline equivalence, and the like. One can, of course, create composite scales from the individually coded items (but for cautions regarding the use of quality scales, see chapter 7, this volume). After delineating the study characteristics of interest, a synthesist should draft items for the coding protocol and circulate it to other research team members or research colleagues, particularly those with substantive knowledge of the research domain being synthesized, for feedback. The protocol should be refined based on this feedback.

Sharon Brown, Sandra Upchurch, and Gayle Acton proposed an alternative approach that derives the coding categories from the studies themselves (2003). Their method begins by listing the characteristics of the studies into a matrix reflecting the general dimensions of interest, such as participant type, variations in the treatment, and the like. This information is then examined to determine how best to categorize studies on these critical dimensions. That is, rather than start with a priori notions of what to code based on the synthesist's ideas about the literature, Brown and colleagues recommended reading through the studies and extracting descriptive information about the essential characteristics of interest. The actual items and categories of the coding protocol are then based on this information. This approach may be particularly helpful in developing items designed to describe the more complex aspects of studies, such as treatment variations and operationalizations of dependent measures. This approach is also better suited to smaller syntheses.

It is good to think of a coding protocol in much the same way as any measure developed for a primary research study. Ideally, it will be valid and reliable—measure what it is intended to measure and do so consistently across studies. Validity and reliability can be enhanced by developing coding items that require as little inference as possible on the part of the coders, yet allows coders to record the requested information in a way that corresponds to how it is typically described in the literature. The amount of inference that a coder needs to make can be reduced by breaking apart the coding of complex study characteristics into a series of focused and simpler decisions. For closed-ended coding items (that is, items with a fixed set of options), the correspondence between the items and manner in which they are described in the literature can be increased by developing the items from a sample of studies. An open-ended question can also be used. With this approach, coders record the relevant information as it appears in the study report and specific coding items or categories are created from an examination of these open-ended responses.

It is important to keep in mind the purpose that each coded construct serves as part of the research synthesis, because it is possible to code too many items. Some constructs may be important to code simply for their descriptive value, providing a basis for effectively summarizing the characteristics of the collection of studies. Other constructs are primarily coded for use as moderators in effect size analyses. The number of possible moderator analyses are often limited unless the meta-analysis has a large number of studies. Thus it is wise to consider carefully the potential value of each item you plan to coded. You can always return to the studies and code an additional item if it becomes necessary.

Although the specific items included in the coding protocol of each research synthesis vary, there are several general categories that most address: report identification, study setting, participants, methodology, treatment or experimental manipulation, dependent measures, and effect size data. I briefly discuss the types of information you might consider for each of these categories.

9.4.1 Types of Information to Code

9.4.1.1 Report Identification Although it is rather mundane, you will want to assign each study an identification code. A complication is that some research studies are reported in multiple manuscripts (such as a journal article and technical report or multiple journal articles), and some written reports present the results from multiple studies or experiments. The primary unit of analysis for a

research synthesis will be an independent research study. As such, the synthesist needs to create a system that allows each manuscript and each research study to be tracked. The coding forms should record basic information about the manuscript, such as the authors' names, type of manuscript (journal article, dissertation, technical report, unpublished manuscript, and so on), year of publication, and year or years of data collection.

9.4.1.2 Study Setting Research is always conducted in a specific context and a synthesist may wish to code aspects of this setting. For example, it may be of value to record the geographic location where the study was conducted, particularly if the synthesis includes studies from different countries or continents. The degree of specificity with regard to geographic location will depend on the research question and nature of the included studies. Also of interest might be the year or years of data collection. In some research areas, such as intervention studies, the research may vary in the degree of naturalness of the research setting. For example, an experimental test of child psychotherapy may occur entirely in a laboratory, such as the research lab of a clinical psychology professor, and rely on participants recruited through newspaper ads, or be set in a clinic and involve participants seeking mental health services (for an example of a meta-analysis that compared the results from children's psychotherapy studies in the lab versus the mental health clinics, see Weisz, Weiss, and Donenberg 1992). Other potentially interesting contextual issues are the academic affiliation of the researcher (such as disciplinary affiliation, in-house researcher or outside academic, and the like) and source of funding, if any. Each of these contextual issues may be related in some way to the findings. For example, researchers with a strong vested interest in the success of the program may report more positive findings than other researchers.

9.4.1.3 Participants Only in a very narrowly focused research synthesis will all of the subject samples be identical on essential characteristics. The coding protocol should therefore include items that address the variations in participant samples that occur across the included studies. A limit on the ability to capture essential subject features is the aggregate nature of the data at the study level and inadequate reporting. For example, age of the study sample is likely to be of interest. Study authors may report this information in a variety of ways, including the age range, the school grade or grades, the mean age, the median age, or they may make no mention of age. The coding protocol needs to be able to accommodate these variations and have clear guidelines for coders on how to characterize the age of the study sample. For example, a synthesist may develop guidelines that indicate that if the mean age is not reported, it should be estimated from the age range or school grade (such as, a sample of kindergarten students in the United States could be coded as having an average age of five and a half). The specific sample characteristics chosen to code will depend on the research question and the nature of the studies included, but other variables to consider are gender mix, racial mix, risk or severity level, any restrictions placed on subject selection (such as, clinical diagnosis), and motivation for participation in the study (such as, course credit, volunteers, court mandated, paid participants, and so on). It is also important to consider whether summary statistics adequately measure characteristic of interest and whether they are potentially biased. The mean age for samples of college students, for example, may be skewed by a small number of older students. Knowing how best to capture sample characteristics will depend on substantive knowledge of the research domain being synthesized.

9.4.1.4 Methodology Research studies included in a research synthesis will often vary in certain methodological features. Research synthesists will want to develop coding items to reflect that variability so that they can assess the influence of method variation on observed results. The importance of methodological features in explaining variability in effects across studies has been well established empirically (Lipsey and Wilson 1993; Shadish and Ragsdale 1996; Weisburd, Lum, and Petrosino 2001; Wilson and Lipsey 2001).

The focus should be on the ways in which the studies in the sample differ methodologically. Things to consider coding include the basic research design, nature of assignment to conditions, subject and experimenter blinding, attrition from baseline (both overall and differential), crossovers, dropouts, other changes to assignment, the nature of the control condition, and sampling methods. Each research domain will have unique method features of interest, and as such the features that should be coded will differ. For example, in conducting a meta-analysis of analog studies of the relationship between alcohol consumption and aggressive behavior, it would be important to code the method used to keep the participants unaware of their experimental condition. In other meta-analyses, coding the type of quasi-experimental design or whether the study used a between-subjects or within-subjects design might be essential. It is also critical to code items specifically related to methodological quality. The issues

involved in assessing methodological quality are addressed in detail elsewhere in this volume (see chapter 7).

9.4.1.5 Treatment or Experimental Manipulation(s) If studies involve a treatment or experimental manipulation, then the synthesist will want to develop coding items to fully capture any differences between the studies in these features. It is useful to have one or more items that capture the general type or essential nature of the treatment or experimental manipulation in addition to more detailed codes. For example, the studies included in our meta-analysis of experimental studies on the effects of alcohol on aggressive behavior used one of two basic types of experimental manipulations, a competitive-reaction time paradigm or a teacher-learner paradigm (Lipsey et al. 1997). An initial coding item captured this distinction and additional items addressed other nuances within each paradigm. As an additional example, a meta-analysis of school-based drug and delinquency prevention programs first categorized the programs into basic types, such as instructional, cognitive-behavioral or behavioral modeling, and the like (Wilson, Gottfredson, and Najaka 2001). Additional coding items were used to evaluate the presence or absence of specific program elements. Other issues to consider are the dose (length) and intensity of treatment, who delivers the treatment, and the integrity of the treatment or experimental manipulation. In my experience meta-analyzing treatment effectiveness research, integrity of treatment is inadequately reported by study authors. However, it may still be of interest to document what is available and if possible examine whether effect size is related to the integrity of the treatment or other experimental manipulation.

9.4.1.6 Dependent Measures Studies will have one or more dependent measures. Research syntheses of correlational research will have at least two measures (either two dependent variables or a dependent and an independent variable) per study but often many more. Not all measures used in a study will necessarily be relevant to the research question. Therefore, the coding manual needs to clearly elaborate a decision rule for determining which measures to code and which to ignore.

For each dependent measure to be coded, a synthesist will want to develop items that capture the essential construct measured and the specific measure used. Note that different studies may operationalize a common construct differently. For example, the construct of depression may be operationalized as scores on the Hamilton Rating Scale for Depression, the Beck Depression Inventory, or a clinician rating. A coding protocol should be designed to capture both the characteristics of the construct and its operationalization. It also may include other information about the measure, such as how the data were collected (such as self-report, observer rating, or clinical interview), psychometric properties (such as reliability and validity), whether the measure was a composite scale or based on a single item, and the scale of the measure (that is, dichotomous, discrete ordinal, or continuous). In many situations it is also essential to know the timing of the measurement relative to the experimental manipulation or treatment (such as baseline, three months, six months, twelve months, and so on). For measures of past behavior, the reporting timeframe would also be considered relevant. Participants in drug prevention studies, to take one example, are often asked to self-report their use of drugs over the past week, past month, and past year.

9.4.1.7 Effect Sizes The heart of a meta-analysis is the effect size, because it encodes the essential research findings from the studies of interest. Often, the data reported in a study need to be reconfigured to compute the effect size of interest. For example, a study may only report data separately for boys and girls, but you might be interested in the overall effect across both genders. In such cases, you will need to properly aggregate the reported data to compute your effect size. Similarly, a study may use a complex factorial design, of which you are only interested in a single factor. As such, you may need to compute the marginal means if they are not reported. In treatment effectiveness research, some primary authors will remove treatment dropouts from the analysis. However, you may be interested in an intent-to-treat analysis and therefore may need to compute a weighted mean between the treatment completers and treatment dropouts to obtain the mean for those assigned to the treatment group. The coding protocol cannot anticipate all of the potential statistical manipulations you may need to make. A copy of these computations should be kept for reference, either saved as a computer file, such as a spreadsheet, or on a sheet of paper attached to the completed coding protocol.

I have found it helpful to code the data on which the effect size is based as well as on the computed effect size. For example, for the standardized mean difference effect size, this would mean recording the means, standard deviations or standard errors, and sample sizes for each condition. Other common data on which this effect size is based include the t-test and sample sizes and the exact p-value for the t-test and sample sizes. If the synthesists also are coding dichotomous outcomes, they would record the frequencies or proportions of successes or failures for each condition. This does not exhaust the possibilities but captures the common configurations for the computation of the standardized mean difference (see Lipsey and Wilson 2001). For the remaining instances based on more complete situations, I recommend simply keeping track of the computation whether electronically or on paper. To assist in data clean-up, it is useful to record the page number of the manuscript where the effect size data can be found. Figure 9.1 shows an example of an effect size coding form for a meta-analysis based on the standardized mean difference. Variations of this form could easily be created for other effect size types, such as the correlation coefficient, odds-ratio, and risk-ratio.

9.4.1.8 Confidence Ratings Synthesists will regularly be frustrated by the inadequate reporting of information in studies. Often this will simply mean they must code an item as "cannot tell" or "missing." However, there are times when information is provided but is in some way inadequate. A study, to take one example, may report the overall sample size but not the sample sizes for the individual groups. Based on the method of constructing the conditions, it may be reasonable to assume that conditions were of roughly equal size. In a meta-analysis of correctional boot camps I conducted with colleagues (MacKenzie, Wilson, and Kider 2001), several studies failed to mention whether the sample was all boys, all girls, or a mix of the two. Given the context, I was reasonably certain that the samples were all boys. A synthesist may wish to track the degree of uncertainty or confidence in the information coded for key items, such as the data on which the effect size is based. Robert Orwin and David Cordray recommended the use of confidence judgment or ratings (1985). For example, coders could rate on a three-point scale their confidence in the data on which the effect size is based, as shown in figure 9.1.

9.4.1.9 Iterative Refinement An important component of good questionnaire development is pilot testing. Similarly, pilot tests of the coding protocol should be conducted on a sample of studies to identify problem items and lack of fit between the coding categories and the characteristics of the studies. Synthesists should not simply start with the studies at the top of the pile or front section of the filing cabinet. These studies are likely to be those that were easy to retrieve or were from the same bibliographic source, such as PsychLit or ERIC. Synthesists want to ensure that the pilot test of the coding protocol is conducted on studies that represent the full range of characteristics of the sample, to the extent possible. Although there is no need to throw out the data coded as

EFFECT SIZE LEVEL CODING FORM

Code this sheet separately for each eligible effect size.

Identifying Information:

1. Study (document) identifier — StudyID |__|__|__|__|
2. Treatment-Control identifier — TxID |__|__|__|__|
3. Outcome (dependent variable) identifier — OutID |__|__|__|__|
4. Effect size identifier — ESID |__|__|__|__|
5. Coder's initials — ESCoder |__|__|__|__|
6. Date coded — ESDate |__/__/__|

Effect Size Related Information:

7. Pre-test, post-test, or follow-up (1 = pre-test; 2 = post-test; 3 = follow-up) — ES_Type |__|
8. Weeks Post-Treatment Measured (code 999 if cannot tell) — ES_Time |__|__|__|
9. Direction of effect (1 = favors treatment; 2 = favors control; 3 = neither) — ESDirect |__|

Effect Size Data—All Effect Sizes:

10. Treatment group sample size — ES_TxN |__|__|__|__|__|
11. Control group sample size — ES_CgN |__|__|__|__|__|

Effect Size Data—Continuous Type Measures:

12. Treatment group mean — ES_TxM |__|__.__|__|
13. Control Group mean — ES_CgM |__|__.__|__|
14. Are the above means adjusted (e.g., ANCOVA adjusted)? (1 = yes, 0 = no) — ES_MAdj |__|
15. Treatment group standard deviation — ES_TxSD |__|__.__|__|
16. Control group standard deviation — ES_CgSD |__|__.__|__|
17. t-value — ES_t |__.__|__|__|

Effect Size Data—Dichotomous Measures:

18. Treatment group; number of failures (recidivators) — ES_TxNf |__|__|__|__|
19. Control group; number failures (recidivators) — ES_CgNf |__|__|__|__|
20. Treatment group; proportion failures — ES_TxPf |__.__|__|
21. Control group; proportion failures — ES_CgPf |__.__|__|
22. Are the above proportions adjusted for pretest variables? (1 = yes; 0 = no) — ES_PAdj |__|
23. Logged odds-ratio — ES_LgOdd |__.__|__|__|
24. Standard error of logged odds-ratio — ES_SELgO |__.__|__|__|
25. Logged odds-ratio adjusted for covariates? (1 = yes; 0 = no) — ES_OAdj |__|

Figure 9.1 Sample Effect-Size Level Coding Form

SOURCE: Author's compilation.

Effect Size Data—Hand Calculated:

26. Hand calculated *d*-type effect size ES_Hand1 |___.___|___|___|

27. Hand calculated standard error of the *d*-type effect size ES_Hand2 |___.___|___|___|

Effect Size Data Location

28. Page number where effect size data found ES_Pg |___|___|___|___|

Effect Size Confidence

29. Confidence in effect size value ES_Conf |___|
    1. Highly Estimated—have N and crude p value only, e.g., $p < .10$, or other limited information
    2. Some Estimation—have complex but complete statistics; some uncertainty about precision of effect size or accuracy of information
    3. No Estimation—have conventional statistical information and am confident in accuracy of information

Figure 9.1 (*Continued*)

part of the pilot testing, synthesists do need to recode any items that were refined or added as a result of the pilot test.

9.4.3 Structure of the Data

A complexity of meta-analytic data is that it is by definition nested or hierarchical. The coding protocol must allow for this complexity. Two general approaches to handling the nested nature of the data can be adapted to any situation: a flat-file approach and a hierarchical or relational file approach. In the flat-file approach, the synthesists specify or know beforehand the nature and extent of the nesting and repeat the effect size variables the necessary number of times. For example, they will have a set of variables for the first effect size of interest, say construct 1, and another set of variables for the second effect size of interest, say construct 2. This works well for a limited and highly structured nesting. In my experience, however, the number of effect sizes of interest per study is typically not known before hand. A hierarchical or relational data structure creates separate data tables for each level of the hierarchy. In its simplest form, this would involve a study level data table with one row per study and a related effect size level data table with one row per effect size. A study identifier links the two data tables. I illustrate each approach.

9.4.3.1 Flat File Approach The flat file approach creates a single data table with one row per study and as many columns as variables or coded items. It is suitable for situations in which the synthesists know the maximum number of effect sizes per study that will be coded and can place each effect size in a distinct category. For example, suppose synthesist are interested in meta-analyzing a collection of studies examining a new teaching method. The outcomes of interest are math and verbal achievement as measured by a well-known standardized achievement test. They might develop a coding protocol with one set of effect size items that addresses the math achievement outcome and another set that addresses the verbal achievement outcome. The advantage of this approach is that all the data are contained within one file. Also, with one row per study, by design, only one effect size per study can be included in any single analysis. The disadvantage is that the flat file approach becomes cumbersome as the number of effect sizes per study increases: wide data tables are difficult to work with. Also, any statistical manipulation of effect sizes, such as the application of the small sample size bias correction to standardized mean difference effect sizes needs to be performed separately for each additional effect size, because each effect size within a study is contained in a different column or variable within the dataset.

9.4.3.2 Hierarchical or Relational File Approach The hierarchical or relational file approach creates separate data tables for each conceptually distinct level of data. Data that can be coded only a single time for a given study, such as the characteristics of the research methods, sampling procedures, study context, and experimental manipulation, are entered into a single study-level data

STUDY_ID	PUBYEAR	MEANAGE	TX_TYPE
001	1992	15.5	2
002	1988	15.4	1
003	2001	14.5	1

STUDY_ID	DV_ID	CONSTRUCT	SCALE
001	01	2	2
001	02	6	1
002	01	2	2
003	01	2	2
003	02	3	1
003	03	6	1

STUDY_ID	ES_ID	DV_ID	FU_MONTHS	TX_N	CG_N	ES
001	02	01	0	44	44	−.39
001	02	01	6	42	40	−.11
001	03	02	0	44	44	.10
001	04	02	6	42	39	.09
002	01	01	0	30	30	.34
002	02	01	12	30	26	.40
003	01	01	6	52	52	.12
003	02	02	6	51	50	.21
003	03	03	6	52	49	.33

Figure 9.2 Hierarchical Data Structure

SOURCE: Author's compilation.

NOTE: Three data tables, one for study level data, one for dependent variable level data, and one for effect size level data. The boxes and arrows show the related data for study 001. The variables are: STUDY_ID = study identifier; PUBYEAR = publication year; MEANAGE = mean age of sample; TX_TYPE = treatment type; DV_ID = dependent variable identifier; CONSTRUCT = construct measured by the dependent variable; SOURCE = source of measure; ES_ID = effect size identifier; FU_MONTHS = months post-intervention; TX_N = treatment sample size; CG_N = control group sample size; ES = effect size.

table. Data that change with each effect size are entered into a separate data table with one row per effect size. This data structure can accommodate any number of effect sizes per study.

In my experience, additional data tables are often useful: a study level, a dependent variable (measurement) level, and an effect-size level. In many research domains, a single dependent variable may be measured at multiple time points, such as at baseline, posttest, and follow-up. Rather than code the features of the dependent variable multiple times for each effect size, one can separate the dependent variable coding items into separate data tables and link that information to the effect-size data table with a study identifier and a dependent variable identifier. This

simplifies coding and eases data cleanup. In a more complex meta-analysis, a separate sample level data table is often required. This allows for the coding of subgroup effects, such as males and females.

This approach is shown in figure 9.2 for a meta-analysis of wilderness challenge programs for juvenile delinquents (Lipsey and Wilson 2001). These studies may each include any number of dependent measures and each may be measured at multiple time points. The coding protocol was divided into three sets of items: those that are at the study level, that is, they do not change for different outcomes; those that describe each dependent variable used in the study; and those that encode each effect size. The effect size data table includes a study

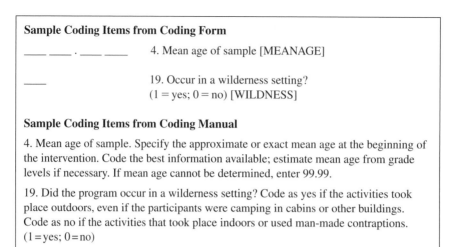

Sample Coding Items from Coding Form

____ ____ . ____ ____   4. Mean age of sample [MEANAGE]

____   19. Occur in a wilderness setting?
(1 = yes; 0 = no) [WILDNESS]

Sample Coding Items from Coding Manual

4. Mean age of sample. Specify the approximate or exact mean age at the beginning of the intervention. Code the best information available; estimate mean age from grade levels if necessary. If mean age cannot be determined, enter 99.99.

19. Did the program occur in a wilderness setting? Code as yes if the activities took place outdoors, even if the participants were camping in cabins or other buildings. Code as no if the activities that took place indoors or used man-made contraptions. (1 = yes; 0 = no)

Figure 9.3 Sample Coding Items

SOURCE: Author's compilation.
NOTE: Variable names are shown in brackets. (Extracted from example used in Lipsey and Wilson, 2001, of a meta-analysis of challenge programs for juvenile delinquents.)

identifier (STUDY_ID) and a dependent variable identifier (DV_ID) that link each effect size row with the appropriate data in both the study level and dependent variable level data tables. This approach also works well for meta-analyses of correlational studies. However, rather than each effect size being associated with a single dependent variable, it will be associated with two. In this way, you can efficiently code all correlations in a correlation matrix by first coding the characteristics of each measured variable and then simply indicate which pair of variables is associated with each correlation.

The hierarchical approach has the advantage of flexibility, simplified coding, and easier data cleanup. The disadvantage is the need to manipulate the data tables prior to analysis. The steps involved are merging the individual data tables into a single flat file, selecting a subset of effect sizes for analysis, and confirming that the selection algorithm yields or results in a single effect size per study. Because the effect size is in a single variable (one column of the data table), it is far easier to create a composite effect size within each study to maintain the one effect size per study restriction. For example, you may wish to perform an analysis based on any achievement effect size within a study, averaging multiple achievement effect sizes if they exist. This can easily be accomplished with a hierarchical data structure.

9.4.4 Coding Forms and Coding Manual

It is often advantageous to develop both a coding manual and coding forms. The former is a detailed documentation of how to code each item and includes notes about decision rules developed as the synthesis team codes studies. A coding form, however, is designed for efficiently recording coded data. Figure 9.3 provides an example of the contents of a code manual for two items and the corresponding items on the coding form. Note that the coding form is decidedly light on information, providing just enough information for trained coders to know where to record relevant study data.

There are several advantages to this approach. First, the synthesists can annotate the coding manual as the team codes studies and that information does not become trapped in the coding form of a single study. Second, coding is more efficient, because coders only need to consult the manual for cases where they are uncertain as to the correct way to code an item for a given study. Finally, coders can condense the coding forms in a matrix layout, as shown in figure 9.4, for efficient coding of dependent measures and effect size data, or any other data items that may need to be coded multiple times for a single study. Note that in figure 9.4, four effect sizes from a single study can be coded on a single page, simplifying the

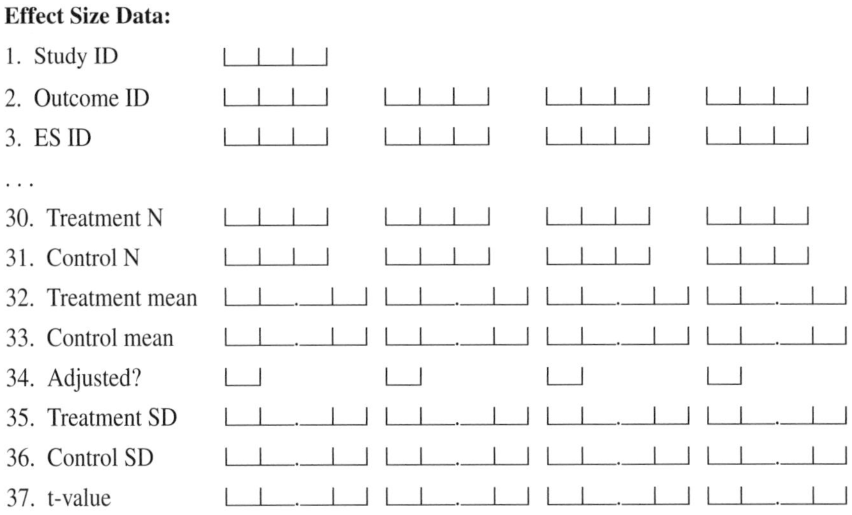

Effect Size Data:

1. Study ID   └┴┴┴┘

2. Outcome ID   └┴┴┴┘   └┴┴┴┘   └┴┴┴┘   └┴┴┴┘

3. ES ID   └┴┴┴┘   └┴┴┴┘   └┴┴┴┘   └┴┴┴┘

. . .

30. Treatment N   └┴┴┴┘   └┴┴┴┘   └┴┴┴┘   └┴┴┴┘

31. Control N   └┴┴┴┘   └┴┴┴┘   └┴┴┴┘   └┴┴┴┘

32. Treatment mean   └┴┴.┴┴┘ └┴┴.┴┴┘ └┴┴.┴┴┘ └┴┴.┴┴┘

33. Control mean   └┴┴.┴┴┘ └┴┴.┴┴┘ └┴┴.┴┴┘ └┴┴.┴┴┘

34. Adjusted?   └┴┘   └┴┘   └┴┘   └┴┘

35. Treatment SD   └┴┴.┴┴┘ └┴┴.┴┴┘ └┴┴.┴┴┘ └┴┴.┴┴┘

36. Control SD   └┴┴.┴┴┘ └┴┴.┴┴┘ └┴┴.┴┴┘ └┴┴.┴┴┘

37. t-value   └┴┴.┴┴┘ └┴┴.┴┴┘ └┴┴.┴┴┘ └┴┴.┴┴┘

Figure 9.4 Matrix Layout

SOURCE: Author's compilation.
NOTE: Layout for a subset of items from an effect-size coding form allowing up to four effect sizes per form.

transfer of data from tables in reports onto the coding form.

9.5 CODING MECHANICS

At a mechanical level, studies can be coded on paper coding forms or directly into the computer. This discussion so far has presumed that paper coding forms are used. I recommend that even if synthesists do develop a database complete with data entry forms, they start with paper coding forms as the template. Paper coding forms are also easier to share with others who request a copy of your coding protocol.

9.5.1 Paper Forms

Paper coding forms, such as the ones shown in figures 9.1 and 9.4, have the advantage of being simple to create and to use. There is also no need to worry about computer backups or database corruption. If the research synthesis involves a small number of studies, such as twenty or fewer, paper coding forms are likely to be a good choice. However, with paper coding forms, coders must enter the data into a computer and manage all the completed paper coding forms. Comparing double-coding for a large synthesis is also more tedious when using paper forms.

When creating paper coding forms, synthesists should always include the variable names that will be used in the statistical software program on the forms. Memory fades quickly and it can be a tedious task reconstructing which variables in a data file corresponded to specific items in a coding protocol. I also recommend that synthesists place the fields for recording the data either down the left or right margin of the coding forms. This has two advantages. First, it allows easy visual inspect of the forms to ensure that each item has been completed. Second, data entry proceeds more smoothly as all data are in a column in either the left or right margin, rather than scattered around the page.

9.5.2 Coding Directly into a Database

Databases provide an alternative to paper coding forms and allow direct coding of studies into a computer, eliminating the need to transfer data from paper forms to computer files. Most any modern database program can be used, such as FileMaker Pro and Microsoft Access. There are several advantages to using a database for study

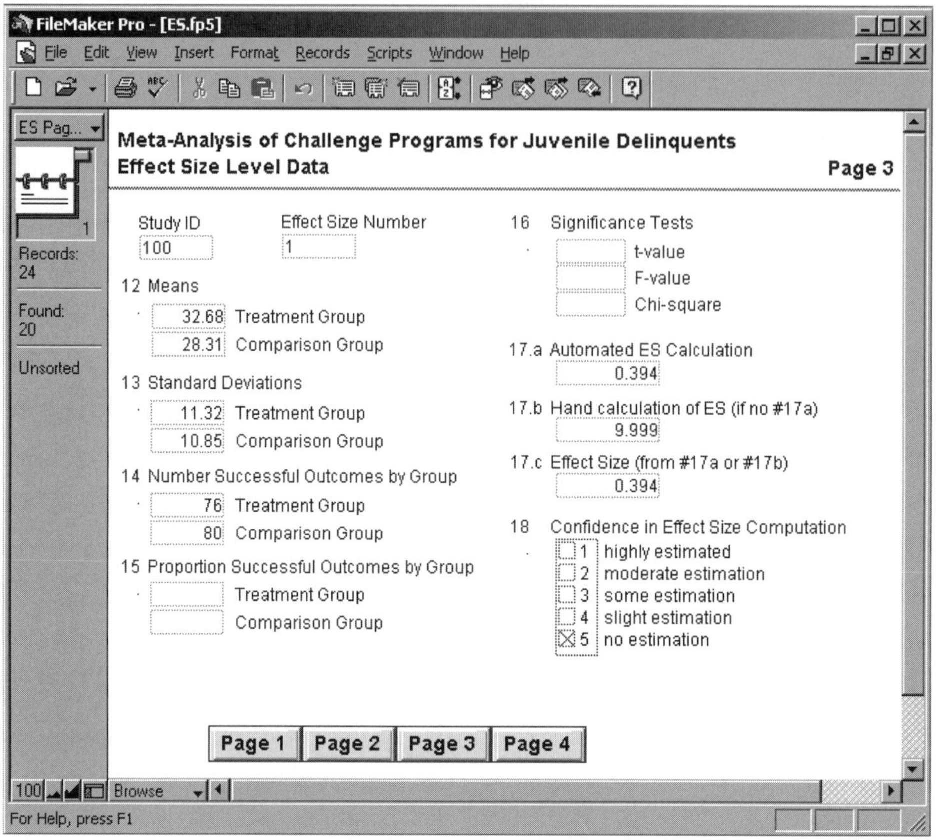

Figure 9.5 Database Coding Form

SOURCE: Author's compilation.

coding. First, synthesists can specify allowable values for any given field (coded item), helping to enhance data accuracy. Second, databases save data as coders work, helping reduce the likelihood of data loss. Third, the data in most databases can easily be exported to any number of statistical software programs. Fourth, the synthesist can create data entry forms that look similar to paper coding forms, providing all of the same information about how to code an item that might have on a paper coding form. A simple example of a coding form in a database program is shown in figure 9.5. Fifth, with a bit more effort, synthesists can have the database compute effect sizes, at least for the more common statistical configurations. Last, data queries can be constructed to assess data consistency, aiding in the process of data cleanup. The disadvantages

of using a database are the time, effort, and technical skills needed to create the database and design the data entry forms. For any reasonably large meta-analysis, however, the result is generally well worth the effort.

9.6 DISADVANTAGES OF A SPREADSHEET

Spreadsheets, such as MS Excel or OpenOffice.org Calc, are extremely useful programs. They are not, however, well suited for systematically coding meta-analytic data. The spreadsheet framework is rows and columns with columns representing different variables and rows representing different studies or effect sizes. This is similar to the typical flat-file data table familiar to anyone accustomed to data analysis, and as such seems a natural fit with the task at hand. However, coding directly into a spreadsheet, however, is different from basic data entry into a database program, where the data are already neatly organized on a survey form or other data instrument. Even in such a case, data entry using a spreadsheet is typically inefficient compared to using a good text editor to create a simple ASCII data file or using a database program. In coding studies, the order of items on a coding form never corresponds to the order of the information in a research write-up. As such, coders generally "bounce around" the coding form, recording information as they find it. In a spreadsheet, this means moving back and forth across the columns used for the different coding items, and the number of columns needed typically far exceeds what can be easily displayed on a single screen. Furthermore, the column headings are cryptic, given limited column width, provide little if any information on how to code an item. For example, it is difficult to display the values associated with a nominal scale in a spreadsheet (such as, 1 = randomized; 2 = quasi-experimental with baseline adjustment; and so on). When using a spreadsheet, it is also easy to inadvertently change rows or columns and overwrite previously coded data.

The disadvantages of the spreadsheet in this context lead me to discourage its use. I recommend that you either use paper coding forms and then enter the data into your favorite statistical or meta-analytic software package (directly or by creating a raw ASCII data file), or that you create a database with data entry forms.

9.7 TRAINING CODERS

Coding studies is a challenging and at times tedious task. Most research syntheses will involve a research team, even if that team consists of only two people. The training of the coders is important to ensure the consistency and accuracy of the information extracted from the studies. Practice coding, regular meetings, coder specialization, and assessment of coder reliability all contribute to coder training and the quality of the final synthesis.

9.7.1 Practice Coding

William Stock recommends a training process that involves eight steps (1994). A central feature of this process is practice coding of a sample of studies, comparing the results among all members of the research team,

modifying coding items if necessary, and repeating this process until consistency across coders is achieved. Stock's eight steps are fairly straightforward:

- An overview of the synthesis is given by the principal investigator.
- Each item on a form and its description in the code book is read and discussed.
- The process for using the forms is described. This is the method chosen to organize forms so that the transition from coding to data entry and data management is facilitated, and should take into account studies that include reports for subsamples, or multiple occasions or measures.
- A sample of five-to-ten studies is chosen to test the forms.
- A study is coded by everyone, with each coder recording how long it takes to code each item. These time data are used to estimate how long it will take to code individual items, entire study reports, and the complete synthesis.
- Coded forms are compared and discrepancies are identified and resolved.
- The forms and code book are revised as necessary.
- Another study is coded and so on. Steps four through eight are repeated until apparent consensus is achieved. (134–35)

This process should be adapted to your needs. A small meta-analysis of five to fifteen studies may abbreviate this process by practice coding only two or three studies and then double-coding all studies to ensure coding reliability.

9.7.2 Regular Meetings

During the coding phase of a meta-analysis, coders should meet regularly to discuss coding difficulties. Discussing difficult coding decisions provides an opportunity to develop a normative understanding of the application of coding rules among all coders. Synthesists should also annotate the coding manual during these meetings to ensure that decisions made about how to handle various study ambiguities are not forgotten and similar studies encountered later in the coding process can be handled in a similar fashion.

9.7.3 Using Specialized Coders

Typically, each coder reads through an entire study and codes all the aspects the synthesis requires. This approach presumes that all members of the research team have the same skills and can accurately complete the entire protocol. Colleagues and research assistants working on the meta-analysis may not, however. Someone with exceptional substantive and theoretical knowledge of the relevant research domain, for example, may not be facile with the computation of effect sizes, particularly when it requires heavy manipulation of the statistical data presented. Similarly, a research assistant with strong quantitative abilities may not have the appropriate background training and experience to code substantive issues that require complex judgments. In these situations, it may be advantageous for coders to specialize, each completing a different section of the coding protocol. Coder specialization has the advantage of exploiting coder strengths to improving coding quality. It can also help address the problem of coder drift (that is, a change in the interpretation of coding items over time), given that each coder will proceed more quickly through the studies, focusing on a smaller set of items. Finally, you can more easily keep effect size coders blind to study authorship and institutional affiliation than coders of more general information. The issue of masking specific study features, such as authorship, is dealt with in more detail in section 9.7.5.

9.7.4 Assessing Coding Reliability

The reliability of coding is critical to the quality of the research synthesis. Both intra- and interrater reliability are important. Intrarater reliability is the consistency of a single coder from occasion to occasion. A common problem in a research synthesis is coder drift: As more studies are coded, items are interpreted differently, resulting in coding inconsistencies. This is most likely to affect items that require judgment on the part of coders, such as an assessment of the similarity of groups at baseline. Purely factual items are less likely to suffer from coder drift. You can assess intrarater reliability by having coders recode studies they coded at some earlier point. Items that show low intrarater agreement might need to be recoded or double-coded to improve accuracy.

Interrater reliability is the consistency of coding between different coders. It is assessed by having at least two different members of the synthesis team code a sample of the studies. As such, in a small meta-analysis, all studies are typically double-coded, whereas a random sample of studies is often selected in larger syntheses. Studies should be coded in a different order by the two coders. Once a sample of studies has been double-coded, the two sets of results are compared. Reliability is often computed as the percentage of agreement across the sample of studies. A weakness of this percentage agreement is that it is affected by the base rate, resulting in artificially high agreement rates for items where most studies share a rating. Other reliability statistics, such as kappa or the intra class correlation, can also be computed and are likely to be more informative (for a fuller discussion, see chapter 10, this volume).

A complication of determining the reliability of coded items stems from the dependent nature of some questions. How the synthesist answers one question often restricts the options for another. William Yeaton and Paul Wortman recommended that reliability estimates take into account this contingent nature of some items (1993). Ignoring this hierarchy of items may result in underestimating the reliability of these contingent items. They therefore recommended that for contingent items, reliability only be assessed when there is agreement on the higher order or more general item. Although this idea has merit, I am unaware of any meta-analyses that have adopted it.

Synthesists have several options for handling an item that shows low coder agreement. First, they can drop the item from any analysis. They can also rewrite the item, possibly breaking it down into multiple simpler judgments, and then code the new item for all of the studies. Finally, they can double-code that item for all studies and resolve any discrepancies either by the senior researcher or through an examination and discussion of the study among all members of the research team.

9.7.5 Masking Coders

In primary research, it is often advisable to have participants unaware of the conditions of assignment. For example, in a randomized clinical trial of a new drug, it is common for participants to be kept uninformed as to whether they are receiving the new drug or a placebo drug. Similarly, keeping raters unaware of conditions, such as medical personnel evaluating a patient's status, is also often advisable. A study may involve both of these methods (a double-blind study) to protect against both sources of potential bias.

There is an analog in meta-analysis. Might not coders be influenced in making quality judgments or other coding decisions about a study by knowledge of who conducted the study, the author's institutional affiliation, or the study findings? Well-known research teams might benefit from a positive bias and research teams at obscure or low status institutions may suffer from a negative bias. Alejandro Jadad and his colleagues examined the effect of masking on the ratings of study quality and found that masked assessment was more consistent than open assessment and resulted in lower quality ratings, suggesting that some coders were affected by knowledge of the authors (1996). In another study, the University of Pennsylvania Meta-Analysis Blinding Study Group randomly assigned pairs of coders to one of two conditions: unaware of author and author affiliation information or aware of such information (Berlin 1997). Of interest was the effect of masking on effect size data. The study found no statistically or clinically meaningful difference in computed effect sizes. Masked assessment may thus be beneficial in assessing study quality but does not appear to matter in coding effect size information. The latter is likely to generalize to the coding of other factual information.

At present, there is not enough empirical evidence to provide clear guidance on the value of masking study authorship and affiliation. Unless coders are already familiar with key studies in research domain of interest, then masking may be beneficial and fairly easy to implement. Clearly, more work is needed in this area.

9.8 COMMON MISTAKES

Novice research synthesists tend to make several mistakes in designing their coding protocols beyond the various recommendations I have discussed already. These errors often stem from a failure to fully plan the analysis of the meta-analytic data. The most common and serious is failure to recognize the implications of the hierarchical nature of meta-analytic data. The creation of a large data file with a single row per effect size without creating codes that allow for the selection of independent subsets of effect sizes for analysis leads to an unmanageable data structure. If synthesists are going to code multiple effect sizes per study, they need to think through how they are going to categorize those effect sizes such that only one effect size (or a composite of multiple effect sizes) is selected for any given analysis. If this is not ensured, the dependency in the results yields biased meta-analytic results, such as standard errors that are too small (for a discussion of this issue and a method for analyzing statistically dependent effect sizes, see chapter 19, this volume).

Another common error in coding is to underestimate the time, effort, and difficulty of coding studies for meta-analysis. Although it may be possible to code some studies in thirty minutes to an hour, spending up to eight hours coding a single study is not unusual. Dissertations and technical reports are often more time consuming than journal articles. Resolving differences between two versions of a study also adds to the coding burden.

Finally, research synthesist should be able to clearly justify why they are coding specific items. Also, in my experience, individuals new to meta-analysis overestimate the level of detail that will be available in the individual studies. During the development phase of the coding protocol, items for which there is rarely adequately information should be eliminated unless they are of enough theoretical interest that it is important to document that particular reporting inadequacy (for a somewhat different take on this issue, see chapter 8, this volume). Also, the synthesist should keep in mind that their ability to use items for moderator analyses will be limited by the size of the meta-analysis. Complex moderator analyses are simply not possible for small meta-analyses, reducing the utility of a large number of variables coded for that purpose.

9.9 CONCLUSION

The goal of this chapter has been to provide guidance on how to achieve a coding protocol that is both replicable and transparent. No matter how fancy the statistical analyses of effect sizes are or impressive the forest plots look, the scientific credibility of a research synthesis rests of the ability of others to scrutinize how the data on which the results of the synthesis were generated. Science progresses not from an authoritarian proclamation of what is known on a topic but rather from a thoughtful, systematic, and empirical taking of stock of the evidence that is open to debate, discussion, and reanalysis. A detailed coding protocol is a key component of this process.

9.10 REFERENCES

Berlin, Jesse A. 1997. "Does Blinding Readers Affect the Results of Meta-Analysis?" *Lancet* 350(9072): 185–86.

Brown, Sharon A., Sandra L. Upchurch, and Gayle J. Acton. 2003. "A Framework for Developing a Coding Scheme for Meta-Analysis." *Western Journal of Nursing Research* 25(2): 205–22.

Jadad, Alejandro R., R. Andrew Moore, Dawn Carroll, Crispin Jenkinson, D. John M. Reynolds, David J. Gavaghan, and Henry J. McQuay. 1996. "Assessing the Quality of Reports of Randomized Clinical Trials: Is Blinding Necessary?" *Controlled Clinical Trials* 17(1): 1–12.

Lipsey, Mark W., and David B. Wilson. 1993. "The Efficacy of Psychological, Educational, and Behavioral Treatment: Confirmation from Meta-Analysis." *American Psychologist* 48(12): 1181–1209.

Lipsey, Mark W., and David B. Wilson. 2001. *Practical Meta-Analysis.* Thousand Oaks, Calif.: Sage.

Lipsey, Mark W., David B. Wilson, Mark A. Cohen, and James H. Derzon. 1997. "Is There a Causal Relationship Between Alcohol Use and Violence? A Synthesis of Evidence." In *Recent Developments in Alcoholism*, vol. XIII, *Alcohol and Violence*, edited by Marc Galanter. New York: Plenum.

MacKenzie, Doris L., David B. Wilson, and Susan Kider. 2001. "Effects of correctional boot camps on offending." *Annals of the American Academy of Political and Social Science* 578: 126–43.

Orwin, Robert G., and David B. Cordray. 1985. "Effects of Deficient Reporting on Meta-Analysis: A Conceptual Framework and Reanalysis." *Psychological Bulletin* 97(1): 134–47.

Shadish, William R., and Kevin Ragsdale. 1996. "Random Versus Nonrandom Assignment in Psychotherapy Experiments: Do You Get the Same Answer?" *Journal of Consulting and Clinical Psychology* 64(6): 1290–1305.

Stock, William A. 1994. "Systematic Coding for Research Synthesis." In *The Handbook of Research Synthesis*, edited by Harris Cooper and Larry V. Hedges. New York: Russell Sage Foundation.

Weisburd, David, Cynthia Lum, and Anthony Petrosino. 2001. "Does Research Design Affect Study Outcomes in Criminal Justice?" *Annals of the American Academy of Social and Political Sciences* 578(1): 50–70.

Weisz, John R., B. Weiss, and Geri R. Donenberg. 1992. "The Lab Versus the Clinic: Effects of Child and Adolescent Psychotherapy." *American Psychologist* 47(12): 1578–85.

Wilson, David B., Denise C. Gottfredson, and Stacy S. Najaka. 2001. "School-Based Prevention of Problem Behaviors: A Meta-Analysis." *Journal of Quantitative Criminology* 17(3): 247–72.

Wilson, David B., and Mark W. Lipsey. 2001. "The Role of Method in Treatment Effectiveness Research: Evidence from Meta-Analysis." *Psychological Methods* 6(4): 413–29.

Yeaton, William H., and Paul M. Wortman. 1993. "On the Reliability of Meta-Analytic Reviews: The Role of Intercoder Agreement." *Evaluation Review* 17(3): 292–309.

10

EVALUATING CODING DECISIONS

ROBERT G. ORWIN AND JACK L. VEVEA

CONTENTS

10.1 INTRODUCTION

Coding is a critical part of research synthesis. It is an attempt to reduce a complex, messy, context-laden, and quantification-resistant reality to a matrix of numbers. Thus it will always remain a challenge to fit the numerical scheme to the reality, and the fit will never be perfect. Systematic strategies for evaluating coding decisions enable the synthesist to control for much of the error inherent in the process. When used in conjunction with other strategies, they can help reduce error as well. This chapter discusses strategies to reduce error as well as those to control for error and suggests further research to advance the theory and practice of this particular aspect of the synthesis process. To set the context, however, it is first useful to describe the sources of error in synthesis coding decisions.

10.2 SOURCES OF ERROR IN CODING DECISIONS

10.2.1 Deficient Reporting in Primary Studies

Reporting deficiencies in original studies presents an obvious problem for the synthesist, to whom the research report is the sole documentation of what was done and what was found. Reporting quality of primary research studies has variously been called "shocking" (Light and Pillemer 1984), "deficient" (Orwin and Cordray 1985), and "appalling" (Oliver 1987).[1] In the worst case, deficient reporting can force the abandonment of a synthesis.[2]

Virtually all write-ups will report some information poorly, but some will be so vague as to obscure what took place entirely. The absence of clear or universally accepted norms undoubtedly contributes to the variation, but there are other factors: different emphases in training, scarcity of journal space, statistical mistakes, and poor writing, for example. The consequences are differences in the completeness, accuracy, and clarity with which empirical research is reported.

Treatment regimens and subject characteristics cannot be accurately transcribed by the coder when inadequately reported by the original author. Similarly, methodological features cannot be coded with certainty when research methods are poorly described. The immediate consequence of coder uncertainty is coder error. One way to address such shortcomings is to agree on conventions for imputing guesses at the values of poorly reported data. For example, in the Mary Smith, Gene Glass, and Thomas Miller analysis, when an investigator was remiss in reporting the length of time the therapist had been practicing, the guessing convention for therapist experience assumed five years (1980). Such a device serves to standardize decisions under uncertainty and therefore to increase intercoder agreement. It is unknown whether it reduces coder error, however, because there is no external way to validate the accuracy of the convention. Furthermore, a guessing convention carries the possibility of bias in addition to error. Unlike pure observational error, which presumably distributes itself around the true value across coders, the convention-generated errors may not balance out, but consistently over- or underestimate the true value. This would happen, for example, if in reality the average PhD therapist had been practicing eight years, rather than five years, as in Smith, Glass, and Miller's guessing convention.

Through its influence on coding accuracy, deficiencies in reporting quality degrade the integrity of later analyses in predictable ways. Errors of observation in dependent variables destabilize parameter estimates and decrease statistical power, whereas errors of observation in independent variables cause parameter estimates to be biased (Kerlinger and Pedhazur 1973). Guessing conventions artificially deflate true variance in coded variables, thereby diminishing the sensitivity of the analysis to detect relationships with other variables. Moreover, any systematic over- and underestimation of true values by guessing conventions exacerbates matters. Specifically, doing so creates a validity problem (for example, therapist experience as coded would not be a valid indicator of therapist experience).

An alternative to guessing conventions that often provides a better solution is to simply code the information as undetermined. When the variable is a categorical one, this has the effect of adding one more category to the possible levels of the moderator (for example, sex of sample predominantly female, predominantly male, mixed, or unknown). In the somewhat rarer case when the problematic variable is continuous, the situation may be handled by dummy coding the availability of the information ($0 =$ not available, $1 =$ available) and adding to the explanatory model both the dummy variable and the product of the dummy and the continuously coded variable. This has the effect of allowing the estimation of the variable's impact on effect size when good information about the variable is available. Such practice is in keeping with the principle of preference for low-inference coding (see chapter 9, this volume).

10.2.2 Ambiguities in the Judgment Process

Coding difficulty reflects the complexity of the item to be coded, not simply the clarity with which it is reported. There is some relationship between reporting quality and the need for judgment calls, in that deficient reporting of study characteristics increases the need for judgment in coding. Many other variables, however, intrinsically require judgment regardless of reporting quality. In fact, more judgment can sometimes be necessary when reporting is thorough, because the coder has more information to weigh.

Numerous variables pose judgment problems for the coder. Consider, for example, treatment integrity (that is, the extent to which the delivered treatment measures up to the intended treatment.) In the psychotherapy literature that Smith, Glass, and Miller synthesized, attrition from treatment (but not from measurement), spotty attendance, failure to meet advertised theoretical requirements, and assorted other implementation problems all potentially degraded treatment integrity (1980). The authors did not

attempt to code treatment integrity per se, though they did code their degree of confidence that the labels placed on therapies by authors described what actually transpired. In a reanalysis, Robert Orwin and David Cordray (1985) attempted to code integrity more globally, but without much success. The following examples (representing actual cases) point up some of the difficulties:

Case 1: The efficacy of ego therapy was tested against a placebo treatment and no-treatment controls. Several participants did not attend every session; only those attending four or more sessions out of seven were posttested.

Case 2: The comparative efficacy of therapist-administered desensitization and self-administered desensitization was tested (relative to various control groups) for reducing public speaking anxiety. The therapists were advanced graduate students in clinical psychology who had been trained in using desensitization but were inexperienced in applying it.

Case 3: The comparative effect of implosive therapy, eclectic verbal therapy, and bibliotherapy was tested for reducing fear of snakes. All eclectic verbal therapy was performed by Jack Vevea, who has published several articles on implosive therapy.

Case 1 exemplifies by far the most common treatment-integrity problem Orwin and Cordray (1985) encountered: the potential dilution of the treatment regimen by nonattendance. Here the coder must judge whether participants with absentee rates up to 43 percent can be said to have received the intended treatment. If not, the coder needs to determine by how much the treatment has degraded, and how this degradation affects the estimated effect size (a question made still more difficult by the authors' failure to report the number of participants with nonattendance problems)? In case 2, the treatment as advertised is potentially degraded by the use of inexperienced treatment providers. The coder must judge whether the lack of practice made a difference and, if so, by how much. In case 3, the treatment provider's motivation to maintain the integrity of the treatment comes into question. The coder must ascertain whether the apparent conflict of interest degraded the treatment and, if so, by how much (for example, uninspired compliance or outright sabotage).

Theoretically, the need for judgment calls in coding such complex constructs could be eliminated by a coding algorithm that considered every possible contingency and provided the coder with explicit instructions in the event of each one, singly and in every combination. The Smith, Glass, and Miller algorithm for internal validity suggests an attempt at this (1980, 63–64). The components of this decision rule represent generally accepted internal validity concerns. Yet, in trying to apply it, Orwin and Cordray frequently found that it failed to accommodate several important contingencies and thus put their own sense of the study's internal validity in contradiction to the score yielded by the algorithm (see Orwin 1985). But the fault is not with the failure of the algorithm to include and instruct on all contingencies, for in practice that would not be possible. With a construct as complex as treatment integrity or internal validity, no amount of preliminary work on the coding algorithm will eliminate the need for judgment calls. Indeed, the contingency instructions are judgment calls themselves, so at best the point of judgment has only been moved, not eliminated.[3] Nevertheless, when it is practical to modify the coding protocol so as to accommodate exceptional contingencies, the practice of striving for low-inference definitions of complex constructs may be advantageous, if only as an aid to coding reliability.

Other examples abound. Although David Terpstra (1981) reported perfect coding reliability in his synthesis of organization development research, R. J. Bullock and Daniel Svyantek's replication (1985) reported numerous problems with both reliability and validity. Specifically, they noted that problems occurred in the coding of the dependent variable, where coding required a great deal of subjectivity. Similarly, Joanmarie McGuire and her colleagues were unable to achieve adequate intercoder agreement on methodological quality despite having methodologically sophisticated coders, written coding instructions, and clearly reported study methods (1985; see also chapter 7, this volume). As noted previously, coding difficulty reflects the complexity of the item to be coded, not just the clarity with which it is reported. One solution to this problem that has gained favor is using low-inference criteria to assess such issues as study quality or implementation fidelity. To the degree that constructs such as study quality or implementation fidelity can be reduced to unambiguous, directly observable criteria, both the validity and the reliability of these measures may be enhanced (see, for example, Valentine and Cooper 2008).

Inherent ambiguities affect more fundamental coding decisions than what values to code, such as what effect sizes to include. Bert Green and Judith Hall observed that synthesists are divided as to whether to use multiple outcomes per group comparison (1984). As shown by Orwin

and Cordray (1985) and Georg Matt (1989), the so-called conceptual redundancy rule—that is, the process used in the original Smith, Glass, and Miller (1980) psychotherapy synthesis for determining which of multiple effect sizes within a given study should be counted in determining overall effect size—can be interpreted quite differently by different coders.[4] In recoding a twenty-five-study sample from the original Smith, Glass, and Miller psychotherapy data, four independent coders extracted 81, 159, 172, and 165 effect sizes, respectively (Smith, Glass, and Miller 1980; Orwin and Cordray 1985; Matt 1989). The corresponding average effect sizes were $d = .90$, .47, .68, and .49. Thus, though the coders were attempting to follow the same decision rule for extracting effect sizes, both the number of effect sizes and the resulting findings varied by a factor of two. Using the same set of printed guidelines on conceptual redundancy, the coders still disagreed substantially. Not surprisingly, coder disagreement on which effect sizes to include led to more discrepant results than coder disagreement on effect size computations once the set of effect sizes had been decided on (Matt 1989). Again, more clarity in the decision rule might have better guided the judgment process, but it is unlikely to have eliminated the need for judgment.

Additional inclusion rules can compensate for the lack of agreement stemming from the use of relatively high inference criteria like the conceptual redundancy rule. In this case, for instance, the synthesis can be restricted to that subset of nonredundant effect sizes on which all or at least most coders agree (Orwin and Cordray 1985). This has the effect of eliminating potentially questionable effect sizes. In still another variation, David Shapiro and Diana Shapiro retained all measures except those permitting only a "relatively imprecise" effect size estimate (1982, 589). These were discarded if, in the coders' view, the study provided more complete data on enough other measures. Matt presents additional rules for selecting effect sizes (1989; see also chapter 8, this volume).

10.2.3 Coder Bias

An additional error source is coder bias (for example, for or against a given therapy). A coder with an agenda is not a good coder, especially for items that require an inference. The ideal coder is totally unbiased and expert in the content area, but such a coder is difficult to find. Some would argue that by definition, it is impossible. Expertise carries the baggage of opinions, and keeping those opinions out of coding decisions—many of which are by definition judgmental—is difficult to do. Ambiguities in the judgment process and coder bias are related in that ambiguity creates a hospitable environment for bias to creep in unnoticed.

Sometimes bias is more blatant. In a synthesis of the effects of school desegregation on black achievement (Wortman and Bryant 1985), a panel of six experts convened by the National Institute of Education independently analyzed the same set of studies, obtaining different results and reaching different conclusions. Panelists excluded studies, substituted alternative control groups, and sought missing information from authors in accordance with their prior beliefs on the effectiveness of desegregation. Even after discussion, the panel "disbanded with their initial views intact" (315).

One approach to reducing bias is to keep coders selectively unaware of information that triggers bias. Thomas Chalmers and his colleagues had two coders independently examine papers in random order, with the methods sections separated from the results (1987).[5] In addition, papers were photocopied in such a way that the coders could not determine their origins. Harold Sacks and his colleagues suggested that this is an ideal way to control for this type of bias, but we note that it is rarely done (1987). In the eighty-six meta-analyses of randomized clinical trials analyzed, none were successfully masked (three showed evidence of attempts). The rationale for such masking is exactly same as in primary studies, to reduce experimenter expectancy effects and related artifacts. It is consistent with the central theme of this book, that research synthesis is a scientific endeavor subject to the same rigorous standards as primary research.

In practice, such masking procedures are difficult to implement, and hence are rarely followed. Two other approaches can help to minimize the impact of coder bias. First, to the degree possible, it is best to identify how issues such as implementation fidelity and study quality will be defined in advance of data collection; this helps prevent creating definitions in such a way that particular individual studies supporting a particular viewpoint will be favored. Second, there is growing agreement that a preference for low-inference coding is helpful with coder bias. To the degree that issues such as study quality can be objectified by criteria like experimenter masking, psychometric properties of measures, impact rating of journal, Carnegie status of primary authors' institutions, and so on, subjectivity is abated, leaving less opportunity for bias (for more discussion of low-inference coding, see chapters 7 and 9, this volume.)

10.2.4 Coder Mistakes

Of course, coders can be unbiased in the sense of holding no views about the likely outcomes of the research being coded, and still make systematic mistakes—systematic in the statistical sense of nonrandom error. The problem is by no means unique to synthesis coding. In their analysis of errors in the extraction of epidemiological data from patient records, Ralph Horwitz and Eunice Yu found that most coding errors occurred because the data extractor simply failed to find information in the medical record (1984). Additional errors were made when information was correctly extracted but the coding criteria were incorrectly applied.

The synthesis coding process is also subject to the same range of simple coder mistakes as any other coding process, including slips of the pencil and keyboard errors. It is particularly vulnerable to the effects of boredom, fatigue, and so on. In a synthesis of any size, many hours are required, often over a period of months, to code a set of studies. The degradation of coder performance under these conditions is discussed further in section 10.4.1.6.

10.3 STRATEGIES TO REDUCE ERROR

Here we discuss nine strategies that potentially reduce error: contacting original investigators, consulting external literature, training coders, pilot testing the coding protocol, revising the coding protocol, possessing substantive expertise, improving primary reporting, using averaged ratings, and seeking coder consensus. Although reducing coding error is distinct from evaluating coding decisions, it is the higher purpose that evaluation of coding decisions serves, and hence merits discussion in this context.

10.3.1 Contacting Original Investigators

An apparent solution to the problem of missing or unclear information is to contact the original investigators in the hope of retrieving or clarifying it. This becomes labor intensive when the number of studies is large, so it is prudent to consider the odds of successful retrieval. The investigators need to be alive; they need to be located; they need to have collected the information in the first place; they need to have kept it; and they need to be willing and able to provide it. Janet Hyde's synthesis of cognitive gender differences is informative here (1981). Rather than trying to estimate effect sizes from incomplete information, she wrote to the authors. Of the fifty-three studies in her database, eighteen lacked the necessary means and standard deviations. Although all eighteen authors were located, only seven responded, and only two were able to supply the information. Furthermore, the successful contact rate may have been atypically high because the topic of cognitive gender differences at that time was relatively young; studies that predated the synthesis by more than fifteen years were rare. Of course, Hyde's final success rate could have been still worse had more esoteric information than means and standard deviations been sought. As Richard Light and David Pillemer noted, the chance of success probably depends quite idiosyncratically on the field, the investigators, and other factors, such as the dates of the studies (1984). One might suppose that the advent of electronic communication and electronic data storage would improve the general success rate of obtaining information through author contact. However, of the nineteen meta-analytic papers published in *Psychological Bulletin* in 2006 that reported on such attempts, eight (57 percent) indicated that the technique failed in some cases; reported response rates were sometimes lower than 50 percent.[6]

10.3.2 Consulting External Literature

A subset of the information of interest to the synthesist is theoretically obtainable from other published sources if omitted or obscured in the reports. For one variable, experimenter affiliation (training), this option was exercised by Smith, Glass, and Miller in their original psychotherapy synthesis, and by others since (1980). When the experimenter's affiliation was not evident from the report, the American Psychological Association directory was consulted. In their reanalysis of the same data, Orwin and Cordray attempted to get the reliability of the outcome measure, when not extractable from the report, from the *Mental Measurements Yearbook* (Orwin and Cordray 1985; Buros 1978). The strategy was unsuccessful, for a number of reasons. Many measures were not included (for example, less-established personality inventories, experimenter-developed convenience scales); when measures were included, a discussion of reliability was frequently omitted; when measures were included and a discussion of reliability included, the range of estimates was sometimes too wide to be useful; and when measures were included, reliability discussed, and an interpretable range of values provided, they were not always generalizable to the population sampled for the psychotherapy

study. Contacting test developers directly would provide more information than consulting the *Mental Measurements Yearbook*, but reliabilities of experimenter-developed convenience scales are not obtainable this way. Nor could the problem of generalizing to different populations be resolved.

The use of external sources may be more successful with other variables. For example, detailed information needed for effect-size calculations may be available in a dissertation, but not in the final, shorter publication of the research. However, the proportion of variables that are potentially retrievable using this strategy will typically be small. For example, of the more than fifty variables coded in the Smith, Glass, and Miller study, experimenter allegiance appeared to be the only variable other than experimenter affiliation that might be deducible through an external published source.

10.3.3 Training Coders

Given the importance and complexity of the coding task, the need for solid coder training as an error-reduction strategy is self-evident. This topic is covered in detail in chapter 9, this volume. The training should include a phase in which coders are made familiar with the project and the coding protocol. Ideally, all coders who will be involved in the project should independently code a subset of studies selected to exemplify particularly challenging coding problems. The process of jointly examining the results sensitizes the coders to the possibility of coder variability and provides a vehicle for training in how to resolve the issues that have been identified as most likely to present difficulties. It may sometimes be practical to merge this training phase with the process of developing and pilot testing the coding protocol. Moreover, if the protocol is revised (see section 10.3.5), the need for appropriate retraining is evident.

10.3.4 Pilot Testing the Coding Protocol

Piloting the coding protocol for a synthesis is no less essential than piloting any treatment or measurement protocol in primary research. First, it supplies coders with direct experience in applying the conventions and decision rules of the process before coding the main sample, which is necessary to minimize learning effects. Second, it assesses whether a coder's basic interpretations of conventions and decision rules are consistent with the synthesist's intent and with the other coders' intent. This is necessary to preclude underestimating attainable levels of interrater reliabilities. Third, it can identify inadequacies in the protocol, such as the need for additional categories for particular variables or additional variables to adequately map the studies.

The concept of pilot testing was taken a step further by Bullock and Svyantek in their reanalysis of a prior synthesis of the organization development literature (1985). After dividing the sixteen years of the study period in half, each author independently coded the first half so as to develop reliability estimates as well as resolve preliminary coding problems. After examining their work, along with each instance of disagreement, the authors attempted to improve interrater reliability by developing more explicit decision rules for the same coding scheme. The more explicit rules were then applied to the studies from the second half of the time period. (The first half was also recoded in accordance with the revised protocol.) Agreement was higher in the second half on six of seven variables tested, sometimes dramatically so (for example, agreement on sample size increased from 70 to 94 percent). It is not clear why the authors halved the sample by period rather than randomly, given that using period introduced a potential confound into the explanation of improvement. That is, the quality of reporting improved during the second half of the study alone could account for much of the improvement in interrater reliability.[7] The distinction between what these authors did from what is typically done is that the preliminary codings were performed on a large enough subset of studies to yield reliable quantitative estimates of agreement. This enabled an empirical validation that agreement had indeed improved on the second half.

10.3.5 Revising the Coding Protocol

On occasion, inadequacies in the coding protocol will not be identified in the pilot testing phase. Indeed, the *Cochrane Handbook of Systematic Reviews* states that it is rare that a coding form does not require modification after piloting (Higgins and Green 2005). It may be, for example, that in the course of coding the fiftieth study, a coder becomes aware of an ambiguity in a categorical coding scheme that requires attention. Although this issue is partly the domain of chapter 9, it may come to the attention of those monitoring the coding process as a result of checks on coder reliability (such as a check for coder drift or a consensus discussion). When a coder modifies the way in which the coding protocol is used, it is important to identify that change and assess whether the modification

is appropriate. If it is deemed appropriate, then the coding protocol is changed, and one must take steps to ensure the uniform (across coders and studies) and retroactive implementation of the change. Following such a practice will ultimately result in higher quality, more consistent coding than would be obtained if coders continued to use the original coding form.

10.3.6 Possessing Substantive Expertise

Substantive expertise will not reduce the need for judgment calls but should increase their accuracy. Numerous authors have stressed the need for substantive expertise, and with good reason: The synthesist who possesses it makes more informed and thoughtful judgments at all levels. Still, scholars with comparable expertise disagree frequently on matters of judgment in the social sciences as elsewhere, particularly when they bring preexisting biases, as noted earlier. Substantive expertise informs judgment, but will not guarantee that the right call was made. Employing low-inference coding protocols will reduce, but not eliminate, the need for expertise. We also note that low-inference codes often require a great deal of expertise to develop.

10.3.7 Improving Primary Reporting

Although the individual synthesist can do nothing about improving primary reporting, social science and medical publications can do something to reduce error in synthesis coding. There is evidence of progress in this regard. There is a recent effort on the part of the American Psychological Association (APA) to encourage full reporting of descriptive statistics and effect sizes (Journal Article Reporting Standards Working Group 2007). A similar policy has been adopted by a number of prominent journals in the medical field, in the form of the Consolidated Standards of Reporting Trials (CONSORT) (see Begg et al. 1996; Moher, Schulz, and Altman 2001).

10.3.8 Using Averaged Ratings

It is a psychometric truism that averages of multiple independent ratings will improve reliability (and therefore reduce error) relative to individual ratings. In principle, then, it is desirable for the synthesist to use such averages whenever possible. Two practical problems limit the applicability of this principle. First, the resources required to double- or triple-code the entire set of studies can be substantial, particularly when the number of studies is large or the coding form is extensive. Second, many variables of interest in a typical synthesis are categorical, and therefore cannot be readily averaged across raters. For example, *therapy modality* from Smith, Glass, and Miller could be coded as individual, group, family, mixed, automated, or other (1980). It is not clear how an mean rating would be possible for such a variable (with at least three coders, a median or a modal rating might make sense).

The first problem can be ameliorated by targeting for multiple coding only those variables known in advance to be both important to the analysis and prone to high error rates. Effect size and internal validity, for example, are two variables that would meet both criteria. The foreknowledge to select the variables can frequently be acquired through a targeted examination of existing research (Cordray and Sonnefeld 1985), as well as through the synthesist's own pilot test. The second problem is more structural, being a property of the categorical measure. A commonsense solution might be, again, to target only those variables that are both important to the analysis and error-prone and have each been coded by three raters. Three raters would permit a majority rule in the event that unanimity was not achieved. If the three coders selected three different responses—that is, there was no majority—the synthesist should probably reconsider the use of that variable altogether.

In circumstances where it is deemed desirable to use averaged codings, the rate of agreement among coders is of less interest than the reliability of the average. Generalizability theory can provide a framework for estimating that reliability (Shavelson and Webb 1991). In principle, a *D* study could estimate the number of raters needed to attain a targeted reliability for the average. In practice, however, it is not common to employ more than two raters; indeed, of the nineteen meta-analytic studies published in *Psychological Bulletin* in 2006, only nine used multiple coders for all studies included in the meta-analysis, and only three double coded effect-size calculations.

10.3.9 Using Coder Consensus

Many meta-analyses are relatively limited in their scope, involving only a few effect sizes and potential explanatory variables. This situation is particularly common for syntheses of medical clinical trials, where the tendency is to focus on narrowly defined questions. Often under such circumstances it is possible to arrange for every study to be coded by two or more independent coders. When that occurs, it is appropriate to have periodic discussions

during which consensus is sought if it is not already present. It may be the case that, where there is disagreement, one coder has noticed a detail the others have missed, and that when the detail is described, all will concur with the minority coder. In that respect, consensus discussions may represent a distinct advantage over strategies such as averaging or majority rule, where the correct minority opinion would be obscured or overruled.

The consensus approach is not without pitfalls. It may be impractical or impossible to implement for projects of large scope. The consensus meeting may present opportunities for systematic coder bias that would not otherwise arise (see section 10.2.3). One particularly insidious form of such bias occurs if there is a tendency for coders to defer to the most senior, who may be the most likely to have a conscious or unconscious agenda for the analysis. Nevertheless, coder consensus can be a highly effective approach. Indeed, it is common practice; of the twelve meta-analyses published in *Psychological Bulletin* in 2006 that reported any measure of rater agreement, half also mentioned that all disagreements were resolved, either by consensus or by appeal to a principal investigator. This makes the meaning of the reported agreement statistics somewhat nebulous, because it does not reflect the coding used in the meta-analyses.

10.4 STRATEGIES TO CONTROL FOR ERROR

It is always better to reduce error than to attempt to control and correct for it after the fact. For example, methods such as the correction for attenuation from measurement error, though useful, tend to overcorrect for error, because the reliability estimates they use in the denominator often are biased downward. Yet strategies to control for error are necessary because, as documented earlier in this chapter, strategies to reduce error can succeed only to a limited degree. Whether because of deficient reporting quality, the limits of coder judgment, or a combination of the two, there will always be a large residual error that cannot be eliminated. The question then becomes how to control for it, and the answer begins with strategies for evaluating coding decisions. Three such strategies are discussed here: interrater reliability assessment, confidence ratings, and sensitivity analysis.

10.4.1 Reliability Assessment

10.4.1.1 Rationale As in primary research applications, the amount of observer error in research synthesis can be at least partially estimated via one or more forms of interrater reliability (IRR). The reanalysis of the Smith, Glass, and Miller psychotherapy data (Orwin and Cordray 1985) suggested that failing to assess IRR and to consider it in subsequent analyses can yield misleading results.

In the original synthesis of psychotherapy outcomes, a sixteen-variable simultaneous multiple regression analysis was used to predict outcomes. Identical regressions were run within three treatment classes and six subclasses, as defined through multidimensional scaling techniques with the assistance of substantive experts (see Smith, Glass and Miller 1980). Orwin and Cordray recoded a sample of studies from the original synthesis, and computed IRRs (1985). They then reran the original regression analyses with corrections for attenuation. Approximately twice per class and twice per subclass, on average, corrections based on one or more reliability estimates caused the sign of an uncorrected predictor's coefficient to reverse.[8] The implication of a sign reversal (when significant) is that interpretation of the uncorrected coefficient would have led to an incorrect conclusion regarding the direction of the predictor's influence on the effect size. In every run in every class and subclass, the reliability correction altered the ranking of the predictors in their capacity to account for variance in effect size.

At least two of the major conclusions of the original study were brought into question by these findings. The first was that the study disconfirmed allegations by critics of psychotherapy that poor-quality research methods have accounted for observed positive outcomes.

This conclusion was based on the trivial amount of observed correlation $(r = .03)$ between internal validity and effect size, which was taken as evidence that design quality had no effect. The reanalyses suggested that unreliability may have so seriously attenuated both the bivariate correlation between the two variables and contribution of internal validity to the regression equations that only an unrealistically large relationship could have been detected. Orwin and Cordray found that the reliability of internal validity may have been as low as .36 (1985).[9]

The second Smith, Glass, and Miller conclusion— bearing another look in light of these reanalyses—was that the outcomes of psychotherapy treatments cannot be very accurately predicted from characteristics of studies. Although the reliability-corrected regressions do not account for enough additional variance in effect size to claim "very accurate" prediction, they do improve the

situation. It would be likely to improve still more with better-specified models, at least to the extent that reporting quality and coder judgment would permit them. The point is that though unreliability alone cannot explain the poor performance of the Smith, Glass, and Miller model, it could be part of a larger process that does.

Unreliability in the primary study's dependent measure attenuates not only relationships with effect size, but the effect-size estimate itself. Under classical test theory assumptions, measurement error has no effect on the means of the treatment and comparison groups, but increases the within-group variance. Hedges and Olkin showed that under that model, the true effect size for a given study is equal to the observed effect size over the square root of the reliability of the dependent measure (1985). If the dependent measure had a reliability of .70, for example, the estimated true effect size would equal the observed effect size times $1/.70^{1/2}$, so the observed effect size would underestimate the true effect size by 16 percent. Error in coding effect-size estimates then exacerbates the problem by increasing the variance of the effect-size distribution at the aggregate level.

Despite wide recognition by writers on research synthesis of the need to assess IRR, in practice addressing coder reliability may still be the exception rather than the rule. For example, only 29 percent of the meta-analyses published in *Psychological Bulletin* from 1986 through 1988 reported a measure of coder reliability (Yeaton and Wortman 1991). A survey of nineteen meta-analytic papers that appeared in *Psychological Bulletin* in 2006 suggests that while the situation has changed for the better, double coding is still far from standard practice. Seven of the nineteen papers presented no information on coding reliability, and only eight reported on the reliability of coding for all variables. Interestingly, only three papers reported that effect-size estimates were double coded. Given the high methodological standards of *Psychological Bulletin* relative to many other social research journals that publish meta-analyses, the field-wide percentage may be significantly lower.

10.4.1.2 Across the Board Versus Item-by-Item Agreement Smith, Glass, and Miller's reliability assessment consisted of the computation of a simple agreement rate across all variables in an abbreviated coding form, which came to 92 percent (1980). According to William Stock and his colleagues, who examined the practice of synthesists regarding interrater reliabilities, subsequent synthesists did much the same thing, if that much (1982). That is, a single agreement rate, falling somewhere between .7 and 1.0, is the extent of what typically gets reported.

There are at least two major problems with this practice. First, it makes little psychometric sense. The coding form is simply a list of items; it is not a multi-item measure of a trait, such that a total-scale reliability would be meaningful. Some items, such as publication date, will have very high interrater agreement, whereas others, such as internal validity, may not. Particularly if reliabilities are to be meaningfully incorporated into subsequent analyses (for example, by correcting correlation matrices for attenuation), it is this variation that needs to be recognized. In their reanalysis of the Smith, Glass, and Miller psychotherapy data (1980), Orwin and Cordray replicated an equivalently high overall agreement rate (1985). However, agreement across individual variables ranged from .24 to 1.00 (and other indices of IRR showed similar variability).

Second, an across-the-board reliability fails to inform the synthesist of specific variables needing refinement or replacement. Thus, an opportunity to improve the process is lost. In sum, the synthesist should assess IRR on an item-by-item basis.

10.4.1.3 Hierarchical Approaches When synthesists do compute item-by-item IRRs (or variable-by-variable IRRs when variables comprise several items), they typically treat the reliability estimate of each item or variable independently. William Yeaton and Paul Wortman suggested an alternative approach (1991). In addition to reinforcing the need to examine item-level reliabilities rather than simply overall reliability, the approach recognizes that item reliabilities are hierarchical, in that the correctness of coding decisions at a given level of the hierarchy is contingent on decisions made at higher levels. In such cases, the code given to a particular variable may restrict the possible codes for other variables. In the hierarchical structure, the IRRs of different variables, as well as the codings themselves, are nonindependent. Existing approaches to IRR in research synthesis do not take this into account.

Using three variables from the original Smith, Glass, and Miller synthesis (1980), Yeaton and Wortman illustrated the hierarchical approach (1991). Each variable can be conceived at a different hierarchical level:

Level 1 (condition): Are these treatment or control group data?

Level 2 (type): What type of treatment or control group is used?

Level 3 (measure): For the type of condition, what measure is reported?

Level 1 identifies the group as psychotherapy or control; level 2 involves a choice of ten general types of therapy and control groups; level 3 consists of four general types of outcome measures used in the calculation of effect size measures. The levels of the hierarchy reflect the contingent nature of the coding decisions; that is, agreement at lower levels is conditional on agreement at higher levels. If the variables are strictly hierarchical, and this is an empirical question in each synthesis, then incorrect coding at one level ensures incorrect coding at all lower levels, and, conversely, correct coding at the lowest level indicates that coding was correct at each higher level. That errors can be made at each level adds to the variability of IRR. Without a hierarchical approach, this variability is obscured and results can be misleading. For example, what does it mean to correctly code type of control group as placebo (level 2), if in fact that group was actually a treatment group (level 1)?

Yeaton and Wortman examined IRRs by level in several syntheses and found that agreement did indeed decrease as level increased, as predicted by their hierarchical model (1991). The results provide yet another argument against an across-the-board agreement rate—that it overestimates true agreement because of the uneven distribution of items across levels (level 1 items tending to be the most numerous).

Although more work is needed on the details of how best to apply a hierarchical approach, the approach does clearly present a more realistic model of the coding process than one that assumes that IRR for each variable is independent. At minimum, synthesists should understand the hierarchical relationships between their own variables prior to interpreting the IRRs calculated on those variables.

10.4.1.4 Specific Indices of Interrater Reliability Laurel Oliver and others have noted that authorities do not agree on the best index of IRR to use in coding syntheses (1987). This presentation does not attempt to resolve all the controversies, but will—it is hoped—provide enough of a foundation for the synthesist who is not a statistician to make informed choices. Five indices are presented: agreement rate, kappa and weighted kappa, delta, intercoder correlation, and intraclass correlation. The discussion of each will include a description (including formulas when possible), a computational illustration, and a discussion of strengths and limitations in the

Table 10.1 Illustrative Data: Ratings of Studies

Study	Coder 1	Coder 2
1	3	2
2	3	1
3	2	2
4	3	2
5	1	1
6	3	1
7	2	2
8	1	1
9	2	2
10	2	1
11	2	2
12	3	3
13	3	1
14	2	1
15	1	1
16	1	1
17	3	3
18	2	2
19	2	2
20	3	1
21	2	1
22	1	1
23	3	2
24	3	3
25	2	2

SOURCE: Authors' compilation.

context of research synthesis. The following section covers the selection, interpretation, and reporting of IRR indices.

Agreement Rate. Percentage agreement, alternately called agreement rate (*AR*), has been the most widely used index of IRR in research synthesis. The formula for *AR* is as follows:

$$AR = \frac{\text{number of observations agreed upon}}{\text{total number of observations}} \quad (10.1)$$

Table 10.1 presents a hypothetical data set that might have been created had two coders independently rated twenty-five studies on a three-point study characteristic (for example, the 1980 Smith, Glass, and Miller internal validity scale). On this particular characteristic, a rating of 1 = low, 2 = medium, and 3 = high. As shown, the coders agreed in fifteen cases out of twenty five, so *AR* = .60.

AR is computationally simple and intuitively interpretable, being basically a batting average. Yet numerous writers on observational measurement have discussed the pitfalls of using it (Cohen 1960; Hartmann 1977; Light 1971; Scott 1955). When variables are categorical (the usual application), the main problem is chance agreement, particularly when response marginals are extreme. For example, the expected agreement between two raters on a yes-no item in which each rater's marginal response rate is 10–90 percent would be 82 percent by chance alone (Hartmann 1977). In other words, these raters could post a respectable (by most standards) interrater reliability simply by guessing without ever having observed an actual case. Extreme marginal response rates are commonplace in many contexts (for example, psychiatric diagnosis of low-prevalence disorders), including research synthesis (see, for example, historical effects in Kulik, Kulik, and Cohen 1979; design validity in Smith 1980). Additional problems arise when marginal response rates differ across raters (Cohen 1960).

When applied to ordinal (as opposed to nominal) categorical variables, *AR* has an additional drawback: the inability to discriminate between degrees of disagreement. In Smith, Glass, and Miller's three-point IQ scale, for example, a low-high interrater pattern indicates greater disagreement than does a low-average or average-high pattern, yet a simple *AR* registers identical disagreement for all three patterns (1980). The situation is taken to the extreme with quantitative variables.

Cohen's Kappa and Weighted Kappa. Various statistics have been proposed for categorical data to improve on *AR*, particularly with regard to removing chance agreement (see Light 1971). Of these, Cohen's kappa (κ) has frequently received high marks (Cohen 1960; Fleiss, Cohen, and Everitt 1969; Hartmann 1977; Light 1971; Shrout, Spitzer, and Fleiss 1987). The parameter κ is defined as the proportion of the best possible improvement over chance that is actually obtained by the raters (Shrout, Spitzer, and Fleiss 1987). The formula for the estimate K of kappa computed from a sample is as follows:

$$K = \frac{P_o - P_e}{1 - P_e}, \quad (10.2)$$

where P_o and P_e are the observed and expected agreement rates, respectively. The observed agreement rate is the proportion of the total count for which there is perfect agreement between raters (that is, the sum of the diagonals in the contingency table divided by the total count). The expected agreement rate is the sum of the expected

Table 10.2 Illustrative Data: Cell Counts and Marginals

		Coder 1			
	Value	1	2	3	Sum
	1	5	3	4	12
Coder 2	2	0	7	3	10
	3	0	0	3	3
	Sum	5	10	10	25

SOURCE: Authors' compilations.

agreement cell probabilities, which are computed exactly as in a chi-square test of association. That is,

$$P_e = \frac{1}{n^2}\sum_{i=1}^{C} n_{i\bullet} n_{\bullet i}, \qquad (10.3)$$

where n is the number of observations, C is the number of response categories, and $n_{i\bullet}$ and $n_{\bullet i}$ are the observed row and column marginals for response i for raters 1 and 2, respectively.

The formula for the estimate of weighted kappa (K_w) is as follows:

$$K_w = 1 - \frac{\sum w_i P_{oi}}{\sum w_i P_{ei}}, \qquad (10.4)$$

where w_i is the disagreement level assigned to cell i, and p_{oi} and p_{ei} are the observed and expected proportions, respectively, in cell i. If regular (unweighted) K is re-expressed as $1 - D_o/D_e$, where D_o and D_e are observed and expected proportion disagreement, it can be seen that K is a special case of K_w, in which all disagreement weights equal 1.

Table 10.2 shows the cell counts and marginals from the illustrative data from table 10.1. Plugging in the values, $P_e = [(12)(5) + (10)(10) + (3)(10)]/(25)^2 = .304$. The proportion of observed agreement, P_o, is $(5 + 7 + 3)/25 = .6$. $K = (.6 - .304)/(1 - .304) = .43$. Thus, chance-corrected agreement in this example is slightly less than half of what it could have been. To compute K_w, the weights (w_i) must be assigned first. Agreement cells are assigned weights of 0. With a three-point scale such as this one, some logical weights would be 2 for the low-high interrater pattern and 1 for the low-medium and medium-high patterns, although other weights are possible of course. Summing across the $w_i p_{oi}$ products for each cell i yields $\sum w_i p_{oi} = .56$. Similarly, summing across the $w_i p_{ei}$ products $\sum w_i p_{ei} = .912$. Then, $K_w = 1 - .56/.912 = .39$. In this case, a relatively modest percentage of disagree-

ments (four of ten) were by two scale points, as one would hope to be the case with a three-point scale. Consequently, K_w is only slightly lower than K. With more scale points, the difference will generally be larger.

As is evident from the formula for K, chance agreement is directly removed from both numerator and denominator. Kappa has other desirable properties:

It is a true reliability statistic, which in large samples is equivalent to the intraclass correlation coefficient (discussed later in this section; see also Fleiss and Cohen 1973).

Because P_e, is computed from observed marginals, no assumption of identical marginals across raters (required with certain earlier statistics) is needed.

It can be generalized to multiple (that is, more than two) raters (Fleiss 1971; Light 1971).

It can be weighted to reflect varying degrees of disagreement (Cohen 1968), for example, for the Smith, Glass, and Miller IQ variable mentioned earlier.

Large-sample standard errors have been derived for both K and K_w (Fleiss, Cohen, and Everitt 1969), thus permitting the use of significance tests and confidence intervals.

It takes on negative values when agreement is less than chance (range is -1 to 1), thus indicating the presence of systematic disagreement as well as agreement.

It can be adapted to stratified reliability designs (Shrout, Spitzer, and Fleiss 1987).

Thus, K is not a single index, but rather a family of indices that can be adapted to various circumstances.

One issue that the analyst must keep in mind is that indices such as K may not be well estimated with the sample sizes available for some meta-analyses. Asymptotic standard errors for K and weighted K are available (Fleiss, Cohen, and Everitt 1969); these may be relatively large unless the number of studies coded is high. It is impossible to provide a simple rule of thumb for a minimum sample size because the standard errors depend partly on the cell proportions. However, it is noteworthy that for the data from table 10.1 (which produced an estimate of $K = .43$), the standard error is 0.12, leading to a 95 percent confidence interval from .18 to .67. For the example provided in the original paper on asymptotic standard errors (Fleiss, Cohen, and Everitt 1969), a sample size of

60 would be required for the standard error of unweighted K to be as low as .10. Hence, for the numbers of studies present in many meta-analyses, there may be very little information about coding reliability.

Although K and K_w resolve most of the problems of AR, potential problems remain if observations are concentrated in only a few cells (Jones et al. 1983). This circumstance increases the probability that a high proportion of scores will be assigned to one or two rating categories while other categories receive small or nonexistent proportions. Allan Jones and his colleagues and others, such as Nancy Burton 1981, have argued that this distribution violates the assumption on which K and K_w are based, reducing the usefulness of the information. In a similar vein, Grover Whitehurst noted that K is sensitive to the degree of disagreement between raters, which is desirable, but also remarkably sensitive to the distribution of ratings, which he argued was not desirable (1984). A highly skewed distribution of ratings will yield high estimates of chance agreement in just those cells in which obtained agreement is also high. This makes for low estimates of true agreement in the face of high levels of obtained agreement and, consequently, low estimates of IRR. As noted earlier, the phenomenon is common in psychiatric diagnosis when prevalence is low and has been termed the base rate problem with K (Carey and Gottesman 1978).

However, others view this as a case of shooting the messenger (see, for example, Shrout, Spitzer, and Fleiss 1987). As they see it, the observation that low base rates decrease K is not an indictment of K, but rather represents the real problem of making distinctions in increasingly homogeneous populations. Similarly, Rebecca Zwick questioned whether the sensitivity of K to the shape of the marginal distributions is necessarily undesirable: if cases are concentrated into a small number of categories, it is less demonstrable that coders can reliably discriminate among *all C* categories, and the IRR coefficient should reflect this (1988). Finally, there is no mathematical necessity for small K values with low base rates (the maximum value of K remains at 1.0 even when the base rate is low) and high Ks have in fact been demonstrated empirically with base rates as low as 2 percent (American Psychiatric Association 1980).

Andrés and Marzo's Delta. One measure of agreement that addresses the base-rate problem of kappa (if, indeed, that is a problem) adapts theory from guessing corrections in multiple-choice testing (Andrés and Marzo 2004). The proposed index ($\hat{\Delta}$) has one somewhat more relaxed assumption than K—nonconcordance rather than independence of raters. In addition to allowing an assessment of the concordance of multiple raters, the index can be computed specifying one rater as a gold standard. It depends, however, on an assumed probability model that describes how raters choose among the options. Moreover, the index lacks a closed-form solution, relying instead on iterative maximum-likelihood estimation (for which software is readily available). Nonetheless, the approach does have the advantage of tending to agree with K in the absence of the base-rate problem, and tending to produce higher values that may be more representative of actual coder agreement when the base-rate problem is present. For the example of table 10.2 (for which unweighted K was .43), the value of delta is .41.

Intercoder Correlation. K and its alternatives were designed for categorical data and are not appropriate for continuous variables, to which we now turn. Numerous statistics have been suggested to assess IRR on continuous variables as well. One of the more popular is the common Pearson correlation coefficient (r), sometimes called the intercoder correlation in this context. Pearson r is as follows:

$$r = \frac{\sum_{i=1}^{n}(X_i - \overline{X})(Y_i - \overline{Y})}{n s_X s_Y}, \qquad (10.5)$$

where (X_i, Y_i) are the n pairs of values, and s_X and s_Y are the standard deviations of the two variables computed using n in the denominator. With more than two coders, this index is generally obtained by calculating r across all pairs of coders rating the phenomenon (Jones et al. 1983). The resulting correlations can then be averaged to determine an overall value.

In our example, $\overline{X}$ and $\overline{Y}$ are 2.2 and 1.64, and s_x and s_y are .75 and .69, respectively, and n is 25. Plugging the values into the Equation 10.5, $r = 5.8 / 12.94 = .45$. In practice, the synthesist will only rarely need to calculate r by hand, as all standard statistical packages and many hand calculators compute it.

The use of Pearson r with observational data to estimate IRRs is analogous to its use in education to estimate test reliabilities when parallel test forms are available (see Stanley 1971). As such, it bears the same relationship to formal reliability theory.[10] The Pearson r also has some drawbacks. First, although it describes the degree to which the scores produced by each coder covary, it says nothing about the degree to which the scores themselves are identical. In principle, this means that coders can produce high r's without actually agreeing on any scores. Conversely, an increase in absolute agreement could actually reduce r

if the disagreement had been part of a consistent disagreement pattern (for example, one rater rating targets consistently lower than the other).[11] If the IRR estimate will be used only to adjust subsequent analyses, this may not be particularly important. If it will be used to diagnose coder consistency in interpreting instructions (for example, instructions for extracting effect sizes), it can be quite important. Second, the between-coders variance is always removed in computing the product-moment formula. Robert Ebel noted that this is especially problematic when comparisons are made among single raw scores assigned to different subjects by different coders—in research synthesis, of course, the subject is the individual study (1951).

Intraclass Correlation. Analogous to the concepts of true score and error in classical reliability theory, the intraclass correlation (r_I) is computed as a ratio of the variance of interest over the sum of the variance of interest plus error. Ebel and others suggested that r_I was preferable to Pearson r as an index of IRR with continuous variables because it permits the researcher to choose whether to include the between-raters variance in the error term (1951).

Like K, r_I is not a single index but a family of indices that permit great flexibility in matching the form of the reliability index to the reliability design used by the synthesist; here we mean design in the sense used in Cronbach's generalizability (G) theory (see Cronbach et al. 1972). Generalizability theory enables the analyst to isolate different sources of variation in the measurement, for example, forms or occasions, and to estimate their magnitude using the analysis of variance (Shavelson and Webb 1991). The reliability design as discussed here is a special case of the one-facet G study, in which coders are the facet (Shrout and Fleiss 1979).

So far, the running example has been discussed in terms of twenty-five studies being coded by two coders, but in fact at least three different reliability designs share this description.

Design 1: Each study is rated by a pair of coders, randomly selected from a larger population of coders (one-way random effects model).

Design 2: A random pair of coders is selected from a larger population, and each coder rates all twenty-five studies (two-way random effects model).

Design 3: All twenty-five studies are rated by each of the same pair of coders, who are the only coders of interest (two-way mixed effects model).

Designs 1 and 2 are common in research synthesis. Although actual random selection of coders is rare, the synthesist's usual intent is that those selected represent, at least in principle, a larger population of coders who might have been selected. Whether explicitly stated or not, synthesists typically present their substantive findings as generalizable across that population. The findings will hold little scientific interest if, say, only a particular subset of graduate students at a particular university can reproduce them. Design 3 will hold in rare instances in which the specific coders selected are the population of interest and can therefore be modeled as a fixed effect. This could happen if, for example, the coders are a specially convened panel of all known experts in the content area (for example, the Wortman and Bryant school desegregation synthesis described in section 2 might have met this criterion).

Each design requires a different form of r_I, based on a different analysis of variance (ANOVA) structure. These forms for the running example are estimated as follows (for a fuller discussion of the statistical models underlying the estimates, see Shrout and Fleiss 1979):

$$r_I(\text{design 1}) = \frac{BMS - WMS}{BMS + WMS}, \qquad (10.6)$$

$$r_I(\text{design 2}) = \frac{BMS - EMS}{BMS + EMS + 2(CMS - EMS)/25}, \qquad (10.7)$$

$$r_I(\text{design 3}) = \frac{BMS - EMS}{BMS + EMS}, \qquad (10.8)$$

where *BMS* and *WMS* are the between-studies and within-study mean squares, respectively, and *CMS* and *EMS* are the between-coders and residual (error) mean square components, resulting from the partitioning of the within-study sum of squares. Note that this partitioning is not possible in design 1; the component representing the coder effect is estimable only when the same pair of coders rated all twenty-five studies.[12]

The next step is the computation of sums of squares and mean squares from the coder data (for computational details, see any standard statistics text, for example, Hays 1994). Table 10.3 shows the various mean squares and their associated degrees of freedom for the data from table 10.1. Plugging the values into equations 10.6, 10.7, and 10.8 yields the following results:

$$r_I (\text{design 1}) = .28,$$

$$r_I (\text{design 2}) = .35,$$

$$r_I (\text{design 3}) = .44.$$

Table 10.3 Analysis of Variance for Illustrative Ratings

Source of Variance	Degrees of Freedom	Mean Squares
Between-studies (BMS)	24	.78
Within-study (EMS)	25	.44
Between-coders (CMS)	1	3.92
Residual (EMS)	24	.30

SOURCE: Authors' compilation.

On average, r_I (design 1) will give smaller values than r_I (design 2) or r_I (design 3) for the same set of data (Shrout and Fleiss 1979).

When its assumptions are met, the r_I family has been mathematically shown to provide appropriate estimates of classical reliability for the IRR case (Lord and Novick 1968). Its flexibility and linkage into G theory also argue in its favor.[13] Perhaps most important, its different variants reinforce the idea that a good IRR assessment requires the synthesist to think through the appropriate IRR design, not just calculate reliability statistics. However, because the r_I is ANOVA-based, it fails to produce coefficients that reflect the consistency of ratings when assumptions of normality are grossly violated (Selvage 1976). Furthermore, like K, it requires substantial between-items variance to show a significant indication of agreement. As with K, some writers consider r_I less useful as an index of IRR when the distributions are concentrated in a small range (for example, Jones et al. 1983). The arguments and counterarguments on this issue are essentially as described for K.

10.4.1.5 Selecting, Interpreting, and Reporting Interrater Reliability Indices With the exception of AR, the indices presented in section 10.4.1.4 have appealing psychometric properties as reliability measures and, when properly applied, can be recommended for use in evaluating coding decisions. The AR was included because of its simplicity and widespread use by research synthesis practitioners, but, for the reasons discussed earlier, it cannot be recommended as highly. These five indices by no means exhaust all those proposed in the literature for assessing IRR. Most of these, such as Finn's r (Whitehurst 1984), Yule's Y (Spitznagel and Helzer 1985), Maxwell's RE (Janes 1979), and Scott's π (Zwick 1988), were proposed to improve on perceived shortcomings of the indices presented here, in particular the r_I, and K family.

Interested readers can explore that literature and draw their own conclusions. They should keep in mind, however, that the prevailing view among statisticians is that, to date, alternatives to r_I and K are not improvements, but in fact propose misleading solutions to misunderstood problems (see, for example, Cicchetti 1985 on Whitehurst 1984; Shrout, Spitzer, and Fleiss 1987 on Spitznagel and Helzer 1985). The synthesist who sticks to the appropriate form of K for categorical variables, and the appropriate model of r_I, for continuous variables, will be on solid ground. Either AR or r can be used to supplement K and r_I for the purpose of computing multiple indices for sensitivity analysis (see section 10.4.3.3). The synthesist who chooses to rely on one of the less common alternatives (such as delta) would be wise to first become intimately familiar with its strengths and weaknesses and should probably anticipate defending the choice. The defense is likely to be viewed more favorably if the alternative approach is used to supplement one of the more usual indexes.

Once an index is selected and computed, the obvious question is how large it should be. The answer is less obvious. For r, .80 is considered adequate by many psychometricians (for example, Nunnally 1978), and at that level correlations between variables are attenuated very little by measurement error. Rules of thumb have also been suggested and used for K and r_I (Fleiss 1981; Cicchetti and Sparrow 1981).

These benchmarks, however, were suggested for evaluating IRR of psychiatric diagnoses, and their appropriateness for synthesis coding decisions is not clear. Indeed, general caution is usually appropriate about statistical rules of thumb that are not tied to a specific context; for that reason, we believe it inappropriate to provide them here (for a similar take on the inappropriateness of context-independent rules of thumb, see chapter 7, this volume).

The issue of how large is large enough is further complicated by distributional variation. The running example showed how the different indices varied for a particular data set, but not how they vary over different conditions. Allan Jones and his colleagues systematically compared the results obtained when these indices are applied to a common set of ratings under various distributional assumptions (1983). Four raters rated job descriptions using a questionnaire designed and widely used (and validated) for that purpose. The majority of ratings were on six-point Likert-type scales with specific anchors for each point; a minority were dichotomous. The K, K_w, AR, r (in

Table 10.4 Estimates of Interrater Agreement for Different Types of Data Distributions

Distributional Conditions	Kappa	Weighted Kappa	Agreement Rate	Average Correlation	Intraclass Correlation
Variations in ratings across jobs and high agreement among raters	.43	.45	.88	.79	.74
Variations in ratings across jobs and low agreement among raters	.01	.04	.16	.13	.05
Little variation in ratings across jobs and high agreement among raters	.04	.04	.77	−.01	−.03

SOURCE: Jones et al. 1983.

the form of average pairwise correlations), and r_1 were computed on each item. Aggregated across items, K and K_w yielded the lowest estimates of IRR, producing median values of .19 and .22, respectively. The AR yielded the highest, with a median value of .63; r and r_1 occupied an intermediate position, with median values of .51 and .39, respectively. For the reasons noted earlier, an overall aggregate agreement index is not particularly meaningful for evaluating coding decisions, but is still useful for illustrating the wide variation among indices.

To examine the effect of the distributions of individual items on the variation across indices, sample items representing three conditions were analyzed: high variation across ratings and high agreement across raters, high variation across ratings and low agreement across raters, and low variation across ratings and high agreement across raters. As shown in table 10.4, the effects can be substantial, in particular under the third condition, where AR registered 77 percent agreement, but the other indices suggested no agreement beyond chance. With moderate to high variance, it makes far less difference which index is used. Therefore, Jones and his colleagues argued, a proper interpretation of the different indices—including whether they are large enough—requires an understanding of the actual distributions of the data (1983).

It should be evident by now that the indices are not directly comparable, even though all but AR have the same range of possible values. It is therefore essential that synthesists report not only their IRR values, but the indices used to compute them. To give the reader the full picture, it would also be wise to include information about the raters' base rates, as William Grove and his colleagues

suggested, particularly when they are very low or very high (1981).

10.4.1.6 Assessing Coder Drift Whatever index of IRR is chosen, the synthesist cannot assume that IRR will remain stable throughout the coding period, particularly when the number of studies to code is large. As Gregg Jackson noted, coder instability arises because many hours are required, often over a period of months, to code a set of studies (1980). When the coding is lengthy, IRR may change over time, and a single assessment may not be adequate.

Orwin and Cordray assessed coder drift in their reanalysis of the 1980 Smith, Glass, and Miller psychotherapy synthesis (1985). Before coding, the twenty-five reports in the main sample were randomly ordered, with coders instructed to adhere to the resulting sequence. It was assumed, given the random ordering, that any trend in IRRs over reports would be attributable to changes in coders over time (for example, practice or boredom).[14] Equating for order permits this trend to be detected and, if desired, removed.

Following the completion of coding, a per case agreement rate was computed, consisting of the number of variables agreed on divided by the number of variables coded.[15] In itself, per case agreement rate is not particularly meaningful because it weights observations on all variables equally and is sensitive to the particular subset of variables (considered here as sampled from a larger domain) constituting the coding form. Change in this indicator over time is meaningful, however, because it speaks to the stability of agreement. Per case agreement rate was plotted against case sequence, and little if any trend was observed. The absence of significant correlation between the two variables substantiated this conclusion. In regression terms, increasing the case sequence by one resulted in a decrease of the agreement rate by less than one-tenth of 1 percent ($b = -.09$). From this exercise it was concluded that further consideration of detrending IRRs was unnecessary in our synthesis. We do not presume, though, to generalize this finding to other syntheses. Coder drift will be more of a concern in some syntheses than in others and needs to be assessed on a case-by-case basis.

The drift assessment could also be integrated into the main IRR assessment with a G theory approach. The one-facet G study with coders as the facet could be expanded to a two-facet G study with time (if measured continuously) or occasions (if measured discretely) as the second facet. This would permit the computation of a single IRR coefficient that captured the variance contributed by each facet, as well as by their interaction.

10.4.2 Confidence Ratings

10.4.2.1 Rationale As a strategy for evaluating coding decisions, IRR is indirect. Directly, it assesses only coder disagreement, which in turn is an indicator of coder uncertainty, but a flawed one. It is desirable, therefore, to seek more direct methods of assessing the accuracy of the numbers gleaned from reports. Tagging each number with a confidence rating is one such method. It is based on the premise that questionable information should not be discarded, and should not be allowed to freely mingle with less questionable information, as is usually done. The former procedure wastes information, which, though flawed, may be the best available on a given variable, whereas the latter injects noise at best and bias at worst into a system already beset by both problems. With confidence ratings, questionable information can be described as such, and both data point and descriptor entered into the database.

In other contexts, confidence ratings and facsimiles (for example, adequacy ratings, certainty ratings) have been around for some time. For example, in their international comparative political parties project, Kenneth Janda constructed an adequacy-confidence scale to evaluate data quality (1970). Similarly, it is not unusual to find interviewer confidence ratings embedded in questionnaires. In the *Addiction Severity Index*, for example, interviewers provided confidence ratings of the accuracy of patient responses (McLellan et al. 1988). Specifically,

they coded whether in their view the information was significantly distorted by the patient's misrepresentation or the patient's inability to understand.

Two early meta-analyses, one that integrates the literature on the effects of psychotherapy (Smith and Glass 1977; Smith, Glass, and Miller 1980) and another devoted to the influence of class size on achievement (Glass and Smith 1979), are used to illustrate how some of the preceding issues can be assessed. For both, the data files were obtained and subjected to reanalysis (Cordray and Orwin 1981; Orwin and Cordray 1985). In examining the primary studies, these authors were careful to note (for some variables) how the data were obtained. For example, in the class size study, the actual number of students enrolled in each class was not always reported directly, so the values recorded in the synthesis were not always based on equally accurate information. In response to this, Glass and Smith included a second variable that scored the perceived accuracy of the numbers recorded. Unfortunately, the accuracy scales were never used in the analysis (Gene Glass, personal communication, August 1981). Similarly, in the psychotherapy study, source of IQ and confidence of treatment classifications were also coded and not used.

The synthesis of class size and achievement compared smaller class and larger class on achievement (Glass and Smith 1979). After assigning each comparison an effect size, the authors regressed effect size on three variables: the size of the smaller class (S), the size of the smaller class squared (S^2), and the difference between the larger class and smaller class ($L - S$). This procedure was done for the total database and for several subdivisions of the data, one of which was well controlled versus poorly controlled studies. Cordray and Orwin reran the regressions as originally reported, except that a dummy variable for overall accuracy was added to the equations (1981). The results are presented in table 10.5. Most remarkable is that accuracy made its only nontrivial contribution in well-controlled studies. A possible explanation is that differential accuracy in the reporting of class sizes is only one of many method factors operating in poorly controlled studies (and in the entire sample, as well-controlled studies are a minority) and that these mask the influence of accuracy by interacting with it in unknown ways or by inflating error variance. In any event, the considerable influence of accuracy in well-controlled studies suggested that it does matter, at least under some circumstances.

In the synthesis of psychotherapy outcomes described earlier, an identical simultaneous multiple regression

Table 10.5 Comparison of R^2 for Original Glass and Smith (1979) Class Size Regression (R_1^2) and Original with Accuracy Added (R_2^2)

	R_1^2	R_2^2	$R_2^2 - R_1^2$
Total sample ($n = 699$)	.1799	.1845	.0046, $F_{(1,694)} = 3.94^*$
Well-controlled studies ($n = 110$)	.3797	.4273	.0476, $F_{(1,105)} = 8.73^*$
Poorly controlled studies ($n = 338$)	.0363	.0369	.0006, $F_{(1,333)} = 0.65$

SOURCE: Cordray and Orwin 1981.
$^*p < .05$

analysis was run within each treatment class and subclass, but with a different orientation. Rather than specifying the treatment in the model and crosscutting by nontreatment characteristics, as with class size, the nontreatment characteristics (diagnosis, method of client solicitation, and so on) were included in the model. Orwin and Cordray recoded twenty variables and found that across items, high confidence proportions ranged from 1.00 for comparison type and location down to .09 for therapist experience (1985; for selection criteria, see Orwin 1983b). Across studies, the proportion of variables coded with certainty or near certainty ranged from 52 to 83 percent. At the opposite pole, the proportion of guesses ranged from 25 to 0 percent.

The effect of confidence on reliabilities was dramatic; for example, the mean AR for high-confidence observations more than doubled that for low-confidence observations (.92 vs. .44). As to the effect of confidence on relationships between variables, 82 percent of the correlations increased in absolute value when observations rated with low and medium confidence were removed. Moreover, all correlations with effect size increased. The special importance of these should be self-evident; the relationships between study characteristics and effect size were a major focus of the original study (and are in research synthesis generally). The absolute values of correlations increased by an average of .15 (.21 − .36), or in relative terms, 71 percent.

Questions posed by these findings include how credible the lack of observed relationship between therapist experience and effect size can be when therapist experience is extracted from only one out of ten reports with confidence. Even for variables averaging considerably higher confidence ratings, the depression of IRRs and consequent attenuation of observed relationships between variables is evident. There are plausible alternative explanations for

individual variables. For example, the outcome measures that are easiest to classify could also be the most reliable. In such circumstance, the outcome type-effect size relationship should be stronger for high confidence observations simply because the less reliable outcome types have been weeded out. An explanation like this can be plausible for individual variables, but cannot account for the generally consistent and nontrivial strengthening of observed relationships throughout the system. Each study in the sample has its own pattern of well-reported and poorly reported variables; there was not a particular subset of studies, presumably different from the rest, that chronically fell out when high confidence observations were isolated. Consequently, any alternative explanation must be individually tailored to a particular relationship. Postulating a separate ad hoc explanation for each relationship would be unparsimonious to say the least.

10.4.2.2 Empirical Distinctness from Reliability It was argued previously that confidence judgments provide a more direct method of evaluating coding decisions than does IRR. It is therefore useful to ask whether this conceptual distinctness is supported by an empirical distinctness. If the information provided by confidence judgments simply duplicated that provided by reliabilities, the case for adopting both strategies in the same synthesis would be weakened.

Table 10.6 presents agreement rates by level of confidence for the complete set of variables ($K = 25$) to which confidence judgments were applied. The table indicates that interrater agreement is neither guaranteed by high confidence nor precluded by low confidence. Yet it also shows that confidence and agreement are associated. Whereas table 10.6 shows the nonduplicativeness of confidence and agreement on a per variable basis, table 10.7 shows this nonduplicativeness across variables. To enable this, variables were first rank ordered by the proportion of

Table 10.6 Agreement Rate by Level of Confidence

	Low	Medium	High
Experimenter affiliation	1.00 (14)	1.00 (37)	.95 (75)
Blinding	.83 (6)	.91 (66)	.93 (44)
Diagnosis	—	1.00 (12)	.99 (114)
Client IQ	1.00 (11)	.18 (74)	1.00 (41)
Client age	1.00 (1)	.68 (50)	.83 (75)
Client source	—	1.00 (10)	.89 (116)
Client assessment	—	.53 (17)	.98 (103)
Therapist assessment	.00 (15)	.67 (18)	.75 (93)
Internal validity	—	.52 (27)	.88 (93)
Treatment mortality	.00 (4)	1.00 (1)	.94 (121)
Comparison mortality	.00 (4)	1.00 (1)	.93 (121)
Comparison type	—	—	1.00 (126)
Control group type	—	.00 (9)	.69 (114)
Experimenter allegiance	.10 (10)	.64 (36)	1.00 (80)
Modality	—	.33 (4)	1.00 (122)
Location	—	—	1.00 (126)
Therapist experience	.55 (55)	.56 (57)	1.00 (11)
Outcome type	.00 (1)	.00 (11)	.92 (114)
Follow-up	.00 (2)	.92 (47)	.83 (75)
Reactivity	.83 (6)	.28 (65)	.94 (47)
Client participation	1.00 (1)	.67 (3)	1.00 (122)
Setting type	—	.40 (15)	.99 (111)
Treatment integrity	.60 (10)	.83 (60)	1.00 (56)
Comparison group contamination	.30 (37)	.32 (50)	1.00 (39)
Outcome Rxx	.13 (150)	.56 (70)	.92 (38)

SOURCE: Orwin 1983b.
NOTE: Selected Variables from the Smith, Glass, and Miller (1980) Psychotherapy Meta-Analysis ($n = 126$) (sample sizes in parenthesis)

observations in which confidence was judged as high. As is clear in the table, the correlation between the two sets of rankings fell far short of perfect, regardless of the reliability estimate selected. In sum, the conceptual distinctness between reliability and confidence is supported by an empirical distinctness.

10.4.2.3 Methods for Assessing Coder Confidence There is no set method for assessing confidence. Orwin and Cordray used a single confidence item for each variable (1985). This is not to imply that confidence is unidimensional; no doubt it is a complex composite of numerous factors. These factors can interact in various ways to affect the coder's ultimate confidence judgment, but Orwin and Cordray did not attempt to spell out rules for handling the many contingencies that arise. Confidence judgments reflected the overall pattern of information as deemed appropriate by the coder.

Alternative schemes are of course possible. One might use two confidence judgments per data point—one rating confidence in the accuracy or completeness of the information as reported and the other rating confidence in the coding interpretation applied to that information. The use of two or more such ratings explicitly recognizes multiple sources of error in the coding process (as described in section 10.2) and makes some attempt to isolate them. More involved schemes, such as Janda's scheme might also be attempted, particularly if information is being sought from multiple sources (1970). As noted earlier, Janda constructed an adequacy-confidence scale to evaluate data quality. The scale was designed to reflect four factors considered important in determining the researchers' belief in the accuracy of the coded variable values: the number of sources providing relevant information for the coding decision, the proportion of agreement to disagreement in the

Table 10.7 Spearman Rank Order Correlations

	r_{RHO}
All Variables ($K = 25$)	
Agreement rate	.71
Variables for Which Kappa was Computed[a] ($K = 20$)	
Agreement rate	.71
Kappa	.62
Variables for Which Intercoder Correlation was Computed ($K = 15$)	
Agreement rate	.81
Intercoder rate	.67
Variables for Which All Three Estimates Were Computed ($K = 10$)	
Agreement rate	.73
Kappa	.79
Intercoder correlation	.66

SOURCE: Authors' compilation.

NOTE: Between Confidence and Interrater Agreement for Selected Variables from the Smith, Glass, and Miller (1980) Psychotherapy Meta-Analysis

information reported by different sources, the degree of discrepancy among sources when disagreement exists, and the credibility attached to the various sources of information. An adequacy-confidence value was then assigned to each recorded variable value.

Along with the question of what specific indicators of confidence to use is the question of how to scale them. Orwin and Cordray pilot tested a five-point scale (1 = low, . . . , 5 = high) for each confidence rating, fashioned after the Smith, Glass, and Miller confidence of treatment classification variable (1985). Analysis of the pilot results revealed that five levels of confidence were not being discriminated. Specifically, the choices of 2 versus 3 and 3 versus 4 seemed to be made arbitrarily. The five categories were then collapsed into three for the main study. In addition, each category was labeled with a verbal descriptor to concretize it and minimize coder drift: 3 = certain or almost certain, 2 = more likely than not, 1 = guess (see Kazdin 1977). Discrepancies were resolved through discussion. The three-point confidence scale was simple yet adequate for its intended purpose—to establish a mechanism for discerning high-quality from lesser-quality information in the conduct of subsequent analyses.

10.4.3 Sensitivity Analysis

10.4.3.1 Rationale Sensitivity analysis can assess robustness and bound uncertainty. It has been a part of research synthesis at least since the original Smith and Glass psychotherapy study (1977). Glass's position of including methodologically flawed studies was attacked by his critics as implicitly advocating the abandonment of critical judgment (for example, Eyesenck 1978). He rebutted these and related charges on multiple grounds, but the most enduring was the argument that meta-analysis does not ignore methodological quality, but rather presents a way of determining empirically whether particular methodological threats systematically influence outcomes. Glass's indicators of quality (for example, the three-point internal validity scale) were crude then and appear even cruder in retrospect, and may not have been at all successful in what they were attempting to do.[16] The principle, however—that of empirically assessing covariance of quality and research findings rather than assuming it a priori—was perfectly sensible and consistent with norms of scientific inquiry. In essence, the question was whether research findings were sensitive to variations in methodological quality. If not, lesser-quality studies can be analyzed along with high-quality studies, with consequent increase in power, generalizability, and so on. The worst-case scenario—that lesser-quality studies produce systematically different results and cannot be used—is no worse than had the synthesist followed the advice of the critics and excluded those studies a priori.

The solutions of Rosenthal and Orwin to the so-called file-drawer problem (that is, publication bias), as well as more recent approaches are in essence types of sensitivity analysis as well (Rosenthal 1978; Orwin 1983a; see chapter 23, this volume; Rothstein, Sutton, and Borenstein 2005). They test the data's sensitivity to the violation of the assumption that the obtained sample of studies is an unbiased sample of the target population. While some level of sensitivity analysis has always been part of research synthesis, much more could have been done. In their synthesis of eighty-six meta-analyses of randomized clinical trials, Harold Sacks and his colleagues found that only fourteen presented any sensitivity analyses, defined as "data showing how the results vary through the use of different assumptions, tests, and criteria" (1987, 452). In the corpus of nineteen meta-analyses published in *Psychological Bulletin* in 2006, fewer than half (eight) discussed sensitivity analyses related to publication bias.

The general issue of sensitivity analysis in research synthesis is discussed in chapter 22, this volume. This chapter focuses on applying the logic of sensitivity analysis to the evaluation of coding decisions.

10.4.3.2 Multiple Ratings of Ambiguous Items The sources of error identified earlier (for example, deficient reporting, ambiguities of judgment) are not randomly dispersed across all variables. The synthesist frequently knows at the outset which variables will be problematic. If not, a well-designed pilot test will identify them. For those variables, multiple ratings should be considered. Like multiple measures in other contexts, multiple ratings help guard against biases stemming from a particular way of assessing the phenomenon (monomethod bias) and can set up sensitivity analyses to determine if the choice of rating makes any difference. Mary Smith's synthesis of sex bias is an example (1980). When an investigator reported only their significant effect, Smith entered an effect size of 0 for the remaining effects. Aware of the pitfalls of this, she alternately used a different procedure and deleted these cases. Neither changed the overall mean effect size by more than a small fraction. In their synthesis of sex differences, Alice Eagly and Linda Carli took the problem of unreported nonsignificant effects a step further (1981). Noting that the majority of these (fifteen of sixteen) tended in the positive (female) direction, they reasoned that setting effect size at 0 would lead to an underestimation of the true mean effect size, whereas deleting them would lead to overestimation. They therefore did both to be confident of at least bracketing the true value. They also used two indicators of researcher's sex (an important variable in this area), overall percentage of male authors and sex of first author. These were found to differ only slightly in their correlations with outcome. Each of these examples illustrates sensitivity analysis directed at specific rival hypotheses. It should be evident that many opportunities exist for thoughtful sensitivity analyses in the evaluation of coding decisions and that these do not require specialized technical expertise; conscientiousness and common sense will frequently suffice.

10.4.3.3 Multiple Measures of Interrater Agreement As described earlier, the choice of IRR index can have a significant effect on the reliability estimate obtained. Although the guidelines in section 10.4.1.5 should narrow the range of choices, they may not uniquely identify the best one. Computing multiple indices is therefore warranted.

In their reanalysis of the Smith, Glass, and Miller psychotherapy data (1980), Orwin and Cordray computed multiple estimates of IRR for each variable (1985). For continuous variables, AR and r were computed (r_I was not computed, but should have been). For ordinal categorical variables, AR, r, and K_w were computed. For nominal categorical variables, AR and K were computed; when nominal variables were dichotomous, r (in the form of phi) was also computed. Four regressions were then run, each using a different set of reliability estimates. The first used the highest estimate available for each variable. Its purpose was to provide a lower bound on the amount of change produced by disattenuation. The second and third runs were successively more liberal (for details, see Orwin and Cordray 1985). A final run, intended as the most liberal credible analysis, disattenuated the criterion variable (effect size) as well as the predictors. The reliability estimates for the four runs are shown in table 10.8.

10.4.3.4 Isolating Questionable Variables Both IRRs and confidence ratings are useful tools for flagging variables that may be inappropriate for use as meta-analytic predictors of effect size. In the Orwin and Cordray study, for example, the therapist experience variable was coded with high confidence in only 9 percent of the studies, was "guessed" in 45 percent, and had an IRR (using Pearson r) of .56 (1985). Such numbers would clearly suggest the exercise of caution in further use of that variable. Conducting analyses with and without it, and comparing the results, would be a logical first step.

With regard to both cases and variables, the finding of significant differences between inclusion and exclusion does not automatically argue for outright exclusion; rather, it alerts the analyst that mindless inclusion is not warranted. As in primary research, the common practices of dropping cases, dropping variables, and working from a missing-data correlation matrix result in loss of information, loss of statistical power and precision, and biased estimates when, as is frequently the case, the occurrence of missing data is nonrandom (Cohen and Cohen 1975). Therefore, more sophisticated approaches are preferable when feasible (for more on the treatment of missing data in research synthesis, see chapter 21, this volume).

10.5 SUGGESTIONS FOR FURTHER RESEARCH

10.5.1 Moving from Observation to Explanation

As Matt concluded, synthesists need to become more aware, and take deliberate account, of the multiple sources of uncertainty in reading, understanding, and interpreting research reports (1989). But, in keeping with the scientific

Table 10.8 Reliability Estimates

Variable	Run 1	Run 2	Run 3	Run 4
Diagnosis: neurotic, phobic, or depressive	.98	.98	.89	.89
Diagnosis: delinquent, felon, or habituée	1.00	1.00	1.00	1.00
Diagnosis: psychotic	1.00	1.00	1.00	1.00
Clients self-presented	.97	.57	.71	.71
Clients solicited	.93	.86	.81	.81
Individual therapy	1.00	1.00	.85	.85
Group therapy	.98	.96	.94	.94
Client IQ	.69	.69	.60	.60
Client age[a]	.99	.99	.91	.91
Therapist experience × neurotic diagnosis	.76	.75	.70	.70
Therapist experience × delinquent diagnosis	1.00	1.00	1.00	1.00
Internal validity	.76	.71	.42	.42
Follow-up time[b]	.99	.99	.95	.95
Outcome type[c]	.87	.70	.76	.76
Reactivity[d]	.57	.56	.57	.57
ES	1.00	1.00	1.00	.78

SOURCE: Orwin and Cordray 1985.
NOTES: Reliability-Corrected Regression Runs on the Smith, Glass, and Miller (1980) Psychotherapy Data
[a]Transformed age $= (\text{age} - 25)(|\text{age} - 25|)^{1/2}$.
[b]Transformed follow-up $= (\text{follow-up})^{1/2}$.
[c]"Other" category removed for purpose of dichotomization.
[d]Transformed reactivity $= (\text{reactivity})^{2.25}$.

endeavor principle, they also need to begin to move beyond observation of these phenomena and toward explanation. For example, further research on evaluating coding decisions in research synthesis should look beyond how to assess IRR toward better classification and understanding of why coders disagree. Evidence from other fields could probably inform some initial hypotheses. Horwitz and Yu conducted detailed analyses of errors in the extraction of epidemiological data from patient records and suggested a rough taxonomy of six error categories (1984). Of the six, four were errors in the actual data extraction, and two were errors in the interpretation of the extracted data. The two interpretation errors were incorrect interpretation of criteria and correct interpretation of criteria, but inconsistent application. Note the similarity between these types of errors and the types of judgment errors described in section 10.2 with regard to coding in research synthesis.

When the coder must make judgment calls, the rationale for each judgment should be documented. In this way, we can begin to formulate a general theory of the coder judgment process. For complex yet critical variables like effect size, interim codes should be recorded. One code might simply indicate what was reported in the original study. A second code might indicate what decision rule was invoked in further interpretation of the data point. A third code—representing the endpoint and the variable of interest to the synthesis (effect size, for example)—would be the number that results from the coder's application of that interpretation. Coder disagreement on the third code would no longer be a black box, but could be readily traced to its source. Coder disagreement could then be broken down to its component parts (that is, the errors could be partitioned by source), facilitating diagnosis much like a failure analysis in engineering or computer programming. The detailed documentation of coding decision rules could be made for examination and reanalysis by other researchers (much like the list of included studies is currently), including any estimation procedures used for missing or incomplete data (for an example, see Bullock and Svyantek 1985). Theoretical models could also be tested and, most important, more informed strategies for reducing and controlling for error rates would begin to emerge.

10.5.2 Assessing Time and Necessity

The coding stage of research synthesis is time-consuming and tedious, and in many instances additional efforts to evaluate coding decisions further lengthen and complicate it. The time and resources of practicing synthesists is limited, and they would likely want answers to two questions before selecting a given strategy: How much time will it take? How necessary is it?

The time issue has not been systematically studied. Orwin and Cordray included a small time-of-task side study and concluded that the marginal cost in time of augmenting the task as they did (double codings plus confidence ratings) was quite small given the number of additions involved and the distinct impact they had on the analytic results (1985). In part, this is because though double coding a study approximately doubles the time required to code it, in an actual synthesis only a sample of studies may be double coded. In a synthesis the size of Smith, Glass, and Miller's, the double coding of twenty-five representative studies would increase the total number of studies coded by only 5 percent. When a particular approach to reliability assessment leads to low occurrence rates or restricted range on key variables—as can happen with Yeaton and Wortman's hierarchical approach (1991)—oversampling techniques can be used. Research that estimated how often oversampling was required would in itself be useful. Other possibilities are easy to envision. In short, this is an area of considerable practical importance that has not been seriously examined to date.

The question of necessity could be studied the same way that methodological artifacts of primary research have been studied. For example, a handful of syntheses have used coder masking techniques, as described in section 10.2. Keeping coders unaware of hypotheses or other information is an additional complication that adds to the cost of the effort and may have unintended side effects. An empirical test of the effect of selective masking on research synthesis coding would be a useful contribution, as such studies have been in primary research, to determine whether the particular artifact that masking is intended to guard against (coder bias) is real or imaginary, large or trivial, and so forth. As was the case with pretest sensitization and Hawthorne effects in the evaluation literature (Cook and Campbell 1979), research may show that some suspected artifacts in synthesis coding are only that—perfectly plausible yet not much in evidence when subjected to empirical scrutiny. In reality, few artifacts will be universally present or absent, but rather will interact with the topic area (for example, coder gender is more likely to matter when the topic is sex differences than when it is acid rain) or with other factors. Research could shed more light on these interactions as well. In sum, additional efforts to assess both the time and need for specific methods of evaluating coding decisions could significantly help the practicing synthesist make cost-effective choices in the synthesis design.

10.6 NOTES

1. It should be noted that significant reporting deficiencies, and their deleterious effect on research synthesis, are not confined to social research areas (see, for example, synthesis work on the effectiveness of coronary artery bypass surgery in Wortman and Yeaton 1985).

2. For example, Oliver (1987) reports on a colleague who attempted to apply a meta-analytic approach to research on senior leadership, but gave up after finding that only 10 percent of the eligible studies contained sufficient data to sustain the synthesis.

3. That detailed criteria for decisions fail to eliminate coder disagreement should not be surprising, even if all known contingencies are incorporated (for an analog from coding patient records in epidemiology, see Horwitz and Yu 1984).

4. Specifically, this rule aims to exclude effect sizes based on redundant outcome measures. A measure was judged to be redundant if it matched another in outcome type, reactivity, follow-up time, *and* magnitude of effect.

5. The Chalmers et al. (1987) review was actually a review of replicate meta-analyses rather than a meta-analysis per se (that is, the individual studies were meta-analyses), but the principle is equally applicable to meta-analysis proper.

6. Here, and elsewhere in the chapter, we rely on meta-analytic work in the most recent complete year of *Psychological Bulletin* as a barometer of current practice.

7. Such improvement is quite plausible. For example, Orwin (1983b) found that measures of reporting quality were positively correlated with publication date in the Smith, Glass, and Miller (1980) psychotherapy data.

8. The rationale for using multiple estimates is discussed in section 10.4.3.3.

9. There is nothing anomalous about the present coders' relative lack of agreement on internal validity; lack of

consensus on ratings of research quality is common-place (for an example within meta-analysis, see Stock et al. 1982).

10. This desirable characteristic has led some writers (for example, Hartmann 1977) to advocate extending the use of r, in the form of the phi coefficient (ϕ), to reliability estimation of dichotomous items: $\phi = (BC - AD)/((A + B)(C + D)(A + C)(B + D))^{1/2}$, where $A, B, C,$ and D represent the frequencies in the first through fourth quadrants of the resulting 2×2 table.

11. This phenomenon is easily demonstrated with the illustrative data; the details are left to the reader as an exercise.

12. The *general* forms of the equations are

$$r_1(\text{design 1}) = \frac{BMS - WMS}{BMS + (k-1)WMS},$$

$$r_1(\text{design 1}) = \frac{BMS - EMS}{BMS + (k-1)EMS + k(CMS - EMS)/n},$$

$$r_1(\text{design 1}) = \frac{BMS - EMS}{BMS + (k-1)EMS},$$

where k is the number of coders rating each study and n is the number of studies.

13. r_1 is closely related to the coefficient of generalizability and index of dependability used in G theory (Shavelson and Webb 1991).

14. Training and piloting were completed before the rater drift assessment began, to avoid picking up obvious training effects.

15. The term *case* is used rather than *study* because a study can have multiple cases (one for each effect size). The term *observation* refers to the value assigned to a particular variable within a case.

16. Chapter 10 shows how the state of the art has evolved (see also U.S. General Accounting Office 1989, appendix II).

10.7 REFERENCES

American Psychiatric Association, Committee on Nomenclature and Statistics. 1980. *Diagnostic and Statistical Manual of Mental Disorders*, 3rd ed. Washington, D.C.: American Psychiatric Association.

Andrés, A.Martín, and P. Fernina Marzo. 2004. "Delta: A New Measure of Agreement Between Two Raters." *British Journal of Mathematical and Statistical Psychology* 57(1): 1–19.

Begg, Colin., Mildred Cho, Susan Eastwood, Richard Horton, David Moher, Ingram Olkin, Roy Pitkin, Drummond Rennie, Kennteh F. Schulz, David Simel, and Donna F. Stroup. 1996. "Improving the Quality of Reporting of Randomised Controlled Trials. The CONSORT Statement." *Journal of the American Medical Association* 276(8): 637–39.

Bullock, R. J., and Daniel J. Svyantek. 1985. "Analyzing Meta-Analysis: Potential Problems, an Unsuccessful Replication, and Evaluation Criteria." *Journal of Applied Psychology* 70(1): 108–15.

Buros, Oscar K. 1978. *Mental Measurements Yearbook*, 8th ed. Highland Park, N.J.: Gryphon Press. Available at http://www.unl.edu/buros/bimm/html/catalog.html.

Burton, Nancy W. 1981. "Estimating Scorer Agreement for Nominal Categorization Systems." *Educational and Psychological Measurement* 41(4): 953–61.

Carey, Gregory, and Irving I. Gottesman. 1978. "Reliability and Validity in Binary Ratings: Areas of Common Misunderstanding in Diagnosis and Symptom Ratings." *Archives of General Psychiatry* 35(12): 1454–59.

Chalmers, Thomas C., Jayne Berrier, Henry S. Sacks, Howard Levin, Dinah Reitman, and Raguraman Nagalingham. 1987. "Meta-Analysis of Clinical Trials as a Scientific Discipline, II. Replicate Variability and Comparison of Studies that Agree and Disagree." *Statistics in Medicine* 6(7): 733–44.

Cicchetti, Domenic V. 1985. "A Critique of Whitehurst's 'Interrater Agreement for Journal Manuscript Reviews': De Omnibus Disputandem Est." *American Psychologist* 40(5): 563–68.

Cicchetti, Dominic V., and Sara S. Sparrow. 1981. "Developing Criteria for Establishing the Interrater Reliability of Specific Items in a Given Inventory." *American Journal of Mental Deficiency* 86: 127–37.

Cohen, Jacob. 1960. "A Coefficient of Agreement for Nominal Scales." *Educational and Psychological Measurement* 20(1): 37–46.

Cohen, Jacob. 1968. "Weighted Kappa: Nominal Scale Agreement with Provision for Scaled Disagreement or Partial Credit." *Psychological Bulletin* 70(4): 213–20.

Cohen, Jacob, and Patricia Cohen. 1975. *Applied Multiple Regression/Correlation Analysis for the Behavioral Sciences.* Hillsdale, N.J.: Lawrence Erlbaum.

Cook, Thomas D., and Donald T. Campbell. 1979. *Quasi-Experimentation: Design and Analysis Issues for Field Settings.* Chicago: Rand McNally.

Cordray, David S., and Robert G. Orwin. 1981. "Technical Evidence Necessary for Quantitative Integration of Research and Evaluation." Paper presented at the joint conference of the International Association of Social Science Information

Services and Technology and the International Federation of Data Organizations. Grenoble, France (September 1981).

Cordray, David S., and L. Joseph Sonnefeld. 1985. "Quantitative synthesis: An actuarial base for planning impact evaluations." In *Utilizing Prior Research in Evaluation Planning*, edited by David S. Cordray. New Directions for Program Evaluation No. 27. San Francisco: Jossey-Bass.

Cronbach, Lee J., G. C. Gleser, H. Nanda, and N. Rajaratnam. 1972. *Dependability of Behavioral Measurements*. New York: John Wiley & Sons.

Eagly, Alice H., and Carli, Linda L. 1981. "Sex of Researchers and Sex-Typed Communications as Determinants of Sex Differences in Influenceability: A Meta-Analysis of Social Influence Studies." *Psychological Bulletin* 90(1): 1–20.

Ebel, Robert L. 1951. "Estimation of the Reliability of Ratings." *Psychometrika* 16(4): 407–24.

Eyesenck, Hans. J. 1978. "An Exercise in Mega-Silliness." *American Psychologist* 33(5): 517.

Fleiss, Jospeh L. 1971. "Measuring Nominal Scale Agreement Among Many Raters." *Psychological Bulletin* 76(5): 378–82.

Fleiss, Joseph L. 1981. *Statistical Methods for Rates and Proportions*, 2nd ed. New York: John Wiley & Sons.

Fleiss, Joseph L., and Jacob Cohen. 1973. "The Equivalence of Weighted Kappa and the Intraclass Correlation Coefficient as Measures of Reliability." *Educational and Psychological Measurement* 33(3): 613–19.

Fleiss, Joseph L., Jacob Cohen, and B. S. Everitt. 1969. "Large Sample Standard Errors of Kappa and Weighted Kappa." *Psychological Bulletin* 72(5): 323–27.

Glass, Gene V., and Mary L. Smith. 1979. "Meta-Analysis of Research on the Relationship of class Size and Achievement." *Educational Evaluation and Policy Analysis* 1(1): 2–16.

Green, Bert F., and Judith A. Hall. 1984. "Quantitative Methods for Literature Reviews." *Annual Review of Psychology* 35: 37–53.

Grove, William M., Nancy C. Andreasen, Patricia McDonald-Scott, Martin B. Keller, and Robert W. Shapiro 1981. "Reliability Studies of Psychiatric Diagnosis: Theory and Practice." *Archives of General Psychiatry* 38(4): 408–13.

Hartmann, Donald P. 1977. "Considerations in the Choice of Interobserver Reliability Estimates." *Journal of Applied Behavior Analysis* 10(1): 103–16.

Hays, William L. 1994. *Statistics*, 5th ed. Fort Worth, Tex.: Harcourt Brace College.

Hedges, Larry V., and Ingram Olkin. 1985. *Statistical Methods for Meta-Analysis*. Orlando, Fl.: Academic Press.

Higgins, Julian P.T., and Sally Green, eds. 2005. *Cochrane Handbook for Systematic Reviews of Interventions*, ver. 4.2.5. Chichester: John Wiley & Sons. http://www.cochrane.org/resources/handbook/hbook.html.

Horwitz, Ralph I., and Eunice C. Yu 1984. "Assessing the Reliability of Epidemiological Data Obtained from Medical Records." *Journal of Chronic Disease* 37(11): 825–31.

Hyde, Janet S. 1981. "How Large Are Cognitive Gender Differences? A Meta-Analysis Using W and D." *American Psychologist* 36(8): 892–901.

Jackson, Gregg B. 1980. "Methods for Integrative Reviews." *Review of Educational Research* 50(3): 438–60.

Janda, Kenneth 1970. "Data Quality Control and Library Research on Political Parties." In *A Handbook of Method in Cultural Anthropology*, edited by R. Narall and R. Cohen. New York: Doubleday.

Janes, Cynthia L. 1979. "Agreement Measurement and the Judgment Process." *Journal of Nervous and Mental Disorders* 167(6): 343–47.

Jones, Allan P., Lee A. Johnson, Mark C. Butler, and Deborah S. Main 1983. "Apples and Oranges: An Empirical Comparison of Commonly Used Indices of Interrater Agreement." *Academy of Management Journal* 26(3): 507–19.

Journal Article Reporting Standards Working Group. 2007. "Reporting Standards for Research in Psychology: Why Do WE Need Them? What Might They Be?" Report to the American Psychological Association Publications & Communications Board. Washington, D.C.: American Psychological Association.

Kazdin, Alan E. 1977. "Artifacts, Bias, and Complexity of Assessment: The ABC's of research." *Journal of Applied Behavior Analysis* 10(1): 141–50.

Kerlinger, Fred N., and Elazar E. Pedhazur. 1973. *Multiple Regression in Behavioral Research*. New York: Holt, Rinehart and Winston.

Kulik, James A., Chen-lin C. Kulik, and Peter A. Cohen. 1979. "Meta-Analysis of Outcome Studies of Keller's Personalized System of Instruction." *American Psychologist* 34(4): 307–18.

Light, Richard J. 1971. "Measures of Response Agreement for Qualitative Data: Some Generalizations and Alternatives." *Psychological Bulletin* 76: 365–77.

Light, Richard J., and David B. Pillemer. 1984. *Summing Up: The Science of Reviewing Research*. Cambridge, Mass.: Harvard University Press.

Lord, Frederic M., and Melvin R. Novick. 1968. *Statistical Theories of Mental Test Scores*. Reading, Mass.: Addison-Wesley.

Matt, Georg E. 1989. "Decision Rules for Selecting Effect Sizes in Meta-Analysis: A Review and Reanalysis of Psychotherapy Outcome Studies." *Psychological Bulletin* 105(1): 106–15.

McGuire, Joanmarie., Gary W. Bates, Beverly J. Dretzke, Julia E. McGivern, Karen L. Rembold, Daniel R. Seobold, Betty Ann M. Turpin, and Joel R. Levin. 1985. "Methodological Quality as a Component of Meta-Analysis." *Educational Psychologist* 20(1): 1–5.

McLellan, A. Thomas, Lester Luborsky, John Cacciola, Jason Griffith, Peggy McGahan, and Charles O'Brien. 1988. *Guide to the Addiction Severity Index: Background, Administration, and Field Testing Results.* Philadelphia, Pa.: Veterans Administration Medical Center.

Moher, David, Kenneth F. Schulz, Douglas G. Altman, and CONSORT Group 2001. "The CONSORT Statement: Revised Recommendations for Improving the Quality of Reports of Parallel-Group Randomised Trials." *Lancet* 357(9263): 1191–94.

Nunnally. Jum C. 1978. *Psychometric Theory* 2nd ed. New York: McGraw-Hill.

Oliver, Laurel W. 1987. "Research Integration for Psychologists: An Overview of Approaches." *Journal of Applied Social Psychology* 17(10): 860–74.

Orwin, Robert G. 1983a. "A Fail-Safe N for Effect Size in Meta-Analysis." *Journal of Educational Statistics* 8(2): 157–59.

———. 1983b. "The Influence of Reporting Quality in Primary Studies on Meta-Analytic Outcomes: A Conceptual Framework and Reanalysis." PhD diss. Northwestern University.

———. 1985. "Obstacles to Using Prior Research and Evaluations." In *Utilizing Prior Research in Evaluation Planning*, edited by D. S. Cordray. New Directions for Program Evaluation No. 27. San Francisco: Jossey-Bass.

Orwin, Robert, and David S. Cordray. 1985. "Effects of Deficient Reporting on Meta-Analysis: A Conceptual Framework and Reanalysis." *Psychological Bulletin,* 97(1): 134–47.

Rothstein, Hannah R., Alexander J. Sutton, and Michael Borenstein, eds. 2005. *Publication Bias in Meta-Analysis: Prevention, Assessment and Adjustments.* Chichester: John Wiley & Sons.

Rosenthal, Robert 1978. "Combining Results of Independent Studies." *Psychological Bulletin* 85: 185–93.

Sacks, Harold S., Jayne Berrier, Dinah Reitman, Vivette A. Ancona-Berk, and Thomas C. Chalmers 1987. "Meta-Analyses of Randomized Controlled Trials." *New England Journal of Medicine* 316(8): 450–55.

Scott, William A. 1955. "Reliability of Content Analysis: The Case of Nominal Scale Coding." *Public Opinion Quarterly* 19(3): 321–25.

Selvage, Rob. 1976. "Comments on the Analysis of Variance Strategy for the Computation of Intraclass Reliability." *Educational and Psychological Measurement* 36(3): 605–9.

Shapiro, David A., and Diana Shapiro. 1982. "Meta-Analysis of Comparative Therapy Outcome Studies: A Replication and Refinement." *Psychological Bulletin* 92(3): 581–604.

Shavelson, Richard J., and Noreen M. Webb 1991. *Generalizability Theory: A Primer.* Newbury Park, Calif.: Sage Publications.

Shrout, Patrick E., and Joseph L. Fleiss. 1979. "Intraclass Correlations: Uses in Assessing Rater Reliability." *Psychological Bulletin* 86(2): 420–28.

Shrout, Patrick E., Spitzer, R. L., and Joseph L. Fleiss. 1987. Quantification of agreement in psychiatric diagnosis revisited. *Archives of General Psychiatry* 44(2): 172–77.

Smith, Mary L. 1980. "Sex Bias in Counseling and Psychotherapy." *Psychological Bulletin* 87: 392–407.

Smith, Mary L., and Gene V. Glass. 1977. "Meta-Analysis of Psychotherapy Outcome Studies." *American Psychologist* 32: 752–60.

Smith, Mary L., Gene V. Glass, and Thomas I. Miller. 1980. *The Benefits of Psychotherapy.* Baltimore, Md.: Johns Hopkins University Press.

Spitznagel, Edward L., and John E. Helzer 1985. "A Proposed Solution to the Base Rate Problem in the Kappa Statistic." *Archives of General Psychiatry* 42(7): 725–28.

Stanley, Julian C. 1971. "Reliability." In *Educational Measurement*, 2d ed., edited by R. L. Thorndike. Washington, D.C.: American Council on Education.

Stock, William A., Morris A. Okun, Marilyn J. Haring, Wendy Miller, Clifford Kenney, and Robert C. Ceurvost. 1982. "Rigor in Data Synthesis: A Case Study of Reliability in Meta-Analysis." *Educational Researcher* 11(6): 10–20.

Terpstra. David E. 1981. "Relationship Between Methodological Rigor and Reported Outcomes in Organization Development Evaluation Research." *Journal of Applied Psychology* 66: 541–43.

U.S. General Accounting Office. 1989. *Prospective Evaluation Methods: The Prospective Evaluation Synthesis.* GAO/PEMD Transfer Paper 10-1-10. Washington, D.C.: Government Printing Office.

Valentine, Jeffrey C., and Harris Cooper. 2008. "A Systematic and Transparent Approach for Assessing the Methodological Quality of Intervention Effectiveness Research: The Study Design and Implementation Assessment Device Study DIAD." *Psychological Methods* 13(2): 130–49.

Whitehurst, Grover J. 1984. "Interrater Agreement for Journal Manuscript Reviews." *American Psychologist* 39(1): 22–28.

Wortman, Paul M., and Fred B. Bryant. 1985. "School Desegregation and Black Achievement: An Integrative Review." *Sociological Methods and Research* 13(3): 289–324.

Wortman. Paul M., and William H. Yeaton. 1985. "Cumulating Quality of Life Results in Controlled Trials of Coronary Artery Bypass Graft Surgery." *Controlled Clinical Trials* 6(4): 289–305.

Yeaton, William H., and Paul M. Wortman. 1991. "On the Reliability of Meta-Analytic Reviews: The Role of Intercoder Agreement." Unpublished manuscript.

Zwick. Rebecca. 1988. "Another Look at Interrater Agreement." *Psychological Bulletin* 103(3): 374–78.

STATISTICALLY DESCRIBING STUDY OUTCOMES

11

VOTE-COUNTING PROCEDURES IN META-ANALYSIS

BRAD J. BUSHMAN
University of Michigan and VU University

MORGAN C. WANG
University of Central Florida

CONTENTS

11.1 INTRODUCTION

As the number of scientific studies continues to grow, it becomes increasingly important to integrate the results from these studies. One simple approach involves counting votes. In the conventional vote-counting procedure, one simply divides studies into three categories: those with significant positive results, those with significant negative results, and those with nonsignificant results. The category containing the most studies is declared the winner. For example, if the majority of studies examining a treatment found significant positive results, then the treatment is considered to have a positive effect.

Many authors consider the conventional vote-counting procedure to be crude, flawed, and worthless (see Friedman 2001; Jewell and McCourt 2000; Lee and Bryk 1989; Mann 1994; Rafaeli-Mor and Steinberg 2002; Saroglou 2002; Warner 2001). Take, for example, the title of one article: "Why Vote-Count Reviews Don't Count" (Friedman 2001). We agree that the conventional vote-counting procedure can be described in these ways. But all vote-counting procedures are not created equal. The vote-counting procedures described in this chapter are far more sophisticated than the conventional procedure. These more sophisticated procedures can have an important place in the meta-analyst's toolbox.

When authors use both vote-counting procedures and effect size procedures with the same data set, they quickly discover that vote-counting procedures are less powerful (see Dochy et al. 2003; Jewell and McCourt 2000; Saroglou 2002). However, vote-counting procedures should never be used as a substitute for effect size procedures. Research synthesists generally have access to four types of information from studies:

- a reported effect size,

- information that can be used to compute an effect size estimate (for example, raw data, means and standard deviations, test statistic values),

- information about whether the hypothesis test found a statistically significant relationship between the independent and dependent variables, and the direction of that relationship (for example, a significant positive mean difference), and

- information about only the direction of the relationship between the independent and dependent variables (for example, a positive[1] mean difference).

These types are rank ordered from most to least in terms of the amount of information they contain (Hedges 1986).[2] Effect size procedures should be used for the studies that contain enough information to compute an effect size estimate (see chapters 12 and 13, this volume). Vote-counting procedures should be used for the studies that do not contain enough information to compute an effect size estimate but do contain information about the direction and the statistical significance of results, or that contain just the direction of results. We recommend that vote-counting procedures never be used alone unless none of the studies contain enough information to compute an effect size estimate. Rather, vote-counting procedures should be used in conjunction with effect size procedures. As we describe in section 11.4, effect size estimates and vote-count estimates can be combined to obtain a more precise overall estimate of the population effect size.

11.2 COUNTING POSITIVE AND SIGNIFICANT POSITIVE RESULTS

Although some studies do not contain enough information to compute an effect size estimate, they may still provide information about the magnitude of the relationship between the independent and dependent variables.[3] Often this information is in the form of a report of the decision yielded by a significance test (for example, a significant positive mean difference) or in the form of the direction of the effect without regard to its statistical significance (for example, a positive mean difference). The first form of information corresponds to whether the test statistic exceeds a conventional critical value at a given significance level, such as $\alpha = .05$. The second form of information corresponds to whether the tests statistic exceeds the rather unconventional level $\alpha = .5$. For the vote-counting procedures described in this chapter, we consider both the $\alpha = .05$ and $\alpha = .5$ significance levels.

11.3 VOTE-COUNTING PROCEDURES

11.3.1 The Conventional Vote-Counting Procedure

Richard Light and Paul Smith were among the first to describe a formal procedure for "taking a vote" of study results:

> All studies which have data on a dependent variable and a specific independent variable of interest are examined.

Three possible outcomes are defined. The relationship between the independent variable and the dependent variable is either significantly positive, significantly negative, or there is no specific relationship in either direction. The number of studies falling into each of these three categories is then simply tallied. If a plurality of studies falls into any one of these three categories, with fewer falling into the other two, the modal category is declared the winner. This modal category is then assumed to give the best estimate of the direction of the true relationship between the independent and dependent variable. (1971, 433)

The conventional vote-counting procedure Light and Smith described has been criticized on several grounds (1971). First, the conventional vote-counting procedure does not incorporate sample size into the vote. As sample size increases, the probability of finding a statistically significant relationship between the independent and dependent variables also increases. Thus, a small study with a nonsignificant result receives as much weight as a large study with a significant result. Second, although the conventional vote-counting procedure allows one to determine which modal category is the winner, it does not allow one to determine what the margin of victory is. If you compare it to a horse race, did the winner barely win by a nose, or did the winner impressively win by a length or more? In other words, the traditional vote-counting procedure does not provide an effect size estimate. Third, the conventional vote-counting procedure has very low power for small to medium effect sizes and sample sizes (Hedges and Olkin 1980). Moreover, for such effect sizes, the power of the conventional vote-counting procedure decreases to zero as the number of studies to be integrated increases (Hedges and Olkin 1980). Unfortunately, many effects and samples in the social sciences are small to medium in size. For example, in the field of social psychology a synthesis of 322 meta-analyses of more than 25,000 studies involving more than 8 million participants found an average correlation of 0.21 (Richard, Bond, and Stokes-Zoota 2003). According to Jacob Cohen (1988), a small correlation is 0.1, a medium correlation is 0.3, and a large correlation is .5. Thus, the 0.21 correlation is between small and medium in size.

11.3.2 The Sign Test

The sign test can be used to test the hypothesis that the effect sizes from a collection of k independent studies

are all zero. The sign test, the oldest nonparametric test (dating back to 1710), is simply the binomial test with probability $\pi = .5$ (Conover 1980, 122). If the best estimate for each effect size is zero, the probability of getting a positive result from a binomial test of sampled effect size estimates equals .5. If the effect size is expected to be greater than zero (as when a treatment has a positive effect), the probability of getting a binomial test result is greater than .5. Thus, the null (H_0) and alternative (H_A) hypotheses are

$$H_0: \pi = .5$$
$$H_A: \pi > .5 \tag{11.1}$$

where π is the proportion of positive results in the population. If one is counting significant positive results rather than positive results, use $\pi = .05$ rather than $\pi = .5$ in (11.1). If U equals the number of positive results in n independent studies, then an estimate of π is $p = U/k$.

A $(1 - \alpha) \times 100$ percent confidence interval for π is (Brown, Cai, and DasGupta 2001)

$$\left(\frac{\left(U + \frac{\kappa^2}{2}\right)}{(n + \kappa^2)} - \frac{(\kappa\sqrt{n})}{(n + \kappa^2)} \times \left(\hat{p}\hat{q} + \frac{\kappa^2}{4n}\right), \right.$$
$$\left. \frac{\left(U + \frac{\kappa^2}{2}\right)}{(n + \kappa^2)} + \frac{(\kappa\sqrt{n})}{(n + \kappa^2)} \times \left(\hat{p}\hat{q} + \frac{\kappa^2}{4n}\right) \right), \tag{11.2}$$

where $\kappa \equiv Z_{\alpha/2} = \Phi^{-1}(1 - \alpha/2)$, $\hat{p} = \frac{U}{n}$, and $\hat{q} = 1 - \hat{p}$. For a 95 percent confidence interval, $\kappa = 1.96$.

Example: Sign Test. In data set II, there are eleven positive results in the nineteen studies (58 percent). The estimate of π is therefore $11/19 = .58$. When $\pi = .5$, it is found from a binomial distribution table that the corresponding area is .0593 (that is, .0417 + .0139 + .0032 + .0005 + .0000) Thus, we would fail to reject the null hypothesis that $\pi = .5$ at the .05 significance level (because the area .0593 is greater than .05). The corresponded 95 percent Wilson confidence interval is (0.456 0.676) Because the confidence interval includes the value .5, we fail to reject the null hypothesis that $\pi = .5$ at the .05 significance level.

This sign test is based on the simple direction of each study result. Like the conventional vote-counting procedure, it does not incorporate sample size in the vote. It also does not provide an effect size estimate. The procedures described in the next section overcome both of these weaknesses.

11.3.3 Confidence Intervals Based on Vote-Counting Procedures

Suppose that the synthesist has a collection of k independent studies that test whether a relationship exists between the independent and dependent variables. The synthesist wants to test whether the effect size for each study, assumed to be the same for each study, is zero. Let $T_1, \ldots, T_k$ be the independent effect size estimators for parameters $\theta_1, \ldots, \theta_k$. If the effect sizes are assumed to be homogeneous, the null and alternative hypotheses are

$$H_0: \theta_1 = \ldots = \theta_k = 0$$
$$H_A: \theta_1 = \ldots = \theta_k = \theta > 0. \tag{11.3}$$

The null hypothesis in (11.3) is rejected if the estimator T of the common effect size θ exceeds the one-sided[4] critical value C_α. The $100(1 - \alpha)$ percent confidence interval for θ is given by

$$T - C_{\alpha/2} S_T \leq \theta \leq T + C_{\alpha/2} S_T, \tag{11.4}$$

where $C_{\alpha/2}$ is the two-sided critical value from the distribution of T at significance level α, and S_T is the standard error of T.

The essential feature of vote-counting procedures is that the values $T_1, \ldots, T_k$ are not observed. If we do not observe T_i, we simply count the number of times T_i exceeded the one-sided critical value C_α. For $\alpha = .5$, we count the number of studies with positive results, and use the null and alternative hypotheses in (11.1). For $\alpha = .05$, we count the number of studies with significant positive results, and use the null and alternative hypotheses

$$H_0: \pi = .05$$
$$H_A: \pi > .05. \tag{11.5}$$

(If $\theta = 0$, the probability of getting a significant positive result equals .05.)

To test the hypotheses in (11.1) and (11.5) we need an estimator of π. The method of maximum likelihood can be used to obtain an estimator of π. For each study there

are two possible outcomes: a success if $T_i > C_{\alpha i}$, or a failure $T_i \leq C_{\alpha i}$. Let X_i be an indicator variable such that

$$X_i = \begin{cases} 1 & \text{if } T_i > C_{\alpha_i} \\ 0 & \text{if } T_i \leq C_{\alpha_i} \end{cases} \tag{11.6}$$

If one counts the number of positive results, the probability π_i is

$$\pi_i = \Pr(X_i = 1) = \Pr(T_i > C_{\alpha = .5, v_i}) \tag{11.7}$$

where v_i are the degrees of freedom for the ith study at the $\alpha = .5$ significance level. If one counts the number of significant positive results, the probability π_i is

$$\pi_i = \Pr(X_i = 1) = \Pr(T_i > C_{\alpha = .05, v_i}) \tag{11.8}$$

If T is approximately normally distributed with mean θ and variance v_i, then the probability π can be expressed as an explicit function of θ,

$$\begin{aligned} \pi_i &= \Pr(T_i > C_{\alpha_i}) \\ &= \Pr[(T_i - \theta)/\sqrt{v_i} > (C_{\alpha_i} - \theta)/\sqrt{v_i}] \\ &= 1 - \Phi[(C_{\alpha_i} - \theta)/\sqrt{v_i}], \end{aligned} \tag{11.9}$$

where $\Phi(x)$ is the standard normal distribution function. Solving (11.9) for θ gives

$$\theta = C_\alpha - \sqrt{v_i}\, \Phi^{-1}(1 - \pi_i). \tag{11.10}$$

Equation 11.10 provides a general large-sample approximate relationship between a proportion and an effect size. It applies to effect sizes that are normally distributed in large samples.

The method of maximum likelihood can be used to obtain an estimator of θ by observing whether $T_i > C_{\alpha_i}$ in each of the k studies. The log-likelihood function is given by

$$L(\theta) = \sum_{i=1}^{k} [x_i \log(\pi_i)] + [(1 - x_i)\log(1 - \pi_i)], \tag{11.11}$$

where x_i is the observed value of X_i (that is 1 or 0) and $\log(x)$ is the natural logarithmic function. The log-likelihood function $L(\theta)$ can be maximized over θ to obtain the maximum likelihood estimator (MLE) $\hat{\theta}$. Generally, however, there is no closed form expression for the MLE $\hat{\theta}$, and $\hat{\theta}$ must be obtained numerically. One way to obtain $\hat{\theta}$ is to compute the log-likelihood function $L(\theta)$ for a coarse grid of θ values and use the maximum value of $L(\theta)$ to determine the region where $\hat{\theta}$ lies. Then a finer grid of θ values can be used to obtain a more precise estimate of $\hat{\theta}$.

When the number of studies combined is large, $\hat{\theta}$ is approximately normally distributed, and the large-sample variance of $\hat{\theta}$ is given by (Hedges and Olkin 1985, 70)

$$\text{Var}(\hat{\theta}) = \left[\sum_{i=1}^{k} \frac{[D_i^{(1)}]^2}{\pi_i(1-\pi_i)} \right]^{-1}, \quad (11.12)$$

where $D_i^{(1)}$ is the first order derivative of π_i evaluated at $\theta = \hat{\theta}$ is

$$D_i^{(1)} = \frac{\partial \pi_i}{\partial \theta} = \frac{S_i(\theta) - \theta S_i'(\theta)}{S_i^2(\theta)\sqrt{2\pi}} \exp\left(-\frac{1}{2}\frac{(C_{\alpha_i}-\theta)^2}{S_i^2(\theta)}\right), \quad (11.13)$$

where $S_i'(\theta) = \partial S_i(\theta)/\partial \theta$, evaluated at $\hat{\theta}$ and the variance v of the effect size estimate can be expressed as a function of the effect size parameter θ via $v = [S(\theta)]^2$.

The upper and lower bounds of a $100(1-\alpha)\%$ confidence interval for the population effect size θ are given by,

$$\hat{\theta} - z_{\alpha/2}\sqrt{\text{Var}(\hat{\theta})} \le \theta \le \hat{\theta} + z_{\alpha/2}\sqrt{\text{Var}(\hat{\theta})}, \quad (11.14)$$

where $z_{\alpha/2}$ is the two-sided critical value from the standard normal distribution at significance level α.

Section 11.3.3.1 describes how to obtain an estimate and a confidence interval for the population standardized mean difference δ. Section 11.3.3.2 illustrates how to obtain an estimate and a confidence interval for the population correlation coefficient ρ.

11.3.3.1 Population Standardized Mean Difference

δ If the independent variable is dichotomous (for example, experimental vs. control group), one common effect size for the ith study is the population standardized mean difference

$$\delta_i = \frac{\mu_i^E - \mu_i^C}{\sigma_i}, \quad i = 1, \ldots, k, \quad (11.15)$$

where μ_i^E and μ_i^C are the respective population means for the experimental and control groups in the ith study, and σ_i is the common population standard deviation in the ith study. The sample estimator of δ_i is

$$d_i = \frac{\overline{Y}_i^E - \overline{Y}_i^C}{S_i}, \quad i = 1, \ldots, k, \quad (11.16)$$

where $\overline{Y}_i^E$ and $\overline{Y}_i^C$ are the respective sample means for the experimental and control groups in the ith study, and S_i is the pooled within-group sample standard deviation in the

ith study. If we assume the effect sizes are homogeneous, we test the hypotheses

$$H_0: \delta_1 = \ldots = \delta_k = 0$$
$$H_A: \delta_1 = \ldots = \delta_k = \delta > 0. \quad (11.17)$$

Suppose that we observe whether the mean difference $\overline{Y}_i^E - \overline{Y}_i^C$ is positive for each study in a collection of k independent studies. If we assume that the effect sizes are homogeneous, then the mean difference $\overline{Y}_i^E - \overline{Y}_i^C$ is approximately normally distributed with mean $\delta\sigma_i$ and variance $\sigma_i^2/\bar{n}_i$, where $\bar{n}_i = n_i^E n_i^C/(n_i^E + n_i^C)$. The probability of a positive result is

$$\pi_i = \text{Pr}(T_i > C_{\alpha=.5, v_i}) = \text{Pr}(\overline{Y}_i^E - \overline{Y}_i^C > 0)$$
$$= 1 - \Phi(-\sqrt{\bar{n}_i}\delta), \quad (11.18)$$

Note that π_i is a function of both δ and $\bar{n}_i$. The log-likelihood function is

$$L(\delta) = \sum_{i=1}^{k} [x_i \log[1 - \Phi(-\sqrt{\bar{n}_i}\delta)]]$$
$$+ [(1-x_i)\log\Phi(-\sqrt{\bar{n}_i}\delta)], \quad (11.19)$$

and the derivative is

$$D_i^{(1)} = \frac{\partial \pi_i}{\partial \delta} = \frac{\sqrt{\bar{n}_i}}{2\pi} \exp\left(-\frac{1}{2}\bar{n}_i\hat{\delta}^2\right). \quad (11.20)$$

The large sample variance of $\hat{\delta}$ is obtained by substituting equation 11.18 for π_i and equation 11.20 for $D_i^{(1)}$ in equation 11.12.

Example: Estimating δ from the Proportion of Positive Results. Suppose we wish to estimate δ based on the proportion of positive results in data set II. Table 11.1 gives the values of n^E, n^C, $\bar{n}$, d and X for the nineteen studies in data set II. Inserting these values into equation 11.19 for values of δ ranging from -0.50 to 0.50 gives the coarse log-likelihood function values on the left side of table 11.2. The coarse grid of δ values reveals that the maximum value of $L(\delta)$ lies between -0.10 and 0.10. To obtain an estimate of δ to two decimal places, we compute the log-likelihood function values between -0.10 and 0.10 in steps of 0.01. The fine grid of δ values is given on the right side of table 11.2. Because the log-likelihood function is largest at $\delta = 0.02$, the MLE is $\hat{\delta} = 0.02$.

The computations for obtaining the large sample variance of $\hat{\delta}$ are given in table 11.3. Substituting these values in equation 11.12 yields the $\text{Var}(\hat{\delta}) = 1/473.734 = 0.00211$.

Table 11.1 Experimental and Control Group Sample Sizes, Standardized Mean Differences, and Indicator Variable Values

Study	n^E	n^C	$\tilde{n} = n_i^E \times n_i^C/(n_i^E + n_i^C)$	d	X
1	77	339	62.748	0.03	1
2	60	198	46.047	0.12	1
3	72	72	36.000	−0.14	0
4	11	22	7.333	1.18	1
5	11	22	7.333	0.26	1
6	129	348	94.113	−0.06	0
7	110	636	93.780	−0.02	0
8	26	99	20.592	−0.32	0
9	75	74	37.248	0.27	1
10	32	32	16.000	0.80	1
11	22	22	11.000	0.54	1
12	43	38	20.173	0.18	1
13	24	24	12.000	−0.02	0
14	19	32	11.922	0.23	1
15	80	79	39.748	−0.18	0
16	72	72	36.000	−0.06	0
17	65	255	51.797	0.30	1
18	233	224	114.206	0.07	1
19	65	67	32.992	−0.07	0

SOURCE: Authors' compilation.

NOTE: n^E = experimental group sample size; n^C = control group sample size; d = sample standardized mean difference given in (15), and X = indicator variable given in (5).

The 95 percent confidence interval for δ given in equation 11.13 is:

$$0.02 - 1.96\sqrt{0.00211} \le \delta \le 0.02 + 1.96\sqrt{0.00211},$$

or simplifying [−0.070 0.110]. Because the confidence interval includes the value 0, the effect does not differ from 0.

11.3.3.2 Population correlation coefficient ρ If both the independent and dependent variables are continuous, the effect size is the population correlation coefficient

$$\rho_i = \frac{\text{Cov}(X_i, Y_i)}{\sigma_X \sigma_Y}, \quad i = 1, \dots, k, \quad (11.21)$$

where $\text{Cov}(X_i, Y_i)$ is the population covariance for variables X and Y in the ith study, and σ_X and σ_Y are the respective population standard deviations for variables X and Y in the ith study. The sample estimator of ρ_i is the Pearson product moment correlation coefficient

Table 11.2 Log-Likelihood Function Values for δ

Coarse Grid		Fine Grid	
δ	$L(\delta)$	δ	$L(\delta)$
−0.50	−52.430	−0.10	−15.259
−0.40	−39.135	−0.09	−14.912
−0.30	−28.409	−0.08	−14.596
−0.20	−20.387	−0.07	−14.311
−0.10	−15.259	−0.06	−14.056
0.00	−13.170	−0.05	−13.831
0.10	−14.082	−0.04	−13.638
0.20	−17.733	−0.03	−13.475
0.30	−23.813	−0.02	−13.343
0.40	−32.149	−0.01	−13.241
0.50	−42.657	0.00	−13.170
		0.01	−13.129
		0.02	−13.118
		0.03	−13.136
		0.04	−13.185
		0.05	−13.263
		0.06	−13.370
		0.07	−13.505
		0.08	−13.669
		0.09	−13.862
		0.10	−14.082

SOURCE: Authors' compilation

NOTE: Based on data from nineteen studies of the effects of teacher expectancy on pupil IQ.

$$r_i = \frac{\sum_{j=1}^{n_i}(X_j - \bar{X})(Y_j - \bar{Y})}{\sqrt{\left[\sum_{j=1}^{n_i}(X_j - \bar{X})^2\right]\left[\sum_{j=1}^{n_i}(Y_j - \bar{Y})^2\right]}}, \quad i = 1, \dots, k. \quad (11.22)$$

If we assume the effect sizes are homogeneous, the hypotheses in (3) can be written as

$$H_0: \rho_1 = \dots = \rho_k = 0 \quad (11.23)$$

$$H_A: \rho_1 = \dots = \rho_k = \rho > 0.$$

Suppose that we observe whether the mean difference r_i is positive for each study in a collection of k independent studies. If we assume that the effect sizes are homogeneous, then r_i is approximately normally distributed with mean ρ and variance $(1-\rho^2)^2/n_i$. The probability of a positive result is

$$\pi_i = \Pr(T_i > C_{\alpha = .5, v_i})$$
$$= \Pr(r_i > 0) = 1 - \Phi[-\sqrt{n_i}\rho/(1-\rho^2)], \quad (11.24)$$

Table 11.3 Computations for Obtaining the Large-Sample Variance of $\hat{\delta}$

Study	p_i	$D_i^{(1)}$	$[D_i^{(1)}]^2/[p_i/(1-p_i)]$
1	0.563	3.121	39.583
2	0.554	2.682	29.119
3	0.548	2.376	22.799
4	0.522	1.079	4.664
5	0.522	1.079	4.664
6	0.577	3.798	59.100
7	0.577	3.792	58.893
8	0.536	1.803	13.070
9	0.549	2.417	23.585
10	0.532	1.591	10.162
11	0.526	1.320	6.992
12	0.536	1.785	12.805
13	0.528	1.379	7.626
14	0.528	1.374	7.576
15	0.550	2.495	25.159
16	0.548	2.376	22.799
17	0.557	2.842	32.728
18	0.585	4.167	71.507
19	0.546	2.276	20.903
			473.734

SOURCE: Authors' compilation.
NOTE: From nineteen studies of the effects of teacher expectancy on pupil IQ.

Table 11.4 Sample Sizes, Correlations, and Indicator Variable Values

Study	n	r	X
1	10	.68	1
2	20	.56	1
3	13	.23	1
4	22	.64	1
5	28	.49	1
6	12	−.04	0
7	12	.49	1
8	36	.33	1
9	19	.58	1
10	12	.18	1
11	36	−.11	0
12	75	.27	1
13	33	.26	1
14	121	.40	1
15	37	.49	1
16	14	.51	1
17	40	.40	1
18	16	.34	1
19	14	.42	1
20	20	.16	1

SOURCE: Authors' compilation.
NOTE: From twenty studies of the relation between student ratings of the instructor and student achievement.
n = sample size; r = Pearson product moment correlation coefficient given in (20); and X = indicator variable given in (5).

Note that π_i is a function of both δ and $\bar{n}_i$. The log-likelihood function is

$$L(\rho) = \sum_{i=1}^{k} \left[x_i \log\left[1 - \Phi\left(-\sqrt{n_i}\,\rho/(1-\rho)^2\right)\right]\right] + \left[(1-x_i)\log\Phi\left(-\sqrt{n_i}\,\rho/(1-\rho)^2\right)\right], \quad (11.25)$$

and the derivative is

$$D_i^{(1)} = \frac{\partial \pi_i}{\partial \rho} = \sqrt{\frac{n_i}{2\pi}}\left[\frac{1+\rho^2}{(1-\rho^2)^2}\right]\exp\left(-\frac{1}{2}\frac{n_i\rho^2}{(1-\rho^2)^2}\right). \quad (11.26)$$

The large sample variance of $\hat{\rho}$ is obtained by substituting equation 11.24 for π_i and equation 11.26 for $D_i^{(1)}$ in equation (12).

Example: Estimating ρ from the Proportion of Positive Results. Suppose we wish to estimate ρ based on the proportion of positive results in data set III. Table 11.4 gives the values of n, r, and X for the 20 studies in data set III. Inserting these values into equation (25) for values of ρ ranging from −0.50 to 0.50 gives the coarse

log-likelihood function values on the left side of table 11.5. The coarse grid of ρ values reveals that the maximum value of $L(\rho)$ lies between 0.10 and 0.30. To obtain an estimate of ρ to two decimal places, we compute the log-likelihood function values between 0.10 and 0.30 in steps of 0.01. The fine grid of ρ values is given on the right side of table 11.5. Because the log-likelihood function is largest at $\rho = 0.24$, the MLE is $\hat{\rho} = 0.24$.

The computations for obtaining the large sample variance of $\hat{\rho}$ are given in table 11.6. Substituting these values in equation 11.12 yields the Var$(\hat{\rho}) = 1/2026.900 = 0.000493$. The 95 percent confidence interval for ρ given in 11.13 is:

$$0.24 - 1.96\sqrt{0.000493} \leq \rho \leq 0.24 + 1.96\sqrt{0.000493},$$

or simplifying [0.196 0.283]. Because the confidence interval excludes the value 0, the correlation is significantly different from 0.

Table 11.5 Log-Likelihood Function Values for δ

Coarse Grid		Fine Grid	
ρ	$L(\rho)$	ρ	$L(\rho)$
-0.50	-159.677	0.10	-8.961
-0.40	-95.780	0.11	-8.640
-0.30	-59.089	0.12	-8.346
-0.20	-36.659	0.13	-8.079
-0.10	-22.568	0.14	-7.837
0.00	-13.863	0.15	-7.621
0.10	-8.961	0.16	-7.430
0.20	-6.900	0.17	-7.263
0.30	-7.108	0.18	-7.119
0.40	-9.552	0.19	-6.998
0.50	-14.990	0.20	-6.900
		0.21	-6.824
		0.22	-6.770
		0.23	-6.738
		0.24	-6.727
		0.25	-6.737
		0.26	-6.769
		0.27	-6.821
		0.28	-6.895
		0.29	-6.991
		0.30	-7.108

SOURCE: Authors' compilation.
NOTE: Based on data from twenty studies of the relation between student ratings of the instructor and student achievement.

Table 11.6 Computations for Obtaining the Large-Sample Variance of $\hat{p}$

Study	p_i	$D_i^{(1)}$	$[D_i^{(1)}]^2/[p_i/(1-p_i)]$
1	0.790	3.413	70.118
2	0.873	3.489	109.552
3	0.821	3.530	84.708
4	0.884	3.430	114.607
5	0.911	3.185	125.273
6	0.811	3.503	80.133
7	0.811	3.503	80.133
8	0.937	2.787	131.041
9	0.867	3.513	106.710
10	0.811	3.503	80.133
11	0.937	2.787	131.041
12	0.986	1.136	95.356
13	0.928	2.940	129.840
14	0.997	0.325	41.491
15	0.939	2.735	131.219
16	0.830	3.547	89.008
17	0.946	2.580	131.152
18	0.846	3.553	96.821
19	0.830	3.547	89.008
20	0.873	3.489	109.552
			2026.900

SOURCE: Authors' compilation.
NOTE: From twenty studies of the relation between student ratings of the instructor and student achievement.

11.4 COMBINING VOTE COUNTS AND EFFECT SIZE ESTIMATES

Effect size procedures should be used for studies that include enough information to compute an effect size estimate. Vote-counting procedures should be used for studies that do not provide enough information to compute an effect size estimate, but do provide information about the direction and/or statistical significance of effects. We have proposed a procedure, called the combined procedure, that combines estimates based on effect size and vote-counting procedures to obtain an overall estimate of the population effect size (see Bushman and Wang 1996).

11.4.1 Population Standardized Mean Difference δ

Out of k independent studies, suppose that the first m_1 studies ($m_1 < k$) report enough information to compute

standardized mean differences, that the next m_2 studies ($m_2 < k$) only report the direction and statistical significance of results, and that the last m_3 studies ($m_3 < k$) only report the direction of results, where $m_1 + m_2 + m_3 = k$. The synthesist can use effect size procedures to obtain an estimator d. of the population standardized mean difference δ using the first m_1 studies (see chapter 14, this volume). The synthesist can also use vote-counting procedures based on the proportion of significant positive results to obtain a maximum likelihood estimator $\hat{\delta}_S$ of δ using the second m_2 studies. Finally, vote-counting procedures based on the proportion of positive results can be used to obtain the maximum likelihood estimator $\hat{\delta}_D$ of δ using the last m_3 studies. (We use the subscript S for the estimate based on studies that provide information about the significance and direction of results, and we use the subscript D for the estimate based on studies that provide information only about the direction of results.) The respective variances for d., $\hat{\delta}_S$, and $\hat{\delta}_D$ are Var(d.), Var($\hat{\delta}_S$), and $\text{Var}(\hat{\delta}_D)$. The combined weighted estimator of δ is given by

$$\hat{\delta}_C = \frac{d./\text{Var}(d.) + \hat{\delta}_S/\text{Var}(\hat{\delta}_S) + \hat{\delta}_D/\text{Var}(\hat{\delta}_D)}{1/\text{Var}(d.) + 1/\text{Var}(\hat{\delta}_S) + 1/\text{Var}(\hat{\delta}_D)}. \quad (11.27)$$

Because $d.$, $\hat{\delta}_S$, and $\hat{\delta}_D$ are consistent estimators of δ, the combined estimator $\hat{\delta}_C$ will also be consistent if: the studies containing each type of information (that is, information used to compute effect size estimates, information about the direction and significance of results, information about only the direction of results) can be considered a random sample from the population of studies containing the same type of information, and m_1, m_2, and m_3 are all large enough and all converge to infinity at the same rate. The variance of the combined estimator is given by

$$\text{Var}(\hat{\delta}_C) = \frac{1}{1/\text{Var}(d.) + 1/\text{Var}(\hat{\delta}_S) + 1/\text{Var}(\hat{\delta}_D)} \quad (11.28)$$

Because $\frac{1}{\text{Var}(d.)} > 0$, $\frac{1}{\text{Var}(\hat{\delta}_S)} > 0$, and $\frac{1}{\text{Var}(\hat{\delta}_D)} > 0$, the combined estimator $\hat{\delta}_C$ is more efficient (in other words, has smaller variance) than $d.$, $\hat{\delta}_S$, and $\hat{\delta}_D$.

The upper and lower bounds of a $100(1-\alpha)$ percent confidence interval for the population standardized mean difference δ are given by

$$\hat{\delta}_C - z_{\alpha/2}\sqrt{\text{Var}(\hat{\delta}_C)} \leq \delta \leq \hat{\delta}_C + z_{\alpha/2}\sqrt{\text{Var}(\hat{\delta}_C)}. \quad (11.29)$$

The confidence interval based on the combined estimator is narrower than the confidence intervals based on the other estimators.

Example: Using the Combined Procedure to Obtain an Estimate and a 95 Percent Confidence Interval for the Population Standardized Mean Difference δ. Suppose we wish to estimate δ based on the results in data set II. Because all studies report effect size estimates, we assumed that study 4, 5, 11, 12, and 15 report only the direction of the study outcome for purpose of illustration. The effect size estimates are $d = 0.048$ and $\hat{\delta}_D = 0.259$. The variances for the estimates are $\text{Var}(d.) = 0.0015$ and $\text{Var}(\hat{\delta}_D) = 0.0077$. Inserting the estimates and variances into equation 11.27 gives the combined estimate

$$\hat{\delta}_C = \frac{d/\text{Var}(d.) + \hat{\delta}_D/\text{Var}(\hat{\delta}_D)}{1/\text{Var}(d.) + 1/\text{Var}(\hat{\delta}_D)}$$
$$= \frac{0.048/0.0015 + 0.259/0.0077}{1/0.0015 + 1/0.0077} = 0.082$$

From equation 11.28, the variance of the combined estimate is

$$\text{Var}(\hat{\delta}_C) = \frac{1}{1/\text{Var}(d.) + 1/\text{Var}(\hat{\delta}_D)}$$
$$= \frac{1}{1/0.0015 + 1/0.0077} = 0.00126$$

From equation 11.29, the 95 percent confidence interval for the combined estimate is $(0.012 - 0.152)$. Note that the confidence interval based on the combined estimate is narrower $(0.152 - 0.012 = 0.140)$ than the confidence intervals based on effect sizes $[0.124 - (-0.290) = 0.153]$ or the confidence interval based on the proportion of positive results $(0.431 - 0.086 = 0.345)$.

11.4.2 Population Correlation Coefficient ρ

Out of k independent studies, suppose that the first m_1 studies $(m_1 < k)$ report enough information to compute sample correlation coefficients, that the next m_2 studies $(m_2 < k)$ only report the direction and statistical significance of results, and that the last m_3 studies $(m_3 < k)$ only report the direction of results, where $m_1 + m_2 + m_3 = k.$. The synthesist can use effect size procedures to obtain an estimator r of the population correlation coefficient ρ using the first m_1 studies (see chapter 14, this volume). The synthesist can use vote-counting procedures based on the proportion of significant positive results to obtain a maximum likelihood estimator $\hat{\rho}_S$ of ρ using the second m_2 studies. Finally, vote-counting procedures based on the proportion of positive results can be used to obtain the maximum likelihood estimator $\hat{\rho}_D$ of ρ using the last m_3 studies. The respective variances for $r.$, $\hat{\rho}_S$, and $\hat{\rho}_D$ are $\text{Var}(r.)$, $\text{Var}(\hat{\rho}_S)$, and $\text{Var}(\hat{\rho}_D)$. The combined weighted estimator of ρ is given by

$$\hat{\rho}_C = \frac{r./\text{Var}(r.) + \hat{\rho}_S/\text{Var}(\hat{\rho}_S) + \hat{\rho}_D/\text{Var}(\hat{\rho}_D)}{1/\text{Var}(r.) + 1/\text{Var}(\hat{\rho}_S) + 1/\text{Var}(\hat{\rho}_D)}. \quad (11.30)$$

As with $\hat{\delta}_C$, the combined estimate $\hat{\rho}_C$ will also be consistent if: the studies containing each type of information can be considered a random sample from the population of studies containing the same type of information, and m_1, m_2, and m_3 are all large enough and all converge to infinity at the same rate. The variance of the combined estimator is given by

$$\text{Var}(\hat{\rho}_C) = \frac{1}{1/\text{Var}(r.) + 1/\text{Var}(\hat{\rho}_S) + 1/\text{Var}(\hat{\rho}_D)} \quad (11.31)$$

Because $\frac{1}{\mathrm{Var}(r_.)} > 0$, $\frac{1}{\mathrm{Var}(\hat{\rho}_S)} > 0$, and $\frac{1}{\mathrm{Var}(\hat{\rho}_D)} > 0$, the combined estimator is more efficient (that is, has smaller variance) than $r_.$, $\hat{\rho}_S$, and $\hat{\rho}_D$.

The upper and lower bounds of a $100(1 - \alpha)$ percent confidence interval for the population correlation coefficient ρ are given by

$$\hat{\rho}_C - z_{\alpha/2}\sqrt{\mathrm{Var}(\hat{\rho}_C)} \leq \rho \leq \hat{\rho}_C + z_{\alpha/2}\sqrt{\mathrm{Var}(\hat{\rho}_C)}. \quad (11.32)$$

The confidence interval based on the combined estimator is narrower than the confidence intervals based on the other estimators.

Example: Using the Combined Procedure to Obtain an Estimate and a 95 Percent Confidence Interval for the Population Correlation Coefficient ρ. Suppose we wish to estimate ρ based on the results in data set III. The effect size estimates are $r = 0.362$ and $\hat{\rho}_D = 0.318$. The variances for the estimates are $\mathrm{Var}(r) = 0.0017$ and $\mathrm{Var}(\hat{\rho}_D) = 0.0127$. Inserting the estimates and variances into equation 11.29 gives the combined estimate

$$\hat{\rho}_C = \frac{r_./\mathrm{Var}(r_.) + \hat{\rho}_D/\mathrm{Var}(\hat{\rho}_D)}{1/\mathrm{Var}(r_.) + 1/\mathrm{Var}(\hat{\rho}_D)}$$
$$= \frac{0.362/0.0017 + 0.318/0.0127}{1/0.0017 + 1/0.0127} = 0.357$$

From equation (31), the variance of the combined estimate is

$$\mathrm{Var}(\hat{\rho}_C) = \frac{1}{1/\mathrm{Var}(r_.) + 1/\mathrm{Var}(\hat{\rho}_D)}$$
$$= \frac{1}{1/0.0017 + 1/0.0127} = 0.0015$$

From equation 11.31, the 95 percent confidence interval for the combined estimate is (0.281 0.432).

Note that the confidence interval based on the combined estimate is narrower ($0.432 - 0.281 = 0.151$) than the confidence intervals based on effect sizes ($0.442 - 0.281 = 0.161$) or the confidence interval based on the proportion of positive results ($0.539 - 0.097 = 0.442$).

11.5 APPLICATIONS OF VOTE-COUNTING PROCEDURES

11.5.1 Publication Bias

It is important to note that vote-counting procedures provide meta-analysts another tool to deal with the problem of publication bias (for an in-depth discussion, see chapter 23, this volume). If studies with significant results are more likely to be published than studies with nonsignificant results, then published studies constitute a biased, unrepresentative sample of the studies in the population. This bias can be reduced if meta-analysts count both positive and negative significant results, as Larry Hedges and Ingram Olkin suggested:

> When both positive and negative significant results are counted, it is possible to dispense with the requirement that the sample available is representative of all studies conducted. Instead, the requirement is that the sample of positive and negative *significant* results is representative of the population of positive and negative *significant* results. If only statistically significant results tend to be published, this requirement is probably more realistic. (1980, 366 [italics in original])

Suppose that a research synthesist wishes to obtain an estimate and confidence interval for θ using the k estimators $T_1, \ldots, T_k$. In this case, because the values $T_1, \ldots, T_k$ are not observed, we count the number of times that $T_i > C_{\alpha/2}$, and the number of times $T_i < -C_{\alpha/2}$. That is, we count the number of *two-sided* tests that found significant positive or significant negative results.

To find a confidence interval for θ, we find the values $[\theta_L, \theta_U]$ that correspond to the upper and lower confidence intervals for π $[\pi_L, \pi_U]$, where π denotes the proportion of the total number of significant results that were positive. That is,

$$\pi = \frac{\pi^+}{\pi^+ + \pi^-}, \quad (11.33)$$

where $\pi^+ = \mathrm{Pr}(T_i > C_{\alpha/2})$ and $\pi^- = \mathrm{Pr}(T_i < -C_{\alpha/2})$. The MLE of π is

$$p = \frac{p^+}{p^+ + p^-}, \quad (11.34)$$

where p^+ is the proportion of significant positive results and p^- is the proportion of significant negative results in the k independent studies.

Larry Hedges and Ingram Olkin produced a table (see table 11.7) for obtaining an estimate and confidence interval for δ given the proportion of the total number of significant standardized mean differences that were positive (1980). Brad Bushman produced a table (see table 11.8) for obtaining an estimate and confidence interval for ρ given the proportion of the total number of significant correlations that were positive (1994). Examples 1 and 2 illustrate how to use tables 11.7 and 11.8, respectively.

Table 11.7 Conditional Probability $p(\delta, n) = p\{t > 0 \mid |t| > C_t\}$

					Population Standardized Mean Difference δ								
n	0.00	0.02	0.04	0.06	0.08	0.10	0.15	0.20	0.25	0.30	0.40	0.50	0.70
2	0.500	0.516	0.531	0.547	0.562	0.577	0.615	0.651	0.685	0.718	0.777	0.827	0.900
4	0.500	0.528	0.556	0.584	0.611	0.638	0.700	0.756	0.805	0.845	0.906	0.945	0.982
6	0.500	0.537	0.573	0.609	0.643	0.676	0.751	0.814	0.863	0.901	0.950	0.976	0.994
8	0.500	0.544	0.586	0.628	0.668	0.706	0.788	0.852	0.899	0.932	0.971	0.988	0.998
10	0.500	0.549	0.598	0.644	0.689	0.729	0.816	0.879	0.923	0.952	0.982	0.993	0.999
12	0.500	0.555	0.608	0.659	0.706	0.750	0.838	0.900	0.940	0.964	0.988	0.996	1.000
14	0.500	0.559	0.617	0.672	0.722	0.767	0.857	0.916	0.952	0.973	0.992	0.998	1.000
16	0.500	0.564	0.625	0.683	0.736	0.783	0.872	0.929	0.961	0.979	0.994	0.998	1.000
18	0.500	0.568	0.633	0.694	0.749	0.796	0.886	0.939	0.968	0.984	0.996	0.999	1.000
20	0.500	0.572	0.640	0.704	0.760	0.809	0.897	0.947	0.974	0.987	0.997	0.999	1.000
22	0.500	0.575	0.647	0.713	0.771	0.820	0.907	0.954	0.978	0.990	0.998	1.000	
24	0.500	0.579	0.654	0.721	0.781	0.830	0.915	0.960	0.982	0.992	0.998		
50	0.500	0.614	0.716	0.800	0.864	0.910	0.970	0.991	0.997	0.999	1.000		
100	0.500	0.659	0.789	0.878	0.933	0.964	0.993	0.999	1.000	1.000			

SOURCE: Hedges and Olkin 1980.
NOTES: Proportions less than .500 correspond to negative values of δ.
Probability that a two-sample t statistic with n subjects per group is positive given that the absolute value of the t statistic exceeds the $\alpha = .05$ critical value.

Example 1: Obtaining an Estimate and Confidence Interval for the Population Standardized Mean Difference δ Based on the Proportion of the Total Number of Significant Mean Differences that Were Positive. Suppose that all the results in data set II are significant; thus we have eleven significant positive results and eight significant negative results. One needs an average sample size for table 11.7. There are several averages to choose from (arithmetic mean, for example). Jean Gibbons, Ingram Olkin, and Milton Sobel recommended using the square mean root (n_{SMR}) as the average sample size because it is not as influenced by extreme values as the arithmetic mean is (1977):

$$n_{SMR} = \left(\frac{\sqrt{n_1} + \cdots + \sqrt{n_k}}{k} \right)^2. \quad (11.35)$$

The square mean root for the nineteen studies was calculated by substituting the mean of the experimental and control group sample sizes in equation 11.35,

$$n_{SMR} = \left[\left(\sqrt{(77 + 339)/2} + \cdots + \sqrt{(65 + 67)/2} \right)/19 \right]^2$$
$$= 83.515, \text{ or } 84.$$

Our estimate of π given by equation 11.34 is $p = 11/(11 + 8) = 0.579$. Linear interpolation in table 11.7

yields the estimate $\hat{\delta} = 0.011$ and the 95 percent confidence interval $[-0.019 \; 0.048]$. Because the confidence interval includes the value zero, we fail to reject the null hypothesis that $\delta = 0$.

Example 2: Obtaining an Estimate and Confidence Interval for the Population Correlation Coefficient ρ Based on the Proportion of the Total Number of Significant correlations that Were Positive. Suppose that all the results in data set III are significant; thus we have eighteen significant positive results and two significant negative results. Using equation 11.35, we obtain the square mean root

$$n_{SMR} = \left[(\sqrt{10} + \cdots + \sqrt{20})/20 \right]^2 = 25.878, \text{ or } 26.$$

Our estimate of π given by equation (34) is $p = 18/(18 + 2) = 0.900$. Linear interpolation in table 11.8 yields the estimate $\hat{\rho} = 0.100$ and the 95 percent confidence interval $[0.058 \; 0.300]$. Because the confidence interval excludes the value zero, we reject the null hypothesis that $\rho = 0$.

11.5.2 Results All in the Same Direction

The method of maximum likelihood cannot be used if the MLE of π (that is, $p = U/k$) is 1 or 0 because there is no

Table 11.8 Conditional Probability $p(\rho, n) = p\{r>0 \mid |t| > C_r\}$

n	0	0.01	0.02	0.03	0.04	0.05	0.06	0.07	0.08	0.09	0.10	0.20	0.30	0.40	0.50	0.60	0.70	0.80	0.90
3	0.500	0.508	0.516	0.524	0.531	0.539	0.547	0.555	0.563	0.570	0.578	0.654	0.725	0.790	0.848	0.897	0.937	0.968	0.989
4	0.500	0.512	0.525	0.537	0.550	0.562	0.574	0.586	0.598	0.610	0.622	0.732	0.823	0.891	0.938	0.968	0.986	0.995	0.999
5	0.500	0.516	0.533	0.549	0.565	0.581	0.597	0.613	0.628	0.644	0.659	0.790	0.883	0.940	0.972	0.988	0.996	0.999	1.000
6	0.500	0.520	0.540	0.559	0.578	0.598	0.617	0.635	0.654	0.671	0.689	0.832	0.920	0.965	0.986	0.995	0.999	1.000	
7	0.500	0.523	0.546	0.568	0.590	0.612	0.634	0.655	0.675	0.695	0.714	0.864	0.943	0.978	0.993	0.998	0.999	1.000	
8	0.500	0.526	0.551	0.576	0.601	0.625	0.649	0.672	0.694	0.715	0.736	0.887	0.958	0.986	0.996	0.999	1.000		
9	0.500	0.528	0.556	0.583	0.610	0.637	0.662	0.687	0.711	0.733	0.755	0.906	0.969	0.991	0.997	0.999	1.000		
10	0.500	0.530	0.560	0.590	0.619	0.647	0.674	0.701	0.725	0.749	0.771	0.920	0.976	0.994	0.998	1.000			
11	0.500	0.532	0.565	0.596	0.627	0.657	0.686	0.713	0.739	0.763	0.786	0.932	0.981	0.995	0.999	1.000			
12	0.500	0.534	0.569	0.602	0.635	0.666	0.696	0.724	0.751	0.776	0.799	0.942	0.985	0.997	0.999	1.000			
13	0.500	0.536	0.572	0.607	0.642	0.674	0.706	0.735	0.762	0.788	0.811	0.950	0.988	0.998	1.000				
14	0.500	0.538	0.576	0.613	0.648	0.682	0.714	0.745	0.773	0.799	0.822	0.956	0.991	0.998	1.000				
15	0.500	0.540	0.579	0.618	0.655	0.690	0.723	0.754	0.782	0.808	0.832	0.962	0.992	0.999	1.000				
16	0.500	0.542	0.582	0.622	0.661	0.697	0.731	0.762	0.791	0.818	0.841	0.966	0.994	0.999	1.000				
17	0.500	0.543	0.586	0.627	0.666	0.704	0.738	0.770	0.800	0.826	0.850	0.970	0.995	0.999	1.000				
18	0.500	0.545	0.589	0.631	0.672	0.710	0.745	0.778	0.807	0.834	0.857	0.974	0.996	0.999	1.000				
19	0.500	0.546	0.591	0.635	0.677	0.716	0.752	0.785	0.815	0.841	0.865	0.977	0.997	1.000					
20	0.500	0.548	0.594	0.639	0.682	0.722	0.759	0.792	0.822	0.848	0.871	0.979	0.997	1.000					
21	0.500	0.549	0.597	0.643	0.687	0.728	0.765	0.798	0.828	0.854	0.877	0.981	0.998	1.000					
22	0.500	0.550	0.600	0.647	0.692	0.733	0.771	0.804	0.834	0.861	0.883	0.983	0.998	1.000					
23	0.500	0.552	0.602	0.651	0.696	0.738	0.776	0.810	0.840	0.866	0.889	0.985	0.998	1.000					
24	0.500	0.553	0.605	0.654	0.701	0.743	0.782	0.816	0.846	0.872	0.894	0.986	0.999	1.000					
25	0.500	0.554	0.607	0.658	0.705	0.748	0.787	0.821	0.851	0.877	0.898	0.988	0.999	1.000					
50	0.500	0.579	0.654	0.723	0.782	0.832	0.872	0.904	0.928	0.947	0.961	0.998	1.000						
100	0.500	0.613	0.715	0.799	0.863	0.909	0.941	0.962	0.976	0.985	0.990	1.000							
200	0.500	0.658	0.788	0.877	0.933	0.964	0.981	0.990	0.995	0.997	0.999	1.000							
400	0.500	0.717	0.866	0.943	0.977	0.991	0.996	0.999	0.999	1.000	1.000	1.000							

SOURCE: Bushman 1994.

NOTES: Probability that a correlation coefficient r from a sample of size n is positive given that the absolute value of r exceeds the $\alpha = .05$ critical value for effect size ρ. Proportions less than .500 correspond to negative values of ρ.

unique value of p. If all the results are in the same direction, we can obtain a Bayes estimate of π (Hedges and Olkin 1985, 300). For example, we can assume that π has a truncated uniform prior distribution. That is, we assume that there is a value of π_0 such that any value of π in the interval $[\pi_0 \; 1]$ is equally likely. The Bayes estimate of π is

$$p = \frac{(k+1)(1-\pi_0^{k+2})}{(k+2)(1-\pi_0^{k+1})} \qquad (11.36)$$

Example: Bayes Estimate for Results All in the Same Direction. Suppose that all the studies in data set II found positive results and that it is reasonable to assume that the value of π is in the interval $[.5 \; 1]$, and that any value in the interval is equally likely. The Bayes estimate given in equation (35) is

$$p = \frac{(19+1)(1-.5^{(19+2)})}{(19+2)(1-.5^{(19+1)})} = 0.952.$$

Unfortunately, one cannot compute a confidence interval for the Bayes estimate because the variance estimator for this Bayes estimate does not exist.

11.5.3 Missing Data

Missing effect size estimates is one of the largest problems facing the practicing meta-analyst. Sometimes research reports do not include enough information (for example, means, standard deviations, statistical tests) to permit the calculation of an effect size estimate. Unfortunately, the proportion of studies with missing effect size estimates in a research synthesis is often quite large, about 25 percent in psychological studies (Bushman and Wang 1995, 1996). While sophisticated missing data methods are available for meta-analysis (see chapter 21, this volume), the most commonly used solutions to the problem of missing effect size estimates are to: omit from the synthesis those studies with missing effect size estimates and analyze only complete cases, set the missing effect size estimates equal to zero, set the missing effect size estimates equal to the mean obtained from studies with effect size estimates, and use the available information in a research report to get a lower limit for the effect size estimate (Rosenthal 1994). Unfortunately, all of these procedures have serious problems that limit their usefulness (Bushman and Wang 1996). Solutions one and three make the unrealistic assumption that missing effect size estimates are missing completely at random. It is well known that nonsignificant effects are less likely to be reported than significant effects (Greenwald 1975). Imputing the value zero (solution one) or the mean (solution three), for missing effect sizes artificially decreases the variance of effect size estimates. Imputing the value zero (solution two) or using the lower limit for an effect size (solution four) underestimates the population effect size.

Our combined procedure (described in section 11.4) has at least five advantages over conventional procedures for dealing with the problem of missing effect size estimates (Bushman and Wang 1996). First, the combined procedure uses all the information available from studies in a meta-analysis; most of the other procedures ignore information about the direction and statistical significance of results. Second, the combined estimator is consistent (that is, if the number of studies k is large, the combined estimator will not underestimate or overestimate the population effect size). Third, the combined estimator is more efficient (that is, has a smaller variance) than either the effect size or vote-counting estimators. Fourth, the variance of the combined estimator is known. All of the conventional approaches for handling missing data either artificially inflate or artificially deflate the variance of the effect size estimates. Fifth, the combined estimator gives weight to all studies proportional to the Fisher information they provide (that is no studies are overweighted).

11.6 CONCLUSIONS

We hope this chapter will improve the bad reputation that vote-counting procedures have gained. It is true that vote-counting procedures are not the method of choice. If studies include enough information to compute effect size estimates, then effect size procedures should be used. Moreover the vote-counting procedures that are currently available are based on the fixed effects model and they do not provide information about the potential heterogeneity of effects across studies. Often, however, not all studies contain enough information to compute an effect size estimate. If the studies with missing effects do provide information about the direction and statistical significance of results (for example, a table of means with no standard deviations or statistical tests) and a fixed effects estimate is desired, vote-counting procedures can and should be used. The meta-analyst can then combine the effect size estimate with the vote-counting estimates to obtain an overall estimate of the population effect size.

11.7 NOTES

1. We use the term *positive* to refer to effects in the predicted direction.
2. Larry Hedges found that estimators based on the third type of data are, at best, only 64 percent as efficient as estimators based on the first type of data (1986). The maximum relative efficiency occurs when the population effect size equals zero for all studies. When the population effect size does not equal zero for all studies, relative efficiency decreases as the population effect size increases and as the sample size increases.
3. Generally the term *independent variable* refers to variables that are directly manipulated by the researcher. We also use the term to refer to variables that are measured by the researcher (for example, participant gender). Although this is not technically correct, it simplifies the discussion considerable.
4. We assume that the synthesist has a priori predictions about the direction of the relationship and is therefore using a one-sided test. For a two-sided test, we let T_i^2 be an estimator of θ_i^2. In general, it is not possible to use a two-sided test statistic T_i^2 to estimate θ if θ can be either positive or negative (Hedges and Olkin 1985, 52).

11.8 REFERENCES

Brown, Lawrence D., T. Tony Cai, and Anirban DasGupta 2001. "Interval Estimation for a Binomial Proportion." *Statistical Science* 16(2): 101–33.

Bushman, Brad J. 1994. "Vote-Counting Procedures in Meta-Analysis." In *The Handbook of Research Synthesis*, edited by Harris Cooper and Larry V. Hedges. New York: Russell Sage Foundation.

Bushman, Brad J., and Morgan C. Wang. 1995. "A Procedure for Combining Sample Correlations and Vote Counts to Obtain an Estimate and a Confidence Interval for the Population Correlation Coefficient." *Psychological Bulletin* 117(3): 530–46.

———. 1996. "A Procedure for Combining Sample Standardized Mean Differences and Vote Counts to Estimate the Population Standardized Mean Difference in Fixed Effects Models." *Psychological Methods* 1(1): 66–80.

Cohen, Jacob. 1988. *Statistical Power Analysis for the Behavioral Sciences,* 2nd ed. New York: Academic Press.

Conover, William J. 1980. *Practical Nonparametric Statistics,* 2nd ed. New York: John Wiley & Sons.

Dochy, Filip, Mien Segers, Piet Van den Bossche, and David Gijbels. 2003. "Effects of Problem-Based Learning: A Meta-Analysis." *Learning and Instruction* 5(13): 533–68.

Friedman, Lee 2001. "Why Vote-Count Reviews Don't Count." *Biological Psychiatry* 49(2): 161–62.

Gibbons, Jean D., Ingram Olkin, and Milton Sobel 1977. *Selecting and Ordering Populations: A New Statistical Methodology.* New York: John Wiley & Sons.

Greenwald, Anthony G. 1975. "Consequences of Prejudice Against the Null Hypothesis." *Psychological Bulletin* 82(1): 1–20.

Hedges, Larry V. 1986. "Estimating Effect Sizes from Vote Counts or Box Score Data." Paper presented at the annual meeting of the American Educational Research Association. San Francisco, Calif. (April 1986).

Hedges, Larry V., and Ingram Olkin. 1980. "Vote-Counting Methods in Research Synthesis." *Psychological Bulletin* 88(2): 359–69.

———. 1986. *Statistical Methods for Meta-Analysis.* New York: Academic Press.

Jewell, George, and Mark E. McCourt 2000. "Pseudoneglect: A Review and Meta-Analysis of Performance Factors in Line Bisection Tasks." *Neuropsychologia* 38: 93–110.

Lee, Valerie E., and Anthony S. Bryk. 1989. "Effects of Single-Sex Schools: Response to Marsh." *Journal of Educational Psychology* 81(4): 647–50.

Light, Richard J., and Paul V. Smith. 1971. "Accumulating Evidence: Procedures for Resolving Contradictions Among Different Research Studies." *Harvard Educational Review* 41(4): 429–71.

Mann, Charles C. 1994. "Can Meta-Analysis Make Policy?" *Science* 266(11): 960–62.

Rafaeli-Mor, Eshkol, and Jennifer Steinberg. 2002. "Self-Complexity and Well-Being: A Review and Research Synthesis." *Personality and Social Psychology Review* 6(1): 31–58.

Richard, F. Daniel, Charles F. Bond, and Juli J. Stokes-Zoota. 2003. "One Hundred Years of Social Psychology Quantitatively Described." *Review of General Psychology* 7(4): 331–63.

Rosenthal, Robert. 1994. "Parametric Measures of Effect Size." In *Handbook of Research Synthesis*, edited by Harris Cooper and Larry V. Hedges. New York: Russell Sage Foundation.

Saroglou, Vassilis. 2002. "Religion and the Five Factors of Personality: A Meta-Analytic Review." *Personality and Individual Differences* 32(1): 15–25.

Warner, James. 2001. Quality of Evidence in Meta-Analysis." *British Journal of Psychiatry* 179(July): 79.

12

EFFECT SIZES FOR CONTINUOUS DATA

MICHAEL BORENSTEIN
Biostat

CONTENTS

12.1 INTRODUCTION

In any meta-analysis, we start with summary data from each study and use it to compute an effect size for the study. An effect size is a number that reflects the magnitude of the relationship between two variables. For example, if a study reports the mean and standard deviation for the treated and control groups, we might compute the standardized mean difference between groups. Or, if a study reports events and nonevents in two groups we might compute an odds ratio. It is these effect sizes that are then compared and combined in the meta-analysis.

Consider figure 12.1, the forest plot of a fictional meta-analysis to assess the impact of an intervention. In this plot, each study is represented by a square, bounded on either side by a confidence interval. The location of each square on the horizontal axis represents the effect size for that study. The confidence interval represents the precision with which the effect size has been estimated, and the size of each square is proportional to the weight that will be assigned to the study when computing the combined effect. This figure also serves as the outline for this chapter, in which I discuss what these items mean and how they are computed. This chapter addresses effect sizes for continuous outcomes such as means and correla-

tions (for effect sizes for binary outcomes, see chapter 13, this volume).

12.1.1 Effect Sizes and Treatment Effects

Meta-analyses in medicine often refer to the effect size as a *treatment effect*, and this term is sometimes assumed to refer to odds ratios, risk ratios, or risk differences, which are common in meta-analyses that deal with medical interventions. Similarly, meta-analyses in the social sciences often refer to the effect size simply as an *effect size*, and this term is sometimes assumed to refer to standardized mean differences or to correlations, which are common in social science meta-analyses.

In fact, though, both the terms effect size and treatment effect can refer to any of these indices, and the distinction between the terms lies not in the index but rather in the nature of the study. The term *effect size* is appropriate when the index is used to quantify the relationship between any two variables or a difference between any two groups. By contrast, the term *treatment effect* is appropriate only for an index used to quantify the impact of a deliberate intervention. Thus, the difference between males and females could be called an *effect size* only, while the difference between treated and control groups could be

Impact of Intervention

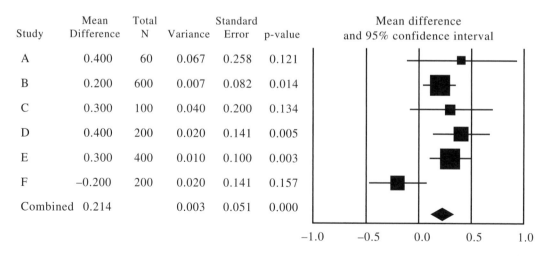

Study	Mean Difference	Total N	Variance	Standard Error	p-value
A	0.400	60	0.067	0.258	0.121
B	0.200	600	0.007	0.082	0.014
C	0.300	100	0.040	0.200	0.134
D	0.400	200	0.020	0.141	0.005
E	0.300	400	0.010	0.100	0.003
F	−0.200	200	0.020	0.141	0.157
Combined	0.214		0.003	0.051	0.000

Figure 12.1 Fictional Meta-Analysis Showing Impact of an Intervention

SOURCE: Author's compilation.

called either an *effect size* or a *treatment effect*. Note, however, that the classification of an index as an *effect size* or a *treatment effect* has no bearing on the computations.

Four major considerations should drive the choice of an effect-size index. The first is that the effect sizes from the different studies should be comparable to one another in the sense that they measure, at least approximately, the same thing. That is, the effect size should not depend on aspects of study design that may vary from study to study (such as sample size or whether covariates are used). The second is that the effect size should be substantively interpretable. This means that researchers in the substantive area of the work represented in the synthesis should find the effect size meaningful. The third is that estimates of the effect size should be computable from the information likely to be reported in published research reports. That is, it should not require reanalysis of the raw data. The fourth is that the effect size should have good technical properties. For example, its sampling distribution should be known so that variances and confidence intervals can be computed.

12.1.2 Effect Sizes Rather than *p*-Values

Reports of primary research typically include the *p*-value corresponding to a test of significance. This *p*-value reflects the likelihood that the sample would have yielded the observed effect (or one more extreme) if the null hypothesis were true.

Researchers often use the *p*-value as a surrogate for the effect size, with a significant *p*-value taken to imply a large effect and a nonsignificant *p*-value taken to imply a trivial effect. In fact, however, while the *p*-value is partly a function of effect size, it is also partly a function of sample size. A *p*-value of 0.01 could reflect a large effect but could also reflect a trivial effect in a large sample. Conversely, a *p*-value of 0.20 could reflect a trivial effect but could also reflect a large effect in a small sample. In figure 12.1, for example, study A has a *p*-value of 0.12 whereas study B has a *p*-value of 0.01, but it is study A that has the larger effect size (0.40 versus 0.20) (see, for example, Borenstein 1994).

In primary studies we can avoid this kind of confusion by reporting the effect size and the precision separately. The former gives us a pure estimate of the effect in the sample, and the latter gives us a range for the effect in the population. Similarly, in a meta-analysis, we need to work with a pure measure of the effect size from each primary study to determine if the effects are consistent, and to compute a combined estimate of the effect across studies. Here, the precision of each effect is used to assign a weight to that effect in these analyses.

12.1.3 Effect-Size Parameters and Sample Estimates of Effect Sizes

Throughout this chapter we make the distinction between an underlying effect-size parameter (denoted by the Greek letter θ) and the sample estimate of that parameter (denoted by T).

If a study had an infinitely large sample size, it would yield an effect size T that was identical to the population parameter θ. In fact, though, sample sizes are finite and so the effect-size estimate T always differs from θ by some amount. The value of T will vary from sample to sample, and the distribution of these values is the sampling distribution of T. Statistical theory allows us to compute the variance of the effect-size estimates.

12.1.4 The Variance of the Effect Size

A dominant factor in variance is the sample size, with larger studies having a smaller variance and yielding a more precise estimate of the effect-size parameter. Using a matched design or including a covariate to reduce the error term will usually lead to a lower variance and more precise estimate. Additionally, the variance of any given effect size is affected by specific factors that vary from one effect-size index to the next.

The sampling distribution of an effect-size estimate T can be expressed as a standard error or as a variance (which is simply the square of the standard error).

When our focus is on the effect size for a single study, we generally work with the standard error of the effect size, which is turn may be used to compute confidence intervals about the effect size. In figure 12.1, for example, study E has four times the sample size of study C (400 versus 100). Its standard error (the square root of the variance) is therefore half as large (0.10 versus 0.20) and its confidence interval half as wide as that of study C. (In this example, all other factors that could affect the precision were held constant).

By contrast, in a meta-analysis we work primarily with the variance rather than the standard error. In a fixed effect analysis, for example, we weight by the inverse variance, or $1/v$. In figure 12.1, study E has four times the sample size of study C (400 versus 100) and its variance is therefore one-fourth as large (0.01 versus 0.04). The square representing study E has four times the area as the one for study C, reflecting the fact that it will be assigned four times as much weight in the analysis.

12.1.5 Calculating Effect-Size Estimates from Reported Information

When researchers have access to a full set of summary data such as means, standard deviations, and sample size for each group, the computation of the effect size and its variance is relatively straightforward. In practice however, researchers will often find themselves working with only partial data. For example, a paper may publish only the p-value and sample size from a test of significance, leaving it to the meta-analyst to back-compute the effect size and variance. For this reason, each of the following sections includes a table that shows how to compute the effect size and variance from some of the more common reporting formats. For additional information on computing effect sizes from partial information, see the resources section at the end of this chapter.

12.2 THE RAW (UNSTANDARDIZED) MEAN DIFFERENCE D

When the outcome is reported on a meaningful scale and all studies in the analysis use the same scale, the meta-analysis can be performed directly on the raw mean difference. The primary advantage of the raw mean difference is that it is intuitively meaningful, either inherently (blood pressure) or because of widespread use (for example, scores on a national achievement test for students).

Consider a study that reports means for two groups (treated and control) and suppose we wish to compare the means of these two groups. Let μ_1 and μ_2 be the true (population) means of the two groups. The population mean difference is defined as

$$\Delta = \mu_1 - \mu_2. \tag{12.1}$$

In this section I show how to estimate Δ from studies that use two independent groups, and from studies that use matched groups or a pre-post design.

12.2.1 Computing D, Independent Groups

We can estimate the mean difference Δ from a study that uses two independent groups as follows. Let $\bar{Y}_1$ and $\bar{Y}_2$ be the sample means of the two independent groups. The sample estimate of Δ is just the sample mean difference, namely

$$D = \bar{Y}_1 - \bar{Y}_2. \tag{12.2}$$

Note that uppercase D is used for the raw mean difference, whereas lowercase d will be used for the standardized mean difference (section 12.3).

Let S_1 and S_2 be the sample standard deviations of the two groups, and n_1 and n_2 be the sample size in the two groups. If we assume that the two population standard deviations are the same (as is assumed in most parametric data analysis techniques), so that $\sigma_1 = \sigma_2 = \sigma$, then the variance of D is

$$v_D = \frac{n_1 + n_2}{n_1 n_2} S^2_{Pooled}, \tag{12.3}$$

where

$$S^2_{Pooled} = \frac{(n_1 - 1)S_1^2 + (n_2 - 1)S_2^2}{n_1 + n_2 - 2}. \tag{12.4}$$

If we don't assume that the two population standard deviations are the same, then the variance of D is

$$v_D = \frac{S_1^2}{N_1} + \frac{S_2^2}{N_2}. \tag{12.5}$$

In either case, the standard error of D is then the square root of v_D,

$$SE_D = \sqrt{v_D}. \tag{12.6}$$

For example, suppose that a study has sample means $\bar{Y}_1 = 103$, $\bar{Y}_2 = 100$, sample standard deviations $S_1 = 5.5$, $S_2 = 4.5$, and sample sizes $n_1 = n_2 = 50$. The raw mean difference D is

$$D = 103 - 100 = 3.000.$$

If we assume that $\sigma_1^2 = \sigma_2^2$ then the pooled standard deviation within groups is

$$S_{Within} = \sqrt{\frac{(50-1) \times 5.5^2 + (50-1) \times 4.5^2}{50 + 50 - 2}} = 5.0249.$$

The variance and standard error of D are given by

$$v_D = \frac{50 + 50}{50 \times 50} \times 5.0249^2 = 1.0100$$

and

$$SE_D = \sqrt{1.0100} = 1.0050.$$

If we do not assume that $\sigma_1^2 = \sigma_2^2$, then the variance and standard error of D are given by

$$v_D = \frac{5.5^2}{50} + \frac{4.5^2}{50} = 1.0100,$$

and

$$SE_D = \sqrt{1.0100} = 1.0050.$$

In this example both formulas yield the same result because $n_1 = n_2$.

12.2.2 Computing *D*, Pre-Post Scores or Matched Groups

We can estimate the mean difference Δ from a study that uses two matched groups as follows. Let $\overline{Y}_1$ and $\overline{Y}_2$ be the sample means of the two groups. The sample estimate of Δ is just the sample mean difference, namely

$$D = \overline{Y}_1 - \overline{Y}_2. \qquad (12.7)$$

However, for the matched study, the variance of D is computed as

$$v_D = \frac{S_{Difference}^2}{n}. \qquad (12.8)$$

In this equation, n is the number of pairs and $S_{Difference}$ is the standard deviation of the paired differences,

$$S_{Difference} = \sqrt{S_1^2 + S_2^2 - 2 \times r \times S_1 \times S_2}, \qquad (12.9)$$

where r is the correlation between siblings in matched pairs. As r moves toward 1.0, the standard error of the paired difference will decrease.

The formulas for matched designs apply to pre-post designs as well. The pre and post means correspond to the means in the matched groups, n is the number of subjects, and r is the correlation between pre-scores and post-scores.

12.2.3 Including Different Study Designs in the Same Analysis

To include data from independent groups and matched groups in the same meta-analysis, one would simply compute the raw mean difference and its variance for each study using the appropriate formulas. The effect size has the same meaning regardless of the study design, and

the variances have been computed to take account of the study design. We can therefore use these effect sizes and variances to perform a meta-analysis without concern for the different designs in the original studies. Note that the phrase *different study designs* is being used here to refer to the fact that one study used independent groups and another used a pre-post design or matched groups. It does not refer to the case where one study was a randomized trial (or quasi-experimental study) and another was observational. The issues involved in combining these different kinds of studies are beyond the scope of this chapter.

For all study designs the direction of the effect ($\overline{Y}_1 - \overline{Y}_2$ or $\overline{Y}_2 - \overline{Y}_1$) is arbitrary, except that the researcher must decide on a convention and then apply this consistently. For example, if a positive difference will indicate that the treated group did better than the control group, then this convention must apply for studies that used independent designs and for studies that used pre-post designs. In some cases it might be necessary to reverse the computed sign of the effect size to ensure that the convention is followed.

12.3 THE STANDARDIZED MEAN DIFFERENCE *d* AND *g*

As noted, the raw mean difference is a useful index when the measure is meaningful, either inherently or because of widespread use. By contrast, when the measure is less well known (for example, a proprietary scale with limited distribution), the use of a raw mean difference has less to recommend it. In any event, the raw mean difference is an option only if all the studies in the meta-analysis use the same scale. If different studies use different instruments, such as different psychological or educational tests, to assess the outcome, then the scale of measurement will differ from study to study and it would not be meaningful to combine raw mean differences.

In such cases, Gene Glass suggested dividing the mean difference in each study by that study's standard deviation to create an index (the standardized mean difference) that would be comparable across studies (1976). The same approach had been suggested earlier by Jacob Cohen in connection with describing the magnitude of effects in statistical power analysis (1969, 1977).

The standardized mean difference can be thought of as being comparable across studies based on either of two arguments (Hedges and Olkin 1985). If the outcome measures in all studies are linear transformations of each other the standardized mean difference can be seen as the mean difference that would have been obtained if all data were transformed to a scale where the standard deviation within-groups was equal to 1.0.

The other argument for comparability of standardized mean differences is that the standardized mean difference is a measure of overlap between distributions. In this telling, the standardized mean difference reflects the difference between the two distributions, how each represents a distinct cluster of scores even if they do not measure exactly the same outcome (see Cohen 1977; Grissom and Kim 2005).

Consider a study that uses independent groups, and suppose we wish to compare their means. Let μ_1 and σ_1 be the true (population) mean and standard deviation of the first group and let μ_2 and σ_2 be the true (population) mean and standard deviation of the other group. If the two population standard deviations are the same (as is assumed in most parametric data analysis techniques), so that $\sigma_1 = \sigma_2 = \sigma$, then the standardized mean difference parameter or population standardized mean difference is defined as

$$\delta = \frac{\mu_1 - \mu_2}{\sigma}. \tag{12.10}$$

In this section I show how to estimate δ from studies that use independent groups, studies that use pre-post or matched group designs, and studies that use analysis of covariance.

12.3.1 Computing *d* and *g*, Independent Groups

We can estimate the standardized mean difference (δ) from studies that use two independent groups as

$$d = \frac{\overline{Y}_1 - \overline{Y}_2}{S_{Within}}. \tag{12.11}$$

In the numerator, $\overline{Y}_1$ and $\overline{Y}_2$ are the sample means in the two groups. In the denominator S_{Within} is the within-groups standard deviation, pooled across groups,

$$S_{Within} = \sqrt{\frac{(n_1 - 1)S_1^2 + (n_2 - 1)S_2^2}{n_1 + n_2 - 2}}, \tag{12.12}$$

where n_1, n_2 are the sample size in the two groups, and S_1, S_2 are the standard deviations in the two groups. The reason that we pool the two sample estimates of the standard deviation is that even if we assume that the underlying population standard deviations are the same (that is

$\sigma_1 = \sigma_2 = \sigma$), it is unlikely that the sample estimates S_1 and S_2 will be identical. By pooling, we obtain a more accurate estimate of their common value.

This index, the *sample estimate* of the standardized mean difference, is often called Cohen's *d* in research synthesis. Some confusion about the terminology has resulted from the fact that the index δ, which Cohen originally proposed as a *population parameter* for describing the size of effects for statistical power analysis is also sometimes called *d*. In this handbook, we use the symbol δ to denote the effect-size parameter and *d* for the sample estimate of that parameter.

The variance of *d* is given (to a very good approximation) by

$$v_d = \frac{n_1 + n_2}{n_1 n_2} + \frac{d^2}{2(n_1 + n_2)}, \tag{12.13}$$

In this equation the first term on the right expresses the uncertainty in the estimate of the mean difference [the numerator in formula 12.11], and the second expresses the uncertainty in the estimate of S_{Within} [the denominator in formula 12.11].

The standard error of *d* is the square root of v_d,

$$SE_d = \sqrt{v_d}. \tag{12.14}$$

It turns out that *d* has a slight bias, tending to overestimate the absolute value of δ in small samples This bias can be removed by a simple correction that yields an unbiased estimate of δ. This estimate is sometimes called Hedges' *g* (Hedges 1981). To convert from *d* to Hedges' *g* we use a correction factor, *J*, defined as

$$J(df) = 1 - \frac{3}{4df - 1}. \tag{12.15}$$

In this expression, *df* is the degrees of freedom used to estimate S_{Within}, which for two independent groups is $n_1 + n_2 - 2$. Then,

$$g = j(df)d, \tag{12.16}$$

$$v_g = [J(df)]^2 v_d, \tag{12.17}$$

and

$$SE_g = \sqrt{v_g}. \tag{12.18}$$

For example, suppose a study has sample means $\overline{Y}_1 = 103$, $\overline{Y}_2 = 100$, sample standard deviations $S_1 = 5.5$,

$S_2 = 4.5$, and sample sizes $n_1 = n_2 = 50$. We would estimate the pooled-within-groups standard deviation as

$$S_{Within} = \sqrt{\frac{(50-1) \times 5.5^2 + (50-1) \times 4.5^2}{50 + 50 - 2}} = 5.0249.$$

Then,

$$d = \frac{103 - 100}{5.0249} = 0.5970,$$

$$v_d = \frac{50 + 50}{50 \times 50} + \frac{0.5970^2}{2(50 + 50)} = 0.0418,$$

and

$$SE_d = \sqrt{0.0418} = 0.2045.$$

The correction factor J, Hedges' g, its variance and standard error are given by

$$J(50 + 50 - 2) = \left(1 - \frac{3}{4 \times 98 - 1}\right) = 0.9923,$$

$$g = 0.9923 \times 0.5970 = 0.5924,$$

$$v_g = 0.9923^2 \times 0.0418 = 0.0412,$$

and

$$SE_g = \sqrt{0.0412} = .2030.$$

The correction factor J is always less than 1.0, and so g will always be less than d in absolute value, and the variance of g will always be less than the variance of d. However, J will be very close to 1.0 unless df is very small (say, less than 10) and so (as in this example) the difference is usually trivial.

Some slightly different expressions for the variance of d (and g) have been given by different authors and even by the same authors at different times. For example, the denominator of the second term of the variance of d is given here as $2(n_1 + n_2)$. This expression is obtained by one method (assuming the n's become large with δ fixed). An alternate derivation (assuming n's become large with $\sqrt{n}\delta$ fixed) leads to a denominator in the second term that is slightly different, namely, $2(n_1 + n_2 - 2)$. Unless n_1 and n_2 are very small, these expressions will be almost identical.

Similarly, the expression given here for the variance of g is J^2 times the variance of d, but many authors ignore the J^2 term because it is so close to unity in most cases. Again, though it is preferable to include this correction factor, including it is likely to make little practical difference.

Table 12.1 provides formulas for computing d and its variance for independent groups, working from information that may be reported in a published paper.

12.3.2 Computing d and g, Pre-Post Scores or Matched Groups

We can estimate the standardized mean difference (δ) from studies that use matched groups or pre-post scores in one group. The formula for the sample estimate of d is

$$d = \frac{\overline{Y}_1 - \overline{Y}_2}{S_{Within}}. \tag{12.19}$$

This is the same formula as for independent groups, 12.11. However, there is an important difference in the mechanism used to compute S_{Within}. When we are working with independent groups, the natural unit of deviation is the standard deviation within groups. Therefore, this value is typically reported, or easily imputed. By contrast, when we are working with matched groups, the natural unit of deviation is the standard deviation of the difference scores, and so this is the value that is likely to be reported. To compute d from this kind of study, we need to impute the standard deviation within groups, which would then serve as the denominator in formula 12.19.

When working with a matched study, the standard deviation within groups can be imputed from the standard deviation of the difference, using

$$S_{Within} = \frac{S_{Difference}}{\sqrt{2(1 - r)}}, \tag{12.20}$$

where r is the correlation between pairs of observations (for example, the correlation between pre-test and post-test). Then we can apply formula 12.19 to compute d. The variance of d is given by

$$v_d = \left(\frac{1}{n} + \frac{d^2}{2n}\right)2(1 - r), \tag{12.21}$$

where n is the number of pairs. The standard error of d is just the square root of v,

$$SE_d = \sqrt{v_d}. \tag{12.22}$$

Because the correlation between pre-test post-test scores is required to impute the standard deviation within groups from the standard deviation of the difference, we must assume that this correlation is known or can be estimated with high precision. Often, the researcher will need to use data from other sources to estimate this correlation.

Table 12.1 Formulas for Computing d in Designs with Independent Groups

Reported	Computation of Needed Quantities	
$\overline{Y}_1, \overline{Y}_2, S_{Pooled}, n_1, n_2$	$d = \dfrac{Y_1 - Y_2}{S_{Pooled}},$	$v = \dfrac{n_1 + n_2}{n_1 n_2} + \dfrac{d^2}{2(n_1 + n_2)}$
t, n_1, n_2	$d = t\sqrt{\dfrac{n_1 + n_2}{n_1 n_2}},$	$v = \dfrac{n_1 + n_2}{n_1 n_2} + \dfrac{d^2}{2(n_1 + n_2)}$
F, n_1, n_2	$d = \pm\sqrt{\dfrac{F(n_1 + n_2)}{n_1 n_2}},$	$v = \dfrac{n_1 + n_2}{n_1 n_2} + \dfrac{d^2}{2(n_1 + n_2)}$
p(one-tailed), n_1, n_2	$d = \pm t^{-1}(p)\sqrt{\dfrac{n_1 + n_2}{n_1 n_2}},$	$v = \dfrac{n_1 + n_2}{n_1 n_2} + \dfrac{d^2}{2(n_1 + n_2)}$
p(two-tailed), n_1, n_2	$d = \pm t^{-1}\left(\dfrac{p}{2}\right)\sqrt{\dfrac{n_1 + n_2}{n_1 n_2}},$	$v = \dfrac{n_1 + n_2}{n_1 n_2} + \dfrac{d^2}{2(n_1 + n_2)}$

SOURCE: Author's compilation.
NOTE: The function $t^{-1}(p)$ is the inverse of the cumulative distribution function of student's t with $n_1 + n_2 - 2$ degrees of freedom. Many computer programs and spreadsheets provide functions that can be used to compute t^{-1}. Assume $n_1 = n_2 = 10$, so that df $= 18$. Then, in Excel™, for example, if the reported p-value is 0.05 (two-tailed) TINV(p, df) = TINV(0.05,18) will return the required value (2.1009). If the reported p-value is 0.05 (one-tailed), TINV$(2p, df)$ = TINV(0.10,18) will return the required value 1.7341. The F in row 3 of the table is the F-statistic from a one-way analysis of variance. In rows 3–5 the sign of d must reflect the direction of the mean difference.

If the correlation is not known precisely, one could work with a range of plausible correlations and use a sensitivity analysis to see how these affect the results.

To compute Hedges' g and associated statistics, we would use formulas 12.15 through 12.18. The degrees of freedom for computing J is $n - 1$, where n is the number of pairs. For example, suppose that a study has pretest and posttest sample means $\overline{Y}_1 = 103$, $\overline{Y}_2 = 100$, sample standard deviation of the difference $S_{Difference} = 5.5$, and sample size $n = 50$ and a correlation between pretest and posttest of $r = 0.7$.

The standard deviation within groups is imputed from the standard deviation of the difference by

$$S_{Within} = \frac{5.5}{\sqrt{2(1 - 0.7)}} = 7.1000.$$

Then d, its variance and standard error are computed as

$$d = \frac{103 - 100}{7.1000} = 0.4225,$$

$$v_d = \left(\frac{1}{50} + \frac{0.4225^2}{2 \times 50}\right)(2(1 - 0.7)) = 0.0131,$$

and

$$SE_d = \sqrt{0.0131} = .1145.$$

The correction factor J, Hedges' g, its variance and standard error are given by

$$J(n - 1) = \left(1 - \frac{3}{4 \times 49 - 1}\right) = 0.9846,$$

$$g = 0.9846 \times 0.4225 = 0.4160,$$

$$v_g = 0.9846^2 \times 0.0131 = 0.0127,$$

and

$$SEg = \sqrt{0.0127} = 0.1127.$$

Table 12.2 provides formulas for computing d and its variance for matched groups, working from information that may be reported in a published paper.

12.3.3 Computing d and g, Analysis of Covariance

We can estimate the standardized mean difference (δ) from studies that use analysis of covariance. The formula for the sample estimate of d is

$$d = \frac{\overline{Y}_1^{Adjusted} - \overline{Y}_2^{Adjusted}}{S_{Within}}. \tag{12.23}$$

Table 12.2 Formulas for Computing *d* in Designs with Paired Groups

Reported	Computation of Needed Quantities
$\overline{Y}_1, \overline{Y}_2, S_{Difference}, r, n$ (number of pairs)	$d = \left(\dfrac{\overline{Y}_1 - \overline{Y}_2}{S_{Difference}}\right)\sqrt{2(1-r)}, \quad v = \left(\dfrac{1}{n} + \dfrac{d^2}{2n}\right)2(1-r)$
t (from paired t-test), r, n	$d = t\sqrt{\dfrac{2(1-r)}{n}}, \quad v = \left(\dfrac{1}{n} + \dfrac{d^2}{2n}\right)2(1-r)$
F (from repeated measures ANOVA), r, n	$d = \pm\sqrt{\dfrac{2F(1-r)}{n}}, \quad v = \left(\dfrac{1}{n} + \dfrac{d^2}{2n}\right)2(1-r)$
p (one-tailed), r, n	$d = \pm t^{-1}(p)\sqrt{\dfrac{2(1-r)}{n}}, \quad v = \left(\dfrac{1}{n} + \dfrac{d^2}{2n}\right)2(1-r)$
p (two-tailed), r, n	$d = \pm t^{-1}\left(\dfrac{p}{2}\right)\sqrt{\dfrac{2(1-r)}{n}}, \quad v = \left(\dfrac{1}{n} + \dfrac{d^2}{2n}\right)2(1-r)$

SOURCE: Author's compilation.

NOTE: The function $t^{-1}(p)$ is the inverse of the cumulative distribution function of Student's t with $n-1$ degrees of freedom. Many computer programs and spreadsheets provide functions that can be used to compute t^{-1}. Assume $n = 19$, so that $df = 18$. Then, in Excel™, for example, if the reported p-value is 0.05 (2-tailed), TINV(p,df) = TINV(0.05,18) will return the required value (2.1009). If the reported p-value is 0.05 (1-tailed), TINV($2p,df$) = TINV(0.10,18) will return the required value 1.7341. The F in row 3 of the table is the F-statistic from a one-way repeated measures analysis of variance. In rows 3–5 the sign of d must reflect the direction of the mean difference.

This is similar to the formula for independent groups (12.11), but with two differences. The first difference is in the numerator, where we use adjusted means rather than raw means. This has no impact on the expected value of the mean difference, but serves to increase precision and possibly removes bias due to pretest differences.

The second is in the mechanism used to compute S_{Within}. When we were working with independent groups, the natural unit of deviation was the unadjusted standard deviation within groups. Therefore, this value is typically reported, or easily imputed. By contrast, the analysis of covariance works with the adjusted standard deviation (typically smaller than the unadjusted value since variance explained by covariates has been removed). Therefore, to compute d from this kind of study, we need to impute the unadjusted standard deviation within groups, which is given by

$$S_{Within} = \frac{S_{Adjusted}}{\sqrt{1-R^2}}, \qquad (12.24)$$

where $S_{Adjusted}$ is the covariate adjusted standard deviation and R is the covariate outcome correlation (or multiple correlation if there is more than one covariate). Note the similarity to (12.20) which we used to impute S_{Within} from

$S_{Difference}$ in matched studies. Equivalently, S_{Within} may be computed as

$$S_{Within} = \sqrt{\frac{MSW_{Adjusted}}{1-R^2}}. \qquad (12.25)$$

where $MSW_{Adjusted}$ is the covariate adjusted mean square within treatment groups.

The variance of d is given by

$$v = \frac{(n_1 + n_2)(1-R^2)}{n_1 n_2} + \frac{d^2}{2(n_1 + n_2)} \qquad (12.26)$$

where n_1 and n_2 are the sample size in each group, and R is the multiple correlation between the covariates and the outcome.

To compute Hedges' g and associated statistics we would use formulas 12.15 through 12.18. The degrees of freedom for computing J is $n_1 + n_2 - 2 - q$, where n_1 and n_2 are the number of subjects in each group, 2 is the number of groups, and q is the number of covariates.

For example, suppose that a study has sample means $\overline{Y}_1 = 103$, $\overline{Y}_2 = 100$, sample standard deviations $S_{Adjusted} = 5.5$, and sample sizes $n_1 = n_2 = 50$. Suppose that we know that the covariate-outcome correlation for the single covariate (so that $q = 1$) is $R = 0.7$.

Table 12.3 Formulas for Computing _d_ in Designs with Independent Groups Using Analysis of Covariance (ANCOVA)

Reported	Computation of Needed Quantities	
$\overline{Y}_1, \overline{Y}_2, S_{Pooled}, n_1, n_2, R, q$	$d = \dfrac{\overline{Y}_1^{\,Adjusted} - \overline{Y}_2^{\,Adjusted}}{S_{Pooled}},$	$v = \dfrac{(n_1 + n_2)(1 - R^2)}{n_1 n_2} + \dfrac{d^2}{2(n_1 + n_2)}$
t (from ANCOVA), n_1, n_2, R, q	$d = t\sqrt{\dfrac{n_1 + n_2}{n_1 n_2}}\sqrt{1 - R^2},$	$v = \dfrac{(n_1 + n_2)(1 - R^2)}{n_1 n_2} + \dfrac{d^2}{2(n_1 + n_2)}$
F (from ANCOVA), n_1, n_2, R, q	$d = \pm\sqrt{\dfrac{F(n_1 + n_2)}{n_1 n_2}}\sqrt{1 - R^2},$	$v = \dfrac{(n_1 + n_2)(1 - R^2)}{n_1 n_2} + \dfrac{d^2}{2(n_1 + n_2)}$
p (one-tailed, from ANCOVA), n_1, n_2, R, q	$d = \pm t^{-1}(p)\sqrt{\dfrac{n_1 + n_2}{n_1 n_2}}\sqrt{1 - R^2},$	$v = \dfrac{(n_1 + n_2)(1 - R^2)}{n_1 n_2} + \dfrac{d^2}{2(n_1 + n_2)}$
p (two-tailed, from ANCOVA), n_1, n_2, R, q	$d = \pm t^{-1}\!\left(\dfrac{p}{2}\right)\sqrt{\dfrac{n_1 + n_2}{n_1 n_2}}\sqrt{1 - R^2},$	$v = \dfrac{(n_1 + n_2)(1 - R^2)}{n_1 n_2} + \dfrac{d^2}{2(n_1 + n_2)}$

SOURCE: Author's compilation.

NOTE: The function $t^{-1}(p)$ is the inverse of the cumulative distribution function of student's t with $n_1 + n_2 - 2 - q$ degrees of freedom, q is the number of covariates, and R is the covariate outcome correlation or multiple correlation. Many computer programs and spreadsheets provide functions that can be used to compute t^{-1}. Assume $n1 = n2 = 11$, and $q = 2$, so that $df = 18$. Then, in Excel™, for example, if the reported p-value is 0.05 (2-tailed), TINV$(p, df) =$ TINV$(0.05, 18)$ will return the required value (2.1009). If the reported p-value is 0.05 (1-tailed), TINV$(2p, df) =$ TINV$(0.10, 18)$ will return the required value 1.7341. The F in row 3 of the table is the F-statistic from a one-way analysis of covariance. In rows 3–5 the sign of d must reflect the direction of the mean difference.

The unadjusted standard deviation within groups is imputed from $S_{Adjusted}$ by

$$S_{Within} = \frac{S_{Adjusted}}{\sqrt{1 - 0.7^2}} = 7.7015.$$

Then d, its variance and standard error are computed as

$$d = \frac{103 - 100}{7.7015} = 0.3895,$$

$$v_d = \frac{(50 + 50)(1 - 0.7^2)}{50 \times 50} + \frac{0.3895^2}{2(50 + 50 - 2 - 1)} = 0.0222,$$

and

$$SE_d = \sqrt{0.0222} = .1490.$$

The correction factor J, Hedges' g, its variance and standard error are given by

$$J(50 + 50 - 2 - 1) = \left(1 - \frac{3}{4 \times 97 - 1}\right) = 0.9922,$$

$$g = 0.9922 \times 0.3895 = 0.3834,$$

$$v_g = 0.9922^2 \times 0.0222 = 0.0219,$$

and

$$S_g = \sqrt{0.0219} = .1480.$$

Table 12.3 provides formulas for computing d and its variance for analysis of covariance, working from information that may be reported in a published paper.

12.3.4 Including Different Study Designs in the Same Analysis

To include data from two or three study designs—independent groups, matched groups, analysis of covariance—in the same meta-analysis, one simply computes the effect size and variance for each study using the appropriate formulas. The effect sizes have the same meaning regardless of the study design, and the variances have been computed to take account of the study design. We can therefore use these effect sizes and variances to perform a meta-analysis without concern for the different designs in the original studies. Note that the phrase _different study designs_ is used here to refer to the fact that one study used independent groups and another used a pre-post design or matched groups, and yet another included covariates. It does not refer to the case where one study was a randomized trial or quasi-experimental design and another was observational. The issues involved in combining these kinds of studies are beyond the scope of this chapter.

For all study designs the direction of the effect $(\overline{Y}_1 - \overline{Y}_2$ or $\overline{Y}_2 - \overline{Y}_1)$ is arbitrary, except that the researcher must decide on a convention and then apply this consistently. For example, if a positive difference will indicate that the treated group did better than the control group, then this convention must apply for studies that used independent designs and for studies that used pre-post designs. It must also apply for all outcome measures. In some cases (for example, if some studies defined outcome as the number of correct answers while others defined outcome as the number of mistakes) it might be necessary to reverse the computed sign of the effect size to ensure that the convention is applied consistently.

12.4 CORRELATION AS AN EFFECT SIZE

The correlation coefficient itself can serve as the effect-size index. The correlation is an intuitive measure that, like δ, has been standardized to take account of different metrics in the original scales. The population parameter is denoted by ρ.

12.4.1 Computing r

The estimate of the correlation parameter ρ is simply the sample correlation coefficient, r. The variance of r is approximately

$$v_r = \frac{(1-r^2)^2}{n-1}, \qquad (12.27)$$

where n is the sample size.

Most meta-analysts do not perform syntheses on the correlation coefficient itself because the variance depends so strongly on the correlation (see, however, chapter 17, this volume; Hunter and Schmidt 2004). Rather, the correlation is converted to the Fisher's z scale (not to be confused with the z-score used with significance tests), and all analyses are performed using the transformed values. The results, such as the combined effect and its confidence interval, would then be converted back to correlations for presentation. This is analogous to the procedure used with odds ratios or risk ratios where all analyses are performed using log transformed values, and then converted back to the original metric.

The transformation from sample correlation r to Fisher's z is given by

$$z = 0.5 \times \ln\left(\frac{1+r}{1-r}\right). \qquad (12.28)$$

The variance of z is (to an excellent approximation)

$$v_z = \frac{1}{n-3}, \qquad (12.29)$$

with standard error

$$SE_z = \sqrt{V_z}. \qquad (12.30)$$

The effect size z and its variance would be used in the analysis, which would yield a combined effect, confidence limits, and so on, in the Fisher's z metric. We could then convert each of these values back to correlation units using

$$r = \frac{e^{2z}-1}{e^{2z}+1}, \qquad (12.31)$$

where $e^x = \exp(x)$ is the exponential (anti-log) function. For example, if a study reports a correlation of 0.50 with a sample size of 100, we would compute

$$z = 0.5 \times \ln\left(\frac{1+0.5}{1-0.5}\right) = 0.5493,$$

$$v_z = \frac{1}{100-3} = 0.0103.$$

and

$$SE_z = \sqrt{0.0103} = 0.1015.$$

To convert the Fisher's z value back to a correlation, we would use

$$r = \frac{e^{2(0.5493)}-1}{e^{2(0.5493)}+1} = 0.5000.$$

12.5 CONVERTING EFFECT SIZES FROM ONE METRIC TO ANOTHER

Earlier I discussed the case where different study designs were used to compute the same effect size. For example, studies that used independent groups and studies that used matched groups were both used to yield estimates of the mean difference D, or studies that used independent groups, studies that used matched groups, and studies that used analysis of covariance were all used to yield estimates of the standardized mean difference, d or g. There is no problem in combining these estimates in a meta-analysis because the effect size has the same meaning in all studies.

Consider, however, the case where some studies report a difference in means, which is used to compute a standardized mean difference. Others report a difference in proportions which is used to compute an odds ratio (see chapter 13, this volume). And others report a correlation. All the studies address the same question, and we want to

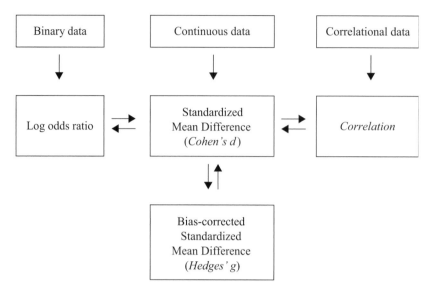

Figure 12.2 Converting Among Effect Sizes

SOURCE: Author's compilation.
NOTE: This schematic outlines the mechanism for incorporating multiple kinds of data in the same meta-analysis. First, each study is used to compute an effect size and variance in its "native" index – log odds ratio for binary data, *d* for continuous data, and *r* for correlational data. Then, we convert all of these indices to a common index, which would be either the log odds ratio, *d*, or *r*. If the final index is *d*, we can move from there to Hedges' *g*. This common index and its variance are then used in the analysis.

include them in one meta-analysis. Unlike the earlier case, we are now dealing with different indices, and we need to convert them to a common index before we can proceed. As in section 12.3.4, we assume that all the studies are randomized trials, all are quasi-experimental, or all are observational.

Here I present formulas for converting between an odds ratio and *d*, or between *d* and *r*. By combining formulas, it is also possible to convert from an odds ratio, via *d*, to *r* (see figure 12.2). In every case the formula for converting the effect size is accompanied by a formula to convert the variance.

When we convert from an odds ratio to a standardized mean difference, we then apply the sampling distribution for *d* to data that were originally binary, despite the fact that the new sampling distribution does not really apply. The same problem holds for any of these conversions. If this approach is therefore less than perfect, it is often better than the alternative of omitting the studies that happened to use an alternate metric. This would involve loss of information, and possibly the systematic loss of infor-

mation. It may also be better than the alternative of separately summarizing studies with effect sizes in different metrics, which leads to not one, but several summaries.

The decision to use these transformations or to exclude some studies will need to be made on a case-by-case basis. For example, if all the studies originally used continuous scales but some had been dichotomized (into pass-fail) for reporting, a good case could be made for converting these effects back to *d* for the meta-analysis. By contrast, if some studies originally used a dichotomous outcome and others used a continuous outcome, the case for combining the two kinds of data would be more tenuous. In either case, a sensitivity analysis to compare the effect size from the two (or more) kinds of studies would be important.

12.5.1 Converting from the Log Odds Ratio to *d*

We can convert from a log odds ratio [ln(*o*)] to the standardized mean difference *d* using

$$d = \frac{\ln(o)\sqrt{3}}{\pi},$$ (12.32)

Table 12.4 Formulas for Computing *r* in Designs with Independent Groups

Reported	Computation of Needed Quantities
r, n	$v_r = \frac{(1-r^2)^2}{n-1}, \quad z = 0.5\ln\left(\frac{1+r}{1-r}\right), \quad v_z = \frac{1}{n-3}$
t, n	$r = \pm\sqrt{\frac{t^2}{t^2+n-2}}, \quad v_r = \frac{(1-r^2)^2}{n-1}, \quad z = 0.5\ln\left(\frac{1+r}{1-r}\right), \quad v_z = \frac{1}{n-3}$
t, r	$n = t^2\left(\frac{1-r^2}{r^2}\right) - 2, \quad v_r = \frac{(1-r^2)^2}{n-1}, \quad z = 0.5\ln\left(\frac{1+r}{1-r}\right), \quad v_z = \frac{1}{n-3}$
p(one-tailed), r	$n = [t^{-1}(p)]^2\left(\frac{1-r^2}{r^2}\right) - 2, \quad v_r = \frac{(1-r^2)^2}{n-1}, \quad z = 0.5\ln\left(\frac{1+r}{1-r}\right), \quad v_z = \frac{1}{n-3}$
p(two-tailed), r	$n = \left[t^{-1}\left(\frac{p}{2}\right)\right]^2\left(\frac{1-r^2}{r^2}\right) - 2, \quad v_r = \frac{(1-r^2)^2}{n-1}, \quad z = 0.5\ln\left(\frac{1+r}{1-r}\right), \quad v_z = \frac{1}{n-3}$

SOURCE: Author's compilation.
NOTE: The function $t^{-1}(p)$ is the inverse of the cumulative distribution function of Student's t with $n-2$ degrees of freedom. Many computer programs and spreadsheets provide functions that can be used to compute t^{-1}. Assume $n = 20$, so that $df = 18$. Then, in Excel™, for example, if the reported p-value is 0.05 (2-tailed), TINV(p, df) = TINV(0.05, 18) will return the required value (2.1009). If the reported p-value is 0.05 (1-tailed), TINV($2p$, df) = TINV(.10, 18) will return the required value 1.7341.

where π is the mathematical constant (approximately 3.14159). The variance of d would then be

$$v_d = \frac{3v_{\ln(o)}}{\pi^2}, \tag{12.33}$$

where $v_{\ln(o)}$ is the variance of the log odds ratio. This method was originally proposed by Victor Hasselblad and Larry Hedges but other variations have also been proposed (Hasselblad and Hedges 1995; see also Sanchez-Meca, Marin-Martinez, and Chacon-Moscoso 2003; Whitehead 2002).

For example, if the log odds ratio were $\ln(o) = 0.9070$ with a variance of $v_{\ln(o)} = 0.0676$, then

$$d = \frac{0.9070\sqrt{3}}{3.1416} = 0.5000,$$

with variance

$$v_d = \frac{3 \times 0.0676}{3.1416^2} = 0.0205.$$

Table 12.4 provides formulas for computing r and its variance from information that may be reported in a published paper.

12.5.2 Converting from *d* to the Log Odds Ratio

We can convert from the standardized mean difference d to the log odds ratio [$\ln(o)$] using

$$\ln(o) = \frac{\pi d}{\sqrt{3}}, \tag{12.34}$$

where π is the mathematical constant (approximately 3.14159). The variance of $\ln(o)$ would then be

$$V_{\ln(o)} = \frac{\pi^2 v_d}{3}. \tag{12.35}$$

For example, if $d = 0.5000$ and $v_d = 0.0205$ then

$$\ln(o) = \frac{3.1415 * 0.5000}{\sqrt{3}} = 0.9069, \tag{12.36}$$

and

$$v_{\ln(o)} = \frac{3.1415^2 * 0.0205}{3} = 0.0674. \tag{12.37}$$

12.5.3 Converting from *r* to *d*

We convert from a correlation (r) to a standardized mean difference (d) using

$$d = \frac{2r}{\sqrt{1-r^2}}, \tag{12.38}$$

and the variance of d computed in this way (converted from r) is

$$v_d = \frac{4v_r}{(1-r^2)^3}. \tag{12.39}$$

For example, if $r = 0.50$ and $v_r = 0.0058$, then

$$d = \frac{2 \times 0.500}{\sqrt{1 - 0.500^2}} = 1.1547,$$

and the variance of d is

$$v_d = \frac{4 \times 0.0058}{(1 - 0.50^2)^3} = 0.0550.$$

12.5.4 Converting from *d* to *r*

We can convert from a standardized mean difference (d) to a correlation (r) using

$$r = \frac{d}{\sqrt{d^2 + a}}, \qquad (12.40)$$

where a is a correction factor for cases where $n_1 \neq n_2$,

$$a = \frac{(n_1 + n_2)^2}{n_1 n_2}. \qquad (12.41)$$

The correction factor a is based on the ratio of n_1 to n_2, rather than the absolute values of these numbers. Therefore, if n_1 *and* n_2 are not known precisely, use $n_1 = n_2$, which will yield $a = 4$. The variance of r computed in this way (converted from d) is

$$v_r = \frac{a^2 v_d}{(d^2 + a)^3}. \qquad (12.42)$$

For example, if $n_1 = n_2$, $d = 1.1547$ and $v_d = 0.0550$, then

$$r = \frac{1.1547}{\sqrt{1.1547^2 + 4}} = 0.5000,$$

and the variance of r converted from d will be

$$v_r = \frac{4^2 \times 0.0550}{(1.1547^2 + 4)^3} = 0.0058.$$

12.6 INTERPRETING EFFECT SIZES

I have addressed the calculation of effect-size estimates and their sampling distributions, a technical exercise which depends on statistical theory. Equally important, however, is the interpretation of effect sizes, a process that requires human judgment and which is not amenable to technical solution (see Cooper 2008).

To understand the substantive implications of an effect size, we need to look at the effect size in context. For example, to judge whether the effect of an intervention is large enough to be important, it may be helpful to compare it with the effects of related interventions that are appropriate for the same population and have similar characteristics (such as cost or complexity). Compendia of effect sizes from various interventions may be helpful in making such comparisons (for example, Lipsey and Wilson 1993).

Alternatively, it may be helpful to compare an effect size to other effect sizes that are well understood conceptually. For example, one might compare an educational effect size to the effect size corresponding to one year of growth. Even here, however, context is critical. For example, one year's growth in achievement from kindergarten to grade one corresponds approximately to a d of 1.0, whereas one year's growth in achievement from grade eleven to grade twelve corresponds to a d of about 0.1.

Technical methods can help with the interpretation of effect sizes in one limited way. They can permit effect sizes computed in one metric (such as a correlation coefficient) to be expressed in another metric (such as a standardized mean difference or an odds ratio). Such re-expression of effect sizes in other metrics can make it easier to compare the effect with other effect sizes that may be known to the analyst and be judged to be relevant (see Rosenthal and Rubin 1979, 1982).

12.7 RESOURCES

For a more extensive discussion of the topics covered in this chapter, see Gene Glass, Barry McGaw, and Mary Smith (1981), Mark Lipsey and David Wilson (2001), Robert Rosenthal, Ralph Rosnow, and Donald Rubin (2000), and Michael Borenstein, Larry Hedges, Julian Higgins, and Hannah Rothstein (2009a, 2009b).

12.8 ACKNOWLEDGMENTS

This work was funded in part by the following grants from the National Institutes of Health: "Combining data types in meta-analysis" (AG021360), "Publication bias in meta-analysis" (AG20052), "Software for meta-regression" (AG024771), from the National Institute on Aging, under the direction of Dr. Sidney Stahl; and "Forest plots for meta-analysis" (DA019280) from the National Institute on Drug Abuse, under the direction of Dr. Thomas Hilton.

My thanks to Julian Higgins for his assistance in clarifying some of the issues discussed in this chapter, to

Steven Tarlow for checking the accuracy of the formulas, and to Hannah Rothstein for helping me to put it all in perspective.

12.9 REFERENCES

Borenstein Michael. 1994. "The Case for Confidence Intervals in Clinical Trials." *Controlled Clinical Trials* 15(5): 411–28.

Borenstein, Michael, Larry V. Hedges, Julian Higgins, and Hannah R. Rothstein. 2009a. *Introduction to Meta-Analysis.* Chichester: John Wiley & Sons.

———. 2009b. *Computing Effect Sizes for Meta-Analysis.* Chichester: John Wiley & Sons.

Cohen, Jacob. 1969. *Statistical Power Analysis for the Behavioral Sciences.* New York: Academic Press.

———. 1988. *Statistical Power Analysis for the Behavioral Sciences*, 2nd ed. New York: Academic Press.

Cooper, Harris. 2008. "The Search for Meaningful Ways to Express the Effects of Interventions." *Child Development Perspectives* 2: 181–86.

Glass, Gene V. 1976. "Primary, Secondary, and Meta-Analysis of Research." *Educational Researcher* 5(10): 3–8.

Glass, Gene V., Barry McGaw, and Mary L. Smith. 1981. *Meta-Analysis in Social Research.* Thousand Oaks, Calif.: Sage Publications.

Grissom, R. J., and J. J. Kim. 2005. *Effect Sizes for Research.* Mahwah, N.J.: Lawrence Erlbaum.

Hasselblad, Victor and Hedges, Larry V. 1995. "Meta-Analysis of Screening and Diagnostic Tests." *Psychological Bulletin* 117(1): 167–78.

Hedges, Larry V. 1981. "Distribution Theory for Glass's Estimator of Effect Size and Related Estimators." *Journal of Educational Statistics* 6(2): 107–28.

Hedges, Larry V., and Ingram Olkin. 1985. *Statistical Models for Meta-Analysis.* New York: Academic Press.

Hunter, John E. and Frank L. Schmidt. 2004. *Methods of Meta-Analysis: Correcting Error and Bias in Research Findings*, 2d ed. Newbury Park, Calif.: Sage Publications.

Lipsey, Mark W., and David B. Wilson. 1993. "The Efficacy of Psychological, Educational, and Behavioral Treatment: Confirmation of Meta-Analysis." *American Psychologist* 48(12): 1181–1209.

———. 2001. *Practical Meta-Analysis.* Thousand Oaks, Calif.: Sage Publications.

Rosenthal, Robert, and Donald B. Rubin. 1979. "A Note on Percentage of Variance Explained as a Measure of Importance of Effects." *Journal of Applied Social Psychology* 9(5): 395–96.

———. 1982. "A Simple, General Purpose Display of Magnitude of Experimental Effect." *Journal of Educational Psychology* 74(2): 166–69.

Rosenthal, Robert, Ralph L. Rosnow, and Donald B. Rubin. 2000. *Contrasts and Effect Sizes in Behavioral Research: A Correlational Approach.* Cambridge: Cambridge University Press.

Sanchez-Meca, Julio., Fulgencio Marin-Martinez, and Salvador Chacon-Moscoso. 2003. "Effect-Size Indices for Dichotomized Outcomes in Meta-Analysis." *Psychological Methods* 8(4): 448–67.

Whitehead, Anne. 2002. *Meta-Analysis of Controlled Clinical Trials (Statistics in Practice).* Chichester: John Wiley & Sons.

13

EFFECT SIZES FOR DICHOTOMOUS DATA

JOSEPH L. FLEISS
Columbia University

JESSE A. BERLIN
*Johnson & Johnson Pharmaceutical
Research and Development*

CONTENTS

13.1 INTRODUCTION

In many studies measurements are made on binary (dichotomous) rather than numerical scales. Examples include studies of attitudes or opinions (the two categories for the response variable being agree or disagree with some statement), case-control studies in epidemiology (the two categories being exposed or not exposed to some hypothesized risk factor), and intervention studies (the two categories being improved or unimproved or, in studies of medical interventions, experiencing a negative event or not). In this chapter we present and analyze four popular measures of association or effect appropriate for categorical data: the difference between two probabilities, the ratio of two probabilities, the phi coefficient, and the odds ratio. Considerations in choosing among the various measures are presented, with an emphasis on the context of research synthesis. The odds ratio is shown to be the measure of choice according to several statistical criteria, and the major portion of the discussion is devoted to methods for making inferences about this measure under a variety of study designs. The chapter wraps up by discussing converting the odds ratio to other measures that have more straightforward substantive interpretations.

Consider a study in which two groups are compared with respect to the frequency of a binary characteristic. Let π_1 and π_2 denote the probabilities in the two underlying populations of a subject being classified into one of the two categories, and let P_1 and P_2 denote the two sample proportions based on samples of sizes n_1 and n_2. Three of the four parameters to be studied in this chapter are the simple difference between the two probabilities,

$$\Delta = \pi_1 - \pi_2; \tag{13.1}$$

the ratio of the two probabilities, or the rate ratio (referred to as the risk ratio or relative risk in the health sciences),

$$RR = \pi_1 / \pi_2; \tag{13.2}$$

and the odds ratio,

$$\omega = (\pi_1/(1-\pi_1))/(\pi_2/(1-\pi_2)). \tag{13.3}$$

The fourth measure, φ (the phi coefficient), is defined later.

The parameters Δ and φ are such that the value 0 indicates no association or no difference. For the other two parameters, RR and ω, the logarithm of the measure is typically analyzed to overcome the awkward features that the value 1 indicates no association or no difference, and that the finite interval from 0 to 1 is available for indexing negative association, but the infinite interval from 1 on up is available for indexing positive association. With the parameters transformed so that the value 0 separates negative from positive association, we present for each the formula for its maximum likelihood estimator, say L, and for its non-null standard error, say SE. (A non-null standard error is one in which no restrictions are imposed on the parameters, in particular, no restrictions that are associated with the null hypothesis.)

We assume throughout that the sample sizes within each individual study are large enough for classical large-sample methods assuming normality to be valid. The quantities L, SE and

$$w = 1/SE^2 \tag{13.4}$$

are then adequate for making the following important inferences about the parameter being analyzed within each individual study, and for pooling a given study's results with the results of others. Let C_α denote the value cutting off the fraction α in the upper tail of the standard normal distribution. The interval $L \pm C_\alpha SE$ is, approximately, a $100(1-2\alpha)$ percent confidence interval for the parameter that L is estimating. As detailed in chapter 14 of this volume (see, in particular, formula 14.14), the given study's contribution to the numerator of the pooled (meta-analytic) estimate of the parameter of interest is, under the model of fixed effects, wL, and its contribution to the denominator is w. (In a so-called fixed-effects model for k studies, the assumption is made that there is interest in only these studies. In a random effects model, the assumption is made that these k studies are a sample from a larger population of studies. We restrict our attention to fixed-effects analyses.)

13.2 THE DIFFERENCE BETWEEN TWO PROBABILITIES

13.2.1 A Critique of the Difference

The straightforward difference, $\Delta = \pi_1 - \pi_2$, is the simplest parameter for estimating the effect on a binary variable of whatever characteristic or intervention distinguishes group 1 from group 2. Its simplicity and its interpretability with respect to numbers of events avoided (or caused) are perhaps its principal virtues, especially in the context of research syntheses. A technical difficulty with Δ is that its range of variation is limited by the magnitudes of π_1 and π_2: the possible values of Δ when π_1 and π_2 are close to 0.5 are greater than when π_1 and π_2 are close to 0 or to 1. If the values of π_1 and π_2 vary across the k studies whose results are being synthesized, the associated values of Δ may also vary. There might then be the appearance of heterogeneity (that is, nonconstancy in the measure of effect) in this scale of measurement due to the mathematical constraints imposed on probabilities rather than to substantive reasons. Empirically, two separate investigations have found that across a wide range of meta-analyses, heterogeneity among studies is most pronounced when using the difference as the measure of effect (Deeks 2002; Engels et al. 2000).

The example in table 13.1 illustrates this phenomenon. The values there are mortality rates from lung cancer in four groups of workers: exposed or not to asbestos in the work place cross-classified by cigarette smoker or not (Hammond, Selikoff, and Seidman 1979). The asbestos–no asbestos difference in lung cancer mortality rates for cigarette smokers is 601.6 to 122.6, or approximately 500 deaths per 100,000 person-years. Given the mortality rate of 58.4 per 100,000 person-years for nonsmokers exposed to asbestos, it is obviously impossible for the asbestos-no asbestos difference between lung cancer mortality rates for nonsmokers to come anywhere near 500 per 100,000 person-years. Heterogeneity—a difference between smokers and nonsmokers in the effect of exposure to asbestos, with effect measured as the difference between two rates—exists, but it is difficult to assign biological meaning to it.

13.2.2 Inference about the Difference

In spite of the possible inappropriateness for a meta-analysis of the difference between two probabilities, the

Table 13.1 Mortality Rates from Lung Cancer (Per 100,000 Person-Years)

Smoker	Exposed to Asbestos	
	Yes	No
Yes	601.6	122.6
No	58.4	11.3

SOURCE: Authors' compilation.

investigator might nevertheless choose to use Δ as the measure of effect. The estimate of the difference is simply

$$D = P_1 - P_2$$

The factor w by which a study's estimated difference is to be weighted in a fixed-effects meta-analysis is equal to the reciprocal of its estimated squared standard error,

$$w = 1/SE^2 = \left(\frac{P_1(1-P_1)}{n_1} + \frac{P_2(1-P_2)}{n_2} \right)^{-1}. \quad (13.5)$$

13.2.3 An Example

Consider as an example a comparative study in which one treatment was applied to a randomly constituted sample of $n_1 = 80$ subjects and the other was applied to a randomly constituted sample of $n_2 = 70$ subjects. If the proportions unimproved are $P_1 = 0.60$ and $P_2 = 0.80$, the value of D is -0.20 and the standard error of D is

$$SE = \left(\frac{0.60 \times 0.40}{80} + \frac{0.80 \times 0.20}{70} \right)^{1/2} = 0.0727.$$

The resulting weighting factor is $w = 1/SE^2 = 189.2$. An approximate 95 percent confidence interval for Δ is $D \pm 1.96 \cdot SE$. In this example, the interval is $-0.20 \pm 1.96 \times 0.0727$, or the interval from -0.34 to -0.06.

13.3 RATIO OF TWO PROBABILITIES

13.3.1 Rate Ratio

The ratio of two probabilities, $RR = \pi_1/\pi_2$, is another popular measure of the effect of an intervention. Its use requires that one be able to distinguish between the two outcomes so that one is in some sense the undesirable one and the other is in some sense the preferred one.

The names given to this measure in the health sciences, the *risk* ratio or relative *risk*, reflect the fact that the measure is not symmetric (that is, $\pi_1/\pi_2 \neq (1 - \pi_1)/(1 - \pi_2)$), but is instead the ratio of the first group's probability of an undesirable outcome to the second's probability. (In fact, in the health sciences, the term *rate* is often reserved for quantities involving the use of person-time, that is, the product of the number of individuals and their length of follow-up. For the remainder of this chapter, the term will be used in its more colloquial sense, or simply a number of events divided by a number of individuals.) *RR* is a natural measure of effect for those investigators accustomed to comparing two probabilities in terms of their proportionate (relative) difference, that is, their ratio. If π_1 is the probability of an untoward outcome in the experimental group and if π_2 is the same in the control or comparison group, the proportionate reduction in the likelihood of such an outcome is

$$\frac{\pi_2 - \pi_1}{\pi_2} = 1 - RR \qquad (13.6)$$

Inferences about a proportionate reduction may therefore be based on inferences about the associated relative risk.

13.3.2 Inferences About the Rate Ratio

RR is estimated simply as

$$rr = P_1/P_2.$$

The range of variation of *rr* is an inconvenient one for drawing inferences. Only the finite interval from 0 to 1 is available for indexing a lower risk in population 1, but the infinite interval from 1 up is, theoretically, available for indexing a higher risk in population 1. A corollary is that the value 1 (rather than the more familiar and convenient value 0) indexes no differential risk. It is standard practice to undo these inconveniences by carrying out one's statistical analysis on the logarithms of the rate ratios, and by transforming point and interval estimates back to the original units by taking antilogarithms of the results. An important by-product is that the sampling distribution of the logarithm of *rr* is more nearly normal than the sampling distribution of *rr*.

The large sample standard error of ln(*rr*), the natural logarithm of *rr*, may be estimated as

$$SE = \left(\frac{1 - P_1}{n_1 P_1} + \frac{1 - P_2}{n_2 P_2} \right)^{1/2} \qquad (13.7)$$

An approximate two-sided $100(1 - \alpha)$ percent confidence interval for *RR* is obtained by taking the antilogarithms of the limits

$$\ln(rr) - C_{\alpha/2} SE \qquad (13.8)$$

and

$$\ln(rr) + C_{\alpha/2} SE. \qquad (13.9)$$

Consider again the numerical example in section 13.2.3: $n_1 = 80$, $P_1 = 0.60$; $n_2 = 70$ and $P_2 = 0.80$. The estimated rate ratio is $rr = 0.60/0.80 = 0.75$, so group 1 is estimated to be at a risk that is 25 percent less than group 2's risk. Suppose that a confidence interval for *RR* is desired. One first obtains the value $\ln(rr) = -0.2877$, and then obtains the value of its estimated standard error,

$$SE = \left(\frac{0.40}{80 \times 0.60} + \frac{0.20}{70 \times 0.80} \right)^{1/2} = 0.0119^{1/2} = 0.1091$$

(the associated weighting factor is $w = 1/SE^2 = 84.0$). An approximate 95 percent confidence interval for ln(*RR*) has as its lower limit

$$\ln(RR_L) = -0.2877 - 1.96 \times 0.1091 = -0.5015$$

and as its upper limit

$$\ln(RR_U) = -0.2877 + 1.96 \times 0.1091 = -0.0739.$$

The resulting interval for *RR* extends from $RR_L = \exp(-0.5015) = 0.61$ to $RR_U = \exp(-0.0739) = 0.93$. The corresponding interval for the proportionate reduction in risk, finally, extends from $1 - RR_U = 0.07$ to $1 - RR_L = 0.39$, or from 7 percent to 39 percent. Notice that the confidence interval is not symmetric about the point estimate of 25 percent.

13.3.3 Problems with the Rate Ratio

Thanks to its connection with the intuitively appealing proportionate reduction in the probability of an undesirable response, the rate ratio continues to be a popular measure of the effect of an intervention in controlled studies. It is also a popular and understandable measure of association in nonexperimental studies. There are at least two technical problems with this measure, however. If the chances are high in the two groups being compared (experimental treatment versus control or exposed versus not exposed) of a subject's experiencing the outcome under study, many values of the rate ratio are mathematically impossible. For example, if the probability π_2 in the comparison group is equal to 0.40, only values for RR in the interval $0 \leq RR \leq 2.5$ are possible. The possibility therefore exists of the appearance emerging of study-to-study heterogeneity in the value of RR only because the studies differ in their values of π_2. As we will see in section 13.5, this kind of constraint does not characterize the odds ratio, a measure related to the rate ratio.

A second difficulty with RR, one that is especially important in epidemiological studies, is that it is not estimable from data collected in retrospective studies. Let E and $\overline{E}$ denote exposure or not to the risk factor under study, and let D and $\overline{D}$ denote the development or not of the disease under study. The rate ratio may then be expressed as the ratio of the two conditional probabilities of developing disease,

$$RR = \frac{P(D|E)}{P(D|\overline{E})}.$$

Studies that rely on random, cross-sectional sampling from the entire population and those that rely on separate random samples from the exposed and unexposed populations permit $P(D|E)$ and $P(D|\overline{E})$, and thus RR, to be estimated. In a retrospective study, subjects who have developed the disease under investigation, as well as subjects who have not, are evaluated with respect to whether or not they had been exposed to the putative risk factor. Such a study permits $P(E|D)$ and $P(E|\overline{D}R)$ to be estimated, but these two probabilities are not sufficient to estimate RR. We will see in Section 5 that, unlike RR, the odds ratio is estimable from all three kinds of studies. Jonathan Deeks presented an in-depth discussion of the various measures of effect described here, including consideration of rate ratio for a harmful, versus a beneficial, outcome from the same study (2002). He noted, for example that the apparent heterogeneity of the rate ratio can be very sensitive to the choice of benefit versus harm as the outcome measure.

13.4 PHI COEFFICIENT

13.4.1 Inference About φ

Consider a series of cross-sectional studies in each of which the correlation coefficient between a pair of binary random variables, X and Y, is the measure of interest. Table 13.2 presents notation for the underlying parameters and table 13.3 presents notation for the observed

Table 13.2 Underlying Probabilities Associated with Two Binary Characteristics

	Y		
X	Positive	Negative	Total
Positive	π_{11}	π_{12}	$\pi_{1\cdot}$
Negative	π_{21}	π_{22}	$\pi_{2\cdot}$
Total	$\pi_{\cdot 1}$	$\pi_{\cdot 2}$	1

SOURCE: Authors' compilation.

Table 13.3 Observed Frequencies in a Study Cross-Classifying Subjects

	Y		
X	Positive	Negative	Total
Positive	n_{11}	n_{12}	$n_{1\cdot}$
Negative	n_{21}	n_{22}	$n_{2\cdot}$
Total	$n_{\cdot 1}$	$n_{\cdot 2}$	$n_{\cdot\cdot}$

SOURCE: Authors' compilation.

frequencies in the 2×2 table cross-classifying subjects' categorizations on the two variables. Let the two levels of X be coded numerically as 0 or 1, and let the same be done for Y. The product-moment correlation coefficient in the population between the two numerically coded variables is equal to

$$\varphi = \frac{\pi_{11}\pi_{22} - \pi_{12}\pi_{21}}{\sqrt{\pi_{1\cdot}\pi_{2\cdot}\pi_{\cdot 1}\pi_{\cdot 2}}} \tag{13.10}$$

and its maximum likelihood estimator (assuming random, cross-sectional sampling) is equal to

$$\hat{\varphi} = \frac{n_{11}n_{22} - n_{12}n_{21}}{\sqrt{n_{1\cdot}n_{2\cdot}n_{\cdot 1}n_{\cdot 2}}}. \tag{13.11}$$

Note that $\hat{\varphi}$ is closely related to the classical chi-square statistic for testing for association in a fourfold table: $\chi^2 = n_{\cdot\cdot}\hat{\varphi}^2$. If numerical values other than 0 or 1 are assigned to the categories of X or Y, the sign of $\hat{\varphi}$ may change but not its absolute value.

The value of $\hat{\varphi}$ for the frequencies in table 13.4 is

$$\hat{\varphi} = \frac{135 \times 10 - 15 \times 40}{\sqrt{150 \times 50 \times 175 \times 25}} = 0.13,$$

a modest association.

Table 13.4 Hypothetical Frequencies in a Fourfold Table

X	Y		Total
	Positive	Negative	
Positive	135	15	150
Negative	40	10	50
Total	175	25	200

SOURCE: Authors' compilation.

Yvonne Bishop, Stephen Fienberg, and Paul Holland ascribe the following formula for the large-sample standard error of $\hat{\varphi}$ to Yule (1975, 381–82):

$$\frac{1}{\sqrt{n_{..}}}\left(1 - \hat{\varphi}^2 + \hat{\varphi}\left(1 + \frac{\hat{\varphi}^2}{2}\right)\frac{(p_{1.} - p_{2.})(p_{.1} - p_{.2})}{\sqrt{p_{1.}p_{.1}p_{2.}p_{.2}}} \right.$$
$$\left. - \frac{3}{4}\hat{\varphi}^2\left[\frac{(p_{1.} - p_{2.})^2}{p_{1.}p_{2.}} + \frac{(p_{.1} - p_{.2})^2}{p_{.1}p_{.2}} \right] \right)^{1/2}. \quad (13.12)$$

where $p_{1.}$ and $p_{2.}$ are the overall (marginal) row probabilities and $p_{.1}$ and $p_{.2}$ are the overall (marginal) column probabilities.

For the data under analysis,

$$SE = \frac{1}{\sqrt{200}}(1.245388)^{1/2} = 0.079 \quad (13.13)$$

13.4.2 Variance Estimation for φ

A straightforward large sample procedure known as the jackknife (Quenouille 1956; Tukey 1958) was used extensively in the past and provided an excellent approximation to the standard error of $\hat{\varphi}$ (as well as the standard errors of other functions of frequencies (Fleiss and Davies 1982) without the necessity for memorizing or programming a formula as complicated as the one in expression 13.12. The jackknife involves reestimating the φ coefficient (or other function) with one unit deleted from the sample and repeating this process for each of the units in the sample, then calculating a mean and variance using these estimates.

With the existence of faster and more powerful computers, methods for variance estimation that are computationally intensive have become possible. In particular, the so-called bootstrap approach has achieved some popularity (Efron and Tibshirani 1993; Davison and Hinkley 1997). Essentially, the bootstrap method involves resam-

pling from the population giving rise to the observed sample and taking an empirical variance estimate, or directly estimating a 95 percent confidence interval from the distribution of resampled estimates. Technically, the steps are as follows: first, sample an observation (with replacement) from the observed data (that is, put the sampled observation into the new dataset, record its value, then put it back so it can be sampled again); second, repeat this n times for a sample of n observations; third, when n observations have been sampled, calculate the statistic of interest (in this case, $\hat{\varphi}$); and, fourth, repeat this entire process a large number of times, say 1,000, generating 1,000 estimates of the value of interest. The bootstrap estimate of φ is just the sample mean of the 1,000 estimates. A variance estimate can then be obtained empirically from the usual sample variance of the 1,000 estimates, or a 95 percent confidence interval can be obtained directly from the distribution of the 1,000 estimates.

Although the bootstrap is technically described in terms of the notion of sampling with replacement, it may be easier to think of the process as generating an infinitely large population of individual subjects, that exactly reflects the distribution of observations in the observed sample. Each bootstrap iteration can then be considered a sample from this population.

13.4.3 Problems with the Phi Coefficient

A number of theoretical problems with the phi coefficient limit its usefulness as a measure of association between two random variables. If the binary random variables X and Y result from dichotomizing one or both of a pair of continuous random variables, the value of φ depends strongly on where the cut points are set (Carroll 1961; Hunter and Schmidt 1990). A second problem is that two studies with populations having different marginal distributions (for example, $\pi_{1.}$ for one study different from $\pi_{1.}$ for another) but otherwise identical conditional probability structures (for example, equal values of $\pi_{11}/\pi_{1.}$ in the two study populations) may have strongly unequal phi coefficients. An untoward consequence of this phenomenon is that there will be the appearance of study to-study heterogeneity (that is, unequal correlations across studies), even though the associations defined in terms of the conditional probabilities are identical.

Finally, the phi coefficient shares with other correlation coefficients the characteristic that it is an invalid measure of association when a method of sampling other

Table 13.5 Hypothetical Fourfold Tables, Problems with Phi Coefficient

X	Y Positive	Y Negative	Total
Second Study			
Positive	45	5	50
Negative	120	30	150
Total	165	35	200
Third Study			
Positive	90	10	100
Negative	80	20	100
Total	170	30	200

SOURCE: Authors' compilation.
NOTE: Data for the original study are in table 13.4.

than cross-sectional (multinomial and simple random sampling are synonyms) is used. The two fourfold tables in table 13.5, plus the original one in table 13.4, illustrate this as well as the previous problem. The associations are identical in the three tables in the sense that the estimated conditional probabilities that Y is positive, given that X is positive, are all equal to 0.90, and the estimated conditional probabilities that Y is positive, given that X is negative, are all equal to 0.80. If these tables summarized the results of three studies in which the numbers of subjects positive on X and negative on X were sampled in the ratio 150:50 in the first study, 50:150 in the second and 100:100 in the third, the problem becomes apparent when the three phi coefficients are calculated. Unlike the other three measures of association studied in this chapter, the values of φ vary across the three studies, from 0.13 to 0.11 to 0.14. The important conclusion is not that the values of φ are nearly equal one to another, but that the values are not identical (for further discussion of the phi coefficient, see Bishop, Fienberg, and Holland 1975, 380–83).

13.5 ODDS RATIO

We have seen that the phi coefficient is estimable only from data collected in a cross-sectional study and that the simple difference and the rate ratio are estimable only from data collected using a cross-sectional or prospective study. The odds ratio, however, is estimable from data collected using any of the three major study designs: cross-sectional, prospective or retrospective. Consider a bivariate population with underlying multinomial probabilities given in table 13.2. The odds ratio, sometimes referred to as the cross-product ratio, associating X and Y is equal to

$$\omega = \frac{\pi_{11}\pi_{22}}{\pi_{12}\pi_{21}}. \tag{13.17}$$

If the observed multinomial frequencies are as displayed in table 13.3, the maximum likelihood estimator of ω is

$$o = \frac{n_{11}n_{22}}{n_{12}n_{21}}. \tag{13.18}$$

Suppose that the study design calls for $n_{1.}$ units to be sampled from the population positive on X, and for $n_{2.}$ units to be sampled from the population negative on X. $P(Y+|X+)$, the conditional probability that Y is positive given that X is positive, is equal to $\pi_{11}/\pi_{1.}$, and the odds for Y being positive, conditional on X being positive, are equal to

$$\text{Odds } (Y+|X+) = P(Y+|X+) / P(Y-|X+)$$
$$= (\pi_{11}/\pi_{1.})/(\pi_{12}/\pi_{1.}) = \pi_{11}/\pi_{12}.$$

Similarly,

$$\text{Odds } (Y+|X-) = P(Y+|X-) / P(Y-|X-)$$
$$= (\pi_{21}/\pi_{2.})/(\pi_{22}/\pi_{2.}) = \pi_{21}/\pi_{22}.$$

The odds ratio is simply the ratio of these two odds values (see equation 13.3),

$$\omega = \frac{Odds(Y+|X+)}{Odds(Y+|X-)} = \frac{\pi_{11}/\pi_{12}}{\pi_{21}/\pi_{22}} = \frac{\pi_{11}\pi_{22}}{\pi_{12}\pi_{21}}. \tag{13.19}$$

The odds ratio is therefore estimable from a study in which prespecified numbers of units positive on X and negative on X are selected for a determination of their status on Y.

A similar analysis shows that the same is true for the odds ratio associated with a study in which prespecified numbers of units positive on Y and negative on Y are selected for a determination of their status on X. Because the latter two study designs correspond to prospective and retrospective sampling, it is clear that the odds ratio is estimable using data from either of these two designs, as well as from a cross-sectional study. Other properties of the odds ratio are discussed in the final subsection of this section.

Unlike the meanings of the measures Δ and RR, the meaning of the odds ratio is not intuitively clear, perhaps because the odds value itself is not. Consider, for example, the quantity

$$\text{Odds}\,(Y+\,|\,X+) = \pi_{11} / \pi_{12}.$$

The odds value is thus the ratio, in the population of individuals who are positive on X, of the proportion who are positive on Y to the proportion who are negative. Given the identities

$$\text{P}(Y+\,|\,X+) = \text{Odds}\,(Y+\,|\,X+) / (1 + \text{Odds}\,(Y+\,|\,X+))$$

and the complementary

$$\text{Odds}\,(Y+\,|\,X+) = P(Y+\,|\,X+) / (1 - P(Y+\,|\,X+)),$$

it is clear that the information present in Odds $(Y+\,|\,X+)$ is identical to the information present in $P(Y+\,|\,X+)$. Nevertheless, the impact of "for every 100 subjects negative on Y (all of whom are positive on X), $100 \times$ Odds $(Y+\,|\,X+)$ is the number expected to be positive on Y" may be very different from that of "out of every 100 subjects (all of whom are positive on X), $100 \times P\,(Y+\,|\,X+)$ is the number expected to be positive on Y." Once one develops an intuitive understanding of an odds value, it quickly becomes obvious that two odds values are naturally contrasted by means of their ratio—the odds ratio.

The functional form of the maximum likelihood estimator of $\hat{\omega}$ is the same for each study design (see equation 13.18). Remarkably, the functional form of the estimated standard error is also the same for each study design. For reasons similar to those given in section 13.3.2, it is customary to perform statistical analyses on $\ln(o)$, the natural logarithm of the sample odds ratio, rather than on o directly. Tosiya Sato, however, obtained a valid confidence interval for the population odds ratio without transforming to logarithms (1990). The large sample standard error of $\ln(o)$, due to Barnet Woolf (1955), is given by the equation

$$SE = \left(\frac{1}{n_{11}} + \frac{1}{n_{12}} + \frac{1}{n_{21}} + \frac{1}{n_{22}} \right)^{1/2} \tag{13.20}$$

for each of the study designs considered here.

13.5.1 A Single Fourfold Table

Thus, for data arrayed as in table 13.3, equations 13.18 and 13.20 provide the point estimate of the odds ratio and the estimated standard error of its logarithm. To reduce the bias caused by one or more small cell frequencies (note that neither $\ln(o)$ nor SE is defined when a cell frequency is equal to zero), authors in the past have recommended that one add 0.5 to each cell frequency before proceeding with the analysis (Gart and Zweifel 1967). For the frequencies in table 13.4, the values of the several statistics calculated with (versus without) the adjustment factor are $o = 2.25$ (versus 2.27), $\ln\,(o) = 0.81$ (versus 0.82), and $SE = 0.4462$ (versus 0.4380). The effect of adjustment is obviously minor in this example. More recently, other authors have called into question the practice of addition of 0.5 to cell counts, suggesting instead that other values may be preferred based on improved statistical properties (Sweeting, Sutton, and Lambert 2004). For example, Michael Sweeting and his colleagues suggest using a function of the reciprocal of the sample size of the opposite treatment arm. That approach reduces the bias that may be associated with the use of a constant correction factor of 0.5. These authors also suggest Bayesian analyses as yet another alternative. In the context of a meta-analysis, the need to add any constant to cells with a zero cell count has also been called into question, as there are alternative methods that yield more accurate estimates with improved confidence intervals (Bradburn et al. 2007; Sweeting, Sutton, and Lambert 2004).

13.5.2 An Alternative Analysis: $(O - E)/V$

In their meta-analysis of randomized clinical trials of a class of drugs known as beta blockers, Salim Yusuf and his colleagues used an approach that has become popular in the health sciences. It may be applied to a single fourfold table as well as to several tables, such as in a study incorporating stratification (1985). Consider the generic table displayed in table 13.3, and select for analysis any of its four cells (the upper left hand cell is traditionally the one analyzed). The observed frequency in the selected cell is

$$O = n_{11}, \tag{13.21}$$

the expected frequency in that cell, under the null hypothesis that X and Y are independent and conditional on the marginal frequencies being held fixed, is

$$E = \frac{n_{1.}n_{.1}}{n_{..}}, \qquad (13.22)$$

and the exact hypergeometric variance of n_{11}, under the same restrictions as above, is

$$V = \frac{n_{1.}n_{2.}n_{.1}n_{.2}}{n_{..}^2(n_{..}-1)}. \qquad (13.23)$$

When ω is close to unity, a good estimator of $\ln(\omega)$ is

$$L = \frac{O-E}{V}, \qquad (13.24)$$

with an estimated standard error of

$$SE = 1/\sqrt{V}. \qquad (13.25)$$

For the frequencies in table 13.4,

$$L = \frac{135-131.25}{4.1222} = 0.9097$$

[which corresponds to an estimated odds ratio of $\exp(0.9097) = 2.48$], with an estimated standard error of

$$SE = 1/\sqrt{4.1222} = 0.49.$$

In this example, what has become known in the health sciences as the Peto method, or sometimes as the Yusuf-Peto method, or as the $(O-E)/V$ method, produces overestimates of the point estimate and its standard error.

The method is capable of producing underestimates as well as overestimates of the underlying parameter. For the frequencies from the two studies summarized in table 13.5, the respective values of L are 0.6892 (yielding 1.99 as the estimated odds ratio) and 0.7804 (yielding 2.18 as the estimated odds ratio). The latter estimate, coming from a study in which one of the two sets of marginal frequencies was uniform, comes closest of all (for the frequencies in tables 13.4 and 13.6) to the correct value of 2.25. Sander Greenland and Alberto Salvan showed that the $(O-E)/V$ method may yield biased results when there is serious imbalance on both margins (1990). Furthermore, they showed that the method may yield biased results when the underlying odds ratio is large, even when there is balance on both margins. These findings have been confirmed in subsequent work (Sweeting et al. 2004; Bradburn et al. 2006). Earlier, Nathan Mantel, Charles Brown, and David Byar (1977) and Joseph Fleiss, Bruce Levin, and Myunghee C. Paik (2003) pointed out flaws in

a measure of association identical to $(O-E)/V$, the standardized difference.

Under many circumstances, given the simplicity and theoretical superiority of the more traditional methods described in the preceding and following subsections of section 5, there might be no compelling reason for the $(O-E)/V$ method to be used. However, more recent simulation studies have demonstrated that, when the margins of the table are balanced, and the odds ratio is not large, the $(O-E)/V$ method performs well, especially relative to use of a constant as a correction factor (Bradburn et al. 2006).

13.5.3 Inference in the Presence of Covariates

Covariates (that is, variables predictive of the outcome measure) are frequently incorporated into the analysis of categorical data. In randomized intervention studies, they are sometimes adjusted for to improve the precision with which key parameters are estimated and to increase the power of significance tests. In nonrandomized studies, they are adjusted for to eliminate the bias caused by confounding, that is, the presence of characteristics associated with both the exposure variable and the outcome variable. There are three major classes of methods for controlling for the effects of covariates: regression adjustment, stratification and matching. Each is considered.

13.5.3.1 Regression Analysis Let Y represent the binary response variable under study ($Y=1$ when the response is positive, $Y=0$ when it is negative), let X represent the binary exposure variable ($X=1$ when the study unit was exposed, $X=0$ when it was not), and let $Z_1, \ldots, Z_p$ represent the values of p covariates. A popular statistical model for representing the conditional probability of a positive response as a function of X and the Z's is the *linear logistic regression model*,

$$\Pi = P(Y=1 \mid X, Z_1, \ldots, Z_p)$$
$$= \frac{1}{1 + e^{-(\alpha + \beta X + \gamma_1 Z_1 + \cdots + \gamma_p Z_p)}}. \qquad (13.26)$$

This model is such that the logit or log odds of π,

$$\mathrm{logit}(\pi) = \ln\frac{\pi}{1-\pi}, \qquad (13.27)$$

is linear in the independent variables:

$$\mathrm{logit}(\pi) = \alpha + \beta X + \gamma_1 Z_1 + \cdots + \gamma_p Z_p \qquad (13.28)$$

(for a thorough examination of logistic regression analysis, see Hosmer and Lemeshow 2000).

The coefficient β is the difference between the log odds for the population coded $X = 1$ and the log odds for the population coded $X = 0$, adjusting for the effects of the p covariates (that is, β is the logarithm of the adjusted odds ratio). The antilogarithm, $\exp(\beta)$, is therefore the odds ratio associating X with Y. Every major package of computer programs has a program for carrying out a linear logistic regression analysis using maximum likelihood methods. The program produces $\hat{\beta}$, the maximum likelihood estimate of the adjusted log odds ratio, and SE, the standard error of $\hat{\beta}$. Inferences about the odds ratio itself may be drawn in the usual ways.

13.5.3.2 Mantel-Haenszel Estimator When the p covariates are categorical (for example, sex, social class, and ethnicity), or are numerical but may be partitioned into a set of convenient and familiar categories (for example, age twenty to thirty-four, thirty-five to forty-nine, and fifty to sixty-four), their effects may be controlled for by stratification. Each subject is assigned to the stratum defined by his or her combination of categories on the p prespecified covariates, and, within each stratum, a fourfold table is constructed cross-classifying subjects by their values on X and Y. Within a study, the strata defined for an analysis will typically be based on the kinds of covariates just described (for example, sex, social class, and ethnicity). Of note, though, in performing a meta-analysis, the covariate of interest is the study, that is, the stratification is made by study and a fourfold table is constructed within each study. Because the detailed formulae for this stratified approach are presented in chapter 14 of this volume (formulas 14.14 and 14.15), only the principles are presented here.

The frequencies in table 13.6 are adapted from a comparative study reported by Peter Armitage (1971, 266). The associated relative frequencies are tabulated in table 13.7. Within each of the three strata, the success rate for the subjects in the experimental group is greater than the success rate for those in the control group. The formula for the summary odds ratio is attributable to Nathan Mantel and William Haenszel (1959). In effect, the stratified estimate is a weighted average of the within-stratum odds ratios, with weights proportional to the inverse of the variance of the stratum-specific odds ratio. (For a meta-analysis, again, the strata are defined by the studies, but the mathematical procedures are identical.) By the formula presented in chapter 14 (see 14.14 or 14.15), the value of the Mantel-Haenszel odds ratio is $o_{MH} = 7.31$.

James Robins, Norman Breslow, and Sander Greenland derived the non-null standard error of $\ln(o_{MH})$, which

Table 13.6 Stratified Comparison of Two Treatments

| | | Outcome | | |
	Treatment	Success	Failure	Total
Stratum 1	Experimental	4	0	4
	Control	0	1	1
	Total	4	1	5
Stratum 2	Experimental	7	4	11
	Control	3	8	11
	Total	10	12	22
Stratum 3	Experimental	1	0	1
	Control	4	9	13
	Total	5	9	14

SOURCE: Authors' compilation.

Table 13.7 Quantities to Calculate the Pooled Log Odds Ratio

Stratum	L_s	SE_s	L_s/SE^2_s	$1/SE^2_s$
1	3.2958	2.2111	0.6741	0.2045
2	1.3981	0.8712	1.8421	1.3175
3	1.8458	1.7304	0.6164	0.3340
Total			3.1326	1.8560

SOURCE: Authors' compilation.

is also presented in chapter 14 as formula 14.16 (1986; see also Robins, Greenland, and Breslow 1986). For the data under analysis (in table 13.6), $o_{MH} = 7.3115$, as above, and $SE = 0.8365$. An approximate 95 percent confidence interval for $\ln(\omega)$ is: $\ln(7.3115) \pm 1.96 \times 0.8365$, or the interval from 0.35 to 3.63. The corresponding interval for ω extends from 1.42 to 37.7.

13.5.3.3 Combining Log Odds Ratios The pooling of log odds ratios across the S strata is a widely used alternative to the Mantel-Haenszel method of pooling. The pooled log odds ratio is explicitly a weighted average of the stratum-specific log odds ratios, with weights equal to the inverse of the variance of the stratum-specific estimate of the log odds ratio.

The reader will note, for the frequencies in table 13.6, that the additive adjustment factor of 0.5, or one of the other options defined by Michael Sweeting, Alex Sutton, and Paul Lambert (2004), must be applied for the log odds ratios and their standard errors to be calculable in strata 1 and 3. Summary values appear in table 13.7. They yield a point estimate for the log odds ratio of $\bar{L}_\bullet = 1.6878$, and a point estimate for the odds ratio of $o_L = \exp(1.6878) = 5.41$. This estimate is appreciably less than the Mantel-Haenszel estimate of $o_{MH} = 7.31$. As all sample sizes increase, the two estimates converge to the same value. The estimated standard error of $\bar{L}_\bullet$ is equal to $SE = 0.7340$. The (grossly) approximate 95 percent confidence interval for ω derived from these values extends from 1.28 to 22.8.

13.5.3.4 Exact Stratified Method The exact method estimates the odds ratio by using permutation methods under the assumption of a conditional hypergeometric response within each stratum (Mehta, Patel, and Gray 1985). It is the stratified extension of Fisher's exact test in which the success probabilities are allowed to vary among strata. The theoretical basis is provided by John Gart (1970), but the usual implementation is provided by network algorithms such as those developed by Cyrus Mehta for the computer program StatXact, as described in the manual for that program (Cytel Software 2003).

13.5.3.5 Comparison of Methods When the cell frequencies within the S strata are all large, the Mantel-Haenszel estimate will be close in value to the estimate obtained by pooling the log odds ratios, and either method may be used (see Fleiss, Levin, and Paik 2003). As seen in the preceding subsection, close agreement may not be the case when some of the cell frequencies are small. In such a circumstance, the Mantel-Haenszel estimator is superior to the log odds ratio estimator (Hauck 1989). Note that there is no need to add 0.5 (or any other constant) to a cell frequency of zero in order for a stratum's frequencies to contribute to the Mantel-Haenszel estimator. In fact, it would be incorrect to add any constant to the cell frequencies when applying the Mantel-Haenszel method. The method must be applied to the cell count values directly.

Others have compared competing estimators of the odds ratio in the case of a stratified study (Agresti 1990, 235–37; Gart 1970; Kleinbaum, Kupper, and Morgenstern 1982, 350–51; McKinlay 1975; Sweeting et al. 2004; Bradburn et al. 2006). There is some consensus that the Mantel-Haenszel procedure is the method of choice, though logistic regression may also be applied. In situations where data are sparse, simulation studies have demonstrated reasonable performance by the logistic regression and Mantel-Haenszel methods (Sweeting et al. 2004; Bradburn et al. 2006). When there are approximately

Table 13.8 Notation for Observed Frequencies Within Typical Matched Set

| Group | Outcome Characteristic | | Total |
	Positive	Negative	
1	a_m	b_m	t_{m1}
2	c_m	d_m	t_{m2}
Total	$a_m + c_m$	$b_m + d_m$	$t_{m\bullet}$

SOURCE: Authors' compilation.

equal numbers of subjects in both treatment arms, and the intervention (or exposure) effects are not large, the stratified version of the one-step (Peto or Yusuf-Peto) method also performs surprisingly well, actually outperforming other methods, including exact methods, in situations in which the data are very sparse (Sweeting, Sutton, and Lambert 2004; Bradburn et al. 2006).

13.5.3.6 Control by Matching Matching, the third and final technique considered for controlling for the effects of covariates, calls for the creation of sets of subjects similar one to another on the covariates. Unlike stratification, in which the total number of strata, S, is specifiable in advance of collecting one's sample, with matching one generally cannot anticipate the number of matched sets one will end up with. Consider a study with $p = 2$ covariates, sex and age, and suppose that the investigators desire close matching on age - say ages no more than one year apart. If the first subject enrolled in the study is, for example, a twelve-year-old male, then the next male between eleven and thirteen, whether from group 1 or group 2, will be matched to the first subject. If one or more later subjects are also males within this age span, they should be added to the matched set that was already constituted, and should not form the bases for new sets. If M represents the total number of matched sets, and if the frequencies within set m are represented as in table 13.8, all that is required for the inclusion of this set in the analysis is that $t_{m1} \geq 1$ and $t_{m2} \geq 1$. The values $t_{m1} = t_{m2} = 1$ for all m correspond to pairwise matching, and the values $t_{m1} = 1$ and $t_{m2} = r$ for all m correspond to $r:1$ matching. Here, such convenient balance is not necessarily assumed. The reader is referred to David Kleinbaum, Lawrence Kupper, and Hal Morgenstern (1982, chap. 18) or to Kenneth Rothman and Sander Greenland (1998, chap. 10) for a discussion of matching as a strategy in study design.

Table 13.9 Results of Study of Association Between Use of Estrogens and Endometrial Cancer

a_m	b_m	c_m	d_m	$t_{m\bullet}$	Matched Sets with Given Pattern
1	0	0	3	4	1
1	0	1	2	4	3
1	0	0	4	5	4
1	0	1	3	5	17
1	0	2	2	5	11
1	0	3	1	5	9
1	0	4	0	5	2
0	1	0	4	5	1
0	1	1	3	5	6
0	1	2	2	5	3
0	1	3	1	5	1
0	1	4	0	5	1

SOURCE: Authors' compilation.

The Mantel-Haenszel estimator of the odds ratio is given by the adaptation of equation 14.15 to the notation in table 13.8:

$$o_{\text{MH, matched}} = \frac{\sum a_m d_m / t_m}{\sum b_m c_m / t_m}. \quad (13.29)$$

Several studies have derived estimators of the standard error of $\ln(o_{\text{MH, matched}})$ in the case of matched sets (Breslow 1981; Connett et al. 1982; Fleiss 1984; Robins, Breslow, and Greenland 1986; Robins, Greenland, and Breslow 1986). Remarkably, formula 14.16, with the appropriate changes in notation, applies to matched sets as well as to stratified samples. The estimated standard error of $\ln(o_{\text{MH, matched}})$ is identical to formula 14.16 for meta-analyses (Robins, Breslow, and Greenland 1986).

The frequencies in table 13.9 have been analyzed by several scholars (Mack et al. 1976; Breslow and Day 1980, 176–82; Fleiss 1984). There were, for example, nine matched sets consisting of one exposed case, no unexposed cases, three exposed controls and one unexposed control. For the frequencies in the table, $o_{\text{MH, matched}} = 5.75$, $\ln(o_{\text{MH, matched}}) = 1.75$, and the estimated standard error of $\ln(o_{\text{MH, matched}})$ is $SE = 0.3780$. The resulting approximate 95 percent confidence interval for the value in the population of the adjusted odds ratio extends from $\exp(1.75 - 1.96 \times 0.3780) = \exp(1.0091) = 2.74$ to $\exp(1.75 + 1.96 \times 0.3780) = \exp(2.4909) = 12.07$.

An important special (and simple) case is that of matched pairs. This analysis simplifies to the following.

Let D represent the number of matched pairs that are discordant in the direction of the case having been exposed and the control not, and let E represent the number of matched pairs that are discordant in the other direction. The Mantel-Haenszel estimator of the underlying odds ratio is simply $o_{\text{MH, matched}} = D/E$, and the estimated standard error of $\ln(o_{\text{MH, matched}})$ simplifies to

$$SE = \left(\frac{D+E}{DE}\right)^{1/2}. \quad (13.30)$$

13.5.4 Reasons for Analyzing the Odds Ratio

The odds ratio is the least intuitively understandable measure of association of those considered in this chapter, but it has a great many practical and theoretical advantages over its competitors. One important practical advantage, that it is estimable from data collected according to a number of different study designs, is demonstrated in section 13.5. A second, illustrated in section 13.5.3.1, is that the logarithm of the odds ratio is the key parameter in a linear logistic regression model, a widely used model for describing the effects of predictor variables on a binary response variable. Other important features of the odds ratio will now be pointed out.

If X represents group membership (coded 0 or 1), and if Y is a binary random variable obtained by dichotomizing a continuous random variable Y^* at the point y, then the odds ratio will be independent of y if Y^* has the cumulative logistic distribution,

$$P(Y^* \leq y^* \mid X) = \frac{1}{1 + e^{-(\alpha + \beta X + \gamma y^*)}}, \quad (13.31)$$

for all y^*. If, instead, a model specified by two cumulative normal distributions with the same variance is assumed, the odds ratio will be nearly independent of y (Edwards 1966; Fleiss 1970).

Many investigators automatically compare two proportions by taking their ratio, rr. When the proportions are small, the odds ratio provides an excellent approximation to the rate ratio. Assume, for example, that $P_1 = 0.09$ and $P_2 = 0.07$. The estimated rate ratio is

$$rr = P_1/P_2 = 0.09/0.07 = 1.286,$$

and the estimated odds ratio is nearly equal to it,

$$o = \frac{P_1(1-P_2)}{P_2(1-P_1)} = \frac{0.09 \times 0.93}{0.07 \times 0.91} = 1.314.$$

A practical consequence is that the odds ratio estimated from a retrospective study—the kind of study from which the rate ratio is not directly estimable—will, in the case of relatively rare events and in the absence of bias, provide an excellent approximation to the rate ratio.

Two theoretically important properties of the odds ratio pertain to underlying probability models. Consider, first, the sampling distribution of the frequencies in table 13.3, conditional on all marginal frequencies being held fixed at their observed values. When X and Y are independent, the sampling distribution of the obtained frequencies is given by the familiar hypergeometric distribution. When X and Y are not independent, however, the sampling distribution, referred to as the *noncentral hypergeometric distribution*, is more complicated. This exact conditional distribution depends on the underlying probabilities only through the odds ratio:

$$\Pr(n_{11},n_{12},n_{21},n_{22}\,|\,n_1,n_2,n_{\cdot1}n_{\cdot2},\omega)$$
$$= \frac{\frac{n_1!}{n_{11}!n_{12}!}\cdot\frac{n_2!}{n_{21}!n_{22}!}\omega^{n_{11}}}{\sum_x \frac{n_1!}{x!(n_1.-x)!}\cdot\frac{n_2!}{(n_{\cdot1}-x)!(n_{\cdot\cdot}-n_1.-n_{\cdot1}+x)!}\omega^x}, \quad (13.32)$$

where the summation is over all permissible integers in the upper left hand cell (Agresti 1990, 66–67).

The odds ratio also plays a key role in *loglinear models*. The simplest such model is for the fourfold table. When the row and column classifications are independent, the linear model for the natural logarithm of the expected frequency in row i and column j is

$$\ln\left(E(n_{ij})\right) = \mu + \alpha_i + \beta_j, \quad (13.33)$$

with the parameters subject to the constraints $\alpha_1 + \alpha_2 = \beta_1 + \beta_2 = 0$. In the general case, dependence is modeled by adding additional terms to the model:

$$\ln\left(E(n_{ij})\right) = \mu + \alpha_i + \beta_j + \gamma_{ij}, \quad (13.34)$$

with $\Sigma\gamma_{1j} = \Sigma\gamma_{2j} = \Sigma\gamma_{i1} = \Sigma\gamma_{i2} = 0$. Let $\gamma = \gamma_{11}$. Thanks to these constraints, $\gamma = -\gamma_{12} = -\gamma_{21} = \gamma_{22}$, and the association parameter γ is related to the odds ratio by the simple equation

$$\gamma = \tfrac{1}{4}\ln(\omega). \quad (13.35)$$

Odds ratios and their logarithms represent descriptors of association in more complicated loglinear models as well (Agresti 1990, chap. 5).

A final property of the odds ratio with great practical importance is that it can assume any value between 0 and ∞, no matter what the values of the marginal frequencies and no matter what the value of one of the two probabilities being compared. As we learned earlier, neither the simple difference nor the rate ratio nor the phi coefficient is generally capable of assuming values throughout its possible range. We saw for the rate ratio, for example, that the parameter value was restricted to the interval $0 \le RR \le 2.5$ when the probability π_2 was equal to 0.40. The odds ratio can assume any nonnegative value, however. Given that

$$\omega = \frac{\pi_1(1-\pi_2)}{\pi_2(1-\pi_1)} = 1.5\frac{\pi_1}{(1-\pi_1)}$$

when $\pi_2 = 0.40$, $\omega = 0$ when $\pi_1 = 0$ and $\omega \to \infty$ when $\pi_1 \to 1$.

Consider, as another example, the frequencies in table 13.4. With the marginal frequencies held fixed, the simple difference may assume values only in the interval $-0.17 \le D \le 0.50$ (and not $-1 \le D \le 1$); the rate ratio with Positive as the untoward outcome may assume values only in the interval $0.83 \le RR \le 2.0$ (and not $0 \le RR \le \infty$); and the phi coefficient may assume values only in the interval $-0.22 \le \varphi \le 0.65$ (and not $-1 \le \varphi \le 1$); but the odds ratio may assume values throughout the interval $0 \le \omega \le \infty$. (To be sure, the rate ratio may assume values throughout the interval $0 \le RR \le \infty$ when Negative represents the untoward outcome.)

The final example brings us back to table 13.1. We saw earlier that there was heterogeneity in the association between exposure to asbestos and mortality from lung cancer, when association was measured by the simple difference between two rates:

$$D_{smokers} = 601.6 - 122.6$$
$$= 479.0 \text{ deaths per } 100,000 \text{ person years}$$

and

$$D_{nonsmokers} = 58.4 - 11.3$$
$$= 47.1 \text{ deaths per } 100,000 \text{ person years.}$$

This appearance of heterogeneous association essentially vanishes when association is measured by the odds ratio:

$$O_{smokers} = \frac{601.6 \times (100,000 - 122.6)}{122.6 \times (100,000 - 601.6)} = 4.93$$

and

$$O_{\text{nonsmokers}} = \frac{58.4 \times (100{,}000 - 11.3)}{11.3 \times (100{,}000 - 58.4)} = 5.17.$$

Whether one is considering smokers or nonsmokers, the odds that a person exposed to asbestos will die of lung cancer are approximately five times those that a person not so exposed will. In fairness, the rate ratio in this example also has a relatively constant value for smokers ($rr = 601.6/122.6 = 4.91$) and for nonsmokers ($rr = 58.4/11.3 = 5.17$), also illustrating the similarity of the values of the odds ratio and the rate ratio for this rare outcome.

As a result, we see that the odds ratio is not prone to the artifactual appearance of interaction across studies due to the influence on other measures of association or effect of varying marginal frequencies or to constraints on one or the other sample proportion. On the basis of this and all of its other positive features, the odds ratio is recommended as the measure of choice for measuring effect or association when the studies contributing to the research synthesis are summarized by fourfold tables.

13.5.5 Conversion to Other Measures

The discussion of the choice of an effect measure is complicated by the mentioned difficulty in the interpretation of the odds ratio. Despite its favorable mathematical properties, the odds ratio remains less than intuitive. In addition, though the odds ratio nicely reproduces the rate ratio (a more intuitive quantity) when the event of interest is uncommon (for example, for the data in table 13.1), it is well known to overestimate the rate ratio when the event of interest is more common (Deeks 1998; Altman, Deeks, and Sackett 1998). The report on the effect of race and sex on physician referrals by Kevin Schulman and his colleagues, illustrates this distortion, which generated some controversy in the medical literature (Schulman et al. 1999; see also, for example, Davidoff 1999). Presented with a series of clinical scenarios portrayed on video recordings involving trained actors, physicians were asked to make a decision as to whether to refer the given patient for further cardiac testing. Whites were referred by 91 percent of physicians, compared with African Americans by 85 percent. The odds ratio of 0.6, meaning the odds of referral for blacks are 60 percent of the odds of referral for whites, is a mathematically valid measure of the effect of race on referrals, but the temptation to think in terms of blacks being 60 percent as likely to be referred clearly needs to be resisted. There was a

similar pattern for referral of women (by 85 percent of physicians) versus men (by 91 percent of physicians). On closer examination of the data, black women were the least likely to be referred, with all others being about as likely as each other to be referred.

It is possible, however, and potentially useful, to perform analyses on the odds ratio scale and then convert the results back into terms of either risk ratios or risk differences (Localio, Margolis, and Berlin 2007). Such an approach would be particularly useful when adjustment for multiple covariates in a logistic regression model is called for in an analysis. In the simplest case, the relationship between the two measures is captured by:

$$rr = \frac{\omega}{(1 - \pi_1)(\omega \pi_1)}. \tag{13.36}$$

One recommended approach, using this relationship, takes the estimate of the proportion in the unexposed or untreated group (π_1) from the data, as the raw risk in the reference group, and takes the estimated odds ratio from a logistic regression model. The upper and lower confidence limits for the rate ratio are calculated using the same relationship, substituting the respective values from the confidence interval for the odds ratio. That approach, known as the method of substitution, has subsequently been criticized for its failure to take into account the variability in the baseline risk of π_1 (McNutt, Hafner, and Xue 1999; McNutt et al. 2003).

It is also possible to estimate adjusted relative risks using alternative models, for example, log-binomial models or Poisson regression models. Variations on these approaches, however, have been demonstrated to suffer from theoretical, as well as practical, limitations (Localio, Margolis, and Berlin 2007). As an alternative, Russell Localio and his colleagues proposed the use of logistic regression models, with adjustment for all relevant covariates, followed by conversion to the *RR* using the following relationship, which follows directly from equation 13.26 (2007):

$$rr = \frac{1 + e^{-\alpha}}{1 + e^{-\alpha - \beta}}. \tag{13.37}$$

Two challenges present themselves in such conversions. One is that equation 13.37 appears to ignore the values of the covariates. In fact, this is a deliberate simplification. From 13.26, it is clear that the value of the risk ratio will depend on both the value of the odds ratio and on the value of the baseline risk. In particular, it is clear that in the logistic model, the risk is specified as depending on the values of the covariates. Thus, even though the value of the odds ratio is constant in the logistic model, the value of the relative rate should depend on the values of the covariates. In 13.37, it is assumed that the covariates have been centered, that is, that the mean values have been subtracted. The value of the relative rate calculated in 13.37 is then the value calculated for the average member of the sample under study. One can extend equation 13.37 to include specified values of the covariates. This would allow, for example, calculating relative risks separately for males and females from a model in which sex is specified as being associated with risk of the outcome.

The second challenge in the conversion from the odds ratio to the relative rate (or risk difference) is variance estimation. As noted, the bootstrap approach is a convenient one in situations where variance estimation is otherwise complicated (Carpenter and Bithell 2000; Efron and Tibshirani 1993; Davison and Hinkley 1997) and this method is suggested by Localio and his colleagues (2007). They also provide a method that doesn't depend on the iterative bootstrap approach and can be used when fitting a single logistic regression model.

The availability of the conversion approach described here makes it all the more compelling to fit models using odds ratios and then to convert the results onto the appropriate scale to facilitate interpretation by substantive or policy experts.

In some meta-analyses, some of the component studies may have reported continuous versions, and others dichotomized versions of the outcome variables. When the majority of meta-analytic studies have maintained the continuous nature of the outcome variable, and only a few others have dichotomized it, a practical strategy would be to transform the odds ratio to the *d* index (described in detail in chapter 12, this volume) with one of the conversion formulae proposed and discussed in a number of papers (for example, Chinn 2000; Haddock, Rindskopf, and Shadish 1998; Hasselblad and Hedges 1995; Sánchez-Meca, Marín-Martínez, and Salvador Chacón-Moscoso 2003; Ulrich and Wirtz 2004). The reverse strategy could be applied when the majority of the studies present odds ratios and only a few include *d* indices.

13.6 ACKNOWLEDGMENTS

Drs. Melissa Begg and Bruce Levin provided valuable criticisms of a version of this chapter written for the previous edition. The original work was supported in part by grant MH45763 from the National Institute of Mental Health. Arthur Fleiss kindly provided permission for Dr. Berlin to modify this chapter for the current edition.

13.7 REFERENCES

Agresti, Alan. 1990. *Categorical Data Analysis*. New York: John Wiley & Sons.

Altman, Douglas G., Jonathan J. Deeks, and David L. Sackett. 1998. "Odds Ratios Should Be Avoided when Events are Common." Letter. *British Medical Journal* 317(7168): 1318.

Armitage, Peter. 1971. *Statistical Methods in Medical Research*. New York: John Wiley & Sons.

Bishop, Yvonne M.M., Stephen E. Fienberg, and Paul W. Holland. 1975. *Discrete Multivariate Analysis: Theory and Practice*. Cambridge, Mass.: MIT Press.

Bradburn, Michael J., Jonathan J. Deeks, A. Russell Localio, Jesse A. Berlin. 2007. "Much Ado About Nothing: A Comparison of the Performance of Meta-Analytical Methods with Rare Events." *Statistics in Medicine* 26(1): 53–77.

Breslow, Norman E. 1981. "Odds Ratio Estimators When the Data are Sparse." *Biometrika* 68(1): 73–84.

Breslow, Norman E., and Nicholas E. Day. 1980. *Statistical Methods in Cancer Research*, vol. 1, *The Analysis of Case-Control Studies*. Lyons: International Agency for Research on Cancer.

Carpenter, James, and John Bithell. 2000. "Bootstrap Confidence Intervals: When, Which, What? A Practice Guide for Medical Statisticians." *Statistics in Medicine* 19(9): 1141–64.

Chinn, Susan. 2000. "A Simple Method for Converting an Odds Ratio to Effect Size for Use in Meta-Analysis." *Statistics in Medicine* 19(22): 3127–31.

Carroll, John B. 1961. "The Nature of the Data, or How to Choose a Correlation Coefficient." *Psychometrika* 26(4): 347–72.

Connett, John, Ejigou, Ayenew, McHugh, Richard, and Breslow, Norman. 1982. "The Precision of the Mantel-Haenszel Estimator in Case-Control Studies with Multiple Matching." *American Journal of Epidemiology* 116(6): 875–77.

Cytel Software Corporation. 2003. *StatXact*: Cambridge, Mass.

Davidoff, Frank. 1999. "Race. Sex, and Physicians' Referral for Cardiac Catheterization." Letter. *New England Journal of Medicine* 341(4): 285–86.

Davison, Anthony C., and David V. Hinkley. 1997. *Bootstrap Methods and Their Application*. Cambridge: Cambridge University Press.

Deeks, Jonathan J. 1998. "When Can Odds Ratio Mislead?" Letter. *British Medical Journal* 317(7166): 1155.

———. 2002. "Issues in the Selection of a Summary Statistic for Meta-Analysis of Clinical Trials with Binary Outcomes." *Statistics in Medicine* 21(11): 1575–1600.

Edwards, John H. 1966. "Some Taxonomic Implications of a Curious Feature of the Bivariate Normal Surface." *British Journal of Preventive and Social Medicine* 20(1): 42–43.

Efron, Bradley, and Rob J. Tibshirani. 1993. *An Introduction to the Bootstrap.* New York: Chapman and Hall.

Engels, Eric A., Christopher H. Schmid, Norma Terrin, Ingram Olkin, and Joseph Lau. 2000. "Heterogeneity and Statistical Significance in Meta-Analysis: An Empirical Study of 125 Meta-Analyses." *Statistics in Medicine* 19(13): 1707–28.

Fleiss, Joseph L. 1970. "On the Asserted Invariance of the Odds Ratio." *British Journal of Preventive and Social Medicine* 24(1): 45–46.

———. 1984. "The Mantel-Haenszel Estimator in Case-Control Studies with Varying Numbers of Controls Matched to Each Case." *American Journal of Epidemiology* 120(1): 1–3.

Fleiss, Joseph L., and Mark Davies. 1982. "Jackknifing Functions of Multinomial Frequencies, with an Application to a Measure of Concordance." *American Journal of Epidemiology* 115(6): 841–45.

Fleiss, Joseph L., Bruce Levin, and Myunghee C. Paik. 2003. *Statistical Methods for Rates and Proportions*, 3rd ed. New York: John Wiley & Sons.

Gart, John J., and James R. Zweifel. 1967. "On the Bias of Various Estimators of the Logit and its Variance, with Application to Quantal Bioassay." *Biometrika* 54(1): 181–87.

Gart, John J. 1970. "Point and Interval Estimation of the Common Odds Ratio in the Combination of 2×2 Tables with Fixed Marginals." *Biometrika* 57(3): 471–75.

Greenland, Sander, and Alberto Salvan. 1990. "Bias in the One-Step Method for Pooling Study Results." *Statistics in Medicine* 9(3): 247–52.

Haddock, C. Keith, David Rindskopf, and William R. Shadish. 1998. "Using Odds Ratios as Effect Sizes for Meta-Analysis of Dichotomous Data: A Primer on Methods and Issues." *Psychological Methods* 3(3): 339–53.

Hammond, E. Cuyler, Irving J. Selikoff, and Herbert Seidman. 1979. "Asbestos Exposure, Cigarette Smoking and Death Rates." *Annals of the New York Academy of Sciences* 330: 473–90.

Hasselblad, Victor, and Larry V. Hedges. 1995. "Meta-Analysis of Screening and Diagnostic Tests." *Psychological Bulletin* 117(10: 167–78.

Hauck, Walter W. 1989. "Odds Ratio Inference from Stratified Samples." *Communications in Statistics* 18A(2): 767–800.

Hosmer, David W., and Stanley Lemeshow. 2004. *Applied Logistic Regression*, 2nd ed. New York: John Wiley & Sons.

Hunter, John E., and Frank L. Schmidt. 1990. "Dichotomization of Continuous Variables: The Implications for Meta-Analysis." *Journal of Applied Psychology* 75: 334–49.

Kleinbaum, David G., Lawrence L. Kupper, and Hal Morgenstern. 1982. *Epidemiologic Research: Principles and Quantitative Methods.* Belmont, Calif.: Lifetime Learning Publications.

Localio, A. Russell, D. J. Margolis, Jesse A. Berlin. 2007. "Relative Risks and Confidence Intervals Were Easily Compared Indirectly from Multivariable Logistic Regression." *Journal of Clinical Epidemiology* 60(9): 74–82.

Mack, Thomas M., Malcolm C. Pike, Brian E. Henderson, R. I. Pfeffer, V. R. Gerkins, B. S. Arthur, and S. E. Brown. 1976. "Estrogens and Endometrial Cancer in a Retirement Community." *New England Journal of Medicine* 294(23): 1262–67.

Mantel, Nathan, Charles Brown, and David P. Byar. 1977. "Tests for Homogeneity of Effect in an Epidemiologic Investigation." *American Journal of Epidemiology* 106(2): 125–29.

Mantel, Nathan, and William Haenszel. 1959. "Statistical Aspects of the Analysis of Data from Retrospective Studies of Disease." *Journal of the National Cancer Institute* 22(4): 719–48.

McKinlay, Sonja M. 1975. "The Effect of Bias on Estimators of Relative Risk for Pair-Matched and Stratified Samples." *Journal of the American Statistical Association* 70(352): 859–64.

McNutt Louise A., Jean P. Hafner, and Xue Xiaonan. 1999. "Correcting the Odds Ratio in Cohort Studies of Common Outcomes." Letter. *Journal of the American Medical Association* 282(6): 529.

McNutt, Louise A., Chuntao Wu, Xiaonan Xue, and Jean P. Hafner. 2003. "Estimating the Relative Risk in Cohort Studies and Clinical Trials of Common Outcomes." *American Journal of Epidemiology* 157(10): 940–43.

Mehta, Cyrus, Nitin R. Patel, and Robert Gray. 1985. "On Computing an Exact Confidence Interval for the Common Odds Ratio in Several 2×2 Contingency Tables." *Journal of the American Statistical Association* 80(392): 969–73.

Quenouille, M. H. 1956. "Notes on Bias in Estimation." *Biometrika* 43(3–4): 353–60.

Robins, James, Norman Breslow, and Sander Greenland. 1986. "Estimators of the Mantel-Haenszel Variance Consistent in

Both Sparse Data and Large-Strata Limiting Models." *Biometrics* 42(2): 11–23.

Robins, James., Sander Greenland, and Norman E. Breslow. 1986. "A General Estimator for the Variance of the Mantel-Haenszel Odds Ratio." *American Journal of Epidemiology* 124(5): 719–23.

Rothman, Kenneth J., and Sander Greenland. 1998. *Modern Epidemiology*, 2nd ed. Philadelphia, Pa.: Lippincott Williams and Wilkins.

Sánchez-Meca, Julio, Fungencio Marín-Martínez, and Salvador Chacón-Moscoso. 2003. "Effect-Size Indices for Dichotomized Outcomes in Meta-Analysis." *Psychological Methods* 8(4): 448–67.

Sato, Tosiya. 1990. "Confidence Limits for the Common Odds Ratio Based on the Asymptotic Distribution of the Mantel-Haenszel Estimator." *Biometrics* 46: 71–80.

Schulman Kevin A., Jesse A. Berlin, William Harless, Jon F. Kerner, Shyrl Sistrunk, Benard J. Gersh et al. 1999. "The Effect of Race and Sex on Physicians' Recommendations for Cardiac Catheterization." *New England Journal of Medicine* 340(8): 618–26.

Sweeting, Michael J., Alex J. Sutton, and Paul C. Lambert. 2004. "What to Add to Nothing? Use and Avoidance of Continuity Corrections in Meta-Analysis of Sparse Data." *Statistics in Medicine* 23(9): 1351–75.

Tukey, John W. 1958. "Bias and Confidence in Not-Quite-Large Samples." *Annals of Mathematical Statistics* 29: 614.

Ulrich, Rolf, and Markus Wirtz. 2004. "On the Correlation of a Naturally and an Artificially Dichotomized Variable." *British Journal of Mathematical and Statistical Psychology* 57(2): 235–51.

Woolf, Barnet. 1955. "On Estimating the Relation Between Blood Group and Disease." *Annals of Human Genetics* 19(4): 251–53.

Yusuf, Salim, Richard Peto, John Lewis, Rory Collins, and Peter Sleight. 1985. "Beta Blockade During and After Myocardial Infarction: An Overview of the Randomized Trials." *Progress in Cardiovascular Diseases* 27(5): 335–71.

PART
VI

STATISTICALLY COMBINING EFFECT SIZES

14

COMBINING ESTIMATES OF EFFECT SIZE

WILLIAM R. SHADISH
University of California, Merced

C. KEITH HADDOCK
University of Missouri, Kansas City

CONTENTS

14.1 INTRODUCTION

In 1896, Sir Almroth Wright—a colleague and mentor of Sir Alexander Fleming, who discovered penicillin—developed a vaccine to protect against typhoid (Susser 1977; see Roberts 1989). The typhoid vaccine was tested in several settings, and on the basis of these tests the vaccine was recommended for routine use in the British army for soldiers at risk for the disease. In that same year, Karl Pearson, the famous biometrician, was asked to examine the empirical evidence bearing on the decision. To do so, he synthesized evidence from five studies reporting data about the relationship between inoculation status and typhoid immunity, and six studies reporting data on inoculation status and fatality among those who contracted the disease. These eleven studies used data from seven independent samples, with four of those being used twice, once in a synthesis of evidence about incidence of typhoid among those inoculated, and once in a synthesis of evidence about death among those contracting typhoid. He computed tetrachoric correlations for each of these eleven cases, and then averaged these correlations (separately for incidence and fatality) to describe average inoculation effectiveness (see table 14.1). In his subsequent report of this research, Karl Pearson concluded that the average correlations were too low to warrant adopting the vaccine, since other accepted vaccines at that time routinely produced correlations at or above .20 to .30: "I think the right conclusion to draw would be not that it was desirable to inoculate the whole army, but that improvement of the serum and method of dosing, with a view to a far higher correlation, should be attempted" (1904b, 1245).

We tell this story for three reasons. First, we discovered this example when first researching this chapter in 1991, and at the time it was the earliest example of what we would now call a meta-analysis. So it is historically interesting, although a few older examples have now been located. Second, we will use the data in table 14.1 to illustrate how to combine results from fourfold tables. Finally, the study illustrates some conceptual and methodological issues involved in combining estimates of effect size across studies, points we return to in the conclusion of this chapter. In many ways, the capacity to combine results across studies is the defining feature of meta-analysis,

so the conceptual and statistical issues involved in computing these combined estimates need careful attention.

14.1.1 Dealing with Differences Among Studies

Combining estimates of effect size from different studies would be easy if studies were perfect replicates of each other—if they made the same methodological choices about such matters as within-study sample size, measures, or design, and if they all investigated exactly the same conceptual issues and constructs. Under the simplest fixed-effects model of a single population effect size θ, all these replicates would differ from each other only by virtue of having used just a sample of observations from the total population, so that observed studies would yield effect sizes that differed from the true population effect size only by sampling error (henceforth we refer to the variance due to sampling error as conditional variance, for reasons explained in section 14.3). An unbiased estimate of the population effect would then be the simple average of observed study effects; and its standard error would allow computation of confidence intervals around that average. Under random-effects models, one would not assume that there is just one population effect size; instead, a distribution of population effect sizes would exist generated by a distribution of possible study realizations (see chapter 3, this volume). Hence, observed outcomes in studies would differ from each other not just because of sampling error, but also because they reflect these true, underlying population differences.

It is rarely (if ever) the case that studies are perfect replicates of each other, however, because they almost always differ from each other in many methodological and substantive ways. Hence, most research syntheses must grapple with certain decisions about how to combine studies that differ from each other in multiple ways. Consider, for example, the eleven samples in table 14.1. The eleven samples differ dramatically in within-study sample size. They also differ in other ways that were extensively and acrimoniously debated by Karl Pearson and Almroth Wright (Susser 1977). For example, both Wright and Pearson recognized that unreliability in inoculation history and typhoid diagnosis attenuated sample correlations. Further, clinical diagnoses of typhoid were less reli-

Table 14.1 Effects of Typhoid Inoculations on Incidence and Fatality

	Inoculated		Not Inoculated			
	N	Cases	N	Cases	r	Odds Ratio
Incidence						
Hospital staffs	297	32	279	75	.373	3.04
Ladysmith's garrison	1,705	35	10,529	1,489	.445	7.86
Methuen's column	2,535	26	10,981	257	.191	2.31
Single regiments	1,207	72	1,285	82	.021	1.07
Army in India	15,384	128	136,360	2,132	.100	1.89
Average					.226	
	Lived	Died	Lived	Died		
Fatality Among Those Contracting Typhoid						
Hospital staffs	30	2	63	12	.307	2.86
Ladysmith garrison	27	8	1,160	329	−.010	.96
Single regiments	63	9	61	21	.300	2.41
Special hospitals	1,088	86	4,453	538	.119	1.53
Various military hospitals	701	63	2,864	510	.194	1.98
Army in India	73	11	1,052	423	.248	2.67
Average					.193	

SOURCE: Adapted with permission from Susser (1977), adapted from Pearson (1904b).

able than autopsy diagnoses, and studies differed in which diagnostic measure they used. One might reasonably consider, therefore, whether and how to take these differences into account.

14.1.2 Weighting Studies

In general, one could deal with such study differences either before or while combining results or subsequent to combining results. Weighting schemes that are applied at the earlier points rest on three assumptions: theory or evidence suggests that studies with some characteristics are more accurate or less biased with respect to the desired inference than studies with other characteristics, the nature and direction of that bias can be estimated prior to combining, and appropriate weights to compensate for the bias can be constructed and justified. Examples of such weighting schemes are described in the following three sections.

14.1.2.1 Variance and Within-Study Sample Size Schemes Everyone who has had an undergraduate statis-

tics course will recognize the principle underlying these schemes. Studies with larger within-study sample sizes will give more accurate estimates of population parameters than studies with smaller sample sizes. They will be more accurate in the sense that the variance of the estimate around the parameter will be smaller, allowing greater confidence in narrowing the region in which the population value falls. Pearson noted this dependence of confidence on within-study sample size in the typhoid inoculation data, stating that "many of the groups . . . are far too small to allow of any definite opinion being formed at all, having regard to the size of the probable error involved" (1904b, 1243; for explanation of probable error, see Cowles 1989). In table 14.1, for example, estimates of typhoid incidence from the army-in-India sample were based on more than 150,000 subjects, more than 200 times as many subjects as the hospital staff sample of less than 600. The large sample used in the army-in-India study greatly reduced the role of sampling error in determining the outcome. Hence, all things being equal, such large-sample studies should be weighted more heavily; doing so would improve Pearson's unweighted estimates. To implement this scheme, one must have access to the within-study sample sizes used in each of the studies, which is almost always available. This is by far the most widely accepted weighting scheme, and the only one to be discussed in detail in this chapter.

This weighting scheme fulfills all three assumptions outlined previously. First, our preference for studies with large within-study sample sizes is based on strong, widely accepted statistical theory. To wit, large-sample studies tend to produce sample means that are closer to the population mean, and estimating that population mean is (often) the desired inference. Second, the nature and direction of uncertainty in studies that have small within-study sample sizes is clear—greater variability of sample estimates symmetrically around the population parameter. Third, statistical theory suggests very specific weights (to be described shortly) that are known to maximize the chances of correctly making the desired inference to the population parameter. We know of only one other weighting scheme that so unambiguously meets these three conditions, which we now describe.

14.1.2.2 Reliability and Validity of Measurement

If studies were perfect replicates, they would measure exactly the same variables. This is rarely the case. As a result, each different measure, both within and between studies, can have different reliability and validity. For example, some studies in table 14.1 used incidence, some used mortality, and others used both measures; and incidence and mortality were measured different ways in different studies. The differential reliability and validity of these measures was a source of controversy between Pearson and Wright. Fortunately, the effects of reliability and validity on assessments of study outcomes are well specified in classical measurement theory. In the simplest case, unreliability of measurement attenuates relationships between two variables, so that univariate correlation coefficients and standardized mean difference statistics are smaller than they would be with more reliable measures. In multivariate cases, the effects of unreliability and invalidity can be more complex (Bollen 1989), but even then they are well known, and specific corrections are available to maximize the chances of making the correct population inference (Hedges and Olkin 1985; Hunter and Schmidt 2004).

However, unlike within-study sample sizes, information about reliability and validity coefficients often will not be available in reports of primary studies. Neither Pearson nor Wright, for example, reported these coefficients in the

articles in which they debated the issues. Sometimes relevant coefficients can be borrowed from test manuals or reports of similar research. Even when no reasonable reliability or validity estimates can be found, however, procedures exist for integrating effect size estimates over studies when some studies are corrected and others are not (Hunter and Schmidt 2004). These and related matters are addressed by Frank Schmidt, Huy Le, and In-Sue Oh elsewhere in this book (see chapter 17, this volume), so we do not discuss them further. However, those procedures can be applied concurrently with the variance-based weighting schemes illustrated in this chapter (Hedges and Olkin 1985).

14.1.2.3 Other Weighting Schemes

No schemes other than those discussed above seem clearly to meet the three conditions previously outlined. Hence, other schemes should be applied with caution and viewed as primarily exploratory in nature (Boissel et al. 1989, see also chapter 7, this volume). Perhaps the most frequent contender for an alternative scheme is weighting by study quality (Chalmers et al. 1981). The latter term, of course, means different things to different people. For example, when treatment effectiveness is an issue, one might weight treatment outcome studies according to aspects of design that, presumably, facilitate causal inference. Of all such design features, random assignment of units to conditions has the strongest justification, so it is worth examining it against our three assumptions. Strong theory buttresses the preference we give to random assignment in treatment outcome studies, so it meets the first criterion. But it does not clearly meet the other two. We know only that estimates from nonrandomized studies may be biased; we rarely know the direction or nature of the bias. In fact, the latter can vary considerably from study to study (Heinsman and Shadish 1996; Lipsey and Wilson 1993; Shadish and Ragsdale 1996). Further, theory does not suggest a very specific weighting scheme that makes it more likely we will obtain a better estimate of the population parameter. We might assign all randomized studies somewhat greater weight than all quasi-experiments. But even this is arguable because it may be that we should assign all-or-none weights to this dichotomy: If quasi-experiments are biased, allowing them any weight at all may bias the combined result in unknown ways. But this dichotomous (0, 1) weighting scheme is not really a weighting scheme at all because applying it is equivalent to excluding all quasi-experiments in the first place. Given our present lack of knowledge about random versus nonrandom assignment related to our second and third conditions for using weighting schemes, we think it is best to examine differences between randomized and quasi-experiments after combining effect sizes rather than before. Other quality measures, such as whether the experimenter is blind to conditions, can be even more equivocal than random assignment and combining such diverse quality measures into quality scales can compound such problems if the scales are not carefully constructed. For example, a recent study applied several different quality scales to the same meta-analytic data set, and found that the overall results varied considerably depending upon which scale was used to do the weighting (Jüni et al. 1999). Because all scales claimed to be measuring the same construct (namely quality), and none is generally acknowledged to be superior, we cannot know which result to believe.

14.1.3 Examining Effects of Study Quality After Combining Results

Weighting schemes that are a function of either within-study sample size or reliability and validity can routinely be used before combining studies. However, we cannot recommend that other weighting schemes be used routinely before combining results: We do not yet know if they generate more accurate inferences, and they can generate less accurate inferences. However, the relationship of effect size to such scales should be freely explored after combining results, and results with and without such weighting can be compared. In fact, such explorations are one of the few ways to generate badly needed information about the nature and direction of biases introduced by the many ways in which studies differ from each other. Methods for such explorations are the topics of other chapters in this book.

14.2 FIXED-EFFECTS MODELS

Suppose that the data to be combined arise from a series of k independent studies, in which the ith study reports one observed effect size T_i, with population effect size θ_i, and variance v_i. Thus, the data to be combined consist of k effect size estimates $T_1, \ldots, T_k$ of parameters $\theta_1, \ldots, \theta_k$, and variances $v_1, \ldots, v_k$. Under the fixed-effects model, we assume $\theta_1 = \ldots = \theta_k = \theta$, a common effect size. Then a general formula for the weighted average effect size over those studies is

$$\overline{T}_{\cdot} = \frac{\sum_{i=1}^{k} w_i T_i}{\sum_{i=1}^{k} w_i}, \qquad (14.1)$$

where w_i is a weight assigned to the ith study that is defined in equation 14.2. Equation 14.1 can be adapted, if desired, to include a quality index for each study, where q_i is the ith study's score on the index, perhaps scaled from 0 to 1:

$$\overline{T}_{\cdot} = \frac{\sum_{i=1}^{k} q_i w_i T_i}{\sum_{i=1}^{k} q_i w_i} .$$

When all observed effect size indicators estimate a single population parameter, as is hypothesized under a fixed-effects model, then $\overline{T}_{\cdot}$ is an unbiased estimate of the population parameter θ. In equation 14.1, T_i may be estimated by any specific effect size statistic, including the standardized mean difference statistic d (for which the relevant population parameter is δ), the correlation coefficient r (population parameter ρ) or its Fisher z transformation (population parameter ζ), the odds ratio o (population parameter ω), the log odds ratio l (population parameter λ), the difference between proportions D (population parameter Δ), or even differences between outcomes measured in their original metric.

The weights that minimize the variance of $\overline{T}_{\cdot}$ are inversely proportional to the conditional variance (the square of the standard error) in each study:

$$w_i = \frac{1}{v_i}. \qquad (14.2)$$

Formulas for the conditional variances, v_i vary for different effect size indices, so we present them separately for each index in later sections. However, they have in common the fact that conditional variance is inversely proportional to within-study sample size—the larger the sample, the smaller the variance, so the more precise the estimate of effect size should be. Hence, larger weights are assigned to effect sizes from studies that have larger within-study sample sizes. Of course, equation 14.2 defines the optimal weights assuming we know the conditional variances, v_i, for each study. In practice we must estimate those variances ($\hat{v}_i$), so we can only estimate these optimal weights ($\hat{w}_i$).

Given the use of weights as defined in 14.2, the average effect size $\overline{T}_{\cdot}$ has conditional variance $v_{\cdot}$, which itself is a function of the conditional variances of each effect size being combined:

$$v_{\cdot} = \frac{1}{\sum_{i=1}^{k} (1/v_i)}. \qquad (14.3)$$

If the quality-weighted version of 14.1 is used to compute the weighted average effect size, where the weights w_i are defined in 14.2, then the conditional variance of the quality-weighted average effect size is

$$v_. = \frac{\sum\limits_{i=1}^{k} q_i^2 w_i}{\left(\sum\limits_{i=1}^{k} q_i w_i\right)^2}.$$

This simplifies to 14.3 if $q_i = 1$ for all i—in effect, not weighting for quality.

The square root of $v_.$ is the standard error of estimate of the average effect size. Multiplying the standard error by an appropriate critical value C_α, and adding and subtracting the resulting product to $\overline{T}_.$, yields the (95 percent) confidence interval (θ_L, θ_U) for θ:

$$\theta_L = \overline{T}_. - C_\alpha(v_.)^{1/2}, \qquad \theta_U = \overline{T}_. + C_\alpha(v_.)^{1/2}. \quad (14.4)$$

C_α is often the unit normal $C_\alpha = 1.96$ for a two-tailed test at $\alpha = .05$; but better Type I error control and better coverage probability of the confidence interval may occur if C_α is Student's t-statistic at $k - 1$ degrees of freedom. In either case, if the confidence interval does not contain zero, we reject the null hypothesis that the population effect size θ is zero. Equivalently, we may test the null hypothesis that $\theta = 0$ with the statistic

$$Z = \frac{|\overline{T}_.|}{(v_.)^{1/2}}, \quad (14.5)$$

where $|\overline{T}_.|$ is the absolute value of the weighted average effect size over studies (given by equation 14.1). Under 14.5, $\overline{T}_.$ differs from zero if Z exceeds 1.96, the 95 percent two-tailed critical value of the standard normal distribution.

The conditional variances used to compute 14.2 and estimates in subsequent equations must be estimated from the available data. Given that those conditional variances are estimated with uncertainty, some have suggested modifying estimates of significance tests and confidence intervals to take into account the uncertainty of estimation (see, for example, Hartung and Knapp 2001). The latter authors have suggested this in the context of random-effects models, but others have suggested its use more generally. We present this procedure later in the chapter under random-effects models, in part because the particular modification involves a weighted estimate of the random-effects variance first introduced in that section as equation 14.20, but—more important—because the mod-

ification assumes that study effect sizes may vary from each other not only due to sampling error and fixed covariates, but also due to potentially random effects. The latter assumption seems counter to the assumptions inherent to fixed-effects models.

A test of the assumption of equation 14.1 that studies do, in fact, share a common population effect size uses the following homogeneity test statistic:

$$Q = \sum_{i=1}^{k} \left[(T_i - \overline{T}_.)^2 / v_i\right] = \sum_{i=1}^{k} w_i (T_i - \overline{T}_.)^2. \quad (14.6)$$

A computationally convenient form of 14.6 is

$$Q = \sum_{i=1}^{k} w_i T_i^2 - \frac{\left(\sum\limits_{i=1}^{k} w_i T_i\right)^2}{\sum\limits_{i=1}^{k} w_i}.$$

If Q exceeds the upper-tail critical value of chi-square at $k - 1$ degrees of freedom, the observed variance in study effect sizes is significantly greater than we would expect by chance if all studies shared a common population effect size. If homogeneity is rejected, $\overline{T}_.$ should not be interpreted as an estimate of a single effect parameter θ that gave rise to the sample observations, but rather simply as describing a mean the of observed effect sizes, or as estimating a mean θ—which may, of course, be of practical interest in a particular research synthesis. If Q is rejected, the researcher may wish to disaggregate study effect sizes by grouping studies into appropriate categories until Q is not rejected within those categories, or may use regression techniques to account for variance among the θ_i. These latter techniques are discussed in subsequent chapters, as are methods for conducting sensitivity analyses that help examine the influence of particular studies on combined effect size estimates and on heterogeneity.

Alternatively, we will see later that one can incorporate this heterogeneity into one's model using random effects rather than fixed-effects models. For example, when the heterogeneity cannot be explained as a function of known study-level covariates (characteristics of studies such as type of treatment or publication status, or some aggregate characteristic of patients within studies such as average age), or when fixed-effects covariate models are not feasible because sufficient information is not reported in primary studies, random-effects models help take heterogeneity into account when estimating the average size of effect and its confidence interval (Hedges 1983; DerSimonian and Laird 1986). Essentially, in the presence of unexplained heterogeneity, the random-effects test is more conservative than the fixed-effects test, for reasons explained later (Berlin et al. 1989). Finally, sometimes a move to random-effects models is suggested for purely theoretical reasons, as when levels of the variable are thought to be sampled from a larger population to which one desires to generalize (Hunter and Schmidt 2004).

In both the fixed- and random-effects cases, however, Q is a diagnostic tool to help researchers know whether they have, to put it in the vernacular, accounted for all the variance in the effect sizes they are studying. Experience has shown that Q is usually rejected in most simple, fixed-effects univariate analyses—for example, when the researcher simply lumps all studies into one category or contrasts one category of studies, such as randomized experiments, with another category of studies, such as quasi-experiments. In such simple category systems, rejection of homogeneity makes eminent theoretical sense. Each simple categorical analysis can be thought of as the researcher's theoretical model about what variables account for the variance in effect sizes. We would rarely expect that just one variable, or even just two or three variables, would be enough to account for all observed variance. The phenomena we study are usually far more complex than that. Often, extensive multivariate analyses are required to model these phenomena successfully. In essence, then, the variables that a researcher uses to categorize or predict effect sizes can be considered to be the model that the researcher has specified about what variables generated study outcome. The Q statistic tells whether that model specification is statistically adequate. Thus, though homogeneity tests are not without problems, they can serve a valuable diagnostic function (Bohning et al. 2002; Boissel et al. 1989; Fleiss and Gross 1991; Spector and Levine 1987).

The power of the Q test for homogeneity has been discussed often (Hardy and Thompson 1998; Harwell 1997; Jackson 2006; Takkouche et al. 1999). Q tests a very general omnibus hypothesis that all possible contrasts are zero. For that omnibus hypothesis, Q is the most powerful unbiased test possible. Larry Hedges and Therese Pigott provided formulas for computing the power of Q (2001). These formulas show that the power of the test based on Q depends on the noncentrality parameter and that the test will have low power whenever the noncentrality parameter is small. The noncentrality parameter is basically k times the ratio of between study variance in effect parameters to sampling error variance. Thus you tend to get low power when the number of studies (k) is small, between study variance is small, and the sampling error vari-

ance is large. In addition the magnitude of Q also depends on which effect size metric is used, presenting a problem for comparing heterogeneity across meta-analyses that use different effect size measures.

In recognition of some of these issues, Q is usefully supplemented by the descriptive statistic:

$$I^2 = 100\% * \left(\frac{Q - (k-1)}{Q} \right).$$

Negative values of I^2 are set to zero. I^2 is the proportion of total variation in the estimates of treatment effects that is due to heterogeneity rather than to chance (Higgins and Thompson 2002; Higgins et al. 2003). Values of I^2 are not affected by the numbers of studies or the effect size metric. Julian Higgins and Simon Thompson suggested the following approximate guidelines for interpreting this statistic: $I^2 = 25\%$ (small heterogeneity), $I^2 = 50\%$ (medium heterogeneity), and $I^2 = 75\%$ (large heterogeneity) (2002).

In addition, Higgins and Thompson presented equations to compute confidence intervals around I^2, whereby homogeneity is rejected if the confidence interval does not include zero, as an alternative to the Q-test for homogeneity (2002). However, the sampling distribution of I^2 is derived from the sampling distribution of Q, so it must have exactly the same power characteristics as Q does, as has been shown empirically (Huedo-Medina et al. 2006). Consequently, we recommend using I^2 as a descriptive statistic without the confidence intervals to supplement Q rather than to replace Q.

In the following subsections, we use the equations just developed to show how to combine standardized mean differences, correlations, various results from fourfold tables, and raw mean differences. These equations hinge on knowing the conditional variance of the effect-size estimates, for use in equations 14.2 and 14.3. Sometimes alternative estimators of conditional variance or of homogeneity are available that might be preferable in some situations to those we use (DerSimonian and Laird 1986; Guo et al. 2004; Hedges and Olkin 1985; Knapp, Biggerstaff, and Hartung 2006; Kulinskaya et al. 2004; Paul and Thedchanamoorthy 1997; Reis, Kirji, and Afifi 1999). The estimates we present should prove satisfactory in most situations, however. In addition, equations 14.1 to 14.6 can usually be used in the meta-analysis of other effect-size indicators as long as a conditional variance for the estimator is known. For example, Stephen Raudenbush and Anthony Bryk presented conditional variances for the meta-analysis of standard deviations and logits (2001), and

Jonathan Blitstein and his colleagues (2005) did the same for the meta-analysis of intraclass correlations.

14.2.1 Fixed Effects—Standardized Mean Differences

Suppose that the effect-size statistic to be combined is the standardized mean difference statistic d:

$$d_i = \frac{\overline{Y}_i^T - \overline{Y}_i^C}{s_i},$$

where $\overline{Y}_i^T$ is the mean of the treatment group in the ith study, $\overline{Y}_i^C$ is the mean of the control group in the ith study, and s_i is the pooled standard deviation of the two groups. A correction for small sample bias can also be applied (Hedges and Olkin 1985, 81). If the underlying data are normal, the conditional variance of d_i is estimated as

$$v_i = \frac{n_i^T + n_i^C}{n_i^T n_i^C} + \frac{d_i^2}{2(n_i^T + n_i^C)}, \qquad (14.7)$$

where n_i^T is the within-study sample size in the treatment group of the ith study, n_i^C is the within-study sample size in the comparison group of the ith study, and d_i estimates the population parameter δ.

Consider how this applies to the nineteen studies of the effects of teacher expectancy on pupil IQ in data set II. Table 14.2 presents details of this computation. (Throughout this chapter, results will vary slightly depending on the number of decimals carried through computations; this can particularly affect odds ratio statistics, especially when some proportions are small). The weighted average effect size $\overline{d}_. = 45.3303/751.2075 = 0.06$ with a variance of $v_. = 1/751.205 = 0.00133$, and standard error $= (1/751.2075)^{1/2} = 0.03649$. We test the significance of this effect size in either of two equivalent ways: by computing the 95 percent confidence interval, which in this case ranges from $\delta_L = [(.06) - 1.96(.036)] = -.01$ to $\delta_U = [(.06) + 1.96(.036)] = .13$, which includes zero in the confidence interval, and the statistic $Z = |\overline{d}_.|/(v_.)^{1/2} = |0.06|/0.036 = 1.65$, which does not exceed the 1.96 critical value at $\alpha = 0.05$. Hence we conclude that teacher expectancy has no significant effect on student IQ. However, homogeneity of effect size is rejected in this data, with the computational version of equation 14.6, yielding $Q = [38.56 - (45.33)^2/751.21] = 35.83$, which exceeds 34.81, the 99 percent critical value of the chi-square distribution for $k-1 = 19 - 1 = 18$ degrees of freedom, so we reject homogeneity of effect sizes at $p < .01$. Similarly, $I^2 = 100\% \times [35.83 - (19 - 1)]/35.83 = 49.76\%$, suggesting that nearly half the

Table 14.2 Computational Details of Standardized Mean Difference Example

Study	d_i	v_i	w_i	$w_i d_i$	$w_i d_i^2$
1	0.03	0.01563	64.000	1.9200	0.0576
2	0.12	0.02161	46.277	5.5532	0.6664
3	−0.14	0.02789	35.856	−5.0199	0.7028
4	1.18	0.13913	7.188	8.4813	10.0080
5	0.26	0.13616	7.344	1.9095	0.4965
6	−0.06	0.01061	94.260	−5.6556	0.3393
7	−0.02	0.01061	94.260	−1.8852	0.0377
8	−0.32	0.04840	20.661	−6.6116	2.1157
9	0.27	0.02690	37.180	10.0387	2.7104
10	0.80	0.06300	15.873	12.6982	10.1586
11	0.54	0.09120	10.964	5.9208	3.1972
12	0.18	0.04973	20.109	3.6196	0.6515
13	−0.02	0.08352	11.973	−0.2395	0.0048
14	0.23	0.08410	11.891	2.7348	0.6290
15	−0.18	0.02528	39.555	−7.1200	1.2816
16	−0.06	0.02789	35.856	−2.1514	0.1291
17	0.30	0.01932	51.757	15.5271	4.6581
18	0.07	0.00884	113.173	7.9221	0.5545
19	−0.07	0.03028	33.029	−2.3121	0.1618
Sum			751.206	45.3300	38.5606

SOURCE: Authors' compilation.

variation in effect sizes is due to heterogeneity. We might therefore assume that other variables could be necessary to explain fully the variance in these effect sizes; subsequent chapters in this book will explore such possibilities.

14.2.2 Fixed Effects—Correlations

14.2.2.1 Fixed Effects—z-Transformed Correlations
For the correlation coefficient r, some authors recommend first converting each correlation by Sir Ronald Fisher's variance stabilizing z transform (1925):

$$z_i = .5\{\ln[(1 + r_i)/(1 - r_i)]\},$$

where $\ln(x)$ is the natural (base e) logarithm of x. The correlation parameter corresponding to z_i is

$$\zeta_i = .5\{\ln[(1 + \rho_i)/(1 - \rho_i)]\}.$$

If the underlying data are bivariate normal, the conditional variance of z_i is

$$v_i = \frac{1}{(n_i - 3)}, \qquad (14.8)$$

where n_i is the within-study sample size of the ith study. Note that this variance depends only on within-study sample size and not on the correlation parameter itself, a desirable property; and when correlations are small ($|r| < .5$), z_i is close to r_i so the variance of r_i is close to the variance of z_i. Using this variance estimate in equations 14.1 through 14.6, an estimate $\bar{z}_.$ and a confidence interval bounded by ζ_L and ζ_U are obtained directly. Interpretation of results is facilitated if $\bar{z}_.$, ζ_L, and ζ_U are converted back again to the metric of a correlation, $\bar{\rho}_.$, via the inverse of the r to z transform given by

$$\rho(z) = (e^{2z} - 1)/(e^{2z} + 1), \tag{14.9}$$

where $e^x = \exp(x)$ is the exponential (antilog) function, yielding both an estimate of ρ and also upper and lower bounds (ρ_L, ρ_U) around that estimate.

As an example, data set III reported correlations from twenty studies between overall student ratings of the instructor and student achievement. We obtain an average z transform of $\bar{z}_. = 201.13/530.00 = 0.38$, and a variance $v_. = (1/530.00) = 0.0019$, the square root of which (0.044) is the standard error. This average effect size is significantly different from zero since $Z = |\bar{z}_.|/(v_.)^{1/2} = 0.38/0.044 = 8.64$, which exceeds the critical value of 1.96 for $\alpha = 0.05$ in the standard normal distribution. The limits of the 95 percent confidence interval are $\zeta_L = [0.38 - 1.96(0.044)] = 0.294$ and $\zeta_U = [0.38 + 1.96(0.044)] = 0.466$. For interpretability, we can use equation (14.9) to transform the zs back into estimates of correlations, because $2\bar{z}_. = 2(0.38) = 0.76$, $\bar{\rho}_. = (e^{0.76} - 1)/(e^{0.76} + 1) = 0.36$, where $e^{.76} \approx 2.14$. Transforming the upper and lower bounds using the same equation yields $\rho_L = 0.287$ and $\rho_U = 0.434$. Hence we conclude that the better the ratings of the instructor, the better is student achievement. Homogeneity of correlations is not rejected [$Q = 97.27 - (201.13)^2/530.00 = 20.94$, $df = 19$, $p = 0.338$]. A Q value this large would be expected between 30 and 35 percent of the time if all the population correlations were identical. In this case, $I^2 = 100\% \times [20.94 - (20 - 1)]/20.94 = 9.26\%$, suggesting very little of the variation in effect sizes is due to heterogeneity.

14.2.2.2 Fixed Effects—Correlation Coefficients Directly Other authors argue that the average z transform is positively biased, so they prefer combining correlations without z transform (see, for example, Hunter and Schmidt 2004). In that case, if the underlying data are bivariate normal, the variance of the correlation coefficient, r, is approximately

$$v_i = (1 - r_i^2)^2/(n_i - 1), \tag{14.10}$$

where r_i is the single study correlation and n_i is the within-study sample size in that study. Note that this variance depends not only on within-study sample size, but also on the size of the original correlation. Use of 14.1 to 14.6 is straightforward. Using data set III, we obtain an estimate of the common correlation of $\bar{r}_. = 337.00/847.180 = 0.40$, with a variance of $\bar{v}_. = 1/847.18 = 0.0012$, yielding a standard error of .0344. This average correlation is significantly different from zero [$Z = |\bar{r}_.|/(v_.)^{1/2} = 0.40/0.035 = 11.43$], with the limits of the 95 percent confidence interval being $\rho_L = 0.40 - (1.96) \times 0.035 = 0.33$ and $\rho_U = 0.3978 + (1.96) \times 0.035 = 0.47$. Homogeneity of effect size is not rejected [$Q = 159.687 - (337.00)^2/847.18 = 25.63$; $df = k - 1 = 19$; $p = .14$]; a value of Q this large would be expected about 15 percent of the time if all the population correlations were identical. Here, $I^2 = 100\% \times [25.63 - (20 - 1)]/25.63 = 25.87\%$, suggesting a quarter of the variation in effect sizes is due to heterogeneity.

Two points should be noted in comparing these two methods of combining correlations. First, both methods yield roughly the same point estimate of the average correlation, a result noted elsewhere (Field 2005; Law 1995). However, few statisticians would advocate the use of untransformed correlations unless sample sizes are very large because standard errors, confidence intervals, and homogeneity tests can be quite different. The present data illustrate this, where the probability associated with the Q test for the untransformed correlations was only about 40 percent as large as the probability associated with the z-transformed correlations. The reason is the poor quality of the normal approximation to the distribution of r, particularly when the population correlation is relatively large. Second, the approximate variances in 14.8 and 14.10 assume bivariate normality to the raw measurements; however, there should be a good deal of robustness of the distribution theory, just as there is for the r distribution. A similar argument applies to the robustness of variance of the standardized mean difference statistic under nonnormality.

14.2.3 Fixed Effects—Fourfold Tables

Fourfold tables report data concerning the relationship between two dichotomous factors—say, a risk factor and a disease or treatment-control status and successful outcome at posttest. The dependent variable in such tables is always either the number of subjects or the percentage of subjects in the cell. Table 14.3 presents such a fourfold table constructed from the data in the first row of table 14.1—the

Table 14.3 Fourfold Table from Pearson's Hospital Staff Incidence Data

| Group | Condition of Interest | | |
	Immune	Diseased	All
Inoculated	$A = 265$	$B = 32$	$M_1 = 297$
Not Inoculated	$C = 204$	$D = 75$	$M_0 = 279$
Total	$N_1 = 469$	$N_0 = 107$	$T = 576$

SOURCE: Authors' compilation.

incidence data for the hospital staff sample. In table 14.3, the risk factor is immunization status and the disease is typhoid incidence. We could construct a similar fourfold table for each row in table 14.1. From such primary data, various measures of effect size can be computed.

One measure is the odds ratio, which may be defined in several equivalent ways. One formula depends only on the four cell sizes A, B, C, and D (see table 14.3);

$$o_i = \frac{AD}{BC}.$$

If any cell size is zero, then 0.5 should be added to all cell sizes for that sample; however, if the total sample size is small, this correction may not work well. This version of the odds ratio is sometimes called the cross-product ratio because it depends on the cross-products of the table cell sizes. Computing this odds ratio based on the data in table 14.3 yields $o_i = 3.04$, suggesting that the odds that inoculated subjects were immune to typhoid were three times the odds for subjects not inoculated. The same odds ratio can also be computed from knowledge of the proportion of subjects in each condition of a study who have the disease:

$$o_i = \frac{p_{i1}(1 - p_{i2})}{p_{i2}(1 - p_{i1})}, \qquad (14.11)$$

where p_{ij} is the proportion of individuals in condition j of study i who have the disease. In the incidence data in table 14.1, for example, if one wants to assess the odds of being immune to typhoid depending on whether the subject was inoculated, p_{11} is the proportion of subjects in the inoculation condition of the hospital staff study who were immune to typhoid ($p_{11} = 265/297 = 0.8923$); and p_{32} is the proportion of subjects in the noninoculated condition of the Methuen's column study who remained immune to typhoid ($p_{32} = 10724/10981 = 0.9766$). Computing the

odds ratio using this formula for the data in table 14.3 yields $o_i = (0.892/1.08)/(0.731/0.269) = 3.04$, the same as before.

From either odds ratio, we may also define the log odds ratio $l_i = \ln(o_i)$, where ln is the natural logarithm and is equal to $l_i = \ln(3.04) = 1.11$ for the data in table 14.3. An advantage of the log odds ratio is that it takes on a value of zero when no relationship exists between the two factors, yielding a similar interpretation of a zero effect size as r and d.

We may also compute the difference between proportions $D_i = p_{i1} - p_{i2}$, which in table 14.3 yields $D_i = 0.892 - 0.731 = 0.161$, interpretable as the actual gains expected in terms of the percentage of patients treated. Rebecca DerSimonian and Nan Laird (1986) and Jesse Berlin and colleagues (1989) discussed the relative merits of these various indicators. Use of these statistics in meta-analysis has been largely limited to medicine and epidemiology, but should apply more widely. For example, psychotherapy researchers often report primary data about percentage success or failure of clients in treatment or control groups. Analyzing the latter data as standardized mean difference statistics or correlations can lead to important errors that the methods presented in this section avoid (Haddock, Rindskopf, and Shadish 1998).

14.2.3.1 Fixed Effects—Differences Between Proportions Some studies report the proportion of the treatment (exposed) group with a condition (a disease, treatment success, status as case or control) and the proportion of the comparison (unexposed) group with that condition. The difference between these proportions, sometimes called a rate difference, is $D_i = p_{i1} - p_{i2}$. The rate difference has a variance of

$$v_i = \frac{p_{i1}(1 - p_{i1})}{n_{i1}} + \frac{p_{i2}(1 - p_{i2})}{n_{i2}}, \qquad (14.12)$$

where all terms are defined as in equation 14.11 and n_{ij} is the number of subjects in condition j of study i. Use of 14.1 to 14.6 is again straightforward.

Using these methods to combine rate differences for the incidence data in table 14.1, we find that the average rate difference $\overline{D}_. = 18811.91/1759249 = .011$, with a standard error of .00075. The significance test $Z = |\overline{D}_.|/(v_.)^{1/2} = 14.67$, which exceeds the 1.96 critical value at $\alpha = 0.05$. The 95 percent confidence interval is bounded by $\Delta_L = 0.011 - 1.96 \times 0.00075 = 0.0095$ and $\Delta_U = 0.011 + 1.96 \times 0.00075 = 0.012$; that is, about 1 percent more of the inoculated soldiers would be expected to be immune to typhoid compared with the noninoculated soldiers.

However, homogeneity of effect size is rejected, with $Q = 763.00 - (18812)^2/1759249 = 561.84$, which exceeds 14.86, the 99.5 percent critical value of the chi-square distribution at $5 - 1 = 4$ df, so we reject homogeneity of rate differences over these five studies at significance level .005. The $I^2 = 100\% \times [561.84 - (5 - 1)]/561.84 = 99.29\%$, suggesting nearly all the variation in effect sizes is due to heterogeneity.

This example points to an interesting corollary problem with the weighting scheme used in this chapter (this problem and its solutions apply equally to combining other effect size indicators such as d or r). The incidence data in table 14.1 come from studies with vastly discrepant within-study sample sizes. In an analysis in which weights depend on within-study sample size, the influence of the one study with over 150,000 subjects can dwarf the influence of the other studies, even though some of their sample sizes—at more than 10,000 subjects—are large enough that these studies may also provide reliable estimates of effect. One might bound weights to some a priori maximum to avoid this problem, but bounding complicates the variance of the weighted mean, which is no longer equation 14.3, and bounding produces a less efficient (more variable) estimator. An alternative is to compute estimates separately for "big" and "other" studies, an option we do not pursue here only because the number of studies in table 14.1 is small. One might also redo analyses excluding one or more studies—say, the study with the largest sample size, or the smallest, or both—to assess stability of results. So, for example, if we eliminate the samples with the largest (army in India) and smallest (hospital staffs) number of subjects, the rate difference increases somewhat to $\overline{D}_. = 0.034$, still significantly different from zero and heterogeneous.

14.2.3.2 Fixed Effects—Log Odds Ratios Joseph Fleiss distinguished two different cases in which we might want to combine odds ratios from multiple fourfold tables (1981). In the first, the number of studies to be combined is small, but the within-study sample sizes per study are large, as with the incidence data in table 14.1. The proper measure of association to be combined is thus the log odds ratio l_i, the variance of which is

$$v_i = \frac{1}{A_i} + \frac{1}{B_i} + \frac{1}{C_i} + \frac{1}{D_i},$$

or, alternatively,

$$v_i = \frac{1}{n_{i1}p_{i1}(1-p_{i1})} + \frac{1}{n_{i2}p_{i2}(1-p_{i2})}, \quad (14.13)$$

where all terms are defined in equation 14.11, and n_{ij} is the number of subjects in condition j of study i. In table 14.1, for example, $n_{11} = 297$ is the number of subjects in the inoculated condition of the hospital staff study; and $n_{23} = 10,981$ is the number of subjects in the noninoculated condition of the Methuen's column study. Using this variance estimate in equations 14.1 through 14.6 is again straightforward. Subsequently, if the researcher wants to report final results in terms of the mean odds ratio over studies, this is done by transforming the average log odds ratio, $\bar{l}_.$, back into an average odds ratio $\bar{o}_. = \text{antilog}(\bar{l}_.)$, and similarly by reporting upper and lower bounds as odds ratios, $\omega_L = \text{antilog}(\lambda_L)$ and $\omega_U = \text{antilog}(\lambda_U)$.

For example, using the incidence data in table 14.1, we find the average log odds ratio $\bar{l}_. = 188.005/230.937 = 0.814$, with variance $\bar{v}_. = 1/230.937 = 0.0043$, yielding a standard error of .066. The significance test $Z = |\bar{l}_.|/(v_.)^{1/2} = 12.33$, which exceeds the 1.96 critical value at $\alpha = 0.05$; and the 95 percent confidence interval is bounded by $\lambda_L = .814 - 1.96 \times 0.066 = 0.685$ and $\lambda_U = 0.814 + 1.96 \times 0.066 = 0.943$. Returning the log odds ratios to their original metrics via the antilog transformation yields a common odds ratio of $\bar{o}_. = \text{antilog}(0.814) = 2.26$ with bounds of $\omega_L = \text{antilog}(0.685) = 1.98$ and $\omega_U = \text{antilog}(0.943) = 2.57$. The common odds ratio of 2.26 suggests that the odds that inoculated soldiers were immune to typhoid were twice the odds for soldiers not inoculated; and this improvement in the odds over groups is statistically significant. However, homogeneity of effect size is rejected, with $Q = 230.317 - (188.005)^2/230.937 = 77.26$, substantially larger than 14.86, the 99.5 percent critical value of the chi-square distribution with $5 - 1 = 4$ df; that is, the odds radios do not seem to be consistent over these five studies. The $I^2 = 100\% \times [77.26 - (5 - 1)]/77.26 = 94.82\%$, again suggesting nearly all the variation in effect sizes is due to heterogeneity.

14.2.3.3 Fixed Effects—Odds Ratios with Mantel-Haenszel The second case that Fleiss distinguishes is when one has many studies to combine, but the within-study sample size in each study is small. In this case, the general logic of equations 14.1 to 14.6 does not dictate analysis. Instead, we compare the Mantel-Haenszel summary estimate of the odds ratio:

$$\overline{o}_{MH} = \frac{\sum_{i=1}^{k} w_i[p_{i1}(1-p_{i2})]}{\sum_{i=1}^{k} w_i[p_{i2}(1-p_{i1})]}, \quad (14.14)$$

where the proportions are as defined for 14.11 and where

$$w_i = \frac{n_{i1}n_{i2}}{n_{i1}+n_{i2}}.$$

A formula that is equivalent to 14.14, but computed from cell sizes as indexed in table 14.3, is

$$\bar{o}_{MH} = \frac{\sum_{i=1}^{k} A_i D_i / T_i}{\sum_{i=1}^{k} B_i C_i / T_i}, \tag{14.15}$$

where T_i is the total number of subjects in the sample. Note that in the case of $k = 1$ (when there is just one study), the T_is cancel from the numerator and denominator of 14.15, which then reduces to the odds ratio defined in 14.11. An estimate of the variance of the log of $\bar{o}_{MH}$ (Robins, Breslow, and Greenland 1986; Robins, Greenland, and Breslow 1986) is

$$v_{\cdot} = \frac{\sum_{i=1}^{k} P_i R_i}{2\left(\sum_{i=1}^{k} R_i\right)^2} + \frac{\sum_{i=1}^{k}(P_i S_i + Q_i R_i)}{2\left(\sum_{i=1}^{k} R_i\right)\left(\sum_{i=1}^{k} S_i\right)} + \frac{\sum_{i=1}^{k} Q_i S_i}{2\left(\sum_{i=1}^{k} S_i\right)^2}, \tag{14.16}$$

where $P_i = (A_i + D_i)/T_i$, $Q_i = (B_i + C_i)/T_i$, $R_i = A_i D_i / T_i$, and $S_i = B_i C_i / T_i$. When $k = 1$, (14.16) reduces to the standard equation for the variance of a log odds ratio in a study as defined in 14.13. Then, the 95 percent confidence interval is bounded by $\omega_L = \text{antilog}[\log(\bar{o}_{MH}) + 1.96(v_{\cdot})^{1/2}]$ and $\omega_U = \text{antilog}[\log(\bar{o}_{MH}) + 1.96(v_{\cdot})^{1/2}]$.

To test the significance of the overall Mantel-Haenszel odds ratio, we can compute the Mantel-Haenszel chi-square statistic:

$$\chi^2_{MH} = \frac{\left\{\left|\sum_{i=1}^{k}[(n_{i1}n_{i2})/(n_{i1}+n_{i2})](p_{i1}-p_{i2})\right| - .5\right\}^2}{\sum_{i=1}^{k}\{[(n_{i1}n_{i2})/(n_{i1}+n_{i2}-1)]p_i(1-p_i)\}}, \tag{14.17}$$

where

$$p_i = \frac{n_{i1}p_{i1} + n_{i2}p_{i2}}{n_{i1}+n_{i2}}.$$

Alternatively, 14.17 can be written in terms of cell frequencies and marginals as

$$\chi^2_{MH} = \frac{\left[\left|\sum_{i=1}^{k}(O_i - E_i)\right| - .5\right]^2}{\sum_{i=1}^{k} V_i},$$

where $O_i = A_i$, $E_i = [(A_i + C_i)/T_i](A_i + B_i)$, and $V_i = E_i[(T_i - (A_i + C_i))/T_i][(T_i - (A_i + B_i))/(T_i - 1)]$. The difference $(O_i - E_i)$ in the numerator is simply the difference "between the observed number of treatment patients who experienced the outcome event, O, and the number expected to have done so under the hypothesis that the treatment is no different from the control, E" (Fleiss and Gross 1991, 131). In the denominator, V_i is the variance of the difference $(O_i - E_i)$, which can also be expressed as the product of the four marginal frequencies divided by $T_i^2(T_i - 1)$, or $[(A_i + C_i)(A_i + B_i)(B_i + D_i)(C_i + D_i)]/[T_i^2(T_i - 1)]$. The result of this chi-square is compared with the critical value of chi-square with 1 df. Homogeneity may be tested using the same test developed for the log odds ratio. However, better homogeneity tests, although computationally complex, are available (Jones et al. 1989; Kuritz, Landis, and Koch 1988; Paul and Donner 1989).

We illustrate the Mantel-Haenszel technique by applying it to the fatality data in table 14.1. The Mantel-Haenszel common odds ratio is $218.934/123.976 = 1.77$; that is, the odds that inoculated subjects who contracted typhoid would survive were almost two times the odds that noninoculated subjects who contracted typhoid would survive. The variance of the log of this ratio is $v_{\cdot} = 63.996/[2(218.934)^2] + 190.273/[2(218.934)(123.976)] + 88.64/[2(123.976)^2] = 0.00706$, and the 95 percent confidence interval is bounded by $\omega_L = \text{antilog}[\log(1.77) - 1.96 \times (0.007)^{1/2}] = 1.50$ and $\omega_U = \text{antilog}[\log(1.77) + 1.96 \times (0.007)^{1/2}] = 2.08$. The Mantel-Haenszel $\chi^2 = (94.96 - .5)^2/191.31 = 46.64$, $p < 0.00001$, suggesting that inoculation significantly increases the odds of survival among those who contracted typhoid. Finally, homogeneity of odds ratios is not rejected over these six studies, $Q = 50.88 - [(78.97)^2/141.46] = 6.80$, $df = 5$, $p = 0.24$; a Q value this large would be expected between 20 and 25 percent of the time if all the population effect sizes were identical. The $I^2 = 100\% \times [6.80 - (6 - 1)]/6.80 = 26.47\%$, suggesting that nearly half the variation in effect sizes is due to heterogeneity.

The results of a Mantel-Haenszel analysis are valid whether one is analyzing many studies with small sample sizes or few studies with large sample sizes; but the results of the log odds ratio analysis in section 14.2.3.2 are valid only in the latter case. Therefore, we can reanalyze the incidence data in table 14.1 using the Mantel-Haenszel techniques and compare the results to the log odds ratio analysis. The Mantel-Haenszel common odds ratio for the incidence data is $537.007/205.823 = 2.61$; that is, the odds that inoculated subjects who contracted typhoid would survive were two and a half times the odds that noninoculated subjects who contracted typhoid would survive. The variance of the log of this ratio is

$$v_. = 125.355/\ [2(537.007)^2] + 459.181/[2(537.007)$$
$$\times (205.823)] + 158.294/\ [2(205.823)^2] = 0.00416,$$

and the 95 percent confidence interval is bounded by $\omega_L =$ antilog[log(2.61) $- 1.96 \times (.004)^{1/2}$] = 2.30 and $\omega_U =$ antilog[log(2.61) $+ 1.96 \times (v.004)^{1/2}$] = 2.96. The Mantel-Haenszel $\chi^2 = (331.185 - .5)^2/462.879 = 236.244$, $p <$ 0.00001, suggesting that inoculation significantly increases the odds of survival among those who contracted typhoid. Finally, homogeneity of odds ratios is rejected over these five studies, $Q = 230.373 - [(188.188)^2/231.173] = 77.18$, which greatly exceeds 14.86, the 99.5 percent critical value of the chi-square distribution at $5 - 1 = 4$ df. Under the Mantel-Haenszel techniques, therefore, the odds ratio was somewhat larger than under the log odds approach; but the variance and Q test results were very similar, and the confidence interval was about as wide under both approaches. The $I^2 = 100\% \times [77.18 - (5 - 1)]/77.18 = 94.82\%$, so that nearly all the variation in effect sizes is due to heterogeneity.

14.2.4 Combining Data in Their Original Metric

On some occasions, the researcher may wish to combine outcomes over studies in which the metric of the outcome variable is exactly the same across studies. Under such circumstances, converting outcomes to the various effect-size indicators might be not only unnecessary but also distracting since the meaning of the original variable may be obscured. An example might be a synthesis of the effects of treatments for obesity compared with control conditions, where the dependent variable is always weight in pounds or kilograms; in this case computing a standardized mean difference score for each study might be less clear than simply reporting outcomes as weight in pounds. Even so, the researcher should still weight each study by a function of sample variance. If T_i is a mean difference between treatment and control group means, $T_i = \overline{Y}_i^T - \overline{Y}_i^C$, then an estimate of the variance of T_i is

$$v_i = \sigma_i^2(1/n_i^T + 1/n_i^C), \qquad (14.18)$$

where n_i^T is the within-study sample size in the treatment group, n_i^C is the within-study sample size for the control group, and σ_i^2 is the assumed common variance (note that this assumes the two variances within each study are ho-

Table 14.4 Effects of Psychological Intervention on Reducing Hospital Length of Stay (in days)

Study	Psychotherapy Groups			Control Group			Mean Difference
	N	Mean Days	SD	N	Mean Days	SD	
1	13	5.00	4.70	13	6.50	3.80	−1.50
2	30	4.90	1.71	50	6.10	2.30	−1.20
3	35	22.50	3.44	35	24.90	10.65	−2.40
4	20	12.50	1.47	20	12.30	1.66	.20
5	10	3.37	.92	10	3.19	.79	.18
6	13	4.90	1.10	14	5.50	.90	−.60
7	9	10.56	1.13	9	12.78	2.05	−2.22
8	8	6.50	.76	8	7.38	1.41	−.88

SOURCE: Adapted from Mumford et al. 1984, table 1.
Includes only randomized trials that reported means and standard deviations.

mogeneous, but different studies need not have the same variances). The latter can be estimated by the pooled within-group variance if the within-study sample sizes and standard deviations of each group are reported:

$$s_{pi}^2 = \left[(n_i^T - 1)(s_i^T)^2 + (n_i^C - 1)(s_i^C)^2\right]/[n_i^T + n_i^C - 2],$$

where $(s_i^T)^2$ is the variance of the treatment group in the ith study, and $(s_i^C)^2$ is the variance of the control group or the comparison treatment in the ith study. The pooled variance can sometimes be estimated other ways. Squaring the pooled standard deviation yields an estimate of the pooled variance. Lacking any estimate of the pooled variance at all, one cannot use 14.1 to 14.6; but then one cannot compute a standardized mean difference, either.

As an example, Emily Mumford and her colleagues synthesized studies of the effects of mental health treatment on reducing costs of medical utilization, the latter measured by the number of days of hospitalization (1984). From their table 1, we extract eight randomized studies comparing treatment groups that received psychological intervention with control groups, which report means, within-study sample sizes, and standard deviations (see table 14.4). The weighted average of the difference between the number of days the treatment groups were hospitalized versus the number of days the control groups were hospitalized was $\overline{T}_. = -14.70/27.24 = -0.54$ days; that is, the treatment group left the hospital about half a day sooner than the control group. The variance of this effect is $v_. = 1/27.24 = 0.037$, yielding a standard error of 0.19. The 95 percent confidence interval is bounded by $T_L = -0.54 - 1.96 \times (0.19)^{1/2} = -0.91$ and $T_U = -0.54 + 1.96 \times (0.19)^{1/2} = -0.17$, and $Z = |\overline{T}_.|/v_.^{1/2} = |-0.54|/0.19 = 2.84$ ($p = 0.0048$), both of which indicate that the effect is significantly different from zero. However, the homogeneity test closely approaches significance, with $Q = [21.85 - (-14.70)^2/27.24] = 13.92$, $df = 7$, $p = 0.053$, suggesting that mean differences may not be similar over studies. This result is supported by the $I^2 = 100\% \times [13.92 - (8 - 1)]/13.92 = 49.71\%$, suggesting that half the variation in effect sizes is due to heterogeneity.

14.3 RANDOM-EFFECTS MODELS

Under a fixed-effects model, the effect-size statistics, T_i, from k studies estimate a population parameter $\theta_1 = \ldots = \theta_k = \theta$. The estimate T_i in any given study differs from the θ due to sampling error, or what we have referred to as conditional variability; that is, because a given study used a sample of subjects from the population, the estimate of T_i computed for that sample will differ somewhat from θ for the population.

Under a random-effects model, θ_i is not fixed, but is itself random and has its own distribution. Hence, total variability of an observed study effect-size estimate v_i^* reflects *both* conditional variation v_i of that effect size around each population θ_i *and* random variation, σ_θ^2, of the individual θ_i around the mean population effect size:

$$v_i^* = \sigma_\theta^2 + v_i. \qquad (14.19)$$

In this equation, we will refer to σ_θ^2 as either the between-studies variance or the variance component (others sometimes refer to this as random effects variance); to v_i as either within-study variance or the conditional variance of T_i (the variance of an observed effect size conditional on θ being fixed at the value of θ_i; others sometimes call this estimation variance); and to v_i^* as the unconditional variance of an observed effect size T_i (others sometimes call this the variance of estimated effects). If the between-studies variance is zero, then the equations of the random-effects model reduce to those of the fixed-effects model, with unconditional variability of an observed effect size [v_i^*] hypothesized to be due entirely to conditional variability [v_i] (to sampling error).

The reader may wonder how to choose between a fixed versus random effects model. No single correct answer to this question exists. Statistically, rejection of homogeneity implies the presence of a variance component estimate that is significantly different from zero, which might, in turn, indicate the presence of random effects to be accounted for. However, this finding is not definitive because the homogeneity test has low power under some circumstances, and because the addition of fixed-effects covariates are sometimes enough to produce a nonsignificant variance component. Some authors argue that random-effects models are preferred on conceptual grounds because they better reflect the inherent uncertainty in meta-analytic inference, and because they reduce to the fixed-effects model when the variance component is zero. Larry Hedges and Jack Vevea noted that the choice of model depends on the inference the researcher wishes to make (1998). The fixed-effects model allows inferences about what the results of the studies in the meta-analysis would have been had they been conducted identically except for using other research participants (that is, except for sampling error). The random-effects model allows inferences about what the results of the studies in the meta-analysis might have been had they been conducted with other participants but also with changes in other study features such as in the treatment, setting, or outcome measures. So the choice between fixed- and random-effects models must be made after careful consideration of the applicability of all these matters to the particular meta-analysis being done.

Once the researcher decides to use a random-effects analysis, a first task is to determine whether the variance component differs significantly from zero and, if it does, then to estimate its magnitude. The significance of the variance component is known if the researcher has already conducted a fixed-effects analysis—if the Q test for homogeneity of effect size (equation 14.6) was rejected, the variance component differs significantly from zero. In that case, the remaining task is to estimate the magnitude of the variance component so that 14.19 can be estimated. If the investigator has not conducted a fixed-effects analysis already, Q must be calculated in addition to estimating the variance component.

Estimating the variance component can be done in many different ways (Viechtbauer 2005). A common method begins with the ordinary (unweighted) sample estimate of the variance of the effect sizes, $T_1, \ldots, T_k$, computed as

$$s^2(T) = \sum_{i=1}^{k}[(T_i - \overline{T})^2/(k-1)], \qquad (14.20)$$

where $\overline{T}$ is the unweighted mean of T_1 through T_k. A computationally convenient form of (14.20) is

$$s^2(T) = \left[\sum_{i=1}^{k} T_i^2 - \left(\sum_{i=1}^{k} T_i\right)^2 \middle/ k\right] \middle/ (k-1).$$

The expected value of $s^2(T)$, that is, the unconditional variance we would expect to be associated with any particular effect size, is

$$E[s^2(T)] = \sigma_\theta^2 + (1/k)\sum_{i=1}^{k} \sigma^2(T_i | \theta_i). \qquad (14.21)$$

The observed variance $s^2(T)$ from equation (14.20) is an unbiased estimate of $E[s^2(T)]$ by definition. To estimate $\sigma^2(T_i | \theta_i)$, we use the v_i from equations 14.7, 14.8, 14.10, 14.12, 14.13, 14.16, or 14.18, depending on which effect-size indicator is being aggregated. With these estimates, equation 14.21 can be solved to obtain an estimate of the variance component:

$$\hat{\sigma}_\theta^2 = s^2(T) - (1/k)\sum_{i=1}^{k} v_i. \qquad (14.22)$$

This unbiased sample estimate of the variance component will sometimes be negative, even though the population variance component must be a positive number. In these cases, it is customary to fix the component to zero.

A problem with equation 14.22 is that it may be negative or zero even though the homogeneity statistic Q rejects the null hypothesis that $\sigma_\theta^2 = 0$. The second method for estimating the variance component avoids that problem. It begins with Q as defined in equation 14.6, which we now take as an estimate if the weighted sample estimate of the unconditional variance $\sigma^2(T_i)$. The expected value of Q is

$$E[Q] = c(\sigma_\theta^2) + (k-1),$$

where

$$c = \sum_{i=1}^{k} w_i - \left[\sum_{i=1}^{k} w_i^2 \middle/ \sum_{i=1}^{k} w_i\right].$$

Solving for σ_θ^2 and substituting Q for its expectation gives another unbiased estimator of the variance component:

$$\hat{\sigma}_\theta^2 = [Q - (k-1)]/c. \qquad (14.23)$$

If homogeneity is rejected, this estimate of the variance component will be positive, since Q must then be bigger than $k-1$.

Lynn Friedman offered the following suggestions for choosing between these two estimators (2000):

- If Q is not rejected and 14.23 is negative, assume $\sigma_\theta^2 = 0$. Equation 14.23 is positive, use the estimate produced by equation 14.23, especially if the studies in the meta-analysis each have small total sample sizes ≤ 40.

- If Q is rejected and equation 14.23 is positive but equation 14.22 is negative, use 14.23 because it is more efficient when σ_θ^2 is small. Equations 14.22 and 14.23 are both positive, use 14.23 if $Q \leq 3(k-1)$, and use 14.22 otherwise.

So equation 14.22 should only be used if Q is rejected and is very large relative to the number of studies in the meta-analysis. Otherwise equation 14.23 should be used.

Neither of the estimators 14.22, 14.23 described above are always optimal (Viechtbauer 2005). In particular, 14.22 tends to be less efficient than 14.23, but 14.23 works best when the conditional variances of each effect size are homogenous, something unlikely to be the case in practice with sets of studies that have notably varying sample sizes. A restricted maximum likelihood estimator described by Wolfgang Viechtbauer (2005) tended to perform better than either 14.22 or 14.23, but must be estimated with an iterative procedure not widely implemented in available software.

When the variance component is significant, one may also compute $\overline{T}_*$, the random effects weighted mean of $T_1, \ldots, T_k$, as an estimate of μ_θ, the average of the random effects in the population; v_*, the variance of $\overline{T}_*$ (the square root of v_* is the standard error of $\overline{T}_*$); and random-effects confidence limits θ_L and θ_U for μ_θ by multiplying the standard error by an appropriate critical value (often, 1.96 at $\alpha = .05$), and adding and subtracting the resulting product from $\overline{T}_*$. In general, the computations follow equations 14.1 to 14.5, except that the following unconditional variance estimate is used in equations 14.2 through 14.5 in place of the conditional variances outlined in the fixed-effects models:

$$v_i^* = \hat{\sigma}_\theta^2 + v_i, \qquad (14.24)$$

where $\hat{\sigma}_\theta^2$ is the variance component estimate yielded by equations 14.22 or 14.23 and v_i is defined by 14.7, 14.8, 14.10, 14.12, 14.13, 14.16, or 14.18, as appropriate. The square root of the variance component describes the standard deviation of the distribution of effect parameters. Multiplying that standard deviation by 1.96 (or an appropriate critical value of t), and adding and subtracting the result from the average random-effect size $\overline{T}_*$, yields the limits of an approximate 95 percent confidence interval. All these random-effects analyses assume that the random effects are normally distributed with constant variance, an assumption that is particularly difficult to assess when the number of studies is small.

In section 14.2, after introducing equation 14.5, we noted that Joachim Hartung and Guido Knapp proposed a modification to random-effects estimates of significance tests and confidence intervals to take into account the uncertainty of sample estimates of the conditional variance and the variance component (2001). Essentially, the modification presents a weighted version of equation 14.20:

$$s^2(T) = \frac{1}{k-1}\sum_{i=1}^{k}\frac{w_i}{w}(T_i - \overline{T})^2,$$

where w is the summation of all the individual w_i. This estimate is then used in equation 14.22 and subsequent computations of significance tests and confidence intervals, both using critical values of t at $df = k-1$ rather than z. Doing so will decrease the power of these tests to reject the null hypothesis, but may result in more accurate coverage of type I and II error rates. Chapter 15 in this volume presents more details of this approach in the context of fixed-effects categorical and regression analyses.

In summary, we can now see why the random effects model usually provides more conservative estimates of the significance of average effect sizes over studies *in the presence of unexplained heterogeneity*. Under both fixed- and random-effects models, significant unexplained heterogeneity results in rejection of Q, because Q is exactly the same under both models. The variance component is an index of how much variance remained unexplained. In the fixed-effects model, this unexplained variance is ignored in the computation of standard errors and confidence intervals. By contrast, in the random-effects model the unexplained heterogeneity represented by the variance component is added to the conditional variance, and the sum of these is then used in computing standard errors and confidence intervals. With more variance, the latter will be usually larger, which often means the random-effects test is usually less likely to reject the null hypothesis than the fixed effects test.

However, the latter is not always true because the estimate of the mean effect under the random-effects model also changes. In some circumstances, for example when the smallest effect sizes are associated with the largest sample sizes, the average effect size can be larger under the random-effects model than under the fixed-effects. Depending on how large this change is, it is theoretically possible to obtain a significant random-effects mean effect size when the fixed effects counterpart is not significant.

Finally, because the weights in equations 14.1 and 14.2 under the random-effects model are computed using the unconditional variance in equation 14.24, these weights will tend toward equality over studies—no matter how unequal their within-study sample size—as study heterogeneity, and thus the variance component, increases.

14.3.1 Random Effects—Standardized Mean Differences

Estimating the conditional variance of d_i with equation 14.7 allows us to solve equations 14.20 through 14.24 to estimate the variance component σ_δ^2. If the variance component is larger than zero, then the weighted average of the random effects is computed using 14.1 with weights as computed by equation 14.2, but where the variance for the denominator of 14.2 is estimated as equation 14.24:

$$v_i^* = \hat{\sigma}_\theta^2 + v_i,$$

where v_i is defined as 14.7. Similarly, the variance of the estimate of μ_δ is given by equation 14.3 but using the estimate of the unconditional variance, v_i^*; using the resulting variance estimate yields the 95 percent confidence intervals and Z test as defined in equations 14.4 and 14.5. If the variance component exceeds zero, the random-effects unconditional variance v_i^* will be larger than the fixed-effects variance v_i, so the variance of the estimate and the confidence intervals will both be larger under random-effects models. Thus, Z will usually be smaller, except for cases like those described at the end of the previous section, where it is theoretically possible for an increase in the random effects mean to offset an increase in the variance enough to increase Z in the random-effects model.

We apply these procedures to data set II, using the data from nineteen studies on the effects of teacher expectancy on student IQ. Equation 14.20 yields a sample estimate of the variance of $s^2(d) = [2.8273 - (3.11)^2/19]/18 = 0.1288$. We then estimate

$$(1/k)\sum_{i=1}^{k}\sigma^2(d_i\,|\,\delta_i)$$

by averaging the v_i as defined by equation 14.7 to obtain

$$(1/k)\sum_{i=1}^{k}v_i = 0.04843,$$

and solve (14.22) for a variance component estimate of

$$\hat{\sigma}_\theta^2 = 0.1288 - 0.04843 = 0.082,$$

which is significantly different from zero because $Q = 35.83$ exceeds 34.81, the 99 percent critical value of the chi-square distribution for $k - 1 = 19 - 1 = 18$ df (note that this is the same Q for homogeneity under a fixed-effects model). Recomputing the aggregate statistics under the random effects model then yields an average population effect size of $\bar{d}_* = 18.22/159.36 = 0.114$, with a standard error $= (1/159.36)^{1/2} = 0.079$.

The limits of the 95 percent confidence interval is

$$\delta_L = 0.114 - 1.96 \times 0.079 = -0.04$$

and

$$\delta_U = 0.114 + 1.96 \times .079 = 0.27,$$

and

$$Z = |\bar{d}_*|/(v_*^{1/2}) = .114|/0.079 = 1.44,$$

which does not exceed 1.96, so we do not reject the hypothesis that the average effect size in the population is zero. These results suggest that though the central tendency over the nineteen studies is rather small at .114, population effect sizes do vary significantly from that tendency. Note the much wider confidence interval relative to the fixed-effects model analysis, reflecting the greater inherent variability of effect sizes under the random-effects model.

Finally, the square root of the variance component in this data is 0.29, which we can take (under the untestable assumption of a normal distribution of effect sizes in the population) as the standard deviation of the population effect sizes. Multiplying it by 1.96 yields a 95 percent confidence interval bounded by -0.45 to 0.67, suggesting that most effect sizes in the population probably fall into this range. Indeed, this mirrors the data in the sample, with all but two study effect sizes falling within this range.

However, in this data set $Q = 35.83 < 3(k - 1)$, so following Friedman's guidelines, we would prefer the weighted variance model of equation 14.23. This yields

$$c = 751.2075 - 47697.39/751.2075 = 687.71,$$

and then the variance component is

$$\hat{\sigma}_\delta^2 = [(35.83 - 18)/687.71] = 0.026,$$

somewhat smaller than the variance component estimated using equation 14.22. Recomputing the aggregated statistics yields an average population effect size of

$$\bar{\delta}_* = 28.68/321.08 = 0.0893,$$

with a standard error $= (1/321.08)^{1/2} = 0.0558$. The limits of the 95 percent confidence interval are $\delta_L = 0.089 - 1.96 \times 0.0558 = -0.020$ and $\delta_U = 0.089 + 1.96 \times 0.0558 = 0.199$, and $Z = |0.0893|/0.0558 = 1.60$.

14.3.2 Random Effects—Correlations

14.3.2.1 Random Effects—z-Transformed Correlations If we follow the z-transform procedure outlined in section 14.2.2.1, then equation 14.8 defines the conditional variance of the individual z statistics, allowing us to solve equations 14.20 to 14.24. We apply this procedure to data set III, which reported correlations from twenty studies between overall student ratings of the instructor and student achievement. Equation 14.20 yields a sample estimate of the variance of

$$s^2(z) = 0.0604.$$

We then estimate $(1/k)\sum_{i=1}^{k}\sigma^2(z_i \mid \zeta_i)$ by averaging the v_i as defined by 14.8 to obtain $(1/k)\sum_{i=1}^{k}v_i = 0.0640$, and solve equation 14.22 for a variance component estimate of $\hat{\sigma}_\zeta^2 = 0.0604 - 0.0640 = -0.0036$, $Q = 20.94$ (the same Q for homogeneity under a fixed effects model), which at $k - 1 = 19$ df is not significant ($p = 0.34$). Because this variance component is negative, we set it equal to zero by convention. However, because Q is not significant, we prefer the weighted variance method (equation 14.23). This yields $c = 530 - 27334/530 = 478.43$, and then the variance component is $\hat{\sigma}_\zeta^2 = [20.94 - 19]/478.43 = 0.0041$, also not significantly different from zero by the same Q test. When the variance component is not significantly different from zero, we will not recompute average effect sizes, standard errors, and confidence intervals; but the reader should recall our previous comments that some analysts would choose to compute all these random effect statistics even though the variance component is not significantly different from zero.

14.3.2.2 Random Effects—Correlation Coefficients Directly If we choose to combine correlation coefficients, the conditional variance of r_i is defined in 14.20 to 14.24. We apply this procedure to data set III. Equation

14.20 yields a sample estimate of the variance of $s^2(r) = (1/19)[3.4884 - (7.28)^2/20] = 0.0441$. Using equation 14.22, we then obtain a variance component of $\hat{\sigma}_\rho^2 = 0.0441 - 0.0374 = 0.0067$, which is not significantly different from zero $Q = 25.631$, $p = 0.14$. Again, because Q is not significant, we prefer the weighted variance method of 14.23, which yields $c = 847.18 - (60444.68/847.18) = 775.83$ and a variance component of $\hat{\sigma}_\rho^2 = (25.631 - 19)/775.83 = 0.0085$, also nonsignificant by the same Q test.

14.3.3 Random Effects—Fourfold Tables

14.3.3.1 Random Effects—Differences Between Proportions In the case of rate differences $D_i = p_{i1} = p_{i2}$, the conditional variance is defined in 14.12. As an example, we apply equations 14.20 to 14.24 to the incidence data in table 14.1. Equation 14.20 yields a sample estimate of the variance of $s^2(D) = (1/4)[0.0408 - (0.3065)^2/5] = 0.0055$. Using equation 14.22, we then obtain a variance component of $\hat{\sigma}_\Delta^2 = 0.0055 - 0.0002 = 0.0053$, which at $k - 1 = 4$ df is significantly different from zero, $Q = 562.296$. Recomputing the aggregate effect size statistics yields an average rate difference $\overline{D}_* = 53.00/912.49 = 0.058$, a variance $v_* = 1/912.49 = 0.0011$, standard error = 0.033, with a 95 percent confidence interval bounded by $\Delta_L = 0.058 - 1.96 \times 0.033 = -0.0069$ and $\Delta_U = 0.058 + 1.96 \times 0.033 = 0.12$.

Using equation 14.23, the weighted variance method yields $c = 393167.45$, which then yields a variance component estimate of $\hat{\sigma}_\Delta^2 = (562.296 - 4)/c = .0014$, also significant by the same Q test. Recomputing the average effect size statistics yield an average difference between proportions of $\overline{D}. = 166.75/3170.42 = .053$, a variance $v. = 1/3170.42 = .0003$, a standard error = .018, and a 95 percent confidence interval bounded by $\Delta_L = .053 - (1.96)(.018) = .018$ and $\Delta_U = .053 + (1.96)(.018) = .087$. The weighted and unweighted methods yield very similar results; however, if we follow Friedman's guidelines (2000), we would prefer the estimate from equation 14.22 because $Q = 562.296 > 3(k - 1)$. This preference makes a difference here because the 95 percent confidence interval includes zero under 14.22 but excludes it under 14.23.

14.3.3.2 Random Effects—Odds and Log Odds Ratios The methods outlined in this section can be used to test random effects models in both cases discussed by Fleiss—combining log odds ratios, and combining odds ratios using Mantel-Haenszel methods. We use equation 14.13 to estimate the conditional variance of the log odds ratio and then use equations 14.20 to 14.24 with

the incidence data in table 14.1 Equation 14.20 yields a sample estimate of variance of $s^2(l) = (1/4)[6.604 - (4.72)^2/5] = 0.537$. Using equation 14.22, we then obtain a variance component estimate of $\hat{\sigma}_\lambda^2 = .537 - 0.032 = 0.505$, which at $k - 1 = 4$ df is significantly different from zero, $Q = 77.2626$. Recomputing the average effect-size statistics yields an average log odds ratio $\overline{l}_* = 8.75/9.32 = 0.94$, a variance $v_* = 1/9.32 = 0.107$, standard error = 0.328, and a 95 percent confidence interval bounded by $\lambda_L = 0.94 - 1.96 \times 0.328) = 0.30$ and $\lambda_U = 0.94 + 1.96 \times 0.328) = 1.58$. As in section 14.2.3, we then reconvert the common log odds ratios into the more interpretable common odds ratio as $\overline{o}_* = $ antilog$(\overline{l}_*) = 2.56$, and $\omega_L = $ antilog $(\lambda_L) = 1.35$ and $\omega_U = $ antilog$(\lambda_U) = 4.85$ are the bounds of the 95 percent confidence interval.

Using equation 14.23, the weighted variance method again yields similar results, with $c = 230.94 - (17566.09/230.94) = 154.88$, which then yields a variance component estimate of $\hat{\sigma}_\lambda^2 = (77.26 - 4)/154.88 = 0.473$, also significant by the same Q test. Here we would again prefer the previous estimate from (14.22) because $Q = 77.26 > 3(k - 1)$. However, to complete the example, recomputing the average effect-size statistics yields an average log odds ratio $\overline{l}_* = 9.30/9.90 = 0.94$, a variance $v_* = 1/9.90 = 1.02$, standard error = 0.319, and a 95 percent confidence interval bounded by $\lambda_L = 0.94 - 1.96 \times 0.319 = 0.32$ and $\lambda_U = 0.94 + 1.96 \times 0.319 = 1.56$. As in section 14.2.3, we then reconvert the common log odds ratio into the more interpretable common odds ratio as $\overline{o}_* = $ antilog$(\overline{l}_*) = 2.56$, and the 95 percent confidence interval is then bounded by $\omega_L = $ antilog$(\lambda_L) = 1.37$ and $\omega_U = $ antilog$(\lambda_U) = 4.73$.

14.3.4 Combining Differences Between Raw Means

Equation 14.18 gives the conditional variance for differences between two independent means. Using that estimate in equations 14.20 to 14.24 yields the random-effects solution. For the data in table 14.4, equation 14.20 yields a sample estimate of the variance of $s^2(T) = (1/7)[15.59 - (8.42)^2/8] = 0.96$. Using equation 14.22, we then obtain a variance component of $\hat{\sigma}_\theta^2 = 0.96 - 1.01 = -0.05$, which by convention we round to zero; since $Q = 13.92$, and $p = 0.053$, however, this is very close to significance. Using equation (14.23), the weighted variance method yields $c = 27.24 - (138.75/27.24) = 22.15$, which then yields $\hat{\sigma}_\theta^2 = (13.92 - 7)/22.15 = 0.31$, which is preferable according the Friedman's rules. Recomputing the average effect-size statistics under the random-effects model for the unweighted variance estimate yields an average mean difference of $\overline{T}_* = -18.41/37.94 = -0.49$ days, with $v_* = 1/37.94 = 0.027$, standard error $= 0.16$, and a 95 percent confidence interval bounded by $\theta_L = -0.49 - 1.96 \times 0.16 = -0.80$ and $\theta_U = -0.49 + 1.96 \times 0.16 = -0.13$. Recomputing these same statistics under the weighted variance model yields $\overline{T}_* = -7.66/11.25 = -0.68$ days, with $v_* = 1/11.25 = 0.09$, standard error $= 0.30$, and a 95 percent confidence interval bounded by $\theta_L = -0.68 - 1.96 \times 0.30 = -1.27$ and $\theta_U = -0.68 + 1.96 \times 0.30 = -0.09$.

14.4 COMPUTER PROGRAMS

Since the first edition of this handbook appeared more than ten years ago, computer programs for doing meta-analysis have proliferated, including those that will do the computations described in this chapter. The programs range from freeware to shareware to commercial products, with the latter including both programs devoted exclusively to meta-analysis and additions to standard comprehensive statistical packages. The first author maintains a web page with links to many such programs, along with a brief description of each (http://faculty.ucmerced.edu/wshadish/Meta-Analysis%20Software.htm).

14.5 CONCLUSIONS

We began this chapter with a discussion of the debate between Pearson and Wright about the effectiveness of the typhoid vaccine. Hence it is fitting to return to this topic as we conclude. If we were to apply the methods outlined in section 14.2.2.1 to the data in table 14.1, we would find that the typhoid vaccine is even less effective than Pearson concluded. For example, the weighted average tetrachoric correlation between immunity and inoculation is $\overline{\rho}_. = 0.132(Z = 56.41)$ and the weighted average tetrachoric correlation between inoculation and survival among those who contract typhoid is $\overline{\rho}_. = 0.146(Z = 17.199)$. Both these averages are smaller than the unweighted averages reported by Pearson of 0.23 and 0.19, respectively. It is ironic in retrospect, then, that Wright was right: subsequent developments showed that the vaccine works very well indeed.

Yet this possibility was not lost on Pearson. Anticipating informally the results of the formal Q tests, which reject homogeneity of effect size for these data ($Q = 1740.36$, and $Q = 76.833$, respectively), Pearson said early in his report that "the material appears to be so heterogeneous, and the results so irregular, that it must be doubtful how much weight is to be attributed to the different results" (1904b, 1243). He went on to outline a host of possible sources of heterogeneity. For example, some of the data came from army units which for a variety of reasons he thought "seem hardly comparable" (1243); for example, "the efficiency of nursing . . . must vary largely from hospital to hospital under military necessities" (1904a, 1663). He also noted that the samples differed in ways that might affect study outcome: "the single regiments were very rapidly changing their environment . . . the Ladysmith garrison least, and Methuen's column more mobile than the Ladysmith garrison, but far less than the single regiments" (1904b, 1244). Because "it is far more easy to be cautious under a constant environment than when the environment is nightly changing," it might be that "the apparent correlation between inoculation and immunity arose from the more cautious and careful men volunteering for inoculation" (1904b, 1244). Thus he called for better-designed "experimental inquiry," using "volunteers, but while keeping a register of all men who volunteered, only to inoculate every second volunteer" (1245); recall that Fisher's work on randomized experiments had not yet been published (1925, 1926).

Wright, too, cited various sources of heterogeneity, noting that the validity of typhoid diagnosis was undoubtedly highly variable over units, partly because "under ordinary circumstances the only practically certain criterion of typhoid fever is to be found in the intestinal lesions which are disclosed by *post-mortem* examination" (1904a, 1490). Such autopsies were not routine in all units, so measurements of fatality in some units were probably more valid than in others. Moreover, autopsy diagnosis is definitive, whereas physician diagnosis of incidence of typhoid was very unreliable. For these and other reasons, Wright questioned the validity of measurement in these studies, chiding Pearson for neglecting "to make inquiry into the trustworthiness of . . . the comparative value of a death-rate and an incidence-rate of the efficacy of an inoculation process" (1904b, 1727).

Methods for exploring such sources of heterogeneity are the topic of the next several chapters of this book. Unfortunately, neither Pearson nor Wright provided more precise information about which studies were associated with which sources of heterogeneity in the data they argued about. Hence, use of the Pearson-Wright example ends here. But the issues are timeless. Pearson and Wright understood them over a century ago; they lacked only the methods to pursue them.

14.6 REFERENCES

Berlin, Jesse A., Nan M. Laird, Henry S. Sacks, and Thomas C. Chalmers. 1989. "A Comparison of Statistical Methods for Combining Event Rates from Clinical Trials." *Statistics in Medicine* 8(2): 141–51.

Blitstein, Jonathan L., David M. Murray, Peter J. Hannan and William R Shadish,. 2005. "Increasing the Degrees of Freedom in Existing Group Randomized Trials: The df_* Approach." *Evaluation Review* 29(3): 241–67.

Bohning, Dankmar, Uwe Malzahn, Ekkehart Dietz, Peter Schlattmann, Chukiat Viwatwongkasem, and Annibale Biggeri. 2002. "Some General Points in Estimating Heterogeneity Variance with the DerSimonian-Laird Estimator." *Biostatistics* 3(4): 445–57.

Boissel, Jean-Pierre, Jean Blanchard, Edouard Panak, Jean-Claude Peyrieux, and Henry Sacks. 1989. "Considerations for the Meta-Analysis of Randomized Clinical Trials." *Controlled Clinical Trials* 10(3): 254–81.

Bollen, Kenneth A. 1989. *Structural Equations with Latent Variables.* New York: John Wiley & Sons.

Chalmers, Thomas C., Henry Smith Jr., Bradley Blackburn, Bernard Silverman, Biruta Schroeder, Dinah Reitman, and Alexander Ambroz. 1981. "A Method for Assessing the Quality of a Randomized Control Trial." *Controlled Clinical Trials* 2(1): 31–49.

Cowlesm Michael. 1989. *Statistics in Psychology: An Historical Perspective.* Hillsdale, N.J.: Lawrence Erlbaum.

DerSimonian, Rebecca, and Nan Laird. 1986. "Meta-Analysis in Clinical Trials." *Controlled Clinical Trials* 7(3): 177–88.

Field, Andy P. 2005. "Is the Meta-Analysis of Correlation Coefficients Accurate when Population Correlations Vary?" *Psychological Methods* 10(4): 444–67.

Fisher, Ronald A. 1925. *Statistical Methods for Research Workers.* Edinburgh: Oliver and Boyd.

Fisher, Ronald A. 1926. "The Arrangement of Field Experiments." *Journal of the Ministry of Agriculture of Great Britain* 33: 505–13.

Fleiss, Joseph L. 1981. *Statistical Methods for Rates and Proportions*, 2nd ed.. New York: John Wiley & Sons.

Fleiss, Joseph L., and Alan J. Gross. 1991. "Meta-Analysis in Epidemiology, with Special Reference to Studies of the Association Between Exposure to Environmental Tobacco Smoke and Lung Cancer: A Critique." *Journal of Clinical Epidemiology* 44(2): 127–39.

Friedman, Lynn. 2000. "Estimators of Random Effects Variance Components in Meta-Analysis." *Journal of Educational and Behavioral Statistics* 25(1): 1–12.

Guo, Jianhua, Yanping Ma, Ning-Zhong Shi, and Tai Shing Lau. 2004. "Testing for Homogeneity of Relative Differences under Inverse Sampling." *Computational Statistics and Data Analysis* 44(4): 513–624.

Haddock, C. Keith, David Rindskopf, and William R. Shadish. 1998. "Using Odds Ratios as Effect Sizes for Meta-Analysis of Dichotomous Data: A primer on Methods and Issues." *Psychological Methods,* 3(3): 339–53.

Hardy, Rebecca J., and Simon G. Thompson. 1998. "Detecting and Describing Heterogeneity in Meta-Analysis." *Statistics in Medicine* 17(8): 841–56.

Hartung, Joachim, and Guido Knapp. 2001. "On Tests of the Overall Treatment Effect in Meta-Analysis with Normally Distributed Responses." *Statistics in Medicine* 20(12): 1771–82.

Harwell, Michael. 1997. "An Empirical Study of Hedges' homogeneity Test." *Psychological Methods* 2(2): 219–31.

Hedges, Larry V. 1983. "A Random Effects Model for Effect Sizes." *Psychological Bulletin* 93(2): 338–95.

Hedges, Larry V., and Ingram Olkin. 1985. *Statistical Methods for Meta-Analysis.* Orlando, Fl.: Academic Press.

Hedges, Larry V., and Therese D Pigott. 2001. "The Power of Statistical Tests in Meta-Analysis." *Psychological Methods* 6(3): 203–17.

Hedges, Larry V., and Jack L. Vevea. 1998. "Fixed- and Random-Effects Models in Meta-Analysis." *Psychological Methods* 4(3): 486–504.

Heinsman, Donna T., and William R. Shadish. 1996. "Assignment Methods in Experimentation: When do Nonrandomized Experiments Approximate the Answers from Randomized Experiments?" *Psychological Methods* 1(2): 154–69.

Higgins, Julian P. T., and Simon G. Thompson. 2002. "Quantifying Heterogeneity in a Meta-Analysis." *Statistics in Medicine* 21(11): 1539–58.

Higgins, Julian P. T., Simon G. Thomson, Jonathan J. Deeks, and Douglas G. Altman. 2003. "Measuring Inconsistency in Meta-Analyses." *British Medical Journal* 327(6 September): 557–60.

Huedo-Medina, Tania B., Julio Sanchez-Meca, J., Fulgencio Marin-Martinez, and Juan Botella. 2006. "Assessing Heterogeneity in Meta-Analysis: Q Statistic or I^2 Index?" *Psychological Methods* 11(2): 193–206.

Hunter, Jack E., and Frank L. Schmidt. 2004. *Methods of Meta-Analysis: Correcting Errors and Bias in Research Findings* 2nd ed. Thousand Oaks, Calif.: Sage Publications.

Jackson, Dan. 2006. "The Power of the Standard Test for the Presence of Heterogeneity in Meta-Analysis." *Statistics in Medicine* 25(15): 2688–99.

Jones, Michael P., Thomas W. O'Gorman, T. W., Jon H. Lemke, J. H., and Robert F. Woolson. 1989. "A Monte Carlo Investigation of Homogeneity Tests of the Odds Ratio under Various Sample Size Configurations." *Biometrics* 45(1): 171–81.

Jüni, Peter, Anne Witschi, Ralph Bloch, and Matthias Egger. 1999. "The Hazards of Scoring the Quality of Clinical Trials for Meta-Analysis." *Journal of the American Medical Association,* 282(11): 1054–60.

Knapp, Guido, Brad J. Biggerstaff and Joachim Hartung. 2006. "Assessing the Amount of Heterogeneity in Random-Effects Meta-Analysis." *Biometrical Journal* 48(2): 271–85.

Kulinskaya, E., M. B. Dollinger, E. Knight, and H. Gao. 2004. "A Welch-type test for homogeneity of contrasts under heteroscedasticity with application to meta-analysis." *Statistics in Medicine* 23(23): 3655–70.

Kuritz. Stephen J., J. Richard Landis, and Gary G. Koch. 1988. "A General Overview of Mantel-Haenszel Methods: Applications and Recent Developments." *Annual Review of Public Health* 9: 123–60.

Law, Kenneth S. 1995. "Use of Fisher's Z in Hunter-Schmidt—Type Meta-Analyses." *Journal of Educational and Behavioral Statistics* 20(3): 287–306.

Lipsey, Mark W., and David B. Wilson. 1993. "The Efficacy of Psychological, Educational, and Behavioral Treatment: Confirmation from Meta-Analysis." *American Psychologist* 48(2): 1181–1209.

Mumford, Emily, Herbert J. Schlesinger, Gene V. Glass, Cathleen Patrick, and Timothy Cuerdon. 1984. "A New Look at Evidence about Reduced Cost of Medical Utilization Following Mental Health Treatment." *American Journal of Psychiatry* 141(10): 1145–58.

Paul, Sudhir, and Alan Donner. 1989. "A Comparison of Tests of Homogeneity of Odds Ratios in k 2 × 2 Tables." *Statistics in Medicine* 8(12): 1455–68.

Paul, Sudhir, and S. Thedchanamoorthy. 1997. "Likelihood-Based Confidence Limits for the Common Odds Ratio." *Journal of Statistical Planning and Inference* 64(1): 83–92.

Pearson, Karl. 1904a. "Antiphoid Inoculation." *British Medical Journal* 2(2296): 1667–68.

———. 1904b. "Report on Certain Enteric Fever Inoculation Statistics." *British Medical Journal* 2: 1243–46.

Raudenbush, Stephen W., and Anthony S. Bryk. 2001. *Hierarchical Linear Models*, 2nd ed. Thousand Oaks, Calif.: Sage Publications.

Reis, Isildinha M., Karim F. Kirji, and Abdelmonem A. Afifi. 1999. "Exact and Asymptotic Tests for Homogeneity in Several 2 × 2 Tables." *Statistics in Medicine* 18(8): 893–906.

Roberts, Royston M. 1989. *Serendipity: Accidental Discoveries in Science.* New York: John Wiley & Sons.

Robins, James, Norman Breslow, and Sander Greenland. 1986. "Estimators of the Mantel-Haenszel Variance Consistent in Both Sparse Data and Large-Strata Limiting Models." *Biometrics* 42(2): 311–23.

Robins, James, Sander Greenland, and Norman Breslow. 1986. "A General Estimator for the Variance of the Mantel-Haenszel Odds Ratio." *American Journal of Epidemiology* 124: 719–23.

Shadish, William R., and Kevin Ragsdale. 1996. "Random Versus Nonrandom Assignment in Psychotherapy Experiments: Do You Get the Same Answer?" *Journal of Consulting and Clinical Psychology* 64(6): 1290–1305.

Spector, Paul E., and Edward L. Levine. 1987. "Meta-Analysis for Integrating Study Outcomes: A Monte Carlo Study of its Susceptibility to Type I and Type II Errors." *Journal of Applied Psychology* 72(1): 3–9.

Susser, Mervyn. 1977. "Judgment and Causal Inference: Criteria in Epidemiologic Studies." *American Journal of Epidemiology* 105(1): 1–15.

Takkouche, Bahi, Carmen Cadarso-Suarez, and Donna Spiegelman. 1999. "Evaluation of Old and New Tests of Heterogeneity in Epidemiological Meta-Analysis." *American Journal of Epidemiology* 150(2): 206–15.

Wright, Almroth E. 1904a. "Antiphoid Inoculation." *British Medical Journal* 2(Jul–Dec): 1489–91.

———. 1904b. "Antiphoid Inoculation." *British Medical Journal* 2(Jul–Dec): 1727.

Viechtbauer, Wolfgang. 2005. "Bias and Efficiency of Meta-Analytic Estimators in the Random-Effects Model." *Journal of Educational and Behavioral Statistics* 30(3): 261–93.

15

ANALYZING EFFECT SIZES: FIXED-EFFECTS MODELS

SPYROS KONSTANTOPOULOS
Boston College

LARRY V. HEDGES
Northwestern University

CONTENTS

15.1 INTRODUCTION

A central question in research synthesis is whether methodological, contextual, or substantive differences in research studies are related to variation in effect-size parameters.

Both fixed- and random-effects statistical methods are available for studying the variation in effects. The choice of which to use is sometimes a contentious issue in both meta-analysis as well as primary analysis of data. The choice of statistical procedures should primarily be determined by the kinds of inference the synthesist wishes to make. Two different inference models are available, sometimes called *conditional* and *unconditional* inference (see Hedges and Vevea 1998). The conditional model attempts to make inference about the relation between covariates and the effect-size parameters in the studies that are observed. In contrast, the unconditional model attempts to make inferences about the relation between covariates and the effect-size parameters in the population of studies from which the observed studies are a representative sample. Fixed-effects statistical procedures are well suited to drawing conditional inferences about the observed studies (see, for example, Hedges and Vevea 1998). Random- or mixed-effects statistical procedures are well suited to drawing unconditional inferences (inferences about the population of studies from which the observed studies are a sample. Fixed-effects statistical procedures may also be a reasonable choice when the number of studies is too small to support the effective use of mixed- or random-effects models.

In this chapter, we present two general classes of fixed-effects models. One class is appropriate when the independent (study characteristic) variables are categorical. This class is analogous to the analysis of variance, but is adapted to the special characteristics of effect-size estimates. The second class is appropriate for either discrete or continuous independent (study characteristic) variables and therefore technically includes the first class as a special case. This second class is analogous to multiple regression analysis for effect sizes. In both cases, we describe the models along with procedures for estimation and hypothesis testing. Although some formulas for hand computation are given, we stress computation using widely available packaged computer programs.

Tests of goodness of fit are given for each fixed-effect model. They test the notion that there is no more variability in the observed effect sizes than would be expected if all (100 percent) of the variation in effect size parameters is explained by the data analysis model. These tests can be conceived as tests of model specification. If a fixed-effects model explains all of the variation in effect-size parameters, the (fixed-effect) model is unquestionably appropriate. Models that are well specified can provide a strong basis for inference about effect sizes in fixed-effects models, but are not essential for inference from them. If differences between studies that lead to differences in effects are not regarded as random (for example, if they are regarded as consequences of purposeful design decisions) then fixed effects methods may be appropriate for the analysis. Similarly fixed-effects analyses are appropriate if the inferences desired are regarded as conditional—applying only to studies like those under examination.

15.2 ANALYSIS OF VARIANCE FOR EFFECT SIZES

One of the most common situations in research synthesis arises when the effect sizes can be sorted into independent groups according to one or more characteristics of the studies generating them. The analytic questions are whether the groups' (average) population effect sizes vary and whether the groups are internally homogeneous, that is, whether the effect sizes vary within the groups. Alternatively, we could describe the situation as one of exploring the relationship between a categorical independent variable (the grouping variable) and effect size.

Many experimental social scientists have encountered the statistical method of analysis of variance (ANOVA) as a way of examining the relationship between categorical independent variables and continuous dependent variables in primary research. Here we describe an analogue to the analysis of variance for effect sizes in the fixed-effect case (for technical details, see Hedges 1982b; Hedges and Olkin 1985).

15.2.1 The One-Factor Model

Situations frequently arise in which we wish to determine whether a particular discrete characteristic of studies is related to effect size. For example, we may want to know if the type of treatment is related to its effect or if all variations of the treatment produce essentially the same effect. The analogue to one-factor analysis of variance for effect sizes is designed to answer just such questions. We present this analysis in somewhat greater detail than its multifactor extensions both because it illustrates the logic of more complex analyses and because it is among the simplest and most frequently used methods for analyzing variation in effect sizes.

15.2.1.1 Notation In the discussion of the one-factor model, we use a notation emphasizing that the independent effect size estimates fall into p groups, defined a priori by the independent (grouping) variable. Suppose that there are p distinct groups of effects with m_1 effects in the first group, m_2 effects in the second group, . . . , and m_p effects in the pth group. Denote the jth effect parameter in the ith group by θ_{ij} and its estimate by T_{ij} with (conditional) variance v_{ij}. That is, T_{ij} estimates θ_{ij} with standard error $\sqrt{v_{ij}}$. In most cases, v_{ij} will actually be an estimated variance that is a function of the within-study sample size and the effect-size estimate in study j. However, unless the within-study sample size is exceptionally small, we can treat v_{ij} as known. Therefore, in the rest of this chapter we assume that v_{ij} is known for each study. The sample data from the collection of studies can be represented as in table 15.1.

15.2.1.2 Means Using the dot notation from the analysis of variance, define the group mean effect estimate for the ith group $\overline{T}_{i\bullet}$ by

$$\overline{T}_{i\bullet} = \frac{\sum_{j=1}^{m_i} w_{ij} T_{ij}}{\sum_{j=1}^{m_i} w_{ij}}, \quad i = 1, \ldots, p \tag{15.1}$$

where the weight w_{ij} is simply the reciprocal of the variance of T_{ij},

$$w_{ij} = 1/v_{ij}. \tag{15.2}$$

The grand weighted mean $\overline{T}_{\bullet\bullet}$ is

$$\overline{T}_{\bullet\bullet} = \frac{\sum_{i=1}^{p} \sum_{j=1}^{m_i} w_{ij} T_{ij}}{\sum_{i=1}^{p} \sum_{j=1}^{m_i} w_{ij}}. \tag{15.3}$$

The grand mean $\overline{T}_{\bullet\bullet}$ could also be seen as the weighted mean of the group means $\overline{T}_{1\bullet}, \ldots, \overline{T}_{p\bullet}$.

$$\overline{T}_{\bullet\bullet} = \frac{\sum_{i=1}^{p} w_{i\bullet} \overline{T}_{i\bullet}}{\sum_{j=1}^{p} w_{i\bullet}},$$

where the weight $w_{i\bullet}$ is just the sum of the weights for the ith group

$$w_{i\bullet} = w_{i1} + \cdots + w_{im_i}.$$

Table 15.1 Effect Size Estimates and Sampling Variances for p Groups of Studies

	Effect Size Estimates	Variances
Group 1		
Study 1	T_{11}	v_{11}
Study 2	T_{12}	v_{12}
.	.	.
.	.	.
Study m_1	T_{1m_1}	v_{1m_1}
Group 2		
Study 1	T_{21}	v_{21}
Study 2	T_{22}	v_{22}
.	.	.
Study m_2	T_{2m_2}	v_{2m_2}
Group p		
Study 1	T_{p1}	v_{p1}
Study 2	T_{p2}	v_{p2}
.	.	.
Study m_p	T_{pm_p}	v_{pm_p}

SOURCE: Authors' compilation.

Thus, $\overline{T}_{i\bullet}$ is simply the weighted mean that would be computed by applying formula (15.1) to the studies in group i and $\overline{T}_{\bullet\bullet}$ is the weighted mean that would be obtained by applying formula (15.1) to all of the studies. If all of the studies in group i estimate a common effect-size parameter $\theta_{i\bullet}$, that is if $\theta_{i1} = \theta_{i2} = \ldots = \theta_{im_i} = \theta_{i\bullet}$, then $\overline{T}_{i\bullet}$ estimates $\theta_{i\bullet}$. If the studies within the ith group do *not* estimate a common effect parameter, then $\overline{T}_{i\bullet}$ estimates the weighted mean of the effect-size parameters θ_{ij} given by

$$\overline{\theta}_{i\bullet} = \frac{\sum_{j=1}^{m_i} w_{ij} \theta_{ij}}{\sum_{j=1}^{m_i} w_{ij}}, \quad i = 1, \ldots, p. \tag{15.4}$$

Similarly, if all of the studies in the collection estimate a common parameter $\overline{\theta}_{\bullet\bullet}$, that is, if

$$\theta_{11} = \ldots = \theta_{1m_1} = \theta_{21} = \ldots = \theta_{pm_p} = \overline{\theta}_{\bullet\bullet},$$

then $\overline{T}_{\bullet\bullet}$ estimates $\overline{\theta}_{\bullet\bullet}$.

If the studies do not all estimate the parameter, then $\overline{T}_{..}$ can be seen as an estimate of a weighted mean $\overline{\theta}_{..}$ of the effect parameters given by

$$\overline{\theta}_{..} = \frac{\sum\limits_{i=1}^{p}\sum\limits_{j=1}^{m_i} w_{ij}\theta_{ij}}{\sum\limits_{i=1}^{p}\sum\limits_{j=1}^{m_i} w_{ij}}. \tag{15.5}$$

Alternatively, $\overline{\theta}_{..}$ can be viewed as a weighted mean of the $\overline{\theta}_{i.}$:

$$\overline{\theta}_{..} = \frac{\sum\limits_{i=1}^{p} w_{i.}\overline{\theta}_{i.}}{\sum\limits_{j=1}^{p} w_{i.}}.$$

where $w_{i.}$ is just the sum of the weights w_{ij} for the ith group as in the alternate expression for $\overline{T}_{..}$ above.

15.2.1.3 Tests of Heterogeneity In the analysis of variance, tests for systematic sources of variance are constructed from sums of squared deviations from means. That the effects attributable to different sources of variance partition the sums of squares leads to the interpretation that the total variation about the grand mean is partitioned into parts that arise from between-group and within-group sources. The analysis of variance for effect sizes has a similar interpretation. The total heterogeneity statistic $Q = Q_T$ (the weighted total sum of squares about the grand mean) is partitioned into a between-groups-of-studies part Q_B (the weighted sum of squares of group means about the grand mean) and a within-groups-of-studies part Q_W (the total of the weighted sum of squares of the individual effect estimates about the respective group means). These statistics Q_B and Q_W yield direct omnibus tests of variation across groups in mean effects and variation within groups of individual effects.

An Omnibus Test for Between-Groups Differences. To test the hypothesis that there is no variation in group mean effect sizes, that is, to test

$$H_0 : \overline{\theta}_{1.} = \overline{\theta}_{2.} = \cdots = \overline{\theta}_{p.},$$

we use the between group heterogeneity statistic Q_B defined by

$$Q_B = \sum_{i=1}^{p} w_{i.}(\overline{T}_{i.} - \overline{T}_{..})^2 \tag{15.6}$$

where $w_{i.}$ is the reciprocal of the variance of $v_{i.}$. Note that Q_B is just the weighted sum of squares of group mean effect sizes about the grand mean effect size. When the

null hypothesis of no variation across group mean effect sizes is true, Q_B has a chi-square distribution with $(p-1)$ degrees of freedom. Hence we test H_0 by comparing the obtained value of Q_B with the upper tail critical values of the chi-square distribution with $(p-1)$ degrees of freedom. If Q_B exceeds the $100(1-\alpha)$ percent point of the chi-square distribution (for example, $C_{.05} = 18.31$ for 10 degrees of freedom and $\alpha = 0.05$), H_0 is rejected at level α and between group differences are significant.

This test is analogous to the omnibus F-test for variation in group means in a one-way analysis of variance in a primary research study. It differs in that Q_B, unlike the F-test, incorporates an estimate of unsystematic error in the form of the weights. Thus no separate error term (such as the mean square within groups as in the typical analysis of variance) is needed and the sum of squares can be used directly as a test statistic.

The test based on equation 15.6 depends entirely on theory that is based on the approximate sampling distributions of the T_{ij}, and on the fact that any differences among the θ_{ij} are not random.

An Omnibus Test for Within-Group Variation in Effects. To test the hypothesis that there is no variation of population effect sizes within the groups of studies, that is to test

$$H_0: \begin{array}{l} \theta_{11} = \ldots = \theta_{1m_1} = \overline{\theta}_{1.} \\ \theta_{21} = \ldots = \theta_{2m_2} = \overline{\theta}_{2.} \\ \quad \cdot \quad\quad\quad \cdot \quad\quad\quad \cdot \\ \quad \cdot \quad\quad\quad \cdot \quad\quad\quad \cdot \\ \quad \cdot \quad\quad\quad \cdot \quad\quad\quad \cdot \\ \theta_{p1} = \ldots = \theta_{pm_p} = \overline{\theta}_{p.} \end{array}$$

use the within-group homogeneity statistic Q_W given by

$$Q_W = \sum_{i=1}^{p}\sum_{j=1}^{m_i} w_{ij}(T_{ij} - \overline{T}_{i.})^2 \tag{15.7}$$

where the w_{ij} are the reciprocals of the v_{ij}, the sampling variances of the T_{ij}. When the null hypothesis of perfect homogeneity of effect size parameters is true, Q_W has a chi-square distribution with $(k-p)$ degrees of freedom where $k = m_1 + m_2 + \ldots + m_p$ is the total number of studies. Therefore within-group homogeneity at significance level α is rejected if the computed value of Q_W exceeds the $100(1-\alpha)$ percent point (the upper tail critical value) of the chi-square distribution with $(k-p)$ degrees of freedom.

Although Q_W provides an overall test of within-group variability in effects it is actually the sum of p separate (and independent) within-group heterogeneity statistics, one for each of the p groups of effects. Thus

$$Q_W = Q_{W_1} + Q_{W_2} + \cdots + Q_{W_p} \qquad (15.8)$$

where each Q_{W_i} is just the heterogeneity statistic Q given in formula 14.6 of chapter 14 in this volume. In the notation used here, Q_{W_i} is given by

$$Q_{W_i} = \sum_{j=1}^{m_i} w_{ij}(T_{ij} - \bar{T}_{i\bullet})^2. \qquad (15.9)$$

These individual within-group statistics are often useful in determining which groups are the major sources of within-group heterogeneity and which groups have relatively homogeneous effects. For example in analyses of the effects of study quality in treatment/control studies, study effect estimates might be placed into two groups: those from quasi-experiments and those from randomized experiments. The effect sizes within the two groups might be quite heterogeneous overall, leading to a large Q_W, but most of that heterogeneity might arise within the group of quasi-experiments so that Q_{W_1} (the statistic for quasi-experiments) would indicate great heterogeneity, but Q_{W_2} (the statistic for randomized experiments) would indicate relative homogeneity (see Hedges 1983a).

If the effect-size parameters within the ith group of studies are homogeneous, that is if $\theta_{i1} = \ldots = \theta_{im_i}$, then Q_{W_i} has the chi-square distribution with $m_i - 1$ degrees of freedom. Thus the test for homogeneity of effects within the ith group at significance level α consists of rejecting the hypothesis of homogeneity if Q_{W_i} exceeds the 100 $(1 - \alpha)$ percent point of the chi-square distribution with $(m_i - 1)$ degrees of freedom.

It is often convenient to summarize the relationships among the heterogeneity statistics via a table analogous to an ANOVA source table (see table 15.2).

Partitioning Heterogeneity. There is a simple relationship among the total homogeneity statistic Q given in formula (15.6) and the between-and within-group fit statistics discussed in this section. This relationship corresponds to the partitioning of the sums of squares in ordinary analysis of variance. That is,

$$Q = Q_B + Q_W.$$

One interpretation is that the total heterogeneity about the mean Q is partitioned into between-group heterogeneity Q_B and within-group heterogeneity Q_W. The ideal is

Table 15.2 Heterogeneity Summary Table

Source	Statistic	Degrees of Freedom
Between groups	Q_{BET}	$p-1$
Within groups		
Within group 1	Q_{W_1}	$m_1 - 1$
Within group 2	Q_{W_2}	$m_2 - 1$
.	.	.
.	.	.
.	.	.
Within group p	Q_{W_p}	$m_p - 1$
Total within groups	Q_W	$k-p$
Overall	Q	$k-1$

SOURCE: Authors' compilation.
NOTE: Here $k = m_1 + m_2 + \cdots + m_p$.

to select independent (grouping) variables that explain variation (heterogeneity) so that most of the total heterogeneity is between-groups and relatively little remains within groups of effects. Of course, the grouping variable must, in principle, be chosen before examination of the effect sizes to ensure that tests for the significance of group effects do not capitalize on chance.

15.2.1.4 Computing the One-Factor Analysis Although Q_B and Q_W can be computed using a computer program for weighted ANOVA, the weighted cell means and their standard errors generally cannot. Computational formulas can greatly simplify direct calculation of Q_B and Q_W as well as the cell means and their standard errors. These formulas are analogous to those used in analyzing variance and enable all of the statistics to be computed in one pass through the data (for example, by a packaged computer program). The formulas are expressed in terms of totals (sums) across cases of the weights, of the weights times the effect estimates, and of the weights times the squared effect estimates. Define

$$TW_i = \sum_{j=1}^{m_i} w_{ij}, \qquad TW_\bullet = \sum_{i=1}^{p} TW_i,$$

$$TWD_i = \sum_{j=1}^{m_i} w_{ij}T_{ij}, \qquad TWD_\bullet = \sum_{i=1}^{p} TWD_i,$$

$$TWDS_i = \sum_{j=1}^{m_i} w_{ij}T_{ij}^2, \qquad TWDS_\bullet = \sum_{i=1}^{p} TWDS_i,$$

where $w_{ij} = 1/v_{ij}$ is just the weight for T_{ij}. Then the overall heterogeneity statistic Q is

$$Q = TWDS_\bullet - (TWD_\bullet)^2/TW_\bullet.$$

Table 15.3 Data for Male-Authorship Example

Study	% Male Authors	Group	# Items	T	v	w	wT	wT^2
1	25	1	2	−.33	.029	34.775	−11.476	3.787
2	25	1	2	.07	.034	29.730	2.081	.146
3	50	2	2	−.30	.022	45.466	−13.640	4.092
4	100	3	38	.35	.016	62.233	21.782	7.624
5	100	3	30	.70	.066	15.077	10.554	7.388
6	100	3	45	.85	.218	4.586	3.898	3.313
7	100	3	45	.40	.045	22.059	8.824	3.529
8	100	3	45	.48	.069	14.580	6.998	3.359
9	100	3	5	.37	.051	19.664	7.275	2.692
10	100	3	5	−.06	.032	31.218	−1.873	0.112

SOURCE: Authors' compilations.

Each of the within-group heterogeneity statistics is given by

$$Q_{W_i} = TWDS_i - (TWD_i)^2/TW_i, \quad i = 1, \ldots, p.$$

The overall within group homogeneity statistic is obtained as $Q_W = Q_{W_1} + Q_{W_2} + \cdots + Q_{W_p}$. The between groups heterogeneity statistic is obtained as $Q_B = Q - Q_W$. The weighted overall mean effects and its variance are

$$\bar{T}_{..} = TWD_. / TW_., \qquad v_{..} = 1/TW_.,$$

and the weighted group means and their variances are

$$v_{i.} = TWD_i / TW_i, \quad i = 1, \ldots, p,$$

and

$$v_{i.} = 1/TW_i, \quad i = 1, \ldots, p.$$

The omnibus test statistics Q, Q_B, and Q_W can also be computed using a weighted analysis of variance program. The grouping variable is used as the factor in the weighted ANOVA, the effect-size estimates are the observations, and the weight given to effect size T_{ij} is just w_{ij}. The weighted between-group or model sum of squares is exactly Q_B, the weighted within-group or residual sum of squares is exactly Q_W, and the corrected total sum of squares is exactly Q.

Example. Alice Eagly and Linda Carli (1981), in data set I, reported a synthesis of ten studies of gender differences in conformity using the so called fictitious norm group paradigm. The effect sizes were standardized mean differences which were classified into three groups on the basis of the percentage of the authors of the research report who were male. Group 1 consisted of two studies with 25 percent male authorship, group 2 of a single study with 50 percent, and group 3 of seven studies with all male authorship. The data are given in table 15.3.

Using formula 14.6 from chapter 14 of this volume), and the sums given in table 15.3, the overall heterogeneity statistic Q is

$$Q = 36.040 - (34.423)^2 / 279.388 = 31.799.$$

Using the sums given in table 15.3, the within-group heterogeneity statistics Q_{W_1}, Q_{W_2}, and Q_{W_3} are

$$Q_{W_1} = 3.933 - (-9.395)^2/64.505 = 2.565,$$
$$Q_{W_2} = 4.092 - (13.640)^2/45.466 = 0.0,$$
$$Q_{W_3} = 28.015 - (57.458)^2/169.417 = 8.528.$$

The overall within-group heterogeneity statistic is therefore

$$Q_W = 2.565 + 0.0 + 8.528 = 11.093.$$

Because 11.093 does not exceed 14.07, the 95 percent point of the chi-square distribution with $10 - 3 = 7$ degrees of freedom, we do not reject the hypothesis that the effect-size parameters are homogeneous within the groups. In fact, a value this large would occur between 10 and 25 percent of the time due to chance even with perfect homogeneity of effect parameters. Thus we conclude that the effect sizes are relatively homogeneous within groups.

The between group heterogeneity statistic is calculated as

$$Q_B = Q - Q_W = 31.799 - 11.093 = 20.706.$$

Because 20.706 exceeds 5.99, the 95 percent point of the chi-square distribution with $3 - 1 = 2$ degrees of freedom, we reject the null hypothesis of no variation in effect size across studies with different proportions of male authors. In other words, the relationship between the percentage of male authors and effect size is statistically significant.

15.2.1.5 Standard Errors of Means The sampling variances $v_1, \ldots, v_p$ of the group mean effect estimates $\bar{T}_1, \ldots, \bar{T}_p$ are given by the reciprocals of the sums of the weights in each group, that is

$$v_{i\cdot} = \frac{1}{\displaystyle\sum_{j=1}^{m_i} w_{ij}}, \quad i = 1, \ldots, p. \tag{15.10}$$

Similarly, the sampling variance $v_{\cdot\cdot}$ of the grand weighted mean is given by the reciprocal of the sum of all the weights or

$$v_{\cdot\cdot} = \frac{1}{\displaystyle\sum_{i=1}^{p}\sum_{j=1}^{m_i} w_{ij}}. \tag{15.11}$$

The standard errors of the group mean effect estimates $\bar{T}_{i\cdot}$ and the grand mean $\bar{T}_{\cdot\cdot}$ are just the square roots of their respective sampling variances.

15.2.1.6 Tests and confidence intervals The group means $\bar{T}_1, \ldots, \bar{T}_p$ are normally distributed about the respective effect-size parameters $\bar{\theta}_1, \ldots, \bar{\theta}_p$ that they estimate. The fact that these means are normally distributed with the variances given in equation 15.10 leads to rather straightforward procedures for constructing tests and confidence intervals. For example, to test whether the ith group mean effect $\bar{\theta}_{i\cdot}$ differs from a predefined constant θ_0 (for example, to test if $\bar{\theta}_{i\cdot} - \theta_0 = 0$) by testing the null hypothesis

$$H_{0i} : \bar{\theta}_{i\cdot} = \theta_0,$$

use the statistic

$$Z_i = \frac{\bar{T}_{i\cdot} - \theta_0}{\sqrt{v_{i\cdot}}} \tag{15.12}$$

Table 15.4 Quantities Used to Compute Male-Authorship Example Analysis

	w	wT	wT^2
Group 1 (25% studies)	64.505	−9.395	3.933
Group 2 (50% studies)	45.466	−13.640	4.092
Group 3 (100% studies)	169.417	57.458	28.015
Over all Groups	279.388	34.423	36.040

SOURCE: Authors' compilation.

and reject H_0 at level α (that is, decide that the effect parameter differs from θ_0) if the absolute value of Z_i exceeds the 100α percent critical value of the standard normal distribution. For example, for a two-sided test that $\bar{\theta}_{i\cdot} = 0$ at $\alpha = 0.05$ level of significance, reject the null hypothesis if the absolute value of Z exceeds 1.96. When there is only one group of studies, this test is identical to that described in chapter 14 of this volume using the statistic Z given in 14.5.

Confidence intervals for the group mean effect $\bar{\theta}_{i\cdot}$ can be computed by multiplying the standard error $\sqrt{v_{i\cdot}}$ by the appropriate two-tailed critical value of the standard normal distribution ($C_\alpha = 1.96$ for $\alpha = 0.05$ and 95 percent confidence intervals) then adding and subtracting this amount from the weighted mean effect size $\bar{T}_{i\cdot}$. Thus the $100(1 - \alpha)$ percent confidence interval for $\bar{\theta}_{i\cdot}$ is given by

$$\bar{T}_{i\cdot} - C_\alpha\sqrt{v_{i\cdot}} \le \bar{\theta}_{i\cdot} \le \bar{T}_{i\cdot} + C_\alpha\sqrt{v_{i\cdot}}. \tag{15.13}$$

Example. We now continue the analysis of the standardized mean differences from studies of gender differences in conformity of data set I. The effect-size estimate T_{ij} for each study, its variance v_{ij}, the weight $w_{ij} = 1/v_{ij}$, $w_{ij} T_{ij}$, and $w_{ij} T_{ij}^2$ (which will be used later) are given in table 15.3. Using the sums for each group from table 15.4 the weighted mean effect sizes for the three classes $\bar{T}_1, \bar{T}_2,$ and $\bar{T}_3$ are given by

$$\bar{T}_{1\cdot} = -9.395/64.505 = -0.15,$$

$$\bar{T}_{2\cdot} = -13.640/45.466 = -0.30,$$

$$\bar{T}_{3\cdot} = 57.458/169.417 = 0.34$$

and the weighted grand mean effect size is

$$\bar{T}_{\cdot\cdot} = 34.423/279.388 = 0.123.$$

The variances $v_{1\cdot}$, $v_{2\cdot}$, and $v_{3\cdot}$ of $\overline{T}_{1\cdot}, \overline{T}_{2\cdot}$, and $\overline{T}_{3\cdot}$ are given by

$$v_{1\cdot} = 1/64.505 = 0.015,$$

$$v_{2\cdot} = 1/45.466 = 0.022,$$

$$v_{3\cdot} = 1/169.417 = 0.006,$$

and the variance $v_{\cdot\cdot}$ of $\overline{T}_{\cdot\cdot}$ is

$$v_{\cdot\cdot} = 1/279.388 = 0.00358.$$

Using formula 15.9 with $C_{.05} = 1.96$, the limits of the 95 percent confidence interval for the group mean parameter $\overline{\theta}_{1\cdot}$ are given by

$$-0.15 \pm 1.96\sqrt{0.015} = -0.15 \pm 0.24.$$

Thus the 95 percent confidence interval for $\overline{\theta}_{1\cdot}$ is given by

$$-0.39 \leq \overline{\theta}_{1\cdot} \leq 0.09.$$

Because this confidence interval contains zero, or alternately, because the test statistic

$$Z = |-0.15|/\sqrt{0.015} < 1.96,$$

we cannot reject the hypothesis that $\overline{\theta}_{1\cdot} = 0$ at the $\alpha = 0.05$ level of significance. Similarly 95 percent confidence intervals for the group mean parameters $\overline{\theta}_{2\cdot}$ and $\overline{\theta}_{3\cdot}$ are given by

$$-0.59 = -0.30 - 1.96\sqrt{0.022}$$

$$\leq \overline{\theta}_{2\cdot} \leq -0.30 + 1.96\sqrt{0.022} = -0.01$$

and

$$0.19 = 0.34 - 1.96\sqrt{0.006}$$

$$\leq \overline{\theta}_{3\cdot} \leq 0.34 + 1.96\sqrt{0.006} = 0.49.$$

Thus we see that the mean effect size for group 2 is significantly less than zero, for group 3 is significantly greater than zero, and for group 1 is not different from zero.

15.2.1.7 Comparisons or Contrasts Among Mean Effects Omnibus tests for differences among group means can reveal that the mean effect parameters are not all the same, but are not useful for revealing the specific pattern of mean differences that might be present. For example, the Q_B statistic might reveal variation in mean effects when the effects were grouped according to type of treatment, but the omnibus statistic gives no insight about which types of treatment (which groups) were associated with the largest effect size. In other cases, the omnibus test statistic may not be significant but we may want to test for a specific a priori difference that the test might not have been powerful enough to detect. In conventional analysis of variance, contrasts or comparisons are used to explore the differences among group means. Contrasts can be used in precisely the same way to examine patterns among group mean effect sizes in meta-analysis. In fact, all of the strategies used for selecting contrasts in ANOVA (such as orthogonal polynomials to estimate trends, Helmert contrasts to discover discrepant groups) are also applicable in meta-analysis.

A contrast parameter is just a linear combination of group means

$$\gamma = c_1\overline{\theta}_{1\cdot} + \cdots + c_p\overline{\theta}_{p\cdot}. \tag{15.14}$$

where the coefficients $c_1, \ldots, c_p$ (called the contrast coefficients) are known constants that satisfy the constraint $c_1 + \ldots + c_p = 0$ and are chosen so that the value of the contrast will reflect a particular comparison or pattern of interest. For example, the coefficients $c_1 = 1$, $c_2 = -1$, $c_3 = \ldots = c_p = 0$ might be chosen so that the value of the contrast is the difference between the mean $\overline{\theta}_{1\cdot}$ of group 1 and the mean $\overline{\theta}_{2\cdot}$ of group 2. Sometimes we refer to a contrast among population means as a population contrast or a contrast parameter to emphasize that it is a function of population parameters and to distinguish it from estimates of the contrast. The contrast parameter specified by coefficients $c_1, \ldots, c_p$ is usually estimated by a sample contrast

$$g = c_1\overline{T}_{1\cdot} + \ldots + c_p\overline{T}_{p\cdot}. \tag{15.15}$$

The estimated contrast g has a normal sampling distribution with variance v_g given by

$$v_g = c_1^2 v_{1\cdot} + \ldots + c_p^2 v_{p\cdot}. \tag{15.16}$$

This notation for contrasts suggests that the contrasts compare group mean effects, but they can in fact be used to compare individual studies (groups consisting of a single study) or a single study with a group mean. All that is required is the appropriate definition of the groups involved.

Confidence Intervals and Tests of Significance. Because the estimated contrast g has a normal distribution with known variance v_g, confidence intervals and tests of statistical significance are relatively easy to construct. Note, however, that just as with contrasts in ordinary analysis of variance, test procedures differ depending on whether the contrasts were planned or were selected using information from the data. Procedures for testing planned comparisons and for constructing nonsimultaneous confidence intervals are given in this section. Procedures for testing post hoc contrasts (contrasts selected using information from the data) and simultaneous confidence intervals are given in the next section.

Planned Comparisons. Confidence intervals for the contrast parameter γ are computed by multiplying the standard error of g, $\sqrt{v_g}$ by the appropriate two-tailed critical value of the standard normal distribution ($C_\alpha = 1.96$ for $\alpha = 0.05$ and 95 percent confidence intervals) and adding and subtracting this amount from the estimated contrast g. Thus the $100(1 - \alpha)$ percent confidence interval for the contrast parameter γ is

$$g - C_\alpha \sqrt{v_g} \le \gamma \le g + C_\alpha \sqrt{v_g}. \qquad (15.17)$$

Alternatively, a (two-sided) test of the null hypothesis that $\gamma = 0$ uses the statistic

$$X^2 = g^2/v_g. \qquad (15.18)$$

If X^2 exceeds the $100(1 - \alpha)$ percent point of the chi-square distribution with one degree of freedom, reject the hypothesis that $\gamma = 0$ and declare the contrast to be significant at the level of significance α.

Post Hoc Contrasts and Simultaneous Tests. Situations often occur where several contrasts among group means are of interest. If several tests are made at the same nominal significance level α, the chance that at least one of the tests will reject (when all of the relevant null hypotheses are true) is generally greater than α and can be considerably greater if the number of tests is large. Similarly, the probability that tests will reject may also be greater than the nominal significance level for contrasts that are selected because they appear to stand out when examining the data. Simultaneous and post hoc testing procedures are designed to address these problems by ensuring that the probability of at least one type I error is controlled at a preset significance level α. Many simultaneous test procedures have been developed (see Miller 1981). We now discuss the application of two of these

procedures to contrasts in meta-analysis (see also Hedges and Olkin 1985).

The simplest simultaneous test procedure is called the Bonferroni method. This exacts a penalty for simultaneous testing by requiring a higher level of significance from each individual contrast for it to be declared significant in the simultaneous test. If a number $L \ge 1$ of contrasts are to be tested simultaneously at level α, the Bonferroni test requires that any contrast be significant at (nonsimultaneous) significance level α/L o be declared significant at level α in the simultaneous analysis. For example, if $L = 5$ contrasts were tested at simultaneous significance level $\alpha = 0.05$, any one of the contrasts would have to be individually significant at the $0.01 = 0.05/5$ level (that is, X^2 would have to exceed 6.635, the 99 percent point of the chi-square distribution with $df = 1$) to be declared significant at $\alpha = 0.05$ level by the simultaneous test.

The Bonferroni method can be used as a post hoc test if the number of contrasts L is chosen as the number that could have been conducted. This procedure works well when all contrasts conducted are chosen from a well-defined class. For example, if there are four groups, there are six possible pairwise contrasts, so the Bonferroni method is applied to any pairwise contrast chosen post hoc by treating it as one of six contrasts examined simultaneously. If number L of comparisons (or possible comparisons) is large, the Bonferroni method can be quite conservative given that it rejects only if a contrast has a very low significance value.

An alternative test procedure, a generalization of the Scheffé method from the analysis of variance, can be used for both post hoc and simultaneous testing (see Hedges and Olkin 1985). It consists of computing the statistic X^2 given in equation 15.18 for each contrast and rejecting the null hypothesis whenever X^2 exceeds the $100(1 - \alpha)$ percent point of the chi-square distribution with L' degrees of freedom, where L' is the smaller of L (the number of contrasts) or $p - 1$ (the number of groups minus one).

When the number of contrasts (or potential contrasts) is small, simultaneous tests based on the Bonferroni method will usually be more powerful. When the number of contrasts is large the Scheffé method will usually be more powerful.

Example. Continuing the analysis of standardized mean difference data on gender difference in conformity, recall that there are three groups of effects: those in which 25 percent, 50 percent, and 100 percent respectively, of the authors are male. To contrast the mean of the effects of group 1 (25 percent male authors) with those of group 3 (100 percent male authors), use the contrast coefficients

$$c_1 = -1.0, \quad c_2 = 0.0, \quad c_3 = 1.0.$$

The value of the contrast estimate g is

$$g = -1.0(-0.15) + 0.0(-0.30) + 1.0(0.34) = 0.49,$$

with an estimated variance of

$$v_g = (-1.0)^2 (0.015) + (0.0)^2 (0.022) + (1.0)^2 (0.006)$$
$$= 0.021.$$

Hence a 95 percent confidence interval for $\gamma = \overline{T}_{3\cdot} - \overline{T}_{1\cdot}$ is

$$0.21 = 0.49 - 1.96\sqrt{0.021} \leq \gamma \leq 0.49 + 1.96\sqrt{0.021}$$
$$= 0.77$$

Because this interval does not contain zero, or alternatively, because $Z = 0.49/\sqrt{0.021} = 3.38$ exceeds 1.96, we reject the hypothesis that $\gamma = 0$ and declare the contrast statistically significant at the $\alpha = 0.05$ level.

Had we picked the contrast above after examining the data, it would have been a post hoc contrast and testing the contrast using a post hoc test would have been appropriate. Specifically, because there are a total of $L = 3$ pairwise contrasts among three group means, we might have carried out an $\alpha = 0.05$ level post hoc test using the Bonferroni method and compared $Z = 3.38$ with 2.39, the $\alpha = 0.05/3 = 0.167$ critical value of the standard normal distribution. Because the obtained value of the test statistic, $Z = 3.39$ exceeds the critical value 2.39, we would still reject the hypothesis that $\gamma = 0$. Alternatively, we might have used the Scheffé method and compared $X^2 = Z^2 = 3.39^2 = 11.49$ with 5.99, the critical value of the chi-square distribution with $p - 1 = 2$ degrees of freedom. Because the obtained value of the test statistic, $X^2 = 11.49$ exceeds the critical value 5.99, we would, using the Scheffé method, still reject the hypothesis that $\gamma = 0$.

15.2.2 Multifactor Models

A multifactor analysis of variance for effect sizes exists and has the same relationship to conventional multifactor analysis of variance as the one-factor analysis for effect sizes has to one-factor analysis of variance. The principal difference is that, because of the weighting involved, multifactor analyses for effect sizes are essentially always unbalanced. The details of the computations are therefore rather involved and the formulas for the omnibus test statistics not intuitively appealing. They are therefore not presented here. Computations of multifactor analyses for effect sizes, like unbalanced analyses of variance in primary research, will almost always be carried out via packaged computer programs. The test statistics for main effects and interactions are the weighted sums of squares, which have chi-square distributions with the conventional degrees of freedom under the corresponding null hypotheses. Procedures for estimating group means and contrasts are exactly the same in the multifactor case as in the one-factor case.

15.2.2.1 Computation of Omnibus Test Statistics
Suppose a two-factor model with two categorical variables. In this case, each of the omnibus test statistics is obtained by computing a weighted analysis of variance on the effect estimates (for example, by using SAS, SPSS, or STATA). Let T_{ijk} be the kth effect estimate at the ith level of the first factor and the jth level of the second factor and let its sampling variance be v_{ijk}. Then the weight w_{ijk} for corresponding to the estimate T_{ijk} is

$$w_{ijk} = 1/v_{ijk},$$

The omnibus test statistics used in the analysis are the weighted sums of squares. When the corresponding null hypotheses are true, these sums have chi-square distributions with the number of degrees of freedom indicated by the computer program. Thus, for a two-factor analysis, the test statistic for the first factor, the second factor, the interaction, and the within-cell homogeneity are the corresponding weighted sums of squares. The mean squares and F-test statistics are not used in this analysis. Similarly, the cell means and standard deviations reported by the computer program are generally not useful. Thus the ANOVA computer program is used exclusively to compute the weighted sums of squares that are the omnibus test statistics. Weighted means for each cell and for each factor level should be computed separately using the methods discussed in connection with the one-factor analysis.

15.2.2.2 Multifactor Models with Three or More Factors
Procedures for multifactor models with more than two factors are analogous to those in two-factor models. Omnibus tests for main effects, interactions, and within cell variation are carried out using the weighted

sums of squares as the test statistics with appropriate degrees of freedom as indicated by the computer program. The interpretation of main effects and interactions is analogous to that in conventional analyses of variance in primary research. As in primary analysis, examining a table of cell means (or cross tab table) before conducting two-way or multiway ANOVA is recommended, because of the possibility of having empty cells.

15.3 MULTIPLE REGRESSION ANALYSIS FOR EFFECT SIZES

It is often desirable to represent the characteristics of research studies by continuously coded variables or by a combination of discrete and continuous variables. In such cases, the synthesist often wants to determine the relationship between these continuous variables and effect size. One very flexible analytic procedure for investigating these relationships is an analogue to multiple regression analysis for effect sizes (see Hedges 1982c, 1983b; Hedges and Olkin 1985). These methods share the generality and ease of use of conventional multiple regression analysis, and like their conventional counterparts they can be viewed as including ANOVA models as a special case. In the recent medical literature, such methods have been called meta-regression.

15.3.1 Models and Notation

Suppose that we have k independent effect size estimates $T_1, \ldots, T_k$ with (estimated) sampling variances $v_1, \ldots, v_k$. The corresponding effect-size parameters are $\theta_1, \ldots, \theta_k$. Suppose also that there are p known predictor variables $X_1, \ldots, X_p$ which are believed to be related to the effects through a linear model of the form

$$\theta_i = \beta_0 + \beta_1 x_{i1} + \ldots + \beta_p x_{ip}, \qquad (15.19)$$

where $x_{i1}, \ldots, x_{ip}$ are the values of the predictor variables $X_1, \ldots, X_p$ for the ith study (that is x_{ij} is the value of X_j for study i), and $\beta_0, \beta_1, \ldots, \beta_p$ are unknown regression coefficients.

15.3.2 Estimation and Significance Tests

Estimation is usually carried out using weighted least squares algorithms. The formulas for estimators and test statistics can be expressed most succinctly in matrix no-

tation (see Hedges and Olkin 1985). The analysis can be conducted using standard computer programs such as SAS, SPSS, or STATA that compute weighted multiple regression analyses. The regression should be run with the effect estimates as the dependent variable and the predictor variables as independent variables with weights defined by the reciprocal of the sampling variances. That is, the weight for T_i is $w_i = 1/v_i$.

Standard computer programs for weighted regression analysis produce the correct (asymptotically efficient) estimates $b_0, b_1, \ldots, b_p$ of the unstandardized regression coefficients $\beta_0, \beta_1, \ldots, \beta_p$. Note that unlike the SPSS computer program, we use the symbols $\beta_0, \beta_1, \ldots, \beta_p$ to refer to the population values of the unstandardized regression coefficients not to the standardized sample regression coefficients. Although these programs give the correct estimates of the regression coefficients, the standard errors and significance values are based on a slightly different model than that used for fixed-effects meta-analysis and are incorrect for the meta-analysis model. Calculating the correct significance tests for individual regression coefficients requires some straightforward hand computations from information given in the computer output.

The correct fixed-effects standard error S_j of the estimated coefficient estimate b_j is simply

$$S_j = SE_j / \sqrt{MS_{ERROR}} \qquad (15.20)$$

where SE_j is the standard error of b_j as given by the computer program and MS_{ERROR} is the error or residual mean square from the analysis of variance for the regression as given by the computer program. Alternatively the correct standard errors of $b_0, b_1, \ldots, b_p$ are the square roots of the diagonal elements of the inverse of the $(X'WX)$ matrix which is sometimes called the inverse of the weighted sum of squares and cross-products matrix. Many regression analysis programs, such as SAS Proc GLM, will print this matrix as an option. The (correct) standard errors obtained from both methods are, of course, identical. The methods are simply alternative ways to compute the same thing.

The regression coefficient estimates (the b_j's) are normally distributed about their respective parameter (the S_j's) values with standard deviations given by the standard errors (the S_j's). A $100(1 - \alpha)$ percent confidence interval for each β_j can thus be obtained by multiplying S_j by the two-tailed critical value C_α of the standard normal distribution (for $\alpha = 0.05$, $C_\alpha = 1.96$) and then adding and subtracting this product from b_j. Thus the $100(1 - \alpha)$ percent confidence interval for β_j is

$$b_j - C_\alpha S_j \leq \beta_j \leq b_j + C_\alpha S_j \qquad (15.21)$$

A two-sided test of the null hypothesis that the regression coefficient is zero,

$$H_0: \beta_j = 0,$$

at significance level α consists of rejecting H_0 if

$$Z_j = |b_j| / S_j \qquad (15.22)$$

exceeds the 100α percent two-tailed critical value of the standard normal distribution.

Example. Consider the example of the standardized mean differences for gender differences in conformity given in data set I. In this analysis, we fit the linear model suggested by Betsy Becker, who explained variation in effect sizes by a predictor variable that was the natural logarithm of the number of items on the conformity measure, see column 4 of table 15.3 (1986). This predictor variable is highly correlated with the percentage of male authors used as a predictor in the example given for the categorical model analysis. Using SAS Proc GLM with effect sizes and weights given in table 15.3, we computed a weighted regression analysis. The estimates of the regression coefficients were $b_0 = -0.323$ for the intercept and $b_1 = 0.210$ for the effect of the number of items. The standard errors of b_0 and b_1 could be computed in either of two ways. The $(X'WX)$ inverse matrix computed by SAS was

$$\begin{pmatrix} 0.01219 & -0.00406 \\ -0.00406 & 0.00191 \end{pmatrix}$$

and hence the standard errors can be computed as

$$S_0 = \sqrt{0.01219} = 0.110,$$
$$S_1 = \sqrt{0.00191} = 0.044.$$

Alternatively, we could have obtained the standard errors by correcting the standard errors printed by the program (which are incorrect for our purposes). The standard errors printed by the SAS program were $SE(b_0) = 0.1151$ and $SE(b_1) = 0.0456$, and the residual mean square from the analysis of variance for the regression was $MS_{ERROR} = 1.086$. Using formula 15.20 gives

$$S_0 = .1151/\sqrt{1.086} = 0.110,$$
$$S_1 = .0456/\sqrt{1.086} = 0.044.$$

A 95 percent confidence interval for the effect β_1 of the number of items using $C_{0.05} = 1.96$, $S_1 = 0.044$, and formula 15.21 is given by $.210 \pm 1.96 \times 0.044$

$$0.12 \leq \beta_1 \leq 0.30.$$

Because the confidence interval does not contain zero, or alternatively, because the statistic $Z_1 = 0.210/0.044 = 4.77$ exceeds 1.96, we reject the hypothesis that there is no relationship between number of items on the conformity measures and effect size. Thus the number of items on the response measure has a statistically significant relationship to effect size.

15.3.3 Omnibus Tests

It is sometimes desirable to test hypotheses about groups or blocks of regression coefficients. For example, stepwise regression strategies may involve entering one block of predictor variables (such as a set of variables reflecting methodological characteristics) and then entering another block of predictor variables (such as a set of variables reflecting treatment characteristics) to see if the second block of variables explains any of the variation in effect size not accounted for by the first block of variables. Formally, we need a test of the hypothesis that all of the regression coefficients for predictor variables in the second block are zero.

Suppose that the a predictor variables $X_1, \ldots, X_a$ have already been entered and we wish to test whether the regression coefficients for a block of b additional predictor variables $X_{a+1}, X_{a+2}, \ldots, X_{a+b}$ are simultaneously zero. That is, we wish to test

$$H_0: \beta_{a+1} = \ldots = \beta_{a+b} = 0.$$

The test statistic is the weighted sum of squares for the addition of this block of variables. It can be obtained directly as the difference in the weighted error sum of squares for the model with a predictors and the weighted error sum of squares of the model with $(a + b)$ predictors.

It can also be computed from the output of the weighted stepwise regression as

$$Q_{CHANGE} = bF_{CHANGE} \, MS_{ERROR} \qquad (15.23)$$

where F_{CHANGE} is the value of the F-test statistic for testing the significance of the addition of the block of b predictor variables and MS_{ERROR} is the weighted error or residual mean square from the analysis of variance for the regression. The test at significance level α consists of rejecting H_0 if Q_{CHANGE} exceeds the $100(1 - \alpha)$ percent point of the chi-square distribution with b degrees of freedom.

If the number k of effects exceeds $(p + 1)$, the number of predictors plus the intercept, then a test of goodness of fit or model specification is possible. The test is formally a test of the null hypothesis that the population effect sizes $\theta_1, \ldots, \theta_k$ are exactly determined by the linear model

$$\theta_i = \beta_0 + \beta_1 x_{i1} + \ldots + \beta_p x_{ip}, \quad i = 1, \ldots, p,$$

versus the alternative that some of the variation in the θ_i's is not fully explained by $X_1, \ldots, X_p$. The test statistic is the weighted residual sum of squares Q_E about the regression line, and the test can be viewed as a test for greater than expected residual variation. This statistic is given in the analysis of variance for the regression and is usually called the error or residual sum of squares on computer printouts. The test at significance level θ consists of rejecting the null hypothesis of model fit if Q_E exceeds the $100(1 - \alpha)$ percent point of the chi-square distribution with $(k - p - 1)$ degrees of freedom.

The tests of homogeneity of effect and tests of homogeneity of effects within groups of independent effects described in connection with the analysis of variance for effect sizes are special cases of the test of model fit given here. That is, the statistic Q_E reduces to the statistic Q given in formula 15.6 when there are no predictor variables, and Q_E reduces to the statistic Q_W given in formula 15.7 when the predictor variables are dummy coded to represent group membership.

Example. Continue the example of the regression analysis of the standardized mean differences for gender differences in conformity, using SAS Proc GLM to compute a weighted regression of effect size on the logarithm of the number of items on the conformity measure. Although we can illustrate the test for the significance of blocks of predictors, there is only one predictor. We start with $a = 0$ predictors and add $b = 1$ predictor variables. The weighted sum of squares for the regression in the analysis

of variance for the regression gives $Q_{CHANGE} = 23.11$. We could also have computed Q_{CHANGE} from the F-test statistic F_{CHANGE} for the R-squared change and the MS_{ERROR} for the analysis of variance for the regression. Here $F_{CHANGE} = 21.29$ and $MS_{ERROR} = 1.086$, so the result using formula 15.23

$$Q_{CHANGE} = 1(21.29)(1.086) = 23.12,$$

is identical to that obtained directly (the two values disagree only because of rounding). Comparing 23.12 to 3.84, the 95 percent point of the chi-square distribution with 1 degree of freedom, we reject the hypothesis that the (single) predictor is unrelated to effect size. This is, of course, the same result obtained by a test for the significance of the regression coefficient.

We also test the goodness of fit of the regression model. The weighted residual sum of squares was computed by SAS Proc GLM as $Q_E = 8.69$. Comparing this value to 15.51, the 95 percent point of the chi-square distribution with $10 - 2 = 8$ degrees of freedom, we see that we cannot reject the fit of the linear model. In fact, chi-square values as large as 8.43 would occur between 25 and 50 percent of the time due to chance if the model fit exactly.

15.3.4 Collinearity

All of the problems that arise in connection with multiple regression analysis can also arise in meta-analysis. Collinearity may degrade the quality of estimates of regression coefficients in primary research studies, wildly influencing their values and increasing their standard errors. The same procedures used to safeguard against excessive collinearity in multiple regression analysis in primary research are useful in meta-analysis. Examination of the correlation matrix of the predictors, and the exclusion of some predictors that are too highly intercorrelated with the others can often be helpful. In some cases, predictor variables derived from critical study characteristics may be too highly correlated for any meaningful analysis using more than a very few predictors. It is important to recognize, however, that collinearity is a limitation of the data (reflecting little information about the independent relations among variables) and not an inadequacy of the statistical method. Highly collinear predictors based on study characteristics thus imply that the studies simply do not have the array of characteristics that might make it possible to ascertain precisely their joint relationship with study effect size. In our example about gender differences in conformity, the correlation coefficient between percent of male authors and number of items is 0.66. The correlation between percent of male authors and log items is even higher, 0.79.

15.4 QUANTIFYING EXPLAINED VARIATION

Although the Q_E statistic (or the Q_W statistic for models with categorical independent variables) is a useful test statistic for assessing whether there is any statistically reliable unexplained variation, it is not a useful descriptive statistic for quantifying the amount of unexplained variation. Quantifying the amount of unexplained variation typically involves imposing a model of randomness on this unexplained variation. This is equivalent to imposing a mixed model on the data to describe unexplained variation. The details of such models are beyond the scope of this chapter (for a discussion, see chapter 16 of this volume).

A descriptive statistic R_B, called the Birge ratio, has long been used in the physical sciences and was recently proposed as a descriptive statistic for quantifying unexplained variation in medical meta-analysis with no covariates, where it was called H^2 (see Birge 1932; Higgins and Thompson 2002). It is the ratio of Q_E (or Q_W) to its degrees of freedom, that is, $R_B = Q_E / (k - p - 1)$, (for a regression model with intercept) or $R_B = Q_W/(k - p)$, (for a categorical model). The expected value of R_B is exactly 1 when the effect parameters are determined exactly by the linear model. When the model does not fit exactly, R_B tends to be larger than one. The Birge ratio has the crude interpretation that it estimates the ratio of the between-study variation in effects to the variation in effects due to (within-study) sampling error. Thus a Birge ratio of 1.5 suggests 50 percent more between-study variation than might be expected given the within-study sampling variance.

The squared multiple correlation between the observed effect sizes and the predictor variables is sometimes used as a descriptive statistic. However, the multiple correlation may be misinterpreted in this context because the maximum value of the population multiple correlation is always less than, and can be much less than, one. The reason is that the squared multiple correlation is a measure of variance (in the observed effect-size estimates) accounted for by the predictors. But there are two sources of variation in the effect-size estimates, between-study (systematic) effects and within-study (nonsystematic or

sampling) effects. We might write this partitioning of variation symbolically as

$$Var[T] = \sigma_\theta^2 + v,$$

where $Var[T]$ is the total variance, σ_θ^2 is the between-study variance in the effect-size parameters and v is the within-study variance (the variance of the sampling errors). Only between-study effects are systematic and therefore only can be explained using predictor variables. Variance due to within-study sampling errors cannot be explained. Consequently, the maximum proportion of variance that could be explained is determined by the proportion of total variance attributable to between-study effects. Thus the maximum possible value of the squared multiple correlation could be expressed (loosely) as

$$\frac{\sigma_\theta^2}{\sigma_\theta^2 + v} = \frac{\sigma_\theta^2}{Var[T]}.$$

Clearly, this ratio can be quite small when the between-studies variance is small relative to the within-study variance. For example, if the between-study variance (component) σ_θ^2 is 50 percent of the (average) within-study variance, the maximum squared multiple correlation would be

$$0.5/(0.5 + 1.0) = 0.33.$$

In this example, predictors that yielded an R^2 of 0.30 would have explained 90 percent of the explainable variance even though they explain only 30 percent of the total variance in effect estimates.

A better measure of explained variance than the conventional R^2 would be based on a comparison of the between-studies variance in the effect-size parameters in a model with no predictors and that in a model with predictors. If $\sigma_{\theta 0}^2$ is the variance in the effect-size parameters in a model with no predictors and $\sigma_{\theta 1}^2$ is the variance in the effect-size parameters in a model with predictors, then the ratio

$$P_{MA}^2 = \frac{\sigma_{\theta 0}^2 - \sigma_{\theta 1}^2}{\sigma_{\theta 0}^2} = 1 - \frac{\sigma_{\theta 1}^2}{\sigma_{\theta 0}^2}$$

of the explained variance to the total variance in the effect-size parameters is an analogue to the usual concept of squared multiple correlation in the meta-analytic context. Such a concept is widely used in the general mixed model in primary data analysis (see Bryk and Raudenbush 1992, 65). The parameter P_{MA}^2 is interpretable as the proportion of explainable variance that is explained in the meta-analytic model.

15.5 CONCLUSION

Fixed-effects approaches to meta-analysis provide a variety of techniques for statistically analyzing effect sizes. These techniques are analogous to fixed-effects statistical methods commonly used in the analysis of primary data—the analysis of variance and multiple regression analysis. Consequently, familiar analysis strategies, such as contrasts from analysis of variance, or coding methods, such as dummy or effect coding from multiple regression analysis, can be used in meta-analysis just as they are in primary analyses. Fixed-effects analyses, however, may not be suitable for all purposes in meta-analysis. One major consideration in the choice to use fixed-effects analyses is the nature of the generalizations desired. Fixed-effects methods are well suited to making conditional inferences about the effects in the studies that have been observed. They are less well suited to making generalizations to populations of studies from which the observed studies are a sample.

15. 6 REFERENCES

Becker, Betsy J. 1986. "Influence Again: An Examination of Reviews and Studies of Gender Differences in Social Influence." In *The Psychology of Gender: Advances Through Meta-Analysis* edited by Janet S. Hyde and Marcia C. Linn. Baltimore, Md.: Johns Hopkins University Press.

Birge, Raymond T. 1932. "The Calculation of Errors by the Method of Least Squares." *Physical Review* 40(2): 207–27.

Bryk, Anthony S., and Steven S. Raudenbush. 1992. *Hierarchical Linear Models*. Thousand Oaks, Calif.: Sage Publications.

Eagly, Alice H., and Linda L. Carli. 1981. "Sex of Researchers and Sex Typed Communication as Determinants of Sex Differences in Influencability: A Meta-Analysis of Social Influence Studies." *Psychological Bulletin* 90(1): 1–20.

Hedges, Larry V. 1982b. "Fitting Categorical Models to Effect Sizes from a Series of Experiments." *Journal of Educational Statistics* 7(2): 119–37.

———. 1982c. "Fitting Continuous Models to Effect Size Data." *Journal of Educational Statistics* 7(4): 245–70.

———. 1983a. "A Random Effects Model for Effect Sizes." *Psychological Bulletin* 93(2): 388–95.

———. 1983b. "Combining Independent Estimators in Research Synthesis." *British Journal of Mathematical and Statistical Psychology* 36: 123–31.

Hedges, Larry V., and Ingram Olkin. 1985. *Statistical Methods for Meta-Analysis*. Orlando, Fl.: Academic Press.

Hedges, Larry V., and Jack L. Vevea. 1998. "Fixed and Random Effects Models in Meta-Analysis." *Psychological Methods* 3(4): 486–504.

Higgins, Julian P.T., and Simon G. Thompson. 2002. "Quantifying Heterogeneity in Meta-Analysis." *Statistics in Medicine* 21(11): 1539–58.

Miller, Rupert. G. Jr. 1981. *Simultaneous Statistical Inference*, 2nd ed. New York: Springer-Verlag.

16

ANALYZING EFFECT SIZES: RANDOM-EFFECTS MODELS

STEPHEN W. RAUDENBUSH
University of Chicago

CONTENTS

16.1 INTRODUCTION

This volume considers the problem of quantitatively summarizing results from a stream of studies, each testing a common hypothesis. In the simplest case, each study yields a single estimate of the impact of some intervention. Such an estimate will deviate from the true effect size as a function of random error because each study uses a finite sample size. What is distinctive about this chapter is that the true effect size itself is regarded as a random variable taking on different values in different studies, based on the belief that differences between the studies generate differences in the true effect sizes. This approach is useful in quantifying the heterogeneity of effects across studies, incorporating such variation into confidence intervals, testing the adequacy of models that explain this variation, and producing accurate estimates of effect size in individual studies.

After discussing the conceptual rationale for the random effects model, this chapter provides a general strategy for answering a series of questions that commonly arise in research synthesis:

1. Does a stream of research produce heterogeneous results? That is, do the true effect sizes vary?

2. If so, how large is this variation?

3. How can we make valid inferences about the average effect size when the true effect sizes vary?

4. Why do study effects vary? Specifically do observable differences between studies in their target populations, measurement approaches, definitions of the treatment, or historical contexts systematically predict the effect sizes?

5. How effective are such models in accounting for effect size variation? Specifically, how much variation in the true effect sizes does each model explain?

6. Given that the effect sizes do indeed vary, what is the best estimate of the effect in each study?

I illustrate how to address these questions by re-analyzing data from a series of experiments on teacher expectancy effects on pupil's cognitive skill. My aim is to illustrate, in a comparatively simple setting, to a broad audience with a minimal background in applied statistics, the conceptual framework that guides analyses using random effects models and the practical steps typically needed to implement that framework.

Although the conceptual framework guiding the analysis is straightforward, a number of technical issues must be addressed satisfactorily to ensure the validity the inferences. To review these issues and recent progress in solving them requires a somewhat more technical presentation. Appendix 16A considers alternative approaches to estimation theory, and appendix 16B considers alternative approaches to uncertainty estimation, that is, the estimation of standard errors, confidence intervals, and hypothesis tests. These appendices together provide reanalyses of the illustrative data under alternative approaches, knowledge of which is essential to those who give technical advice to analysts.

16.2 RATIONALE

A skilled researcher wanting to assess the generalizability of findings from a single study will try to select a sample that represents the target population. The researcher will examine interaction effects—by asking, for example, whether a new medical treatment's impact depends on the age of the patient, whether the effectiveness of a new method of instruction depends on student aptitude, whether men are as responsive as women to nonverbal communication. If such interactions are null—if treatment effects do not depend on the age, aptitude, or sex of the participant—a finding is viewed as generalizing across these characteristics. If, on the other hand, such interactions are significant, the researcher can clarify the limits of generalization.

Inevitably, however, the capacity of a single study to clarify the generalizability of a finding is limited. Of necessity, many conditions that might affect a finding will be constant in any given study. The manner in which the treatment is implemented, the geographic or cultural setting, the instrumentation used, and the era during which an experiment is implemented are all potential moderators of study findings. Such moderators can rarely be examined in the context of a single study because these factors rarely vary within a single study.

Across a series of replicated or similar studies, however, such moderators will naturally tend to vary. For example, later in this chapter we reanalyze data from nineteen experiments testing the effects of teacher expectancy on pupil IQ. Several factors widely believed to influence results in this type of study varied across those studies: the timing and duration of the treatment, the conditions of pupil testing, the age of the students, the year the study was conducted, its geographic location and administrative context, the characteristics of the teachers expected to implement the treatment, and the socioeconomic status

of the communities providing setting. For any synthesis, a long list of possible moderators of effect size can be enumerated, many of which cannot be ascertained even by the most careful reading of each study's report.

The multiplicity of potential moderators underlies the concept of a study's true effect size as random. Indeed, the concept of randomness in nature arises from a belief that the outcome of a process cannot be predicted in advance, precisely because the sources of influence on the outcome are both numerous and unidentifiable. Yet two distinctly different statistical rationales—classical and Bayesian—provide quite different justifications for the random effects approach.

16.2.1 The Classical View

We may conceive of the set of studies in a synthesis as constituting a random sample from a larger population of studies. For some syntheses, this assumption is literally true. For example, because Robert Rosenthal and Donald Rubin located far more studies of interpersonal expectancy effects than they could code in any detail with available resources, they drew a random sample of studies for intensive analysis (1978).

One might also view the set of studies under synthesis as constituting a realization from a universe of possible studies—studies that realistically could have been conducted or that might be conducted in the future. Because one wishes to generalize to this larger universe of possible implementations of treatment or observations of a relationship, one conceives of the true effect size of each available study as one item from a random sample of all possible effect sizes. This notion of a hypothetical universe of possible observations is standard in measurement theory. Lee Cronbach and his colleagues put it this way: "A behavioral measurement is a sample from the measurements that might have been made, and interest attaches to the obtained score only because it is representative of the whole collection or universe" (1972, 18).

Once the set of studies under investigation is conceived as a random sample, it is clear that a meta-analysis can readily be viewed as a survey having a two-stage sampling procedure. At the first stage, one obtains a random sample of studies from a larger population of studies and regards these studies as varying in their true effects. At the second stage, within each study, one then obtains a random sample of subjects from a population of subjects.[1]

The two-stage sampling design implies a model with two components of variance. The first component (random effects variance) represents the variance that arises as a result of the sampling of studies, each having a unique effect size. The second component (estimation variance) arises because, for any given study, the effect size estimate is based on a finite sample of subjects.[2]

16.2.2 The Bayesian View

The Bayesian view avoids the specification of any sampling mechanism as a justification for the random effects model. Instead, the random variation of the true effect sizes represents the investigator's lack of knowledge about the process that generates them. Imagine that one could have observed the true effect size for a very large number, k, of related studies. A Bayesian might say that there studies are exchangeable (De Finetti 1964) in that she or he has a priori no reason to believe that one study's effect size should be any larger than any other study's effect size. The Bayesian investigator represents his or her uncertainty about the heterogeneity of the effect sizes with a prior distribution having a central tendency and a dispersion.

Just as the classical statistician might envision a synthesis as a two-stage sampling process, the Bayesian statistician envisions two components of uncertainty. At one level, there is the uncertainty that a particular true effect size will be near the prior mean, that is, the statistician's a priori belief about of the central tendency of all true effect sizes. The degree of uncertainty about the true effect size is expressed quantitatively as a prior variance. At the next level, there is uncertainty that an estimated effect for any given study will be near that study's true effect size. This uncertainty is expressed as a within-study variance—the variability of the estimated effect size around the true effect size for each study.

The notion that the true effect sizes are exchangeable is functionally equivalent to the classical statistician's assumption that they constitute a random sample from a population of studies. However, exchangeability as a basis for a random effects model can be invoked without appealing to the notion of a random sampling mechanism. The Bayesian statistician's way of thinking about the variance of the estimated effect size around the true effect size is generally identical to the thinking of the classical statistician.

Suppose, however, that there is a priori reason to believe that some studies will have larger effects than others. For example, a treatment may have a longer duration in one set of studies than in another, and one may hypothesize that longer durations are associated with larger effect sizes. The exchangeability assumption would then be unwarranted. Both Bayesian and classical statisticians would then be inclined to incorporate such information into the model, as I illustrate in section 16.3.2.

Our discussion reveals that classical and Bayes approaches to random effects meta-analysis assign different definitions to quantities like the mean and variance of the true random effects, but use those quantities in functionally similar ways. In interpreting the results presented here, you the reader are free to use either the classical conception (that the variance of the computed effect sizes represents a two-stage sampling process) or the Bayesian interpretation (that the variance of the computed effects reflects two sources of the investigator's uncertainty). I say more about the Bayesian approach in section 16.6.

16.3 THE MODEL

I represent the random effects model as a two-level hierarchical linear model (Raudenbush and Bryk 1985; 2002, chapter 7). At the first, or micro level, the estimated effect size for study i varies randomly around the true effect size. This might be regarded as a measurement model in which the estimated effect size measures the true effect size with and error. At the second, or macro level, the true effect sizes vary from study to study as a function of observed study characteristics plus a random effect representing unobserved sources of heterogeneity of effect size. The model can be expanded at both levels: the first level can specify multiple effect sizes, each of which is explained by study characteristics at level 2 (Kalaian and Raudenbush 1996; see chapter 9, this volume). This chapter focuses on the case of a single effect size per study, however.

16.3.1 Level-1 Model

Suppose, then, that we have calculated an effect size estimate T_i of the true effect size θ_i for each of k studies, $i = 1, \ldots, k$. The true effect size θ_i might be a mean difference, a standardized mean difference, a correlation (possibly transformed using the Fisher transformation), the natural logarithm of the odds ratio, a log-event rate, or some other meaningful indicator of effect magnitude. The estimate T_i is the corresponding sample statistic, sometimes corrected for bias. Chapters 12 and 13 of this volume present a number of useful measures of effect and their corresponding sample estimators and sampling variances.

If the sample size per study is moderate to large, assuming that T_i is normally distributed about θ_i with known variance v_i is often reasonable. The form of v_i depends on the effect indicator of interest, but generally diminishes at a rate proportional to $1/n_i$, where n_i is the sample size in study i. We therefore write the level-1 model as

$$T_i = \theta_i + e_i, \quad e_i \sim N(0, v_i), \qquad (16.1)$$

where we assume that the errors of estimation e_i are statistically independent, each with a mean of zero and known variance v_i. The normality assumption for e_i is often justifiable based on the central limit theorem (compare Casella and Berger 2001) as long as the sample size in each study is not too small.

16.3.2 Level-2 Model

We now formulate a prediction model for the true effects as depending on a set of study characteristics plus error:

$$\theta_i = \beta_0 + \beta_1 X_{i1} + \beta_1 X_{i2} + \cdots + \beta_p X_{ip} + u_i,$$
$$u_i \sim N(0, \sigma_\theta^2) \qquad (16.2)$$

where β_0 is the model intercept;

$X_{i1}, \ldots, X_{ip}$ are coded characteristics of studies hypothesized to predict the study effect size θ_i;

$\beta_1, \ldots, \beta_p$ are regression coefficients capturing the association between study characteristics and effect sizes; and

u_i is the random effect of study i, that is, the deviation of the study i's true effect size from the value predicted on the basis of the model. Each random effect, u_i, is assumed independent of all other effect sizes with the mean of zero and variance σ_θ^2.

16.3.3 Mixed Model

If we combine the two models by substituting the expression for θ_i in (16.2) into (16.1), we have a mixed effects linear model

$$T_i = \beta_0 + \beta_1 X_{i1} + \beta_1 X_{i2} + \cdots + \beta_p X_{ip} + u_i + e_i$$
$$u_i + e_i \sim N(0, v_i^*), \qquad (16.3)$$

where $v_i^* = \sigma_\theta^2 + v_i$ is the total variance in the observed effect size T_i.

Equation 16.3 is identical to the fixed effects regression model with one exception, the addition of the random effect u_i. Under the fixed-effects specification, the study characteristics $X_{i1}, \ldots, X_{ip}$ are presumed to account completely for the variation in the true effect sizes so that $\sigma_\theta^2 = 0$ and therefore $v_i^* = v_i$. In contrast, the random effects specification allows for the possibility that part of the variability in these true effects is unexplainable by the model. The term *mixed effects linear model* emphasizes that the model has both fixed effects $\beta_1, \ldots, \beta_p$ and random effects $u_i, i = 1, \ldots, k$.

16.4 DATA

In their landmark experiment, Robert Rosenthal and Linda Jacobson tested all of the children in a single elementary school and, ostensibly based on their results, informed teachers that certain children could be expected to make dramatic intellectual growth in the near future (1968). In fact, these expected "bloomers" had been assigned at random to the experimental high expectancy condition. The remaining children made up the control group. The experimenters found that, by the end of the year, the experimental children had gained significantly more than the control children in IQ, leading to the conclusion that the teachers had treated the experimental children in ways that fulfilled the expectation of rapid cognitive growth, a demonstration of the self-fulfilling prophecy in education. Practitioners, policymakers, and many advocates of educational reform hailed these results as showing that unequal teacher expectations were creating inequality in the cognitive skills of the nation's children (Ryan 1971). Naturally, these findings drew the attention of skeptical social scientists, some of whom challenged the study's procedures and findings, leading to a series of re-analyses and ultimately to a number of attempts to replicate those findings in new experiments. I was able to locate nineteen experimental studies of the hypothesis that teacher expectations influence pupil IQ (see Raudenbush 1984; Raudenbush and Bryk 1985). The data from that synthesis are reproduced in table 16.1 (see table 16.1).

The data from each study produced an estimate $T_i = (\bar{Y}_{Ei} - \bar{Y}_{Ci})/S_i$ of the unknown true effect size $\theta_i = (\mu_{Ei} - \mu_{Ci})/\sigma_i$ where $\bar{Y}_{Ei}, \bar{Y}_{Ci}$ are the sample mean outcomes of the experimental and control groups, respectively, within study i; μ_{Ei}, μ_{Ci} are the corresponding study-specific population means; S_i is the pooled, within-treatment sample standard deviation, and σ_i is the corresponding population

Table 16.1 Effect of Teacher Expectancy on Pupil IQ

Study	T_i = effect size estimate	v_i = sampling variance of T_i	X_i = Weeks of Prior Teacher Student Contact
1	0.03	0.015625	2
2	0.12	0.021609	3
3	−0.14	0.027889	3
4	1.18	0.139129	0
5	0.26	0.136161	0
6	−0.06	0.010609	3
7	−0.02	0.010609	3
8	−0.32	0.048400	3
9	0.27	0.026896	0
10	0.87	0.063001	1
11	0.54	0.091204	0
12	0.18	0.049729	0
13	−0.02	0.083521	1
14	0.23	0.084100	2
15	−0.18	0.025281	3
16	−0.06	0.027889	3
17	0.30	0.019321	1
18	0.07	0.008836	2
19	−0.07	0.030276	3

SOURCE: Raudenbush and Bryk 1985.

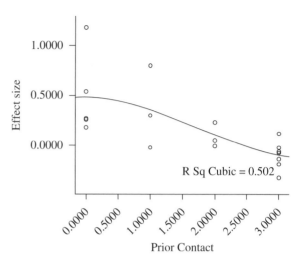

Figure 16.1 Estimated Effect Size T_i as a Function of X_i

SOURCE: Author's compilation.

standard deviation. Following Larry Hedges (1981), we define the sampling variance of T_i as an estimate of θ_i to be closely approximated by $v_i = (n_{iE} + n_{iC})/(n_{iE}n_{iC}) + \theta_i^2/[2(n_{iE} + n_{iC})]$ where we substitute a consistent estimate (for example T_i) for the unknown value of θ_i.

Early commentary on teacher-expectancy experiments suggested that the experimental treatment is difficult to implement if teachers and their students know each other well at the time of the experiment. Essentially, the treatment depends on deception: researchers attempt to convince teachers that certain children can be expected to show dramatic intellectual growth. In fact, those designated are assigned at random to this high expectancy condition, so that if the deception is successful, the only difference between experimental and control children is in the minds of the teachers. Early studies provided anecdotal evidence that the deception will fail if the teachers have had enough previous contact to form their own judgments of the children's capabilities. In those cases, teachers will rely more heavily on their own experience with the children than on test information supplied by outside experts.

For these reasons, we hypothesize that the length of time that teachers and children knew each other at the time of the experiment will be negatively related to a study's effect size. Table 16.1 therefore lists $X_i = 0, 1, 2,$ or 3, depending on whether prior teacher-pupil contact occurred for less than one, one, two, or more than two weeks before the experiment.

A simple visual examination of the data provides quite striking support for this hypothesis. Figure 16.1 shows a scatter plot of the estimated effect size T_i as a function of X_i = weeks of prior teacher-student contact. I have superimposed a cubic spline to summarize this relationship, revealing that the relationship is essentially linear.

16.5 THE LOGIC OF INFERENCE

Let us now consider how to apply the logic of the random effects model 16.3 to answer the six questions listed at the beginning of this chapter and using the teacher expectancy data (table 16.1) for illustration. The results discussed below are summarized in table 16.2.

16.5.1 Do the Effect Sizes Vary?

According to the general model (see equation 16.3), the true effect sizes vary for two reasons. There are predictable differences between studies, based on study charac-

Table 16.2 Statistical Inferences for Three Models

Parameter	Homogenous Effects Model[a]	Simple Random-Effects Model[b]	Mixed-Effects Model[c]
Intercept			
$\hat{\beta}_0$	0.060	0.083	0.407
$SE(\hat{\beta}_0)$	(0.036)	(0.052)	(0.087)
t-ratio	1.65	1.62	4.68
Coefficient for X = Weeks			
$\hat{\beta}_1$			−0.157
$SE(\hat{\beta}_0)$			(0.036)
t-ratio			−4.90
Random-Effects Variance, σ_θ^2			
$\hat{\sigma}_\theta^2$		0.019	0.000
Q		35.83	16.57
df		18	18
p-value		(p < .01)	(p > .50)
Study-Specific Effects, θ			
min, max(T_i)[d]	(−0.14, 1.18)	(−0.14, 1.18)	(−0.14, 1.18)
min, max(θ_i^*)		(−0.03, 0.25)	(−0.07, 0.41)
95% P.V. Interval for (θ)		(−0.19, 0.35)	(−0.07, 0.41)

SOURCE: Authors' compilation.
NOTE: The homogenous model was estimated by maximum likelihood; the simple random-effects model and the mixed-effects model were estimated using restricted maximum likelihood.
[a] $\theta_i = \beta_0$
[b] $\theta_i = \beta_0 + u_i$ $\qquad Var(u_i) = \sigma_\theta^2$
[c] $\theta_i = \beta_0 + \beta_1 X_i + u_i$ $\qquad Var(u_i|X_i) = \sigma_\theta^2$
[d] In all cases $T_i = \theta_i + e_i$ $\qquad Var(e_i) = v_i$

teristics X; and unpredictable differences based on study-specific random effects, u. Therefore, if every regression coefficient β and every random effect u are zero, all studies will have the same true effect size ($\theta_i = \beta_0$ for all i), and the only variation in the estimated effect sizes will arise from the sampling errors, e. This is the case of homogenous effect sizes. Equation 16.3 thus reduces to

$$T_i = \beta_0 + e_i, \quad e_i \sim N(0, v_i). \qquad (16.4)$$

Every effect size T_i is thus an unbiased estimate of the common effect size β_0, having known variance v_i. The best estimate[3] of β_0 is then

$$\hat{\beta}_0 = \sum_{i=1}^{k} v_i^{-1} T_i \bigg/ \sum_{i=1}^{k} v_i^{-1}. \qquad (16.5)$$

The reciprocals of the variances v_i are known as precisions, and model 16.5 is a precision-weighted average. The precision of model 16.5 is the sum of the precisions,

that is, $Precision(\hat{\beta}_0) = \sum_{i=1}^{k} v_i^{-1}$, so that the sampling variance of 16.5 is

$$Var(\hat{\beta}_0) = \left(\sum_{i=1}^{k} v_i^{-1} \right)^{-1}. \qquad (16.6)$$

Table 16.2 shows the results, $\hat{\beta}_0 = .060$, $se = \sqrt{Var(\hat{\beta}_0)} = .036$. Under the assumption of homogeneity of effect size, we can test the null hypothesis $H_0: \beta_0 = 0$. Under this hypothesis, all study effect sizes are null. We therefore compute

$$z = \hat{\beta}_0 / \sqrt{Var(\hat{\beta}_0)} = 1.65, \qquad (16.7)$$

providing modest evidence against the null hypothesis. A 95 percent confidence interval is

$$\hat{\beta}_0 \pm 1.96 \sqrt{Var(\hat{\beta}_0)} = (-0.011, 0.132).$$

Such an inference is based, however, on the assumption that the effect sizes are homogenous. We need to test this assumption to discover whether the effect sizes vary and, if so, why.

Under the homogeneous model, $\sigma_\theta^2 = 0$, and the sum of squared differences between the effect size estimate T_i and their estimated mean $\hat{\beta}_0$, each divided by v_i, will be distributed as a central chi-squared random variable with $k - 1 = 18$ degrees of freedom. However, if $\sigma_\theta^2 > 0$, the denominators v_i will be too small, suggesting rejection of the null hypothesis $\sigma_\theta^2 = 0$. Thus, following Hedges and Olkin (1983), we can compute the chi-square test of homogeneity

$$Q_0 = \sum_{i=1}^{k} v_i^{-1}(T_i - \hat{\beta}_0)^2 = 35.83, \quad df = 18, \quad p = .007, \quad (16.8)$$

suggesting rejection of the null hypothesis, implying that the true effect sizes do indeed vary across studies, that is $\sigma_\theta^2 > 0$. The next logical question is how large this variation is.

16.5.2 How Large is the Variation in True Effect Sizes?

16.5.2.1. A Variation to Signal Ratio One simple indicator of the extent to which studies vary is the simple ratio (Higgins and Thompson 2002):

$$H = Q_0/(k - 1) = 35.83/18 = 1.99 \quad (16.9)$$

Under the null hypothesis, H has an expected value of unity. Our value of H can be therefore regarded as suggesting that the variation in the estimated effect sizes T_i, $i = 1, \ldots, k$ is about twice what would be expected under the null hypothesis. A nice feature of this statistic is that it can be used to compare the heterogeneity of effect sizes from research syntheses that use different measures of effect. For example, one synthesis might use an odds ratio to describe the effect of a treatment on a binary outcome, but another study, like ours, might use as standardized mean difference as a measure of effect of a treatment on a continuous outcome. The analyst can meaningfully compare H statistics from the two studies.

Although useful, H as a measure of heterogeneity of effect size depends on the design of the studies as well as the magnitude of the differences in the true effect sizes. In particular, for any value of σ_θ^2, H will tend to increase as the sample sizes within studies grow. Thus, H does not characterize how different the true effect sizes are from each other.

A more direct measure of heterogeneity is a good estimate of σ_θ^2 itself. We therefore expand the simple model 16.4 to include study specific random effects, yielding

$$T_i = \beta_0 + u_i + e_i, \, e_i \sim N(0, v_i^*), \quad v_i^* = \sigma_\theta^2 + v_i, \quad (16.10)$$

a special case of the general model 16.3 with $\beta_1, \ldots, \beta_p$ set to zero. According to this model, none of the variation in the true random effects is predictable. Instead, all of the variation is summarized by a single parameter, σ_θ^2. This model has been labeled the simple random effects model (Hedges 1983) to distinguish it from the mixed random effects model 16.3.

Our point estimates, based on restricted maximum likelihood estimation,[4] are

$$\hat{\beta}_0 = 0.084, \quad \hat{\sigma}_\theta^2 = 0.019. \quad (16.11)$$

From this simple result (see table 16.2), we can compute two useful descriptive statistics to characterize the magnitude of the heterogeneity in the true effect sizes.

16.5.2.2 Plausible Value Intervals Under our model, the true effect sizes are normally distributed with estimated mean .084 and standard deviation $\sigma_\theta = \sqrt{0.019} = 0.138$. Under this estimated model, 95 percent of the true effect sizes would lie (approximately) between the two numbers:

$$\hat{\beta}_0 \pm 1.96 * \hat{\sigma}_\theta = .083 \pm 1.96 * 0.138.$$
$$= (-0.187, 0.353). \quad (16.12)$$

We thus estimate that about 95 percent of the true effect sizes lie between -0.187 and 0.353. Stephen Raudenbush and Anthony Bryk called this the "95 percent plausible value interval" for θ (Raudenbush and Bryk 2002). We will revise this estimate after fitting an explanatory model. Our interval suggests that the effect sizes are mostly positive though small negative effect sizes are possible as well.

16.5.2.3 Intra-Class Correlation Julian Higgins and Simone Thompson suggested several variance ratios to represent the degree of heterogeneity in a set of estimated effect sizes (2002). In a similar vein, Raudenbush and Bryk (1985) proposed a measure of reliability with which study i's effect size is measured:

$$\hat{\lambda}_i = \hat{\sigma}_\theta^2 / (\hat{\sigma}_\theta^2 + v_i), \quad (16.13)$$

closely related to the intraclass correlation coefficient. Note that this reliability varies from study to study as a function of v_i. An overall measure of reliability is

$$\hat{\lambda} = \sum_{i=1}^{k} \hat{\lambda}_i / k = .37. \qquad (16.14)$$

Thus about 37 percent of the variation in the estimated effect sizes reflects variation in the true effect sizes. This result cautions analysts not to equate variation in the observed effect sizes with true effect size heterogeneity. The statistic $\hat{\lambda}$ is similar to H in that it can also be useful in comparing syntheses using different outcome metrics. However, like H, it will increase with study sample sizes and therefore should not be regarded as pure measure of effect size variation.

16.5.3 How Can We Make Valid Inferences About the Average Effect Size When the True Effect Sizes Vary?

Recall that fitting model 16.5 provided a point estimate of $\hat{\beta}_0 = .060$ under the assumption of constant true effect sizes ($\sigma_\theta^2 = 0$). In contrast, we estimated $\hat{\beta}_0 = 0.083$ under the assumption $\sigma_\theta^2 = 0.019$. The formula for computing this estimate is

$$\hat{\beta}_0 = \sum_{i=1}^{k} \hat{v}_i^{*-1} T_i \bigg/ \sum_{i=1}^{k} \hat{v}_i^{*-1}, \quad \hat{v}_i^* = \hat{\sigma}_\theta^2 + v_i. \qquad (16.15)$$

The latter estimate is more credible because the evidence strongly suggested $\sigma_\theta^2 > 0$, in which case the estimate based on the assumption $\sigma_\theta^2 = 0$, though consistent, is not efficient. Thus, our finding that $\sigma_\theta^2 > 0$ encouraged us in our case to modestly increase the point estimate of the average true effect size.

Perhaps more consequentially, under the assumption $\sigma_\theta^2 = 0$, model 16.7 provided a test of the null hypothesis $H_0: \beta_0 = 0$ with $t = 1.65$. The problem is that the standard error used in this test is likely too small because it fails to reflect the extra variation in the effect size estimates when $\sigma_\theta^2 > 0$, as the evidence suggests is the case for our data.

Note that estimate 16.15 is a precision-weighted average where the estimated precisions are the reciprocals of the estimated study specific variances $\hat{v}_i^* = \hat{\sigma}_\theta^2 + v_i$. The estimated precision of this estimate is (approximately) sum of the estimated precisions, $Precison(\hat{\beta}_0) = \sum_{i=1}^{k} v_i^{*-1}$, so that the estimated sampling variance of (16.15) is

$$\text{Vâr}(\hat{\beta}_0) = \left(\sum_{i=1}^{k} v_i^{*-1} \right)^{-1}. \qquad (16.16)$$

Table 16.2 shows the results, $\hat{\beta}_0 = .083$, $se = \sqrt{V\hat{a}r(\beta_0)} = 0.052$. To test the null hypothesis $H_0: \beta_0 = 0$, we therefore compute

$$t = \hat{\beta}_0 / \sqrt{V\hat{a}r(\hat{\beta}_0)} = 0.083/0.052 = 1.62, \qquad (16.17)$$

This result gives very slightly weaker evidence against the null hypothesis than obtained when we assumed $\sigma_\theta^2 > 0$. Note that the 95 percent confidence interval for the average effect size is now

$$\hat{\beta}_0 \pm 1.96 * \sqrt{V\hat{a}r(\hat{\beta}_0)} = .083 \pm 1.96 * 0.052$$
$$= (-0.019, 0.185). \qquad (16.18)$$

a bit wider than the 95 percent confidence interval based on the assumption $\sigma_\theta^2 > 0$ (see Table 16.2).

16.5.4 Why Do Study Effects Vary?

Having discovered significant variability in the true effect sizes, the question naturally arises about what characteristics of studies might explain this variation. As mentioned in section 16.3, theoretical considerations suggested that the experimental treatment would not take hold if teachers knew their students well before researchers provided them false information about which children were most likely to exhibit rapid cognitive growth. This reasoning supports an essentially a priori hypothesis that weeks of contact between teachers and children before the experimental manipulation would negatively predict the true effect size. Recall that figure 16.1 provided visual empirical evidence in support of this hypothesis. To test this hypothesis, we pose the model

$$T_i = \beta_0 + \beta_1 X_i + u_i + e_i \quad u_i + e_i \sim N(0, v_i^*), \qquad (16.19)$$

where $X_i = 0, 1, 2$ indicating the corresponding number of weeks and $X_i = 3$ indicating three or more weeks; and $v_i^* = \sigma_\theta^2 + v_i$ where $\sigma_\theta^2 = Var(\theta | X)$ is now the conditional variance of the true effect sizes controlling for the covariate X = weeks of prior contact. Based on estimation theory under restricted maximum likelihood (see section 16A), we find

$$\hat{\beta}_0 = 0.407, \quad se = 0.087, \quad t = 4.68$$
$$\hat{\beta}_1 = -0.157, \quad se = 0.036, \quad t = -4.90 \qquad (16.20)$$
$$\hat{\sigma}_\theta^2 = 0.000.$$

These results (see table 16.2) provide quite strong evidence supporting the hypothesis that weeks of prior contact are negatively related with effect sizes. The model suggests that, when $X = 0$, the expected effect size is 0.41 in standard deviation units. For each week of additional contact, we can expect the study effect size to diminish by about 0.16 standard deviation units until week $X = 3$, after which the effect size remains near zero.

16.5.5 How Effective Are Such Models in Accounting for Effect Size Variation?

Our point estimate of the conditional variance of the true effect sizes given X is 0.000 rounded to three decimal points. The inference that this variance is near zero is reinforced by testing the null hypothesis H_0: $\sigma_\theta^2 = 0$ by means of the test of conditional homogeneity

$$Q_1 = \sum_{i=1}^{k} v_i^{-1}(T_i - \hat{\beta}_0 - \hat{\beta}_1 X_i)^2 = , \, df = 17, \quad p > .50.$$

$$(16.21)$$

where $\hat{\beta}_0, \hat{\beta}_1$ are estimated based on the assumption $\sigma_\theta^2 = 0$. The evidence comfortably supports retention of the null hypothesis that, within studies having the same value of X, the true effect sizes are invariant.

We can compare the conditional variance of the true effect sizes to the total variance of the effect sizes to compute an approximate variance explained statistic:

$$\frac{\hat{\sigma}_\theta^2(total) - \hat{\sigma}_\theta^2(given\ X)}{\hat{\sigma}_\theta^2(total)} = \frac{0.019 - 0.000}{0.019} = 1.000 \quad (16.22)$$

suggesting in our case that essentially all of the variance in the true effect sizes is explained by the model. The reader should be cautioned that this estimate is uncertain as the null hypothesis $\sigma_\theta^2 = 0$, though retained, is not known to be true. Nevertheless, the results suggest that X quite effectively explains the random effects variance.

16.5.6 What Is the Best Estimate of the Effect for Each Study?

The estimate T_i is an unbiased estimate of θ_i for each study i with precision equal to the reciprocal of the variance v_i. Although certainly a reasonable estimate, T_i would typically be inaccurate when v_i is large, reflecting that study i used a small sample size. Our analysis suggests two alternatives.

16.5.6.1 Unconditional Shrinkage Estimation A plausible alternative is the estimated average value of θ_i, that is β_0, based on the unconditional model 16.10. In particular, if σ_θ^2 were known to be small, every θ_i would likely be near β_0, indicating that $\hat{\beta}_0$ might be a much better estimate of each θ_i, than T_i is. Of course, $\hat{\beta}_0$ would be a bad choice as an estimator of each θ_i if σ_θ^2 were known to be large, particularly if T_i is highly reliable. A compromise estimator is the conditional (posterior) mean

$$E(\theta_i \mid T_i, \hat{\beta}_0, \hat{\sigma}_\theta^2) = \theta_i^* = \hat{\lambda}_i T_i + (1 - \hat{\lambda}_i)\hat{\beta}_0, \quad (16.23)$$

where $\hat{\lambda}_i$ is the estimated reliability of T_i as an estimate of θ_i (defined in equation 16.13). As $\hat{\lambda}_i$ approaches its maximum of 1.0, all of the weight is assigned to the highly reliable study-specific estimate T_i. However, if study i's information is not particularly reliable, more weight is assigned to the central tendency $\hat{\beta}_0$. An alternative form θ_i^* is also instructive about its properties. Simple algebra shows that

$$\theta_i^* = \frac{v_i^{-1} T_i + \hat{\sigma}_\theta^{-2} \hat{\beta}_0}{v_i^{-1} + \hat{\sigma}_\theta^{-2}}, \quad (16.24)$$

a precision-weighted average. We can regard $\hat{\sigma}_\theta^{-2}$ as the estimated precision of β_0 viewed as an estimator of θ_i. If the precision of the data from study i is high, that is, if v_i^{-1} is large, θ_i^* places large weight on the study-specific estimate T_i. However, if the true effect sizes are concentrated about their central tendency, $\hat{\sigma}_\theta^{-2}$ will be large and the central tendency $\hat{\beta}_0$ will be accorded large weight in composing θ_i^*. Because θ_i^* tends to be closer to the central tendency than T_i does, θ_i^* is known as a shrinkage estimator because it shrinks outlying values of T_i toward the mean (Morris 1983). Such estimators are routinely computed and can be saved in a file using standard software packages for hierarchical linear models.

This shrinkage can readily be seen in table 16.2. We see that the estimates T_i range between -0.14 and 1.18, seeming to imply that modest negative effects of teacher expectancy (as low as -0.14) and extraordinarily large effects (as large as 1.18 standard deviation units) are found in these nineteen studies. We know, however, that much of this variation is attributable to noise. Recall that the mean value of $\hat{\lambda}_i$ is 0.37, suggesting that only about 37 percent of the variation in the estimates T reflects variation in the underlying true effect size θ. In particular, study 4 provided the estimate $T_4 = 1.18$. However, that study had the largest sampling variance in the entire data set (note $v_4 = 0.139$, and see table 16.1). We can speculate

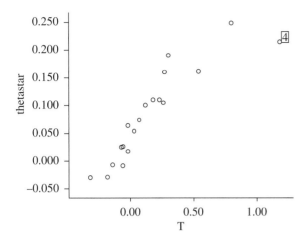

Figure 16.2 Empirical Bayes Estimates θ_i^* Against Study-Specific Estimates T_i (horizontal axis).

SOURCE: Author's compilation.

NOTES: Note that study 4 provided an unusually large study-specific estimate, $T_4 = 1.18$. However, the empirical Bayes estimate $\theta_4^* = 0.21$ is much smaller, reflecting the fact that T_4 is quite unreliable.

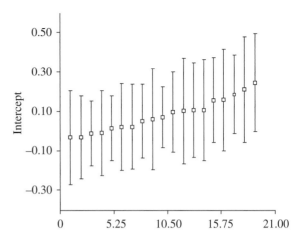

Figure 16.3 Bayes 95 percent Posterior Confidence Intervals, Simple Random-Effects Model

SOURCE: Author's compilation.

that this apparent outlying positive effect size reflects noise.

The empirical Bayes estimates, $\theta_i^*, i = 1, \ldots, k$, having experienced shrinkage to the mean proportional to their unreliability, are much more concentrated, displaying a range of -0.03 to 0.25, suggesting that most of the effects are small and positive. These estimates are plotted against the study-specific estimates in figure 16.2. The experience of study 4 is instructive. Although we found $T_4 = 1.18$, we see that $\theta_4^* = .21$ (case 4 is highlighted in figure 16.2). The seemingly outlying value of T_4 has been severely shrunk toward the mean. This result is not surprising when we inspect the reliability of T_4, that is, $\hat{\lambda}_4 = \hat{\sigma}_\theta^2/(\hat{\sigma}_\theta^2 + v_4) = 0.019/(0.019 + 0.139) = 0.12$, the lowest in the data set.

How much uncertainty is associated with the empirical Bayes point estimates θ_i^* of the effect size? Recall from equation 16.24 that θ_i^* is a precision-weighted average. The precision of this estimate is (approximately) the sum of the precisions, that is, $\mathrm{Precision}(\theta_i|T_i,\hat{\beta}_0, \hat{\sigma}_\theta^{-2}) = v_i^{-1} + \hat{\sigma}_\theta^{-2}$ so that the approximate posterior variance is

$$\mathrm{Var}(\theta_i \mid T_i, \hat{\beta}_0, \hat{\sigma}_\theta^2) = (v_i^{-1} + \hat{\sigma}_\theta^{-2})^{-1}. \qquad (16.25)$$

Figure 16.3 displays the 95 percent posterior confidence intervals for the nineteen effect sizes, rank ordered by

their central values. The widest interval, not surprisingly, is associated with study 4.

The intervals in figure 16.3 are based on the assumption that the parameter σ_θ^2 is equal to its estimated value, $\hat{\sigma}_\theta^2$ and that $\beta_0 = \hat{\beta}_0$. Indeed the term *empirical Bayes* is rooted in the idea that the parameters of the prior distribution of the true effect size θ_i are estimated empirically from the data. Basing θ_i^* on the approximation $\sigma_\theta^2 = \hat{\sigma}_\theta^2$ is a reasonable approximation when the number of studies, k, is large. In our case, with $k = 19$, this approximation is under some doubt. Raudenbush and Bryk re-estimated the posterior distribution of θ_i using a fully Bayesian approach that takes into account the uncertainty in the data about σ_θ^2 (2002, chapter 13; for an extensive discussion of this issue, see Seltzer 1993).[5]

The reader may have noted that the range of the empirical Bayes estimates (from -0.03 to 0.25) is narrower than the 95 percent plausible value interval (see table 16.2) for θ_i given by $(-0.23, 0.41)$. This is a typical result: the empirical Bayes estimates are shrinkage estimators. These shrinkage estimators reduce the expected mean-squared error of the estimates but in doing so they tend to overshrink in the sense that the ensemble of empirical Bayes estimates $\theta_i^*, i = 1, \ldots, k$, will generally be more condensed than the distribution of the true effect sizes $\theta_i, i = 1, \ldots, k$. This bias toward the mean implies θ_i^* will underestimate truly large positive values of θ_i and overestimate truly large negative values of θ_i. One suffers some bias to obtain small posterior variance leading to a small mean-squared error. That is, $E[(\theta_i^* - \theta_i)^2|\theta_i] = Bias^2 (\theta_i^*|\theta_i) + Var(\theta_i^*|\theta_i)$ will be small because the penalty associated with the bias will be offset by the reduction in variance. Indeed, Raudenbush (1988) showed that the ratio of $E[(\theta_i^* - \theta_i)^2|\theta_i]$ to the mean-squared error of the conventional estimate, that is $E[(T_i - \theta_i)^2|\theta_i]$ is, on average, approximately λ_i. The lower the reliability of T_i, then, the greater is the benefit of using the empirical Bayes estimator θ_i^*.

16.5.6.2 Conditional Shrinkage Estimation The empirical Bayes estimators discussed so far and plotted in figures 16.2 and 16.3 are based on the model $\theta_i = \beta_0 + u_i$, $u_i \sim N(0, \sigma_\theta^2)$. This model assumes that the true effect sizes are exchangeable, that is, we have no prior information to suspect that any one effect size is larger than any other. This assumption of exchangeability is refutable based on earlier theory and data in figure 16.1, suggesting that the duration of contact between teachers and children prior to the experiment will negatively predict the true effect size. This reasoning encouraged us to estimate the conditional model $\theta_i = \beta_0 + \beta_1 X_i + u_i$, $u_i \sim N(0, \sigma_\theta^2)$, where $X_i =$ weeks of prior contact, and the random effects u_i, $i = 1, \ldots, k$ are now conditionally exchangeable. That is, once we take weeks of previous contact into account, the random effects are exchangeable. The results, shown in table 16.2, provided strong support for including X_i in the model. This finding suggests an alternative and presumably superior empirical Bayes estimate of each study's effect size:

$$E(\theta_i | T_i, X_i, \hat{\beta}_0, \hat{\sigma}_\theta^2) = \theta_i^*$$
$$= \hat{\lambda}_i T_i + (1 - \hat{\lambda}_i)(\hat{\beta}_0 + \hat{\beta}_1 X_i). \quad (16.26)$$

Equation 16.27 shrinks outlying values of T_i not toward the overall mean but rather toward the predicted value $\hat{\beta}_0 + \hat{\beta}_1 X_i$. The shrinkage will be large when the reliability $\hat{\lambda}_i = \hat{\sigma}_\theta^2/(\hat{\sigma}_\theta^2 + v_i)$ is small. Recall that, after controlling for X_i, we found $\hat{\sigma}_\theta^2 = 0.000$. Our reliability is now conditional; that is, $\hat{\lambda}_i$ is now the reliability with which we can discriminate between studies that share the same number of $X_i = weeks$ of prior contact. The evidence suggests that such discrimination is not possible based on the data, given that the study effect sizes within levels of X_i appear essentially homogeneous.

The result, then, is $\theta_i^* = \hat{\beta}_0 + \hat{\beta}_1 X_i$. The 95 percent posterior confidence intervals for these conditional shrinkage estimators are plotted in figure 16.4. All studies having the same value of X_i share the same value of θ_i^* because our

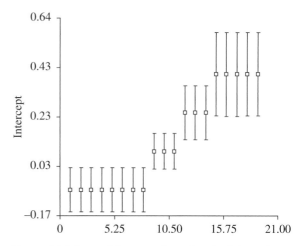

Figure 16.4 Bayes 95 percent Posterior Confidence Intervals, Mixed-Effects Model

SOURCE: Author's compilation.

estimate of the posterior variance, given the regression coefficients, is now $Var(\theta_i|T_i, X_i, \hat{\beta}_0, \hat{\beta}_1, \hat{\sigma}_\theta^2) = (v_i^{-1} + \hat{\sigma}_\theta^{-2})^{-1} = \hat{\sigma}_\theta^2/(1 + \hat{\sigma}_\theta^2 v_i^{-1}) = 0.000$ when $\hat{\sigma}_\theta^2 = 0.000$. This result also has implications for our belief about the range of plausible effect sizes. We see in table 16.2 that the empirical Bayes estimates range from -0.07 (when $X = 3$) to 0.41 (when $X = 0$). The 95 percent plausible value interval is the same, that is $(-0.07, 0,41)$.[6]

16.6 SUMMARY AND CONCLUSIONS

16.6.1 Advantages of the Approach

This chapter has described several potential advantages in using a random effects approach to quantitative research synthesis. First, the random effects conceptualization is consistent with standard scientific aims of generalization: studies under synthesis can be viewed as representative of a larger population or universe of implementations of a treatment or observations of a relationship.

Second, the random effects approach allows a parsimonious summary of results when the number of studies is very large. In such large-sample syntheses, it may be impossible to formulate fully adequate explanatory models for effects of study characteristics on study outcomes. The random effects approach enables the investigator to represent the unexplained variation in the true effect sizes with a single parameter, σ_θ^2, and to make summary statements about the range of likely effects in a particular area of research even when the sources of discrepancies in study results are poorly understood. In addition, re-estimating the random effects variance after adding explanatory variables makes a kind of percentage of variance explained criterion for assessing the adequacy of the explanatory models possible.

Third, the statistical methods described in this chapter allow the investigator to draw inferences about average effect sizes that account for the extra uncertainty that arises when the studies under synthesis are heterogeneous in their true effects. This uncertainty arises because, even despite concerted attempts at replication, study contexts, treatments, and procedures will inevitably vary in many ways that may influence results.

Fourth, the random effects approach facilitates improved estimation of the effect of each study using empirical Bayes or Bayes shrinkage. To the extent each study's estimated effect size lacks reliability, one can borrow information from the entire ensemble of studies to gain a more accurate estimate for each study.

16.6.2 Threats to Valid Statistical Inference

The problem is that the advantages of the random effects conceptualization are purchased at the price of two extra assumptions. First, the approach we have presented treats the random effects variance σ_θ^2 as if it were known, though, in fact, it must be estimated from the data. Second, to allow for hypothesis testing, we have assumed that the random effects are normally distributed with constant variance given the predictors in the between-study model.

16.6.2.1 Uncertainly About the Variance In standard practice and in the methods presented here, estimation of the overall mean effect size or regression coefficients are weighted estimates and the weights depend on the unknown variance of the random effects. The truth is that this variance is uncertain, especially in small sample syntheses, so that the weights are in reality random variables that inject uncertainty into inferences. Conventional methods of inference do not account for this source of uncertainty, leading to confidence intervals for regression coefficients and individual study effect sizes that are too short. This problem is most likely to arise when k is small and when the sampling variances v_i are highly variable. Donald Rubin described this problem and approached to addressing it in the context of parallel randomized experiments (1981). Raudenbush and Bryk described a

procedure for sensitivity analysis in research synthesis that allows the investigator to gauge the likely consequence of uncertainty about the variance for inferences about means and regression coefficients in effect size analyses (1985). In appendix 16B.2, I review an approach proposed by Gudio Knapp and Jocham Hartung that appears to offer help in insuring accurate standard errors (2003).

16.6.2.2 Failure of Parametric Assumptions Standard practice requires the assumption that the random effects are normally distributed with constant variance. This assumption is difficult to assess when the number of studies is small. It should be emphasized that this assumption is in addition to the usual assumption that the error of T_i as an estimate of θ_i is normal. Hedges investigated this latter assumption intensively (1981). Michael Seltzer developed a robust estimation procedure that allows the analyst to assume that the random effects are t-distributed rather than normal (1993). As the degrees of freedom of the t increase, the results converge to those based on normality assumption. The robust estimates are resistant to influence by outliers among the random effects. Appendix 16B.3 describes and illustrated a Huber-White robust estimator that is useful in ensuring that standard errors, tests, and confidence intervals are not sensitive to parametric assumptions. However, this approach works best when k is moderately large.

Alternative methods of point estimation have characteristic strengths and limitations. Maximum likelihood, either full or restricted, requires distributional assumptions about the random effects but provides efficient estimates in large samples when these assumptions are met. The method of moments requires assumptions only about the mean and variance of the random effects. However, the estimators, while consistent, are not efficient. I illustrate their application in appendix 16A.

16.6.2.3 Problems of Model Misspecification and Capitalizing on Chance As in any linear model analysis, estimates of regression coefficients will be biased if confounding variables—exogenous predictors related both to the explanatory variables in the model and to the random effects—are not included in the model. This is the classic problem of underfitting. The opposite problem is overfitting, when the investigator includes a large number of candidate explanatory variables as predictors, some of which may seem to improve prediction strictly as a result of chance. The best solution to both problems seems to be to develop strong a priori theory about study characteristics that might affect outcomes and include as many studies as possible. John Hunter and Frank Schmidt have warned especially against the overfitting that occurs when investigators carry out an extensive fishing expedition for explanatory variables even when there may be little true variance in the effect sizes (1990). In the absence of strong a priori theory, they advocate simple summaries using random effects approach where the key parameters are the mean and variance of the true effects sizes. Raudenbush described a method for deriving a priori hypotheses by carefully reading discussion sections of early studies to find speculations testable on the basis of later studies (1983).

16.6.2.4 Multiple Effect Sizes Most studies use several outcome variables. Such variables will tend to be correlated, and syntheses that fail to account for the resulting dependency may commit statistical errors (see Hedges and Olkin 1985, chapter 19; Rosenthal and Rubin 1986; Kalaian and Raudenbush 1996).

16.6.2.5 Inferences About Particular Effect Sizes In some cases, the investigator may wish to conceive of the effect sizes as random and yet make inferences about particular effects. For example, it may be of interest to assess the largest effect size in a series of studies or to identify for more intensive investigation effect sizes that are somehow unusual. The need to make such inferences, however, poses a problem. On the one hand, each study's data may yield an imprecise based on chance alone. On the other, the grand mean effect or the value predicted on the basis of a regression model may produce a poor indicator of the effects of any particular study. Rebecca DerSimonian and Nan Laird (1986), Donald Rubin (1981), and Stephen Raudenbush and Anthony Bryk (1985) have all described methods for combing data from all studies to strengthen inferences about the effects in each particular study to create a reasonably accurate picture of the effects of a collection of studies. I illustrated these ideas in section 16.5.6.

16A APPENDIX A: ALTERNATIVE APPROACHES TO POINT ESTIMATION

All the results presented so far (see table 16.2) were based on restricted maximum likelihood estimation of the between-study variance, σ_θ^2, with weighted least squares estimation of the regression coefficients β_0, β_1, and empirical Bayes estimation of the study-specific effect sizes θ_i, $i = 1, \ldots, k$. Many would agree that this is a reasonable approach. However, there are alternative approaches, and each has merits as well as vulnerabilities. In this and the next appendix, I consider and compare the alternatives. They differ along two dimensions. First, I consider three approaches to point estimation: full maximum likelihood, restricted maximum likelihood, and the method of moments. I then review three alternatives for computing uncertainty estimates such as standard errors, hypothesis tests, and confidence intervals.

This appendix presents general results for point estimation and illustrates them with application to the teacher expectancy data. Tables 16.3 and 16.4 collect these results in a concise form for the simple random effects model $T = \beta_0 + u_i + e_i$.

16A.1 Full Maximum Likelihood

Full maximum likelihood (FMLE) is based on the joint likelihood of the regression coefficients and the variance σ_θ^2. It ensures large-sample efficiency of these inferences based on distributional assumptions about the true effect sizes. The key criticism is that estimates of σ_θ^2 do not reflect uncertainty about the unknown regression coefficients (the βs). The estimates of σ_θ^2 will therefore be negatively biased. Note that the FMLE estimates of σ_θ^2 are smaller than those provided by other methods in table 16.3. This bias may lead to overly liberal hypothesis tests and confidence intervals for the regression parameters. A counterargument is that FMLE provides estimates of σ_θ^2 that have smaller mean-squared error than other approaches do, as long as the number of predictors in the model, $p + 1$, is not too large. FMLE achieves efficiency by reducing sampling variance at the expense of a negative bias that diminishes rapidly as k increases. One FMLE benefit is that any two nested models can be compared using the likelihood ratio statistic. However, the likelihood ratio test is a large-sample approach and there are some questions about its validity in our case with $k = 19$.

To simplify the presentation, we express the general model (equation 16.3) in matrix notation:

$$T_i = \mathbf{X}_i^T \boldsymbol{\beta} + u_i + e_i, \quad u_i \sim N(0, \sigma_\theta^2), \quad e_i \sim N(0, v_i) \quad (16.27)$$

where $\mathbf{X}_i = (1, X_{i1}, \ldots, X_{ip})^T$ is a column vector having dimension $(p + 1)$ of known characteristics of study i used as predictors of the effect size; $\boldsymbol{\beta} = (\beta_0, \ldots, \beta_p)^T$ is a column vector of dimension $p + 1$ containing unknown regression coefficients; $\mathbf{T} = (T_1, \ldots, T_k)^T$ is a column vector of dimension k containing the estimated effect sizes; and $v_i^* = v_i + \sigma_\theta^2$ is the variance of T_i.

Table 16.3 Statistical Inferences for β_0

Method of Estimation	$\hat{\sigma}_\theta^2$	$\hat{\beta}_0$	SE	$t = \hat{\beta}_0/(SE)$	95% Confidence Interval for β_0
Fixed Effects MLE					
Conventional[a]	0.000[d]	0.060	0.036	1.65	$(-0.016, 0.136)$
Quasi-F[b]	0.000	0.060	0.051	1.17	$(-0.047, 0.167)$
Huber-White[c]	0.000	0.060	0.040	1.51	$(-0.024, 0.136)$
Full MLE					
Conventional[a]	0.012	0.078	0.048	1.64	$(-0.016, 0.136)$
Quasi-F[b]	0.012	0.078	0.061	1.27	$(-0.047, 0.167)$
Huber-White[c]	0.012	0.078	0.048	1.62	$(-0.024, 0.136)$
Restricted MLE					
Conventional[a]	0.018	0.083	0.052	1.62	$(-0.025, 0.193)$
Quasi-F[b]	0.018	0.083	0.063	1.32	$(-0.048, 0.216)$
Huber-White	0.018	0.083	0.050	1.68	$(-0.021, 0.189)$
Method of Moments					
Conventional[a]	0.026	0.089	0.056	1.60	$(-0.028, 0.206)$
Quasi-F[b]	0.026	0.089	0.064	1.40	$(-0.044, 0.222)$
Huber-White[c]	0.026	0.089	0.052	1.71	$(-0.020, 0.199)$

SOURCE: Authors' compilation.

NOTES: For the simple random effects model using three methods of point estimation and three methods of estimating uncertainty.

[a]Conventional $V\hat{a}r(\hat{\beta}_0) = \left(\sum_{i=1}^{k} (\hat{\sigma}_\theta^2 + v_i)^{-1} \right)^{-1}$.

[b]Quasi-F approach is equivalent to assuming $V\hat{a}r(\hat{\beta}_0) = q = \dfrac{\sum_{i=1}^{k} (\hat{\sigma}_\theta^2 + v_i)^{-1}(T_i - \hat{\beta}_0)^2}{(k-1)\sum_{i=1}^{k}(\hat{\sigma}_\theta^2 + v_i)^{-1}}$.

[c]Huber-White $V\hat{a}r(\hat{\beta}_0) = \dfrac{\sum_{i=1}^{k} (\hat{\sigma}_\theta^2 + v_i)^{-2}(T_i - \hat{\beta}_0)^2}{\left(\sum_{i=1}^{k}(\hat{\sigma}_\theta^2 + v_i)^{-1} \right)^2}$.

[d]For the fixed effects model, $\sigma_\theta^2 = 0$ by assumption.

The log-likelihood for FMLE is the density of the observed effect sizes regarded as a function of the parameters given the sample:

$$L_F(\boldsymbol{\beta}, \sigma_\theta^2; T) = \text{constant} - \tfrac{1}{2}\sum_{i=1}^{k} \ln(v_i^{*-1})$$
$$- \tfrac{1}{2}\sum_{i=1}^{k} v_i^{*-1}(T_i - \mathbf{X}_i^T\boldsymbol{\beta})^2. \quad (16.28)$$

The likelihood is maximized by solving

$$\frac{dL_F}{d\boldsymbol{\beta}} = \sum_{i=1}^{k} v_i^{*-1}\mathbf{X}_i(T_i - \mathbf{X}_i^T\boldsymbol{\beta}) = 0 \quad (16.29)$$

$$\frac{dL_F}{d\sigma_\theta^2} = -\tfrac{1}{2}\sum_{i=1}^{k} v_i^{*-1} + \tfrac{1}{2}\sum_{i=1}^{k} v_i^{*-2}(T_i - \mathbf{X}_i^T\boldsymbol{\beta})^2 = 0. \quad (16.30)$$

The resulting estimating equations are

$$\hat{\boldsymbol{\beta}} = \left(\sum_{i=1}^{k} v_i^{*-1}\mathbf{X}_i\mathbf{X}_i^T \right)^{-1} \sum_{i=1}^{k} v_i^{*-1}\mathbf{X}_i T_i \quad (16.31)$$

and

$$\hat{\sigma}_\theta^2 = \sum_{i=1}^{k} v_i^{*-2}[(T_i - \mathbf{X}_i^T\boldsymbol{\beta})^2 - v_i] / \sum_{i=1}^{k} v_i^{*-2}. \quad (16.32)$$

These estimating equations have the interesting form of weighted least squares estimates. The weights for estimating the regression coefficients are the precisions v_i^{*-1}, the reciprocals of $E\lfloor(T_i - \mathbf{X}_i^T\boldsymbol{\beta})^2\rfloor$, while the weights for estimating the variances are v_i^{*-2}, inversely proportional to $Var\lfloor(T_i - \mathbf{X}_i^T\boldsymbol{\beta})^2\rfloor$.

Table 16.4 Statistical Inferences for σ_θ^2

Method of Estimation	$\hat\sigma_\theta^2$	Reliability, λ[a]	95% Plausible Value Interval for θ_i[b]
Full MLE	0.012	.29	$(-0.14, 0.30)$
Restricted MLE	0.018	.37	$(-0.19, 0.35)$
Method of Moments	0.026	.51	$(-0.23, 0.41)$

SOURCE: Authors' compilation.
NOTES: For the simple random effects model using three methods of point estimation.

[a]$\lambda = \sum\limits_{i=1}^{19}\hat\lambda_i, \hat\lambda_i = \hat\tau^2/(\hat\sigma_\theta^2 + v_i)$

[b]95% Plausible value interval for $\theta_i = \beta_0 \pm 1.96*\hat\sigma_\theta$.

Of course, the two equations cannot be solved in closed form. However, it is easy to write a small program that begins with an initial estimate of $\boldsymbol\beta$, for example using weighted least squares with σ_θ^2 set to zero in computing the weights, so that the weight v_i^{*-1} is set initially to v_i^{-1}. These estimates are then used to produce an initial estimate of σ_θ^2 using weights v_i^{*-2} set initially to v_i^{-2}. This initial estimate of σ_θ^2 is then used in constructing new weights with which to re-estimate $\boldsymbol\beta$ and σ_θ^2 using 16.31 and 16.32. This procedure continues until convergence. Alternatively, standard statistical programs for mixed models that allow known but varying level-1 variances can be used to estimate this model. I used HLM version 6.0 for the computation of the results for FMLE shown in table 16.3.

For the simple random effects model (table 16.3), the estimating equations 16.31 and 16.32 reduce to

$$\hat\beta_0 = \sum_{i=1}^{k}v_i^{*-1}T_i \Big/ \sum_{i=1}^{k}v_i^{*-1} \qquad (16.33)$$

and

$$\hat\sigma_\theta^2 = \sum_{i=1}^{k}v_i^{*-2}[(T_i - \beta_0)^2 - v_i] \Big/ \sum_{i=1}^{k}v_i^{*-2}. \qquad (16.34)$$

16A.2 Restricted Maximum Likelihood

Restricted maximum likelihood (RMLE) is based on the likelihood of the variance σ_θ^2 given the data, with the regression coefficients removed through integration (Demptser, Rubin, and Tsutakawa 1981).[7] The advantage of RMLE versus FMLE is that RMLE estimates of σ_θ^2 take into account the uncertainty about $\boldsymbol\beta$, yielding nearly un-biased estimates and possibly improving the standard errors, hypothesis tests, and intervals for $\boldsymbol\beta$. However, the RMLE estimates of σ_θ^2 will be slightly less efficient than those based on FMLE unless the number of parameters, $p + 1$, in $\boldsymbol\beta$ is comparatively large.

The log-likelihood for RMLE is the density of the observed effect sizes regarded as a function of σ_θ^2 alone, given the sample (Raudenbush and Bryk 1985):

$$L_F(\sigma_\theta^2;\mathrm{T}) = \text{constant} - \tfrac{1}{2}\sum_{i=1}^{k}\ln(v_i^{*-1})$$

$$- \tfrac{1}{2}\ln\left(\sum_{i=1}^{k}v_i^{*-1}\mathbf{X}_i\mathbf{X}_i^T\right)$$

$$- \tfrac{1}{2}\sum_{i=1}^{k}v_i^{*-1}(T_i - \mathbf{X}_i^T\hat{\boldsymbol\beta}). \qquad (16.35)$$

The log-likelihood is maximized by solving

$$\frac{dL_R}{d\sigma_\theta^2} = -\tfrac{1}{2}\sum_{i=1}^{k}v_i^{*-1} + \tfrac{1}{2}\left(\sum_{i=1}^{k}v_i^{*-1}\mathbf{X}_i\mathbf{X}_i^T\right)^{-1}$$

$$+ \tfrac{1}{2}\sum_{i=1}^{k}v_i^{*-2}(T_i - \mathbf{X}_i^T\hat{\boldsymbol\beta})^2 = 0. \qquad (16.36)$$

The resulting estimating equation is

$$\hat\sigma_\theta^2 = \frac{\sum\limits_{i=1}^{k}v_i^{*-2}[(T_i - \mathbf{X}_i^T\hat{\boldsymbol\beta})^2 - v_i] + tr\left[\left(\sum\limits_{i=1}^{k}v_i^{*-1}\mathbf{X}_i\mathbf{X}_i^T\right)^{-1}\sum\limits_{i=1}^{k}v_i^{*-2}\mathbf{X}_i\mathbf{X}_i^T\right]}{\sum\limits_{i=1}^{k}v_i^{*-2}}.$$

$$(16.37)$$

Note that this estimating equation is equivalent to that for the case of FMLE (equation 16.32) except for the second term in the numerator of 16.37. It is instructive to compare the form of these in the balanced case, that is $v_i = v$ for all i. In that case, the FMLE estimating equation becomes

$$\hat\sigma_\theta^2 = \frac{\sum\limits_{i=1}^{k}(T_i - \mathbf{X}_i^T\beta)^2 - v}{k}$$

and the RMLE estimating equation becomes

$$\sigma_\theta^2 = \frac{\sum\limits_{i=1}^{k}(T_i - \mathbf{X}_i^T\hat{\boldsymbol\beta})^2 - v}{k - p - 1}$$

where $\hat{\boldsymbol\beta} = (\Sigma\mathbf{X}_i\mathbf{X}_i^T)^{-1}\Sigma\mathbf{X}_iT_i$. Note that the denominator of the RMLE estimator is $k-p-1$ rather than k, as it is in the case of the FMLE estimator, revealing how RMLE accounts for the uncertainty associated with having to estimate $\boldsymbol{\beta}$. These equations for the balanced case are solvable in closed form.

In the typical case of unbalanced data, the RMLE equation (16.37) for the new σ_θ^2 depends on the current estimate σ_θ^2 because σ_θ^2 is needed for the weights. This problem can easily be solved using an iterative program very similar to that used for FMLE. However, I used HLM version 6.0 for the computation of the results for FMLE shown in table 16.3.

For the simple random effects model, the estimating equation for σ_θ^2 reduces to

$$\hat{\sigma}_\theta^2 = \frac{\sum_{i=1}^{k} v_i^{*-2}[(T_i - \hat{\beta}_0)^2 - v_i]}{\sum_{i=1}^{k} v_i^{*-2}} + \left(\sum_{i=1}^{k} v_i^{*-1}\right)^{-1}. \qquad (16.38)$$

where $\hat{\beta}_0 = \sum_{i=1}^{k} v_i^{*-1} T_i \Big/ \sum_{i=1}^{k} v_i^{*-1}$.

16A.3 Method of Moments

The key benefit of a likelihood-based approach is that estimates will be efficient and hypothesis tests powerful in large samples. To obtain these benefits, however, one must assume a distribution for the random effects u_i, $i = 1, \ldots, k$. We have assumed normality of these random effects throughout.

A popular alternative approach is the method of moments (MOM), which requires assumptions about the first two moments of u_i but not the probability density. For this purpose, we assume that u_i, $i = 1, \ldots, k$ are mutually independent with zero means and common variance σ_θ^2. The approach will ensure consistent but not necessarily efficient estimates of the parameters.

MOM uses a two-step approach. One begins with a comparatively simple regression to estimate $\boldsymbol{\beta}$. One then equates the residual sum of squares from this regression to its expected value and solves for σ_θ^2. This estimate of σ_θ^2 is then used to construct weights $v_i^{*-1} = (\sigma_\theta^2 + v_i)^{-1}$ to be used in a second weighted least squares regression. I considered two approaches to the first step, the computation of the simple regression: ordinary least squares (OLS" and weighted least squares using the known v_i^{-1} as weights (WLS, known weights).

16A.3.1 MOM Using OLS Regression at Stage 1
The OLS estimates for the general model are

$$\hat{\boldsymbol{\beta}}_{OLS} = \left(\sum_{i=1}^{k} \mathbf{X}_i \mathbf{X}_i^T\right)^{-1} \sum_{i=1}^{k} \mathbf{X}_i T_i^T, \qquad (16.39)$$

and the expected residual sum of squares is

$$E(RSS) = E \sum_{i=1}^{k} (T_i - \mathbf{X}_i^T \hat{\boldsymbol{\beta}}_{OLS})^2$$
$$= \sum_{i=1}^{k} v_i - tr(S) + (k - p - 1)\,\sigma_\theta^2 \qquad (16.40)$$

yielding the estimating equation

$$\hat{\sigma}_\theta^2 = \frac{RSS - \sum_{i=1}^{k} v_i - tr(S)}{k - p - 1}, \qquad (16.41)$$

where

$$S = \left[\left(\sum_{i=1}^{k} \mathbf{X}_i \mathbf{X}_i^T\right)^{-1} \sum_{i=1}^{k} v_i \mathbf{X}_i \mathbf{X}_i^T\right].$$

In the case of the simple random effects model, 16.41 reduces to

$$\hat{\sigma}_\theta^2 = \frac{\sum_{i=1}^{k} (T_i - \bar{T})^2}{k - 1} - \bar{v}$$

where $\bar{T} = \sum_{i=1}^{k} T_i / k$ and $\bar{v} = \sum_{i=1}^{k} v_i / k$.

16A.3.2 MOM Using WLS with Known Weights
The WLS estimates with weights v_i^{-1} are

$$\hat{\boldsymbol{\beta}}_{WLS} = \left(\sum_{i=1}^{k} v_i^{-1} \mathbf{X}_i \mathbf{X}_i^T\right)^{-1} \sum_{i=1}^{k} v_i^{-1} \mathbf{X}_i T_i^T, \qquad (16.42)$$

and the expected weighted residual sum of squares is

$$E(WRSS) = E \sum_{i=1}^{k} v_i^{-1} \left(T_i - \mathbf{X}_i^T \hat{\boldsymbol{\beta}}_{WLS}\right)^2 = \sigma_\theta^2\, tr(M) + k - p,$$

yielding the estimating equation

$$\hat{\sigma}_\theta^2 = \frac{WRSS - (k - p - 1)}{tr(M)} \qquad (16.43)$$

where

$$tr(M) = \sum_{i=1}^{k} v_i^{-1} - tr\left[\left(\sum_{i=1}^{k} v_i^{-1} \mathbf{X}_i \mathbf{X}_i^T\right)^{-1} \sum_{i=1}^{k} v_i^{-2} \mathbf{X}_i \mathbf{X}_i^T\right].$$

In the case of the simple random effects model, 16.43 reduces to

$$\hat{\sigma}_\theta^2 = \frac{\sum_{i=1}^{k} v_i^{-1}(T_i - \bar{T})^2 - (k - 1)}{tr(M)}$$

where

$$tr(M) = \sum_{i=1}^{k} v_i^{-1} - \sum_{i=1}^{k} v_i^{-2} \Bigg/ \sum_{i=1}^{k} v_i^{-1} \text{ and } \overline{T} = \sum_{i=1}^{k} v_i^{-1} T_i \Bigg/ \sum_{i=1}^{k} v_i^{-1}.$$

16A.4 Results

The results for the simple random effects model based on the three approaches FMLE, RMLE, and MOM are provided in table 16.3. We focus on the results for the simple random effects model, for which there is clear evidence against the null hypothesis H_0: $\sigma_\theta^2 = 0$. Recall that when X_i = weeks of prior contact is included in the model, we can comfortably retain H_0: $\sigma_\theta^2 = 0$ and the differences among FMLE, RMLE, and MOM are no longer of great interest.

For MOM, the OLS approach produced an unrealistically large mean effect size as well as an unrealistically large estimate of σ_θ^2. This problem seems to reflect the fact that the illustrative data have widely varying sampling variances v_i and some outlying values of T_i associated with studies having large v_i. In this setting, the OLS results, which weight each study's results equally in computing the initial estimates, appear suboptimal. As a result, the MOM results in table 16.3 are those based on WLS with known weights, v_i^{-1}. This experience suggests caution in using the OLS approach despite its presence in the literature (for example, Hedges 1983).

I focus on the conventional approach to estimating standard errors because the two other approaches (Quasi-F and Huber-White) will be considered in appendix 16B.

As expected, the point estimate of σ_θ^2 is larger under RMLE ($\hat{\sigma}_\theta^2 = 0.019$) than under FMLE ($\hat{\sigma}_\theta^2 = 0.012$). As a result, the conventional tests of the null hypothesis H_0: $\beta_0 = 0$ are a bit more conservative under RMLE than under FMLE, and the conventional 95 percent confidence intervals are a bit wider under RMLE than under FMLE.

MOM based on WLS with known weights produces a larger point estimate of σ_θ^2 ($\hat{\sigma}_\theta^2 = 0.026$) than either RMLE or FMLE and therefore produces somewhat more conservative inferences about β_0.

The main conclusion to be drawn from the comparison is that the conventional inferences about β_0 are quite similar across FMLE, RMLE, and MOM for these data. What differs across the three methods are the inferences about σ_θ^2 itself· and, relatedly, about the range of plausible values of the true effect sizes θ_i. These differences are apparent in table 16.4. Note the wider 95 percent plausible value interval for θ_i under MOM than under RMLE and

the wider plausible value interval for RMLE than under FMLE. The reliabilities differ as well. Under MOM, about 51 percent of the variation in the observed effect sizes appears to reflect variation in the true effect sizes. The corresponding values for RMLE and FMLE are 37 percent and 29 percent, respectively.

16B APPENDIX B: ALTERNATIVE APPROACHES TO UNCERTAINTY ESTIMATION

This appendix describes the conventional approach to computing hypothesis tests and confidence intervals based on large-sample theory. It then reviews two alternative approaches: a quasi-F statistic and a Huber-White robust variance estimator. I illustrate application of these in the case of the simple random effects model. The results are presented in table 16.3.

16B.1 Conventional Approach

Inferences about the central tendency of the true effect sizes, captured in the regression coefficients $\boldsymbol{\beta}$, are conventionally based on the large-sample variance of $\hat{\boldsymbol{\beta}}$. As $k \to \infty$, the estimated weights $\hat{v}_i^{*-1} = (v_i + \hat{\sigma}_\theta^2)^{-1}$ converge to the true weights $v_i^{*-1} = (v_i + \hat{\sigma}_\theta^2)^{-1}$, and the estimated variance of the regression coefficients converges to

$$Var(\hat{\boldsymbol{\beta}} \mid \hat{\sigma}_\theta^2 = \sigma_\theta^2) = \left(\sum_{i=1}^{k} v_i^{*-1} \mathbf{X}_i \mathbf{X}_i^T \right)^{-1}.$$

The conventional inferences then treat the square roots of the diagonal elements of $Var(\hat{\boldsymbol{\beta}} \mid \hat{\sigma}_\theta^2 = \sigma_\theta^2)$ as known standard errors in computing hypothesis tests and confidence intervals.

For example, in the case of the simple random effects model, section 16.5 used the ratio $\hat{\beta}_0 / \sqrt{Var(\hat{\beta}_0 \mid \hat{\sigma}_\theta^2 = \sigma_\theta^2)}$ to test the null hypothesis H_0: $\beta_0 = 0$, where $Var(\beta_0 \mid \hat{\sigma}_\theta^2 = \sigma_\theta^2) = \left(\sum_{i=1}^{k} v_i^{*-1} \right)^{-1}$. One difficulty with this approach is that, when k is modest, the assumption that $\hat{v}_i^{*-1} = (v_i + \hat{\sigma}_\theta^2)^{-1}$ have converged to the true weights $v_i^{*-1} = (v_i + \sigma_\theta^2)^{-1}$ and therefore that $Var(\hat{\boldsymbol{\beta}} \mid \hat{\sigma}_\theta^2 = \sigma_\theta^2) = Var(\hat{\boldsymbol{\beta}})$ will be difficult to defend, leading to overly liberal tests H_0: $\beta_0 = 0$ and confidence intervals for β_0.

16B.2 A Quasi-F Statistic

To cope with this problem, Jocham Hartung and Guido Knapp proposed an alternative test for the simple random effects model and extended this approach to the case where there is a single between-study predictor X (Hartung and Knapp 2004; Knapp and Hartung 2003). They provided evidence that under certain simulated conditions, their approach produced more accurate tests and intervals than does the conventional approach.

The approach appears to be based on what might be called a quasi-F statistic. For the simple random effects model we compute

$$F^*(1, k - p - 1) = \hat{\beta}_0^2 / \hat{q}, \qquad (16.44)$$

where

$$\hat{q} = \frac{\sum_{i=1}^{k} \hat{v}_i^{*-1}(T_i - \hat{\beta}_0)^2 \backslash \sum_{i=1}^{k} \hat{v}_i^{*-1}}{k - 1} \qquad (16.45)$$

with

$$\hat{\beta}_0 = \sum_{i=1}^{k} \hat{v}_i^{*-1} T_i \bigg/ \sum_{i=1}^{k} \hat{v}_i^{*-1}.$$

We can motivate this approach as follows. We know that, if the weights v_i^{*-1} were known, the estimate

$$\hat{\beta}_0 = \sum_{i=1}^{k} v_i^{*-1} T_i \bigg/ \sum_{i=1}^{k} v_i^{*-1}$$

would have variance

$$Var(\hat{\beta}_0) = 1 \bigg/ \sum_{i=1}^{k} v_i^{*-1}.$$

Therefore the statistic

$$(\hat{\beta}_0 - \beta_0)^2 \sum_{i=1}^{k} v_i^{*-1}$$

would be distributed as chi-square with one degree of freedom. Moreover, the weighted sum of squared residuals

$$q = \sum_{i=1}^{k} v_i^{*-1}(T_i - \hat{\beta}_0)^2$$

would be distributed as chi-square with $k-1$ degrees of freedom. Hence the ratio

$$F(1, k-1) = \frac{(\hat{\beta}_0 - \beta_0)^2 \sum_{i=1}^{k} v_i^{*-1}}{\sum_{i=1}^{k} v_i^{*-1}(T_i - \hat{\beta}_0)^2} \qquad (16.46)$$

would be distributed as a central F statistic with degrees of freedom $(1, k-1)$. Equation 16.46 is identical to equation

16.44 with β_0 set to 0 and $v_i^* = \hat{v}_i^*$. We can therefore regard 16.44 as a quasi-F, that is, an approximation to F. The likely benefit of this approach stems from the fact that the empirical weighted sum of squared residuals strongly influence the test, and these are somewhat robust to inaccurate estimates of σ_θ^2.

The approach can be generalized to any linear hypothesis. Define $H_0: \mathbf{C}^T \boldsymbol{\beta} = 0$, where $\mathbf{C}$ is a $p+1$ by c contrast matrix, c being the number of linear contrasts of interest among the $p+1$ regression parameters. The Knapp-Hartung approach defines the test

$$F^*(c, k - p - 1)$$
$$= \frac{(\hat{\boldsymbol{\beta}} - \boldsymbol{\beta})^T \mathbf{C} \bigg[\mathbf{C}^T \bigg(\sum_{i=1}^{k} \hat{v}_i^{*-1} \mathbf{X}_i \mathbf{X}_i^T \bigg)^{-1} \mathbf{C} \bigg]^{-1} (\hat{\boldsymbol{\beta}} - \boldsymbol{\beta})^{-1}/c}{\sum_{i=1}^{k} \hat{v}_i^{*-1}(T_i - \mathbf{X}_i \hat{\boldsymbol{\beta}})^2/(k - p - 1)}. \qquad (16.47)$$

The results of that approach (table 16.3) show that, regardless of the method of point estimation, the quasi-F produces significantly more conservative inferences (smaller t-ratios, wider confidence intervals) than does the conventional approach. These may be more realistic than the conventional inferences.

16B.3 Huber-White Variances

The conventional approach and the quasi-F approach rely on parametric assumptions about the distribution of the random effects. In particular, in deriving tests and confidence intervals, we have assumed that these are normally distributed with constant variance τ^2. For example, we have assumed that the random effects variance is constant at every $X =$ weeks of prior contact, an assumption that may be false. The Huber-White approach produces standard errors that do not rely on these assumptions (White 1980). Consider the regression of T on $\mathbf{X}$ using any known weights w_i. The estimator $\hat{\boldsymbol{\beta}} = \bigg(\sum_{i=1}^{k} w_i \mathbf{X}_i \mathbf{X}_i^T \bigg)^{-1} \sum_{i=1}^{k} w_i \mathbf{X}_i T_i^T$

has variance

$$Var(\hat{\boldsymbol{\beta}}) = \bigg(\sum_{i=1}^{k} w_i \mathbf{X}_i \mathbf{X}_i^T \bigg)^{-1} \sum_{i=1}^{k} w_i^2 \mathbf{X}_i Var(T_i) \mathbf{X}_i^T \bigg(\sum_{i=1}^{k} w_i \mathbf{X}_i \mathbf{X}_i^T \bigg)^{-1} \qquad (16.48)$$

regardless of the true form of $Var(T_i)$. If in fact

$$w_i^{-1} = Var(T_i) = v_i^*,$$

(16.48) will reduce to the conventional expression for the variance, that is

$$Var(\hat{\boldsymbol{\beta}}) = \left(\sum_{i=1}^{k} v_i^* \mathbf{X}_i \mathbf{X}_i^T\right)^{-1}.$$

The Huber-White approach avoids this step, substituting the empirical estimator $V\hat{a}r(T_i) = (T_i - \mathbf{X}_i^T\hat{\boldsymbol{\beta}})^2$ thus avoiding assumptions about the variance of the random effects, yielding then the expression

$$V\hat{a}r_{HW}(\hat{\boldsymbol{\beta}})$$

$$= \left(\sum_{i=1}^{k} w_i \mathbf{X}_i \mathbf{X}_i^T\right)^{-1} \sum_{i=1}^{k} w_i^2 \mathbf{X}_i (T_i - \mathbf{X}_i^T\hat{\boldsymbol{\beta}})^2 \, \mathbf{X}_i^T \left(\sum_{i=1}^{k} w_i \mathbf{X}_i \mathbf{X}_i^T\right)^{-1}.$$

(16.49)

This expression is familiar in the literature on generalized estimating equations as the sandwich estimator, where the conventional expression for the variance is the bread on each side of 16.49 (Liang and Zeger 1986).

The results, shown in table 16.3, suggest that the Huber-White and conventional inferences are very similar for our data, regardless of the method of point estimation. Apparently, parametric assumptions about the variance of the random effects are not strongly affecting our inferences. A problem with the Huber-White approach is that the theory behind it requires large k. Some simulation studies show that these inferences often are sound even for modest k, but more research is needed (Cheong, Raudenbush, and Fotiu 2000).

16.7 NOTES

1. Of course, many studies are not actually based on random samples of subjects from well-defined populations. Nevertheless, the investigators use standard inferential statistical procedures. By doing so, each investigator is implicitly viewing his or her sample as a random sample from a hypothetical population or universe of possible observations that supply the target of any generalizations that arise from the statistical inferences.
2. At either stage, it is conceivable to view the population as either finite or infinite. Of course, it is hard to imagine that an infinite number of studies have been conducted on a particular topic. Nevertheless, in practice, populations at both levels are viewed as infinite. This simplifies analysis because one need not use finite population correction factors. Perhaps more important, the view of populations as infinite is consistent with the notion of generalizing to a universe of possible observations (Cronbach et al. 1972) and is therefore consistent with standard scientific aims.
3. The estimate is "best" in the sense that it has minimum variance within the class of unbiased estimators.
4. See section 16.6 for a discussion and illustration of alternative estimation procedures.
5. In the present case, we find that the likelihood for σ_θ^2 is quite symmetric about the mode, so that the empirical Bayes inferences and the fully Bayes inferences are quite similar (for more discussion, see Rubin 1981; Morris 1983).
6. Of course σ_θ^2 is not known to be zero. Nonzero values of σ_θ^2 are plausible. We are also conditioning on point estimates of $\hat{\beta}_0$, $\hat{\beta}_1$ in computing these plausible value intervals and posterior variances. A fully Bayesian approach would reflect the uncertainty about σ_θ^2 as well as β_0, β_1 in the posterior intervals for θ_i (Raudenbush and Bryk 2002, chapter 13).
7. Let $f_F(\mathbf{T}|\boldsymbol{\beta},\sigma_\theta^2)$ be the density of the data (equivalent to the FMLE likelihood). Formulate a prior distribution $p(\boldsymbol{\beta}|\boldsymbol{\gamma},\boldsymbol{\Gamma})$ where $\boldsymbol{\gamma}$ is the prior mean of $\boldsymbol{\beta}$ and $\boldsymbol{\Gamma}$ is the prior variance-covariance matrix. Then

$$\lim_{(\Gamma^{-1}\to 0)} f_R(\mathbf{T}|\sigma_\theta^2)$$
$$= \lim_{(\Gamma^{-1}\to 0)} \int f_F(Y|\boldsymbol{\beta},\sigma_\theta^2)p(\boldsymbol{\beta}|\boldsymbol{\gamma},\boldsymbol{\Gamma})d\boldsymbol{\beta}.$$

The idea is that one is treating $\boldsymbol{\gamma}$ in a fully Bayesian fashion but conditioning on fixed σ_θ^2. However, there is essentially no prior information about $\boldsymbol{\gamma}$ so that its prior precision $\boldsymbol{\Gamma}^{-1}$ is essentially null. Equation 16.36 involves the log density of the data given only the variance σ_θ^2, averaging over all possible values of $\boldsymbol{\gamma}$ so that inferences about σ_θ^2 fully take into account the uncertainty about $\boldsymbol{\beta}$. Alternatively, Raudenbush and Bryk provided a classical interpretation in which $f_R(\mathbf{T}|\sigma_\theta^2) = f_F(Y|\boldsymbol{\beta},\sigma_\theta^2)/g((\hat{\boldsymbol{\beta}}|\sigma_\theta^2)$, where $\hat{\boldsymbol{\beta}}$ is given by equation 16.31 (1985).

16.8 REFERENCES

Casella, George, and Roger Berger. 1990. *Statistical Inference*, 2nd ed. Pacific Grove, Calif.: Wadsworth.

Cheong, Yuk F., Randall P. Fotiu, and Stephen W. Raudenbush. 2001. "Efficiency, and Robustness of Alternative Estimators for 2-, and 3- Level Models: The Case of NAEP." *Journal of Educational, and Behavioral Statistics* 26(4): 411–29.

Cronbach, Lee J., Goldine C. Gleser, Harinder Nanda, and Nageswari Rajaratnam. 1972. *The Dependability of Behavioral Measurements: Theory of Generalizability for Scores, and Profiles*. New York: John Wiley & Sons.

De Finetti, Bruno. 1964. "Foresight: Its Logical Laws, Its Subjective Sources." In *Studies in Subjective Probability*, edited by E.E. Kyburg Jr. and H.E. Smokler. New York: John Wiley & Sons.

Dempster, Arthur P., Donald B. Rubin, and Robert K. Tsutakawa. 1981. "Estimation in Covariance Components Models." *Journal of the American Statistical Association* 76(374): 341–53.

DerSimonian, Rebecca, and Nan M. Laird. 1986. "Meta-Analysis in Clinical Trials." *Controlled Clinical Trials* 7(3): 177–88.

Hartung, Jocham., and Guido Knapp. 2001. "On Tests of the Overall Treatment Effect in Meta-Analysis with Normally Distributed Responses." *Statistics in Medicine* 20(12): 1771–82.

Hedges, Larry V. 1981. "Distribution Theory for Glass's Estimator of Effect Size, and Related Estimators." *Journal of Educational Statistics* 6(2): 107.

———. 1983. "A Random Effects Model for Effect Size." *Psychological Bulletin* 93: 388–95.

Hedges, Larry V., and Inrgam Olkin. 1983. "Regression Models in Research Synthesis." *American Statistician* 37: 137–40.

Higgins, Julian.T., and Simon G. Thompson. 2002. "Quantifying Heterogeneity in Meta-Analysis." *Statistics in Medicine* 21(11): 1539–58.

Hunter, John E., and Frank L. Schmidt. 1990. *Methods of Meta-Analysis: Correcting Error, and Bias in Research Findings*. Beverly Hills, Calif.: Sage Publications.

Kalaian, Hripsime, and Stephen W. Raudenbush. 1996. "A Multivariate Mixed Linear Model for Meta-Analysis." *Psychological Methods* 1(3): 227–35.

Knapp, Guido, and Jocham Hartung. 2003. "Improved Tests for a Random Effects Meta-Regression with a Single Covariate." *Statistics in Medicine* 22(17): 2693–2710.

Liang, Kung-Yee, and Scott L. Zeger. 1986. "Longitudinal Data Analysis Using Generalized Linear Models." *Biometrika* 73(1): 13–22.

Morris, Carl N. 1983. "Parametric Empirical Bayes Inference: Theory, and Applications. *Journal of the American Statistical Association* 78(381): 47–65.

Raudenbush, Stephen W. 1983. "Utilizing Controversy as a Source of Hypotheses in Meta-Analysis: The Case of Teacher Expectancy on Pupil IQ." *Evaluation Studies Review Annual*. Beverly Hills, Calif.: Sage Publications.

———. 1984. "Magnitude of Teacher Expectancy Effects on Pupil IQ as a Function of the Credibility of Expectancy Induction: A Synthesis of Findings from 18 Experiments." *Journal of Educational Psychology* 76(1): 85–97.

———. 1988. "Educational Applications of Hierarchical Linear Models." *Journal of Educational, and Behavioral Statistics* 13(2): 85–116.

Raudenbush, Stephen W., and Anthony S. Bryk. 1985. "Empirical Bayes Meta-Analysis." *Journal of Educational, and Behavioral Statistics* 10(2): 241–69.

———. 2002. *Hierarchical Linear Models: Applications, and Data Analysis Methods*. Thousand Oaks, Calif.: Sage Publications.

Rosenthal, Robert, and Linda Jacobson. 1968. *Pygmalion in the Classroom*. New York: Holt, Rinehart, and Winston.

Rosenthal, Robert, and Donald B. Rubin. 1978. "Interpersonal Expectancy Effects: The First 345 Studies." *Behavioral, and Brain Sciences* 3: 377–86.

Rubin, Donald B. 1981. "Estimation in Parallel Randomized Experiments." *Journal of Educational, and Behavioral Statistics* 6(4): 337–401.

Ryan, William. 1971. *Blaming the Victim*. New York: Vintage.

Seltzer, Michael. 1993. "Sensitivity Analysis for Fixed Effects in the Hierarchical Model: A Gibbs Sampling Approach." *Journal of Educational, and Behavioral Statistics* 18(3): 207–35.

White, Halbert. 1980. "A Heteroskedasticity-Consistent Covariance Matrix Estimator, and a direct Test for Heteroskedasticity." *Econometrica* 48(4): 817–38.

17

CORRECTING FOR THE DISTORTING EFFECTS OF STUDY ARTIFACTS IN META-ANALYSIS

FRANK L. SCHMIDT
University of Iowa

HUY LE
University of Central Florida

IN-SUE OH
University of Iowa

CONTENTS

17.1 ARTIFACTS THAT DISTORT OBSERVED STUDY RESULTS

Every study has imperfections, many of which bias the results. In some cases we can define precisely what a methodologically ideal study would be like, and thus say that the effect size obtained from any real study will differ to some extent from the value that would have been obtained had the study been methodologically perfect. Although it is important to estimate and eliminate bias in individual studies, it is even more important to remove such errors in research syntheses such as meta-analyses.

Some authors have argued that meta-analysts should not correct for study imperfections because the purpose of meta-analysis is to provide only a description of study findings, not an estimate of what would have been found in methodologically ideal studies. However, the errors and biases that stem from study imperfections are artifactual: they stem from imperfections in our research methods, not from the underlying relationships that are of scientific interest (Rubin 1990). Thus, scientific questions are better addressed by estimates of the results that would have been observed had studies been free of methodological biases (Cook et al. 1992; Hunter and Schmidt 2004, 30–32; Rubin 1990; Schmidt 1992). For example, in correlation research the results most relevant to evaluation of a scientific theory are those that would be obtained from a study using an infinitely large sample from the relevant population (that is, the population itself) and measures of the independent and dependent variables that are free of measurement error and are perfectly construct valid. Such a study would be expected to provide an exact estimate of the relation between constructs in the population of interest; such an estimate is maximally relevant to the testing and evaluation of scientific theories, and to theory construction. Thus corrections for biases and other errors in study findings due to study imperfections, which we call artifacts, are essential to developing valid cumulative knowledge. The increasing use of estimates from meta-analysis as input into causal modeling procedures (see, for example, Colquitt, LePine, and Noe 2002; Becker and Schram 1994) further underlines the importance of efforts to ensure that meta-analysis findings are free of correctable bias and distortion. In the absence of such corrections, the results of path analyses and other causal modeling procedures are biased in complex and often unpredictable ways.

Most artifacts we are concerned with have been studied in the field of psychometrics. The goal is to develop methods of calibrating each artifact and correcting for its effects. The procedures for correcting for these artifacts can be complex, but software for applying them is available (Schmidt and Le 2004). The procedures summarized here are more fully detailed in another work, where they are presented for both the correlation coefficient and the standardized mean difference (d value statistic) (Hunter and Schmidt 2004, chapters 2, 3). This chapter focuses on applying them to correlations. The procedures described here can also be applied to other effect size statistics such as odds ratios and related risk statistics, but these applications have not been fully explicated.

17.1.1 Unsystematic Artifacts

Some artifacts produce a systematic effect on the study effect size and some cause unsystematic (random) effects. Even within a single study, it is sometimes possible to correct for a systematic effect, though it usually requires special information to do so. Unsystematic effects usually cannot be corrected in single studies and sometimes may not be correctable even at the level of meta-analysis. The two major unsystematic artifacts are sampling error and data errors.

17.1.1.1 Sampling Errors It is not possible to correct for the effect of sampling error in a single study. The confidence interval gives an idea of the potential size of the sampling error, but the magnitude of the sampling error in any one study is unknown and hence cannot be corrected. However, the effects of sampling error can be greatly reduced or eliminated in meta-analysis if the number of studies (k) is large enough to produce a large total sample size, because sampling errors are random and average out across studies. If the total sample size in the meta-analysis is not large, one can still correct for the effects of sampling error, though the correction is less precise and some smaller amount of sampling error will remain in the final meta-analysis results (see Hunter and Schmidt 2004, chapter 9). A meta-analysis that corrects only for sampling error and ignores other artifacts is often called a bare-bones meta-analysis.

17.1.1.2 Data Errors and Outliers Bad data in meta-analysis stem from a variety of errors in handling data. Primary data used in a study may be erroneous due to transcription errors, coding errors, and so on; the initial results of the analysis of the primary data in a particular study may be incorrect due to computational errors, transcriptional errors, computer program errors, and so on; the study results as published may have errors caused by transcriptional error by the investigator, by a typist, or by a printer; or a meta-analyst may miscopy a result or make a computational error. Data errors are apparently very common (Guilliksen 1986; Tukey 1960). Sometimes such errors can be detected and eliminated using outlier analysis, but outlier analysis can be problematic in meta-analysis because it is often impossible to distinguish between data errors and large sampling errors, and deletion of data with large sampling errors can bias corrections for sampling error (Hunter and Schmidt 2004, 196–97).

17.1.2 Systematic Artifacts

Many artifacts have a systematic influence on study effect size parameters and their estimates. If such an effect can be quantified, often there is an algebraic formula for the effect of the artifact. Most algebraic formulas can be inverted, producing a correction formula. The resulting correction removes the bias created by the artifact and estimates the effect size that would have been obtained had the researcher carried out a study without the corresponding methodological limitation.

Correction for an artifact requires knowledge about the size of the effect of that artifact. Correction for each new artifact usually requires at least one new piece of information. For example, to correct for the effects of random error of measurement in the dependent variable, we need to know the reliability of the dependent variable in the primary studies. Many primary studies do not present information on the artifacts in the study, but often this information (for example, scale reliability) is available from other sources. Even when artifact information is presented in the study, it is not always of the required type. For example, in correcting for the influence of measurement error, it is important to use the appropriate type of reliability coefficient. Use of an inappropriate coefficient will lead to a correction that is at least somewhat erroneous (Hunter and Schmidt 2004, 99–103), usually an undercorrection.

There are at least ten systematic artifacts that can be corrected if the artifact information is available. The correction can be made within each study individually if the information is available for all, or nearly all, studies individually. If so, then the meta-analysis is performed on these corrected values, as described in section 17.3 of this chapter. If not, the correction can be made at the meta-analysis level if the distribution of artifact values across studies can be estimated, as described in section 17.4.

As pointed out elsewhere in this volume, study effect sizes can be expressed in a variety of ways, with the two most frequently used indices being the correlation coefficient and the standardized mean difference (d value and variations thereof). For ease of explication, artifact effects and corrections are discussed in this chapter in terms of correlations. The same principles apply to standardized mean differences, though it is often more difficult to make appropriate corrections for artifacts affecting the independent variable in true experiments (see Hunter and Schmidt 2004, chapters 6, 7, and 8).

17.1.2.1 Single Artifacts Most artifacts attenuate the population correlation ρ. The amount of attenuation depends on the artifact. For each artifact, it is possible to present a conceptual definition that makes it possible to quantify the influence of the artifact on the observed effect size. For example, the reliability of the dependent variable calibrates the extent to which there is random error of measurement in the measure of the dependent variable. The reliability, and hence the artifact parameter that determines the influence of measurement error on effect size, can be empirically estimated. Journal editors should require authors to furnish those artifact values but often do not. Most of the artifacts cause a systematic attenuation of the correlation; that is, the study correlation is lower than the actual correlation by some amount. This attenuation is usually most easily expressed as a product in which the actual correlation is multiplied by an artifact multiplier, usually denoted a.

We denote the actual (unattenuated) population correlation by ρ and the (attenuated) study population correlation by ρ_o. Because we cannot conduct the study perfectly (that is, without measurement error), this study imperfection systematically biases the actual correlation parameter downward. Thus the study correlation ρ_o is smaller than the actual correlation ρ.

We denote by a_i the artifact value for the study expressed in the form of a multiplier. If the artifact parameter is expressed by a multiplier a_i, then

$$\rho_o = a_i\rho, \qquad (17.1)$$

where a_i is some fraction, $0 < a_i < 1$. The size of a_i depends on the artifact—the greater the error, the smaller the value of a_i. In the developments that follow, these artifacts are described as they occur in correlation studies. However, each artifact has a direct analogue in experimental studies (for more detail on these analogues, see Hunter and Schmidt 2004, chapters 6–8).

Attenuation artifacts and the corresponding multiplier are as follows:

1. Random error of measurement in dependent variable Y:

$$a_1 = \sqrt{r_{YY}},$$

   where r_{YY} is the reliability of the measure of Y. Example: $r_{YY} = 0.49$, implies $a_1 = 0.70$, $\rho_o = 0.70\rho$, a 30 percent reduction.

2. Random error of measurement in independent variable X:

$$a_2 = \sqrt{r_{XX}},$$

   where r_{XX} is the reliability of the measure of X. Example: $r_{XX} = 0.81$, implies $a_2 = 0.90$, $\rho_o = 0.90\rho$, a 10 percent reduction.

3. Artificial dichotomization of continuous dependent variable split into proportions p and q:

$$a_3 = \text{biserial constant} = \phi(c)/\sqrt{(pq)},$$

   where $\phi(x) = e^{-x^2/2}/\sqrt{2\pi}$ is the unit normal density function and where c is the unit normal distribution cut point corresponding to a split of p. That is, $c = \phi^{-1}(p)$, where $\phi(x)$ is the unit normal cumulative distribution function (Hunter and Schmidt 1990).
   Example: Y is split at the median, $p = q = 0.5$, $a_3 = 0.80$, $\rho_o = 0.80\rho$, a 20 percent reduction.

4. Artificial dichotomization of continuous independent variable split into proportions p and q:

$$a_4 = \text{biserial constant} = \phi(c)/\sqrt{(pq)},$$

   where c is the unit normal distribution cut point corresponding to a split of p and $\phi(x)$ is the unit normal density function. That is, $c = \phi^{-1}(p)$ (Hunter and Schmidt 1990).
   Example: X is split such that $p = 0.9$ and $q = 0.1$. Then $a_4 = 0.60$, $\rho_o = 0.60\rho$, a 40 percent reduction.

5. Imperfect construct validity of the dependent variable Y. Construct validity is the correlation of the dependent variable measure with the actual dependent variable construct:

$$a_5 = \text{the construct validity of } Y.$$

Example: supervisor ratings of job performance, $a_5 = 0.72$ (Viswesvaran, Schmidt, and Ones 1996); mean interrater reliability of supervisor rating is 0.52; the correlation between the supervisor rating and job performance true scores is $\sqrt{0.52} = 0.72$ (see also Rothstein 1990).

$$\rho_o = 0.72\rho, \text{ a 28 percent reduction.}$$

6. Imperfect construct validity of the independent variable X. Construct validity is defined as in (5) above:

$$a_6 = \text{the construct validity of } X.$$

Example: use of a perceptual speed measure to measure general cognitive ability, $a_6 = 0.65$ (the true score correlation between perceptual speed and general cognitive ability is 0.65).

$$\rho_o = 0.65\rho, \text{ a 35 percent reduction.}$$

7. Range restriction on the independent variable X. Range restriction results from systematic exclusion of certain scores on X from the sample compared with the relevant (or reference) population:

   a_7 depends on the standard deviation (SD) ratio, $u_X = (SD_X \text{ study population})/(SD_X \text{ reference population})$.

For example, the average value of u_X among employees for general cognitive ability has been found to be 0.67.

   Range restriction can be either direct, explicit truncation of the X distribution, or indirect, partial or incomplete truncation of the X distribution, usually resulting from selection on some variable correlated with X. Volunteer bias in study subjects, a form of self-selection can produce indirect range restriction. Most range restriction in real data is indirect (Hunter, Schmidt, and Le 2006). The order in which corrections are made for range restriction and measurement error differs for direct and indirect range restriction. A major development since the first edition of this handbook is the discovery of practical methods of correcting for indirect range restriction, both in individual studies and in meta-analysis (Hunter and Schmidt 2004, chapter 5; Hunter, Schmidt, and Le 2006). These methods are included in the Frank Schmidt and Huy Le (2004) software. Correcting for direct range restriction when the restriction has actually been indirect results in estimates that are downwardly biased, typically by about 25 percent.

Complication: For both types of range restriction, the size of the multiplier depends on the size of ρ.

For direct range restriction the formula is:

$$a_7 = u_X / \sqrt{(u_X^2 \rho^2 + 1 - \rho^2)}.$$

Example: For $\rho = 0.20$ and $u_X = 0.67$, $a_7 = 0.68$. $\rho_o = 0.68\rho$, a 32 percent reduction.

For indirect range restriction the formula is:

$$a_7 = u_T / \sqrt{(u_T^2 \rho^2 + 1 - \rho^2)}$$

where u_T is the ratio of restricted to unrestricted true score standard deviations (computed using either equation 3.16 in Hunter and Schmidt 2004, 106; or equation 22 in Hunter, Schmidt, and Le 2006):

$$u_T^2 = \{u_X^2 - (1 - r_{XX_a})\} / r_{XX_a}.$$

Note that in this equation, r_{XX_a} is the reliability of the independent variable X estimated in the unrestricted group. John Hunter, Schmidt, and Le used subscript a to denote values estimated in the applicant (unrestricted) group, and subscript i for the incumbent (restricted) group (2006). We use the same notation in this chapter.

Example: For $\rho = 0.20$ and $u_T = 0.56$, $a_7 = 0.57$.

$$\rho_o = 0.57\rho, \text{ a 43 percent reduction.}$$

8. Range restriction on the dependent variable Y. Range restriction results from systematic exclusion of certain scores on Y from the sample in comparison to the relevant population:

a_8 depends on the *SD* ratio,
$u_Y = (SD_Y \text{ study population})/$
$(SD_Y \text{ reference population})$.

Example: Some workers are fired early for poor performance and are hence underrepresented in the incumbent population. Assume that exactly the bottom 20 percent of performers are fired, a condition of direct range restriction. Then from normal curve calculations, $u_Y = 0.83$.

Complication: The size of the multiplier depends on the size of ρ.

$$a_8 = u_Y / \sqrt{(u_Y^2 \rho^2 + 1 - \rho^2)}.$$

Example: For $\rho = 0.20$ and $u_Y = 0.83$, $a_8 = 0.84$, $\rho_o = 0.84\rho$, a 16 percent reduction.

Note: Correction of the same correlation for range restriction on both the independent and dependent variables is complicated and requires special formulas (for a discussion of this problem, see Hunter and Schmidt 2004, 39–41).

9. Bias in the correlation coefficient. The correlation has a small negative bias:

$$a_9 = 1 - (1 - \rho^2)/(2N - 2).$$

For sample sizes of twenty or more, bias is smaller than rounding error. Thus, bias is usually trivial in size, as illustrated in the following example.

Example: For $\rho = 0.20$ and $N = 68$, $a_9 = 0.9997$, $\rho_o = 0.9997\rho$, a .03 percent reduction.

10. Study-caused variation (covariate-caused confounds).

Example: Concurrent validation studies conducted on ability tests evaluate the job performance of workers who vary in job experience, whereas applicants all start with zero job experience. Job experience correlates with job performance, even holding ability constant.

Solution: Use partial correlation to remove the effects of unwanted variation in experience.

Specific case: Study done in new plant with very low mean job experience (for example, mean = 2 years). Correlation of experience with ability is zero. Correlation of job experience with job performance is 0.50. Comparison of the partial correlation to the zero order correlation shows that $a_{10} = \sqrt{1 - .50^2} = 0.87$. $\rho_o = 0.87\rho$, a 13 percent reduction.

17.1.2.2 Multiple Artifacts Suppose that the study correlation is affected by several artifacts with parameters, $a_1, a_2, a_3, \cdots$. The first artifact reduces the actual correlation from ρ to

$$\rho_{o1} = a_1 \rho.$$

The second artifact reduces that correlation to

$$\rho_{o2} = a_2 \rho_{o1} = a_2(a_1 \rho) = a_1 a_2 \rho.$$

The third artifact reduces that correlation to

$$\rho_{o3} = a_3\rho_{o2} = a_3(a_1a_2\rho) = a_1a_2a_3\rho, \text{ and so on.}$$

Thus, the joint effect of the artifacts is to multiply the population correlation by all the multipliers. For m artifacts

$$\rho_{om} = (a_1a_2a_3 \cdots a_m)\rho$$

or

$$\rho_o = A\rho, \qquad (17.2)$$

where A is the compound artifact multiplier equal to the product of the individual artifact multipliers

$$A = a_1a_2a_3 \cdots a_m.$$

17.1.2.3 A Numerical Illustration We now illustrate the impact of some of these artifacts using actual data from a large program of studies of the validity of a personnel selection test. The quantitative impact is large. Observed correlations can be less than half the size of the estimated population correlations based on perfect measures and computed on the relevant (unrestricted reference) population. We give two illustrations: the impact on the average correlation between general cognitive ability and job performance ratings and the impact of variation in artifacts across studies.

A meta-analysis of 425 validation studies conducted by the U.S. Employment Service shows that for medium-complexity jobs, the average applicant population correlation between true scores on general cognitive ability and true scores job performance ratings is 0.73 (Hunter, Schmidt, and Le 2006).[1] We now show how this value is reduced by study artifacts to an observed mean correlation of 0.27.

The Ideal Study. Ideally, each worker would serve under a population of judges (raters), so that idiosyncrasy of judgment could be eliminated by averaging ratings across judges. Let P = consensus (average) rating of performance by a population of judges, and let A = actual cognitive ability. The correlation r_{AP} would then be computed on an extremely large sample of applicants hired at random from the applicant population. Hence, there would be no unreliability in either measure, no range restriction, and virtually no sampling error. The Hunter, Schmidt, and Le meta-analysis indicates that the obtained correlation would be 0.73 (2006) .

The Actual Study. In the actual study, the independent variable X = score on an imperfect test of general cognitive ability, and dependent variable Y = rating by one immediate supervisor. The correlation r_{XY} is computed on a small sample of range-restricted workers (incumbents) hired by the company based on information available at the time of hire.

Impact of Restriction in Range. Range restriction biases the correlation downward. All studies by the U.S. Employment Service were conducted in settings in which the General Aptitude Test Battery (GATB) had not been used to select workers, so this range restriction is indirect. When range restriction is indirect, its attenuating effects occur before the attenuating effects of measurement error occur (Hunter, Schmidt, and Le 2006). The average extent of restriction in range observed for the GATB was found to be (Hunter 1980):

$$u_X = (SD_X \text{ incumbent population})/ \\ (SD_X \text{ applicant population}) = 0.67.$$

This observed range restriction u_X translates into a true score u_T value of 0.56 based on equation 3.16 in Hunter and Schmidt (2004) or equation 22 in Hunter, Schmidt, and Le (2006) presented in the previous section (this calculation is based on the reliability of the GATB in the unrestricted population, r_{XX_i}, which is 0.81). Further using the equation provided earlier in section 17.1.2.1, we obtain the value of $a_7 = .70$, a 30 percent reduction.

Impact of Measurement Error. The value of r_{XX}, the reliability of the GATB, is 0.81 in the unrestricted population, but in the restricted group this value is reduced to 0.58 (computed using equation 27 in Hunter, Schmidt, and Le 2006: $r_{XX_i} = 1 - (1 - r_{XX_a})/u_X^2$).

Consequently, the attenuation multiplier due to measurement error in the independent variable is $a_2 = \sqrt{0.58} = 0.76$. The value of r_{YY}, the reliability of the dependent variable (supervisory ratings of job performance), is .50 in the restricted group (Viswesvaran, Ones, and Schmidt 1996), which translates into the attenuation multiplier of 0.71 ($a_1 = \sqrt{0.50} = 0.71$).

Taken together, the combined attenuating effect due to indirect range restriction and measurement error in both the independent and dependent variable measures is:

$$A = a_1a_2a_7 = (0.71)(0.76)(0.70) = 0.38.$$

Hence the expected value of the observed r is:

$$r_{XY} = Ar_{AP} = a_1a_2a_7r_{AP} \\ = (0.71)(0.76)(0.70)(0.73) = 0.27.[2]$$

The total impact of study limitations was to reduce the mean population correlation from 0.73 to 0.27, a reduction of 63 percent.

Now we present numerical examples of the effects of variation in artifacts across studies. Again, here we assume a correlation of 0.73 in the applicant population for perfectly measured variables. Even if there was no true variation in this value across studies (for example, employers), variation in artifact values would produce substantial variation in observed correlations even in the absence of sampling error.

Indirect Range Restriction. None of the firms whose data were available for the study used the GATB in hiring. Different firms used a wide variety of hiring methods, including cognitive ability tests other than the GATB. Thus, range restriction on the GATB is indirect. Suppose that the composite hiring dimension used (symbolized as S) correlated on average 0.90 with GATB true scores (that is, $\rho_{ST} = 0.90$, see Hunter, Schmidt, and Le 2006, figure 1). Further assume that some firms are very selective, accepting only the top 5 percent on their composite ($u_S = 0.37$; calculated based on the procedure presented in Schmidt, Hunter, and Urry 1976), while some others are relatively lenient, selecting half of the applicants ($u_S = 0.60$). Explicit selection on S results in range restrictions on true score T of the test (X), which effect varies from $u_T = 0.55$ (when $u_S = 0.37$) to $u_T = 0.69$ (when $u_S = 0.60$). These calculations are based on equation 18 of Hunter, Schmidt, and Le (2006):

$$u_T^2 = \rho_{ST_a}^2 u_S^2 - \rho_{ST_a}^2 + 1.$$

This creates an attenuating effect ranging from $a_7 = 0.69$ (31 percent reduction) to $a_7 = 0.82$ (18 percent reduction). The correlations would then vary from $r_{XY} = 0.51$ to $r_{XY} = 0.60$.

Predictor Reliability. Suppose that each study is conducted using either a long or a short ability test. Assume that the reliability for the long test is $r_{XX} = 0.81$ and that for the short test is 0.49 in the unrestricted population. Due to indirect range restriction, reliabilities of the tests in the samples will be reduced. The following equation (based on equations 25 and 26 in Hunter, Schmidt, and Le 2006) allows calculation of the restricted reliability (r_{XX_i}) from the unrestricted reliability (r_{XX_a}) and the range restriction ratio on T (u_T):

$$r_{XX_i} = \frac{u_T^2 r_{XX_a}}{u_T^2 r_{XX_a} + 1 - r_{XX_a}}$$

Based on the equation, observed reliabilities for the long test will vary from to 0.56 (when $u_T = 0.55$) to 0.67 (when $u_T = 0.69$); for the short test, reliabilities will be 0.22 (when $u_T = 0.55$) and 0.32 (when $u_T = 0.69$). The corresponding correlations would be:

1. High range restriction ratio, long test: $r_{XY} = \sqrt{0.67}$ $(0.60) = 0.49$,

2. High range restriction ratio, short test: $r_{XY} = \sqrt{0.32}$ $(0.60) = 0.33$,

3. Low range restriction ratio, long test: $r_{XY} = \sqrt{0.56}$ $(0.51) = 0.38$,

4. Low range restriction ratio, short test: $r_{XY} = \sqrt{0.22}$ $(0.51) = 0.24$.

Criterion reliability: one rater versus two raters. Suppose that in some studies one supervisor rates job performance and in other studies there are two raters. Inter-rater reliability is 0.50 for one rater and (by the Spearman-Browne formula) is 0.67 for ratings based on the average of two raters. Criterion reliability would then be either $r_{YY} = 0.50$ or $r_{YY} = 0.67$. Consequently, the observed correlations would be:

1. High range restriction ratio, long test, two raters: $r_{XY} = \sqrt{0.67}(0.49) = 0.40$,

2. High range restriction ratio, long test, one rater: $r_{XY} = \sqrt{0.50}(0.49) = 0.35$,

3. High range restriction ratio, short test, two raters: $r_{XY} = \sqrt{0.67}(0.33) = 0.27$,

4. High range restriction ratio, short test, one rater: $r_{XY} = \sqrt{0.50}(0.33) = 0.23$,

5. Low range restriction ratio, long test, two raters: $r_{XY} = \sqrt{0.67}(0.38) = 0.31$,

6. Low range restriction ratio, long test, one rater: $r_{XY} = \sqrt{0.50}(0.38) = 0.27$,

7. Low range restriction ratio, short test, two raters: $r_{XY} = \sqrt{0.67}(0.24) = 0.20$,

8. Low range restriction ratio, short test, one rater: $r_{XY} = \sqrt{0.50}(0.24) = 0.17$.

Thus, variation in artifacts produces variation in study population correlations. Instead of one population correlation of 0.73, we have a distribution of attenuated population correlations: 0.17, 0.20, 0.23, 0.27, 0.27, 0.31, 0.35 and 0.40. Each of these values would be the population correlation underlying a particular study. To that value random sampling error would then be added to yield the correlation observed in the study. In this example, we have assumed a single underlying population value of 0.73. If there were variation in population correlations prior to the introduction of these artifacts, that variance would be increased because the process illustrated here applies to each value of the population correlation.

17.2. CORRECTING FOR ATTENUATION-INDUCED BIASES

17.2.1 The Population Correlation: Attenuation and Disattenuation

The population correlation can be exactly corrected for the effect of any artifact. The exactness of the correction follows from the absence of sampling error. Because

$$\rho_o = A\rho,$$

we can reverse the equation algebraically to obtain

$$\rho = \rho_o/A. \quad (17.3)$$

Correcting the attenuated population correlation produces the value the correlation would have had if it had been possible to conduct the study without the methodological limitations produced by the artifacts. To divide by a fraction is to increase the value. That is, if artifacts reduce the study population correlation, then the corresponding disattenuated (corrected) correlation must be larger than the observed correlation.

17.2.2 The Sample Correlation

The sample correlation can be corrected using the same formula as for the population correlation. This eliminates the systematic error in the sample correlation, but it does not eliminate the sampling error. In fact, sampling error is increased by the correction.

The sample study correlation relates to the (attenuated) population correlation by

$$r_o = \rho_o + e, \quad (17.4)$$

where e is the sampling error in r_o (the observed correlation). To within a close approximation, the average error is zero (Hedges 1989) and the sampling error variance is

$$Var(e) = (1 - \rho_o^2)^2/(N-1). \quad (17.5)$$

The corrected sample correlation is

$$r_c = r_o/A, \quad (17.6)$$

where A is the compound artifact multiplier. The sampling error in the corrected correlation is related to the population correlation by

$$r_c = r_o/A = (\rho_o + e)/A \quad (17.7)$$
$$= (\rho_o/A) + (e/A)$$
$$= \rho + e'.$$

That is, the corrected correlation differs from the actual effect size correlation by only sampling error e', where the new sampling error e' is given by

$$e' = e/A. \quad (17.8)$$

Because A is less than 1, the sampling error e' is larger than the sampling error e. This can be seen in the sampling error variance

$$Var(e') = Var(e)/A^2. \quad (17.9)$$

However, because the average error e is essentially zero, the average error e' is also essentially zero (for a fuller discussion, see Hunter and Schmidt 2004, chapter 3).

17.3 META-ANALYSIS OF CORRECTED CORRELATIONS AND SOFTWARE

The meta-analysis methods described in this section and in section 17.4 are all based on random effects models (see chapter 16 in this volume). These procedures are implemented in the Schmidt and Le (2004) Windows-based software and have been shown in simulation studies to be accurate (for example, see Field 2005; Hall and Brannick 2002; Law, Schmidt, and Hunter 1994a; 1994b; Schulze 2004).

If study artifacts are reported for each study, then for each study, we have three numbers. For study i we have

r_i = the ith study correlation;

A_i = the compound artifact multiplier for study i; and

N_i = the sample size for study i.

We then compute for each study the disattenuated correlation: r_{ci} = the disattenuated correlation for study i. (If some of the studies do not provide some of the artifact information, the usual practice is to fill in this missing information with average values from the other studies.) Two meta-analyses can be computed: one on the biased (attenuated) study correlations (the bare-bones meta-analysis) and one on the corrected (unbiased) correlations.

17.3.1 The Mean Corrected Correlation

As both Frank Schmidt and John Hunter (1977) and Larry Hedges and Ingram Olkin (1985) have noted, large-sample studies contain more information than small-sample studies and thus should be given more weight. So studies are often weighted by sample size (or the inverse of their sampling error variance, which is nearly equivalent). For corrected correlations, Hunter and Schmidt recommended a more complicated weighting formula that takes into account the other artifact values for the study—for example, the more measurement error there is in the measures of the variables, the less the information there is in the study (2004, chapter 3). Thus, a high-reliability study should be given more weight than a low-reliability study. Hedges and Olkin advocated a similar weighting procedure (1985, 135–36).

Hunter and Schmidt (2004) recommended that the weight for study i should be

$$w_i = N_i A_i^2, \qquad (17.10)$$

where A_i is the compound artifact multiplier for study i.

The average correlation can be written

$$Ave(r) = \Sigma w_i r_i / \Sigma w_i, \qquad (17.11)$$

where

$w_i = 1$ for the unweighted average;

$w_i = N_i$ for the sample size weighted average; and

$w_i = N_i A_i^2$ for the full artifact weighted average (applied to corrected correlations).

If the number of studies were infinite (so that sampling error would be completely eliminated), the resulting mean would be the same regardless of which weights were used. But for a finite number of studies, meta-analysis does not totally eliminate sampling error; there is still some sampling error left in the mean correlation. Use of the full artifact weights described here minimizes sampling error in the mean corrected correlation.

17.3.2 Corrected Versus Uncorrected Correlations

In some research domains, the artifact values for most individual studies are not presented in those studies. As a result, some published meta-analyses do not correct for artifacts. Failure to correct means that the mean uncorrected correlation will be downwardly biased as an estimate of the actual (unattenuated or construct level) correlation. The amount of bias in a meta-analysis of uncorrected correlations will depend on the extent of error caused by artifacts in the average study. This average extent of systematic error is measured by the average compound multiplier $Ave(A)$.

To a close statistical approximation, the mean corrected correlation $Ave(r_c)$ relates to the mean uncorrected correlation $Ave(r)$ in much the same way as does an individual corrected correlation. Just as for a single study

$$r_c = r/A, \qquad (17.12)$$

so to a close approximation we have for a set of studies

$$Ave(r_c) = Ave(r)/Ave(A). \qquad (17.13)$$

Thus, to a close approximation, the difference in findings of an analysis that does not correct for artifacts and one that does is the difference between the uncorrected mean correlation $Ave(r)$ and the corrected mean correlation $Ave(r_c)$ (Hunter and Schmidt 2004, chapter 4).

17.3.3 Variance of Corrected Correlations: Procedure

The variance of observed correlations greatly overstates the variance of population correlations. This is true for corrected correlations as well as for uncorrected correlations. From the fact that the corrected correlation is

$$r_{ci} = \rho_i + e_i', \qquad (17.14)$$

where r_{ci} and ρ_i are the corrected sample and population correlations, respectively, and e_i' is the sampling error, we have the decomposition of variance

$$Var(r_c) = Var(\rho) + Var(e').$$

Thus, by subtraction, we have an estimate of the desired variance

$$Var(\rho) = Var(r_c) - Var(e'). \qquad (17.15)$$

The variance of study corrected correlations is the weighted squared deviation of the ith correlation from the mean correlation. If we denote the average corrected correlation by $\bar{r}_c$ then

$$\bar{r}_c = Ave(r_c) = \sum w_i r_{ci} / \sum w_i, \qquad (17.16)$$

$$Var(r_c) = \sum w_i (r_{ci} - \bar{r}_c)^2 / \sum w_i. \qquad (17.17)$$

The sampling error variance is computed by averaging the sampling error variances of the individual studies. The error variance of the individual study depends on the size of the uncorrected population correlation. To estimate that number, we first compute the average uncorrected correlation $\bar{r}$.

$$\bar{r} = Ave(r) = \sum w_i r_i / \sum w_i$$

where the w_i are the sample sizes N_i.

For study i, we have

$$Var(e_i) = v_i = (1 - \bar{r}^2)^2 / (N_i - 1), \qquad (17.18)$$

and

$$Var(e_i') = v_i' = Var(e_i) / A_i^2 = v_i / A_i^2.$$

For simplicity, denote the study sampling error variance $Var(e_i')$ by v_i'. The weighted average error variance for the meta-analysis is the average

$$Var(e') = \sum w_i v_i' / \sum w_i \qquad (17.19)$$

where the $w_i = N_i A_i^2$.

Procedure. The specific computational procedure involves six steps:

1. Given for each study, r_i = uncorrected correlation, A_i = compound artifact multiplier, and N_i = sample size.

2. Compute for each study, r_{ci} = corrected correlation, and w_i = the proper weight to be given to r_{ci}.

3. To estimate the effect of sampling error, compute the average uncorrected correlation $\bar{r}$. This is done using weights $w_i = N_i$.

4. For each study compute the sampling error variance: v_i' = the sampling error variance.

5. The meta-analysis of disattenuated correlations has four steps:
   a. Compute the mean corrected correlation using weights $w_i = N_i A_i^2$: Mean corrected correlation = $Ave(r_c)$.
   b. Compute the variance of corrected correlations using $w_i = N_i A_i^2$: Variance of corrected correlations = $Var(r_c)$.
   c. Compute the sampling error variance $Var(e')$ by averaging the individual study sampling error variances:

$$Var(e') = Ave(v_i'). \qquad (17.20)$$

   d. Now compute the estimate of the variance of population correlations by correcting for sampling error: $Var(\rho) = Var(r_c) - Var(e')$.

6. The final fundamental estimates (the average and the standard deviation (SD) of ρ) are

$$Ave(\rho) = Ave(r_c) \qquad (17.21)$$

and

$$SD_\rho = \sqrt{Var(\rho)}. \qquad (17.22)$$

As mentioned earlier, software is available for conducting these calculations (Schmidt and Le 2004). A procedure similar to the one described here has been presented by Nambury Raju and his colleagues (1991; for simplified examples of application of this approach to meta-analysis see Hunter and Schmidt 2004, chapter 3; for published meta-analyses of this sort, see Carlson et al. 1999; Judge et al. 2001; and Rothstein et al. 1990).

As an example, in the Kevin Carlson et al. study (1999), a previously developed weighted biodata form was correlated with promotion or advancement (with years of experience controlled) for 7,334 managers in twenty-four organizations. Thus, there were twenty-four studies, with a mean N per study of 306. The reliability of the dependent variable—rate of advancement or promotion rate—was estimated at 0.90. The standard deviation (SD) of the independent variable was computed in each organization, and the SD of applicants (that is, the unrestricted SD) was known, allowing each correlation to be corrected for range variation. In this meta-analysis, there was no between—studies variation in correlations due to variation in measurement error in the independent variable-because the same biodata scale was used in all twenty-four studies. Also, in this meta-analysis, the interest was in the effectiveness of this particular biodata scale in predicting managerial advancement. Hence, mean ρ was not corrected for unreliability in the independent variable.

The results were as follows:

$$Ave(r_i) = 0.48,$$

$$Ave(\rho) = 0.53,$$

$$Var(\rho) = Var(r_c) - Var(e')$$

$$= 0.00462 - 0.00230 = 0.00232$$

and

$$SD_\rho = \sqrt{Var(\rho)} = 0.048.$$

Thus, the mean operational validity of this scale across organization was estimated as 0.53, with a standard deviation of 0.048. After correcting for measurement error, range variation, and sampling error, there is only a small apparent variability in correlation across organizations. If we assume a normal distribution for ρ, the value at the tenth percentile is $0.53 - 1.28 \times 0.048 = 0.47$. Thus, the conclusion is that the correlation is at least 0.47 in 90 percent of these (and comparable) organizations. The value at the ninetieth percentile is 0.59, yielding an 80 percent credibility interval of 0.47 to 0.59, indicating that an estimated 80 percent of population values of validity lie in this range. Although the computation procedures used are not described in this chapter, confidence intervals can be placed around the mean validity estimate. In this case the 95 percent confidence interval for the mean is 0.51 to 0.55. The reader should note that confidence intervals and credibility intervals are different and serve different purposes (Hunter and Schmidt 2004, 205–7). Confidence intervals refer only to the estimate of the mean, while credibility intervals are based on the estimated distribution of all of the population correlations. Hence confidence intervals are based on the estimated standard error of the mean, and credibility intervals are based on the estimated standard deviation of the population correlations.

17.4 ARTIFACT DISTRIBUTION META-ANALYSIS AND SOFTWARE

In most contemporary research domains, the artifact values are not provided in many of the studies. Instead, artifact values are presented only in a subset of studies, usually a different but overlapping subset for each individual artifact. Meta-analysis can be conducted in such domains, though the procedures are more complex.

Simplified examples of application of artifact distribution meta-analysis can be found in Hunter and Schmidt (2004, chapter 4). Many published meta-analyses have been based on these artifact distribution meta-analysis methods (see Hunter and Schmidt 2004, chapters 1 and 4). A subset of these have been conducted on correlation coefficients representing the validities of various kinds of predictors of job performance—usually tests, but also interviews, ratings of education and job experience, assessment centers, and others. The implications of the findings of these meta-analyses for personnel selection practices have been quite profound (see Schmidt, Hunter, and Pearlman 1980; Pearlman, Schmidt, and Hunter 1980; Schmidt and Hunter 1981; Schmidt et al. 1985; Schmidt, Hunter, and Raju 1988; Schmidt, Ones, and Hunter 1992; Schmidt and Hunter 1998, 2003; Schmidt, Oh, and Le 2006; McDaniel, Schmidt, and Hunter 1988; Le and Schmidt 2006; McDaniel et al. 1994). Artifact distribution meta-analyses have also been conducted in a variety of other research areas, such as role conflict, leadership, effects of goal setting, and work-family conflict. More than 100 such nonselection meta-analyses have appeared in the literature to date.

The artifact distribution meta-analysis procedures described in this section are implemented in the Schmidt and Le software (2004), though the methods used in these programs to estimate the standard deviation of the population corrected correlations (discussed in section 17.4.2) are slightly different from those described here. Although the methods used in the Schmidt-Le programs for this purpose are slightly more accurate, as shown in simulation studies, than those described in the remainder of this chapter (Hunter and Schmidt 2004, chapter 4), they are also much more complex—in fact, too complex to describe easily in a chapter of this sort. Similar artifact distribution methods for meta-analysis have been presented by John Callender and Hobart Osburn (1980) and Nambury Raju and Michael Burke (1983). In all these methods, the key assumption in considering artifact distributions is independence of artifact values across artifacts. This assumption is plausible for the known artifacts in research domains that have been examined. The basis for independence is the fact that the resource limitations that produce problems with one artifact (for example, range restriction) are generally different and hence independent of those that produce problems with another (for example, measurement error in scales used) (Hunter and Schmidt

2004, chapter 4). Artifact values are also assumed to be independent of the true score correlation ρ_i. In a computer simulation study, Nambury Raju and his colleagues found that violation of these independence assumptions has minimal effect on meta-analysis results unless the artifacts are correlated with the ρ_i, a seemingly unlikely event (1998).

We use the following notation for the correlations associated with the ith study:

ρ_i = the true (unattenuated) study population correlation;

r_{ci} = the study sample corrected correlation that can be computed if artifact information is available for the study so that corrections can be made;

r_{oi} = uncorrected (observed) study sample correlation; and

ρ_{oi} = uncorrected (attenuated) study population correlation.

In the previous section (section 3), we assumed that artifact information is available for every (or nearly every) study individually. Thus, an estimate r_{ci} of the true correlation ρ_i can be computed for each study and meta-analysis can be conducted on these estimates. In this section, we assume that artifact information is missing for many or most studies. However, we assume that the distribution (or at least the mean and variance) of artifact values can be estimated for each artifact. The meta-analysis then proceeds in two steps:

- A bare bones meta-analysis is conducted, yielding estimates of the mean and standard deviation of attenuated study population correlations. A bare bones meta-analysis is one that corrects only for sampling error.

- The mean and standard deviation from the bare bones meta-analysis are then corrected for the effects of artifacts other than sampling error.

17.4.1 Mean of the Corrected Correlations

The attenuated study population correlation ρ_{oi} is related to the actual study population correlation ρ_i by the formula

$$\rho_{oi} = A_i \rho_i,$$

where A_i = the compound artifact multiplier for study i (which is unknown for most studies).

The sample attenuated correlation r_{oi} for each study is related to the attenuated population correlation for that study by

$$r_{oi} = \rho_{oi} + e_{oi},$$

where e_{oi} = the sampling error in study i (which is unknown).

17.4.1.1 Meta-Analysis of Attenuated Correlations
The meta-analysis uses the additivity of means to produce

$$Ave(r_{oi}) = Ave(\rho_{oi} + e_{oi}) = Ave(\rho_{oi}) + Ave(e_{oi}).$$

If the number of studies is large, the average sampling error will tend to zero and hence

$$Ave(r_{oi}) = Ave(\rho_{oi}) + 0 = Ave(\rho_{oi}).$$

Thus, the bare-bones estimate of the mean attenuated study population correlation is the expected mean attenuated study sample correlation.

17.4.1.2 Correcting the Mean Correlation
The attenuated population correlation for study i is related to the disattenuated correlation for study i by $\rho_{oi} = A_i \rho_i$, where A_i = the compound artifact multiplier for study i.

Thus, the mean attenuated correlation is given by

$$Ave(\rho_{oi}) = Ave(A_i \rho_i). \qquad (17.23)$$

Because we assume that artifact values are independent of the size of the true correlation, the average of the product is the product of the averages:

$$Ave(A_i \rho_i) = Ave(A_i) Ave(\rho_i). \qquad (17.24)$$

Hence, the average attenuated correlation is related to the average disattenuated correlation by $Ave(\rho_{oi}) = Ave(A_i) Ave(\rho_i)$, where $Ave(A_i)$ = the *average* compound multiplier across studies.

Note that we need not know all the individual study artifact multipliers; we need only know the average. If the average multiplier is known, then the corrected mean correlation is

$$Ave(\rho_i) = Ave(r_{oi})/Ave(A_i). \qquad (17.25)$$

17.4.1.3 The Mean Compound Multiplier
To estimate the average compound multiplier, it is sufficient to

be able to estimate the average for each single artifact multiplier separately. This follows from the independence of artifacts. To avoid double subscripts, let us denote the separate artifact multipliers by $a, b, c, \ldots$ The compound multiplier A is then given by the product

$$A = abc \ldots$$

Because of the independence of artifacts, the average product is the product of averages:

$$Ave(A_i) = Ave(a)Ave(b)Ave(c) \ldots \quad (17.26)$$

Thus, the steps in estimating the compound multiplier are as follows:

1. Consider the separate artifacts.
   a. Consider the first artifact a. For each study that includes a measurement of the artifact magnitude, denote the value a_i. Average those values to produce $Ave(a_i)$ = average of attenuation multiplier for first artifact.
   b. Consider the second artifact b. For each study that includes a measurement of the artifact magnitude, denote the value b_i. Average those values to produce $Ave(b_i)$ = average of attenuation multiplier for second artifact.
   c. Similarly, consider the other separate artifacts c, d, and so on, that produce estimates of the averages $Ave(c_i)$, $Ave(d_i)$ . . . .

The accuracy of these averages depends on the assumption that the available artifacts are a reasonably representative sample of all artifacts (for a discussion of this assumption, see Hunter and Schmidt 2004). Note that even if this assumption is not fully met, results will still tend to be more accurate than those from a meta-analysis that does not correct for artifacts.

2. Compute the product

$$Ave(A_i) = Ave(a)Ave(b)Ave(c)Ave(d) \ldots \quad (17.27)$$

17.4.2 Correcting the Standard Deviation

The method described in this section for estimating the standard deviation of the population true score correlations is the multiplicative method (Hunter and Schmidt 2004, chapter 4). This approach was first introduced by

Callender and Osburn (1980) and is the least complicated method to present. Other methods have also been proposed and used, however. Nambury Raju and Burke (1983) and John Hunter, Schmidt, and Le (2006) have presented Taylor Series-based methods for use under conditions of direct and indirect range restriction, respectively. Still another procedure is the interactive procedure (Schmidt, Gast-Rosenberg, and Hunter 1980; Hunter and Schmidt 2004, chapter 4); simulation studies have suggested that the interactive procedure (with certain refinements) is slightly more accurate than other procedures (Law, Schmidt, and Hunter 1994a, 1993b; Le and Schmidt 2006), and so it is the procedure incorporated in the Schmidt and Le software (2004). However, by the usual standards of social science research, all three procedures are accurate.

The meta-analysis of uncorrected correlations provides an estimate of the variance of attenuated study population correlations. However, these study population correlations are themselves uncorrected; that is, they are biased downward by the effects of artifacts. Furthermore, the variation in artifact levels across studies causes the study correlations to be attenuated by different amounts in different studies. This produces variation in the size of the study correlations that could be mistaken for variation due to a real moderator variable, as we saw in our numerical example in section 17.1.2.3. Thus the variance of population study correlations computed from a meta-analysis of uncorrected correlations is affected in two different ways. The systematic artifact-induced reduction in the magnitude of the study correlations tends to decrease variability, while at the same time variation in artifact magnitude tends to increase variability across studies. Both sources of influence must be removed to accurately estimate the standard deviation of the disattenuated correlations across studies.

Let us begin with notation. The study correlation free of study artifacts (the disattenuated correlation) is denoted ρ_i, and the compound artifact attenuation factor for study i is denoted A_i. The attenuated study correlation, ρ_{oi} is computed from the disattenuated study correlation by

$$\rho_{oi} = A_i \rho_i. \quad (17.28)$$

The study sample correlation r_{oi} departs from the disattenuated study population correlation ρ_{oi} by sampling error e_i defined by

$$r_{oi} = \rho_{oi} + e_i = A_i \rho_i + e. \quad (17.29)$$

Consider now a bare-bones meta-analysis on the un-corrected correlations. We know that the variance of sample correlations is the variance of population correlations added to the sampling error variance. That is,

$$Var(r_{oi}) = Var(\rho_{oi}) + Var(e_i). \qquad (17.30)$$

Because the sampling error variance can be computed by statistical formula, we can subtract it to yield

$$Var(\rho_{oi}) = Var(r_{oi}) - Var(e_i). \qquad (17.31)$$

That is, the meta-analysis of uncorrected correlations produces an estimate of the variance of attenuated study population correlations, the actual study correlations after they have been reduced in magnitude by the study imperfections.

At the end of a meta-analysis of uncorrected (attenuated) correlations, we have the variance of attenuated study population correlations $Var(\rho_{oi})$, but we want the variance of actual disattenuated correlations $Var(\rho_i)$. The relationship between them is

$$Var(\rho_{oi}) = Var(A_i\rho_i). \qquad (17.32)$$

We assume that A_i and ρ_i are independent. A formula for the variance of this product is given in Hunter and Schmidt (2004, chapter 4). Here we simply use this formula. Let us denote the average disattenuated study correlation by $\bar{\rho}$ and denote the average compound attenuation factor by $\bar{A}$. The variance of the attenuated correlations is given by

$$Var(A_i\rho_i) = \bar{A}^2 Var(\rho_i) + \bar{\rho}^2 Var(A_i)$$
$$+ Var(\rho_i)Var(A_i). \qquad (17.33)$$

Because the third term on the right is negligibly small, to a close approximation

$$Var(A_i\rho_i) = \bar{A}^2 Var(\rho_i) + \bar{\rho}^2 Var(A_i). \qquad (17.34)$$

We can then rearrange this equation algebraically to obtain the desired equation for the variance of actual study correlations free of artifact effects:

$$Var(\rho_i) = [Var(A_i\rho_i) - \bar{\rho}^2 Var(A_i)]/\bar{A}^2. \quad (17.35)$$

That is, starting from the meta-analysis of uncorrected correlations, we have

$$Var(\rho_i) = [Var(\rho_{oi}) - \bar{\rho}^2 Var(A_i)]/\bar{A}^2. \quad (17.36)$$

The right-hand side of equation 17.36 has four numbers:

$Var(\rho_{oi})$: the population correlation variance from the meta-analysis of uncorrected correlations, estimated using equation 17.31;

$\bar{\rho}$: the mean of disattenuated study population correlations, estimated using equation 17.25;

$\bar{A}$: the mean compound attenuation factor, estimated from equation 17.26; and

$Var(A_i)$: the variance of the compound attenuation factor. This quantity has not yet been estimated.

17.4.2.1 Variance of the Artifact Multiplier How do we compute the variance of the compound attenuation factor, A_i? We are given the distribution of each component attenuation factor, which must be combined to produce the variance of the compound attenuation factor. The key to this computation lies in two facts: (a) that the compound attenuation factor is the product of the component attenuation factors and (b) that the attenuation factors are assumed to be independent. That is, because $A_i = a_i b_i c_i \ldots$, the variance of A_i is

$$Var(A_i) = Var(a_i b_i c_i \ldots). \qquad (17.37)$$

The variance of the compound attenuation factor is the variance of the product of independent component attenuation factors. The formula for the variance of this product is derived in Hunter and Schmidt (2004, chapter 4, 148–49). Here we simply use the result.

For each separate artifact, we have a mean and a standard deviation for that component attenuation factor. From the mean and the standard deviation, we can compute the coefficient of variation, which is the standard deviation divided by the mean. For our purposes here, we need a symbol for the squared coefficient of variation:

$$cv = [SD/Mean]^2. \qquad (17.38)$$

For each artifact, we now compute the squared coefficient of variation. For the first artifact attenuation factor a, we compute

$$cv_1 = Var(a)/[Ave(a)]^2. \qquad (17.39)$$

For the second artifact attenuation factor b, we compute

$$cv_2 = Var(b)/[Ave(b)]^2. \qquad (17.40)$$

For the third artifact attenuation factor c, we compute

$$cv_3 = Var(c)/[Ave(c)]^2, \qquad (17.41)$$

and so on. Thus, we compute a squared coefficient of variation for each artifact. These are then summed to form a total

$$CVT = cv_1 + cv_2 + cv_3 + \ldots \qquad (17.42)$$

Recalling that $\bar{A}$ denotes the mean compound attenuation factor, we write the formula for the variance of the compound attenuation factor (to a close statistical approximation) as the product

$$Var(A_i) = \bar{A}^2 CVT. \qquad (17.43)$$

We now have all the elements needed to estimate the variance in actual study correlations $Var(\rho_i)$. The final formula is

$$Var(\rho_i) = [Var(\rho_{oi}) - \bar{\rho}^2 Var(A_i)]/\bar{A}^2$$
$$= [Var(\rho_{oi}) - \bar{\rho}^2 \bar{A}^2 CVT]/\bar{A}^2. \qquad (17.44)$$

The square root of this value is the SD_ρ. Hence we now have two main results of the meta-analysis: the mean of the corrected correlations, from equation 17.25; and the standard deviation of the correction correlations, from equation 17.44. Using these values, we can again compute credibility intervals around the mean corrected correlation, as illustrated at the end of section 17.3.3. Also, using the methods described in Hunter and Schmidt (2004, 205–7) we can compute confidence intervals around the mean corrected correlation.

For data sets for which the meta-analysis methods described in section 17.3 can be applied (that is, correlations can be corrected individually), the artifact distribution methods described in this section can also be applied. When this is done, the results are virtually identical, as expected. Computer simulation studies also indicate that artifact distribution meta-analysis methods are quite accurate (for example, see Le and Schmidt 2006).

17.4.2.2 Decomposition of the Variance of Correlations Inherent in the derivation in section 17.4.2.1 is a decomposition of the variance of uncorrected (observed) correlations. We now present that decomposition:

$$Var(r_{oi}) = Var(\rho_{oi}) + Var(e_i),$$
$$Var(\rho_{oi}) = Var(A_i \rho_i) = \bar{A}^2 Var(\rho_i) + \bar{\rho}^2 Var(A_i), \quad (17.45)$$

and

$$Var(A_i) = \bar{A}^2 CVT. \qquad (17.46)$$

That is,

$$Var(r_{oi}) = \bar{A}^2 Var(\rho_i) + \bar{\rho}^2 \bar{A}^2 CVT + Var(e_i) \qquad (17.47)$$
$$= S1 + S2 + S3.$$

where $S1$ is the variance in uncorrected correlations produced by the variation in actual unattenuated effect size correlations, $S2$ is the variance in uncorrected correlations produced by the variation in artifact levels and $S3$ is the variance in uncorrected correlations produced by sampling error.

In this decomposition, the term $S1$ contains the estimated variance of effect size correlations. This estimated variance is corrected for those artifacts that were corrected in the meta-analysis. For reasons of feasibility, this does not usually include all the artifacts that affect the study value. Thus, $S1$ is an upper-bound estimate of the component of variance in uncorrected correlations due to real variation in the strength of the relationship and not due to artifacts of the study design. To the extent that there are uncorrected artifacts, $S1$ will overestimate the real variation; it may even greatly overestimate that variation. As long as there are uncorrected artifacts, there will be artifactual variation in study correlations produced by variation in those uncorrected artifacts.

17.5 SUMMARY

The methods presented in this chapter for correcting meta-analysis results for sampling error and biases in individual study results might appear complicated, but the reader should be aware that more elaborated and extended descriptions of these procedures are presented in Hunter and Schmidt (2004) and that Windows-based software (Schmidt and Le 2004) is available to apply these methods in meta-analysis. It is also important to bear in mind that without these corrections, for biases and sampling error the results of meta-analysis do not estimate the construct-level population relationships that are of greatest scientific and theoretical interest (Rubin 1990; Schmidt and Hunter 2004, chapters 1, 14). Hence, in meta-analysis, these corrections are essential to the development of valid cumulative knowledge about relations between constructs. This is especially important in light of recent developments concerning the use of meta-analytic results in the testing of causal models. Meta-analytic results are increasingly being used as input to path analyses and other causal modeling techniques. Without appropriate corrections for measurement error, range restriction, and other artifacts that bias meta-analysis results, these causal modeling procedures will produce erroneous results.

17.6 NOTES

1. Table 3 in Hunter, Schmidt, and Le (2006) shows the value of operational validity for medium complexity job is 0.66, which was obtained from the true score correlation of 0.73.
2. The value here is slightly different from .26 shown in Hunter, Schmidt, and Le (2006) due to rounding of values beyond the second decimal place.

17.7 REFERENCES

Becker, Betsy, and Christine Schram. 1994. "Examining Explanatory Models Through Research Synthesis." In *The Handbook of Research Synthesis*, edited by Harris Cooper and Larry Hedges. New York: Russell Sage Foundation.

Callender, John, and Hobart Osburn. 1980. "Development and Test of a New Model for Validity Generalization." *Journal of Applied Psychology* 65(5): 543–58.

Carlson, Kevin, Steve Scullen, Frank Schmidt, Hannah Rothstein, and Frank Erwin. 1999. "Generalizable Biographical Data Validity: Is Multi-Organizational Development and Keying Necessary?" *Personnel Psychology* 52(3): 731–56.

Colquitt, Jason, Jeffrey LePine, and Raymond Noe. 2002. "Toward an Integrative Theory of Training Motivation: A Meta-Analytic Path Analysis of 20 Years of Research." *Journal of Applied Psychology* 85(5): 678–707.

Cook, Thomas, Harris Cooper, David Cordray, Heidi Hartman, Larry Hedges, Richard Light, Thomas Louis, and Fredrick Mosteller. 1992. *Meta-Analysis for Explanation: A Casebook*. New York: Russell Sage Foundation.

Field, Andy. 2005. "Is the Meta-Analysis of Correlation Coefficients Accurate when Population Correlations Vary?" *Psychological Methods* 10(4): 444–67.

Gulliksen, Harold. 1986. "The Increasing Importance of Mathematics in Psychological Research (Part 3)." *The Score* 9: 1–5.

Hall, Steven, and Michael Brannick. 2002. "Comparison of Two Random Effects Methods of Meta-Analysis." *Journal of Applied Psychology* 87(2): 377–89.

Hedges, Larry V. 1989. "An Unbiased Correction for Sampling Error in Validity Generalization Studies." *Journal of Applied Psychology* 74(3): 469–77.

Hedges, Larry, and Ingram Olkin. 1985. *Statistical Methods for Meta-Analysis*. Orlando, Fl.: Academic Press.

Hunter, John. 1980. *Validity Generalization for 12,000 Jobs: An Application of Synthetic Validity and Validity Generalization to the General Aptitude Test Battery (GATB)*. Washington, D.C.: U.S. Department of Labor.

Hunter, John, and Frank Schmidt. 1990. "Dichotomization of Continuous Variables: The Implications for Meta-Analysis." *Journal of Applied Psychology* 75(3): 334–49.

———. 2004. *Methods of Meta-Analysis: Correcting Error and Bias in Research Findings*, 2nd ed. Thousand Oaks, Calif.: Sage Publications.

Hunter, John, Frank Schmidt, and Huy Le. 2006. "Implications of Direct and Indirect Range Restriction for Meta-Analysis Methods and Findings." *Journal of Applied Psychology* 91(3): 594–612.

Judge, Timothy, Carl Thoresen, Joyce Bono, and Gregory Patton. 2001. "The Job Satisfaction-Job Performance Relationship: A Qualitative and Quantitative Review." *Psychological Bulletin* 127(3): 376–407.

Law, Kenneth, Frank Schmidt, and John Hunter. 1994a. "Nonlinearity of Range Corrections in Meta-Analysis: A Test of an Improved Procedure." *Journal of Applied Psychology* 79(3): 425–38.

Law, Kenneth, Frank Schmidt, and John Hunter 1994b. "A Test of Two Refinements in Meta-Analysis Procedures." *Journal of Applied Psychology* 79(6): 978–86.

Le, Huy, and Frank Schmidt 2006. "Correcting for Indirect Range Restriction in Meta-Analysis: Testing a New Analytic Procedure." *Psychological Methods* 11(4): 416–38.

McDaniel, Michael, Debra Whetzel, Frank Schmidt, and Steven Maurer. 1994. "The Validity of Employment Interviews: A Comprehensive Review and Meta-Analysis." *Journal of Applied Psychology* 79(4): 599–616.

McDaniel, Michael, Frank Schmidt, and John Hunter 1988. "A Meta-Analysis of the Validity of Methods for Rating Training and Experience in Personnel Selection." *Personnel Psychology* 41(2): 283–314.

Pearlman, Kenneth, Frank Schmidt, and John Hunter 1980. "Validity Generalization Results for Tests Used to Predict Job Proficiency and Training Criteria in Clerical Occupations." *Journal of Applied Psychology* 65(4): 373–407.

Raju, Nambury, and Michael Burke. 1983. "Two New Procedures for Studying Validity Generalization." *Journal of Applied Psychology* 68(3): 382–95.

Raju, Nambury, Michael Burke, Jacques Normand, and George Langlois. 1991. "A New Meta-Analysis Approach." *Journal of Applied Psychology* 76(3): 432–46.

Raju, Nambury, Tobin Anselmi, Jodi Goodman, and Adrian Thomas. 1998. "The Effects of Correlated Artifacts and True Validity on the Accuracy of Parameter Estimation in Validity Generalization." *Personnel Psychology* 51(2): 453–65.

Rothstein, Hannah, Frank Schmidt, Frank Erwin, William Owens, and Paul Sparks. 1990. "Biographical Data in Employment Selection: Can Validities Be Made Generalizable?" *Journal of Applied Psychology* 75(2): 175–84.

Rothstein, Hannah. 1990. "Interrater Reliability of Job Performance Ratings: Growth to Asymptote with Increasing Opportunity to Observe." *Journal of Applied Psychology* 75(3): 322–27.

Rubin, Donald B. 1990. "A New Perspective on Meta-Analysis." In *The Future of Meta-Analysis*, edited by Kenneth Wachter and Miron Straf. New York: Russell Sage Foundation.

Schmidt, Frank. 1992. "What Do Data Really Mean? Research Findings, Meta-Analysis, and Cumulative Knowledge in Psychology." *American Psychologist* 47(10): 1173–81.

Schmidt, Frank, and Huy Le. 2004. *Software for the Hunter-Schmidt Meta-Analysis Methods*. Iowa City: University of Iowa, Department of Management and Organizations.

Schmidt, Frank, and John Hunter. 1977. "Development of a General Solution to the Problem of Validity Generalization." *Journal of Applied Psychology* 62(5): 529–40.

———. 1981. "Employment Testing: Old Theories and New Research Findings." *American Psychologist* 36(10): 1128–37.

———. 1998. "The Validity and Utility of Selection Methods in Personnel Psychology: Practical and Theoretical Implications of 85 Years of Research Findings." *Psychological Bulletin* 124(2): 262–74.

———. 2003. "History, Development, Evolution, and Impact of Validity Generalization and Meta-Analysis Methods 1975–2001." In *Validity Generalization: A Critical Review*, edited by Kevin Murphy. Mahwah, N.J.: Lawrence Erlbaum.

Schmidt, Frank, Deniz Ones, and John Hunter. 1992. "Personnel Selection." *Annual Review of Psychology* 43: 627–70.

Schmidt, Frank, Ilene Gast-Rosenberg, and John Hunter 1980. "Validity Generalization Results for Computer Programmers." *Journal of Applied Psychology* 65(6): 643–61.

Schmidt, Frank, In-Sue Oh, and Huy Le. 2006. "Increasing the Accuracy of Corrections for Range Restriction: Implications for Selection Procedure Validities and Other Research Results." *Personnel Psychology* 59(2): 281–305.

Schmidt, Frank, John Hunter, and Kenneth Pearlman. 1980. "Task Difference and Validity of Aptitude Tests in selection: A Red Herring." *Journal of Applied Psychology* 66(2): 166–85.

Schmidt, Frank, John Hunter, and Nambury Raju. 1988. "Validity Generalization and Situational Specificity: A Second Look at the 75% Rule and the Fisher *z* Transformation." *Journal of Applied Psychology* 73(4): 665–72.

Schmidt, Frank, John Hunter, Kenneth Pearlman, and Hannah Hirsh. 1985. "Forty Questions about Validity Generalization and Meta-Analysis." *Personnel Psychology* 38(4): 697–798.

Schulze, Ralf. 2004. *Meta-Analysis: A Comparison of Approaches*. Cambridge, Mass.: Hogrefe and Huber.

Tukey, John. 1960. "A Survey of Sampling from Contaminated Distributions." In *Contributions to Probability and Statistics*, edited by Ingram Olkin. Stanford, Calif.: Stanford University Press.

Viswesvaran, Chockalingam, Deniz Ones, and Frank Schmidt. 1996. "Comparative Analysis of the Reliability of Job Performance Ratings." *Journal of Applied Psychology* 81(5): 557–74.

PART
VII

SPECIAL STATISTICAL ISSUES AND PROBLEMS

18

EFFECT SIZES IN NESTED DESIGNS

LARRY V. HEDGES
Northwestern University

CONTENTS

18.1 INTRODUCTION

Studies with nested designs are frequently used to evaluate the effects of social treatments, such as interventions, products, or technologies in education or public health. One common nested design assigns entire sites—often classrooms, schools, clinics, or communities—to the same treatment group, with different sites assigned to different treatments. Experiments with designs of this type are also called *group randomized* or *cluster randomized* designs because sites such as schools or communities correspond to statistical clusters. In experimental design terminology, these designs are designs involving clusters as nested factors. Nested factors are groupings of individuals that occur only within one treatment condition, such as schools or communities in designs that assign whole schools or communities to treatments.

Assigning entire groups to receive the same treatments is sometimes done for the convenience of the investigator because it is easier to provide the same treatment to all individuals in a group. In other situations, group assignment is used to prevent contamination of one treatment group by elements of treatments assigned to others. In this case, physical separation of groups (for example, different schools or communities) minimizes contamination. In other situations, the treatment is an administrative or group-based manipulation, making it conceptually difficult to imagine how the treatment could be assigned at an individual level. For example, treatments that involve changing teacher behavior cannot easily be assigned to individual students, nor can administrative or procedural reforms in a school or clinic easily be assigned to individuals within a school or clinic.

Although the analysis of cluster randomized designs has received considerable attention in the statistical literature (for example, Donner and Klar 2000; Raudenbush and Bryk 2002), the problem of representation of the results of cluster randomized trials (and the corresponding quasi-experiments) in the form of effect sizes and combining them across studies in meta-analyses have received less attention. Brenda Rooney and David Murray called attention to the problem of effect-size estimation in cluster randomized trials and suggested that neither the conventional estimates of effect sizes nor the formulas for their standard errors were appropriate in this situation (1996). In particular, standard errors of effect-size estimates will be profoundly underestimated if conventional formulas are used. Allan Donner and Neil Klar suggested that corrections for the effects of clustering should be introduced in meta-analyses of cluster randomized experiments (2002). Larry Hedges studied the problem of effect sizes in cluster randomized designs and computed their sampling distributions, including their standard errors (2007, in press).

The standardized mean difference, which is widely used as an effect-size index in educational and psychological research, is defined as the ratio of a difference between treatment and control group means to a standard deviation. In designs where there is no nesting, that is, where there is no statistical clustering, the notion of standardized mean difference is often unambiguous. In designs with two or more levels of nesting, such as cluster randomized trials, several possible standard deviations could be used to compute standardized mean differences, so the concept of this effect-size measure is ambiguous.

This chapter has two goals. One is to provide definitions of effect sizes that may be appropriate for cluster randomized trials. The second is to provide methods to estimate these effect sizes (and obtain standard errors for them) from statistics that are typically given in reports of research (that is, without a reanalysis of the raw data). To ease the exposition, I consider studies with two levels of nesting (for example, students within schools or individuals within communities) first, then consider effect sizes in studies with three levels of nesting (for example, students within classrooms within schools or individuals within plants within companies).

18.2 DESIGNS WITH TWO LEVELS

The most widely used designs, or at least the designs most widely acknowledged in analyses, are designs with one level of nesting—so-called two-level designs. In such designs, individuals are sampled by first sampling existing groups of individuals, such as classrooms, schools, communities, or hospitals, and then individuals within the groups In such designs whole groups are assigned to treatments.

18.2.1 Model and Notation

To adequately define effect sizes, we need to first describe the notation used in this section. Let Y_{ij}^T ($i = 1, \ldots, m^T$; $j = 1, \ldots, n_i^T$) and Y_{ij}^C ($i = 1, \ldots, m^C$; $j = 1, \ldots, n_i^C$) be the jth observation in the ith cluster in the treatment and control groups respectively, so that there are m^T clusters in the treatment group and m^C clusters in the control group, and a total of $M = m^T + m^C$. Thus the sample size is

$$N^T = \sum_{i=1}^{m^T} n_i^T$$

in the treatment group,

$$N^C = \sum_{i=1}^{m^C} n_i^C$$

in the control group, and the total sample size is $N = N^T + N^C$.

Let $\overline{Y}_{i\bullet}^T$ ($i = 1, \ldots, m^T$) and $\overline{Y}_{i\bullet}^C$ ($i = 1, \ldots, m^C$) be the means of the ith cluster in the treatment and control groups, respectively, and let $\overline{Y}_{\bullet\bullet}^T$ and $\overline{Y}_{\bullet\bullet}^C$ be the overall (grand) means in the treatment and control groups, respectively. Define the (pooled) within-cluster sample variance S_W^2 via

$$S_W^2 = \frac{\sum_{i=1}^{m^T} \sum_{j=1}^{n_i^T} \left(Y_{ij}^T - \overline{Y}_{i\bullet}^T\right)^2 + \sum_{i=1}^{m^C} \sum_{j=1}^{n_i^C} \left(Y_{ij}^C - \overline{Y}_{i\bullet}^C\right)^2}{N - M}$$

and the total pooled within-treatment group variance S_T^2 via

$$S_T^2 = \frac{\sum_{i=1}^{m^T} \sum_{j=1}^{n_i^T} \left(Y_{ij}^T - \overline{Y}_{\bullet\bullet}^T\right)^2 + \sum_{i=1}^{m^C} \sum_{j=1}^{n_i^C} \left(Y_{ij}^C - \overline{Y}_{\bullet\bullet}^C\right)^2}{N - 2}.$$

Let S_B be the pooled within treatment-groups standard deviation of the cluster means given by

$$S_B^2 = \frac{\sum_{i=1}^{m^T} \left(\overline{Y}_{i\bullet}^T - \overline{Y}_{*\bullet}^T\right)^2 + \sum_{i=1}^{m^C} \left(\overline{Y}_{i\bullet}^C - \overline{Y}_{*\bullet}^C\right)^2}{m^T + m^C - 2},$$

where $\overline{Y}_{*\bullet}^T$ is the (unweighted) mean of the m^T cluster means in the treatment group, and $\overline{Y}_{*\bullet}^C$ is the (unweighted) mean of the m^C cluster means in the control group. That is,

$$\overline{Y}_{*\bullet}^T = \frac{1}{m^T} \sum_{i=1}^{m^T} \overline{Y}_{i\bullet}^T$$

and

$$\overline{Y}_{*\bullet}^C = \frac{1}{m^C} \sum_{i=1}^{m^C} \overline{Y}_{i\bullet}^C$$

Note that when cluster sample sizes are unequal, $\overline{Y}_{*\bullet}^T$ need not equal $\overline{Y}_{\bullet\bullet}^T$, the grand mean of the treatment group and $\overline{Y}_{*\bullet}^C$ need not equal $\overline{Y}_{\bullet\bullet}^C$, the grand mean of the control group. However, when cluster sample sizes are all equal, $\overline{Y}_{*\bullet}^T = \overline{Y}_{\bullet\bullet}^T$ and $\overline{Y}_{*\bullet}^C = \overline{Y}_{\bullet\bullet}^C$.

Suppose that observations within the treatment and control group clusters are normally distributed about cluster means μ_i^T and μ_i^C with a common within-cluster variance σ_W^2. That is

$$Y_{ij}^T \sim N(\mu_i^T, \sigma_W^2), \quad i = 1, \ldots, m^T; \; j = 1, \ldots, n_i^T$$

and

$$Y_{ij}^C \sim N(\mu_i^C, \sigma_W^2), \quad i = 1, \ldots, m^C; \; j = 1, \ldots, n_i^C.$$

Suppose further that the clusters have random effects (for example they are considered a sample from a population of clusters) so that the cluster means themselves have a normal sampling distribution with means $\mu_\bullet^T$ and $\mu_\bullet^C$ and common variance σ_B^2. That is

$$\mu_i^T \sim N(\mu_\bullet^T, \sigma_B^2), \quad i = 1, \ldots, m^T$$

and

$$\mu_i^C \sim N(\mu_\bullet^C, \sigma_B^2), \quad i = 1, \ldots, m^C.$$

Note that in this formulation, σ_B^2 represents true variation of the population means of clusters over and above the variation in sample means that would be expected from variation in the sampling of observations into clusters.

These assumptions correspond to the usual assumptions that would be made in the analysis of a multi-site trial by a hierarchical linear models analysis, an analysis of variance (with treatment as a fixed effect and cluster as a nested random effect), or a *t*-test using the cluster means in treatment and control group as the unit of analysis.

In principle there are three different within-treatment group standard deviations, σ_B, σ_W, and σ_T, the latter defined by

$$\sigma_T^2 = \sigma_B^2 + \sigma_W^2.$$

In most educational data when clusters are schools, σ_B^2 is considerably smaller than σ_W^2. Obviously, if the between cluster variance σ_B^2 is small, then σ_T^2 will be very similar to σ_W^2.

18.2.1.1 Intraclass Correlation with One Level of Nesting
A parameter that summarizes the relationship between the three variances is called the intraclass correlation ρ, which is defined by

$$\rho = \frac{\sigma_B^2}{\sigma_B^2 + \sigma_W^2} = \frac{\sigma_B^2}{\sigma_T^2}. \tag{18.1}$$

The intraclass correlation ρ can be used to obtain one of these variances from any of the others, given that $\sigma_B^2 = \rho\sigma_T^2$ and $\sigma_W^2 = (1-\rho)\sigma_T^2 = (1-\rho)\sigma_B^2/\rho$.

The intraclass correlation describes the degree of clustering in the data. Later in this section, I show that the variance of effect-size estimates in nested designs depends on the intraclass correlation, and how the intraclass correlation can be used to relate different possible effect sizes that can be computed in multilevel designs to one another.

18.2.1.2 Primary Analyses
Many analyses reported in the literature fail to take the effects of clustering into account or do so but by analyzing cluster means as the raw data. In either case, the treatment effect (the mean difference between treatment and control groups), the standard deviations S_B, S_W, and S_T, and the sample sizes may be reported (or deduced from information that is reported). In other cases the analysis reported may have taken clustering into account by treating clusters as random effects. Such an analysis (for example using the program HLM, SAS Proc Mixed, or the STATA routine XT-Mixed) usually yields direct estimates of the treatment effect b and the variance components $\hat{\sigma}_B^2$ and $\hat{\sigma}_W^2$ and their standard errors. This information can be used to calculate both the effect size and its approximate standard error. Let $\hat{\sigma}_B^2$ and $\hat{\sigma}_W^2$ be the estimates of σ_B^2 and σ_W^2 and let $V(b)$, $V(\hat{\sigma}_B^2)$ and $V(\hat{\sigma}_W^2)$ be their variances (the square of their standard errors). Generally, $V(\hat{\sigma}_W^2)$ depends primarily on the number of individuals and $V(\hat{\sigma}_B^2)$ on the number of clusters. Because the number of individuals typically greatly exceeds the number of clusters, $V(\hat{\sigma}_W^2)$ will be so much smaller than $V(\hat{\sigma}_B^2)$ that $V(\hat{\sigma}_W^2)$ can be considered negligible.

18.2.2 Effect Sizes in Two-Level Designs

In designs with one level of nesting (two-level designs), three possible standardized mean difference parameters correspond to the three different standard deviations. The choice of one of these effect sizes should be determined on the basis of the inference of interest to the researcher. If the effect-size measures are to be used in meta-analysis, an important inference goal may be to estimate parameters comparable with those that can be estimated in other studies. In such cases, the standard deviation may be chosen to be the same kind of standard deviation used in the effect sizes of other studies to which the study will be compared. I focus on three effect sizes that seem likely to be the most widely used.

If $\sigma_W \neq 0$ (hence $\rho \neq 1$), one effect-size parameter has the form

$$\delta_W = \frac{\mu_{\cdot\cdot}^T - \mu_{\cdot\cdot}^C}{\sigma_W}. \tag{18.2}$$

This effect size might be of interest, for example, in a meta-analysis where the other studies to which the current study is compared are typically single site studies. In such studies δ_W may (implicitly) be the effect size estimated and hence δ_W might be the effect size most comparable with that in other studies.

A second effect-size parameter is of the form

$$\delta_T = \frac{\mu_{\cdot\cdot}^T - \mu_{\cdot\cdot}^C}{\sigma_T}. \tag{18.2}$$

This effect size might be of interest in a meta-analysis where the other studies are multisite studies or sample from a broader population but do not include clusters in the sampling design. This would typically imply that they used an individual, rather than a cluster, assignment strategy. In such cases, δ_T might be the most comparable with the effect sizes in other studies.

If $\sigma_B \neq 0$ (and hence $\rho \neq 0$), a third possible effect-size parameter would be

$$\delta_B = \frac{\mu_{\cdot\cdot}^T - \mu_{\cdot\cdot}^C}{\sigma_B}. \tag{18.4}$$

This effect size is less likely to be of general interest, but it might be in cases where the treatment effect is defined at the level of clusters and the individual observations are of interest because the average defines an aggregate property. For example, Anthony Bryk and Barbara Schneider used the concept of trust in schools as an aggregate property that is measured by taking means of reports from individuals (2000). The effect size δ_B might be the appropriate effect size to quantify the effect of an intervention designed to improve school trust. The effect size δ_B may also be of interest in a meta-analysis where the studies being compared are typically multisite studies that have been analyzed using cluster means as the unit of analysis.

Note, however, that though δ_W and δ_T will often be similar in size, δ_B will typically be much larger (in the same population) than either δ_W or δ_T, because σ_B is typically considerably smaller than σ_W and σ_T. Note also that if all of the effect sizes are defined (that is, if $0 < \rho < 1$), and ρ is known, any one of these effect sizes may be obtained from any of the others. In particular, both δ_W and δ_T can be obtained from δ_B and ρ because

$$\delta_W = \delta_B \sqrt{\frac{\rho}{1-\rho}} = \frac{\delta_T}{\sqrt{1-\rho}} \qquad (18.5)$$

and

$$\delta_T = \delta_B \sqrt{\rho} = \delta_W \sqrt{1-\rho}. \qquad (18.6)$$

18.2.3 Estimates of Effect Sizes in Two-Level Designs

Formulas for effect-size estimates, and particularly their variances, are considerably simpler when sample sizes are equal than when they are unequal. Many studies are planned with the intention of having equal sample sizes in each cluster because this is simpler and maximizes statistical power. Although the eventual data collected may not have equal sample sizes, sizes are often rather close to being so. As a practical matter, exact cluster sizes are frequently not reported, so the research synthesist often has access only to the average sample sizes and thus may have to proceed as if the sizes are equal. Unless they are profoundly unequal, proceeding as if they are equal to some reasonable compromise value (such as the harmonic mean) will yield reasonably accurate results. In the sections that follow, I present results first when clusters are assumed to have equal sample size n, that is, when $n_i^T = n$, $i = 1, \ldots, m^T$ and $n_i^C = n$, $i = 1, \ldots, m^C$. In this case, $N^T = nm^T$, $N^C = nm^C$, and $N = N^T + N^C = n(m^T + m^C) = nM$.

18.2.3.1 Estimating δ_W
We start with estimation of δ_W, which is the most straightforward. If $\rho \neq 1$ (so that $\sigma_W \neq 0$ and δ_W is defined), the estimate

$$d_W = \frac{\overline{Y}_{..}^T - \overline{Y}_{..}^C}{S_W} \qquad (18.7)$$

is a consistent estimator of δ_W, which is approximately normally distributed about δ_W. When cluster sample sizes are equal the (estimated) variance of d_W is

$$v_W = \left(\frac{N^T + N^C}{N^T N^C}\right)\left(\frac{1+(n-1)\rho}{1-\rho}\right) + \frac{d_W^2}{2(N-M)}. \qquad (18.8)$$

Note that the presence of the factor $(1-\rho)$ in the denominator of the first term is possible because δ_W is defined only if $\rho \neq 1$.

Note that if $\rho = 0$ so that there is no clustering, equation 18.8 reduces to the variance of a mean difference divided by a standard deviation with $(N-M)$ degrees of freedom (see chapter 12, this volume). The leading term of the variance in equation 18.8 arises from uncertainty in the mean difference. Note that it is $[1+(n-1)\rho]/(1-\rho)$ as large as would be expected if there were no clustering in the sample, that is, if $\rho = 0$. Thus $[1+(n-1)\rho]/(1-\rho)$ is a kind of variance inflation factor for the variance of the effect-size estimate d_W.

When cluster sizes are not equal, the estimator d_W of δ_W is the same as in the case of equal cluster sample sizes, but the (estimated) variance of the estimator becomes

$$v_W = \left(\frac{N^T + N^C}{N^T N^C}\right)\left(\frac{1+(\tilde{n}-1)\rho}{1-\rho}\right) + \frac{d_W^2}{2(N-M)}, \qquad (18.9)$$

where

$$\tilde{n} = \frac{N^C \sum_{i=1}^{m^T} (n_i^T)^2}{N^T N} + \frac{N^T \sum_{i=1}^{m^C} (n_i^C)^2}{N^C N}. \qquad (18.10)$$

When all of the n_i^T and n_i^C are equal to n, $\tilde{n} = n$ and equation 18.9 reduces to equation 18.8.

When the analysis properly accounts for clustering by treating the clusters as having random effects, and an estimate b of the treatment effect, its variance $V(b)$, and an estimate $\hat{\sigma}_W^2$ of the variance component σ_W^2 is available (for example, from an HLM analysis), the estimate of δ_W is

$$d_W = \frac{b}{\hat{\sigma}_W}. \qquad (18.11)$$

The approximate variance of the estimate given in equation 18.11 is

$$v_W = \frac{V(b)}{\hat{\sigma}_W^2}. \qquad (18.12)$$

18.2.3.2 Estimating δ_B Estimating the other δ_B and δ_T is less intuitive than estimating δ_W. For example, one might expect that

$$\frac{\bar{Y}_{\cdot\cdot}^T - \bar{Y}_{\cdot\cdot}^C}{S_B},$$

would be that natural estimator of δ_B, but this is not the case. The reason is that S_B is not a pure estimate of σ_B, because it is inflated by the within-cluster variability. In particular, the expected value of S_B^2 is

$$\sigma_B^2 + \frac{\sigma_W^2}{n}.$$

Whenever $\sigma_B \neq 0$ (so that $\rho \neq 0$ and δ_B is defined), the intraclass correlation may be used to obtain an estimate of δ_B using S_B. When all the cluster sample sizes are equal to n, a consistent estimate of δ_B is

$$d_B = \frac{\bar{Y}_{\cdot\cdot}^T - \bar{Y}_{\cdot\cdot}^C}{S_B}\sqrt{\frac{1+(n-1)\rho}{n\rho}}, \qquad (18.13)$$

which is normally distributed in large samples with (estimated) variance

$$v_B = \left(\frac{m^T + m^C}{m^T m^C}\right)\left(\frac{1+(n-1)\rho}{n\rho}\right) + \frac{[1+(n-1)\rho]d_B^2}{2n\rho(M-2)}. \qquad (18.14)$$

Note that the presence of ρ in the denominators of the variance terms is possible since δ_B is only defined if $\rho \neq 0$.

The variance in equation 18.14 is $[1+(n-1)\rho]/n\rho$ as large as the variance of the standardized mean difference computed from an analysis using cluster means as the unit of analysis, that is, applying the usual formula for the variance of the standardized difference between the means of the cluster means in the treatment and control group. Thus $[1+(n-1)\rho]/n\rho$ is a kind of variance inflation factor for the variance of effect-size estimates like d_B.

When the cluster sizes are not equal, an estimator of δ_B that is a generalization of equation 18.13 is given by

$$d_B = \left(\frac{\bar{Y}_{\cdot\cdot}^T - \bar{Y}_{\cdot\cdot}^C}{S_B}\right)\sqrt{\frac{1+(\bar{n}_B-1)\rho}{\bar{n}_B\rho}}, \qquad (18.15)$$

where

$$\bar{n}_B = \left(\frac{(m^T-1)\bar{n}_I^T + (m^C-1)\bar{n}_I^C}{M-2}\right)^{-1},$$

$$\bar{n}_I^T = \frac{1}{m^T}\sum_{i=1}^{m^T}(1/n_i^T),$$

and

$$\bar{n}_I^C = \frac{1}{m^C}\sum_{i=1}^{m^C}(1/n_i^C).$$

When cluster sample sizes are unequal, the variance of d_B is approximately

$$v_B = \left(\frac{m^T + m^C}{m^T m^C}\right)\left(\frac{1+(\bar{n}_B-1)\rho}{\bar{n}_B\rho}\right) + \frac{\bar{n}_B C d_B^2}{2(M-2)^2\rho[1+(\bar{n}_B-1)\rho]}, \qquad (18.16)$$

where

$$\bar{n}_B = \left(\frac{m^C\bar{n}_I^T + m^T\bar{n}_I^C}{M}\right)^{-1},$$

$$C = (M-2)\rho^2 + 2[(m^T-1)\bar{n}_I^T + (m^C-1)\bar{n}_I^C]\rho(1-\rho) + [(m^T-2)\bar{n}_I^{T2} + (m^C-2)\bar{n}_I^{C2} + (\bar{n}_I^T)^2 + (\bar{n}_I^C)^2](1-\rho)^2,$$

$$\bar{n}_I^{T2} = \frac{1}{m^T}\sum_{i=1}^{m^T}(1/n_i^T)^2,$$

and

$$\bar{n}_I^{C2} = \frac{1}{m^C}\sum_{i=1}^{m^C}(1/n_i^C)^2.$$

Note that when the n_i^T and n_i^C are all equal to n,

$$n_B = n, \quad \tilde{n}_B = n,$$
$$n_I^T = \bar{n}_I^C = 1/n, \quad \bar{n}_I^{T2} = \bar{n}_I^{C2} = 1/n^2,$$

and

$$C = \frac{(M-2)[1+(n-1)\rho]^2}{n^2},$$

so that 18.15 reduces to 18.13 and 18.16 reduces to 18.14. When cluster sample sizes are unequal, substituting $\bar{n}_B$ for n in equations 18.13 and 18.14 would give results that are quite close to the exact values.

When the analysis properly accounts for clustering by treating the clusters as having random effects, and an estimate b of the treatment effect, its variance $V(b)$, and an estimate $\hat{\sigma}_B^2$ of the variance component σ_B^2 is available (for example, from an HLM analysis), the estimate of δ_B is

$$d_B = \frac{b}{\hat{\sigma}_B}. \qquad (18.17)$$

The approximate variance of the estimate given in equation 18.17 is

$$v_B = \frac{V(b)}{\hat{\sigma}_B^2} + \frac{b^2 V(\hat{\sigma}_B^2)}{4\hat{\sigma}_B^6}. \tag{18.18}$$

18.2.3.3 Estimating δ_T An estimate of δ_T can also be obtained using the intraclass correlation. When all the cluster sample sizes are equal to n, an estimator of δ_T is

$$d_T = \left(\frac{\bar{Y}_{\cdots}^T - \bar{Y}_{\cdots}^C}{S_T}\right)\sqrt{1 - \frac{2(n-1)\rho}{N-2}}, \tag{18.19}$$

which is normally distributed in large samples with (an estimated) variance of

$$v_T = \left(\frac{N^T + N^C}{N^T N^C}\right)(1 + (n-1)\rho) + d_T^2\left(\frac{(N-2)(1-\rho)^2 + n(N-2n)\rho^2 + 2(N-2n)\rho(1-\rho)}{2(N-2)[(N-2) - 2(n-1)\rho]}\right). \tag{18.20}$$

Note that if $\rho = 0$ and there is no clustering, d_T reduces to the conventional standardized mean difference and equation 18.20 reduces to the usual expression for the variance of the standardized mean difference (see chapter 12, this volume).

The leading term of the variance in equation 18.20 arises from uncertainty in the mean difference. Note that this leading term is $[1 + (n-1)\rho]$ as large as would be expected if there were no clustering in the sample (that is if $\rho = 0$). The expression $[1 + (n-1)\rho]$ is the variance inflation factor mentioned by Allan Donner and colleagues (1981) and the design effect mentioned by Leslie Kish (1965) for the variance of means in clustered samples and also corresponds to a variance inflation factor for the effect-size estimates like d_T.

When cluster sample sizes are unequal, the estimator of δ_T becomes

$$d_T = \left(\frac{\bar{Y}_{\cdots}^T - \bar{Y}_{\cdots}^C}{S_T}\right)\sqrt{1 - \left(\frac{(N - n_U^T m^T - n_U^C m^C) + n_U^T + n_U^C - 2}{N-2}\right)\rho}, \tag{18.21}$$

where

$$n_U^T = \frac{(N^T)^2 - \sum_{i=1}^{m^T} (n_i^T)^2}{N^T(m^T - 1)},$$

and

$$n_U^C = \frac{(N^C)^2 - \sum_{i=1}^{m^C} (n_i^C)^2}{N^C(m^C - 1)}.$$

When all of the n_i^T and n_i^C are equal to n, $n_U^T = n_U^C = n$ and 18.21 reduces to 18.19.

The variance of d_T when cluster samples sizes are unequal is somewhat more complex. It is given by

$$v_T = \left(\frac{N^T + N^C}{N^T N^C}\right)(1 + (\tilde{n} - 1)\rho) + \frac{[(N-2)(1-\rho)^2 + A\rho^2 + 2B\rho(1-\rho)]d_T^2}{2(N-2)[(N-2) - \rho(N-2-B)]}, \tag{18.22}$$

where the auxiliary constants A and B are defined by $A = A^T + A^C$,

$$A^T = \frac{(N^T)^2 \sum_{i=1}^{m^T} (n_i^T)^2 + \left(\sum_{i=1}^{m^T} (n_i^T)^2\right)^2 - 2N^T \sum_{i=1}^{m^T} (n_i^T)^3}{(N^T)^2},$$

$$A^C = \frac{(N^C)^2 \sum_{i=1}^{m^C} (n_i^C)^2 + \left(\sum_{i=1}^{m^C} (n_i^C)^2\right)^2 - 2N^C \sum_{i=1}^{m^C} (n_i^C)^3}{(N^C)^2},$$

and

$$B = n_U^T(m^T - 1) + n_U^C(m^C - 1).$$

When n_i^T and n_i^C are equal to n, $A = n(N - 2n)$, $n_U^T = n_U^C = n$ and $B = (N - 2n)$ so that equation 18.22 reduces to 18.20. These expressions suggest that if cluster sample sizes are unequal, substituting the average of n_U^T and n_U^C into equations 18.19 and 18.20 would give results that are quite close to the exact values.

When the analysis has been carried out using an analytic method that properly accounts for clustering by treating clusters as random effects, the estimate of δ_T is

$$d_T = \frac{b}{\sqrt{\hat{\sigma}_B^2 + \hat{\sigma}_W^2}}. \tag{18.23}$$

Assuming that $V(\hat{\sigma}_W^2)$ is negligible and that $\hat{\sigma}_B^2$ is independent of the estimated treatment effect b, the approximate variance of the estimate 18.23 is

$$v_T = \frac{V(b)}{\hat{\sigma}_B^2 + \hat{\sigma}_W^2} + \frac{b^2 V(\hat{\sigma}_B^2)}{4(\hat{\sigma}_B^2 + \hat{\sigma}_W^2)^3}. \tag{18.24}$$

18.2.3.4 Confidence Intervals for δ_W, δ_B, and δ_T
The results here can also be used to compute confidence intervals for effect sizes. If δ is any one of the effect sizes mentioned, d is a corresponding estimate, and v_d is the estimated variance of d, then a $100(1 - \alpha)$ percent confidence interval for δ based on d and v_d is given by

$$d - c_{\alpha/2} v_d \leq \delta \leq d + c_{\alpha/2} v_d, \tag{18.25}$$

where $c_{\alpha/2}$ is the $100(1 - \alpha/2)$ percent point of the standard normal distribution (for example, 1.96 for $\alpha/2 = 0.05/2 = 0.025$).

18.2.4 Applications in Two-Level Designs

This chapter is intended to be useful in deciding what effect sizes might be desirable to use in studies with nested designs. It has described ways to compute effect-size estimates and their variances from data that may be reported. I illustrate applications in this section.

18.2.4.1 Effect sizes for individuals The results in this chapter can be used to produce effect-size estimates and their variances from studies that report (incorrectly analyzed) experiments as if there were no nested factors. The required means, standard deviations, and sample sizes can usually be extracted from what may be reported.

Suppose it is decided that the effect size δ_T is appropriate because most other studies both assign and sample individually from a clustered population. Suppose that the data are analyzed by ignoring clustering, then the test statistic is likely to be either

$$t = \sqrt{\frac{N^T N^C}{N^T + N^C}} \left(\frac{\bar{Y}^T_{\cdot\cdot} - \bar{Y}^C_{\cdot\cdot}}{S_T} \right)$$

or $F = t^2$. Either can be solved for

$$\left(\frac{\bar{Y}^T_{\cdot\cdot} - \bar{Y}^C_{\cdot\cdot}}{S_T} \right),$$

which can then be inserted into equation 18.19 along with ρ to obtain d_T. This estimate (d_T) of δ_T can then be inserted into equation 18.20 to obtain v_T, an estimate of the variance of d_T.

Alternatively, suppose it is decided that the effect size δ_W is appropriate because most other studies involve only a single site. We may begin by computing d_T and v_T as before. Because we want an estimate of δ_W, not δ_T, we use the fact given in equation 18.5 that

$$\delta_W = \delta_T / \sqrt{1 - \rho}$$

and therefore

$$d_T / \sqrt{1 - \rho} \qquad (18.26)$$

is an estimate of δ_W with a variance of

$$v_T / (1 - \rho). \qquad (18.27)$$

Example. An evaluation of the connected mathematics curriculum reported by Ridgway, and colleagues compared the achievement of $m^T = 18$ classrooms of sixth grade students who used connected mathematics with that of $m^C = 9$ classrooms in a comparison group that did not use connected mathematics (2002). In this quasi-experimental design the clusters were classrooms. The cluster sizes were not identical but the average cluster size in the treatment groups was $N^T/m^T = 338/18 = 18.8$ and N^C/m^C $162/18 = 18$ in the control group. The exact sizes of all the clusters were not reported, but here I treat the cluster sizes as if they were equal and choose $n = 18$ as a slightly conservative sample size. The mean difference between treatment and control groups is $\bar{Y}^T_{\cdot\cdot} - \bar{Y}^C_{\cdot\cdot} = 1.9$, the pooled within-groups standard deviation $S_T = 12.37$. This evaluation involved sites in all regions of the country and it was intended to be nationally representative. Ridgway and colleagues did not give an estimate of the intraclass correlation based on their sample. Larry Hedges and Eric Hedberg have provided an estimate of the grade 6 intraclass correlation in mathematics achievement for the nation as a whole, based on a national probability sample, of 0.264 with a standard error of 0.019 (2007). For this example, I assume that the intraclass correlation is identical to that estimate, namely $\rho = 0.264$.

Suppose that the analysis ignored clustering and compared the mean of all of the students in the treatment with the mean of all of the students in the control group. This leads to a value of the standardized mean difference of

$$\frac{\bar{Y}^T_{\cdot\cdot} - \bar{Y}^C_{\cdot\cdot}}{S_T} = 0.1536,$$

which is not an estimate of any of the three effect sizes considered here. If an estimate of the effect size δ_T is desired, and we had imputed an intraclass correlation of $\rho = 0.264$, then we use equation 18.19 to obtain $d_T = (0.1536)(0.9907) = 0.1522$.

The effect-size estimate is very close to the original standardized mean difference even though the amount of clustering in this case is not small. However this amount of clustering has a substantial effect on the variance of the effect-size estimate. The variance of the standardized mean difference ignoring clustering is

$$\frac{324 + 162}{324 * 162} + \frac{0.1531^2}{2(324 + 162 - 2)} = 0.009259.$$

However, computing the variance of d_T using equation 18.20 with $\rho = 0.264$, we obtain a variance estimate of 0.050865, which is 549 percent of the variance ignoring clustering. A 95 percent confidence interval for δ_T is given by

$$-0.2899 = 0.1522 - 1.96\sqrt{0.050865}$$
$$\leq \delta_T \leq 0.1522 + 1.96\sqrt{0.050865} = 0.5942.$$

If clustering had been ignored, the confidence interval for the population effect size would have been -0.0350 to 0.3422.

If we wanted to estimate δ_W, then an estimate of δ_W given by expression 18.26 is

$$\frac{0.1522}{\sqrt{1 - 0.264}} = 0.1774,$$

with variance given by expression 18.27 as

$$0.050865/(1 - 0.264) = 0.06911,$$

and a 95 percent confidence interval for δ_W based on 18.25 would be

$$-0.3379 = 0.1774 - 1.96\sqrt{0.06911}$$
$$\leq \delta_W \leq 0.1774 + 1.96\sqrt{0.06911} = 0.6926.$$

18.2.4.2 Effect when Clusters Are the Unit of Analysis
The results in this chapter can also be used to obtain different effect size estimates when the data have been analyzed with the cluster mean as the unit of analysis. In such cases, the researcher might report a t-test or a one way analysis of variance carried out on cluster means, but we wish to estimate δ_T. In this case, the test statistic reported will either be

$$t = \sqrt{\frac{m^T m^C}{m^T + m^C}} \left(\frac{\bar{Y}^T_{..} - \bar{Y}^C_{..}}{S_B} \right)$$

or $F = t^2$. Either can be solved for

$$\left(\frac{\bar{Y}^T_{..} - \bar{Y}^C_{..}}{S_B} \right),$$

which can then be inserted into equation 18.13 along with ρ to obtain d_B. This estimate of d_B can then be inserted into equation 18.14 to obtain v_B, an estimate of the variance of d_B. Because we want an estimate of δ_T, not δ_B, we use the fact given in equation 18.6 that

$$\delta_T = \delta_B \sqrt{\rho}$$

and therefore

$$d_B \sqrt{\rho} \qquad (18.28)$$

is an estimate of δ_T with a variance of

$$\rho v_B. \qquad (18.29)$$

Alternatively, suppose it is decided that the effect size δ_W is the desired effect size. We may begin by computing d_B and v_B as before. Because we want an estimate of δ_W, not δ_B, we use the fact given in equation 18.5 that

$$\delta_W = \delta_B \sqrt{\rho/(1-\rho)}$$

and therefore

$$d_W = d_B \sqrt{\rho/(1-\rho)} \qquad (18.30)$$

is an estimate of δ_W with a variance of

$$\rho v_B/(1-\rho). \qquad (18.31)$$

18.3 DESIGNS WITH THREE LEVELS

The methods described in section 18.2 are appropriate for computing effect-size estimates and their variances from designs that involve two stages of sampling (one nested factor or so-called two-level models). Many social experiments and quasi-experiments, however, assign treatments to high level clusters such as schools or communities when lower levels of clustering (two nested factors or so called three-level models) are also available. For example, schools are frequently units of assignment in experiments, but students are nested within classrooms that are in turn nested within schools. Most studies in education that sample students by first sampling schools must take into account this nesting. A similar situation arises when individuals are nested within institutions nested within communities, but communities are assigned to public health interventions.

Although I use the terms *schools* and *classrooms* to characterize the stages of clustering, this is merely a matter of convenient and readily understandable terminology. These results apply equally well to any situation in which there is a three-stage sampling design and treatments are assigned to clusters (for example, schools).

18.3.1 Model and Notation

As in section 18.2.1 to adequately define effect sizes and their variances, I need to first describe the notation. Let

Y_{ijk}^T $(i = 1, \ldots, m^T; j = 1, \ldots, p_i^T; k = 1, \ldots, n_{ij}^T)$ and Y_{ijk}^C $(i = 1, \ldots, m^C; j = 1, \ldots, p_i^C; k = 1, \ldots, n_{ij}^C)$ be the kth observation in the jth classroom in the ith school in the treatment and control groups respectively. Thus, in the treatment group, there are m^T schools, the ith school has p_i^T classrooms, and the jth classroom in the ith school has n_{ij}^T observations. Similarly, in the control group, there are m^C schools, the ith school has p_i^C classrooms, and the jth classroom in the ith school has n_{ij}^C observations. Thus there is a total of $M = m^T + m^C$ schools, a total of

$$P = \sum_{i=1}^{m^T} p_i^T + \sum_{i=1}^{m^C} p_i^C$$

classrooms, and a total of

$$N = N^T + N^C = \sum_{i=1}^{m^T} \sum_{j=1}^{p_i^T} n_{ij}^T + \sum_{i=1}^{m^C} \sum_{j=1}^{p_i^C} n_{ij}^C$$

observations overall.

Let $\overline{Y}_{i\bullet\bullet}^T$ $(i = 1, \ldots, m^T)$ and $\overline{Y}_{i\bullet\bullet}^C$ $(i = 1, \ldots, m^C)$ be the means of the ith school in the treatment and control groups, respectively, let $\overline{Y}_{ij\bullet}^T$ $(i = 1, \ldots, m^T; j = 1, \ldots, p_i^T)$ and $\overline{Y}_{ij\bullet}^C$ $(i = 1, \ldots, m^C; j = 1, \ldots, p_i^C)$ be the means of the jth class in the ith school in the treatment and control groups, respectively, and let $\overline{Y}_{\bullet\bullet\bullet}^T$ and $\overline{Y}_{\bullet\bullet\bullet}^C$ be the overall means in the treatment and control groups, respectively. Define the (pooled) within treatment groups variance S_{WT}^2 via

$$S_{WT}^2 = \frac{\sum_{i=1}^{m^T} \sum_{j=1}^{p_i^T} \sum_{k=1}^{n_{ij}^T} \left(Y_{ijk}^T - \overline{Y}_{\bullet\bullet\bullet}^T\right)^2 + \sum_{i=1}^{m^C} \sum_{j=1}^{p_i^C} \sum_{k=1}^{n_{ij}^C} \left(Y_{ijk}^C - \overline{Y}_{\bullet\bullet\bullet}^C\right)^2}{N - 2},$$

the pooled within-schools variance S_{WS}^2 via

$$S_{WS}^2 = \frac{\sum_{i=1}^{m^T} \sum_{j=1}^{p_i^T} \sum_{k=1}^{n_{ij}^T} \left(Y_{ijk}^T - \overline{Y}_{i\bullet\bullet}^T\right)^2 + \sum_{i=1}^{m^C} \sum_{j=1}^{p_i^C} \sum_{k=1}^{n_{ij}^C} \left(Y_{ijk}^C - \overline{Y}_{i\bullet\bullet}^C\right)^2}{N - M},$$

and the (pooled) within-classroom sample variance S_{WC}^2 via

$$S_{WC}^2 = \frac{\sum_{i=1}^{m^T} \sum_{j=1}^{p_i^T} \sum_{k=1}^{n_{ij}^T} \left(Y_{ijk}^T - \overline{Y}_{ij\bullet}^T\right)^2 + \sum_{i=1}^{m^C} \sum_{j=1}^{p_i^C} \sum_{k=1}^{n_{ij}^C} \left(Y_{ijk}^C - \overline{Y}_{ij\bullet}^C\right)^2}{N - P}.$$

Define the between schools but within treatment groups variance S_{BS}^2 via

$$S_{BS}^2 = \frac{\sum_{j=1}^{m^T} \left(\overline{Y}_{i\bullet\bullet}^T - \overline{Y}_{\bullet\bullet\bullet}^T\right)^2 + \sum_{j=1}^{m^C} \left(\overline{Y}_{i\bullet\bullet}^C - \overline{Y}_{\bullet\bullet\bullet}^C\right)^2}{M - 2},$$

and the (pooled) between classrooms but within treatment groups and schools variance S_{BC}^2 via

$$S_{BC}^2 = \frac{\sum_{i=1}^{m^T} \sum_{j=1}^{p_i^T} \left(\overline{Y}_{ij\bullet}^T - \overline{Y}_{i\bullet\bullet}^T\right)^2 + \sum_{i=1}^{m^C} \sum_{j=1}^{p_i^C} \left(\overline{Y}_{ij\bullet}^C - \overline{Y}_{i\bullet\bullet}^C\right)^2}{P - M}.$$

Suppose that observations within the jth classroom in the ith school within the treatment and control group groups are normally distributed about classroom means μ_{ij}^T and μ_{ij}^C with a common within-classroom variance σ_{WC}^2. That is

$$Y_{ijk}^T \sim N\left(\mu_{ij}^T, \sigma_{WC}^2\right), \quad i = 1, \ldots, m^T; \quad j = 1, \ldots, p_i^T;$$
$$k = 1, \ldots, n_{ij}^T$$

and

$$Y_{ijk}^C \sim N\left(\mu_{ij}^C, \sigma_{WC}^2\right), \quad i = 1, \ldots, m^C; \quad j = 1, \ldots, p_i^C;$$
$$k = 1, \ldots, n_{ij}^C.$$

Suppose further that the classroom means are random effects (a sample from a population of means, for example) so that the class means themselves have a normal distribution about the school means $\mu_{i\bullet}^T$ and $\mu_{i\bullet}^C$ and common variance σ_{BC}^2. That is,

$$\mu_{ij}^T \sim N\left(\mu_{i\bullet}^T, \sigma_{BC}^2\right), \quad i = 1, \ldots, m^T; \quad j = 1, \ldots, p_i^T$$

and

$$\mu_{ij}^C \sim N\left(\mu_{i\bullet}^C, \sigma_{BC}^2\right), \quad i = 1, \ldots, m^C; \quad j = 1, \ldots, p_i^C.$$

Finally, suppose that the school means $\mu_{i\bullet}^T$ and $\mu_{i\bullet}^C$ are also normally distributed about the treatment and control group means $\mu_{\bullet\bullet}^T$ and $\mu_{\bullet\bullet}^C$ with common variance σ_{BS}^2. That is,

$$\mu_{i\bullet}^T \sim N\left(\mu_{\bullet\bullet}^T, \sigma_{BS}^2\right), \quad i = 1, \ldots, m^T$$

and

$$\mu_{i\bullet}^C \sim N\left(\mu_{\bullet\bullet}^C, \sigma_{BS}^2\right), \quad i = 1, \ldots, m^C.$$

Note that in this formulation, σ_{BC}^2 represents true variation of the population means of classrooms over and above the variation in sample means that would be expected from variation in the sampling of observations into classroom. Similarly, σ_{BS}^2 represents the true variation in school means, over and above the variation that would be expected from that attributable to the sampling of observations into schools.

These assumptions correspond to the usual assumptions that would be made in the analysis of a multisite trial by a three-level hierarchical linear models analysis, an analysis of variance (with treatment as a fixed effect and schools and classrooms as nested random effects), or a t-test using the school means in treatment and control group as the unit of analysis.

18.3.1.1 Intraclass Correlations In principle, there are several within-treatment group variances in a three-level design. I have already defined the within-classroom, between-classroom, and between-school, variances σ_{WC}^2, σ_{BC}^2, and σ_{BS}^2. There is also the total variance within treatment groups σ_{WT}^2 defined by

$$\sigma_{WT}^2 = \sigma_{BS}^2 + \sigma_{BC}^2 + \sigma_{WC}^2.$$

In most educational achievement data when clusters are schools, σ_{BS}^2 and σ_{BC}^2 are considerably smaller than σ_{WC}^2. Obviously, if the between school and classroom variances σ_{BS}^2 and σ_{BC}^2 are small, then σ_{WT}^2 will be very similar to σ_{WC}^2.

In two-level models, such as those with schools and students as levels, we saw in section 18.2.1.1 that the relation between variances associated with the two levels was characterized by an index called the intraclass correlation. In three-level models, two indices are needed to characterize the relationship between these variances. Define the school-level intraclass correlation ρ_S by

$$\rho_S = \frac{\sigma_{BS}^2}{\sigma_{BS}^2 + \sigma_{BC}^2 + \sigma_{WC}^2} = \frac{\sigma_{BS}^2}{\sigma_{WT}^2}. \quad (18.32)$$

Similarly, define the classroom level intraclass correlation ρ_C by

$$\rho_C = \frac{\sigma_{BC}^2}{\sigma_{BS}^2 + \sigma_{BC}^2 + \sigma_{WC}^2} = \frac{\sigma_{BC}^2}{\sigma_{WT}^2}. \quad (18.33)$$

These intraclass correlations can be used to obtain one of these variances from any of the others, given that $\sigma_{BS}^2 = \rho_S \sigma_{WT}^2$, $\sigma_{BC}^2 = \rho_C \sigma_{WT}^2$, and $\sigma_{WC}^2 = (1 - \rho_S - \rho_C)\sigma_{WT}^2$. Note that if $\sigma_{BS} = 0$, then $\rho_S = 0$ and ρ_C reduces to ρ given in (18.1) and that if $\sigma_{BC} = 0$, $\rho_C = 0$ and ρ_S reduces to ρ given in (18.1).

In designs with two levels of nesting, the two intraclass correlations describe the degree of clustering, just as the single intraclass correlation describes the degree of clustering in designs with one level of nesting. The variance of effect-size estimates depends on the intraclass correlations in designs with two levels of nesting, just as the variance of effect-size estimates depends on the (single) intraclass correlation in designs with one level of nesting.

18.3.1.2 Primary Analyses Many analyses reported in the literature fail to take the effects of clustering into account or take clustering into account by analyzing cluster means as the raw data. In either case, the treatment effect (the mean difference between treatment and control groups), the standard deviations S_{BS}, S_{BS}, S_{WC}, and S_{WT}, and the sample sizes may be reported (or deduced from information that is reported). In other cases, the analysis reported may have taken clustering into account by treating clusters as having random effects. Such an analysis—for example, using the program HLM, SAS Proc Mixed, or the STATA routine XTMixed—usually yields direct estimates of the estimated treatment effect b and the variance components σ_{BS}^2, σ_{BC}^2 and σ_{WC}^2 and their standard errors. This information can be used to calculate both the effect size and its approximate standard error. Let $\hat{\sigma}_{BS}^2$, $\hat{\sigma}_{BC}^2$, and $\hat{\sigma}_{WC}^2$ be the estimates of σ_{BS}^2, σ_{BC}^2 and σ_{WC}^2 and let $V(b)$, $V(\hat{\sigma}_{BS}^2)$, $V(\hat{\sigma}_{BC}^2)$, and $V(\hat{\sigma}_{WC}^2)$ be their variances (the square of their standard errors). Generally, $V(\hat{\sigma}_{WC}^2)$ depends primarily on the number of individuals, whereas $V(\hat{\sigma}_{BS}^2)$ and $V(\hat{\sigma}_{BC}^2)$ depend on the number of schools and classrooms, respectively. Because the number of individuals typically greatly exceeds the number of schools and classrooms, $V(\hat{\sigma}_{WC}^2)$ will be so much smaller than either $V(\hat{\sigma}_{BS}^2)$ and $V(\hat{\sigma}_{BC}^2)$ that $V(\hat{\sigma}_{WC}^2)$ can be considered negligible in comparison.

18.3.2 Effect Sizes

In three-level designs, several standardized mean differences are possible because at least five standard deviations can be used. The choice should be determined on the basis of the inference of interest to the researcher. If the effect-size measures are to be used in meta-analysis, an important inference goal may be to estimate parameters comparable with those that can be estimated in other studies. In such cases, the standard deviation chosen might be the same kind used in those to which the study will be compared. Three effect sizes seem likely to be the most widely used, but two others might be useful in some circumstances.

One effect size standardizes the mean difference by the total standard deviation within-treatment groups. It is in the form

$$\delta_{WT} = \frac{\mu_{..}^T - \mu_{..}^C}{\sigma_{WT}} = \frac{\mu_{..}^T - \mu_{..}^C}{\sqrt{\sigma_{BS}^2 + \sigma_{BC}^2 + \sigma_{WC}^2}}. \quad (18.34)$$

The effect size δ_{WT} might be of interest in a meta-analysis where the other studies have three-level designs or sample from a broader population but do not include schools or classes in the sampling design. This would typically imply that they used an individual, rather than a cluster, assignment strategy. In such cases, δ_{WT} might be the most comparable effect size.

If $\sigma_{BC}^2 + \sigma_{WC}^2 \neq 0$ (and hence $\rho_S \neq 1$), a second effect size can be defined that standardizes the mean difference by the standard deviation within-schools. It is in the form

$$\delta_{WS} = \frac{\mu_{..}^T - \mu_{..}^C}{\sqrt{\sigma_{BC}^2 + \sigma_{WC}^2}} = \frac{\delta_{WT}}{\sqrt{1 - \rho_S}}. \quad (18.35)$$

The effect size δ_{WS} might be of interest in a situation where most of the studies of interest had two level designs involving multiple schools. In such studies δ_{WS} may (implicitly or explicitly) be the effect size estimated and hence be most comparable with the effect-size estimates of the other studies.

If $\sigma_{WC} \neq 0$ (and hence $\rho_S + \rho_C \neq 1$), a third effect size can be defined that standardizes the mean difference by the within-classroom standard deviation. It has the form

$$\delta_{WC} = \frac{\mu_{..}^T - \mu_{..}^C}{\sigma_{WC}} = \frac{\delta_{WT}}{\sqrt{1 - \rho_S - \rho_C}} = \frac{\delta_{WS}\sqrt{1 - \rho_S}}{\sqrt{1 - \rho_S - \rho_C}}. \quad (18.36)$$

The effect size δ_{WC} might be of interest, for example, in a meta-analysis where the other studies to which the current study is compared are typically studies with a single school. In such studies, δ_{WC} may (implicitly) be the effect size estimated and thus δ_{WC} might be the most comparable effect size.

Two other effect sizes that standardize the mean difference by between classroom and between school standard deviations may be of interest, though less frequently. If $\sigma_{BS} \neq 0$ (and hence $\rho_S \neq 0$), a fourth possible effect size would be

$$\delta_{BS} = \frac{\mu_{..}^T - \mu_{..}^C}{\sigma_{BS}} = \frac{\delta_{WT}}{\sqrt{\rho_S}}. \quad (18.37)$$

The effect size δ_{BS} may also be of interest in a meta-analysis where the studies being compared are typically multilevel studies that have been analyzed by using school means as the unit of analysis. This is less likely to

be of general interest, but might be of interest when the outcome variable on which the treatment effect is defined is conceptually at the level of schools and the individual observations are of interest only because the average defines an aggregate property.

If $\sigma_{BC} \neq 0$ (and hence $\rho_C \neq 0$), a fifth possible effect size would be

$$\delta_{BC} = \frac{\mu_{..}^T - \mu_{..}^C}{\sigma_{BC}} = \frac{\delta_{WT}}{\sqrt{\rho_C}}. \quad (18.38)$$

The effect size δ_{BC} may also be of interest in a meta-analysis when the studies being compared are typically multilevel studies analyzed using classroom means as the unit, or when the outcome is conceptually defined as a classroom-level property. It might be of interest when the outcome variable and the individual observations are of interest only because the classroom average defines an aggregate property.

Because ρ_S and ρ_C will often be rather small, δ_{WS} and δ_{WC} will often be similar in size to δ_{WT}. For the same reason, δ_{BS} and δ_{BC} will typically be much larger in the same population than δ_{WS}, δ_{WC}, or δ_{WT}. Note also that if all of the effect sizes are defined, that is, if $0 < \rho_S < 1, 0 < \rho_C < 1$, and $\rho_S + \rho_C < 1$, any one of these effect sizes may be obtained from any of the others. In particular, equations 18.34 through 18.38 can be solved for δ_{WT} in terms of any of the other effect sizes, ρ_S, and ρ_C. These same equations, in turn, can be used to define any of the five effect sizes in terms of δ_{WT}.

18.3.3 Estimates of Effect Sizes in Three-Level Designs

As in two-level designs, formulas for estimates and their variances are much simpler for balanced designs when all cluster sample sizes are equal. Here I present estimates of the effect sizes and their approximate sampling distributions, first for when all the schools have the same number (p) of classrooms—that is, when $p_i^T = p, i = 1, \ldots, m^T$ and $p_i^C = p, i = 1, \ldots, m^C$—and all the classroom sample sizes are equal to n—that is, when $n_{ij}^T = n, i = 1, \ldots, m^T$; $j = 1, \ldots, p$ and $n_{ij}^C = n, i = 1, \ldots, m^C$; $j = 1, \ldots, p$. In this case, $N^T = npm^T$, $N^C = npm^C$, and $N = N^T + N^C = np(m^T + m^C) = nM$.

18.3.3.1 Estimating δ_{WC} We start with estimation of δ_{WC}, which is the most straightforward. If $\rho_S + \rho_C \neq 1$ (so that $\sigma_{WC} \neq 0$), then δ_{WC} is defined then the estimate

$$d_{WC} = \frac{\overline{Y}_{...}^T - \overline{Y}_{...}^C}{S_{WC}} \quad (18.39)$$

is a consistent estimator of δ_{WC} that is approximately normally distributed with mean δ_{WC}. If cluster sizes are equal the estimated variance of d_{WC} is

$$v_{WC} = \frac{1 + (pn-1)\rho_S + (n-1)\rho_C}{\tilde{m}pn(1-\rho_S-\rho_C)} + \frac{d_{WC}^2}{2(N-Mp)}, \quad (18.40)$$

where

$$\tilde{m} = \frac{m^T m^C}{m^T + m^C}.$$

The leading term of the variance in equation 18.40 arises from uncertainty in the mean difference. It is $[1 + (pn-1)\rho_S + (n-1)\rho_C]/(1-\rho_S-\rho_C)]$ as large as would be expected if there were no clustering in the sample, that is, if $\rho_S = \rho_C = 0$. Thus $[1 + (pn-1)\rho_S + (n-1)\rho_C]/(1-\rho_S-\rho_C)]$ is a kind of variance inflation factor for the variance of the effect-size estimate d_{WC}.

Note that if either $\rho_S = 0$ or $\rho_C = 0$, there is no clustering at one level and that the three-level design logically reduces to a two-level design. If $\rho_S = 0$, the expression 18.39 for d_{WC} and the expression 18.40 for its variance reduce to equations 18.7 and 18.8 in section 18.2.3 for both the corresponding effect-size estimate d_W and its variance in a two-level design, with clusters of size n and intraclass correlation $\rho = \rho_C$. Similarly, if $\rho_C = 0$, expressions 18.39 and 18.40 also reduce to equations 18.7 and 18.8 of section 18.2.3, but with cluster size np and intraclass correlation $\rho = \rho_S$. If $\rho_S = \rho_C = 0$ so that there is no clustering by classroom or by school, the design is logically equivalent to a one-level design. In this case, the expression 18.39 for d_{WT} and expression 18.40 for its variance reduce to the usual expressions for the standardized mean difference and its variance (see, for example, chapter 12, this volume).

When cluster sample sizes are unequal, the estimate d_{WC} of δ_{WC} is identical to that in equation 18.39, but the estimated variance is given by

$$v_{WC} = \frac{1 + (p_U-1)\rho_S + (n_U-1)\rho_C}{\tilde{N}(1-\rho_S-\rho_C)} + \frac{d_{WC}^2}{2(N-P)}, \quad (18.41)$$

where

$$p_U = \frac{N^C \sum_{i=1}^{m^T} \left(\sum_{j=1}^{p_i^T} n_{ij}^T \right)^2}{NN^T} + \frac{N^T \sum_{i=1}^{m^C} \left(\sum_{j=1}^{p_i^C} n_{ij}^C \right)^2}{NN^C},$$

$$n_U = \frac{N^C \sum_{i=1}^{m^T} \sum_{j=1}^{p_i^T} (n_{ij}^T)^2}{NN^T} + \frac{N^T \sum_{i=1}^{m^C} \sum_{j=1}^{p_i^C} (n_{ij}^C)^2}{NN^C},$$

$$P = \sum_{i=1}^{m^T} p_i^T + \sum_{i=1}^{m^C} p_i^C$$

is the total number of classrooms, and $\tilde{N} = N^T N^C/(N^T + N^C)$. Note that if all the n_i^T and n_i^C are equal to n, then $n_U = n$, and that if all the p_i^T and p_i^C are equal to p, then $p_U = pn$, and $P = Mp$, so that equation 18.41 reduces to equation 18.40.

When the analysis properly accounts for clustering by treating the clusters as having random effects, and an estimate b of the treatment effect and $\hat{\sigma}_{WC}^2$ of the variance component σ_{WC}^2 is available (for example, from an HLM analysis), the estimate of δ_{WC} is

$$d_{WC} = \frac{b}{\hat{\sigma}_{WC}}. \quad (18.42)$$

The approximate variance of the estimate given in 18.42 is

$$v_{WC} = \frac{V(b)}{\hat{\sigma}_{WC}^2}. \quad (18.43)$$

18.3.3.2 Estimating δ_{WS} If $\rho_S \neq 1$, then δ_{WS} is defined an estimate of δ_{WS} can also be obtained from the mean difference and S_{WS} if the intraclass correlations ρ_S and ρ_C are known or can be imputed. When all the cluster sample sizes are equal, a consistent estimator of δ_{WS} is

$$d_{WS} = \left(\frac{\overline{Y}_{\cdots}^T - \overline{Y}_{\cdots}^C}{S_{WS}} \right) \sqrt{1 - \frac{(n-1)\rho_C}{(1-\rho_S)(np-1)}}, \quad (18.44)$$

and the estimate d_{WS} is normally distributed in large samples with mean δ_{WS} variance

$$v_{WS} = \frac{1 + (pn-1)\rho_S + (n-1)\rho_C}{\tilde{m}pn(1-\rho_S)}$$
$$+ d_{WS}^2 \left(\frac{(pn-1)\overline{\rho}^2 + 2n(p-1)\overline{\rho}\rho_C + n^2(p-1)\rho_C^2}{2M[(pn-1)^2(1-\rho_S)^2 - (pn-1)(n-1)\rho_C(1-\rho_S)]} \right), \quad (18.45)$$

where $\overline{\rho} = (1-\rho_S-\rho_C)$.

The leading term of the variance in equation 18.45 arises from uncertainty in the mean difference. Note that it is $[1 + (pn-1)\rho_S + (n-1)\rho_C]/(1-\rho_S)]$ as large as would be expected if there were no clustering in the sample, that is, if $\rho_S = \rho_C = 0$. Thus the term $[1 + (pn-1)\rho_S + (n-1)\rho_C]/(1-\rho_S)]$ is a kind of variance inflation factor for the variance of the effect-size estimate d_{WS}.

If $\rho_C = 0$ so that there is no clustering effect by classrooms, the three-level design is logically equivalent to a two-level design. In this case, the expressions 18.44 for d_{WS} and 18.45 for its variance reduce to equations 18.7 and 18.8 in section 18.2.3 for the corresponding effect-size estimate (called there d_W) and its variance in a two-level design, with clusters of size pn and intraclass correlation

$\rho = \rho_S$. If $\rho_S = \rho_C = 0$ so that there is no clustering by classroom or by school, the design is logically equivalent to a one-level design. In this case, the expression 18.44 for d_{WS} and expression 18.45 for its variance reduce to those for the standardized mean difference and its variance in a one-level design (see chapter 12, this volume).

When cluster sample sizes are unequal, an estimate d_{WS} of δ_{WS} is

$$d_{WS} = \left(\frac{\bar{Y}^T_{...} - \bar{Y}^C_{...}}{S_{WS}} \right) \sqrt{1 - \frac{(\tilde{n}-1)\rho_C}{(1-\rho_S)(N_S-1)}}, \qquad (18.46)$$

where

$$\tilde{n} = \frac{\sum\limits_{i=1}^{M^T} \left(\sum\limits_{j=1}^{p^T_i} (n^T_{ij})^2 \Big| n^T_{i+} \right) + \sum\limits_{i=1}^{M^C} \left(\sum\limits_{j=1}^{p^C_i} (n^C_{ij})^2 \Big| n^C_{i+} \right)}{M^T + M^C} \qquad (18.47)$$

$$n^T_{i+} = \sum\limits_{j=1}^{p^T_i} n^T_{ij},$$

$$n^C_{i+} = \sum\limits_{j=1}^{p^C_i} n^C_{ij},$$

and $N_S = N/M$ is the average sample size per school. Note that if all of the n^T_{ij} and n^C_{ij} are equal to n and all the p^T_j and p^C_j are equal to p, then $\tilde{n} = n$, $N_S = pn$, and (18.46) reduces to 18.45.

The variance of d_{WS} in the unequal sample size case is rather complex. It is given exactly in Hedges (in press), but a slight overestimate—that is, a slightly conservative estimate—of the variance of d_{WS} is

$$v_{WS} \approx \frac{1 + (p_U - 1)\rho_S + (n_U - 1)\rho_C}{\tilde{N}(1 - \rho_S)} + \frac{d^2_{WS}}{2(P-M)}. \qquad (18.48)$$

When the analysis properly accounts for clustering by treating the clusters as having random effects, and an estimate b of the treatment effect, $\hat{\sigma}^2_{WC}$ of the variance component σ^2_{WC}, and $\hat{\sigma}^2_{BC}$ of the variance component σ^2_{BC} are available—for example, from an HLM analysis, the estimate of δ_{WS} is

$$d_{WS} = \frac{b}{\sqrt{\hat{\sigma}^2_{BC} + \hat{\sigma}^2_{WC}}}. \qquad (18.49)$$

Assuming that $V(\hat{\sigma}^2_{WC})$ is negligible and that $\hat{\sigma}^2_{BC}$ is independent of the estimated treatment effect b, the approximate variance of the estimate in (18.49) is

$$v_{WS} = \frac{V(b)}{\hat{\sigma}^2_{BC} + \hat{\sigma}^2_{WC}} + \frac{b^2 V(\hat{\sigma}^2_{BC})}{4(\hat{\sigma}^2_{BC} + \hat{\sigma}^2_{WC})^2}. \qquad (18.50)$$

18.3.3.3 Estimating δ_{WT} An estimate of δ_{WT} can also be obtained from the mean difference and S_{WT} if the intra-

class correlations ρ_S and ρ_C are known or can be imputed. When cluster sample sizes are equal, a consistent estimator of δ_{WT} is

$$d_{WT} = \left(\frac{\bar{Y}^T_{...} - \bar{Y}^C_{...}}{S_{WT}} \right) \sqrt{1 - \frac{2(pn-1)\rho_S + 2(n-1)\rho_C}{N-2}}, \qquad (18.51)$$

which is normally distributed in large samples with mean δ_{WT} and variance

$$v_{WT}$$
$$= \frac{1 + (pn-1)\rho_S + (n-1)\rho_C}{\tilde{m}pn}$$
$$+ d^2_{WT} \left(\frac{pn\hat{N}\rho^2_S + n\check{N}\rho^2_C + (N-2)\bar{\rho}^2 + 2n\hat{N}\rho_S\rho_C + 2\hat{N}\rho_S\bar{\rho} + 2\check{N}\rho_C\bar{\rho}}{2(N-2)[(N-2) - 2(pn-1)\rho_S - 2(n-1)\rho_C]} \right), \qquad (18.52)$$

where $\hat{N} = (N - pn)$, $\check{N} = (N - 2n)$, and $\bar{\rho} = 1 - \rho_S - \rho_C$.

The leading term of the variance in equation 18.52 arises from uncertainty in the mean difference. It is $[1 + (np-1)\rho_S + (n-1)\rho_C]$ as large as would be expected if there were no clustering in the sample, that is, if $\rho_S = \rho_C = 0$. The term $[1 + (np-1)\rho_S + (n-1)\rho_C]$ is a generalization to two stages of clustering of the variance inflation factor mentioned by Allan Donner and colleagues (1981) for the variance of means in clustered samples (with one stage of clustering) and corresponds to a variance inflation factor for the effect-size estimate d_{WT}.

If $\rho_C = 0$ so that there is no clustering by classrooms, or if $\rho_S = 0$ so that there is no clustering by schools, the three-level design is logically equivalent to a two-level design. If $\rho_S = 0$, the expression 18.51 for d_{WT} and expression 18.52 for its variance reduce to equations 18.19 and 18.20 in section 18.2.3.3 for the corresponding effect-size estimate d_T and its variance in a two-level design, with clusters of size n and intraclass correlation $\rho = \rho_C$. Similarly, if $\rho_C = 0$, expressions 18.51 and 18.52 reduce to equations 18.19 and 18.20 of section 18.2.3.3 for the corresponding effect size in a two-level design, with cluster size np and intraclass correlation $\rho = \rho_S$.

When school or classroom sample sizes are unequal, a consistent estimate d_{WT} of δ_{WT} is

$$d_{WT} = \left(\frac{\bar{Y}^T_{...} - \bar{Y}^C_{...}}{S_{WT}} \right) \sqrt{1 - \frac{2(p_U-1)\rho_S + 2(n_U-1)\rho_C}{N-2}}, \qquad (18.53)$$

where p_U and n_U are given above. Note that if all of the n^T_{ij} and n^C_{ij} are equal to n and all the p^T_j and p^C_j are equal to p, equation 18.53 is reduced to 18.51.

The exact expression for the variance of d_{WT} is complex when sample sizes are unequal (see Hedges, in press). A slight overestimate, that is, a slightly conservative estimate, of the variance of d_{WT} is

$$v_{WT} \approx \frac{1 + (p_U - 1)\rho_S + (n_U - 1)\rho_C}{\bar{N}} + \frac{d^2_{WT}}{2(M-2)}. \quad (18.54)$$

When the analysis properly accounts for clustering by treating the clusters as having random effects, and an estimate b of the treatment effect, $\hat{\sigma}^2_{BS}$ of the variance component σ^2_{BS}, $\hat{\sigma}^2_{BC}$ of the variance component σ^2_{BC}, and $\hat{\sigma}^2_{WC}$ of the variance component σ^2_{WC} are available (for example, from an HLM analysis), the estimate of δ_{WS} is

$$d_{WT} = \frac{b}{\sqrt{\hat{\sigma}^2_{BS} + \hat{\sigma}^2_{BC} + \hat{\sigma}^2_{WC}}}. \quad (18.55)$$

Assuming that $V(\hat{\sigma}^2_{WC})$ is negligible and that $\hat{\sigma}^2_{BS}$ and $\hat{\sigma}^2_{BC}$ are independent of each other and the estimated treatment effect b, The approximate variance of the estimate given in 18.55 is

$$v_{WT} = \frac{V(b)}{\hat{\sigma}^2_{BS} + \hat{\sigma}^2_{BC} + \hat{\sigma}^2_{WC}} + \frac{b^2 [V(\hat{\sigma}^2_{BS}) + V(\hat{\sigma}^2_{BC})]}{4(\hat{\sigma}^2_{BS} + \hat{\sigma}^2_{BC} + \hat{\sigma}^2_{WC})^3}. \quad (18.56)$$

18.3.3.4 Estimating δ_{BC} If $\rho_C \neq 0$ (so that δ_{BC} is defined), S_{BC} may be used, along with the intraclass correlations ρ_S and ρ_C, to obtain an estimate of δ_{BC}. When all of the cluster sample sizes are equal,

$$d_{BC} = \frac{\bar{Y}^T_{\cdots} - \bar{Y}^C_{\cdots}}{S_{BC}} \sqrt{\frac{1 - \rho_S + (n-1)\rho_C}{n\rho_C}} \quad (18.57)$$

is a consistent estimate of δ_{BC} that is normally distributed in large samples with mean δ_{BC} and variance

$$v_{BC} = \frac{1 - \rho_S + (n-1)\rho_C}{\tilde{m}pn\rho_C} + \frac{[1 - \rho_S + (n-1)\rho_C]d^2_{BC}}{2M(p-1)n\rho_C}. \quad (18.58)$$

The presence of ρ_C in the denominator of the estimator and its variance is possible because the effect-size parameter exists only if $\rho_C \neq 0$.

The variance in equation 18.58 is $[1 - \rho_S + (n-1)\rho_C]/n\rho_C]$ as large as that of the standardized mean difference computed from an analysis using class means as the unit of analysis when $\rho_S = \rho_C = 0$, that is, computing the first factor on the right hand side of equation 18.58. Thus $[1 - \rho_S + (n-1)\rho_C]/n\rho_C]$ is a kind of variance inflation factor for the variance of effect-size estimates like d_{BC}.

Note that if $\rho_S = 0$ so that there is no clustering by school, the expression 18.57 for d_{BC} and expression 18.58 for its variance reduce to those given in equations 18.13

and 18.14 in section 18.2.3.2 for the corresponding effect-size estimate (called there d_B) and its variance in a two-level design with clusters of size n and intraclass correlation $\rho = \rho_C$.

18.3.3.5 Estimating δ_{BS} If $\rho_S \neq 0$ (so that δ_{BS} is defined), S_{BS} may be used, along with the intraclass correlations ρ_S and ρ_C, to obtain an estimate of δ_{BS}. When all the cluster sample sizes are equal,

$$d_{BS} = \frac{\bar{Y}^T_{\cdots} - \bar{Y}^C_{\cdots}}{S_{BS}} \sqrt{\frac{1 - (pn-1)\rho_S + (n-1)\rho_C}{pn\rho_S}} \quad (18.59)$$

is a consistent estimate of δ_{BS} that is normally distributed in large samples with mean δ_{BS} and (estimated) variance

$$v_{BS} = \frac{1 + (pn-1)\rho_S + (n-1)\rho_C}{\tilde{m}pn\rho_S} + \frac{[1 + (pn-1)\rho_S + (n-1)\rho_C]d^2_{BS}}{2(M-2)pn\rho_S}. \quad (18.60)$$

The presence of ρ_S in the denominator of the estimator and its variance is possible since the effect-size parameter exists only if $\rho_S \neq 0$.

The variance in equation 18.60 is $[1 + (pn-1)\rho_S + (n-1)\rho_C]/pn\rho_S]$ as large as the variance of the standardized mean difference computed from an analysis using school means as the unit of analysis when $\rho_S = \rho_C = 0$. Thus $[1 - \rho_S + (n-1)\rho_C]/pn\rho_S]$ is a kind of variance inflation factor for the variance of effect-size estimates like d_{BS}.

Note that if $\rho_C = 0$ so that there is no clustering by classroom, expressions 18.59 and 18.60 reduce to the corresponding estimate (called there d_B) in equation 18.13 and its variance in equation 18.14 of section 18.2.3.2 in a two-level design, with cluster size np and intraclass correlation $\rho = \rho_S$.

18.3.3.6 Unequal Sample Sizes when Using d_{BC} and d_{BS} There is more than one way to generalize the estimator d_{BC} to unequal school and classroom sample sizes. One is to use the means of the classroom means in the treatment and control group in the numerator and standard deviation of the classroom means in the denominator. This corresponds to using the classroom means as the unit of analysis. Another possibility is to use the grand means in the treatment and control groups. Similarly, there are multiple possibilities for the denominator, such as the mean square between classrooms or the standard deviation of the classroom means. When school and classroom sample sizes are identical, all these approaches are equivalent in that the effect-size estimates are identical. When the cluster sample sizes are not identical, the resulting estimators (and their sampling distributions) are also not the same.

Similarly, there is more than one way to generalize the estimator d_{BS} to unequal school and classroom sample sizes. One possibility is to use the means of the school means in the treatment and control group in the numerator and their standard deviation in the denominator. This in effect uses the school means as the unit of analysis. Another possibility for the numerator is to use the grand means in the treatment and control groups. Similarly, there are multiple possibilities for the denominator, such as the mean square between schools or the standard deviation of school means. When school and classroom sample sizes are identical, all these approaches are equivalent in that the effect-size estimates are identical. When the school and classroom sample sizes are not identical, the resulting estimators and their sampling distributions are of course not the same.

Because there are so many possibilities, because some of them are complex, and because the information necessary to use them—means, standard deviations, and sample sizes for each classroom and school—is frequently not reported in research reports, I do not include the estimates or their sampling distributions here. A sensible approach if these estimators are needed is to compute the variance estimates for the equal sample size cases, substituting the equal sample size formulas.

18.3.3.7 Confidence Intervals for δ_{WC}, δ_{WS}, δ_{WT}, δ_{BC}, and δ_{BS} Confidence intervals for any of the effect sizes described in section 18.3 can be obtained from the corresponding estimate using the method described in section 18.2.3.4. That is, if δ is any one of the effect sizes mentioned, d is a corresponding estimate, and v_d is the estimated variance of d, then a $100(1 - \alpha)$ percent confidence interval for δ based on d and v_d is given by (18.25).

18.3.4 Applications at Three Levels

This chapter should be useful in deciding what effect sizes are desirable in a three-level experiment or quasi-experiment. It should also be useful for finding ways to compute effect-size estimates and their variances from data that may be reported.

18.3.4.1 Effect Sizes When Individuals Are the Unit of Analysis The results in section 18.3 can be used to produce effect-size estimates and their variances from

studies that incorrectly analyze three level designs as if there were no clustering in the sampling design. The required means, standard deviations, and sample sizes cannot always be extracted from what may be reported, but often it is possible to extract the information to compute at least one of the effect sizes discussed here.

Suppose it is decided that the effect size δ_{WT} is appropriate because most other studies both assign and sample individually from a clustered population. Suppose that the data in a study are analyzed by ignoring clustering. The test statistic is then likely to be either

$$t = \sqrt{\frac{N^T N^C}{N^T + N^C}} \left(\frac{\overline{Y}^T_{\cdots} - \overline{Y}^C_{\cdots}}{S_{WT}} \right)$$

or $F = t^2$. Either can be solved for

$$\left(\frac{\overline{Y}^T_{\cdots} - \overline{Y}^C_{\cdots}}{S_{WT}} \right),$$

which can then be inserted into equation 18.51 or 18.53 along with ρ_S and ρ_C to obtain d_{WT}. This estimate of d_{WT} can then be inserted into equation 18.52 or 18.54 to obtain v_{WT}, an estimate of the variance of d_{WT}.

Alternatively, suppose it is decided that the effect size δ_{WC} is appropriate because most other studies involve only a single site. We may begin by computing d_{WT} and v_{WT} as before. Because we want an estimate of δ_{WC}, not δ_{WT}, we use the fact given in equation (18.36) that

$$\delta_{WC} = \frac{\delta_{WT}}{\sqrt{1 - \rho_S - \rho_C}}$$

and therefore

$$\frac{d_{WT}}{\sqrt{1 - \rho_S - \rho_C}} \tag{18.61}$$

is an estimate of δ_{WC} with a variance of

$$\frac{v_{WT}}{\sqrt{1 - \rho_S - \rho_C}}. \tag{18.62}$$

Similarly, we might decide that δ_{WS} was a more appropriate effects size because most other studies involve schools that have several classrooms, and nesting of students within classrooms is not taken into account in most of them. We may begin by computing d_{WT} and v_{WT} as before. Because we want an estimate of δ_{WS}, not δ_{WT}, we use the fact given in equation 18.35 that

$$\delta_{WS} = \frac{\delta_{WT}}{\sqrt{1 - \rho_S}}$$

and therefore

$$\frac{d_{WT}}{\sqrt{1-\rho_S}} \qquad (18.63)$$

is an estimate of δ_{WS} with a variance of

$$\frac{v_{WT}}{\sqrt{1-\rho_S}}. \qquad (18.64)$$

If we decided that δ_{BC} or δ_{BS} was a more appropriate effect size, then we could still begin by computing d_{WT} and v_{WT}, but then solve 18.38 for δ_{BC}, namely

$$\delta_{BC} = \delta_{WT}/\sqrt{\rho_C}$$

to obtain

$$d_{WT}/\sqrt{\rho_C} \qquad (18.65)$$

as an estimate of δ_{BC} with variance

$$v_{WT}/\rho_C \qquad (18.66)$$

or solve equation 18.37 for δ_{BS}, namely

$$\delta_{BS} = \delta_{WT}/\sqrt{\rho_S}$$

to obtain

$$d_{WT}/\sqrt{\rho_S} \qquad (18.67)$$

as an estimate of δ_{BS} with variance v_{WT}/ρ_S.

18.4 CAN A LEVEL OF CLUSTERING SAFELY BE IGNORED?

One might be tempted to think that accounting for clustering at the highest level, schools, for example, would be enough to account for most of the effects of clustering at both levels. Indeed, the belief among some researchers that only the highest level of clustering matters in analysis, and that lower levels can safely be omitted if they are not explicitly part of the analysis, is widely held. Another sometimes expressed is that any level of the design can be ignored if it is not explicitly included in the analysis. It is clear from 18.39, 18.44, and 18.51 that the effect-size estimates d_{WC}, d_{WS}, and d_{WT} do not depend strongly on the intraclass correlation, at least within plausible ranges of intraclass correlations likely to be encountered in social and educational research. The analytic results presented here, however, show that in situations where there is a plausible amount of clustering at two levels, omitting either level from variance computations can lead to substantial biases in the variance of effect-size estimators.

Table 18.1 shows several calculations of the variance of the effect-size estimate d_{WT} and its variance in a balanced design with $m = 10$, $n = 20$, $\rho_S = 0.15$, $\rho_C = 0.10$, and $d = (\overline{T}_{...}^T - \overline{T}_{...}^C)/S_{WT} = 0.15$ (fairly typical values). Omitting the classroom (lowest level) of clustering leads to underestimating the variance when the number of classrooms per school is small, underestimation that becomes less serious as the number of classrooms per school increases. Omitting it from variance computations leads to

Table 18.1 Variance Effect of Omitting Levels on the Variance of Effect Size d_{WT}

			Variance Computed				Estimated Variance / True Variance			
p	d_{WT}	v_{WT}	Ignoring Classes	Ignoring Schools	Ignoring Classes $\rho = \rho_S + \rho_C$	Ignoring Classes $\rho = \rho_S + \rho_C/p$	Ignoring Classes	Ignoring Schools	Ignoring Classes $\rho = \rho_S + \rho_C$	Ignoring Classes $\rho = \rho_S + \rho_C/p$
1	0.1482	0.0576	0.0385	0.0290	0.0576	0.0576	0.670	0.504	1.000	1.000
2	0.1485	0.0438	0.0343	0.0145	0.0538	0.0440	0.783	0.332	1.229	1.006
3	0.1486	0.0392	0.0329	0.0097	0.0525	0.0394	0.838	0.247	1.341	1.006
4	0.1487	0.0369	0.0321	0.0073	0.0519	0.0371	0.871	0.197	1.407	1.005
5	0.1487	0.0355	0.0317	0.0058	0.0515	0.0357	0.893	0.163	1.451	1.004
8	0.1488	0.0335	0.0311	0.0036	0.0510	0.0336	0.929	0.108	1.524	1.003
16	0.1488	0.0317	0.0305	0.0018	0.0505	0.0318	0.963	0.057	1.591	1.002

SOURCE: Author's compilation.
NOTE: These calculations assume $n = 20$, $\rho_S = 0.15$, $\rho_C = 0.10$, and $d_{WT} = 0.15$

underestimating the variance of d_{WT} by 22 percent when $p = 2$, but by only 11 percent when $p = 5$. Because the number of classrooms per school is typically small in educational experiments, we judge that variance is likely to be seriously underestimated when the classroom level is ignored in computing variances of effect-size estimates in three-level designs.

The table makes it clear that omitting the school level of clustering leads to larger underestimates of variance when the number of classrooms per school is small. These misestimates become even more serious when the number of classrooms becomes larger. Omitting the school level of clustering from variance computations leads to underestimating the variance of d_{WT} by about 67 percent when $p = 2$, but by 84 percent when $p = 5$.

One possible substitute for computing variances using three levels is to compute the variances assuming two levels, but use a composite intraclass correlation that "includes" both between-school and between-class components. Table 18.1 shows that using $\rho_S + \rho_C$ in place of the school-level intraclass correlation performs poorly, leading to overestimates of the variance of from 23 percent (for $p = 2$) to 45 percent (for $p = 5$). Using $\rho_S + \rho_C/p$ in place of the school-level intraclass correlation performs quite well, however.

18.5 SOURCES OF INTRACLASS CORRELATION DATA

Intraclass correlations are needed for the methods described here, but are often not reported, particularly when these effect sizes are calculated retrospectively in meta-analyses. However, because plausible values of ρ are essential for power and sample size computations in planning cluster randomized experiments, there have been systematic efforts to obtain information about reasonable values of ρ in realistic situations. Some information comes from cluster randomized trials that have been conducted. David Murray and Jonathan Blitstein, for example, reported a summary of intraclass correlations obtained from seventeen articles on cluster randomized trials in psychology and public health (2003). Murray, Sherri Varnell, and Blitstein gave references to fourteen studies that provided data on intraclass correlations for health related outcomes (2004). Other information on reasonable values of ρ comes from sample surveys that use clustered sampling designs. For example, Martin Gulliford, Obioha Ukoumunne, and Susan Chinn (1999) and Vijay Verma and Than Le (1996) have both presented values of intraclass correlations based

on surveys of health outcomes. Larry Hedges and Eric Hedberg presented a compendium of intraclass correlations computed from surveys of academic achievement that used national probability samples at various grade levels (2007). This compendium provides national values for intraclass correlations as well as values for regions of the country and subsets of regions differing in level of urbanicity.

Although most of these surveys have focused in intraclass correlations in two-level designs, they do provide some empirical information and guidelines about patterns of intraclass correlations. The National Assessment of Educational Progress (NAEP) uses a sampling design that makes possible the computation of both school and classroom level intraclass correlations (our ρ_S and ρ_C). Spyros Konstantopoulos and Larry Hedges used NAEP to estimate school and classroom level intraclass correlations in reading and mathematics achievement (2007). The values of school-level intraclass correlations (ρ_S values) are consistent with those obtained with those Barbara Nye, Hedges, and Konstantopoulos obtained for a large experiment involving seventy-one school districts in Tennessee (2000). In both cases, the classroom level intraclass correlations (ρ_C values) are consistently about two-thirds the size of the school-level intraclass correlations (ρ_S values).

In most situations, exact values of intraclass correlation parameters will not be available, so effect sizes and their variances computed from results in this paper will involve imputing values for intraclass correlations. In such cases, it can be helpful to consider a range of values. Similarly, it may be useful to ask how large the intraclass correlation would need to be to result in a nonsignificant effect size. In some cases, studies may actually report enough information, between and within-cluster variance component estimates, to estimate the intraclass correlation. Even in cases where a study provides an estimate of the intraclass correlations, such an estimate is likely to have large sampling uncertainty unless the study has hundreds of clusters. Thus, even in this case, it is wise to consider a range of possible intraclass correlation values. The exact range used should depend on the amount of clustering plausible in the particular setting and outcomes involved. For example, in educational settings where the clusters are schools and academic achievement is the outcome, the data reported in Hedges and Hedberg suggested that it might be wise to consider a range of school-level intraclass correlations from $\rho = 0.10$ to 0.30 (2007).

18.6 CONCLUSIONS

The clustered sampling structure of nested designs leads to different possible effect-size parameters and has an impact on both effect-size estimates and their variances. Methods have been presented for using intraclass correlations to compute various effect-size estimates and their variances in nested designs. It is important to put the impact of clustering on meta-analysis in perspective. The results of this chapter show that, though the effects of clustering on the variance of effect-size estimates can be profound, the effects of clustering on the value of the estimates themselves is often modest. Failing to account for clustering in computing effect-size estimates will therefore typically introduce little bias. Failure to account for clustering in the computation of variances, however, can lead to serious underestimation of these variances. The underestimation of variance can have two consequences. First, it will lead to studies whose variance has been underestimated receiving larger weight than they should. Second, it will lead to underestimating the variance of combined effects. However, if only a small proportion of the studies in a meta-analysis involve nested designs, ignoring clustering in these studies may have only a modest effect on the overall outcome of the meta-analysis. On the other hand, if a large proportion of the studies involve clustering, or if the studies that do involve clustering receive a large proportion of the total weight, then ignoring clustering could have a substantial impact on the overall results, particularly on the variance of the estimated overall effect size.

18.7 REFERENCES

Bryk, Anthony, and Barbara Schneider. 2000. *Trust in Schools*. New York: Russell Sage Foundation.

Donner, Allan, Nicolas Birkett, and Carol Buck. 1981. "Randomization by Cluster" *American Journal of Epidemiology* 114(6): 906–14.

Donner, Allan, and Neil Klar. 2000. *Design and Analysis of Cluster Randomization Trials in Health Research*. London: Arnold.

———. 2002. "Issues in the Meta-Analysis of Meta-Analysis of Cluster Randomized Trials." *Statistics in Medicine* 21(19): 2971–80.

Guilliford, Martin C., Obioha C. Ukoumunne, and Susan Chinn. 1999. "Components of Variance and Intraclass Correlations for the Design of Community-Based Surveys and Intervention Studies: Data from the Health Survey for England 1994." *American Journal of Epidemiology* 149(9): 876–83.

Hedges, Larry V. 1981. "Distribution Theory for Glass's Estimator of Effect Size and Related Estimators." *Journal of Educational Statistics* 6(2): 107–28.

———. 2007. "Effect Sizes in Cluster Randomized Designs." *Journal of Educational and Behavioral Statistics* 32(4): 341–70.

———. In press. "Effect Sizes in Designs with Two Levels of Nesting." *Journal of Educational and Behavioral Statistics*.

Hedges, Larry V., and Eric C. Hedberg. 2007. "Intraclass Correlations for Planning Group Randomized Experiments in Education." *Educational Evaluation and Policy Analysis* 29(1): 60–87.

Kish, Leslie. 1965. *Survey Sampling*. New York: John Wiley & Sons.

Klar, Neil, and Allan Donner. 2001. "Current and Future Challenges in the Design and Analysis of Cluster Randomization Trials." *Statistics in Medicine* 20(24): 3729–40.

Konstantopoulos, Spyros, and Larry V. Hedges. 2007. "How Large an Effect Can We Expect from School Reforms?" *Teachers College Record* 110(8): 1613–40.

Murray, David M., and Jonathan L. Blitstein. 2003. "Methods to Reduce the Impact of Intraclass Correlation in Group-Randomized Trials." *Evaluation Review* 27(1): 79–103.

Murray, David M., Sherri P. Varnell, and Jonathan L. Blitstein. 2004. "Design and Analysis of Group-Randomized Trials: A Review of Recent Methodological Developments." *American Journal of Public Health* 94(3): 423–32.

Nye, Barbara, Larry V. Hedges, and Spyros Konstantopoulos. 2000. "The Effects of Small Classes on Achievement: The Results of the Tennessee Class Size Experiment." *American Educational Research Journal* 37(1): 123–51.

Raudenbush, Stephen W., and Anthony S. Bryk. 2002. *Hierarchical Linear Models*. Thousand Oaks, Calif.: Sage Publications.

Rooney, Brenda L., and David M. Murray. 1996. "A Meta-Analysis of Smoking Prevention Programs after Adjustment for Errors in the Unit of Analysis." *Health Education Quarterly* 23(1): 48–64.

Verma, Vijay, and Than Le. 1996. "An Analysis of Sampling Errors for Demographic and Health Surveys." *International Statistical Review* 64(3): 265–94.

19

STOCHASTICALLY DEPENDENT EFFECT SIZES

LEON J. GLESER
University of Pittsburgh

INGRAM OLKIN
Stanford University

CONTENTS

19.1 INTRODUCTION

Much of the literature on meta-analysis deals with analyzing effect sizes obtained from k independent studies in each of which a single treatment is compared with a control (or with a standard treatment). Because the studies are statistically independent, so are the effect sizes.

Studies, however, are not always so simple. For example, some may compare multiple variants of a type of treatment against a common control. Thus, in a study of the beneficial effects of exercise on blood pressure, independent groups of subjects may each be assigned one of several types of exercise: running for twenty minutes daily, running for forty minutes daily, running every other day, brisk walking, and so on. Each of these exercise groups is to be compared with a common sedentary control group. In consequence, such a study will yield more than one exercise versus control effect size. Because the effect sizes share a common control group, the estimates of these effect sizes will be correlated. Studies of this kind are called *multiple-treatment studies.*

In other studies, the single-treatment, single-control paradigm may be followed, but multiple measures will be used as endpoints for each subject. Thus, in comparing exercise and lack of exercise on subjects' health, measurements of systolic blood pressure, diastolic blood pressure, pulse rate, cholesterol concentration, and so on, may be taken for each subject. Similarly, studies of the use of carbon dioxide for storage of apples can include measures of flavor, appearance, firmness, and resistance to disease. A treatment versus control effect-size estimate may be calculated for each endpoint measure. Because measures on a common subject are likely to be correlated, corresponding estimated effect sizes for these measures will be correlated within studies. Studies of this type are called *multiple-endpoint studies* (for further discussions of multiple-endpoint studies, see Gleser and Olkin 1994; Raudenbush, Becker, and Kalaian 1988; Timm 1999).

A special, but common, kind of multiple-endpoint study is that in which the measures (endpoints) used are subscales of a psychological test. For study-to-study comparisons, or to have a single effect size for treatment versus control, we may want to combine the effect sizes obtained from the subscales into an overall effect size. Because subscales have differing accuracies, it is well known that weighted averages of such effect sizes are required. Weighting by inverses of the variances of the estimated subscale effect sizes is appropriate when these effect sizes are independent, but may not produce the most precise estimates when the effect sizes are correlated.

In each of these above situations, possible dependency among the estimated effect sizes needs to be accounted for in the analysis. To do so, additional information has to be obtained from the various studies. For example, in the multiple-endpoint studies, dependence among the endpoint measures leads to dependence between the corresponding estimated effect sizes, and values for between-measures correlations will thus be needed for any analysis. Fortunately, as will be seen, in most cases this is all the extra information that will be needed. When the studies themselves fail to provide this information, the correlations can often be imputed from test manuals (when the measures are subscales of a test, for example) or from published literature on the measures used.

When dealing with dependent estimated effect sizes, we need formulas for the covariances or correlations. Note that the dependency between estimated effect sizes in multiple-endpoint studies is intrinsic to such studies, arising from the relationships between the measures used, whereas the dependency between estimated effect sizes in multiple-treatment studies is an artifact of the design (the use of a common control). Consequently, formulas for the covariances between estimated effect sizes differ between the two types of studies, necessitating separate treatment of each type. On the other hand, the variances of the estimated effect sizes have the same form in both types of study—namely, that obtained from considering each effect size in isolation (see chapters 15 and 16, this volume). Recall that such variances depend on the true effect size, the sample sizes for treatment and control, and (possibly) the treatment-to-control variance ratio (when the variance of a given measurement is assumed to be affected by the treatment). As is often the case in analyses in other chapters in this volume, the results obtained are large sample (within studies) normality approximations based on use of the central limit theorem.

19.2 MULTIPLE-TREATMENT STUDIES: PROPORTIONS

Meta-analytic methods have been developed for determining the effectiveness of a single treatment versus a control or standard by combining such comparisons over a number of studies. When the endpoint is dichotomous (such as death or survival, agree or disagree, succeed or fail), effectiveness is typically measured by several alternative metrics: namely, differences in risks (proportions),

Table 19.1 Proportions of Subjects Exhibiting Heart Disease for a Control and Three Therapies.

| | Control | | T_1 | | T_2 | | T_3 | |
Study	n	p	n	p	n	p	n	p
1	1000	0.0400	4000	0.0250	4000	0.0375	—	—
2	200	0.0500	400	0.0375	—	—	—	—
3	2000	0.0750	1000	0.0400	1000	0.0800	1000	0.0500

SOURCE: Authors' compilation.
NOTE: Here, n denotes the sample size and p the proportion.

log risk ratios, log odds ratios, and arc sine square root differences. In large studies, more than one treatment may be involved, each treatment being compared to the control. This is particularly true of pharmaceutical studies, in which the effects of several drugs or drug doses are compared. Because investigators may have different goals or different resources, different studies may involve different treatments, and not all treatments of interest may be used in any particular study. When a meta-analytic synthesis of all studies that involve the treatments of interest to the researcher is later attempted, the facts that effect sizes within studies are correlated, and that some studies may be missing one or more treatments, need to be accounted for in the statistical analysis.

19.2.1 An Example and a Regression Model

Suppose that one is interested in combining the results of several studies of the effectiveness of one or more of three hypertension therapies in preventing heart disease in males or females considered at risk because of excessively high blood-pressure readings. For specificity, suppose that the therapies are T_1 = use of a new beta-blocker, T_2 = use of a drug that reduces low-density cholesterol, and T_3 = use of a calcium channel blocker. Suppose that the endpoint is the occurrence (or nonoccurrence) of coronary heart disease. In every study, the effects of these therapies are compared to that of using only a diuretic (the control). Although the data for convenience of exposition are hypothetical, they are illustrative of studies that have appeared in this area of medical research.

That all therapies are compared to the same control in each study suggests that the effect sizes for these therapies will be positively correlated. If the control rate of

occurrence of heart disease is high, it will not take much for all therapies in that study to show a strong improvement over the control, whereas when the control rate in a study is low, even the best therapies are likely to only show a weak effect. In an alternative context, in education, if students in a school are very good, it is hard to show much improvement, whereas if the students in a school have poor skills, then substantial improvement is possible. Given that this is the case, information about the effect size for therapy T_1 can be used to predict the effect size for any other therapy (for example, T_2) in the same study, and vice versa.

To illustrate the model, suppose that a thorough search of published and unpublished studies using various registers yields three studies. Although all of these studies use a similar control, each study need not include all the treatments. The hypothetical data in table 19.1—in which study 1 used therapies T_1 and T_2, study 2 used only therapy T_1, and study 3 involved all three therapies—make this clear.

For our illustration, estimated effect sizes will be differences between the proportion of the control group who experience an occurrence of heart disease and the proportion of those treated who have an occurrence. Later we discuss alternative metrics to define effect sizes. In this case, the effect size for each study is:

$$d_T = p_C - p_T. \qquad (19.1)$$

(Because we expect the treatments to lower the rate of heart disease, in defining the effect size, we have subtracted the treatment proportion from the control proportion in order to avoid a large number of minus signs in the estimates.) The effect-size estimates for each study are

Table 19.2 Effect Sizes (Differences of Proportions) from Table 1

Study	Treatment		
	T_1	T_2	T_3
1	0.0150	0.0025	—
2	0.0125	—	—
3	0.0350	−0.0050	0.0250

SOURCE: Authors' compilation.

shown in table 19.2. The dashes in tables 19.1 and 19.2 indicate where particular therapies are not present in a study.

The estimated variance of a difference of proportions $p_C - p_T$ is

$$\hat{\text{var}}(p_C - p_T) = \frac{p_C(1 - p_C)}{n_C} + \frac{p_T(1 - p_T)}{n_T}, \qquad (19.2)$$

where n_C and n_T are the number of subjects in the control and therapy groups, respectively. The estimated covariance between any two such differences $p_C - p_{T_1}$ and $p_C - p_{T_2}$ is equal to the estimated variance of the common control proportion p_C, namely,

$$\hat{\text{cov}}(p_C - p_{T_1}, p_C - p_{T_2}) = \frac{p_C(1 - p_C)}{n_C}. \qquad (19.3)$$

Notice that the variance and covariance formulas are the same regardless of whether we define effect sizes as $p_C - p_T$ or $p_T - p_C$, as long as the same definition is used for all treatments and all studies.

Thus, the following are the estimated covariance matrices for the three studies. Note that a covariance matrix for a study gives estimated variances and covariances only for the effect sizes observed in that study.

Study 1: $\hat{\boldsymbol{\Psi}}_1 = 10^{-9} \begin{pmatrix} 44,494 & 38,400 \\ 38,400 & 47,423 \end{pmatrix}$,

Study 2: $\hat{\boldsymbol{\Psi}}_2 = 10^{-9}(327,734)$,

Study 3: $\hat{\boldsymbol{\Psi}}_3 = 10^{-9} \begin{pmatrix} 73,088 & 34,688 & 34,688 \\ 34,688 & 108,288 & 34,688 \\ 34,688 & 34,688 & 82,188 \end{pmatrix}$. (19.4)

In study 1, we might be interested in comparing the effect sizes d_{T_1} and d_{T_2} of treatments T_1 and T_2. If we ig-

nore the covariance between d_{T_1} and d_{T_2}, we would obtain the variance of the effect-size difference $d_{T_1} - d_{T_2}$ by summing the variances of d_{T_1} and d_{T_2} obtaining

$$\hat{\text{var}}(d_{T_1} - d_{T_2}) = \hat{\text{var}}(d_{T_1}) + \hat{\text{var}}(d_{T_2})$$
$$= \frac{p_{T_1}(1 - p_{T_1})}{n_{T_1}} + \frac{p_{T_2}(1 - p_{T_2})}{n_{T_2}} + 2\frac{p_C(1 - p_C)}{n_C}.$$

That this variance would be too large can be seen by noting that

$$d_{T_1} - d_{T_2} = (p_C - p_{T_1}) - (p_C - p_{T_2}) = p_{T_2} - p_{T_1},$$

so that actually

$$\hat{\text{var}}(d_{T_1} - d_{T_2}) = \hat{\text{var}}(p_{T_1} - p_{T_2}) = \hat{\text{var}}(p_{T_1}) + \hat{\text{var}}(p_{T_2})$$
$$= \frac{p_{T_1}(1 - p_{T_1})}{n_{T_1}} + \frac{p_{T_2}(1 - p_{T_2})}{n_{T_2}}.$$

In our hypothetical data, the difference between these two variance estimates is (relatively) quite large:

$$\hat{\text{var}}(d_{T_1} - d_{T_2}) = 10^{-9}(44,494 + 47,423) = 10^{-9}(91,917)$$

versus

$$\hat{\text{var}}(d_{T_1} - d_{T_2}) = \frac{(.0250)(.9750)}{4000} + \frac{(.0375)(.9625)}{4000}$$
$$= 10^{-9}(15,117)$$

indicating that in such a comparison of effect sizes it would be a major error to ignore the covariance between the effect-size estimates. A large-sample confidence interval for the difference of the two population effect sizes, for example, would be excessively conservative if the covariance between the effect-size estimates d_{T_1} and d_{T_2} were ignored, leading to the use of an overestimate of the variance of $d_{T_1} - d_{T_2}$. On the other hand, if we were interested in averaging the two effect sizes to create a summary effect size for drug therapy, we would underestimate the variance of $(d_{T_1} + d_{T_2})/2$ if we failed to take account of the covariance between d_{T_1} and d_{T_2}.

The existence of covariance (dependence) between effect sizes within studies means that even when a given study has failed to observe a given treatment, the effect sizes for other treatments that were observed and compared to the common control can provide information about any given effect size. To clarify this point, let us display the estimated effect sizes in a six-dimensional column vector d, where the first two coordinates of d are the effect sizes from study 1, the next coordinate is the effect size from study 2 and the remaining two coordinates are effect sizes from study 3. Thus,

$$d = (0.0150, 0.0025, 0.0125, 0.0350, -0.0050, 0.0250)'.$$

Let δ_{T_j} denote the population effect size for treatment T_j, $j = 1,2,3$. Then we can write the following regression model for the vector $\boldsymbol{\delta} = (\delta_{T_1}, \delta_{T_2}, \delta_{T_3})'$ of population effect sizes:

$$d = X\boldsymbol{\delta} + \boldsymbol{\varepsilon}, \qquad (19.5)$$

where $\boldsymbol{\varepsilon}$ is a random vector of the same dimension as d having mean vector 0 and variance-covariance matrix $\hat{\boldsymbol{\Psi}}$ of block diagonal form:

$$\hat{\boldsymbol{\Psi}} = \begin{pmatrix} \hat{\boldsymbol{\Psi}}_1 & 0 & 0 \\ 0 & \hat{\boldsymbol{\Psi}}_2 & 0 \\ 0 & 0 & \hat{\boldsymbol{\Psi}}_3 \end{pmatrix},$$

in which the matrices $\hat{\boldsymbol{\Psi}}_i$, $i = 1,2,3$ are those displayed in equation 19.4. In the regression model 19.5, the design matrix is

$$X = \begin{pmatrix} 1 & 0 & 0 \\ 0 & 1 & 0 \\ 1 & 0 & 0 \\ 1 & 0 & 0 \\ 0 & 1 & 0 \\ 0 & 0 & 1 \end{pmatrix}.$$

Note that the first two rows of X pertain to study 1, the next row to study 2, and the remaining three rows to study 3. Using weighted least squares, we can form the best linear unbiased estimator

$$\hat{\boldsymbol{\delta}} = (X'\hat{\boldsymbol{\Psi}}^{-1}X)^{-1}X'\hat{\boldsymbol{\Psi}}^{-1}d = (0.0200, 0.0043, 0.0211)'$$
$$(19.6)$$

of $\boldsymbol{\delta} = (\delta_{T_1}, \delta_{T_2}, \delta_{T_3})'$. The estimated variance-covariance matrix of the estimator $\hat{\boldsymbol{\delta}}$ is

$$C = \begin{pmatrix} c_{11} & c_{12} & c_{13} \\ c_{21} & c_{22} & c_{23} \\ c_{31} & c_{32} & c_{33} \end{pmatrix} = (X'\boldsymbol{\Psi}^{-1}X)^{-1}$$

$$= 10^{-9} \begin{pmatrix} 24{,}612 & 19{,}954 & 13{,}323 \\ 19{,}954 & 28{,}538 & 13{,}255 \\ 13{,}323 & 13{,}255 & 69{,}807 \end{pmatrix}. \qquad (19.7)$$

Three points are worth noting here. First, observe that even though treatment T_3 appeared only in study 3, the estimator of the effect size for this treatment that we have obtained by regression using equation 19.6 differs from the effect size for treatment T_3 given for study 3 in table 19.2 (0.0211 versus 0.0250). Note also that the estimated variance of the estimate of the effect size for treatment 3 that is given in equation 19.7, namely 10^{-9} (69,807), is noticeably smaller than the estimated variance of the effect size for this treatment in study 3, namely 10^{-9} (82,188). This indicates that information from the estimated effect sizes of other treatments has been used in estimating the effect size of treatment T_3.

Second, suppose that we ignore the covariances among estimated effect sizes within studies and use the usual weighted combination of estimated effect sizes for a treatment over studies to estimate the population effect size for that treatment. That this estimate will differ from the one given by equation 19.6 has already been shown for treatment T_3. For another example, consider treatment T_1. The usual weighted estimate of the population effect size for treatment T_1, where the weights are inversely proportional to the estimated variances, is:

$$\tilde{\delta}_{T_1}$$
$$= \frac{(44{,}494)^{-1}(0.0150) + (327{,}734)^{-1}(0.0125) + (73{,}088)^{-1}(0.0350)}{(44{,}494)^{-1} + (327{,}734)^{-1} + (73{,}088)^{-1}}$$
$$= .0218,$$

which has (estimated) variance

$$\text{vâr}(\tilde{\delta}) = 10^{-9}[(44{,}494)^{-1} + (327{,}734)^{-1} + (73{,}088)^{-1}]^{-1}$$
$$= 10^{-9}(25{,}505).$$

Compare this to the estimate $\hat{\delta}_{T_1} = 0.0200$ and estimated variance

$$c_{11} = \text{vâr}(\hat{\delta}_{T_1}) = 10^{-9}(24{,}612)$$

obtained from fitting the model 19.5, taking account of the covariances among estimated effect sizes in the studies. Again, the variance of $\hat{\delta}_{T_i}$ is smaller than the variance of $\hat{\delta}_{T_i}$.

Last, note that the regression model 19.5 assumes that the population effect sizes for a given treatment are the same for all studies—that is, that effect sizes for each treatment are homogeneous over studies. Of course, this need not be the case. However, we can test this hypothesis by computing the test statistic

$$Q = (d - X\hat{\delta})'\hat{\Psi}^{-1}(d - X\hat{\delta}) = d'\hat{\Psi}^{-1}d - \hat{\delta}'X'\hat{\Psi}^{-1}X\hat{\delta}$$
$$= 37.608 - 30.417 = 17.191. \tag{19.8}$$

The statistic Q has a chi-squared distribution with degrees of freedom equal to

$$(\text{dimension of } d) - (\text{dimension of } \delta) = 6 - 3 = 3$$

under the null hypothesis of homogeneous (over studies) effect sizes. The observed value of Q in equation 19.8 is between the 90th and 95th percentiles of the chi-squared distribution with three degrees of freedom, so that we can reject the null hypothesis of homogeneous effect sizes at the 0.10 level of significance, but not at the 0.05 level. This result casts some doubt on whether the population effect sizes are homogenous over studies, and consequently we will delay illustrating how to construct confidence intervals for, or make other inferences about, the elements of the vector δ in the model 19.5 until we consider the alternative effect-size metric of the log odds ratio.

19.2.2 Alternative Effect Sizes

Differences between proportions are not the only way that investigators measure differences between rates or proportions. Other metrics that are used to compare rates or proportions are: log odds ratios, log ratios of proportions, differences between the arc sine square root transforms of proportions (the variance-stabilizing metric).

Odds Ratio and Log Odds Ratio The *log odds ratio* between a treatment group T and a statistically independent control group C is

$$d_T = \log\left[\frac{p_C(1-p_T)}{(1-p_C)p_T}\right]$$
$$= \log[p_C/(1-p_C)] - \log[p_T/(1-p_T)]. \tag{19.9}$$

An estimate of the large-sample variance of the log odds ratio d_T is

$$\text{vâr}(d_T) = \frac{1}{n_C p_C(1-p_C)} + \frac{1}{n_T p_T(1-p_T)}, \tag{19.10}$$

where n_c and n_T are the sample sizes of the observed control and treatment groups, respectively, and p_c and p_T are the observed control and treatment proportions. An estimate of the large-sample covariance of estimated log odds ratios d_{T_1} and d_{T_2} of treatment groups T_1 and T_2 versus a common control group C is

$$\text{côv}(d_{T_1}, d_{T_2}) = \frac{1}{n_C p_C(1-p_C)}. \tag{19.11}$$

A regression model of the form given in section 19.2.1 can now be used to estimate the common effect sizes for the treatments. For the data in table 19.1, the vector d of log odds ratio effect sizes is

$$d = (0.4855, 0.0671, 0.3008, 0.6657, -0.0700, 0.4321)',$$

and the estimated large-sample covariance matrix of d is

$$\hat{\Psi} = \begin{pmatrix} \hat{\Psi}_1 & 0 & 0 \\ 0 & \hat{\Psi}_2 & 0 \\ 0 & 0 & \hat{\Psi}_3 \end{pmatrix},$$

where $\hat{\Psi}_i$ now denotes the matrix of estimated large-sample variances and covariances among the estimated log odds ratios for study i, i = 1,2,3; namely,

$$\text{Study 1: } \hat{\Psi}_1 = \begin{pmatrix} 0.03630 & 0.02604 \\ 0.02604 & 0.03297 \end{pmatrix},$$

$$\text{Study 2: } \hat{\Psi}_2 = 0.17453,$$

$$\text{Study 3: } \hat{\Psi}_3 = \begin{pmatrix} 0.03325 & 0.00721 & 0.00721 \\ 0.00721 & 0.02079 & 0.00721 \\ 0.00721 & 0.00721 & 0.02826 \end{pmatrix}.$$

Using the model

$$d = X\delta + \varepsilon,$$

where X is the same design matrix used in section 19.2.1, δ is the vector of common population log odds ratios for the treatments versus the common control, and ε is a random vector having mean vector $\mathbf{0}$ and covariance matrix $\hat{\boldsymbol{\Psi}}$, the best linear unbiased estimator of $\boldsymbol{\delta}$ is

$$\hat{\boldsymbol{\delta}} = (\hat{\delta}_{T_1}, \hat{\delta}_{T_2}, \hat{\delta}_{T_3})' = [X'\hat{\boldsymbol{\Psi}}^{-1}X]^{-1}X'\hat{\boldsymbol{\Psi}}^{-1}d$$
$$= (0.5099, 0.0044, 0.4301)'$$

whose estimated covariance matrix is

$$C = \begin{pmatrix} c_{11} & c_{12} & c_{13} \\ c_{21} & c_{22} & c_{23} \\ c_{31} & c_{32} & c_{33} \end{pmatrix} = [X'\hat{\boldsymbol{\Psi}}^{-1}X]^{-1}$$

$$= \begin{pmatrix} 0.01412 & 0.00712 & 0.00425 \\ 0.00712 & 0.01178 & 0.00455 \\ 0.00425 & 0.00455 & 0.02703 \end{pmatrix}.$$

As before, a test of the model $d = X\boldsymbol{\delta} + \varepsilon$ of homogeneous (over studies) log odds-ratio effect sizes yields a test statistic

$$Q = d'\hat{\boldsymbol{\Psi}}^{-1}d - \hat{\boldsymbol{\delta}}'X'\hat{\boldsymbol{\Psi}}^{-1}X\hat{\boldsymbol{\delta}} = 33.177 - 31.120 = 2.057$$

that is a little smaller than the median of the chi-squared distribution with 3 degrees of freedom (the distribution of Q under the null hypothesis); that is, the p-value for the null hypothesis is greater than 0.50. Consequently, for this effect-size metric for comparing treatment and control proportions, the model of homogeneous effect sizes appears to fit the data. In such a case, we might be interested in forming confidence intervals for the population effect sizes δ_{T_j}, j = 1, 2, 3. These intervals have the form

$$(\hat{\delta}_{T_j} - z\sqrt{\hat{\text{var}}(\hat{\delta}_{T_j})}, \hat{\delta}_{T_j} + z\sqrt{\hat{\text{var}}(\hat{\delta}_{T_j})}) = (\hat{\delta}_{T_j} - z\sqrt{c_{jj}}, \hat{\delta}_{T_j} + z\sqrt{c_{jj}}),$$
(19.12)

where z is a percentile of the standard normal distribution. If we are interested in only one of the population effect sizes, and not in comparison or contrast to the other population effect sizes, and if we wanted a 95 percent confidence interval for this population effect size, then z needs to be the 97.5th percentile of the standard normal distribution, that is, $z = 1.96$. On the other hand, if we are interested in joint or simultaneous confidence intervals for the population effect sizes, such that the probability that all three confidence intervals contain their respective

population parameters is 0.95, then conservative *simultaneous 95 percent confidence intervals* for the population effect sizes can be obtained using the Boole-Bonferroni inequality (Hochberg and Tamhane 1987; Hsu 1996; Miller 1981). In this case, z would be the $100 - (1/3)\,5 = 98.33$rd percentile of the standard normal distribution, that is, $z = 2.128$. For our present example, a 95 percent confidence interval for the population effect size for treatment T_1 is

$$(0.5099 - 1.96\sqrt{0.01412},\ 0.5099 + 1.96\sqrt{0.01412})$$
$$= (0.277, 0.743).$$

Because this interval contains 0, we cannot reject the null hypothesis at a 5 percent level of significance that the log odds-ratio effect size for treatment T_1 equals 0, or equivalently that the odds ratio for treatment T_1 versus the control group equals 1.

Simultaneous 95 percent confidence intervals for the log odds-ratio effect sizes for treatments T_1, T_2, and T_3 are

$$T_1: (0.5099 - 2.128\sqrt{0.01412}, 0.5099 + 2.128\sqrt{0.01412})$$
$$= (0.257, 0.763),$$

$$T_2: (0.0044 - 2.128\sqrt{0.01178}, 0.0044 + 2.128\sqrt{0.01178})$$
$$= (-0.227, 0.235),$$

$$T_3: (0.4301 - 2.128\sqrt{0.02703}, 0.4301 + 2.128\sqrt{0.02703})$$
$$= (0.080, 0.780).$$

Note that the interval for the log odds-ratio effect size for treatment T_3 does not contain 0. Thus we can reject the null hypothesis that this population effect size is zero. Because we used simultaneous confidence intervals for the log odds-ratio effect sizes, the fact that we looked at all three such effect sizes before finding one or more significantly different from zero has been accounted for. That is, this method of testing the individual effect sizes controls at 0.05 the probability that at least one effect size is declared to be different from zero by the data when, in fact, all three population effect sizes equal zero.

Confidence intervals for log odds-ratios of the form 19.12, whether individual confidence intervals or simultaneous confidence intervals, yield corresponding confidence intervals for the odds ratios

$$\omega_{T_i} = p_C(1 - p_{T_i})/(1 - p_C)p_{T_i} \quad i = 1, 2, 3,$$

by simply exponentiating the endpoints of the confidence intervals. For example, the individual 95 percent confidence interval (0.277, 0.743) for the log odds-ratio effect size for treatment T_1 yields the 95 percent confidence interval

$$(e^{0.277}, e^{0.743}) = (1.319, 2.102)$$

for ω_T.

Ratios and Log Ratios Rather than taking ratios of the odds ratios $p_C/(1-p_C)$ and $p_T/(1-p_T)$ for the control and treatment groups, respectively, investigators often find the simple ratio p_C/p_T to be more meaningful. This ratio can be converted to a difference

$$d_T = \log(p_C/p_T) = \log p_C - \log p_T \qquad (19.13)$$

and can be treated by the methodology of this chapter. The estimated large-sample variance of a log ratio (of proportions) d_T is

$$\text{vâr}(d_T) = \frac{1-p_C}{n_C p_C} + \frac{1-p_T}{n_T p_T}, \qquad (19.14)$$

and the large-sample covariance between two log ratios from the same study is

$$\text{côv}(d_{T_1}, d_{T_2}) = \frac{1-p_C}{n_C p_C}. \qquad (19.15)$$

The following are the vector d of effect sizes and the estimated large-sample covariance matrix $\hat{\Psi}$ of d:

$$d = (0.4700,\ 0.0645,\ 0.2877,\ 0.6286,\ -0.0645,\ 0.4055)'$$

and

$$\hat{\Psi} = \begin{pmatrix} \hat{\Psi}_1 & 0 & 0 \\ 0 & \hat{\Psi}_2 & 0 \\ 0 & 0 & \hat{\Psi}_3 \end{pmatrix},$$

where

$$\text{Study 1: } \hat{\Psi}_1 = \begin{pmatrix} 0.03375 & 0.02400 \\ 0.02400 & 0.03042 \end{pmatrix},$$

$$\text{Study 2: } \hat{\Psi}_2 = 0.15917,$$

$$\text{Study 3: } \hat{\Psi}_3 = \begin{pmatrix} 0.03017 & 0.00617 & 0.00617 \\ 0.00617 & 0.01767 & 0.00617 \\ 0.00617 & 0.00617 & 0.02517 \end{pmatrix}.$$

Note that for the data of table 19.1, there is very little difference between the effect sizes based on log odds ratios and the effect sizes for log ratios, and also little difference between their respective estimated covariance matrices. This is not surprising considering that all sample proportions p for the controls and treatments in the three studies are small, between 0.025 and 0.080, and thus p and $p/(1-p)$ are nearly equal. The statistical analysis of log ratio (of proportions) effect sizes for this example follows the same steps as for log odds ratios and thus yields similar estimates, test of fit of the model of homogenous (over studies) population effect sizes, and confidence intervals for population effect sizes as those already given for log odds ratios.

Differences of Arc Sine Square Root Transformations of Proportions Still another metric used to define effect sizes for proportions is based on the *variance-stabilizing transformation* $2 \arcsin(\sqrt{p})$, measured in radians. For a sample proportion p based on a large sample size n, the variance of $2 \arcsin(\sqrt{p})$ is $1/n$. Thus, if the estimated effect size for comparing a treatment group T and an independent control group C is defined to be

$$d_T = 2 \arcsin \sqrt{p_C} - 2 \arcsin \sqrt{p_T},$$

then the large-sample variance of d_T is $\text{var}(d_T) = \frac{1}{n_C} + \frac{1}{n_T}$, independent of the proportions p_C and p_T. (This independence of the values of the proportions is particularly important when one or both of the proportions are equal to zero.) Similarly, the large-sample covariance between the estimated effect sizes for two treatment groups T_1 and T_2 versus a common independent control group is

$$\text{cov}(d_{T_1}, d_{T_2}) = \frac{1}{n_C}.$$

The regression analysis now proceeds in the same steps as with the log odds ratio effect sizes or the difference of proportion effect sizes. For the data in table 19.1,

$$d = (0.0852,\ 0.0130,\ 0.0613,\ 0.1521,\ -0.0187,\ 0.1038)',$$

$$\hat{\Psi}_1 = \begin{pmatrix} 0.00125 & 0.00100 \\ 0.00100 & 0.00125 \end{pmatrix}, \quad \hat{\Psi}_2 = 0.00750,$$

$$\hat{\boldsymbol{\Psi}}_3 = \begin{pmatrix} 0.00150 & 0.00050 & 0.00050 \\ 0.00050 & 0.00150 & 0.00050 \\ 0.00050 & 0.00050 & 0.00150 \end{pmatrix},$$

so that

$$\hat{\boldsymbol{\delta}} = (0.1010, \quad 0.0102, \quad 0.0982)',$$

and

$$\hat{cov}(\hat{\delta}) = \boldsymbol{C} = \begin{pmatrix} c_{11} & c_{12} & c_{13} \\ c_{21} & c_{22} & c_{23} \\ c_{31} & c_{32} & c_{33} \end{pmatrix}$$

$$= \begin{pmatrix} 0.00058 & 0.00040 & 0.00024 \\ 0.00040 & 0.00061 & 0.00025 \\ 0.00024 & 0.00025 & 0.00137 \end{pmatrix}.$$

Further, $Q = 36.069 - 31.805 = 4.264$ is between the 75th and 80th percentiles of the chi-squared distribution with three degrees of freedom, so that the p-value for the null hypothesis of homogeneity of population effect sizes over studies is between 0.20 and 0.25. (The precise p-value associated with the value of Q here is 0.234.) This is extremely weak evidence against the null hypothesis, so we might go on to construct individual or simultaneous confidence intervals for the population effect sizes. For example, a 90 percent individual confidence interval for $\delta_{T_i} = 2 \arcsin \sqrt{p_C} - 2 \arcsin \sqrt{p_{T_i}}$ is

$$(0.1010 - 1.645 \sqrt{0.00058}, \, 0.1010 + 1.645 \sqrt{0.00058})$$

$$= (0.061, 0.141).$$

The results show that one can come to different conclusions about treatment effects relative to a control, depending on which metric is used to define an effect size for proportions. This is already known for a single treatment versus a control, so it is not surprising that the choice of the metric also affects inferences in multiple treatment studies. No matter the metric, it is important to take account of the fact that effect sizes will be statistically dependent in multiple treatment studies (for a discussion of characteristics of these metrics, see chapter 13, this volume).

19.3 MULTIPLE TREATMENT STUDIES: STANDARDIZED MEAN DIFFERENCES

A common way of defining effect sizes when the response variable is quantitative is by the difference between the means of the response for the treatment and for the control, standardized by the standard deviation of the response variable (assumed to be the same for both treatment and control groups). Letting μ_T and μ_C be the population means of the response variable for the treatment and control groups, respectively, and letting σ be the common population standard deviation for both groups, the effect size for treatment T relative to the control C is defined as

$$\delta_T = \frac{\mu_T - \mu_C}{\sigma}. \tag{19.16}$$

In a study in which n_T observations are assigned to treatment T and n_C observations are assigned to the control C, and in which the sample means for treatment and control are $\bar{y}_T$ and $\bar{y}_C$, respectively, and the pooled sample standard deviation is s, the usual estimate of δ_T is

$$d_T = \frac{\bar{y}_T - \bar{y}_C}{s}. \tag{19.17}$$

If the study is a multiple treatment study, with treatments $T_j, j = 1, 2, \ldots, p$, and a common control, the effect-size estimates d_{T_j} will be statistically dependent because they have $\bar{y}_C$ and s in common. If the sample sizes $n_{T_j}, j = 1, 2, \ldots, p$, and n_c are large enough for the central limit theorem approximation to hold, then the effect-size estimates will be approximately jointly normally distributed with the mean of d_{T_j} being $\delta_{T_j}, j = 1, \ldots, p$, the (estimated) large-sample variance of d_{T_j} being

$$\hat{var}(d_{T_j}) = \frac{1}{n_{T_j}} + \frac{1}{n_C} + \frac{d_{T_j}^2}{2n_{total}}, \quad j = 1, \ldots, p, \tag{19.18}$$

and the (estimated) large-sample covariance between d_{T_j} and $d_{T_{j^*}}$ being

$$\hat{cov}(d_{T_j}, d_{T_{j^*}}) = \frac{1}{n_C} + \frac{d_{T_j} d_{T_{j^*}}}{2n_{total}}, \quad j \neq j^*, \tag{19.19}$$

where $n_{total} = n_C + \sum_{j=1}^{k} n_{T_j}$ is the total sample size of the study. In distributional approximations concerning the effect sizes, the estimated large-sample variances and covariances are treated as if they were the true large-sample variances and covariances of the estimated effect sizes.

Comment 1. The large-sample variances and covariances estimated in equations 19.18 and 19.19 are derived under the assumption that the responses are normally distributed. If the distribution is not normal, the large-sample variances and covariances of treatment effect estimates will involve the measures of skewness and kurtosis of the distribution, and these must be determined (or estimated) if $\text{var}(d_{T_j})$ and $\text{cov}(d_{T_j}, d_{T_{j'}})$ are to be estimated.

Comment 2. Because the treatments may affect the variances of the responses as well as their means, some meta-analysts define the population effect size δ_T as the difference in population means between treatment and control divided by the standard deviation of the control population. They will then estimate this effect size by dividing the difference in sample means between treatment and control by the sample standard deviation s_C of the control group. If one can nevertheless assume that the population response variances are the same for the control and all treatments, then to account for the use of s_C in place of s in the definition of d_T one need only replace n_{total} by n_C in the formulas 19.18 and 19.19. If one cannot assume that the treatment and control population variances are the same, estimating the large-sample variances and covariances will require either the sample treatment and control variances from the individual studies or information about the population variances of the treatment and control groups (Gleser and Olkin 1994, 346; see also 2000).

Comment 3. The definition of effect size in equation 19.16 is the usual one. If we expect control means to exceed treatment means, as in section 19.2, then to avoid a large number of negative estimates we can instead define the effect size of a treatment by subtracting the mean of the treatment population from the mean of the control population—thus, $\delta_T = (\mu_C - \mu_T)/\sigma$. As long as this is done consistently over all studies and over all treatments, the formulas 19.18 and 19.19 for the estimated variances and covariances of the effect sizes remain the same.

When several multiple treatment studies are the focus of a meta-analysis, and particularly when not all treatments appear in all studies, the regression model of section 19.2 can be used to estimate treatment effects, assuming that such effects are common to all studies (each treatment has homogeneous effect sizes across studies). Also, the statistic Q introduced in section 19.2 can be used to test the assumption of homogeneity of effect sizes. We illustrate such an analysis in the example that follows. Again, the example is hypothetical, though it is inspired by meta-analyses in the literature. The context is

Table 19.3 Summary Data for Four Studies of Exercise and Systolic Blood Pressure

	Control		Intensive Exercise		Light Exercise		Pooled Standard
	$\bar{y}_c$	n_c	$\bar{y}_{T_i}$	n_{T_i}	$\bar{y}_{T_i}$	n_{T_i}	s
Dev. Study							
1	−1.36	25	7.87	25	4.35	22	4.2593
2	0.98	40	9.32	38	—	—	2.8831
3	1.17	50	8.08	50	—	—	3.1764
4	0.45	30	7.44	30	5.34	30	2.9344

SOURCE: Authors' compilation.

Table 19.4 Estimated Effect Sizes for the Data in Table 19.3 (Computed Using Equation 19.17).

	Intensive Exercise	Light Exercise
Study		
1	2.1670	1.3406
2	2,8927	—
3	2.1754	—
4	2.3821	1.6664

SOURCE: Authors' compilation.

medical, but many similar examples are available in the social sciences.

19.3.1 Treatment Effects, An Example

Suppose that a survey of all randomized studies of changes in systolic blood pressure due to exercise had found four such studies and identified two types of treatment used in those studies: intensive exercise, such as running for thirty minutes; light exercise, such as walking for twenty minutes. All four studies had considered intensive exercise in comparison to an independent nonexercise control group, but only two had included a group who did light exercise. The studies were not large, but the sample sizes were large enough that the sample means could be regarded as having approximately normal distributions. Table 19.3 contains summary data—sample sizes, sample means, pooled sample standard deviations—for the four studies. The response variable was the reduction in systolic blood pressure after a three-month treatment. Table 19.4 presents the estimated effect sizes for each treatment in each study.

As in section 19.2, we create a vector d of estimated effect sizes:

$$d = (2.1670, \ 1.3406, \ 2.8927, \ 2.1754, \ 2.3821, \ 1.6664)',$$

where the first two elements in d pertain to study 1, the next element pertains to study 2, the next to study 3, and the last two elements in d are from study 4. A model of homogeneous effect sizes (over studies) yields the following regression model for d:

$$d = X\delta + \varepsilon, \tag{19.20}$$

where $\delta = (\delta_{T_1}, \delta_{T_2})'$ is the vector of population effect sizes for treatments T_1 (intensive exercise) and T_2 (light exercise), the design matrix for the regression is

$$X = \begin{pmatrix} 1 & 0 \\ 0 & 1 \\ 1 & 0 \\ 1 & 0 \\ 1 & 0 \\ 0 & 1 \end{pmatrix},$$

and ε has a multivariate normal distribution with vector of means equal to 0 and (estimated) covariance matrix

$$\hat{\Psi} = \begin{pmatrix} \hat{\Psi}_1 & 0 & 0 & 0 \\ 0 & \hat{\Psi}_2 & 0 & 0 \\ 0 & 0 & \hat{\Psi}_3 & 0 \\ 0 & 0 & 0 & \hat{\Psi}_4 \end{pmatrix}.$$

Here, making use of equations 19.18 and 19.19,

$$\hat{\Psi}_1 = \begin{pmatrix} 0.11261 & 0.06017 \\ 0.06017 & 0.09794 \end{pmatrix}, \quad \hat{\Psi}_2 = 0.10496,$$

$$\hat{\Psi}_3 = 0.06366, \quad \hat{\Psi}_4 = \begin{pmatrix} 0.09819 & 0.05539 \\ 0.05539 & 0.08209 \end{pmatrix},$$

Fitting the regression model 19.20 by weighted least squares as in 19.6, we find that the estimate of the vector of common, cross-study population effect sizes for treatments T_1 and T_2 is

$$\hat{\delta} = (\hat{\delta}_{T_1}, \hat{\delta}_{T_2})' = [X'\hat{\Psi}^{-1}X]^{-1}X'\hat{\Psi}^{-1}d = (2.374, \ 1.570)'$$

and the estimated covariance matrix of $\hat{\delta}$ is

$$C = \begin{pmatrix} c_{11} & c_{12} \\ c_{21} & c_{22} \end{pmatrix} = [X'\hat{\Psi}^{-1}X]^{-1} = \begin{pmatrix} 0.02257 & 0.01244 \\ 0.01244 & 0.03554 \end{pmatrix}. \tag{19.21}$$

Proceeding to test the goodness-of-fit of the model 19.20, that is, of homogeneity of population effect sizes for treatments over the studies, we find that

$$Q = d'\hat{\Psi}^{-1}d - \hat{\delta}'X'\hat{\Psi}^{-1}X\hat{\delta} = 256.110 - 252.165 = 3.945.$$

Under the null hypothesis 19.20, Q has a chi-squared distribution with degrees of freedom = dimension of d − dimension of $\delta = 6 - 2 = 4$.

Because the observed value of Q is the 59th percentile of the chi-squared distribution with 4 degrees of freedom, the model 19.20 appears to be an adequate fit to the data.

19.3.2 Test for Equality of Effect Sizes and Confidence Intervals for Linear Combinations of Effect Sizes

A natural question is whether the population effect sizes for the treatments are equal; if not, we might want to construct a confidence interval for the difference $\delta_{T_1} - \delta_{T_2}$. It is better to construct the confidence interval first and then use this to test whether $\delta_{T_1} = \delta_{T_2}$. Note that

$$\mathrm{v\hat{a}r}(\hat{\delta}_{T_1} - \hat{\delta}_{T_2}) = \mathrm{v\hat{a}r}(\hat{\delta}_{T_1}) + \mathrm{v\hat{a}r}(\hat{\delta}_{T_2}) - 2\mathrm{c\hat{o}v}(\hat{\delta}_{T_1}, \hat{\delta}_{T_2})$$
$$= c_{11} + c_{22} - 2c_{12}.$$

Thus, a 95 percent approximate confidence interval for $\delta_{T_1} - \delta_{T_2}$ is

$$\hat{\delta}_{T_1} - \hat{\delta}_{T_2} \pm 1.96\sqrt{(c_{11} + c_{22} - 2c_{12})}$$

or

$$2.374 - 1.570 \pm 1.96\sqrt{[0.02257 + 0.03554 - 2(0.01244)]},$$

which yields the interval (0.447, 1.161). Because this interval does not contain 0, we can reject the null hypothesis that $\delta_{T_1} = \delta_{T_2}$ at level of significance 0.05. Based on this hypothetical analysis, intensive exercise is more successful than light exercise at reducing systolic blood pressure by between 0.447 and 1.161 standard deviations.

An alternative way of performing the test of H_0: $\delta_{T_1} = \delta_{T_2}$ would be to fit the regression model

$$d = (1,1,1,1,1,1)' \eta + \varepsilon,$$
$$\eta = \text{common value of } \delta_{T_1} \text{ and } \delta_{T_2}, \quad (19.22)$$

using weighted least squares and test this model against the alternative hypothesis 19.20. To do so, we compute the test statistic

$L = $ lack of fit of model (19.22) $-$ lack of fit of
        model (19.20)

$= [d'\hat{\boldsymbol{\Psi}}^{-1}d - \hat{\eta}^2[(1,1,\ldots,1)\hat{\boldsymbol{\Psi}}^{-1}(1,1,\ldots,1)']$

$\quad - [d'\hat{\boldsymbol{\Psi}}^{-1}d - \hat{\boldsymbol{\delta}}'X'\hat{\boldsymbol{\Psi}}^{-1}X\hat{\boldsymbol{\delta}}]$

$= \hat{\boldsymbol{\delta}}'X'\hat{\boldsymbol{\Psi}}^{-1}X\hat{\boldsymbol{\delta}} - \hat{\eta}^2[(1,1,\ldots,1)\hat{\boldsymbol{\Psi}}^{-1}(1,1,\ldots,1)']$

$= 252.165 - 232.705 = 19.460,$

which, under the null hypothesis H_0: $\delta_{T_1} = \delta_{T_2}$, has a chi-squared distribution with degrees of freedom equal to dimension of $\boldsymbol{\delta}$ $-$ dimension of $\eta = 2 - 1 = 1$. The p-value of L for the null hypothesis is essentially 0, so that we can reject the null hypothesis of equal effect sizes at virtually any level of significance.

As in section 19.2, we can construct 95 percent simultaneous confidence intervals for δ_{T_1} and δ_{T_2} using the Boole-Bonferroni inequality. However, we might also be interested in confidence intervals for $\delta_{T_1} - \delta_{T_2}$ and $(\delta_{T_1} + \delta_{T_2})/2$. To obtain simultaneous $100(1-\gamma)$ percent confidence intervals for these quantities, and for all other linear combinations $a\delta_{T_1} + b\delta_{T_2}$ of δ_{T_1} and δ_{T_2}, we form the intervals

$$a\hat{\delta}_{T_1} + b\hat{\delta}_{T_2} \pm \sqrt{\chi^2_{2;1-\gamma}} \sqrt{(a^2 c_{11} + 2abc_{12} + b^2 c_{22})}, \quad (19.23)$$

where $C = (c_{jj*}) = \hat{\text{cov}}(\hat{\boldsymbol{\delta}})$ is the estimated covariance matrix of $\hat{\boldsymbol{\delta}} = (\hat{\delta}_{T_1}, \hat{\delta}_{T_2})'$ given by 19.21 and $\chi^2_{2;1-\gamma}$ is the $100(1-\gamma)$ percentile of the chi-squared distribution with 2 degrees of freedom. The probability that all intervals of the form 19.23, for all choices of the constants a and b, will simultaneously cover the corresponding linear combinations $a\delta_{T_1} + b\delta_{T_2}$ of the population effect sizes is $1-\gamma$. To achieve simultaneous coverage of the confidence intervals for all linear combinations of δ_{T_1} and δ_{T_2},

we must pay a price in terms of the lengths of the confidence intervals. Thus, individual 95 percent confidence intervals for δ_{T_1} and δ_{T_2} are

$$\delta_{T_1}: 2.374 \pm 1.96\sqrt{0.02257} \quad \text{or} \quad 2.374 \pm 0.294,$$
$$\delta_{T_2}: 1.570 \pm 1.96\sqrt{0.03554} \quad \text{or} \quad 1.570 \pm 0.369,$$

and simultaneous 95 percent confidence intervals for δ_{T_1} and δ_{T_2} using the Boole-Bonferroni inequality are

$$\delta_{T_1}: 2.374 \pm 2.2414\sqrt{0.02257} \quad \text{or} \quad 2.374 \pm 0.337,$$
$$\delta_{T_2}: 1.570 \pm 2.2414\sqrt{0.03554} \quad \text{or} \quad 1.570 \pm 0.423.$$

In comparison, using the simultaneous confidence interval procedure 19.23 would yield

$$\delta_{T_1}: 2.374 \pm 2.4477\sqrt{0.02257} \quad \text{or} \quad 2.374 \pm 0.368,$$
$$\delta_{T_2}: 1.570 \pm 2.4477\sqrt{0.03554} \quad \text{or} \quad 1.570 \pm 0.461.$$

The difference of interval lengths is noticeable even here, where the number of effect sizes being considered is small.

19.4 MULTIPLE-ENDPOINT STUDIES: STANDARDIZED MEAN DIFFERENCES

Multiple-endpoint studies involve a treatment group T and an independent control group C. (In some studies, another treatment may be used in place of a control.) On every subject in a study, p variables (endpoints) $Y_1, Y_2, \ldots, Y_p$ are measured.

For both the treatment (T) group and the control (C) group, it is assumed that the joint distribution of $Y_1, Y_2, \ldots, Y_p$ is multivariate normal with a common matrix $\boldsymbol{\Sigma} = (\sigma_{jj*})$ of population variances and covariances, but possibly different vectors $\boldsymbol{\mu}_C = (\mu_{C1}, \mu_{C2}, \ldots, \mu_{Cp})'$, $\boldsymbol{\mu}_T = (\mu_{T1}, \mu_{T2}, \ldots, \mu_{Tp})'$, of population means.

The population effect size for the jth endpoint (variable) Y_j for a multiple-endpoint study is defined as

$$\delta_j = \frac{\mu_{Cj} - \mu_{Tj}}{\sqrt{\sigma_{jj}}}, \quad j = 1, \ldots, p, \quad (19.24)$$

where again we take differences as control minus treatment because we expect the means corresponding to the control to be larger than the means corresponding to the treatments, and we want to avoid too many minus signs in the estimated effect sizes. Let $\boldsymbol{\delta} = (\delta_1, \ldots, \delta_p)'$ denote the population effect-size vector for such a study. The usual estimate of δ_j would be

$$d_j = \frac{\overline{Y}_{Cj} - \overline{Y}_{Tj}}{\sqrt{s_{jj}}}, \qquad (19.25)$$

where $\overline{Y}_{Cj}$, $\overline{Y}_{Tj}$ are the sample means for the j-th endpoint from the control group and treatment group, respectively, and s_{jj} is the pooled sample variance for the j-th endpoint. Because the measurements $Y_{C1}, Y_{C2}, \ldots, Y_{Cp}$ of the endpoints for any subject in the control group are dependent, as are those of the endpoints for any subject in the treatment group, so will the differences of means $\overline{Y}_{Cj} - \overline{Y}_{Tj}$ and the pooled sample variances s_{jj}, $j = 1, \ldots, p$. The effect-size estimates $d_1, d_2, \ldots, d_p$ will in turn also be dependent. The covariances or correlations among the $d_j's$ will similarly depend on the covariances or correlations among the endpoint measurements for subjects in the control or treatment groups.

Unfortunately, not all studies will report sample covariances or correlations. This forces us to impute values for the population covariances or correlations from other sources, either other studies or, when the endpoints are subscores of a psychological test, from published test manuals. It is most common for sample or population correlations to be reported, rather than covariances, so we will give formulas for the large-sample variances and covariances of the $d_j's$ in terms of correlations among the endpoint measures.

Under the assumption that the endpoint measures for the control and treatment groups have a common matrix of variances and covariances, and that no observations are missing, the large-sample variance of d_j can be estimated by

$$\hat{\Psi}_{jj} = \text{vâr}(d_j) = \frac{1}{n_T} + \frac{1}{n_C} + \frac{d_j^2}{2(n_T + n_C)} \qquad (19.26)$$

as already noted in section 19.3, equation 19.18. The large-sample covariance between d_j and d_{j*} can be estimated by

$$\hat{\Psi}_{jj*} = \left(\frac{1}{n_T} + \frac{1}{n_C}\right)\hat{\rho}_{jj*} + \left(\frac{d_j d_{j*}}{2(n_T + n_C)}\right)\hat{\rho}_{jj*}^2, \quad j \neq j*, \quad (19.27)$$

where $\hat{\rho}_{jj*}$ is an estimate of the correlation between Y_j and Y_{j*}. When the sample covariance matrix of the endpoint measures for the study is published, $\hat{\rho}_{jj*}$ can be taken to be the sample correlation, r_{jj*}. Otherwise, $\hat{\rho}_{jj*}$ will have to be imputed from other available information, for example, from the sample correlation for endpoints j and $j*$ taken from another study. As before, each study will have its own vector $d = (d_1, \ldots, d_p)'$ of effect-size estimates and corresponding estimated large-sample covariance matrix. The vectors from the k studies can be combined into one

large kp-dimensional vector of estimated effect sizes, and the estimated covariance matrices will be diagonal blocks of a combined $kp \times kp$ estimated covariance matrix. This can be done even if not all studies measure all p possible endpoints. In this case, the dimension $p(i)$ of the row vector $d(i)$ of estimated effect sizes for the endpoints measured in the i-th study will depend on the study index i, as will the dimension of the $p(i) \times p(i)$ estimated large-sample covariance matrix $\hat{\Psi}(i)$ of $d(i)$, for $i = 1, \ldots, k$. In general, then, the combined estimated effect-size column vector $d = (d(1), \ldots, d(k))'$ will have dimension $p* \times 1$, and the corresponding block-diagonal estimated large-sample covariance matrix

$$\hat{\Psi} = \begin{pmatrix} \hat{\Psi}(1) & 0 & 0 \\ 0 & \ddots & 0 \\ 0 & 0 & \hat{\Psi}(k) \end{pmatrix},$$

will have dimension $p* \times p*$, respectively, where $p* = \sum_{i=1}^{k} p(i)$.

We can now write the model of homogeneous (over studies) effect sizes as the regression model

$$d = X\delta + \varepsilon, \qquad (19.28)$$

where ε has a $p*$-dimensional normal distribution with mean vector 0 and covariance matrix $\hat{\Psi}$, and where the design matrix X has elements x_{ij} which equal 1 if endpoint j is measured in study i, and otherwise equal 0. As before, the vector δ of common (over studies) population effect sizes for the p endpoints is estimated by weighted least squares

$$\hat{\delta} = (\hat{\delta}_1, \ldots, \hat{\delta}_p)' = (X'\hat{\Psi}^{-1}X)^{-1}X'\hat{\Psi}^{-1}d, \qquad (19.29)$$

and has estimated covariance matrix

$$C = (c_{jj*}) = (X'\hat{\Psi}^{-1}X)^{-1}. \qquad (19.30)$$

A test of the model of homogeneous effect sizes 19.28 against general alternatives can be based on the test statistic

$$Q = (d - X\hat{\delta})'\hat{\Psi}^{-1}(d - X\hat{\delta}) = d'\hat{\Psi}^{-1}d - \hat{\delta}'X'\hat{\Psi}^{-1}X\hat{\delta} \qquad (19.31)$$

whose null distribution is chi-squared with degrees of freedom equal to the dimension $p*$ of d minus the dimension

Table 19.5 Effect of Drill Coaching on Differences Between Post-Test and Pre-Test Achievement Scores

| | | Means | | | | | Pooled Standard Dev. | | |
| | Control | | | Treatment | | | | | |
School	n	Math	Reading	n	Math	Reading	Math	Reading	Correlation ρ
1	22	2.3	2.5	24	10.3	6.6	8.2	7.3	0.55
2	21	2.4	1.3	21	9.7	3.1	8.3	8.9	0.43
3	26	2.5	2.4	23	8.7	3.7	8.5	8.3	0.57
4	18	3.3	1.7	18	7.5	8.5	7.7	9.8	0.66
5	38	1.1	2.0	36	2.2	2.1	9.1	10.4	0.51
6	42	2.8	2.1	42	3.8	1.4	9.6	7.9	0.59
7	39	1.7	0.6	38	1.8	3.9	9.2	10.2	0.49

SOURCE: Authors' compilation.

p of δ. When all p end points are measured in each of the k studies, the degrees of freedom of Q under the null hypothesis 19.28 is equal to $p(k-1)$. An example illustrating these calculations appears in the following section (19.4.1).

Comment. Our discussion has dealt with situations where certain endpoints are not measured in a study, so that all observations on this endpoint can be regarded as missing. Because such "missingness" is a part of the design of the study, with the decision as to what endpoints to use presumably made before the data were obtained, this type of situation is not the kind of missing data of concern in the literature (see, for example, Little and Rubin 1987; Little and Schenker 1994). It is, of course, not unusual for observations not to be obtained on a subset of subjects in a study. If such observations are missing because measurements are missing, bias in the statistical inference is possible. The meta-analyst can generally do little about such missing data because original data are seldom accessible. Whatever methodology used by a study to account for its missing data will thus have to be accepted without adjustment. This is a potential problem, but a few such missing values are not likely to seriously affect the values of estimates or test statistics in the large-sample situations where these methods can be applied.

19.4.1 Missing Values, An Example

Consider the following hypothetical situation, which is based loosely on published meta-analyses that deal with

the possible effects of coaching on SAT-Math and SAT-Verbal scores (Becker 1990; Powers and Rock 1998, 1999). A group of seven community colleges have instituted summer programs for high school graduates planning to attend college. One of the stated goals of the summer program is to improve the students' skills in mathematics and reading comprehension. There is discussion as to whether coaching by drill (the treatment) is effective for this purpose, as opposed to no coaching (the control). Consequently, each student enrolled for the summer program is randomly assigned to either coaching or non-coaching sessions. At the beginning of the summer session, every student is tested in mathematics and reading comprehension. Later, their scores in mathematics and reading comprehension on a posttest are recorded. There are $p = 2$ endpoints measured on each subject—namely, posttest minus pretest scores in mathematics and in reading comprehension. Sample sizes, sample means, and pooled sample variances and correlations are given for these measurements in table 19.5. Note that the sample sizes are not the same, even within a particular summer program at a particular community college, because of dropouts. Only those students who completed the summer session and actually took the posttest are included in the data. All students therefore had measurements on both endpoints.

Using equations 19.25 through 19.27, we obtain table 19.6, which for each school gives the estimated effect sizes for mathematics and reading comprehension attrib-

Table 19.6 Estimated Effect Sizes for Drill Coaching and Their Estimated Covariance Matrices

	Effect Sizes		Estimated Variance and Covariances of the Effect Size Estimates		
	d_{math}	$d_{reading}$	$\hat{v}ar(d_{math})$	$\hat{v}ar(d_{reading})$	$\hat{c}ov(d_{math}, d_{reading})$
School					
1	0.976	0.562	0.0975	0.0906	0.0497
2	0.880	0.202	0.1045	0.0957	0.0413
3	0.729	0.157	0.0874	0.0822	0.0471
4	0.545	0.694	0.1152	0.1177	0.0756
5	0.121	0.010	0.0542	0.0541	0.0276
6	0.104	−0.089	0.0477	0.0477	0.0281
7	0.011	0.324	0.0520	0.0526	0.0255

SOURCE: Authors' compilation.

utable to drill coaching, and also the (estimated) large-sample matrix of variances and covariances for the estimated effect sizes.

Note that there is considerable variation (over schools) in the estimated effect sizes for both mathematics and reading comprehension in table 19.6. Using the test statistic Q for testing the null hypothesis 19.28 of homogeneous population effect sizes for both mathematics and reading comprehension, we find that $Q = 19.62$. This value is the 92.5th percentile of the chi-squared distribution with $p(k-1) = 2(7-1) = 12$ degrees of freedom (the null hypothesis distribution of Q). This result casts doubt on the model of homogeneity of effect sizes. One might look for explanations of the differences in effect sizes. It is the conjectured explanations of heterogeneity of effect sizes, supported by, or at least not contradicting, the data and pointing to directions for future study, that are often the greatest contribution of a meta-analysis to scientific research.

19.4.2 Regression Models and Confidence Intervals

Suppose it is known that schools 1 through 4 and schools 5 through 7 in the example of section 19.4.1 draw students from different strata of the population, or that they used different methods for applying the treatment (the drill). It might then be conjectured that effect sizes for mathematics and reading comprehension are homogeneous within each group of schools, but differ between the groups. This conjecture corresponds to a regression model of the form

$$d = U\theta + \varepsilon, \qquad (19.32)$$

where

$$d = (0.976, 0.562, 0.880, 0.202, \ldots, 0.011, 0.324)'$$

is the fourteen-dimensional column vector of mathematics and reading comprehension effect sizes (in that order) for schools 1 through 7, $\theta = (\theta_{1,math}, \theta_{1,reading}, \theta_{2,math}, \theta_{2,reading})'$, and ε has approximately a multivariate normal distribution with mean 0 and (estimated) block diagonal variance-covariance matrix

$$\hat{\Psi} = \begin{pmatrix} \hat{\Psi}_1 & \cdots & 0 \\ \vdots & \ddots & \vdots \\ 0 & \vdots & \hat{\Psi}_7 \end{pmatrix} \equiv \text{Diag}(\hat{\Psi}_1, \ldots, \hat{\Psi}_7),$$

whose diagonal 2×2 blocks are given in table 6. For example,

$$\hat{\Psi}_3 = \begin{pmatrix} 0.0874 & 0.0471 \\ 0.0471 & 0.0822 \end{pmatrix}$$

is the estimated variance-covariance matrix of the estimated effect sizes from school 3 and can be found in the last three columns of the row for school 3 in table 19.6.

The design matrix U is

$$U = \begin{pmatrix} I & 0 \\ I & 0 \\ I & 0 \\ I & 0 \\ 0 & I \\ 0 & I \\ 0 & I \end{pmatrix},$$

where I is the two-dimensional identity matrix and 0 is a 2×2 matrix of zeroes. Note that U is a 14×4 matrix. To test the null hypothesis 19.32, we use the test statistic

$$\begin{aligned} Q &= \min_{\theta}(d - U\theta)' \hat{\Psi}^{-1}(d - U\theta) \\ &= d'\hat{\Psi}^{-1}d - \hat{\theta}'U'\hat{\Psi}^{-1}U\hat{\theta}, \end{aligned} \quad (19.33)$$

where $\hat{\theta}$ is the (optimal) weighted least squares estimator

$$\hat{\theta} = (U'\hat{\Psi}^{-1}U)^{-1}U'\hat{\Psi}^{-1}d. \quad (19.34)$$

Under the null hypothesis 19.32, Q has an approximate chi-squared distribution with degrees of freedom = dimension of d − dimension of $\theta = 14 - 4 = 10$. The value of Q for the data in table 6 is $Q = 8.087$, which is the 45th percentile of the chi-squared distribution with 10 degrees of freedom. There is thus not enough evidence in the data to reject the model of two homogeneous groups of schools.

If the individuals analyzing the effect-size data decide to use the model 19.32 of two homogeneous groups of schools and then summarize their findings, they might want to form confidence intervals for the components of $\theta = (\theta_{1,math}, \theta_{1,reading}, \theta_{2,math}, \theta_{2,reading})'$ or for such linear combinations of these components as $\theta_{1,math} - \theta_{2,math}$. The large-sample (estimated) covariance matrix of the estimator 19.34 is given by

$$C = (c_{ij}) = (U'\hat{\Psi}^{-1}U)^{-1}. \quad (19.35)$$

One can now proceed to construct confidence intervals using the methods described in section 19.2. For example, an individual 95 percent confidence interval for the common mathematics population effect size $\theta_{1,math}$ in schools 1 through 4 is

$$\hat{\theta}_{1,math} \pm 1.96\sqrt{c_{11}};$$

Simultaneous 95 percent confidence intervals for all linear combinations

$$a'\theta = a_1\theta_{1,math} + a_2\theta_{1,reading} + a_3\theta_{2,math} + a_4\theta_{2,reading}$$

of the components of θ are given by

$$a'\hat{\theta} \pm \sqrt{\chi^2_{4;0.95}}\sqrt{a'Ca}, \quad \text{for every vector } a.$$

Here, $\chi^2_{4;0.95}$ is the 95th percentile of the chi-squared distribution with 4 degrees of freedom and equals 9.488.

Another possible model to explain the heterogeneity of effect sizes observable in table 19.6 would relate the effect sizes to a quantitative characteristic W of the schools whose values for schools 1 through 7 are $w_1, \ldots, w_7$. For example, the characteristic W might be the length of time spent in drill per session. A model that posits separate linear relationships between W and the population effect sizes for mathematics and for reading would be

$$d = W\beta + \varepsilon, \quad (19.36)$$

where d and ε are the same vectors appearing in equation 19.32, but now the elements of the 4×1 vector $\beta = (\beta_{0,math}, \beta_{1,math}, \beta_{0,reading}, \beta_{1,reading})'$ are the intercepts (subscripted 0) and slopes (subscripted 1) of the linear relationship between the effect sizes and the predictor W. The 14×4 design matrix W has the form

$$W = \begin{pmatrix} 1 & w_1 & 0 & 0 \\ 0 & 0 & 1 & w_1 \\ 1 & w_2 & 0 & 0 \\ 0 & 0 & 1 & w_2 \\ \vdots & \vdots & \vdots & \vdots \\ 1 & w_7 & 0 & 0 \\ 0 & 0 & 1 & w_7 \end{pmatrix}.$$

We can fit and test the model 19.36 by the weighted least squares methods already described for the model 19.32. Thus, a test of the model 19.36 is based on the statistic

$$\begin{aligned} Q &= \min_{\beta}(d - W\beta)'\hat{\Psi}^{-1}(d - W\beta) \\ &= d'\hat{\Psi}^{-1}d - \hat{\beta}'W'\hat{\Psi}^{-1}W\hat{\beta}, \end{aligned} \quad (19.37)$$

where

$$\hat{\beta} = (W'\hat{\Psi}^{-1}W)^{-1}W'\hat{\Psi}^{-1}d, \quad (19.38)$$

and the null distribution of Q is the chi-squared distribution with degrees of freedom equal to the difference between the dimensions of d and of $\boldsymbol{\beta}$. Here, that difference equals $14 - 4 = 10$. Linear combinations of the elements of $\boldsymbol{\beta}$ that we might want to estimate are

$$(1)\, \beta_{0,math} + (w)\beta_{1,math}, \quad (1)\beta_{0,reading} + (w)\beta_{1,reading},$$

which are the population effect sizes for drill for the mathematics and reading comprehension parts, respectively, of the achievement test when the characteristic W of a school has the value w. Simultaneous 95 percent confidence intervals (over all values of w) for these population effect sizes are

$$(1)\hat{\beta}_{0,math} + (w)\hat{\beta}_{1,math} \pm \sqrt{\chi^2_{4:0.95}}\sqrt{c_{11} + w^2 c_{22} + 2wc_{12}},$$

$$(1)\hat{\beta}_{0,reading} + (w)\hat{\beta}_{1,reading} \pm \sqrt{\chi^2_{4:0.95}}\sqrt{c_{33} + w^2 c_{44} + 2wc_{34}},$$

where

$$C = (c_{jj*}) = (W'\hat{\boldsymbol{\Psi}}^{-1}W)^{-1} \qquad (19.39)$$

is the (estimated) large-sample covariance matrix of $\hat{\boldsymbol{\beta}}$ and we have already noted that the value of $\chi^2_{4:0.95}$ is 9.488.

Comment. In the example of section 19.4.1, all endpoints are measured on all subjects. In such situations, the general formulas in 19.29 through 19.31, and in 19.33 through 19.35, can be greatly simplified. Let d_i be the column vector of estimated effect sizes from study i, $i = 1$, $2, \ldots, k$. Recall that $\hat{\boldsymbol{\Psi}}_i$ represents the estimated large-sample variance-covariance matrix of d_i. Then the formula 19.29 for the estimate $\hat{\boldsymbol{\delta}}$ of the vector of common population effect sizes is

$$\boldsymbol{\delta} = (\hat{\boldsymbol{\Psi}}_1^{-1} + \hat{\boldsymbol{\Psi}}_2^{-1} + \ldots + \hat{\boldsymbol{\Psi}}_k^{-1})^{-1}$$
$$\times (\hat{\boldsymbol{\Psi}}_1^{-1}d_1 + \hat{\boldsymbol{\Psi}}_2^{-1}d_2 + \ldots + \hat{\boldsymbol{\Psi}}_k^{-1}d_k), \qquad (19.40)$$

which can be viewed as the vector-matrix generalization of the optimal weighted linear combination of scalar effect sizes (see chapter 14, this volume)—that is, the effect-size vectors are weighted by the inverses of their variance-covariance estimates. Similarly, the formula 19.30 for the estimated covariance matrix of $\hat{\boldsymbol{\delta}}$ simplifies to

$$C = (c_{jj*}) = (\hat{\boldsymbol{\Psi}}_1^{-1} + \hat{\boldsymbol{\Psi}}_2^{-1} + \ldots + \hat{\boldsymbol{\Psi}}_k^{-1})^{-1}, \qquad (19.41)$$

and the goodness-of-fit test statistic Q for the model 19.28 becomes

$$Q = d_1'\hat{\boldsymbol{\Psi}}_1^{-1}d_1 + d_2'\hat{\boldsymbol{\Psi}}_2^{-1}d_2 + \ldots + d_k'\hat{\boldsymbol{\Psi}}_k^{-1}d_k$$
$$- \hat{\boldsymbol{\delta}}'(\hat{\boldsymbol{\Psi}}_1^{-1} + \hat{\boldsymbol{\Psi}}_2^{-1} + \ldots + \hat{\boldsymbol{\Psi}}_k^{-1})\hat{\boldsymbol{\delta}}. \qquad (19.42)$$

In the case of model 19.32, one can use formulas 19.40 and 19.41 to estimate the vectors of population effect size for each group based on the effect-size vector d_i and estimated large-sample covariance matrix $\hat{\boldsymbol{\Psi}}_i$ from that group. These estimates give the same result as 19.34. The estimated covariance matrix of such estimators is block diagonal, with the covariance matrix for the estimated effect-size vector of group 1 as the first diagonal block and the covariance matrix for the estimated effect-size vector of group 2 as the other diagonal block.

19.4.3 When Univariate Approaches Suffice

Let δ_{ij} be the effect size of the j-th endpoint in the i-th study and let d_{ij} represent its estimate based on the data in the i-th study, $i = 1, \ldots, k; j = 1, \ldots, p$. Further, let $\Psi_{i:jj}$ be the large-sample variance of d_{ij} and let $\hat{\Psi}_{i:jj}$ be its estimate. These quantities are the same regardless of whether j is the only treatment or endpoint measure considered in study i, or is part of a multiple-treatment or multiple-endpoint study. The temptation thus exists to ignore correlations between the effect-size estimates and to assume that they are independent. In this case, one can view the effect-size estimates d_{ij} as arising from kp independent studies of the treatment versus control type and use univariate methods to compare the corresponding effect sizes δ_{ij} either across studies (over i) or across treatments or endpoints (over j).

Using univariate methods to compare effect sizes δ_{ij} across studies for a single treatment or single endpoint j is a frequently used approach, and provides correct confidence intervals and has correct stated probabilities of error or non-coverage. Even so, such procedures may not be optimally accurate or efficient because they do not make use of statistical information about the errors of estimation $d_{ij} - \delta_{ij}$ in the other effect sizes. As we showed in section 19.2, multivariate procedures take account of such information and thus are generally more statistically efficient.

Examples of across-study inferences for a specified treatment or endpoint j that can be validly performed using univariate methods include

1. tests of the null hypothesis $H_{oj}: \delta_{1j} = \delta_{2j} = \ldots = \delta_{kj}$ of homogeneity of effect sizes over studies, and follow-up multiple comparisons;

2. tests of the null hypothesis that an assumed common effect size (over studies) is equal to 0, or confidence intervals for this common effect size;

3. if effect sizes over studies are not believed to be homogeneous, goodness of-fit tests of alternative models relating such effect sizes (for the endpoint or treatment j) to certain study characteristics (covariates).

Even so, when more than one treatment or endpoint is being considered, the meta-analyst should be aware that any of the inferential procedures 1, 2 or 3 for one endpoint or treatment j can be statistically related to the results of one of these procedures done for another endpoint or treatment j^*. Consequently, it should not be surprising to see clusters of significant tests across treatments or endpoints.

In contrast, using univariate methods to compare effect sizes δ_{ij} over endpoints or treatments within studies can lead to serious error. We saw this in section 19.2, where treating effect sizes as being independent (as is assumed in univariate methods) led to underestimates or overestimates of the variances of certain linear combinations of the effect sizes.

19.5 CONCLUSION

The main goal of this chapter is to introduce two common situations in meta-analysis in which several population effect sizes are estimated in each study, leading to statistical dependence among the effect size estimates. In one case, several treatments may be compared to a common control in each study (multiple treatment studies). The dependence arises because each estimated effect size in a given study contains a contribution from the same control. In the other case (multiple endpoint studies), a treatment and a control may be compared on multiple endpoints. Here the dependence arises because of the dependence of the endpoints when measured on the same subject. In these cases, and in general whenever there is statistical dependency among estimated effect sizes, ignoring such dependencies can lead to biases or lack of efficiency in statistical inferences, as illustrated in section 19.2.1.

One methodology that can be applied to account for statistical dependency in effect sizes was illustrated in sections 19.2 through 19.4—namely, the use of a regression model and weighted least squares estimation of its parameters, followed by tests of fit of the model (or of submodels) and construction of individual or simultaneous confidence intervals for these parameters. The main requirement is that the vector d of all estimated effect sizes in all studies considered has an approximate multivariate normal distribution with mean vector equal to the vector of corresponding population effect sizes δ and a covariance matrix Ψ that can be estimated accurately enough that its estimator $\hat{\Psi}$ can be treated as if it were the known true value of Ψ. That this is possible in the special cases we considered follows from standard large-sample statistical theory. It should be noted, however, that in this theory the number p^* of estimated effect sizes—that is, the dimension of d—is assumed to be fixed and the sample sizes within studies are assumed to be large relative to p^*. This type of approximation for meta-analysis was originally discussed in Hedges and Olkin (1985).

Although the difference in sample proportions discussed in section 19.2.1 is an unbiased estimator of its population value, this is not the case for the other effect sizes discussed in section 19.2 or for those in sections 19.3 and 19.4. Similarly, again with the exception of the difference in sample proportions, the estimates of variances and covariances of effect sizes are not unbiased. Unbiased estimates can be found (Hedges and Olkin 1985). However, for the theory underlying the regression methodology given in this chapter to hold, it would be more important to modify effect-size estimates and the estimates of their variances and covariances to improve the normal distributional approximation. Some univariate studies on the goodness of the normality approximation for various transformations of standardized differences of sample means are given in Hedges and Olkin (1985). These may give insights for the multivariate case.

As illustrated in sections 19.2 and 19.3, one of the advantages of the regression approach is that it enables the meta-analyst to treat the often-encountered situation in which not all population effect sizes of interest are observed in all studies. This meant in sections 19.2 and 19.3 that not all treatments were observed in all studies. In section 19.4, it meant that not all endpoints were measured in all studies.

The applicability of regression models of the form $d = X\delta + \varepsilon$, where ε has an (approximate) normal distribution with (approximated, estimated) covariance matrix $\hat{\Psi}$, extends beyond endpoints based on proportions or standardized mean differences as treated in sections 19.2 through 19.4. For example, this approach can be applied to correlations ρ between behavioral measures X and Y compared under no treatment (control) and under a variety of interventions (multiple treatments) when the correlations are obtained from independent samples of subjects. In this case, one possible effect size is the difference $d_T = r_T - r_C$ between the sample correlation r_C from the control context and the sample correlation r_T under a treatment context. Assume that X and Y are jointly normal and that sample sizes are greater than fifteen, so that d_T has approximately a normal distribution with (estimated) variance

$$\hat{\mathrm{var}}(d_T) = \frac{(1 - r_T^2)^2}{n_T} + \frac{(1 - r_C^2)^2}{n_C}.$$

Further, the estimated effect sizes d_{T_1} and d_{T_2} for treatments T_1 and T_2 within the same study have (estimated) covariance

$$\hat{\mathrm{cov}}(d_{T1}, d_{T2}) = \frac{(1 - r_C^2)^2}{n_C}.$$

Putting all of the estimated effect sizes into a column vector d and using the cited information (plus the independence of the studies) to construct the (estimated) covariance matrix $\hat{\Psi}$ of d, now puts the meta-analyst in a position to estimate, compare, and test models for the population effect sizes. As noted, this can be done even if not all treatments are contained in every study.

An alternative metric for correlations is Fisher's z-transformation,

$$z = (1/2)\log((1 + r)/(1 - r)),$$

whose large sample distribution is normal with mean $\zeta = (1/2)\log((1 + \rho)/(1 - \rho))$ and variance $1/n$, where n is the sample size. If we are in the independent correlations context of the previous paragraph, but now define the estimated effect size for treatment context T as the difference

$$d_T = z_T - z_C = (1/2)\log((1 + r_T)/(1 - r_T))$$
$$- (1/2)\log((1 + r_C)/(1 - r_C))$$

between the z-transformations of the correlations obtained under context T and the control context C, then the large-sample distribution of d_T is normal with mean $\delta_T = \zeta_T - \zeta_C$ and variance $\mathrm{var}(d_T) = 1/n_T + 1/n_C$, and the large-sample covariance between two such estimated effect sizes d_{T_1} and d_{T_2} is $\mathrm{cov}(d_{T_1}, d_{T_2}) = 1/n_C$. These results, as before, allow us to make inferences concerning the population effect sizes.

If we are interested in comparing correlations between a response Y and various covariates $X_1, \ldots, X_p$ measured on the same individual under a treatment and a control (as in multiple endpoint studies), we can determine large-sample variances and covariances of the estimated effect sizes $d_T = r_T - r_C$. However, constructing estimates of the large-sample covariance matrix Ψ of the vector d of estimated effect sizes in this situation is more difficult. It is especially so if studies neglect to report sample correlations between the covariates, because naïve plug-in estimates of this matrix need not be positive definite.

Still another example of the application of the regression methodology involves multiple endpoint studies where the responses are categorical (qualitative). Here, two multinomial distributions are compared—one under the control and the other under the treatment. If the effect sizes (differences in proportions between control and treatment for the various categories) equal 0, then this is equivalent to saying that the response is independent of the treatment assignment. The original test for this hypothesis was the classic chi-squared test of independence. It can be applied to situations that are the combination of multiple treatment and multiple endpoint studies, and simultaneous confidence intervals can be given for the effect sizes that are compatible with the test of independence (Goodman 1964). The test in its classical form. however, can be applied only to cases where every study involves every treatment. When this is not the case, the methodology given in this chapter is still applicable and provides similar inference (see Trikalinos and Olkin 2007).

In sections 19.2 and 19.4, we exclusively discussed situations where the considered studies are assumed to be statistically independent. However, more than one study that a meta-analyst includes can involve the same group of subjects, or can come from the same laboratories (investigators). This creates the possibility of statistical dependencies between estimated effect sizes across studies. If good estimates of the covariances can be obtained, the regression methodology illustrated in sections 19.2 through 19.4 can be applied.

Finally, a word should be said about computation. If values of the vector d of estimated effect sizes, the design matrix X of the regression model, and the covariance matrix $\hat{\Psi}$ are input by the meta-analyst, many computer software programs for regression (for example, SAS, SPSS, Stata) can find weighted least squares estimates of the parameter vector δ of the regression model $d = X\delta + \varepsilon$ and will provide the value of the goodness-of-fit statistic $Q = \min_\delta (d - X\delta)' \hat{\Psi}^{-1}(d - X\delta)$ for that model (usually under the label *sum of squares error*). Some will also provide individual or simultaneous confidence intervals for (linear combinations of) the elements of δ. The calculations given in this chapter, however, were done using only the vector-matrix computation capabilities of the statistical software Minitab. One could also use Matlab, its open-source equivalent Octave, or any other software capable of doing vector-matrix computations. Difficulties in using any such computer software may arise when the total number of effect sizes under consideration is large (over 200), primarily due to the need for inversion of the covariance matrix $\hat{\Psi}$. It is thus worth noting that when the studies under consideration are independent, $\hat{\Psi}$ is block diagonal, and the inverse of this matrix can be found using inverses of its diagonal blocks, thus greatly reducing the dimensionality of the computation.

19.6 ACKNOWLEDGMENT

The second author was supported in part by the National Science Foundation and the Evidence-Based Medicine Center of Excellence, Pfizer, Inc.

19.7 REFERENCES

Becker, Betsy J. 1990. "Coaching for the Scholastic Aptitude Test: Further synthesis and appraisal." *Review of Educational Research* 60(3): 373–417.

Gleser, Leon J., and Ingram Olkin. 1994. "Stochastically Dependent Effect Sizes." In *The Handbook of Research Synthesis*, edited by Harris Cooper and Larrry V. Hedges. New York: Russell Sage Foundation.

———. 2000. " Meta-Analysis for 2×2 Tables with Multiple Treatment Groups." In *Meta-Analysis in Medicine and Health Policy*, edited by Donald Berry and Darlene Stangl. New York: Marcel Dekker.

Goodman, Leo A. 1964. "Simultaneous Confidence Intervals for Contrasts Among Multinomial Populations." *Annals of Mathematical Statistics* 35(2): 716–25.

Hedges, Larry V., and Ingram Olkin. 1985. *Statistical Models for Meta-Analysis.* New York: Academic Press.

Hochberg, Yoav, and Ajit C. Tamhane. 1987. *Multiple Comparison Procedures.* New York: John Wiley & Sons.

Hsu, Jason C. 1996. *Multiple Comparisons. Theory and Methods.* New York: Chapman and Hall.

Little, Richard J.A., and Donald B. Rubin. 1987. *Statistical Analysis With Missing Data.* New York: John Wiley & Sons.

Little, Richard J.A., and Nathaniel Schenker. 1994. "Missing Data." In *Handbook for Statistical Modeling in the Social and Behavioral Sciences*, edited by Gerhard Arminger, Clifford C. Clogg, and Mike E. Sobel. New York: Plenum.

Miller, Rupert G. Jr. 1981. *Simultaneous Statistical Inference*, 2d ed. New York: Springer-Verlag.

Powers, Donald E., and Donald A. Rock. 1998. "Effects of Coaching on SAT I: Reasoning Scores." *College Board Research Report 98–6.* New York: College Board.

———. 1999. "Effects of Coaching on SAT I: Reasoning Test Scores." *Journal of Educational Measurement* 36(2): 93–111.

Raudenbush, Stephen W., Betsy J. Becker, and Hripsime A. Kalaian. 1988. "Modeling multivariate effect sizes." *Psychological Bulletin* 103(11): 111–20.

Timm, Neil H. 1999. "Testing Multivariate Effect Sizes in Multiple-Endpoint Studies." *Multivariate Behavioral Research* 34(4): 457–65.

Trikalinos, Thomas A., and Ingram Olkin. 2008. "A Method for the Meta-Analysis of Mutually Exclusive Outcomes." *Statistics in Medicine* 27(21): 4279–4300.

20

MODEL-BASED META-ANALYSIS

BETSY JANE BECKER
Florida State University

CONTENTS

20.1 WHAT ARE MODEL-BASED RESEARCH SYNTHESES?

In this chapter I introduce model-based meta-analysis and update what is known about research syntheses aimed at examining models and questions more complex than those addressed in typical bivariate meta-analyses. I begin with the idea of model-based (or model-driven) meta-analysis and the related concept of linked meta-analysis, and describe the benefits (and limitations) of model-based meta-analysis. In discussing how to do a model-based meta-analysis, I pay particular attention to the practical concerns that differ from those relevant to a typical bivariate meta-analysis. To close, I provide a brief summary of research that has been done on methods for model-driven meta-analysis since the first edition of this volume was published.

20.1.1 Model Driven Meta-Analysis and Linked Meta-Analysis

The term *meta-analysis* was coined by Gene Glass to mean the analysis of analyses (1976). Originally, Glass applied the idea to sets of results arising from series of independent studies that had examined the same (or very similar) research questions. Many meta-analyses have looked at relatively straightforward questions, such as whether a particular treatment or intervention has positive effects, or whether a particular predictor relates to an outcome. However, linked or model-driven meta-analysis techniques can be used to analyze more complex chains of events such as the prediction of behaviors based on sets of precursor variables.

Suppose a synthesist wanted to draw on the accumulated literature to understand the prediction of sport performance from three aspects of anxiety. For now, let us refer to the anxiety measures as X_1 through X_3 and the outcome as Y. The synthesist might want to use meta-analysis to estimate a model such as the one shown in figure 20.1. I will use the term *model* to mean "a set of postulated interrelationships among constructs or variables" (Becker and Schram 1994, 358). This model might be represented in a single study as $\hat{Y}_j = b_0 + b_1 X_{1j} + b_2 X_{2j} + b_3 X_{3j}$ for person j. Model-based meta-analysis provides one way of summarizing studies that could inform us about this model.

Suppose now that the synthesist thinks that the Xs may also influence each other, and wants to examine whether X_1 and X_2 affect Y by way of the mediating effect of X_3, as

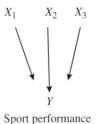

Figure 20.1 Prediction of Sport Performance
SOURCE: Author's compilation.

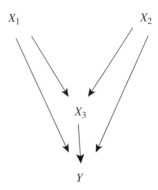

Figure 20.2 X_3 Plays a Mediating Role in Sport Performance
SOURCE: Author's compilation.

well as directly. We can also examine a model like the one shown in figure 20.2 using model-based meta-analysis.

The ideas I cover in this chapter have been called by two different names—linked meta-analysis (Lipsey 1997) and model-driven meta-analysis (Becker 2001). The estimation and data-analysis methods used have been variously referred to as estimating linear models (Becker 1992b, 1995), meta-analytic structural equation modeling, called MASEM by Mike Cheung and Wai Chan (2005) and MA-SEM by Carolyn Furlow and Natasha Beretvas (2005), and two-stage structural equation modeling (TSSEM) (Cheung and Chan 2005).

The term model-driven meta-analysis has been used to refer to meta-analyses designed to examine a particular theoretical model. Although linked meta-analysis can also be used to address models, in its simplest form the term simply means to link a set of meta-analyses on a sequence of relationships. Linked or model-driven meta-analyses go further than traditional meta-analyses by asking about chains of connections among predictors and outcomes (does A lead to B *and* B lead to C?). Linked meta-analyses have been used to look at relationships over time (for example, Mark Lipsey and James Derzon's 1998 work on the prediction of serious delinquency), and to ask about relationships of a set of predictors to an outcome or set of outcomes (do A and B lead to C?). Early applications of the latter sort include my examination of science achievement outcomes and Sharon Brown and Larry Hedges' analysis of metabolic control in diabetics (Becker 1992a; Brown and Hedges 1994). Linked meta-analysis or model-driven meta-analysis is therefore more extensive and more complicated than a typical meta-analysis, but can yield several benefits that go beyond those of a more ordinary synthesis. Of course, with the added complexity come additional caveats; they will be addressed below.

20.1.2 Average Correlation Matrices as a Basis for Inference

When multiple studies have contributed information about a relationship between variables, the statistical techniques of meta-analysis allow estimation of the strength of the relationship between variables (on average) and the amount of variation across studies in relationship strength, often associated with the random-effects model (Hedges and Vevea 1998).

For a linked or model-driven meta-analysis, an average correlation matrix among all relevant variables is estimated under either a fixed-effects model where all studies are presumed to arise from one population, or a random-effects model, which presumes the existence of a population of study effects. Ideally, of course, one would like to compute this average matrix from many large, representative studies that include all the variables of interest, measured with a high degree of specificity. In reality, the synthesist will likely be faced with a group of diverse studies examining different subsets of variables relating proposed predictors to the outcome or outcomes of interest. For this reason it is likely that the random-effects model will apply. In the random-effects case, estimates of variation across studies are incorporated into the uncertainty of the mean correlations, allowing the estimates of the average correlations to be generalized to a broader range of situations, but also adding to their uncertainty (and widening associated confidence intervals). In addi-

tion a test of the appropriateness of the fixed-effects model for the entire correlation matrix can be computed (see, for example, Becker 2000).

Let us again consider the example above with three predictors of sport performance. The first step in the model-based meta-analysis would be to gather estimates of the correlations in the matrix $\mathbf{R}$,

$$\mathbf{R} = \begin{bmatrix} 1 & r_{12} & r_{13} & r_{1Y} \\ r_{12} & 1 & r_{23} & r_{2Y} \\ r_{13} & r_{23} & 1 & r_{3Y} \\ r_{1Y} & r_{2Y} & r_{3Y} & 1 \end{bmatrix},$$

where r_{ab} is the correlation between X_a and X_b for a and $b = 1$ to 3, and r_{aY} is the correlation of X_a with Y, for $a = 1$ to 3. I use the subscript i to represent the ith study, so $\mathbf{R}_i$ represents the correlation matrix from study i. The index p represents the total number of variables in the matrix (here $p = 4$). Therefore any study may have up to $p^* = p(p-1)/2$ unique correlations. For this example $p^* = 6$. Thus each study's matrix $\mathbf{R}_i$ can also contain up to $m_i = p^*$ unique elements. $\mathbf{R}_i$ would contain $m_i < p^*$ elements if one or more variables are not examined in study i or if some correlations are not reported. Also, the rows and columns of $\mathbf{R}_i$ can be arranged in any order, though it is often convenient to place the most important or ultimate outcome in either the first or last position. In the example below, the correlations involving the sport performance outcome appear in the first row and column. From the collection of $\mathbf{R}_i$ matrices the synthesist would then estimate an average correlation matrix, say $\overline{\mathbf{R}}$.

I will give more detail on how this average can be obtained, but for now I introduce one more item—the correlation vector. Suppose we list the unique elements of the matrix $\mathbf{R}$ in a vector, $\mathbf{r}_i$. Then $\mathbf{r}_i = (r_{i12}, r_{i13}, r_{i1Y}, r_{i23}, r_{i2Y}, r_{i3Y})'$ contains the same information as is shown in the display of the square matrix $\mathbf{R}_i$ above. On occasion it may be useful to refer to the entries $(r_{i12}, r_{i13}, r_{i1Y}, r_{i23}, r_{i2Y}, r_{i3Y})$ as r_{i1} through r_{ip^*}, or more generally as $r_{i\alpha}$, for $\alpha = 1$ to p^*, in which case the elements of $\mathbf{r}_i$ are indexed by just two subscripts, and in our example $\mathbf{r}_i = (r_{i1}, r_{i2}, r_{i3}, r_{i4}, r_{i5}, r_{i6})'$.

20.1.3 The Standardized Multiple-Regression Model

For now let us assume the synthesist has an estimate of $\overline{\mathbf{R}}$. The next step in model-driven meta-analysis is to use the matrix of average correlations to estimate an equation or a series of regression equations relating the predictors X_1 through X_3 to Y. These equations provide a picture of the relationships among the variables that takes their co-variation into account (Becker 2000; Becker and Schram 1994) and this is where the possibility of examining partial relations, indirect effects, and mediating effects is manifest.

If the synthesist wanted to examine the model shown in figure 20.2, two regression equations would be obtained. One relates X_1 and X_2 to X_3, and the other relates X_1 through X_3 to Y. Because the primary studies may use different measures of these four variables, and those measures may be on different scales, it may not be possible to estimate raw regression equations or to directly combine regressions of X_3 or Y on the relevant predictors. (Later I briefly discuss issues related to the synthesis of regression equations.) However, the synthesist can estimate a standardized regression model or path model using the estimated mean matrix $\overline{\mathbf{R}}$ (and its variance co-variance matrix).

Recent promising work by Soyeon Ahn has also begun to examine the possibility of using meta-analytic correlation data to estimate latent variable models (Ahn 2007; Thum and Ahn 2007). Adam Hafdahl has also examined factor analytic models based on mean correlations (2001). In this chapter, however, I focus only on path models among observed variables.

20.1.4 Slope Coefficients, Standard Errors, and Tests

As with any standardized regression equation in primary research or in the context of structural equation modeling, the model-based estimation procedure produces a set of standardized slopes and standard errors for those slopes. Individual slopes can be examined to determine whether each differs from zero (that is, the synthesist can test whether each $\beta_j^* = 0$), which indicates a statistically significant contribution of the tested predictor to the relevant outcome. An overall, or omnibus, test of whether all standardized slopes equal zero is also available, thus the synthesist can test the hypothesis that $\beta_j^* = 0$ for all j.

Because standardized regression models express predictor–outcome relationships in terms of standard-deviation units of change in the variables, comparisons of the relative importance of different predictors are also possible. As is true for ordinary least squares (OLS) regression slopes in primary studies, the slopes themselves can be interrelated, and appropriate tests account for the

covariation between slopes in making comparisons of the relative importance of slopes.

Finally, comparisons of slopes from models computed for different samples (from different sets of studies) are available. Thus it is possible to compare, say, models for males and females or for other subgroups of participants—for example, old versus young participants, or experts versus novices.

20.2 WHAT WE CAN LEARN FROM MODELS, BUT NOT FROM BIVARIATE META-ANALYSES

Unlike typical bivariate meta-analyses, linked or model-driven meta-analyses address how sets of predictors relate to an outcome (or outcomes) of interest. In a linked meta-analysis we may examine partial relations, because we will model or statistically control for additional variables, and it is also possible to examine indirect relations and mediator effects. I provide examples of how each of these sorts of effects has been examined in existing linked or model-driven meta-analyses.

20.2.1 Partial Effects

Most meta-analyses of associations take one predictor at a time, and ask if the predictor X relates to the outcome Y. In the real world, however, things are usually not so simple. Consider the issue of the effectiveness of alternate routes to teacher certification. In many studies, this is addressed by a simple comparison: teachers trained in traditional teacher-education programs are compared on some outcome to teachers who have gone through an alternate certification program. An example of such a design is found in a study by John Beery (1962). However, prospective teachers who seek alternate certification are often older than those going though the traditional four-year college programs in teacher education. To make a fair comparison of the effectiveness of alternate-routes programs, one should control for these anticipated differences in teacher age. This control can be exerted through design variations, such as by selecting teachers of the same age, or by statistical means, such as by adding a measure of teacher age to a model predicting the outcome of interest. Studies of alternate routes have used various design controls and also included different statistical control variables to partial out unwanted factors (see, for example, Goldhaber and Brewer 1999).

Model-driven meta-analysis can incorporate such additional variables that need to be controlled into a more

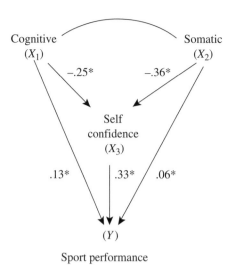

Figure 20.3 Prediction of Sport Performance from Components of Anxiety

source: Author's compilation.

complex model than is possible in a traditional bivariate meta-analysis. A specific example from an application in sport psychology shows this benefit. Lynette Craft and her colleagues examined the role of a three-part measure of anxiety in the prediction of sport performance (2003). The Competitive Sport Anxiety Index (CSAI) includes measures of cognitive aspects of anxiety, denoted as X_1, and somatic anxiety, or physical symptoms, denoted as X_2, as well as a measure of self confidence, denoted as X_3 (Martens, Vealey, and Burton 1990). An important question concerns whether these three factors contribute significantly to sport performance, as shown in figure 20.3.

Estimates of the bivariate correlations of the three aspects of anxiety to performance show mixed results. Although self confidence correlates significantly with performance across studies ($\bar{r}_{.3Y} = .21, p < .001$), the other anxiety subscores showed virtually no zero-order relation to performance, on average ($\bar{r}_{.1Y} = .01, \bar{r}_{.2Y} = -.03$, both not significant). By modeling all three components together, as shown in figure 20.3, Craft and her colleagues found that though the bivariate relationships of cognitive and somatic anxiety to performance were not significant, when the three components were considered together in a regression model, all the subscales showed overall relations to sport performance, even under a random-effects model. The effect of somatic anxiety on performance,

controlling for other CSAI components, was weak, and just barely significant at the traditional .05 level, with $z = 1.961, p = .05$. Thus when one controls for the role of self confidence, other aspects of anxiety appear to play a role in performance.

20.2.2 Indirect Effects

A second major benefit of the use of model-driven or linked meta-analyses is the ability to examine indirect effects. In some situations, predictors may have an impact on an outcome through the force of an intervening or mediating variable. Such a predictor might have no direct effect on the outcome, but have influence only through the third mediating factor. The analysis of indirect relationships is a key aspect of primary-study analyses using structural equation models (see, for example, Kaplan 2000). In many areas of psychology and education, complex models have been posited for a variety of outcomes. It makes sense that synthesist might want to consider such models in the context of meta-analysis.

Model-driven meta-analyses can provide this same benefit by examining indirect relationships among the predictors outlined in theoretical models of the outcome of interest. An example comes from a model-driven synthesis of the predictors of child outcomes in divorcing families (Whiteside and Becker 2000). An important consideration in child custody decisions when parents divorce is the extent of parental visitation for the noncustodial parent. Often when the mother is granted custody, a decision must be made about the extent of father visitation and contact with the child. Curiously and counterintuitively, narrative syntheses of the influence of the extent of father visitation on child adjustment outcomes showed only a weak relationship to child outcomes such as internalizing and externalizing behaviors (see, for example, Hodges 1991).

Indeed when Mary Whiteside and I examined the bivariate associations of measures of father-child contact with adjustment outcomes in their meta-analysis, only the quality of the father-child relationship showed a significant association with child outcome variables. Specifically, a good father-child relationship was related to higher levels of cognitive skills for the child (a good outcome), and to lower scores on externalizing symptoms and internalizing symptoms (also sensible, because high internalizing and externalizing symptoms reflect problems for the child). Measures of father visitation, and

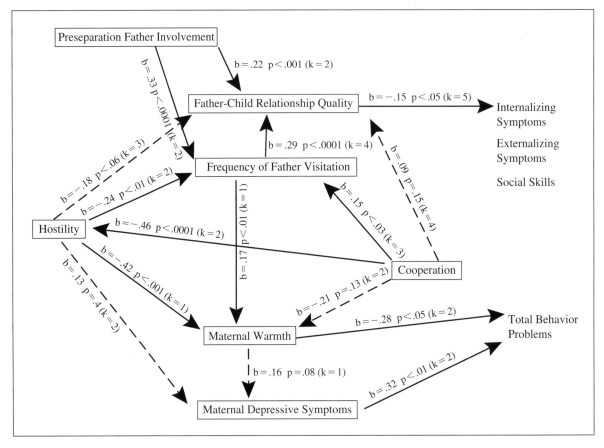

Figure 20.4 Model of Child Outcomes in Divorcing Families
SOURCE: Whiteside and Becker 2000.

pre-separation and current levels of father involvement, showed nonsignificant relationships with child outcomes.

However, when the father-child contact variables were examined in the context of a more realistic multivariate model for child outcomes, Whiteside and I found that extent of father contact showed a significant indirect relationship, through the variable of father-child relationship quality. Figure 20.4 shows the empirical model. Both pre-separation father involvement and extent of father visitation showed positive indirect impacts on child outcomes. Whiteside and I interpreted this finding to mean that when a child has a good relationship with their father, more father contact has a good influence on psychological outcomes. In addition, children with higher levels of involvement with their father before parental separation,

and who had good relationships with their fathers, also showed more positive outcomes. These results would not have been found had the indirect relationships specified in the model-driven synthesis not been examined.

20.2.2.1 Mediator Effects Because it has the potential to model indirect effects, model-driven meta-analysis also allows for mediating effects to be tested. One example of an application where the roles of mediators were critical concerns the effects of risk and protective factors for substance abuse (Collins, Johnson, and Becker 2007). In this synthesis of community-based interventions for substance-abuse prevention, David Collins, Knowlton Johnson, and I were interested in the question of whether risk and protective factors, such as friends' use of drugs and parental views of drug use, mediated the impact of preventive interventions on a variety of substance-abuse outcomes. This was accomplished by using a variation of the approach proposed by Charles Judd and David Kenny (1981). One could also test for mediation in other ways, using variations on the approaches described by David MacKinnon and his colleagues (2002).

Collins and his colleagues examined three models for each of six prevalence outcomes. The first were intervention effect models that had prevalence of substance abuse as the outcome, and used as predictors three intervention variables (intervention, year, and their interaction) plus covariates such as gender and ethnicity. In these models, a negative slope for the intervention by year interaction indicated that the treatment was associated with a reduction in the prevalence of substance abuse between the baseline year and the post-intervention year. If a prevalence outcome showed a significant negative slope for the intervention by year interaction in the first model, the outcome had the potential to show a mediating effect of the risk and protective factors.

The second set of models had as outcomes each of twelve risk and protective factors, with the same predictors as previously mentioned. The authors identified the specific risk and protective factors that could be mediating the intervention effects by finding those with significant treatment effects for the intervention.

Finally, a third set of mediation models again had the prevalence behaviors as outcomes, but included not only the intervention variables and covariates, but also all the risk and protective factors viewed as outcomes in the second set of models. The mediating effect of a risk or protective factor was apparent if a significant intervention by year interaction (from the first set of models) became nonsignificant when the risk and protective factors were added as predictors in mediation models. Mediating effects were found for three outcomes: sustained cigarette use, alcohol use, and binge drinking. The two risk factors identified as potential mediators were friends' drug use and perceived availability of drugs. Again in this example, mediating effects could not have been analyzed in a typical meta-analytic examination of the pairwise relationships among risk and protective factors and the outcomes.

20.2.3 Model Comparisons

Other analyses that are possible in model-based meta-analyses are comparisons of models based on moderator variables. A synthesist may want to ask, for instance, whether the same predictive relationships hold for two or more different groups. Whiteside and I wanted to know whether the same predictors were important for child outcomes for young boys and young girls in divorcing families (2000). Craft and her colleagues coded a variety of aspects of the athletes and sports examined in the studies of anxiety (2003). Team and individual sports were compared, as were elite athletes, European club athletes, college athletes, and college physical education students. Two other moderators were type of skill (open versus closed) and time of administration of the anxiety scale (in terms of minutes prior to performance).

Complications can arise when moderator variables are examined. Most common is that estimating the same models examined for the whole collection of studies may not always be possible for subsets of studies. This occurs when some variables have not been examined in the subsets of studies. When Whiteside and I selected samples where results had been reported separately by gender, we found that several variables had not been studied for the separate-gender samples. Four correlations that had been examined for mixed gender samples had not been studied in the separate-gender groups. This meant some of the models that were estimated for the full set of seventeen studies could not be examined for the five samples of boys and five samples of girls.

20.2.4 We Can Learn What Has Not Been Studied

A final benefit of linked meta-analysis based on a specific model is that it can enable the synthesist to identify relationships that have not been studied, or have not been studied extensively. Absence of data means that a typically bivariate synthesis must be abandoned, because connections between the pair of variables are necessary to support any analysis. In the context of more-complex linked models however, the absence of data on a particular relationship might restrict the complexity of some analyses, but does not imply that no analyses can be done. In addition, such missing data can point out possible areas for future research.

Two examples of how model-driven linked meta-analyses help to identify gaps in the literature can be drawn from previous syntheses. In a synthesis to examine the prediction of science achievement for males and females, I wanted to examine the extent to which an elaborate model for school achievement behaviors proposed by Jacquelynne Eccles and her colleagues had been examined by others in the context of science achievement (Becker

1992a; model in Eccles-Parsons 1984). Admittedly, the literature I gathered had not aimed to examine this model, but it still seemed reasonable to ask whether evidence was available in the literature to support or refute the model. I found that most of the connections between the variables specified in Eccles' model had not been investigated at all, yet other connections, such as the relationship between aptitudes and science achievement, were well studied. In the synthesis of prediction of child outcomes in divorcing families, Whiteside and I found no correlations for maternal restrictiveness with two other predictors—restrictiveness had not been studied with coparent hostility and maternal depressive symptoms (2000). For this reason, the two pairs of predictors (restrictiveness and hostility, and restrictiveness and maternal depression) could not be used together to predict any outcome.

Even if all relationships have been studied, a model-driven meta-analysis can identify when some relationships are better understood than others. In the analysis of Craft and colleagues, each relationship between an aspect of anxiety and sport performance was represented by at least fifty-five correlation values. However, the intercorrelations between the subscales of the CSAI were not reported in all of the studies examined. These relationships were observed between twenty-two and twenty-five times, depending on the pair of subscales examined.

20.2.5 Limitations of Linked Meta-Analysis

As does any statistical analysis technique, model-driven or linked meta-analysis has its limitations. Of particular concern is the issue of missing data. When a particular relationship has not been studied, one cannot even estimate the degree of relatedness between the variables of interest. The positive aspect is that this information may provide future directions for research. Missing data can also, however, lead to analytical problems. Fortunately, some advances have been made in dealing with missing data in the context of correlation matrices (Furlow and Beretvas 2005); these will be discussed briefly in the next section of the chapter.

A second concern is that the results of a model-driven meta-analysis may be fractionated and not apply well to any particular population. In theory, the correlations relevant to one relationship can arise from a completely different set of studies than those that provide results for other relationships, though this is rare. Thus the connection between variables A and C can be based on different samples (or populations) from the samples examining variables B and C. This is a particular concern for relations based on only a few studies. Using random-effects models and testing for the consistency of results across studies can help address this concern by providing estimates that apply broadly and by incorporating uncertainty to account for between-studies variation in the strength of relationships under study.

20.3 HOW TO CONDUCT A MODEL-BASED SYNTHESIS

Many of the tasks required in conducting a model-driven meta-analysis are essentially the same as those required in doing a typical meta-analysis. Because Schram and I went into great detail about some of these tasks, I do not examine all of their ideas here (Becker and Schram 1994).

20.3.1 Problem Formulation

Problem formulation begins the process, and here the synthesist often begins with a model like the ones just shown—with key model components specified. These components can be used in computerized database searches. In a study of sport performance, for example, a particular instrument had been used to study anxiety—the Competitive State Anxiety Index—so searches used that specific name to locate studies. Most studies presented correlations of each subscore with performance, which was the main focus of the simple model in figure 20.3. In more complex problems, like the study of impacts on children of divorce, it is unlikely that all studies will report correlations among all variables. In fact none of the studies Whiteside and I gathered reported on all of the correlations among the fourteen variables studied (eleven of which are shown in figure 20.4). Whiteside and I did set inclusion rules (2000); one of these was that every study in the synthesis must have reported a correlation involving some outcome with either father-child contact or the coparenting relationship (measured as hostility or cooperation). These two variables were chosen because they can be influenced by the courts in divorce custody decisions.

20.3.2 Data Collection

For a linked or model-driven meta-analysis, an average correlation matrix is estimated. Eventually a regression

	Performance (Y)	Cognitive anxiety	Somatic anxiety	Self confidence
Performance (Y)	1.0	.10	.31	−.17
Cognitive anxiety	C1	1.0	NR	NR
Somatic anxiety	C2	C4	1.0	NR
Self confidence	C3	C5	C6	1.0

Figure 20.5 Form for Recording Values of Correlations

SOURCE: Author's compilation.

model or path model of relationships among the variables is as well. What this means for model-driven meta-analysis is that the synthesist searches for studies presenting correlations for sets of variables—as many of the variables of interest as can be found. Searching strategies often involve systematic pairings of relevant terms. For example, in the meta-analysis of outcomes for young children in divorcing families, Whiteside and I required that each study include either a measure of father's contact with the child or a measure of the quality of interactions between the parents. Hostility or cooperation were often measured. Clearly, a measure of some child outcome also had to be available, and zero-order correlations among these and other variables had to be either reported or available from the primary study author or authors. Such search requirements often lead to intersection searches. For instance, the search parameters (*hostility or cooperation*) AND (*anxiety or internalizing behaviors or depression or externalizing behaviors*) should identify studies that would meet one of the conditions required of all studies in the Whiteside and Becker synthesis.

This does not mean, however, that only studies with correlations among all variables would be included—if this requirement were set, no one would ever be able to complete a model-driven meta-analysis. Instead, each study likely will contribute a part of the information in the correlation matrix. Some studies will examine more of the correlations than others and some will report all

desired correlations. This also leads to an increased importance of testing for consistency in the full correlation matrix across studies, and of examining the influence of studies that contribute many correlations to the analysis, for instance, by way of sensitivity analysis.

20.3.3 Data Management

Because one goal of a model-driven meta-analysis is to compute a correlation matrix, it is best to lay out the contents of that matrix early in the data-collection process so that when correlations are found in studies they can be recorded in a systematic way. When the data are analyzed, they will be read in the format of a typical data record—that is, they will appear in a line or row of a data matrix, not in the form of a square correlation matrix. It is sometimes easier, however, to record the data on a form that looks like the real matrix, and then to enter the data from that form into a file. Thus, for the study of anxiety, one might make a form such that shown in figure 20.5, which shows below the diagonal a variable name for each correlation (these are ordered as the correlations appear in the matrix) and shows above the diagonal the values of the correlations for our study 17 (Edwards and Hardy 1996). When a correlation does not appear, the entry above the diagonal may be left blank, but it may be preferable to note its absence with a mark, such as NR for *not reported*. When primary study authors have omitted values that were not significant, one could enter NS so that

Table 20.1 Relationships Between CSAI Subscales and Sport Behavior

			Variable Names and Corresponding Correlations					
			Cognitive/ Performance	Somatic/ Performance	Self Confidence/ Performance	Cognitive/ Somatic	Cognitive/ Self Confidence	Somatic/ Self Confidence
ID	n_i	Type of sport	C1 r_{i1Y}	C2 r_{i2Y}	C3 r_{i3Y}	C4 r_{i12}	C5 r_{i13}	C6 r_{i23}
1	142	2	−.55	−.48	.66	.47	−.38	−.46
3	37	2	.53	−.12	.03	.52	−.48	−.40
6	16	1	.44	.46	9.00	.67	9.00	9.00
10	14	2	−.39	−.17	.19	.21	−.54	−.43
17	45	2	.10	.31	−.17	9.00	9.00	9.00
22	100	2	.23	.08	.51	.45	−.29	−.44
26	51	1	−.52	−.43	.16	.57	−.18	−.26
28	128	1	.14	.02	.13	.56	−.53	−.27
36	70	1	−.01	−.16	.42	.62	−.46	−.54
38	30	2	−.27	−.13	.15	.63	−.68	−.71

SOURCE: Craft et al. 2003.

later on a count could be obtained of the omitted nonsignificant results as a crude index of the possibility of publication bias. Often, however, it is not possible to determine why a result has not been reported. In the case shown in figure 20.5, only the correlations of the anxiety scales with the outcome were reported. This is a fairly common pattern for unreported data. Also, for practical reasons, this form is arranged so the matrix has the correlations involving the performance outcome in the first row. These are the most frequently reported correlations and, because they are also the first three values, can easily be typed first when creating the data file, to be followed by three missing-data codes if the predictor correlations are not reported.

The one other variable the synthesist must record is the sample size. When primary studies have used pairwise deletion, the rs for different relations in a matrix may be based on different sample sizes. In such cases, I generally record the smallest sample size shown, as a conservative choice of the single value to use for all relationships in that study. In most meta-analyses, the synthesist will also record values of other coded variables that are of interest as moderators or simply as descriptors of the primary studies. There are no particular requirements for those additional variables, and each relationship in a matrix or each variable may have its own moderator variables. For

example, type of performance outcome might be relevant mainly to the correlations involving the X-Y relations in our example.

Thus for each case the data will include, at a minimum, a study identification code, the sample size and up to p^* (here, six) correlations. For the example, I have also included a variable representing whether the sport behavior examined was a team (Type = 1) or individual (Type = 2) sport. Table 20.1 shows the data for ten studies drawn from the Craft et al. meta-analysis (2003). The record for a study shows a value of 9.00 when a correlation is missing. The fifth line of table 20.1 shows the data from study 17, also displayed in figure 20.5.

Finally, one additional type of information will eventually be needed in the multivariate approach to synthesizing correlation matrices. Specifically, a set of indicator variables is needed to identify which correlations are reported in each study. Because the analysis methods rely on representations of the study data in vectors of correlations, it is perhaps simplest to describe the indicators in terms of matrices. Matrices are also used in the example. If data are stored in the form of one record per correlation, then a set of p^* variables can be created to serve as indicators for the relationships under study.

Let us consider studies with IDs 3 and 6 in table 20.1. Study 3 has provided estimates of all six correlations among the CSAI subscales and a performance outcome. We can represent the results of study 3 with a 6×1 vector $\mathbf{r}_3$ and the 6×6 indicator matrix $\mathbf{X}_3$, such that $\mathbf{r}_3 = \mathbf{X}_3 \boldsymbol{\rho} + \mathbf{e}_3$ with

$$
\mathbf{r}_3 = \begin{bmatrix} .53 \\ -.12 \\ .03 \\ .52 \\ -.48 \\ -.40 \end{bmatrix}, \quad \mathbf{X}_3 = \begin{bmatrix} 1 & 0 & 0 & 0 & 0 & 0 \\ 0 & 1 & 0 & 0 & 0 & 0 \\ 0 & 0 & 1 & 0 & 0 & 0 \\ 0 & 0 & 0 & 1 & 0 & 0 \\ 0 & 0 & 0 & 0 & 1 & 0 \\ 0 & 0 & 0 & 0 & 0 & 1 \end{bmatrix}, \text{ and } \boldsymbol{\rho} = \begin{bmatrix} \rho_1 \\ \rho_2 \\ \rho_3 \\ \rho_4 \\ \rho_5 \\ \rho_6 \end{bmatrix},
$$

where $\mathbf{e}_3$ represents the deviation of the observed values in $\mathbf{r}_3$ from the unknown population values in ρ. Here $\mathbf{X}_3$ is an identity matrix because study 3 has six correlations and the columns of X represent each of the six correlations being studied. Study 6 does not appear to have used a measure of self-confidence, however, and reports only three correlations: for the first, second, and fourth relationships in the correlation matrix. Thus the data from study 6 are represented by

$$
\mathbf{r}_6 = \begin{bmatrix} .44 \\ .46 \\ .67 \end{bmatrix} \quad \text{and} \quad \mathbf{X}_6 = \begin{bmatrix} 1 & 0 & 0 & 0 & 0 & 0 \\ 0 & 1 & 0 & 0 & 0 & 0 \\ 0 & 0 & 0 & 1 & 0 & 0 \end{bmatrix}.
$$

The indicator matrix still contains six columns to allow for the six correlations we hope to summarize across all studies.

20.3.4 Analysis and Interpretation

The synthesist typically begins a model-driven analysis by estimating a correlation matrix across studies. The task is to find patterns that are supported across studies and to discuss potential explanations for any important variability in the correlations of interest across studies. Another goal is to explicitly identify missing (unstudied) relationships or less-well-studied relationships to suggest fruitful areas for future research. Moderator analyses, such as comparing results for team versus individual sports, may also be conducted. Before describing these steps in the process of data analysis, I present a little additional notation.

20.3.4.1 Distribution of the Correlation Vector r

The methods presented for the synthesis of correlation matrices are based on generalized least (GLS) methods for synthesis, designed after the approach laid out by Stephen Raudenbush, Betsy Becker, and Hripsime Kalaian (1988) and Becker and Schram (1994). These methods require that we have variances and covariances for the correlations we summarize. Ingram Olkin and Minoru Siotani showed that in large samples of size n_i, the correlation vector $\mathbf{r}_i$ is approximately normally distributed with mean vector $\boldsymbol{\rho}_i$ and variance-covariance matrix $\boldsymbol{\Sigma}_i$, where the elements of $\boldsymbol{\Sigma}_i$ are defined by $\sigma_{i\alpha\gamma}$ (1976).

Recall that in section 20.1.2 we had simplified the vector notation for $\mathbf{r}_i = (r_{i12}, r_{i13}, r_{ist}, \ldots, r_{i(p-1)p})$, by using a single subscript in place of the pair of indices s, t. Here, to distinguish among the p variables in each study, we must return to the full subscripting, because the covariance of r_{ist} with r_{iuv} involves all of the cross-correlations $(r_{ist}, r_{isu}, r_{isv}, r_{itu}, r_{itv},$ and $r_{iuv})$. Formulas for the variances and covariances are given by Olkin and Siotani (1976, 238). Specifically

$$
\sigma_{i\alpha\alpha} = \text{Var}(r_{i\alpha}) = \text{Var}(r_{ist}) = (1 - \rho_{ist}^2)^2/n_i, \qquad (20.1)
$$

and

$$
\sigma_{i\alpha\gamma} = \text{Cov}(r_{i\alpha}, r_{i\gamma}) = \text{Cov}(r_{ist}, r_{iuv}).
$$

The covariance $\sigma_{i\alpha\gamma}$ can be expressed by noting that if $r_{i\alpha} = r_{ist}$, the correlation between the sth and tth variables in study i, $r_{i\gamma} = r_{iuv}$, and ρ_{ist} and ρ_{iuv} are the corresponding population values, then

$$
\begin{aligned}
\text{Cov}(r_{ist}, r_{iuv}) = [&0.5\, \rho_{ist}\, \rho_{iuv}\, (\rho_{isu} + \rho_{isv} + \rho_{itu} + \rho_{itv}) \\
&+ \rho_{isu}\, \rho_{itv} + \rho_{isv}\, \rho_{itu} - (\rho_{ist}\, \rho_{isu}\, \rho_{isv} \\
&+ \rho_{its}\, \rho_{itu}\, \rho_{itv} + \rho_{ius}\, \rho_{iut}\, \rho_{iuv} \\
&+ \rho_{ivs}\, \rho_{ivt}\, \rho_{ivu})]/n_i
\end{aligned} \qquad (20.2)
$$

for $s, t, u,$ and $v = 1$ to p. Typically $\sigma_{i\alpha\alpha}$ and $\sigma_{i\alpha\gamma}$ are estimated by substituting the corresponding sample estimates for the parameters in equations 20.1 and 20.2. Also sometimes the denominators in equations 20.1 and 20.2 are computed as $(n_i - 1)$ rather than n_i.

In the data analysis using GLS methods, these variances and covariances must be computed. However research has indicated that substituting individual study correlations into formulas 1 and 2 is not the best approach (Becker and Fahrbach 1994; Furlow and Beretvas 2005).

In short, I recommend computing these variances and co-variances using values of the mean correlations across studies.

However, another issue is whether to use raw correlations for the data analysis or to transform them using Sir Ronald Fisher's z transformation (1921). Fisher's transformation is

$$z = 0.5 \log\left[(1 + r)/(1 - r)\right], \qquad (20.3)$$

which is variance-stabilizing. Thus the variance of $z_{i\alpha}$ is usually taken to be $\Psi_{i\alpha\alpha} = n_i^{-1}$ or $(n_i - 3)^{-1}$, which does not involve the parameter $\rho_{i\alpha}$ or any function of $\rho_{i\alpha}$. The covariance between two z values, however, is even more complicated than the covariance in 2, and is given by

$$\Psi_{i\alpha\gamma} = \mathrm{Cov}(z_{i\alpha}, z_{i\gamma}) = \sigma_{i\alpha\gamma}/[(1 - \rho_{i\alpha}^2)(1 - \rho_{i\gamma}^2)]. \quad (20.4)$$

These elements form the covariance matrix for the z values from study i, Ψ_i. Work by myself and Kyle Fahrbach (1994) and by Carolyn Furlow and Natasha Beretvas (2005) also supports the use of the Fisher transformation both in terms of estimation of the mean correlation matrix, and to improve the behavior of associated test statistics, which are described shortly. For much of this presentation, I offer formulas for tests and estimates based on the use of Fisher's z transformation.

These formulas apply under the fixed-effects model, that is, when one can assume that all k correlation matrices arise from a single population. If that is unreasonable, either on theoretical grounds or because tests of homogeneity are significant, then one should augment the within-study uncertainty quantified in Ψ_i by the addition of between-studies variation. A number of approaches exist for estimating between-studies variation in the univariate context, but little work has been done in the multivariate context (see, for example, chapter 18, this volume; Sidik and Jonkman 2005). My own practical experience, based on applications to a variety of data sets, is that estimating between-studies variances is relatively simple, but estimating between-studies covariance components can be problematic. For instance, particularly with small numbers of studies, one can obtain between-studies correlations above 1. If one has a set of studies with complete data and can use, for instance, a multivariate hierarchical modeling approach these problems may be resolved (Kalaian and Raudenbush 1996). It is rare, however, to have complete data when summarizing correlation matrices. My recommendation is thus to add only the variance

components when adopting a random-effects model, until further research or improved software provide better estimators.

Example. I consider the first study from the used by from Craft and her colleagues (2003). Study 1 has reported all six correlation values, shown here in square as well as in column form and transformed into the z metric, specifically

$$\mathbf{R}_1 = \begin{bmatrix} 1 & -.55 & -.48 & .66 \\ -.55 & 1 & .47 & -.38 \\ -.48 & .47 & 1 & -.46 \\ .66 & -.38 & -.46 & 1 \end{bmatrix} \text{ or}$$

$$\mathbf{r}_1 = \begin{bmatrix} -.55 \\ -.48 \\ .66 \\ .47 \\ -.38 \\ -.46 \end{bmatrix} \text{ and } \mathbf{z}_1 = \begin{bmatrix} -.618 \\ -.523 \\ .793 \\ .510 \\ -.400 \\ -.497 \end{bmatrix}.$$

The covariance of $\mathbf{r}_1$, computed using sample-size weighted mean correlations ($\bar{\mathbf{r}}_{\boldsymbol{\cdot}} = (-.074 \quad -.127 \quad .324 \quad .523 \quad -.416 \quad -.414)'$), is

$$\mathrm{Cov}(\mathbf{r}_1) = \begin{bmatrix} .0070 & .0036 & -.0025 & -.0005 & .0018 & .0009 \\ .0036 & .0068 & -.0025 & -.0002 & .0008 & .0017 \\ -.0025 & -.0025 & .0056 & .0001 & -.0000 & -.0003 \\ -.0005 & -.0002 & .0001 & .0037 & -.0013 & -.0013 \\ .0018 & .0008 & -.0000 & -.0013 & .0048 & .0022 \\ .0009 & .0017 & -.0003 & -.0013 & .0022 & .0048 \end{bmatrix},$$

and the covariance of $\mathbf{z}_1$ computed using the same mean correlations is

$$\mathrm{Cov}(\mathbf{z}_1) = \begin{bmatrix} .0072 & .0037 & -.0029 & -.0008 & .0022 & .0011 \\ .0037 & .0072 & -.0029 & -.0003 & .0010 & .0022 \\ -.0029 & -.0029 & .0072 & .0001 & -.0001 & -.0004 \\ -.0008 & -.0003 & .0001 & .0072 & -.0022 & -.0023 \\ .0022 & .0010 & -.0001 & -.0022 & .0072 & .0033 \\ .0011 & .0022 & -.0004 & -.0023 & .0033 & .0072 \end{bmatrix}.$$

If a study reports fewer than p^* correlations the $\mathrm{Cov}(\mathbf{r}_i)$ and $\mathrm{Cov}(\mathbf{z}_i)$ matrices will be reduced in size accordingly.

20.3.4.2 Mean Correlation Matrix Under Fixed Effects If we denote the Fisher transformation of the population correlation ρ using the Greek letter zeta (ζ), we have $\zeta = 0.5 \log[(1+\rho)/(1-\rho)]$. Generalized least squares methods provide the estimator

$$\hat{\zeta} = (\mathbf{X}'\hat{\mathbf{\Psi}}^{-1}\mathbf{X})^{-1}\mathbf{X}'\hat{\mathbf{\Psi}}^{-1}\mathbf{z}, \qquad (20.5)$$

where $\mathbf{X}$ is a stack of $k\,p^* \times p^*$ indicator matrices, $\mathbf{z}$ is the vector of z-transformed correlations, and $\hat{\mathbf{\Psi}}$ is a blockwise diagonal matrix with the matrices $\hat{\mathbf{\Psi}}_i = \mathrm{Cov}(z_i)$ on its diagonal. The variance of $\hat{\zeta}$ is

$$\mathrm{Var}(\hat{\zeta}) = (\mathbf{X}'\hat{\mathbf{\Psi}}^{-1}\mathbf{X})^{-1}. \qquad (20.6)$$

The elements of $\hat{\zeta}$ are returned to the correlation metric via the inverse Fisher transformation to yield the estimator $\tilde{\rho}$ of ρ, given by $\tilde{\rho} = (\tilde{\rho}_\alpha)$, where $\tilde{\rho}_\alpha = [\exp(2\hat{\zeta}_\alpha)-1]/[\exp(2\hat{\zeta}_\alpha)+1]$, for $\alpha = 1$ to p^*. Also, because the transformation that puts the covariance matrix in equation 20.6 back on the correlation scale is approximate, it is typical to compute confidence intervals in the z metric, and to back-transform the endpoints of each confidence interval or CI to the correlation scale. Any tests of differences among elements of the correlation matrix would be done in the z metric, though these are not common.

Parallel methods can be applied to untransformed correlations, with

$$\hat{\rho} = (\mathbf{X}'\hat{\mathbf{\Sigma}}^{-1}\mathbf{X})^{-1}\mathbf{X}'\hat{\mathbf{\Sigma}}^{-1}\mathbf{r}, \qquad (20.7)$$

where $\mathbf{X}$ is a stack of $k\,p^* \times p^*$ indicator matrices, $\mathbf{r}$ is the vector of correlations, and $\hat{\mathbf{\Sigma}}$ is a blockwise diagonal matrix with the matrices $\hat{\mathbf{\Sigma}}_i$ on its diagonal. The variance of $\hat{\rho}$ is

$$\mathrm{Var}(\hat{\rho}) = (\mathbf{X}'\hat{\mathbf{\Sigma}}^{-1}\mathbf{X})^{-1}. \qquad (20.8)$$

These estimators have been extensively studied and other estimation methods have been proposed. Ingram Olkin and I have shown that the generalized least squares estimators given here are also maximum likelihood estimators (Becker and Olkin 2008). We also provide an estimator that is the matrix closest to a set of correlation matrices, in terms of a general minimum distance function. Meng-Jia Wu has investigated a method based on factored likelihoods and the sweep operator that appears to perform well in many conditions (2006). However, the limitation of Wu's approach is that if any data are missing, the studies must have a nested structure, such that the data missing in any study must be a subset of what is missing in other studies. If a study had no values for r_1 and r_2, and another study were missing r_1, r_2 and r_3, the data would be nested, but if the second study instead were missing r_1 and r_3, the data would not be nested and the method could not be applied.

Example. For the ten studies in our example, the mean correlation matrix obtained from analysis of the correlations is

$$\hat{\rho} = \begin{bmatrix} 1 & -.074 & -.127 & .317 \\ -.074 & 1 & .523 & -.415 \\ -.127 & .523 & 1 & -.405 \\ .317 & -.415 & -.405 & 1 \end{bmatrix},$$

and the mean based on the Fisher z transformed correlations is

$$\tilde{\rho} = \begin{bmatrix} 1 & -.092 & -.141 & .364 \\ -.092 & 1 & .586 & -.449 \\ -.141 & .586 & 1 & -.438 \\ .364 & -.449 & -.438 & 1 \end{bmatrix}.$$

The first rows of these matrices contain the correlations of the CSAI subscales with sport behavior. Clearly, the last entries in that row, which represent the average correlation for self confidence with sport behavior, are the largest. As we will see shortly, however, these values should not be interpreted because the fixed-effects model is not appropriate for these data.

20.3.4.3 Test of Homogeneity, with H_{01}: $\rho_1 = \ldots = \rho_k$ or $\zeta_1 = \ldots = \zeta_k$

In a typical univariate meta-analysis, one of the first hypotheses to be tested is whether all studies appear to be drawn from a single population. This test can conceptually be generalized to the case of correlation matrices, and here the meta-analyst may want to test whether all k correlation matrices appear to be equal or homogeneous. A test of the hypothesis of homogeneity of the k population correlation matrices is a test of H_{01}: $\rho_1 = \ldots = \rho_k$. Similar logic leads to a test of the hypothesis $\zeta_1 = \ldots = \zeta_k$, which is logically equivalent to the hypothesis H_{01} that all the ρ_i are equal. The recommended test, which Fahrbach and I found to have a distribution much closer to the theoretical chi-square distribution than a similar test based on raw correlations (Becker Fahrbach 1994), uses the statistic

$$Q_E^Z = \mathbf{z}'[\hat{\mathbf{\Psi}}^{-1} - \hat{\mathbf{\Psi}}^{-1}\mathbf{X}(\mathbf{X}'\hat{\mathbf{\Psi}}^{-1}\mathbf{X})^{-1}\mathbf{X}'\hat{\mathbf{\Psi}}^{-1}]\mathbf{z}. \qquad (20.9)$$

When H_{01} is true, Q_E^Z has approximately a chi-square distribution with $(k-1)p^*$ degrees of freedom if all studies report all correlation values. If some data are missing the degrees of freedom equal the total number of observed correlations across studies minus p^*.

Mike Cheung and Wai Chan investigated the chi-square test of homogeneity based on raw correlations, and found as I and Fahrbach did that the analogue to Q_E^Z computed using correlations had a rejection rate that was well above the nominal rate (Cheung and Chan 2005; Becker and Fahrbach 1994). Cheung and Chan proposed using a Bonferroni adjusted test which draws on the individual tests of homogeneity (see chapter 14, this volume) for the p^* relationships represented in the correlation matrix. We reject the hypothesis H_{01}: $\mathbf{p}_i = \mathbf{p}$ if at least one of the p^* individual homogeneity tests is significant at level α/p^*. Cheung and Chan refer to this as the BA1 (Bonferroni At least 1) rule. Thus, for example, in the Craft et al. data with $p^* = 6$ correlations, the adjusted significance level for an $\alpha = .05$ test would be $.05/6 = .0083$. Cheung and Chan examined tests of the homogeneity of raw correlations, but this adjustment can also be applied to homogeneity tests for the Fisher z transformed rs.

Example. For the ten studies in our example there are fifty-four reported correlations ($k \times p^* = 10 \times 6 = 60$ were possible, but six are missing) and the value of $Q_E^Z = 209.18$ with $df = 54 - 6 = 48$. This is significant with $p < .001$. In addition, we can consider the six individual homogeneity tests for the six relationships in the matrix. According to the BA1 rule, at least one test must be significant at the .0083 level. In fact, all three of the homogeneity tests of the CSAI subscales with sport performance are highly significant (with p values well below .0001), and the three sets of correlations among the subscales are each homogeneous.

20.3.4.4 Estimating Between-Studies Variation If a random-effects model seems sensible, either on principle or because the hypothesis of homogeneity just described has been rejected, the meta-analyst will need to estimate the variance across studies in each set of correlations, say τ_α^2 for the αth correlation, with $\alpha = 1$ to p^*. A variety of estimators are described elsewhere; the key issue is how to incorporate the estimates of variation into the analysis. A straightforward approach is to estimate the diagonal matrix $\mathbf{T}^Z = \text{diag}(\tau_1^2, \tau_2^2, \ldots, \tau_{p^*}^2)$ and add $\hat{\mathbf{T}}^Z$ to each of the $\hat{\mathbf{\Psi}}_i$, so that $\hat{\mathbf{\Psi}}_i^{RE} = \hat{\mathbf{\Psi}}_i + \hat{\mathbf{T}}^Z$. Then $\hat{\mathbf{\Psi}}^{RE}$ (the estimate of the random-effects covariance matrix with $\hat{\mathbf{\Psi}}_i^{RE}$ matrices on its diagonal) would be used in place of $\hat{\mathbf{\Psi}}$ in formulas 20.5 and 20.6, and in formulas 20.10 and 20.11. When analyses are conducted on raw correlations, variance components are estimated in the r metric (I will call the

matrix $\mathbf{T}^R$) and then the estimate $\hat{\mathbf{T}}^R$ is added to each $\hat{\mathbf{\Sigma}}_i$ to get $\hat{\mathbf{\Sigma}}_i^{RE} = \hat{\mathbf{\Sigma}}_i + \hat{\mathbf{T}}^R$, then $\hat{\mathbf{\Sigma}}^{RE}$ would then be used in place of $\hat{\mathbf{\Sigma}}$ in further estimates, as shown in equations 20.12 and 20.13 below.

Example. For our example a simple method of moments estimator is computed for each set of transformed correlations (see chapter 16 this volume). Based on the homogeneity tests, we expect the τ_α^2 estimates to be nonzero for the correlations of CSAI scales with the outcome, but the three estimates for the CSAI intercorrelations to be very close to zero. The values in the z metric are shown in the diagonal matrix

$$\hat{\mathbf{T}}^Z = \begin{bmatrix} .159 & 0 & 0 & 0 & 0 & 0 \\ 0 & .075 & 0 & 0 & 0 & 0 \\ 0 & 0 & .096 & 0 & 0 & 0 \\ 0 & 0 & 0 & 0 & 0 & 0 \\ 0 & 0 & 0 & 0 & .013 & 0 \\ 0 & 0 & 0 & 0 & 0 & .012 \end{bmatrix},$$

and indeed the values are as expected, with one value—the variance component for the correlations between cognitive and somatic anxiety—estimated at just below zero. Thus this value was set to 0. Also, the first three entries are larger than the last two. To interpret the values, one can take the square root of each diagonal element to get the estimated standard deviation of the population of Fisher z values for the relationship in question. For instance, the ζ_i values for correlations of cognitive anxiety with sport behavior have a standard deviation of $\hat{\tau}_1 = \sqrt{0.159} = 0.40$. If the true average correlation (or ζ) were zero, then roughly 95 percent of the population z values would be in the interval $-.781$ to $.781$ and 95 percent of the population correlations would fall between $-.65$ and $.65$. This range is analogous to the plausible values range computed for hierarchical linear model parameters at level 2 (for example, Raudenbush and Bryk 2002). This is quite a wide spread in population correlation values.

20.3.4.5 Random-Effects Mean Correlation If the meta-analyst believes that a random-effects model is more appropriate for the data, then, as just noted, the value of $\hat{\mathbf{\Psi}}^{RE}$—the variance including between-studies differences—would be used in place of $\hat{\mathbf{\Psi}}$ in (5). Then the random-effects mean $\hat{\boldsymbol{\zeta}}^{RE}$ would be computed as

$$\hat{\boldsymbol{\zeta}}^{RE} = (\mathbf{X}'[\hat{\mathbf{\Psi}}^{RE}]^{-1} \mathbf{X})^{-1} \mathbf{X}'[\hat{\mathbf{\Psi}}^{RE}]^{-1} \mathbf{z}, \qquad (20.10)$$

where $\mathbf{X}$ is still the stack of $k\,p^* \times p^*$ indicator matrices, $\mathbf{z}$ is the vector of z-transformed correlations, and $\hat{\boldsymbol{\Psi}}^{\mathrm{RE}}$ is a blockwise diagonal matrix with the matrices $\hat{\boldsymbol{\Psi}}_i^{\mathrm{RE}}$ on its diagonal. As for the fixed-effects estimate, the values in $\hat{\zeta}^{\mathrm{RE}}$ would be transformed to the correlation metric for reporting. The variance of $\hat{\zeta}^{\mathrm{RE}}$ is

$$\mathrm{Var}(\hat{\zeta}^{\mathrm{RE}}) = (\mathbf{X}'[\hat{\boldsymbol{\Psi}}^{\mathrm{RE}}]^{-1}\,\mathbf{X})^{-1}. \qquad (20.11)$$

Typically the variances obtained using formula 20.11 will be larger than those using 20.6.

Similarly, we can estimate the mean in the correlation metric, as

$$\hat{\boldsymbol{\rho}}^{\mathrm{RE}} = (\mathbf{X}'[\hat{\boldsymbol{\Sigma}}^{\mathrm{RE}}]^{-1}\,\mathbf{X})^{-1}\,\mathbf{X}'[\hat{\boldsymbol{\Sigma}}^{\mathrm{RE}}]^{-1}\,\mathbf{r}, \qquad (20.12)$$

where $\mathbf{X}$ is the stack of $k\,p^* \times p^*$ indicator matrices, B is the vector of correlations, and $\hat{\boldsymbol{\Sigma}}^{\mathrm{RE}}$ is the variance of $\mathbf{r}$ under random effects. The variance of $\hat{\boldsymbol{\rho}}^{\mathrm{RE}}$ is

$$\mathrm{Var}(\hat{\boldsymbol{\rho}}^{\mathrm{RE}}) = (\mathbf{X}'[\hat{\boldsymbol{\Sigma}}^{\mathrm{RE}}]^{-1}\,\mathbf{X})^{-1}. \qquad (20.13)$$

Example. In the data from our example, the first three correlations showed significant variation. Now, under the random-effects model, they are slightly smaller than under the fixed-effects assumptions. The random-effects mean, computed using z values then transformed, is

$$\tilde{\boldsymbol{\rho}}^{\mathrm{RE}} = \begin{bmatrix} 1 & -.040 & -.075 & .252 \\ -.041 & 1 & .548 & -.454 \\ -.075 & .548 & 1 & -.409 \\ .252 & -.454 & -.409 & 1 \end{bmatrix}.$$

The mean in the correlation metric is quite similar:

$$\hat{\boldsymbol{\rho}}^{\mathrm{RE}} = \begin{bmatrix} 1 & -.041 & -.082 & .241 \\ -.041 & 1 & .523 & -.417 \\ -.082 & .523 & 1 & -.418 \\ .241 & -.417 & -.418 & 1 \end{bmatrix}.$$

The variance of $\hat{\boldsymbol{\rho}}^{\mathrm{RE}}$ (used below) is

$$\mathrm{Var}(\boldsymbol{\rho}^{\mathrm{RE}}) = \begin{bmatrix} .0339 & .0033 & -.0017 & 0 & .0002 & .0001 \\ .0033 & .0295 & -.0011 & 0 & .0001 & .0002 \\ -.0017 & -.0011 & .0087 & 0 & .0001 & .0001 \\ 0 & 0 & 0 & .0023 & -.0004 & -.0011 \\ .0002 & .0001 & .0001 & -.0004 & .0303 & .0022 \\ .0001 & .0002 & .0001 & -.0011 & .0022 & .0048 \end{bmatrix}.$$

Values shown as 0 were not exactly zero, but were smaller than 0.0001.

20.3.4.6 Test of No Association, with H_{02}: $\boldsymbol{\rho} = \mathbf{0}$ Synthesists often want to test whether any of the correlations in the matrix $\boldsymbol{\rho}$ are nonzero, that is, to test H_{02}: $\boldsymbol{\rho} = \mathbf{0}$. GLS theory provides such a test for the correlation vector. That test, however, has been shown to reject H_{01} up to twice as frequently as it should for the nominal alpha (Becker and Fahrbach 1994). If we are working with Fisher z transformed data, we can test the hypothesis that $\boldsymbol{\zeta} = \mathbf{0}$, which is logically equivalent to the hypothesis H_{02}: $\boldsymbol{\rho} = \mathbf{0}$ because $\boldsymbol{\zeta} = \mathbf{0}$ implies $\boldsymbol{\rho} = \mathbf{0}$. The test uses the statistic

$$Q_{\mathrm{B}}^{Z} = \hat{\zeta}'\,(\mathbf{X}'\,\hat{\boldsymbol{\Psi}}^{-1}\,\mathbf{X})^{-1}\,\hat{\zeta}. \qquad (20.14)$$

When the null hypothesis is true, Q_{B}^{Z} has approximately a chi-square distribution with p^* degrees of freedom. I and Fahrbach found that the rejection rate of Q_{B}^{Z} for $\alpha = .05$ was never more than .07, but was usually within .01 of $\alpha = .05$, and when the within–study sample size was 250, dropped below .05 in some cases (1994). Rejecting H_{02} implies that at least one element of the matrix $\boldsymbol{\rho}$ is nonzero, but does not mean that all elements are nonzero.

A similar test for a subset of correlations can be obtained by selecting a submatrix of m values from $\hat{\zeta}$ and using the corresponding submatrices of $\mathbf{X}$ and $\hat{\boldsymbol{\Psi}}$ to compute Q_{B}^{Z} for the subset of correlations.

If one has adopted a random-effects model on the basis of theory or because of rejecting either of the homogeneity test criteria, then one would substitute $\hat{\boldsymbol{\Psi}}^{\mathrm{RE}}$ for $\hat{\boldsymbol{\Psi}}$ in 20.14 and use the random-effects mean $\hat{\zeta}^{\mathrm{RE}}$ in place of $\hat{\zeta}$. Typically, the test computed under these conditions will be smaller than the value defined in 20.14. Effects are thus less likely to be judged significant in the context of the random-effects model.

Example. Under both the fixed and random-effects models, the value of Q_{B}^{Z} is quite large. Under the more appropriate random-effects model, $Q_{\mathrm{B}}^{Z} = 402.44$ with $df = 6$, so at least one population correlation differs from 0. However, individual random-effects tests of the significance of the first two correlations, of cognitive and somatic anxiety with performance, are not significant—$z = -0.29$ for the cognitive anxiety-performance correlation and $z = -0.76$ for somatic anxiety with performance.

20.3.4.7 Estimating Linear Models In many model driven meta-analyses, synthesists will also want to estimate path models such as those shown in figures 20.3 and 20.4. We draw on either $\hat{\boldsymbol{\rho}}^{\mathrm{RE}}$ and its variance or $\tilde{\boldsymbol{\rho}}^{\mathrm{RE}}$ and its variance for our computations. Let us consider a model to be estimated under random-effects assumptions. Here I use $\hat{\rho}^{RE}$, the mean based on correlations, to avoid using the approximate inverse variance transformation for $\mathrm{Var}(\hat{\rho}^{RE})$. We begin by arranging the mean correlations in $\hat{\rho}^{RE}$ into a square form, which we call $\bar{\mathbf{R}}$. The matrix $\bar{\mathbf{R}}$ is partitioned so that the mean correlations of the predictors with the outcome of interest are put into a submatrix we call $\bar{\mathbf{R}}_{XY}$ and the intercorrelations among the predictors are put into $\bar{\mathbf{R}}_{XX}$. Different subsets of $\bar{\mathbf{R}}$ are used depending on the model or models to be estimated.

Suppose that we are working with the 4×4 matrix of means from the Craft et al. data, and the matrix is arranged so that correlations involving sport performance are in the first row and column. We would partition the matrix to estimate a model with performance as the outcome, as follows. $\bar{\mathbf{R}}_{XY}$ would be the vector of three values $(-.040, -.075, .252)$ and $\bar{\mathbf{R}}_{XX}$ is the lower square matrix:

$$
\bar{\mathbf{R}} = \begin{bmatrix}
1 & -.040 & -.075 & .252 \\
-.040 & 1 & .548 & -.454 \\
-.075 & .548 & 1 & -.409 \\
.252 & -.454 & -.409 & 1
\end{bmatrix}
$$

$$
= \begin{bmatrix}
1 & \bar{\mathbf{R}}_{XY} \\
\hline
\bar{\mathbf{R}}_{YX} & \bar{\mathbf{R}}_{XX}
\end{bmatrix}.
$$

The product $\mathbf{b}^* = \bar{\mathbf{R}}_{XY}^{-1}\,\bar{\mathbf{R}}_{XY}$ is then computed to obtain the standardized regression slopes in the structural model. Schram and I provided further details about the estimation of variances and tests associated with such models (Becker and Schram 1994). The variance of $\mathbf{b}^*$ is obtained as a function of $\mathrm{Var}(\bar{\mathbf{R}}) = \mathrm{Var}(\hat{\rho}^{RE})$ that depends on the specific model being estimated.

Cheung and Chan noted that multilevel structural equation modeling programs can be used if there is no interest in obtaining the mean correlation matrix as an intermediate step (2005). These analyses, however, assume that a fixed-effects model applies. If that is not likely to be true, using such a model may not be advisable. Also their approach does not appear to incorporate the variances of the mean correlations, thus the standard errors from their approach may be underestimated.

Example. I have estimated the path diagram shown in figure 20.3 for the example data on anxiety and sport behavior, under random-effects assumptions. For this subset of the original Craft et al. data, only self confidence

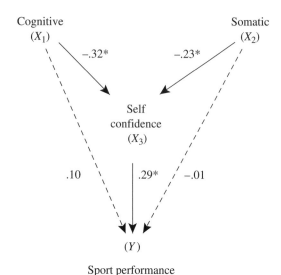

Figure 20.6 Random-Effects Path Coefficients for Prediction of Sport Performance

SOURCE: Author's compilation.

appears to be a significant direct predictor of sport behavior under the random-effects model. Dashed lines represent paths that were tested but found not significant. However, both cognitive and somatic aspects of anxiety appear to have indirect effects on self confidence. Both are significant predictors of self confidence, even under the random-effects model.

In general, when one presents more information than just the path coefficients and some indication of their significance on path diagrams, the images get very "busy". Thus it is not a good idea to add standard errors and the like. In the original Craft et al. meta-analysis, tables of slope coefficients and their standard errors, and z tests computed as the ratios of these quantities, were presented (2003). This information can also be presented in text, such as by discussion of the slopes. In describing this model, the meta-analyst might mention that a one standard-deviation shift in self confidence has nearly a third of a standard deviation impact on sport performance (with a slope of .29, $SE = .115$, and $z = 2.43$, $p = .011$). Other slopes would also be discussed similarly.

20.3.4.8 Moderator Analyses A last step one might take in analyzing a model in a meta-analysis is to examine categorical moderator variables. To do this, separate models are estimated for the groups of interest and compared. For the example data set, the moderator I examine

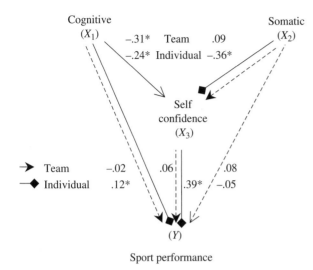

Figure 20.7 Random-Effects Path Coefficients by Type of Sport

SOURCE: Author's compilation.

is the type of sport activity—specifically team sports (studies with Type = 1 in table 20.1) versus individual sports (Type = 2). Four team-sport studies and six individual-sport studies are in the example data. Figure 20.7 shows the slopes from the two samples. The arrow endpoints distinguish differences in significance of relations for the two groups, with curved arrows for team sports, square point arrows for individual sports, and open arrow points when both groups show similar relationships. Again, the dashed lines represent paths with coefficients that do not differ from zero.

Even though the data set is relatively small, some differences are apparent in figure 20.7. Overall, it seems that the anxiety measures do not predict sport performance well for team sports, whereas most aspects of anxiety are important for individual sports, with either direct or indirect effects or both. I discuss only one comparison. The contribution of self confidence to performance appears to be quite different for the two types of sports. Because the studies are independent, we can construct a test of differences in the slopes as

$$z = \frac{(b_T^* - b_I^*)}{\sqrt{Var(b_T^*) + Var(b_I^*)}} = \frac{(.06 - (.39))}{\sqrt{.0049 + .0023}}$$
$$= \frac{-.33}{.085} = -3.89,$$

which in large samples is approximately normal, and thus can be compared to the critical values of the standard normal distribution. For a two-sided test at the .05 level, this difference would clearly be significant, leading us to decide to reject the hypothesis that $\beta_T^* = \beta_I^*$ for this relationship. Self-confidence appears to be more important to performance for individual sports than team sports. The slope representing its contribution to performance also differs from zero for the individual-sport studies, but not for team sports.

20.4 SUMMARY AND FUTURE POSSIBILITIES

Although the steps and data required to implement a model driven meta-analysis are definitely more complex and involved than those for a typical univariate meta-analysis, if the questions of interest are suitable, the benefits of the model-driven analyses are well worth the effort. I have illustrated the strengths of model based analyses for examining such complexities as partial relationships, mediating variables, and indirect effects. The small data set used for the examples did not allow for a good illustration of the ways model-driven meta-analysis can help to identify areas in need of further exploration, but some of the examples drawn from the literature make clear that even when data are sparse, much can be learned.

There is still work to be done in this area, in spite of the good number of dissertations that have addressed questions about estimation methods and the impact of missing data in this complicated realm. The development of tests of indirect effects as well as overall goodness of fit tests would be of value, though Cheung and Chan argued that one can get such tests by using multilevel structural equation modeling packages—at least when the constraints of those packages, such as belief in the fixed-effects model, are reasonable (2005).

Finally, the synthesist may ask whether the model-driven meta-analysis approach of estimating a regression model from a summary of correlation matrices is preferable to a direct synthesis of regression models. In a separate study, a colleague and I have outlined some of the complications that arise when summarizing sets of regression models (Becker and Wu 2007). Some of those complications are avoided with the model-driven meta-analysis approach. In particular, to summarize raw regression coefficients, one must be sure that the predictors and outcome are either measured using the same scale across studies, or that the slopes can be transformed to be on a common standardized scale. Because correlations are scale-free, this issue is skirted in model-driven meta-analysis.

More problematic is that technically, for raw slopes or even standardized slopes to be comparable, the same set of predictors should appear in all of the regression models to be combined. This is almost never the case, given that researchers typically build on and add to models already appearing in the literature, thus making exact replications very rare. Wu and I have shown the conditions under which a summary of slopes from identical models will be equivalent to the same regression model computed from the pooled primary data (Becker and Wu 2007). It is unclear, however, how close the summary of slopes and the pooled analysis results would be if the models were not identical across studies. One might conjecture that the summary result will be closer to the raw-data analysis the less correlated the predictors are, but this problem is still under study and thus making strong statements is not yet advisable.

One last development related to the issue of combining regressions addresses the question of whether one could directly summarize an index of partial relationship based on a regression analysis. Ariel Aloe and I have proposed using the semi-partial correlation for this purpose (in press). This index can be computed from information typically presented in multiple regression studies (specifically, from a t test of the slope and the multiple R^2 for the equation), and does not require the full correlation matrix among predictors to be reported. Its one drawback is that when studies use different sets of predictors, as discussed earlier, the index estimates different partial relationships across studies, and its interpretation may depend on the particular set of predictors examined in each study. Although work on this index is in its early stages, it holds some promise for summaries of regression studies that do not report zero order correlations.

20.5 ACKNOWLEDGMENTS

This work was supported by grant REC-0634013 from the National Science Foundation to Betsy Becker and Ingram Olkin. Any opinions, findings, and conclusions or recommendations expressed in this material are those of the author(s) and do not necessarily reflect the views of the National Science Foundation.

20.6 REFERENCES

Ahn, Soyeon. 2008. "Application of Model-Driven Meta-Analysis and Latent Variable Framework in Synthesizing Studies Using Diverse Measures." Ph.D. diss., Michigan State University.

Aloe, Ariel M., and Betsy J. Becker. In press. "Advances in Combining Regression Results." In *The SAGE Handbook of Methodological Innovation*, edited by Malcolm Williams and Paul Vogt. London: Sage Publications.

Becker, Betsy J. 1992a. "Models of Science Achievement: Factors Affecting Male and Female Performance In School Science." In *Meta-Analysis for Explanation: A Casebook*, edited by Thomas D. Cook, Harris M. Cooper, David S. Cordray, Larry V. Hedges, Heidi Hartmann, Richard J. Light, Thomas A. Louis, and Frederick Mosteller. New York: Russell Sage Foundation.

———. 1992b. "Using Results from Replicated Studies to Estimate Linear Models." *Journal of Educational Statistics* 17(4): 341–62.

———. 1995. "Corrections to 'Using Results from Replicated Studies to Estimate Linear Models.'" *Journal of Educational Statistics* 20(1): 100–2.

———. 2000. "Multivariate Meta-analysis." In *Handbook of Applied Multivariate Statistics and Mathematical Modeling*, edited by Howard E. A. Tinsley and Steven D. Brown. San Diego, Calif.: Academic Press.

———. 2001. "Examining Theoretical Models Through Research Synthesis: The Benefits of Model-Driven Meta-Analysis." *Evaluation and the Health Professions* 24(2): 190–217.

Becker, Betsy J., and Kyle R. Fahrbach. 1994. "A Comparison of Approaches to the Synthesis of Correlation Matrices." Paper presented at the annual meeting of the American Educational Research Association. New Orleans, La. (1994).

Becker, Betsy J., and Ingram Olkin. 2008. "Combining Correlation Matrices from Multiple Studies." Manuscript under review for *Journal of Educational and Behavioral Statistics*.

Becker, Betsy J., and Christine M. Schram. 1994. "Examining Explanatory Models Through Research Synthesis." In *The Handbook of Research Synthesis*, edited by Harris M. Cooper and Larry V. Hedges. New York: Russell Sage Foundation.

Becker, Betsy J., and Meng-Jia Wu. 2007. "The Synthesis of Regression Slopes in Meta-Analysis." *Statistical Science* 22 (3): 414–29.

Beery, John R. 1962. "Does Professional Preparation Make a Difference?" *Journal of Teacher Education* 13(4): 386–95.

Brown, Sharon A., and Larry V. Hedges. 1994. "Predicting Metabolic Control in Diabetes: A Pilot Study Using Meta-Analysis to Estimate a Linear Model." *Nursing Research* 43(6): 362–68.

Cheung, Mike W. L., and Wai Chan. 2005. "Meta-Analytic Structural Equation Modeling: A Two-Stage Approach." *Psychological Methods* 10(1): 40–64.

Collins, David, Knowlton W. Johnson, and Betsy J. Becker. 2007. "A Meta-Analysis of Effects of Community Coalitions Implementing Science-Based Substance Abuse Prevention Interventions." *Substance Use and Misuse* 42(6): 985–1007.

Craft, Lynette L., T. Michelle Magyar, Betsy J. Becker, and Debra L. Feltz. 2003. "The Relationship Between the Competitive State Anxiety Index-2 and Athletic Performance: A Meta-analysis." *Journal of Sport and Exercise Psychology* 25(1): 44–65.

Eccles-Parsons, Jacquelynne. 1984. "Sex Differences in Mathematics Participation." In *Advances in Achievement and Motivation*, edited by Marjorie W. Steinkamp and Martin L. Maehr. Greenwich, Conn.: JAI Press.

Edwards, Tara, and Lew Hardy. 1996. "The Interactive Effects of Intensity and Direction of Cognitive and Somatic Anxiety and Self-Confidence Upon Performance." *Journal of Sport and Exercise Psychology* 18(3): 296–312.

Fisher, Ronald A. 1921. "On the 'Probable Error' of a Coefficient of Correlation Deduced from a Small Sample." *Metron* 1: 1–32.

Furlow, Carolyn F., and S. Natasha Beretvas. 2005. "Meta-Analytic Methods of Pooling Correlation Matrices for Structural Equation Modeling Under Different Patterns of Missing Data." *Psychological Methods* 10(2): 227–254.

Glass, Gene V. 1976. "Primary, Secondary, and Meta-analysis of Research." *Educational Researcher* 5(10): 3–8.

Goldhaber, Dan D., and Dominic J. Brewer. 1999. "Teacher Licensing and Student Achievement." In *Better Teachers, Better Schools*, edited by Marci Kanstoroom and Chester F. J. Finn. Washington, D.C.: Thomas B. Fordham Foundation.

Hafdahl, Adam R. 2001. "Multivariate Meta-Analysis for Exploratory Factor Analytic Research." PhD diss., The University of North Carolina at Chapel Hill.

Hedges, Larry V., and Jack L. Vevea. 1998. "Fixed- and Random-Effects Models in Meta-Analysis." *Psychological Methods* 3(4): 486–504.

Hodges, William F. 1991. *Interventions for Children of Divorce*, 3rd ed. New York: John Wiley & Sons.

Judd, Charles M., and David A. Kenny. 1981. "Process Evaluation: Estimating Mediation in Treatment Evaluations." *Evaluation Review* 5(5): 602–19.

Kalaian, Hripsime A., and Stephen W. Raudenbush. 1996. "A Multivariate Mixed Linear Model for Meta-Analysis." *Psychological Methods* 1(3): 227–35.

Kaplan, David. 2000. *Structural Equation Modeling: Foundations and Extensions*. Newbury Park, Calif.: Sage Publications.

Lipsey, Mark W. 1997. "Using Linked Meta-Analysis to Build Policy Models." In *Meta-Analysis of Drug Abuse Prevention Programs*, edited by William J. Bukoski. NIDA Research Monograph 170. Rockville, Md.: National Institute of Drug Abuse.

Lipsey, Mark W., and James H. Derzon. 1998. "Predictors of Violent or Serious Delinquency in Adolescence and Early Adulthood: A Synthesis of Longitudinal Research." In *Serious and Violent Juvenile Offenders: Risk Factors and Successful Interventions*, edited by Rolf Loeber and David P. Farrington. Thousand Oaks, Calif.: Sage Publications.

MacKinnon, David P., Chondra M. Lockwood, Jeanne M. Hoffman, Stephen G. West, and Virgil Sheets. 2002. "A Comparison of Methods to Test Mediation and Other Intervening Variable Effects." *Psychological Methods* 7(1): 83–104.

Martens, Rainer, Robin S. Vealey, and Damon Burton, eds. 1990. *Competitive Anxiety in Sport*. Champaign, Ill.: Human Kinetics.

Olkin, Ingram, and Minoru Siotani. 1976. "Asymptotic Distribution of Functions of a Correlation Matrix." In *Essays in Probability and Statistics*, edited by S. Ikeda et al. Tokyo: Shinko Tsusho Co., Ltd.

Raudenbush, Stephen W., and Anthony S. Bryk. 2002. *Hierarchical Linear Models: Applications and Data Analysis Methods*. Thousand Oaks, Calif.: Sage Publications.

Raudenbush, Stephen W., Betsy Jane Becker, and Hripsime Kalaian. 1988. "Modeling Multivariate Effect Sizes." *Psychological Bulletin* 103(1): 111–20.

Sidik, Kurek, and Jeffrey N. Jonkman. 2005. "Simple Heterogeneity Variance Estimation for Meta-Analysis." *Applied Statistics* 54(2): 367–84.

Thum, Yeow Meng, and Soyeon Ahn. 2007. "Challenges of Meta-Analysis from the Standpoint of Latent Variable Framework." Paper presented at the 7th Annual Campbell Collaboration Colloquium. London (May 14–17, 2007).

Whiteside, Mary F., and Betsy J. Becker. 2000. "The Young Child's Post-Divorce Adjustment: A Meta-Analysis with Implications For Parenting Arrangements." *Journal of Family Psychology* 14(1): 1–22.

Wu, Meng-Jia. 2006. "Methods of Meta-analyzing Regression Studies: Applications of Generalized Least Squares and Factored Likelihoods." PhD diss., Michigan State University.

PART

VIII

DATA INTERPRETATION

21

HANDLING MISSING DATA

THERESE D. PIGOTT
Loyola University, Chicago

CONTENTS

21.1 INTRODUCTION

This chapter discusses what researchers can do when studies are missing the information needed for meta-analysis. Despite careful evaluation of coding decisions, researchers will find that studies in a research synthesis invariably differ in the types and quality of the information reported. Here I examine the types of missing data that occur in a research synthesis, and discuss strategies synthesists can use when faced with missing data.

Problems caused by missing data can never be entirely alleviated. As such, the first strategy for addressing missing data should be to contact the study authors. Doing so remains a viable strategy in some cases. When it fails, the next strategy is to use statistical missing data methods to check the sensitivity of results to different assumptions about the distribution of the hypothetically complete data and the reasons for the missing data. We can think of these methods as a trade-off—if observations are missing in the data, then the cost is the need to make assumptions about the data and the mechanism that causes the missing data, assumptions that are difficult to verify in practice. Synthesists can feel more confident about results that are robust to different assumptions about the missing data. On the other hand, results that differ depending on the missing data assumptions may indicate that the sample of studies do not provide enough evidence for robust inference. Joseph Schafer and John Graham pointed out that the main goal of statistical methods for missing data is not to recover or estimate the missing values but to make valid inferences about a population of interest (2002). They thus noted that the missing data method is embedded in the particular model or testing procedure the analyst is using. My goal in this chapter is to introduce methods for estimating effect size models when either effect sizes or predictors in the model are missing from primary studies.

21.2 TYPES OF MISSING DATA

Synthesists encounter missing data in three major areas: missing studies, missing effect sizes, and missing study descriptor variables. Although the reasons for missing observations on any of these three areas vary, each type of missing data presents difficulties for the synthesists.

21.2.1 Missing Studies

A number of mechanisms lead to studies missing in a research synthesis. Researchers in both medicine and in the social sciences have documented the bias in published literature toward statistically significant results (for example, Rosenthal 1979; Hemminki 1980; Smith 1980; Begg and Berlin 1988). In this case, studies that find nonsignificant statistical results are less likely to appear in the published literature (see chapters 6 and chapter 23, this volume). Another reason studies may be missing in a synthesis is lack of accessibility; some studies are unpublished reports that are not identifiable through commonly used search engines or are not easily accessed by synthesists. For example, Matthias Egger and George Davey (1997) and Peter Jüni et al. (2002) demonstrated that studies published in languages other than English may have different results than those published in English.

Researchers undertaking a comprehensive synthesis expend significant effort identifying and obtaining unpublished studies to maintain the representative nature of the sample for the synthesis. Strategies for preventing, assessing, and adjusting for publication bias are examined in chapter 23 of this volume, and by Hannah Rothstein, Alexander Sutton, and Michael Borenstein (2005). Because publication bias is treated more thoroughly elsewhere, I focus here on missing data within studies, and not on methods for adjusting analyses for publication bias.

21.2.2 Missing Effect Sizes

A common problem in research syntheses is missing information for computing an effect size. For example, An-Wen Chan and colleagues found that 50 to 65 percent of the outcomes in a set of 102 published clinical trials were incompletely reported (2004). Missing effect sizes occur when studies are missing the descriptive statistics needed to compute an effect size and the effect size's standard error. The typical information needed includes simple descriptive statistics such as the means, standard deviations, and sample sizes for standardized mean differences, the correlation and sample size for correlations, or the frequencies in a two by two table for odds ratios.

In some studies, the effect sizes are missing because the primary researcher did not report enough information to compute an effect size. Chan and colleagues referred to these studies as including incompletely reported outcomes (2004), making a distinction between various categories of incompletely reported outcomes. Studies with partially reported outcomes usually include the value of the effect size and may report the sample size or p-value, or both. Studies with qualitatively reported outcomes may include only the p-value, with or without the sample size. In these two cases of incompletely reported outcomes, an estimated effect size could be computed based on the p-value and sample sizes. William Shadish, Leslie Robinson, and Congxiao Lu's effect size calculator computes effect sizes from more than forty types of data, including p-values, sample sizes, and statistical tests (1997). As journals begin requiring the use of effect sizes in reporting primary study results, problems with computing effect sizes may decrease.

The more difficult missing outcome category is Chan et al.'s unreported outcomes, where no information is given, leading to a missing effect size. Missing outcomes in any statistical analysis, and missing effect sizes in a research synthesis pose a difficult problem. When a measure of outcome is missing for a case in any statistical analysis, the case provides limited information for estimating statistical models and making statistical inferences.

One primary reason for missing effect sizes relates to the effect size itself. A primary research study may not provide descriptive statistics for an outcome that did not yield a statistically significant result. Chan and his colleagues found that the odds of an outcome being completely reported, with enough information to compute an effect size, was twice as high if that outcome was statistically significant (2004). As discussed later in the chapter, when outcomes are missing because of their actual values, the options for analysis require strong assumptions and more complex statistical procedures.

Incomplete reporting on outcomes can occur at the level of the primary outcome for a study, or in subgroup analyses within a study (Hahn et al. 2000). If a study does not include information on the primary outcome—the primary focus of the meta-analysis—then the study is missing an effect size. However, a synthesist may also be interested in subgroup analyses in a given study. For example, Hahn and colleagues found that in a set of studies on the use of routine malaria prophylaxis in pregnant women living in high-risk areas, only one study compared the benefits of treatment among first-time pregnant mothers versus mothers who had a prior pregnancy. They referred to this issue as *selective reporting within a study*. Though the researchers talked about this problem as missing effect sizes within studies, it can also be conceptualized as a problem of a missing descriptor variable— only one study in the Hahn et al. example reports on whether participants were pregnant for the first time. This chapter treats selective reporting of outcomes in a study as a missing descriptor variable problem instead of as a missing outcome. In some cases, the primary researcher may not have collected information on this predictor. In others, however, the researcher may have collected the information, but not reported the results. Paula Williamson and her colleagues found that one reason for not reporting subgroup analyses (or, alternatively, descriptor variables) is that the results for these analyses are not statistically significant (2005). As will be seen later, if descriptor variables are missing because they are not statistically related to the outcome within a study, the synthesist must use missing data methods that require stronger assumptions and more complex analyses.

21.2.3 Missing Descriptor Variables

A major task in the data evaluation phase of a meta-analysis involves coding aspects of a study's design. Because most studies in a given area are not strict replications of earlier work, studies differ not only in their research design, but also in their data collection strategies and reporting practices. Studies may be missing particular descriptor variables because certain information was not collected or was not reported. Missing descriptor variables are an inherent problem in meta-analysis; not every study will collect the same data. In the Seokyung Hahn et al. paper, not all studies reported (or collected) information on whether the participants had had a prior pregnancy (2000). If the synthesist is interested in modeling how the effect of prophylactic malaria treatment differs as a function of whether a woman is pregnant for the first time, those studies not reporting the number of previous pregnancies are missing a key predictor.

As I discuss later, statistical methods for missing data methods require particular assumptions about why variables are missing. Thus it is important to explore possible reasons for missing descriptor variables in a primary study.

In one sense, missing data on study descriptors arise because of differences synthesists and primary researchers place on the importance of particular information. Synthesists may take a wider perspective on an area of research and primary authors may have many constraints in both designing and reporting study results.

For example, disciplinary practices in the field may dictate how primary authors collect, measure, and report certain information. Robert Orwin and David Cordray used the term *macrolevel reporting* to refer to practices in a given research area that influence how constructs are defined and reported (1985). For example, Selcuk Sirin found several measures of socioeconomic status reported in his meta-analysis of the relationship between academic achievement and socioeconomic status (2005). These measures included parent scores on Hollingshead's occupational status scale, parental income, parent education level, free lunch eligibility in school, and composites of these measures. Differences in types of socioeconomic status (SES) measures reported could derive from traditional ways disciplines have reported SES. In this case, differences in reporting between primary studies may relate to the disciplinary practices in the primary author's field.

Primary authors also differ from each other in writing style and thoroughness of reporting. Orwin and Cordray referred to individual differences of primary authors as microlevel reporting quality (1985). In this case, whether a given descriptor is reported across a set of studies may be random given that it depends on individual writing preferences and practices.

Primary authors may also be constrained by a particular journal's publication practices, and thus do not report on information a synthesist considers important. Individual journal reporting constraints would likely result in descriptors missing randomly rather than missing because their values were not acceptable.

Another reason study descriptors may be missing from primary study reports relates to bias for reporting only statistically significant results. If, for example, gender does not have a significant relationship to a given outcome (such as SES), then a primary author may not report the gender distribution of the sample. Williamson and her colleagues provided a discussion of this problem, raising the possibility that the likelihood of a study reporting a given descriptor variable is related to the statistical significance

of the relationship between the variable and the outcome (2005).

Researchers may also avoid reporting study descriptors when those values are not generally acceptable. For example, if a researcher obtains a racially homogeneous sample, the researcher may hesitate to report fully the actual distribution of ethnicity in the sample. This type of selective reporting may also occur when reporting in a study is influenced by a desire to "clean up" the results. For example, attrition information is missing in a published study to present a more positive picture. In both of these cases, the values of the study descriptor influence whether the researcher reports those values.

21.3 REASONS FOR MISSING DATA IN META-ANALYSIS

Statisticians have been studying methods for handling missing data for many years, especially in the area of survey research. Since the publication of Roderick Little and Donald Rubin's seminal book (1987), statisticians have developed many more sophisticated frameworks and analysis strategies. These strategies all rely on specific assumptions the analyst is willing to make about the reasons for missing data, and about the joint distribution of all variables in the data. Rubin first introduced a framework for describing the response mechanism, the reasons why data may be missing in a given research study (1976). This framework describes data as missing completely at random, missing at random, and not missing at random.

21.3.1 Missing Completely at Random

We refer to data as missing completely at random (MCAR) when the missing data mechanism results in observations missing randomly across the data set. The classic example is the missing pattern created when ink is accidentally spilled on a paper copy of a spreadsheet. Although we are no longer plagued by ink spills, other mechanisms could be imagined to operate resulting in MCAR data. For example, microlevel reporting practices may cause primary authors to differ from each other about the information presented in a study, and the result of those differences could be values of a particular variable missing in a completely random way. When data are MCAR, we assume that the values of the missing variable are not related to the probability that they are missing, and that the values of other variables in the data set are also not related to the probability the variable has missing values. More important, when data are MCAR, the studies with completely observed data are a random sample of the studies originally identified for the synthesis.

21.3.2 Missing at Random

We refer to data as missing at random (MAR) when the probability of missing a value on a variable is not related to the missing value, but may be to other completely observed variables in the data set. For example, Sirin reported on a number of moderator analyses in his meta-analysis of the relationship between SES and academic achievement (2005). One moderator examined is type of components of measured SES. These include parental education, parental occupation, parental income, and eligibility for free or reduced lunch. A second is source of the information on SES, whether parent, student or a secondary source. Imagine if all studies report the source of information on SES, but not all report the component used. It is likely that the source of information on SES is highly related to the component of SES measured in the study. Students are much less likely to report income, but may be more likely to report on parental education or occupation. Secondary sources of income are usually derived from eligibility for free or reduced lunch. In this case, the component of SES is MAR because we can assume that the likelihood of observing any given component of SES depends on the completely observed value, source of SES information. Note that MAR refers to what is formally called the response mechanism. This is generally unobservable, unless a researcher can gather direct information from participants about why they did not respond to a given survey question, or from primary authors about why they did not collect a particular variable. The assumption of MAR does not depend on measuring variables without error; it specifies only the assumed relationship between the likelihood of observing a particular variable given another completely observed variable in the data set.

21.3.3 Not Missing at Random

We refer to data as not missing at random (NMAR) when the probability of observing a given value for a variable is related to the missing value itself. For example, effect sizes are not missing at random when they are not reported in a study because they are not statistically significant (see Chan et al. 2004; Williamson et al. 2005). In this example, data are not missing at random due to a censoring mechanism which results in certain values having a higher probability of being missing than other values. Meta-analysts have developed methods to handle censored effect size data in the special case of publication bias (for example, Hedges and Vevea 1996; Vevea and Woods 2005; chapter 23, this volume).

Missing study descriptors could also be NMAR. Primary authors whose research participants are racially homogeneous may not report fully on the distribution of ethnicities among participants as a way to provide a more positive picture of their study. As with the effect size example, the actual ethnic distribution of the samples are related their probability of being missing.

Synthesists rarely have direct evidence about the reasons for missing data and need to assume data are MCAR, MAR, or NMAR if they are to choose an appropriate analysis that will not lead to inferential errors, or else to conduct sensitivity analyses to different assumptions about the reasons for the missing data. In the rest of the chapter, I discuss the options available to meta-analysts when faced with various forms of missing data.

21.4 COMMONLY USED METHODS FOR MISSING DATA IN META-ANALYSIS

Before Little and Rubin's (1987) work, most researchers used one of three strategies to handle missing data: using only cases with all variables completely observed (listwise deletion), using available cases that have particular pairs of variables observed (pairwise deletion), or replacing missing values in a given variable with a single value such as the mean for the complete cases (single value imputation). The performance of these methods depends on the assumptions about why the data are missing. In general, the methods produce unbiased estimates only when data are missing completely at random. The main problem is that the standard errors do not accurately reflect the fact that variables have missing observations.

21.4.1 Complete-Case Analysis

In complete-case analysis, the researcher uses only those cases with all variables fully observed. This procedure, also known as listwise deletion, is usually the default procedure for many statistical computer packages. When some cases are missing values of a particular variable, only cases observing all the variables in the analysis are used. When the missing data are missing completely at random, the complete cases can be considered a random sample from the originally identified set of cases. Thus, if a synthesist can make the assumption that values are in fact missing completely at random, using only complete cases will produce unbiased results.

In meta-analysis as in other studies, using only complete cases can seriously limit the number of observations available for the analysis. Losing cases decreases the power of the analysis and ignores the information contained in the incomplete cases (Kim and Curry 1977; Little and Rubin 1987).

When data are NMAR or MAR, complete case analysis yields biased estimates because the complete cases cannot be considered representative of the original sample. With NMAR data, the complete cases observe only part of the distribution of a particular variable. With MAR data, the complete cases are also not a random sample from a particular variable. With MAR data, though, the incomplete cases still provide information because the variables that are completely observed across the data set are related to the probability of missing a particular variable.

21.4.1.1 Complete Case Analysis Example Table 21.1 presents a subset of studies from a meta-analysis examining the effects of oral anticoagulant therapy for patients with coronary artery disease (Anand and Yusuf 1999). To illustrate the use of missing data methods with MCAR and MAR data, I used two methods for deleting the age of the study in the total data set. For MCAR data, I randomly deleted twelve values (35 percent) of study age from the data. For MAR data, I randomly deleted studies depending on the value of the intensity of the dose of anticoagulant. For studies using a high dose of anticoagulant, studies had a ten in twenty-two (45 percent) chance of missing a value. Studies with a moderate dose of anticoagulants had a one in nine (11 percent) chance of missing a value, and studies with a low dose had a one in three (33 percent) chance of missing a value of study age.

Table 21.2 compares the means and standard deviations between the log-odds ratio, the intensity of the dose, and the age of the study when using complete case analysis on MCAR or MAR data. The MCAR estimates for the two predictors, intensity of dose and age of study, are closer to the true mean than the MAR estimates. This finding is reversed for the estimates of the mean log-odds ratio, where the MAR estimates are closer to the true values. The standard deviation of the MCAR estimate of the log-odds ratio mean is larger than the MAR estimate, resulting in a more conservative estimate. Table 21.3 com-

pares the correlations between these three variables using complete case analysis. The MAR estimates for the correlation of the log-odds ratio with the two predictor variables are closer to the true values than the MCAR estimates. For the correlation of the intensity of dose with study age, the MCAR and MAR estimates are similar to each other and smaller than the true value.

21.4.2 Available Case Analysis

An available case analysis, also called pairwise analysis, estimates parameters using as much data as possible. In other words, if three variables in a data set are missing 0 percent, 10 percent, and 40 percent of their values, respectively, then the correlation between the first and third variable would be estimated using the 60 percent of cases that observe both variables, and that between the first and second variable would use the 90 percent. An entirely different set of cases would provide the estimate of the correlation of the second and third variables given the potential of having between 40 percent and 50 percent of these cases missing both variables.

This simple example illustrates the drawback of using available case analysis: each correlation in the variance-covariance matrix estimated using available cases could be based on different subsets of the original data set. If data are MCAR, these subsets are individually representative of the original data, and available case analysis provides unbiased estimates. If data are MAR, however, these subsets are not individually representative of the original data and will produce biased estimates.

Much of the early research on methods for missing data focuses on the performance of available case analysis versus complete case analysis (for example, Glasser 1964; Haitovsky 1968; Kim and Curry 1977). Kyle Fahrbach examined the research on available case analysis and concludes that such methods provide more efficient estimators than complete case analysis when correlations between two independent variables are moderate, that is, around 0.6 (2001). This view, however, is not shared by all who have examined this literature (for example, Allison 2002).

One statistical problem that could arise from the use of available cases under any form of missing data is a nonpositive definite variance-covariance matrix, that is, a variance-covariance matrix that cannot be inverted to obtain the estimates of slopes for a regression model. One reason for this problem is that different subsets of studies are used to compute the elements of the variance-covariance matrix. Further, Paul Allison pointed out that a more

Table 21.1 Oral Anticoagulant Therapy Data

ID	Odds Ratio	Log-Odds Ratio	Variance of Log-Odds Ratio	Intensity of Dose	Year of Publication	Study Age
1.00	20.49	3.02	2.21	High	1960	39.00[2]
2.00	0.16	−1.84	0.80	High	1960	39.00[2]
3.00	1.27	0.24	0.16	High	1961	38.00[1]
4.00	0.86	0.15	0.07	High	1961	38.00[2]
5.00	0.62	0.47	0.07	High	1964	35.00[2]
6.00	0.68	0.38	0.28	High	1964	35.00[1]
7.00	0.68	−0.38	0.18	High	1966	33.00[2]
8.00	0.62	−0.47	0.22	High	1967	32.00[1]
9.00	0.78	−0.25	0.07	High	1967	32.00[2]
10.00	1.25	0.22	0.10	High	1969	30.00[1,2]
11.00	0.43	0.84	0.08	High	1969	30.00[1,2]
12.00	0.71	0.35	0.04	High	1980	19.00[2]
13.00	0.72	0.33	0.02	High	1990	9.00
14.00	0.89	0.12	0.01	High	1994	5.00
15.00	0.16	−1.81	0.82	High	1969	30.00[1]
16.00	0.65	0.43	0.02	High	1974	25.00
17.00	1.20	0.18	0.06	High	1980	19.00
18.00	1.48	0.39	0.07	High	1980	19.00
19.00	0.41	0.90	0.41	High	1993	6.00
20.00	0.52	0.65	0.29	High	1996	3.00[2]
21.00	4.15	1.42	2.74	High	1990	9.00
22.00	0.87	0.14	4.06	High	1990	9.00[1]
23.00	1.04	0.04	1.36	Moderate	1982	17.00[1]
24.00	0.71	0.35	0.35	Moderate	1981	18.00[1,2]
25.00	0.92	0.08	0.03	Moderate	1982	17.00
26.00	0.94	0.06	0.03	Moderate	1969	30.00
27.00	0.65	0.43	0.07	Moderate	1964	35.00
28.00	0.31	−1.16	0.93	Moderate	1986	13.00
29.00	0.47	0.75	0.98	Moderate	1982	17.00[1]
30.00	0.45	0.81	0.60	Moderate	1998	1.00
31.00	1.04	0.04	0.82	Moderate	1994	5.00[1]
32.00	0.70	0.35	0.70	Low	1998	1.00
33.00	0.71	0.34	0.06	Low	1997	2.00[2]
34.00	1.17	0.15	0.02	Low	1997	2.00[1]

SOURCE: Author's compilation.
[1] Value deleted for MCAR data example
[2] Value deleted for MAR data example

difficult problem in the application of available case analysis concerns the computation of standard errors of available case estimates (2002). At issue is the correct sample size when computing standard errors, given that each parameter could be estimated with a different subset of data. Some of the standard errors could be based on the whole data set, and others on the subset of studies that observe a particular variable or pair of variables. Numerous statistical computing packages calculate available case analysis, but how standard errors are computed differs widely.

Although available case analysis is easy to understand and implement, consensus in the literature is scant about the conditions where available case analysis outperforms complete case analysis when data are MCAR. Previous research on available case analysis does point to the size of the correlations between variables in the data set as

Table 21.2 Descriptive Statistics, Complete Case Analysis

Variable	N	Mean	SD
Log-odds ratio			
Full data	34	−0.239	0.826
MCAR data	22	−0.185	0.938
MAR data	22	−0.262	0.634
Intensity of dose			
Full data	34	2.559	0.660
MCAR data	22	2.591	0.666
MAR data	22	2.454	0.671
Study age			
Full data	34	20.353	13.112
MCAR data	22	19.500	13.780
MAR data	22	17.000	12.107

SOURCE: Author's compilation.

Table 21.4 Descriptive Statistics, Available Case Analysis

Variable	N	Mean	SD
Log-odds ratio			
Full data	34	−0.239	0.826
MCAR data	34	−0.239	0.826
MAR data	34	−0.239	0.826
Intensity of dose			
Full data	34	2.559	0.660
MCAR data	34	2.559	0.660
MAR data	34	2.559	0.660
Study age			
Full data	34	20.353	13.112
MCAR data	22	19.500	13.780
MAR data	22	17.000	12.107

SOURCE: Author's compilation.

Table 21.3 Correlations, Complete Case Analysis

Correlation	Intensity of Dose & Log-Odds Ratio	Study Age & Log-Odds Ratio	Study Age & Intensity of Dose
Full data	0.058	0.049	0.498
MCAR data	0.172	0.158	0.381
MAR data	0.052	−0.121	0.369

SOURCE: Author's compilation.

critical to the efficiency of this strategy, but this same research does not reach consensus about the optimal size of these correlations.

21.4.2.1. Available Case Analysis Example Table 21.4 provides the means and standard deviations using available case analysis; table 21.5 provides the correlations. The entries in the two are a mix between the complete case values for the study age, and the entire data set of values for the log-odds ratio and the intensity of dose. When the data are MAR, both the mean and the standard deviation of study age are smaller than the true values even though fewer cases are used for these estimates. The estimate of the mean of study age from the MCAR data, however, is close to the true value. The standard deviation of study age is slightly larger, reflecting the smaller sample size. The correlations in table 21.5 are a mixture of

the true values and the complete case values from table 21.3. The correlations involving study age are the same as in table 21.3, and the correlation between the log-odds ratio and intensity of dose is the true value for all three analyses. If we added a second variable with missing observations, we could have correlations based on several subsets of the data set.

21.4.3 Single-Value Imputation with the Complete Case Mean

When values are missing in a meta-analysis (or any statistical analysis), many researchers replace the missing value with a reasonable value, such as the mean for the cases that observed the variable. Little and Rubin referred to this strategy as single-value imputation. Researchers commonly use different strategies to fill in missing values (1987). One method fills in the complete case mean, and the other uses regression with the complete cases to estimate predicted values for missing observations given the observed values in a particular case.

21.4.3.1 Imputing the Complete Case Mean Replacing the missing values in a variable with the complete case mean of the variable is also referred to as unconditional mean imputation. When we substitute a single value for all the missing values, the estimate of the variance of that variable is decreased. The estimated variance thus does not reflect the true uncertainty in the variable. Instead, the smaller variance wrongly indicates more certainty

Table 21.5 Correlations, Available Case Analysis

Correlation	Intensity of Dose & Log-Odds Ratio	Study Age & Log-Odds Ratio	Study Age & Intensity of Dose
Full data	0.058 (N = 34)	0.049 (N = 34)	0.498 (N = 34)
MCAR data	0.058 (N = 34)	0.158 (N = 22)	0.381 (N = 22)
MAR data	0.058 (N = 34)	−0.121 (N = 22)	0.369 (N = 22)

SOURCE: Author's compilation.

Table 21.6 Descriptive Statistics, Mean Imputation

Variable	N	Mean	SD
Log-odds ratio			
Full data	34	−0.239	0.826
MCAR data	34	−0.239	0.826
MAR data	34	−0.239	0.826
Intensity of dose			
Full data	34	2.559	0.660
MCAR data	34	2.559	0.660
MAR data	34	2.559	0.660
Study age			
Full data	34	20.353	13.112
MCAR data	34	19.500	10.992
MAR data	34	17.00	9.658

SOURCE: Author's compilation.

Table 21.7 Correlations, Mean Imputation

Correlation	Intensity of Dose and Log-Odds Ratio	Study Age and Log-Odds Ratio	Intensity of Dose and Study Age
Full data	0.058	0.049	0.498
MCAR data	0.058	0.144	0.307
MAR data	0.058	−0.074	0.299

SOURCE: Author's compilation.

about the value. These biases are compounded when the biased variances are used to estimate models of effect size. There are no circumstances under which imputation of the complete case mean leads to unbiased results of the variances of variables with missing data.

21.4.3.1.1 Complete Case Mean Imputation Example Tables 21.6 and 21.7 present the results using single-value imputation with the complete case mean. The means for study age under MCAR and MAR data are the same as in the complete case analysis, but the standard deviations are much smaller than the true values. The estimated correlation between study age and intensity of dose is also more biased than estimates from listwise or pairwise deletion. Because we are missing more of the values for high doses, the correlation between intensity of dose and study age is decreased when replacing missing values of study age with the mean study age.

21.4.3.2 Single-Value Imputation with Conditional Means A single-value imputation method that provides less biased results with missing data was first suggested by S. F. Buck (1960). Instead of replacing each missing value with the complete case mean, each missing value is replaced with the predicted value from a regression model using the variables observed in that particular case as predictors and the missing variable as the outcome. This method is also referred to as conditional mean imputation or as regression imputation. For each pattern of missing data, the cases with complete data on the variables in the pattern are used to estimate regressions using the observed variables to predict the missing values. The result is that each missing value is replaced by a predicted value from a regression using the values of the observed variables in that case. When data are MCAR, each of the subsets used to estimate prediction equations are representative of the original sample. This method results in more variation than in unconditional mean imputation because the missing values are replaced with those that depend on the regression equation. However, the standard errors using Buck's method are still too small because the missing values are replaced with predicted values that lie directly on the regression line used to impute the values. In other words, Buck's method yields values predicted exactly by the regression equation without error.

Little and Rubin presented the form of the bias for Buck's method and suggest corrections to the estimated variances to account for the bias (1987). If we have two variables, Y_1 and Y_2, and Y_2 has missing observations, then the form of the bias using Buck's method to fill in values for Y_2 is given by

$$(n - n^{(2)})(n - 1)^{-1}\sigma_{22.1},$$

where n is the sample size, $n^{(2)}$ is the number of cases that observe Y_2, and $\sigma_{22.1}$ is the residual variance from the regression of Y_2 on Y_1. Little and Rubin provide the more general form of the bias with more than two variables.

To date, there has not been extensive research on the performance of Buck's method to other more complex methods for missing data in meta-analysis. One advantage to the method with the corrections Little and Rubin suggested is that the standard errors of estimates reflect the uncertainty in the data, which more simple methods do not. Buck's method with corrections to the standard errors could thus lead to more conservative and less biased estimates than complete case and available case methods.

SPSS Missing Value Analysis offers regression imputation as an option for filling-in missing observations (Hill 1997). The program does not implement Little and Rubin's 1987 corrections to obtain unbiased variance estimates. Instead, the researcher has the option to add a random variate to the imputed value, thus increasing the variance among imputed observations. An analysis adding a random variate to the imputed value may be an option to consider in a sensitivity analysis of the effects of missing data on a review's results.

21.4.3.2.1 Regression Imputation Example Tables 21.8 and 21.9 present the results using single-value imputation with the conditional mean. In this analysis, estimates are directly from a regression imputation with no random variate added. Standard errors are thus not adjusted and in turn underestimate the variation. In both the MCAR and MAR case, the correlation between the intensity of the dose and study age is closer than other analyses to the true value. The estimate of the mean study age from MAR data is only slightly larger than in earlier analyses.

21.4.4 Missing Effect Sizes

Missing effect sizes pose a more difficult problem in meta-analysis. When any outcome is missing for a case in a statistical analysis, that case can add little information to the estimation of the model. For example, in a regression analysis, a case missing the outcome variable does not provide information about how the predictor variables

Table 21.8 Descriptive Statistics, Regression Imputation

Variable	N	Mean	SD
Log-odds ratio			
Full data	34	−0.239	0.826
MCAR data	34	−0.239	0.826
MAR data	34	−0.239	0.826
Intensity of dose			
Full data	34	2.559	0.660
MCAR data	34	2.559	0.660
MAR data	34	2.559	0.660
Study age			
Full data	34	20.353	13.112
MCAR data	34	19.181	11.359
MAR data	34	17.648	10.129

SOURCE: Author's compilation.

Table 21.9 Correlations, Regression Imputation

Correlation	Intensity of Dose and Log-odds Ratio	Study Age and Log-odds Ratio	Intensity of Dose and Study Age
Full data	0.058	0.049	0.498
MCAR data	0.058	0.128	0.445
MAR data	0.058	−0.193	0.430

SOURCE: Author's compilation.

are related to the outcome. Even when the assumption of MCAR holds for effect sizes, the studies that do not report on the outcome contribute little to estimation procedures. With MCAR missing effect sizes, complete case analysis will provide the most unbiased results. When outcomes are missing, pairwise deletion does not provide any more information than listwise deletion because the relationship of interest—that between the outcome and predictors—can only be estimated in the cases that observe an effect size.

In the case of MAR and NMAR missing effect sizes, the only other option available is single-value imputation. Many meta-analysts have suggested replacing missing effect sizes with zero. The rationale is that because missing effect sizes are likely to be not statistically significant, using zero as a substitute provides only a conservative estimate of overall mean effect size. As noted in the earlier discussion of single-value imputation, replacing missing effect sizes with zero results in estimates of variance that are too small and tests of homogeneity that also do not reflect the true variation among effect sizes.

Using regression imputation would seem to provide somewhat better estimates. However, because the missing effect sizes are replaced with predictions from a regression of effect sizes on observed variables, the imputed effect sizes will be consistent with the relationships between the effect sizes and predictors that exist in the complete cases. In other words, the imputed effect sizes will fit exactly the linear model of effect size that is estimated by the complete cases. Any analyses that subsequently use the regression imputed effect sizes will not adequately reflect that missing data causes more uncertainty than the standard errors and fit of the effect size model indicate. Any single imputation method for missing effect sizes will thus make the models appear to have a better fit to the data than actually exists. When effect sizes are MAR or NMAR, none of the simple methods will provide adequate estimates of effect size models.

If a synthesist cannot derive estimates of effect size from other information reported in a study, even the methods described later in this chapter may not provide adequate estimates for MAR or NMAR effect sizes. Unfortunately, missing effect sizes are likely NMAR, missing because their values do not reach statistical significance. In this case, methods for censored data such as those James Heckman (1976) and Takeshi Amemiya (1984) described could be used to estimate effect size models. The details of these analyses are beyond the scope of this text, and require specialized estimation techniques.

The reviewer faced with missing effect sizes has few options. Dropping the cases with missing effect sizes will likely bias the estimated mean effect size upward, given the likelihood that the missing effect sizes are nonsignificant. Replacing missing effect sizes with a value of zero may seem a sensible solution, but this strategy will tend to underestimate the variance among effect sizes. Given that missing effect sizes tend to be small, reviewers may wish to conduct sensitivity analyses similar to those described for publication bias (see Hedges and Vevea 1996; Vevea and Woods 2005; chapter 23, this volume).

21.4.5 Summary of Simple Methods for Missing Data

When missing predictors are MCAR, complete case analysis, available case analysis, and conditional mean impu-

tation can produce unbiased results. The cost of using these methods lies in the estimation of standard errors. For complete case analysis, the standard errors will be larger than those from the hypothetically complete data. In available case and conditional mean imputation, the standard errors will be too small, though those from conditional mean imputation can be adjusted. When data are MAR or NMAR, none of the simpler methods produce unbiased results.

21.5 MODEL-BASED METHODS FOR MISSING DATA

When data are MCAR, we have a number of simple options for analysis. When data are MAR, however, these simple methods do not produce unbiased estimates. The simple methods share a characteristic—with the exception of regression imputation, they do not make any assumptions about the distribution of the data, and thus do not use the information available in the cases with missing values. For example, in complete case analysis, all of the cases with missing values are deleted. Available case analysis uses pairs of variables to compute correlations, but does not use any information about the multivariate relationships in the data. Little and Rubin described two methods based on a model for the data (1987). Each of these strategies assumes a form for the conditional distribution of the observed data given the missing data. These methods are maximum likelihood using the EM (expectation—maximization) algorithm and multiple imputation. These two methods are standard missing data methods in statistical analysis. The next section provides a brief overview of these two methods and describes how the methods can be applied to missing data in meta-analysis.

21.5.1 Maximum Likelihood Methods Using the EM Algorithm

A full explanation of maximum likelihood methods with missing data is not possible in this chapter (for more information, see Little and Rubin 1987; Schafer 1997). Simply put, maximum likelihood methods for missing data follow the general rules for computing maximum likelihood estimators: given the data likelihood, estimates for target parameters are found that maximize the value of that likelihood. Using maximum likelihood methods for missing data yields estimates for the means and variance-covariance matrix for all the variables in the data set. If we assume that the probability of missing observations does not depend on the missing values as in MCAR or MAR data (also called an ignorable response mechanism), then the likelihood of the hypothetically complete data can be factored into two likelihoods—one for the observed data, and one for the missing data conditional on the observed data. Maximizing this latter likelihood is difficult. Unlike the mean of a normal distribution, the maximum likelihood estimates with missing data do not have a closed form solution. That is, we cannot specify an equation for the estimates of important parameters that maximize the likelihood. Instead, the solution must be found either using numerical techniques, such as the Newton-Raphson method to find the maximum of the likelihood or the EM algorithm (Dempster, Laird, and Rubin 1977).

The EM algorithm starts with a data likelihood for the hypothetically complete data. For many situations with missing data, we assume these are distributed as a multivariate normal distribution. The normal distribution has many properties that simplify the estimation procedures. In a multivariate normal distribution, the maximum likelihood estimates are computed using the sums of squares and the sum of squared cross products of the variables— the sufficient statistics for the maximum likelihood estimators. The algorithm starts with guesses for the means and covariance matrix of the multivariate normal distribution. The E-step provides estimates of the values of the sufficient statistics given the current guesses for the means and covariance matrix. These estimates are derived from the fact that the marginal distributions of a multivariate normal are also normally distributed. For each pattern of missing data, we can find a conditional mean and variance matrix given the observed variables in that case. The conditional means and variances are computed from regressions of variables with missing values on variables with completely observed variables. These conditional means and variances are then used to compute the sums and sums of cross products of the variables in the data.

Once these sums are computed, they are used in the M-step, as if they were they were the real values from the hypothetically complete data, and the algorithm computes new values of the means and covariance matrix. The algorithm goes back and forth between these two steps until the estimates do not change much from one iteration to the next. Once a researcher has maximum likelihood estimates, he or she can use the means and covariance matrix to fit models to the data such as a linear regression.

When using the EM algorithm, the researcher does not obtain individual estimates of missing observations but instead maximum likelihood estimates of important data parameters—the means and variance-covariance matrix. This can then be used to estimate the coefficients of a regression model.

Several commonly used computing packages implement EM. These include the SPSS Missing Value Analysis module (Hill 1997). One remaining drawback of maximum likelihood methods for missing data is the computation of standard errors. SPSS Missing Value Analysis does not provide standard errors as an option. Instead, the analyst needs to use numerical methods for finding the second derivative of a function, and compute these standard errors by hand.

21.5.1.1 Maximum Likelihood Methods As described, obtaining maximum likelihood estimates with missing data requires two important assumptions: that the missing data are either MCAR or MAR (also referred to as ignorable response mechanisms), and that the hypothetically complete data have a particular distribution such as multivariate normal. Though we may not have direct evidence about the missing data mechanism, we can use the methods discussed here to assess the sensitivity of results to differing assumptions about the missing data mechanism. Roderick Little provided a test of MCAR data that is implemented in SPSS Missing Values Analysis (Little 1988; Hill 1997). This test may provide a synthesist with evidence that data are MCAR, but does not provide evidence about any other form of missing data.

For any analysis with missing data, the decision to assume the data are MAR is a logical, not an empirical one, and is based on the synthesist's understanding and experience in the field of study (Schafer 1997). We can never have direct information about why data are missing in a meta-analysis. We can, however, increase our odds that the MAR assumption holds by gathering a rich data set with several completely observed study descriptor variables. The more information we have in a data set, the better the model of the conditional distribution of the missing observations, given the completely observed variables. Thorough and complete coding of a study may provide descriptor variables that are related to the missing values, and thus can improve the estimation of the means and variance-covariance matrix when using the EM algorithm. In the earlier Hahn et al. example, if the synthesists coded the age of the pregnant woman, the information, which may be or may not correlated with whether the woman had previous pregnancies, can improve the maximum likelihood estimates by providing more information about how variables relate in the data (2000).

The second assumption requires the analyst to assume a distribution for the hypothetically complete data to specify the form of the likelihood. For meta-analysis applications, one choice is the multivariate normal distribution. In this case, the joint distribution of the effect sizes and the study descriptors are considered multivariate normal. The advantage is that the conditional distribution of any study descriptor is univariate normal conditional on the other variables in the data.

The assumption of multivariate normal can seem problematic in meta-analysis given that many study descriptor variables are categorical, such as primary author's discipline of study or publication status of a study. Schafer pointed out, however, that even with deviations from multivariate normality, the normal model is useful (1997). If particular variables in the data deviate from normality, they can usually be transformed to normalize the distribution. In addition, if some of the variables are categorical but completely observed, the multivariate normal model can be used if the analyst can make the assumption that within each of the subsets of data defined by the categorical variable, the data are multivariate normal, and that the parameters we want to estimate are relevant to the conditional distribution within each of the defined subsets. For example, say that in the Hahn et al. (2000) data we are interested in examining how the effectiveness of prophylactic malaria treatment varies as a function of the country where the study took place and whether the woman was pregnant for the first time. If the synthesis has complete information on the country where the study took place, we need to assume that the variables collected are distributed as a multivariate normal distribution. In other words, the multivariate normal model will apply to missing data situations in meta-analysis where the categorical variables in the model are completely observed. We can assume that the missing variables have a normal distribution conditional on these categorical ones.

One difficulty in meta-analysis that does cause a problem with the multivariate normal assumption is the distribution of effect sizes. The variance of estimates of effect size depends on the sample size of the study, and thus varies across studies. In Therese Pigott's application of maximum likelihood to missing data in meta-analysis, the likelihood weighted all variables in the data by the variance of the study (2001). This strategy leads to the assumption that studies with small sample sizes not only have effect sizes with larger standard errors, but also have measured study descriptor variables with a precision inversely proportional to the study's sample size. Although

Table 21.10 Descriptive Statistics, Maximum Likelihood Methods

Variable	Mean	SD
Log-odds ratio		
Full data	−0.239	0.826
MAR data	−0.239	0.813
Intensity of dose		
Full data	2.559	0.660
MAR data	2.559	0.650
Study age		
Full data	20.353	13.112
MAR data	17.649	11.885

SOURCE: Author's compilation.

the distribution of effect sizes depends on the sample sizes in the study, synthesists do not assume that because a study uses a small sample size, that a small study will estimate the age of participants with less precision than one with a larger sample. Kyle Fahrbach suggested instead a form of the likelihood that includes estimating the variance component for the effect size, treating the study descriptor variables as fixed, and thus not needing to weight all variables by the variance of the effect size (2001). Unfortunately, Fahrbach's application of maximum likelihood is not implemented in any current software package. If we ignore the weighting of effect sizes, the EM results can be obtained through a number of packaged computer programs, such as SPSS Missing Value Analysis (SPSS 2007) or the NORM, freeware available from Joseph Schafer (1999). More research is needed on the size of the bias that may result when using the EM algorithm for missing data in meta-analysis without applying the weights for effect sizes.

21.5.1.2 Example of Maximum Likelihood Methods Tables 21.10 and 21.11 provide a comparison of the means, standard deviations, and correlations from all of the data with the estimates from maximum likelihood using the EM algorithm. No weighting was used to generate estimates. In this simple example, the estimates are similar to those from regression imputation but closer to the true values than regression imputation. Because EM does not provide standard errors for the estimates, we cannot test these estimates for statistical significance.

21.5.1.3 Other Applications of Maximum Likelihood Methods Other researchers than Fahrbach have applied maximum likelihood methods for missing data to

Table 21.11 Correlations, Maximum Likelihood Methods

Correlation	Intensity of Dose and Log-odds Ratio	Study Age and Log-odds Ratio	Intensity of Dose and Study Age
Full data	0.058	0.049	0.498
MAR data	0.058	−0.162	0.361

SOURCE: Author's compilation.

particular issues in meta-analysis (2001). Meng-Jia Wu, for example, uses maximum likelihood methods with a monotone pattern of missing data to obtain estimates for a regression model of effect size (2006). In Wu's example, a number of studies present the results of a regression analysis examining the relationship between teacher quality and many other variables. The missing data issue arises because the studies differ in the number of variables in their models. When the missing data pattern is monotone—that is, when there is a block of studies that observe variables 1–2, another block that observe variables 1–4, another block that observe variables 1–6, and so on—the maximum likelihood estimates are more easily computed than in general patterns. Wu demonstrates how to obtain a set of synthesized regression coefficients for a set of studies when not all studies observe the same study descriptors using maximum likelihood with the EM algorithm.

Betsy Becker also uses the EM algorithm to obtain a synthesized correlation matrix across a set of studies (1994). This correlation matrix can then be used to estimate explanatory models such as structural equation models. Maximum likelihood methods with the EM algorithm can be applied in many situations where missing data occurs, including in meta-analysis. The difficulty with applying these methods to research synthesis centers on posing the appropriate distribution for the multivariate data set. More research in this area, as well as work on accessible computer programs to compute results, should provide more options for reviewers struggling with missing data issues.

21.5.2 Multiple Imputation

A second method for missing data, also based on the distribution of the data, is multiple imputation. As the name implies, the method imputes values for the missing data. There are two major differences, however, between single-value imputation methods and multiple imputation. With multiple imputation, several possible values are estimated for each missing observation, resulting in several complete data sets. The imputed values are derived by simulating random draws from the probability distribution of the variables with missing observations using Markov chain Monte Carlo techniques. These distributions are based on an assumption of multivariate normality. Once we obtain a number of random draws, the complete data sets created from these draws can then be analyzed using the methods the researcher intended to use before discovering missing data. Thus, unlike maximum likelihood methods using the EM algorithm, each missing observation is filled in with multiple values. The analyst then has a number of complete data sets analyzed separately using standard methods. The result is a series of estimated results for each completed data set. These multiple values allow standard errors to be estimated using the variation among the series of results.

Like maximum likelihood with the EM algorithm, multiple imputation consists of two steps. The first involves generating possible values for the missing observations, and the second involves inference with multiply imputed estimates. The more difficult is generating possible values for the missing observations. The goal is to obtain a random sample from the distribution of missing values given the observed values. When the data are MAR or MCAR, the distribution of missing values given the observed values does not depend on the values of the missing observations, as discussed earlier. The idea is to draw a random sample from the distribution of missing observations given the observed variables, a complex distribution even when we assume multivariate normality.

One method is data augmentation, first discussed by Martin Tanner and Wing Wong (1987). Space considerations do not allow a full explanation (for example, Rubin 1987, 1996; Schafer 1997). In sum, though, the logic is similar to that of the EM algorithm. If we know the means and covariance matrix of the hypothetically complete data, we can use properties of the multivariate normal distribution and Bayes theorem to define the distribution of missing values, given the observed variables. Tanner and Wong provide an iterative method using an augmented distribution of the missing variables. Usually, the imputed values are randomly drawn from the probability distribution at a particular iteration number (for example, at the 100th, 200th, and 300th iterations beyond convergence).

Schafer provided a number of practical guidelines for generating multiple imputations (1997).

Researchers typically draw five different values from the distribution of missing values given the observed variables resulting in five completed data sets (for a rationale for the number of imputations needed, see Schafer 1997). After obtaining the series of completed data sets, the synthesist analyzes the data as originally planned and obtains a set of estimates for target parameters within each completed data set. For example, each set provides a value for the mean of a variable. These estimates are then combined to obtain a multiply imputed estimate and its standard error.

As Paul Allison pointed out, one problem with multiple imputation is that meta-analysts using the same data set will arrive at different values of the multiply imputed estimates because each synthesist will obtain different random draws in creating their completed data sets (2002). This flexibility, however, could also be an asset. Sophisticated users could pose a number of distributions for the missing data apart from multivariate normal to test sensitivity of results to assumptions about the response mechanism.

21.5.2.1 Multiple Imputation and Missing Data In some respects, multiple imputation methods require less specialized knowledge than maximum likelihood. One advantage of multiple imputation is the ability to use the originally planned analysis on the data. Once the synthesist obtains the completed data sets, the analysis can proceed with standard techniques. Multiple imputation also provides standard errors for estimation that are easily computed and that accurately reflect the uncertainty in the data when observations are missing.

As in maximum likelihood methods, the likelihood for meta-analysis data is complicated by effect sizes not being identically distributed, but instead having variances that depend on the sample size in each study. Fahrbach's suggested likelihood distribution should be examined for use in multiple imputation methods as well (2001).

Multiple imputation is now more widely available in statistical computing packages. SAS, to take one example, uses multiple imputation procedures with multivariate normal data (Yuan 2000). The examples in this chapter were computed with Schafer's NORM program, freeware for conducting multiple imputation with multivariate normal data (1999).

21.5.2.2 Example of Multiple Imputation Table 21.12 provides the original data set along with the values for each completed data set after random draws from the distribution of missing values. The imputations for miss-

ing study age range from 0 to 42. Table 21.13 provides the estimates of the means, standard error of the mean, and correlations for each imputed data set, and then for the multiply imputed estimates. The estimate for the mean study age is the same as in maximum likelihood using the EM algorithm. As Schafer discussed, the multiply imputed estimates will approach the maximum likelihood estimates as the number of imputation increases (1997). The standard error of the mean of study age is 2.67 and is larger than the true value, 2.25, correctly reflecting the added uncertainty in the estimate due to missing values.

21.6 RECOMMENDATIONS

Missing data will occur in meta-analysis whenever a range of studies and researchers focus on a given area. Unlike primary researchers, research synthesists cannot avoid missing data by working hard to collect all data from all subjects—in this case, the subjects are studies inherently different in their methods and reporting quality.

What recommendations can we make for practice? As illustrated in this chapter, missing data methods have as a cost the necessity of making assumptions about the reasons for the missing data and the joint distribution of the complete data set. When a researcher has little missing data relative to the number of studies retrieved, the bias introduced by assuming MCAR and using complete cases will also be relatively small. With MCAR data, complete case analysis has the potential of providing unbiased results, and does not overestimate the precision of the estimates.

In meta-analysis, however, missing data across several variables could lead to fewer studies that observe all variables relative to the number of studies gathered for synthesis. A large, rich meta-analysis data set with many study descriptors has more potential than a smaller one to support both the assumption of MAR data and of multivariate normality. With MAR data, the options are using maximum likelihood methods with the EM algorithm or multiple imputation. The major benefit of these over more commonly used methods is that they are based on a model for the data and use as much information as possible from the data set.

In applying these methods to meta-analysis, each technique has its own set of difficulties. Maximum likelihood methods, for example, are now implemented in popular statistical software, but do not provide standard errors of the estimates and does not allow the weighting needed for effect size data.

Table 21.12 Oral Anticoagulant Therapy Data

ID	Log-odds ratio	Intensity of dose	Study age	Imputations				
1.00	3.02	High	39.00[1]	21	17	2	21	6
2.00	−1.84	High	39.00[1]	10	20	25	44	24
3.00	0.24	High	38.00					
4.00	0.15	High	38.00[1]	3	48	20	37	19
5.00	0.47	High	35.00[1]	12	6	30	19	17
6.00	0.38	High	35.00					
7.00	−0.38	High	33.00[1]	25	56	16	12	1
8.00	−0.47	High	32.00					
9.00	−0.25	High	32.00[1]	25	34	34	26	20
10.00	0.22	High	30.00[1]	19	28	16	34	17
11.00	0.84	High	30.00[1]	16	17	9	4	3
12.00	0.35	High	19.00[1]	9	0	23	0	11
13.00	0.33	High	9.00					
14.00	0.12	High	5.00					
15.00	−1.81	High	30.00					
16.00	0.43	High	25.00					
17.00	0.18	High	19.00					
18.00	0.39	High	19.00					
19.00	0.90	High	6.00					
20.00	0.65	High	3.00[1]	26	42	14	35	29
21.00	1.42	High	9.00					
22.00	0.14	High	9.00					
23.00	0.04	Moderate	17.00					
24.00	0.35	Moderate	18.00[1]	18	9	20	24	12
25.00	0.08	Moderate	17.00					
26.00	0.06	Moderate	30.00					
27.00	0.43	Moderate	35.00					
28.00	−1.16	Moderate	13.00					
29.00	0.75	Moderate	17.00					
30.00	0.81	Moderate	1.00					
31.00	0.04	Moderate	5.00					
32.00	0.35	Low	1.00					
33.00	0.34	Low	2.00[1]	5	23	5	0	13
34.00	0.15	Low	2.00					

SOURCE: Author's compilation.
[1] Value deleted for MAR data example

Multiple imputation is the more flexible of the two options, but is not widely implemented in statistical packages. The SAS program has a procedure for both multiple imputation and for combing the results across completed data sets. Although the SPSS Missing Values program claims that multiple imputation is possible, the program does not implement the multiple imputation procedures as outlined by Donald Rubin (1987) and Joseph Schafer (1997). Instead, the manual recommends running the program several times using regression imputation with estimates augmented by a random deviate. Research is needed on the behavior of this procedure compared with the data augmentation procedures Schafer described. After the researcher obtains multiple imputations, the procedures to compute multiply imputed estimates are straightforward.

When applying any statistical method, with or without missing data, the synthesist is obligated to discuss the

Table 21.13 Correlations, Maximum Likelihood Methods

	Study Age		Correlations	
	Mean	SE (Mean)	Study Age and Log-Odds Ratio	Study Age and Intensity of Dose
Imputation 1	16.56	1.84	0.01	0.34
Imputation 2	19.82	2.47	−0.07	0.32
Imputation 3	17.29	1.91	−0.26	0.36
Imputation 4	18.53	2.25	−0.10	0.40
Imputation 5	16.06	1.86	−0.20	0.26
Multiple Imputation	17.65	2.67	−0.12	0.34
Full data	20.35	2.25	0.05	0.50

SOURCE: Author's compilation.

assumptions made in the analysis. In the case of methods for missing data, we cannot provide empirical evidence about the truth of the assumptions. We can, however, fully disclose the assumptions of each analysis and compare the results to examine the sensitivity of our findings.

21.7 REFERENCES

Allison, Paul D. 2002. *Missing Data.* Thousand Oaks, Calif.: Sage Publications.

American Psychological Association. 2005. *Publication Manual of the American Psychological Association*, 5th ed. Washington, D.C.: APA.

Amemiya, Takeshi. 1984. "Tobit Models: A Survey." *Journal of Econometrics* 24(1–2): 3–61.

Anand, Sonia S. and Salim Yusuf. 1999. "Oral Anticoagulant Therapy In Patients With Coronary Artery Disease." *Journal of the American Medical Association* 282(21): 2058–67.

Becker, Betsy J. 1992. "Using Results From Replicated Studies To Estimate Linear Models." *Journal of Educational Statistics* 17(3): 341–62.

Begg, Colin B., and Jesse A. Berlin. 1988. "Publication Bias: A Problem in Interpreting Medical Data (with Discussion)." *Journal of the Royal Statistical Society Series A* 151(2): 419–63.

Buck, S. F. 1960. "A Method of Estimation of Missing Values in Multivariate Data Suitable for Use with an Electronic Computer." *Journal of the Royal Statistical Society Series B* 22(2): 302–3.

Chan, An-Wen, Asbjørn Hróbjartsson, Mette T. Haahr, Peter C. Gøtzsche, and Douglas G. Altman. 2004. "Empirical Evidence for Selective Reporting of Outcomes in Randomized Trials." *Journal of the American Medical Association* 291(20): 2457–65.

Dempster, Arthur P., Nan M. Laird, and Donald B. Rubin. 1977. "Maximum Likelihood from Incomplete Data via the EM Algorithm." *Journal of the Royal Statistical Society Series B* 39(1): 1–38.

Egger, Matthias, and George Davey. 1998. "Meta-Analysis: Bias in Location and Selection of Studies. *British Medical Journal* 316(7124): 61–6.

Fahrbach, Kyle R. 2001. "An Investigation of Methods for Mixed-Model Meta-Analysis in the Presence of Missing Data." PhD diss. Michigan State University.

Glasser, Marvin. 1964. "Linear Regression Analysis with Missing Observations among the Independent Variables." *Journal of the American Statistical Association* 59(307): 834–44.

Hahn, Seokyung, Paula R. Williamson, Jane L. Hutton, Paul Garner, and E. Victor Flynn. 2000. "Assessing the Potential for Bias in Meta-Analysis Due to Selective Reporting of Subgroup Analyses within Studies." *Statistics in Medicine* 19(24): 3325–36.

Haitovsky, Yoel. 1968. "Missing Data in Regression Analysis." *Journal of the Royal Statistical Society Series B* 30(1): 67–82.

Heckman, James. 1976. "The Common Structure of Statistical Models of Truncation, Sample Selection and Limited Dependent Variables, and a Simple Estimator for Such Models." *Annals of Economic and Social Measurement* 5(4): 475–92.

Hedges, Larry V., and Jack L. Vevea. 1996. "Estimating Effect Size under Publication Bias: Small Sample Properties and Robustness of a Random Effects Selection Model." *Journal of Educational and Behavioral Statistics* 21(4): 299–332.

Hemminki, E. 1980. "Study of Information Submitted by Drug Companies to Licensing Authorities." *British Medical Journal* 280(6217): 833–36.

Hill, MaryAnn. 1997. *SPSS Missing Value Analysis 7.5.* Chicago: SPSS Inc.

Jüni, Peter, Franziska Holenstein, Jonathon Sterne, Christopher Bartlett, and Matthias Egger. 2002. "Direction and Impact of Language Bias in Meta-Analyses of Controlled trials: Empirical Study. *International Journal of Epidemiology* 31(1): 115–23.

Kim, Jae-On, and James Curry. 1977. "The Treatment of Missing Data in Multivariate Analysis." *Sociological Methods and Research* 6(2): 215–40.

Little, Roderick J. A. 1988. "A Test of Missing Completely at Random for Multivariate Data with Missing Values." *Journal of the American Statistical Association* 83(404): 1198–1202.

Little, Roderick J. A., and Donald B. Rubin. 1987. *Statistical analysis with Missing Data.* New York: John Wiley & Sons.

Orwin, Robert G., and David S. Cordray. 1985. "Effects of Deficient Reporting on Meta-Analysis: A Conceptual Framework and Reanalysis." *Psychological Bulletin* 97(1): 134–47.

Pigott, Therese D. 2001. "Missing Predictors in Models of Effect Size." *Evaluation & the Health Professions* 24(3): 277–307.

Rosenthal, Robert. 1979. "The 'File Drawer Problem' and Tolerance for Null Results." *Psychological Bulletin* 86(3): 638–41.

Rothstein, Hannah R., Alexander J. Sutton, and Michael Borenstein. 2005. *Publication Bias in Meta-Analysis: Prevention, Assessment and Adjustments.* West Sussex: John Wiley & Sons.

Rubin, Donald B. 1976. "Inference and Missing Data." *Biometrika* 63(3): 581–92.

Rubin, Donald B. 1987. *Multiple Imputation for Nonresponse in Surveys.* New York: John Wiley & Sons.

Rubin, Donald B. 1996. "Multiple Imputation after 18+ Years." *Journal of the American Statistical Association* 91(434): 473–89.

Schafer, Joseph L. 1997. *Analysis of Incomplete Multivariate Data.* New York: Chapman & Hall.

Schafer, Joseph L. 1999. *NORM: Multiple Imputation Of Incomplete Multivariate Data under a Normal Model*, version 2. http://www.stat.psu.edu/~jls/misoftwa.html#win

Schafer, Joseph L. and John W. Graham. 2002. "Missing Data: Our View of the State of the Art." *Psychological Methods* 7(2): 147–77.

Shadish, William R., Leslie Robinson, and *Congxiao* Lu. 1997. *Effect Size Calculator.* St. Paul, Minn.: Assessment Systems Corporation. http://www.assess.com.

Sirin, Selcuk R. 2005. "Socioeconomic Status and Academic Achievement: A Meta-Analytic Review of Research." *Review of Educational Research* 75(3): 417–53.

Smith, Mary L. 1980. "Publication Bias and Meta-Analysis." *Evaluation in Education* 4: 22–24.

SPSS. 2007. *SPSS Missing Value Analysis.* http://www.spss.com/missing_value.

Tanner, Martin A., and Wing H. Wong. 1987. "The Calculation of Posterior Distributions by Data Augmentation." *Journal of the American Statistical Association* 82(398): 528–40.

Vevea, Jack L and Carol M. Woods. 2005. "Publication Bias in Research Synthesis: Sensitivity Analysis Using A Priori Weight Functions." *Psychological Methods* 10(4): 428–43.

Willamson, Paula R., Carrol Gamble, Douglas G. Altman, and J. L. Hutton. 2005. "Outcome Selection Bias in Meta-Analysis." *Statistical Methods in Medical Research* 14(5): 515–24.

Wu, Meng-Jia. 2006. "Methods of Meta-analyzing Regression Studies: Applications of Generalized Least Squares and Factored Likelihoods." PhD diss., Michigan State University.

Yuan, Yang C. 2000. "Multiple Imputation for Missing Data: Concepts and New Development." http://support.sas.com/rnd/app/papers/multipleimputation.pdf

22

SENSITIVITY ANALYSIS AND DIAGNOSTICS

JOEL B. GREENHOUSE
Carnegie Mellon University

SATISH IYENGAR
University of Pittsburgh

CONTENTS

22.1 INTRODUCTION

At every step in a research synthesis, decisions are made that can affect the conclusions and inferences drawn from that analysis. Sometimes a decision is easy to defend, such as one to omit a poor quality study from the meta-analysis based on prespecified exclusion criteria, or to use a weighted average instead of an unweighted one to estimate an effect size. At other times, the decision is less convincing, such as that to use only the published literature, to omit a study with an unusual effect size estimate, or to use a fixed effects instead of a random effects model. When the basis for a decision is tenuous, it is important to check whether reasonable alternatives appreciably affect the conclusions. In other words, it is important to check how sensitive the conclusions are to the method of analysis or to changes in the data. Sensitivity analysis is a systematic approach to address the question, "What happens if some aspect of the data or the analysis is changed?"

When meta-analysis first came on the scene, it was greeted with considerable skepticism. Over time, with advances in methodology and with the experience due to increased use, meta-analysis has become an indispensable research tool for synthesizing evidence about questions of importance, not only in the social sciences but also in the biomedical and public health sciences (Gaver et al. 1992). Given the prominent role that meta-analysis now plays in informing decisions that have wide public health and public policy consequences, the need for sensitivity analyses to assess the robustness of the conclusions of a meta-analysis has never been greater.

A sensitivity analysis need not be quantitative. For instance, the first step in a meta-analysis is to formulate the research question. Although a quantitative sensitivity analysis is not appropriate here, it is, nevertheless, useful to entertain variations of the research question before proceeding further, that is, to ask the "what if " question. Because different questions require different types of data and data analytic methods, it is important to state clearly the aims of the meta-analysis at the outset to collect the relevant data and to explain why certain questions (and not others) were addressed. Or, at another step decisions are made concerning the identification and retrieval of studies. Here the decision is which studies to include. Should only peer-reviewed, published studies be included? What about conference proceedings or dissertations? Should only randomized controlled studies be considered? These are decisions that a meta-analyst must make early on that can affect the generalizability of the

conclusions. We view the process of asking these questions as part of sensitivity analysis because it encourages the meta-analyst to reflect on decisions and to consider the consequences of those decisions in terms of the subsequent interpretation and generalizablility of the results. Our focus in this chapter is primarily on quantitative methods for sensitivity analysis. It is important, however, to keep in mind that these early decisions about the nature of the research question and identifying and retrieving the literature can have a profound impact on the usefulness of the meta-analysis.

We will use a rather broad brush to discuss sensitivity analysis. Because, as we have suggested, sensitivity analysis is potentially applicable at every step of a meta-analysis, we touch on many topics covered in greater detail elsewhere in this volume. Our aim here is to provide a unified approach to the assessment of the robustness, or cogency, of the conclusions of a research synthesis. The key element to this approach is simply the recognition of the importance of asking the "what if" question. Although a meta-analysis need not concern itself with each issue discussed here, our point of view, and the technical tools we describe in this chapter should be useful generally. Many of these tools come from the extensive statistical literature on exploratory data analysis. Some, though, have been developed largely in the context of meta-analysis itself, such as those for assessing the effects of publication bias.

To illustrate some of the quantitative approaches, we use two example data sets throughout this chapter. The first is a subset of studies from a larger study of the effectiveness of open education programs (Hedges, Giaconia, and Gage 1981). The objective was to examine the effects of open education on student outcomes by comparing students in experimental open classroom schools with those from traditional schools. Specifically, in table 22.1, we look at the studies presented in table 8 of Larry Hedges and Ingram Olkin (1985, 25) consisting of the randomized experiments and other well-controlled studies of the effects of open education on self-concept. The table presents the students' grade levels, the sample size for each group, a weight for each study, and an unbiased estimate, T, of the standardized mean difference for each study. The second data set that we will use when we discuss random effects models comes from table 7 of Hedges and Olkin (1985, 24) on the effects of open education on student independence and self-reliance. These data are presented in table 22.2. Unbiased estimates of the standardized mean differences are shown for the seven studies there, all of which have the same sample sizes. No

Table 22.1 Effects of Open Education on Student Self-Concept

Study	Grade	Open Education n^T	Traditional School n^C	Weights w_i	Standardized Mean Difference T
1	4–6	100	180	0.115	0.100
2	4–6	131	138	0.120	−0.162
3	4–6	40	40	0.036	−0.090
4	4–6	40	40	0.036	−0.049
5	4–6	97	47	0.057	−0.046
6	K–3	28	61	0.034	−0.010
7	K–3	60	55	0.051	−0.431
8	4–6	72	102	0.076	−0.261
9	4–6	87	45	0.053	0.134
10	K–3	80	49	0.054	0.019
11	K–3	79	55	0.058	0.175
12	4–6	40	109	0.052	0.056
13	4–6	36	93	0.046	0.045
14	K–3	9	18	0.011	0.103
15	K–3	14	16	0.013	0.121
16	4–6	21	22	0.019	−0.482
17	4–6	133	124	0.115	0.290
18	K–3	83	45	0.052	0.342

SOURCE: Authors' compilation.

Table 22.2 Effects of Open Education on Student Independence and Self-Reliance

Study	$n_E = n_C$	T
1	30	0.699
2	30	0.091
3	30	−0.058
4	30	−0.079
5	30	−0.235
6	30	−0.494
7	30	−0.587

SOURCE: Authors' compilation.

covariate information, such as grade level, was given in Hedges and Olkin (1985) for this data set.

22.2 RETRIEVAL OF LITERATURE

The discussions in part III of this volume clearly demonstrate that retrieving papers that describe relevant studies

for a research synthesis is not as straightforward as it might seem. Several questions arise at the outset of this time-consuming yet essential component of a meta-analysis. Should a single person search the literature, or should (time and resources permitting) at least two people conduct independent searches? Should the search concentrate on the published literature only (that is, from peer-reviewed journals), or should it attempt to retrieve all possible studies? Which time period should be covered by the search? The first two questions deal with the concern that the studies missed by a search may arrive at qualitatively different conclusions than those retrieved. The third question is more subtle: it is possible that within a period of time studies yield homogeneous results, but that different periods show considerable heterogeneity. For instance, initial enthusiastically positive reports may be followed by a sober period of null, or even negative results. Geoffrey Borman and Jeffrey Grigg also discuss the use of graphical displays to see how research results in a variety of fields change with the passage of time (chapter 26, this volume).

Using more than one investigator to search the literature is itself a type of sensitivity analysis. It is especially useful when the central question in a research synthesis spans several disciplines and the identification of relevant studies requires subject-matter expertise in these disciplines (see, for example, Greenhouse et al. 1990). It is intuitively clear that if the overlap between the searches is large, we would be more confident that a thorough retrieval had been done. (A formal approach for assessing the completeness of a literature search might involve capture-recapture calculations of population size estimation; see, for example, Fienberg, Johnson, and Junker 1999). Clearly, the more comprehensive a search, the better, for then the meta-analysis can use more of the available data. However, a thorough search may not be feasible because of various practical constraints. To decide whether a search limited, for example, to the published literature can lead to biased results, it is worthwhile to do a small search of the unpublished literature (for example, dissertations, conference proceedings, and abstracts) and compare the two (for a more complete discussion, see chapters 2 and 3 of this volume; Wachter and Straf 1990, chapter 15).

When a substantial group of studies is retrieved, and data from them gleaned, it is useful to investigate the relationship between study characteristics and study results. Some of the questions that can be addressed using the exploratory data analytic tools are discussed in section 22.3. Do published and unpublished studies yield different results? If the studies are rated according to quality, do the lower quality studies show greater variability or systematic bias? When two disciplines touch on the same issue, how do the results of studies compare? When these issues arise early in the literature search, it is possible to alter the focus of the retrieval process to ensure a more comprehensive database.

22.3 EXPLORATORY DATA ANALYSIS

Once the literature has been searched and studies have been retrieved, the next step is to analyze effect size. Formal statistical procedures for determining the significance of an effect size summary usually are based on assumptions about the probabilistic process generating the observations. Violations of these assumptions can have an important impact on the validity of the conclusions. Here we discuss relatively simple methods for investigating features of a data set and the validity of the statistical assumptions. We consider these methods to be a part of a sensitivity analysis because they help reveal features of the data that could have a large impact on the choice of

more formal statistical procedures and on the interpretation of their results.

The approach described, known as exploratory data analysis, is based on the premise that the more that is known about the data, the more effectively it can be used to develop, test, and refine theory. Methods should be relatively easy to do with or without a computer, quick to use so that a data set may be explored from different points of view, and robust in that they are not adversely influenced by such misleading phenomena as extreme cases, measurement errors, or unique cases that need special attention. Two methods discussed, the stem-and-leaf plot and the box plot, are graphical methods that satisfy these requirements. They are powerful tools for developing insights and hypotheses about the data in a meta-analysis.

22.3.1 Stem-and-Leaf Plot

In a stem-and-leaf plot, the data values are sorted into numerical order and brought together quickly and efficiently in a graphic display. The stem-and-leaf plot uses all of the data and illustrates the shape of a distribution, that is, shows whether it is symmetric or skewed, how many peaks it has, and whether it has outliers (atypical extreme values) or gaps within the distribution. Figure 22.1 shows the stem-and-leaf plot for the distribution of the standardized mean differences, T, for the 18 studies given in table 22.1. Here, for ease of exposition, we have rounded the values of the standardized mean differences in the last column of table 22.1 to two significant digits. The first significant digit of T is designated the *stem* and the second digit the *leaf*. For example, the rounded standardized mean difference for study 2 is -0.16. Therefore, the stem is -1 and the leaf is 6. The possible stem values are listed vertically in increasing order from top (negative stems) to bottom (positive stems) and a vertical line is drawn to the right of the stems. The leaves are recorded directly from table 22.1 onto the lines corresponding to their stem value to the right of the vertical line. Within each stem, the leaves should be arranged in increasing order away from the stem.

In figure 22.1 we can begin to see some important features of the distribution of T.

1. Typical values. It is not difficult to locate by eye the typical values or central location of the distribution of T by finding the stem or stems where most of the observations fall. For the distribution in figure 22.1,

−4	3 8
−3	
−2	6
−1	6
−0	1 4 4 9
0	1 4 5
1	0 0 2 3 7
2	9
3	4

Figure 22.1 Stem-and-Leaf Plot of Standardized Mean Differences.

SOURCE: Authors' compilation.
NOTE: For studies reported in table 22.1 ($n = 18$; leaf unit $= 0.010$).

we see that the center of the distribution seems to be in the stems corresponding to values of T between −0.01 to 0.17.

2. Shape. Does the distribution have one peak or several peaks? Is it approximately symmetric or is it skewed in one direction? We see that the distribution in figure 22.1 may have two peaks and seems to have a longer lower tail, that is, skewed toward negative values.

3. Gaps and atypical values. The stem-and-leaf plot is useful for identifying atypical observations, which are usually highlighted by gaps in the distribution. Note, for example, the gap between study #8 with a T of −0.26 and the next two studies, study #7 with a T of −0.43, and study #16 with a T of −0.49, suggesting that these latter two studies might be unusual observations.

For generalizations and modifications of the stem-and-leaf plot for other sorts of data configurations, see Lambert Koopmans (1987, 9–22).

At this stage of a meta-analysis, the stem-and-leaf plot is useful in helping the data analyst become more familiar with the sample distribution of the effect size. From the point of view of sensitivity analysis, we are particularly interested in identifying unusual or outlying studies, deciding whether the assumptions of more formal statistical procedures for making inferences about the population effect size are satisfied, and investigating whether there are important differences in the sample distributions of effect size for different values or levels of an explanatory variable. The latter is particularly interesting in furthering an understanding of how studies differ. The stem-and-

leaf plot and the box plot, described in the next section, are both useful exploratory data analytic techniques for carrying out these aspects of a sensitivity analysis.

22.3.2 Box Plots

As noted earlier, major features of a distribution include the central location of the data, how spread out the data are, the shape of the distribution, and the presence of any unusual observations, called outliers. The box plot is a visual display of numerical summaries, based in part on sample percentiles, that highlights such major features of the data. The box plot provides a clear picture of where the middle of the distribution lies, how spread out the middle is, the shape of the distribution, and identifies outliers in the distribution. A box plot differs from a stem-and-leaf plot in several ways. The box plot is constructed from numerical summaries of the distribution and does not simply display every observation as the stem-and-leaf plot does. It provides a much clearer picture of the tails of the distribution and is more useful for identifying outliers than the stem-and-leaf plot. Sample percentiles are used to construct the box plot because percentiles, such as the median (50th percentile), are robust measures of location. The mean, by contrast, is not a robust measure of central location because it is influenced by atypical observations, or outliers.

Figure 22.2 presents a box plot of the sample distribution of T. We describe the features of this box plot. The box itself is drawn from the first quartile ($Q_1 = -0.109$) to the third quartile ($Q_3 = 0.127$). The plus sign inside the box denotes the median ($m = 0.032$). The difference or distance between the third quartile and the first quartile ($Q_3 - Q_1 = 0.127 - (-0.109) = .236$) is called the interquartile range (IQR) and is a robust measure of variability. The IQR is the length of the box in figure 22.2. The dotted lines going in either direction from the ends of the box (that is, away from Q_1 and Q_1 respectively) denote the tails of the distribution. The two asterisks in figure 22.2 denote outliers, −0.491 and −0.434, and are so defined because they are located a distance more than one and a half times the IQR from the nearest quartile. The dotted lines extend to an observed data point within one and a half times the IQR from each respective quartile to establish regions far enough from the center of the distribution to highlight observations that may be outliers (see Koopmans 1987, 51–57). Note that the sample mean of T is −0.008 and is smaller than the median because of the inclusion of the two negative atypical values.

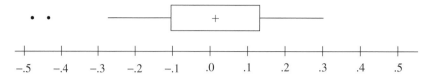

Figure 22.2 Box Plot of Standardized Mean Differences.

SOURCE: Authors' compilation.
NOTE: For studies reported in table 22.1

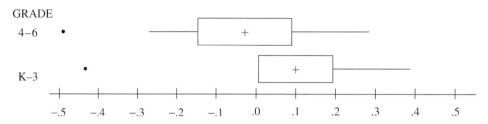

Figure 22.3 Comparative Box Plots of Distribution of Standardized Mean Differences.

SOURCE: Authors' compilation.
NOTE: For studies of grades K–3 versus grade 4–6

From the viewpoint of sensitivity analysis, awareness of unusual features of the data, such as outliers or skewness, can guide the analyst with respect to the choice of statistical methods for synthesis, and alert the analyst to the potential for erroneous conclusions based on a few influential observations. Furthermore, it is often the aim of a research synthesis to compare the sample distribution of effect size at different values of an explanatory variable, that is, to characterize how studies differ and why. Box plots are particularly useful for comparing distributions because they so clearly highlight important features of the data. For example, we might be interested in seeing whether the effects of open education on student self-concept is different for students in grades K–3 versus grades four through six. The investigation of such questions using exploratory data analysis techniques is a form of sensitivity analysis. As an illustration, we present in figure 22.3, a pair of box plots displaying the distribution of T in each grade range. Table 22.3 gives the relevant numerical summary measures for each grade level. A comparison of the two distributions is revealing:

1. Central location. The sample distribution of T for the studies of children in grades K–3 is shifted to the right relative to the distribution for the studies of

Table 22.3 Numerical Summary Measures for Distribution of T

	K–3	4–6	Total
n	7	11	18
median	0.103	−0.046	0.032
mean	0.046	−0.043	−0.008
Q_1	−0.010	−0.162	−0.109
Q_3	0.176	0.100	0.127
IQR	0.186	0.262	0.236

SOURCE: Authors' compilation.

children in grades four through six, suggesting a potentially greater effect of open education on self-concept in the younger students. Even though the median effect size for the studies of children in grades four through six is smaller than that for those of children in grades K–3, note that the two distributions have significant overlap.

2. Spread. Judging by the IQRs for each distribution, that is, the length of the boxes in figure 22.3, the spread of the sample distribution of T is a little larger for the studies of children in grades four

through six, 0.262, versus 0.186 for the studies of children in grades K–3.

3. Shape. The shape of the middle 50 percent of the sample distribution of T for the studies of children in grades K–3 is skewed to the left. Note that the median is closer to Q_3 than to Q_1. Yet, we see that the right tail is longer than the left. The shape of the middle 50 percent of the distribution for the studies of children in grades four through six is approximately symmetric, though the median is a little closer to Q_1 than to Q_3, and the right tail is a little longer than the left, suggesting a slight skewness to the right.

4. Outliers. We see that each distribution has an outlier. The solid circle in the box plot for the studies of children in grades K–3 indicates that the outlier in this group of studies, study 5, is very far away from Q1, whereas the outlier in the group of studies for children in grades four through six is not quite as far out.

Having identified the outliers in each group, decisions about how to handle these potentially anomalous studies in subsequent analyses are necessary. Nevertheless, because of the use of robust measures of central location in the box plots displayed in figure 22.3, we have a relatively good feeling for the amount of overlap of the distributions for the two grade ranges. As a result, for example, we might be wary of a single summary measure of effect size obtained by pooling across the two grade ranges. In summary, informal exploratory analyses from the perspective of sensitivity analysis have been useful in highlighting features of the data that we need to be aware of, sources of heterogeneity, and situations where pooling across a stratification variable might be inappropriate.

22.4 COMBINING EFFECT SIZE ESTIMATES

Based on the exploratory analysis, we have a better understanding of what the interesting features of the data set are, the effect on the research synthesis of studies that are outliers, and issues related to distributional assumptions. In the next step, decisions must be made about how to combine and summarize the effect size estimates. For example, if we assume a fixed effects model, the implication is that studies in our research synthesis represent a random sample from the same population with a common but unknown population effect. If we assume a random effects model, are we then explicitly assuming a proba-

bility distribution of effect sizes? Is there evidence from the data to support a fixed or random effects model? Should we use a weighted or unweighted summary measure of effect size? Should we use parametric or nonparametric procedures for assessing the statistical significance of our combined results? In this section we frame some of these questions as issues in sensitivity analysis.

22.4.1 Fixed Effects

Part III of this volume contains descriptions of effect size estimates from a single study, and methods of combining them across studies (see also Hedges and Olkin 1985). In practice, when there is a choice between two analyses or between two summary measures, it is prudent to do both analyses to see how the conclusions differ and to understand why they differ (Greenhouse et al. 1990). For example, the average of the standardized mean difference for the studies of the effect of open education on student self-concept given in table 22.1 is $\overline{T} = -.008$ with an approximate 95 percent confidence interval of $(-0.111, 0.023)$. Because zero is contained in this interval, we would conclude that there is no statistically significant difference between the effect of open education and traditional education on student self-concept. It is more common to consider the effect of different sample sizes in the primary studies by calculating a weighted average of the standardized mean difference, $\overline{T}_w$ using the weights given in table 22.1 where

$$w_i = \tilde{n}_i \left/ \sum_{i=1}^{k} \tilde{n}_i \right. \quad \text{and} \quad \tilde{n}_i = (n_i^T \times n_i^C) \left/ (n_i^T + n_i^C) \right.$$

From table 22.1 we find the weighted average of the standardized mean difference to be

$$\overline{T}_w = w_1 \times T_1 + \cdots + w_k \times T_k = 0.010.$$

with an approximate 95 percent confidence interval of $(-0.073, 0.093)$. We see that our conclusions about the population effect size in this case do not depend on whether we use a weighted or unweighted estimate.

At this stage, it is of interest to see whether any individual study or groups of studies have a large influence on the effect size estimate. For example, recall that in section 22.3 we identified several studies as potential outliers. One relatively simple method for this is to hold one study out, calculate the average effect size by combining the remaining studies, and compare the results with the overall combined estimate. Here we are interested in assessing how sensitive the combined measure of the effect

Table 22.4 The "Leave One Out" for the Studies in Table 22.1

Study Left Out i	T	Unweighted $\overline{T}(i)$	Weighted $\overline{T}_w(i)$
1	0.100	−0.015	−0.001
2	−0.162	0.001	0.030
3	−0.091	−0.003	0.013
4	−0.049	−0.006	0.012
5	−0.046	−0.006	0.013
6	−0.010	−0.008	0.011
7	−0.434	0.017	0.032
8	−0.262	0.008	0.030
9	0.135	−0.017	0.003
10	0.019	−0.010	0.009
11	0.176	−0.019	0.000
12	0.056	−0.012	0.007
13	0.045	−0.011	0.008
14	0.106	−0.015	0.009
15	0.124	−0.016	0.009
16	−0.491	0.020	0.020
17	0.291	−0.026	−0.023
18	0.344	−0.029	−0.008
median	0.032	−0.011	0.009

SOURCE: Authors' compilation.

size is to any one particular study. In other words, how large is the influence of each study on the overall estimate of effect size? Let $\overline{T}(i)$ be the unweighted average standardized mean difference computed from all the studies except study i. Table 22.4 presents the values of $\overline{T}(i)$ for each study in table 22.1. We see that the effect of leaving study 7 or 16 out of the unweighted average of the standardized mean difference, respectively, is to result in an increase in the summary estimate. Because we identified these two studies earlier as outliers, this result is not surprising. We also see from table 22.4 that studies 17 and 18 have the largest values of T among all of the studies and that the effect of leaving each of these studies out of the combined estimate, respectively, is to result in a decrease in the average standardized mean difference. In summary, the results in this analysis are as we might expect: remove a study that has a smaller-than-the-mean effect size, and the overall summary measure goes up.

In table 22.4, we also report the analysis of leaving one study out using the weighted average, $\overline{T}_w(i)$. We see similar results for the analysis with the unweighted average.

However, in addition, we now see that studies 2 and 8 are fairly influential studies, respectively. Because they are larger studies, removing them one at a time from the analysis causes the weighted average effect size to increase. Computer software is available to facilitate the implementation of a leave-one-out analysis (see, for example, Borenstein and Rothstein 1999).

Leaving out more than one study at a time can become very complicated. For instance, if we leave out all pairs, all triplets, and so on, the number of cases to consider would be too large to give any useful information. It is better to identify beforehand covariates that are of interest, and leave out studies according to levels of the covariates. In our example, the only covariate provided is the grade level (K–3 and four through six). Leaving out one level of that covariate is thus equivalent to the box plot analysis in table 22.3 and figure 22.3. Finally, we note that the leave-one-out method can be used for other combining procedures. For instance, the maximum likelihood estimate (MLE) is common when we have a specific model for the effect sizes. In that case, the sensitivity of each case can be assessed as shown above; of course, the computational burden involved will be much greater.

22.4.2 Random Effects

In the previous sections, the covariate information (grade level) was useful in accounting for some of the variability among the effect size estimates. There are other times, though, when such attempts fail to find "consistent relationships between study characteristics and effect size" (Hedges and Olkin, 1985, 90). In such cases, random effects models provide an alternative framework for combining effect sizes. These models regard the effect sizes themselves as random variables, sampled from a distribution of possible effect sizes. Chapter 20 of this volume and chapter 9 of Hedges and Olkin (1985) provide discussions on the distinction between random and fixed effect models, the conceptual bases underlying random effects models, and their application to research synthesis. The discussion that follows assumes familiarity with those chapters. The development of methods for sensitivity analysis for random effects models are an important, open research area. In fact, Hedges and Olkin did not refer to random effects models in their chapter on diagnostic procedures for research synthesis (1985). In this section, we briefly outline the most commonly used random effects model and the use of sensitivity analysis for random effects models in meta-analysis.

Table 22.5 Comparison of Fixed Effects Analysis to Random Effects Analysis

			Fixed effects			
Table	Study	k	$\hat{\delta}$	$SE(\hat{\delta})$	95% c.i. for δ
22.2	Independence	7	−0.095	0.099	(−0.288, 0.099)
22.1 K–3	Self-Concept	7	0.034	0.081	(−0.125, 0.192)
22.1 4–6	Self-Concept	11	0.001	0.050	(−0.096, 0.099)

			Random effects				
Table	Study	k	$\hat{\Delta}$	$SE(\hat{\Delta})$	95% c.i. for Δ	$\hat{\sigma}^2_{\Delta}$	χ^2 for $H_0{:}\hat{\sigma}^2_{\Delta}=0$
22.2	Independence	7	−0.095	0.161	(−0.411, 0.220)	0.113	$\chi^2_6 = 15.50$
22.1 K–3	Self-Concept	7	0.037	0.066	(−0.093, 0.167)	0.0*	$\chi^2_6 = 9.56$
22.1 4–6	Self-Concept	11	0.004	0.057	(−0.116, 0.107)	0.007	$\chi^2_{10} = 13.96$

SOURCE: Authors' compilation.
*The unbiased estimate, −0.012, is inadmissible.

In the random effects approach, the true effect sizes $\{\theta_i{:}i = 1, \ldots, k\}$ are considered realizations of a random variable Θ. That is, they are a random sample from a population of effect sizes, which is governed by some distribution with mean μ_θ and variance τ^2, a measure of the between-study variability. Given sample estimates, T_i, of θ_i for $i = 1, \ldots, k$, the analysis proceeds by estimating these parameters or performing tests on them. For instance, a common starting point is the test of the hypothesis $\tau^2 = 0$, a test of the homogeneity of the effect sizes across the k studies: $\theta_1 = \cdots = \theta_k = \theta$. Note that this common effect size, θ, in the fixed-effects model has a different interpretation than μ: Even if μ is positive, there may be realizations θ_i that are negative (for more discussion, see Hedges and Olkin 1985; DerSimonian and Laird 1986; Higgins and Thompson 2002).

An important example of sensitivity analysis is the comparison of the results of fixed and random effects models for a particular data set. The output from a fixed effects analysis includes an estimate of the common effect size, $\hat{\theta}$, and its estimated standard error, SE. The output from a random effects analysis includes an estimate of the mean of the distribution of effect sizes, $\hat{\mu}_\theta$, its estimated standard error, $SE(\hat{\mu})$, an estimate of the variance of the distribution of random effects τ^2, and the value of the chi-squared test of the hypothesis of complete homogeneity. A confidence interval for the mean effect size, μ, is also provided (see Hedges and Olkin 1985).

Table 22.5 contains the results of the two analyses for the data in tables 22.1 and 22.2. Recall that table 22.1 consists of studies measuring the effects of open education on self-concept and includes grade level as a covariate. Table 22.2 consists of studies measuring the effects of open education on student independence and self-reliance, but providing no covariate information. For table 22.2, the estimates of (common or mean) effect size are virtually the same ($\hat{\theta} = -0.095$, $\hat{\mu}_\theta = -0.095$), however, the SE for the random effects model, 0.161 is greater than that for the fixed effects model, 0.099. Next, the estimate $\hat{\tau}^2 = 0.113$ is significantly different from zero, in that the corresponding chi-square value is 15.50, which exceeds the 5 percent critical value of 12.59 for six degrees of freedom. Because this test of complete homogeneity rejects, the conservative (in the sense of having wider confidence intervals) random effects analysis is deemed more valid. In this case, the substantive conclusions may well be the same under both models, given that both confidence intervals contain zero. On the other hand, when the two methods differ, it is important to investigate why they differ and to explicitly recognize that the conclusions are sensitive to the modeling assumption.

The rest of table 22.5 compares the two analyses for the data in table 22.1. In section 22.3, we saw that grade level, the one covariate provided, helped to explain variability among the effect-size estimates. Thus, neither the fixed nor the random effects models should be applied to these data without modification. We present these data to illustrate several points. First, we can compare the two on subsets of the data defined by the levels of the covariate (in this case, K–3, four through six). Next, making inferences about τ^2 can be difficult, because a large number of primary studies are needed, and strong distributional assumptions must be satisfied before using the procedures Hedges and Olkin described (1985). In fact, for the K–3 data, the unbiased estimate of τ^2 is negative. Although it is tempting to interpret this as supporting the hypothesis that $\tau^2 = 0$ and then to use the fixed effects analysis, more research is needed to find better methods of estimating τ^2 for such cases (for further discussion, see chapter 20, this volume; Hill 1965.) Of course, the methods from exploratory data analysis discussed earlier are also useful for sensitivity analysis. For instance, the leave-one-out method can be readily used in a random effects context to informally investigate the influence of studies.

Random effects models typically assume the normal distribution for the effect sizes (for continuous data), so one concern is how sensitive the analysis is to this assumption. The two main steps in addressing this concern are to detect departures from normality of the random effects, and then to assess the impact of non-normality on the combined estimates of interest. The weighted normal plot is a graphical procedure designed to check the assumption of normality of the random effects in linear models (Dempster and Ryan 1985). John Carlin used this diagnostic in his study combining the log of odds ratios from several 2×2 tables (1992). If the normality of the random effects is found to be inappropriate, several new problems arise. For instance, the usual parameters (such as μ) are no longer easily interpretable. For example, if the distribution of effect sizes is sufficiently skewed, the mean effect size can be positive (indicating that on average an efficacious result), yet for more than half of all the effect sizes can be negative. One way to assess the sensitivity of the analysis to this kind of departure from the normal assumption is to use skewed densities, such as the gamma or log-normal families. Carrying out this kind of sensitivity analysis is computationally intensive.

22.4.3 Bayesian Hierarchical Models

Bayesian methods have emerged as powerful tools for analyzing data in many complex settings, including meta-analysis. Situations in which Bayesian methods are especially valuable are those where: information comes from many similar sources, which are of interest individually

and also need to be combined, so that hierarchical models may be constructed; ancillary information about a question of interest is available, for example, either from expert judgment or from previous studies, which represents the cumulative evidence about the question of interest that needs to be systematically incorporated with current evidence; modeling assumptions need to be examined systematically, or models are sufficiently complicated that inferences about quantities of interest require substantial computation. The Bayesian hierarchical model integrates the fixed-effect and random-effect models discussed earlier into one framework and provides a unified modeling approach to meta-analysis. The theory and applications of this approach to the problems of research synthesis with illustrative examples can be found in the literature (see, for example, Hedges 1998; Normand 1999; Raudenbush and Bryk 2002; Stangl and Berry 2000; Sutton et al. 2000). The statistical software package WinBUGS is available for implementing this approach (Gilks, Thomas, and Spiegelhalter 1994). Here we briefly review the major elements of the Bayesian approach for meta-analysis with a special emphasis on the role of sensitivity analysis.

As before, assume that there are k independent studies available for a research synthesis, and let T_i denote the study-specific effect from study i ($i = 1, \ldots, k$). It is possible that the differences in the T_i are not only due to experimental error but also due to actual differences between the studies. If explicit information on the source of heterogeneity is available, it may be possible to account for study-level differences in the modeling, but in the absence of such information a useful and simple way to express study-level heterogeneity is with a multistage or hierarchical model. Let σ_i^2 denote the within-study component of variance associated with each outcome measure T_i. The first stage of the model relates the observed outcome measure T_i to the underlying study-specific effect in the ith study, θ_i. At the second stage of the model, the θ_i's are in turn related to the overall effect μ in the population from which all the studies are assumed to have been sampled, and τ^2 is the between-study component of variance or the variance of the effects in the population. The model, presented in two equivalent forms, is

$$\text{\textit{Stage I}:} \quad \tau_i | \theta_i, \sigma_i^2 \sim N(\theta_i, \sigma_i^2) \quad \left| \begin{array}{l} T_i = \theta_i + \varepsilon_i \\ \\ \varepsilon_i \sim N(0, \sigma_i^2) \end{array} \right.$$

$$\text{\textit{Stage II}:} \quad \theta_i | \mu, \tau^2 \sim N(\mu, \tau^2) \quad \left| \begin{array}{l} \theta_i = \mu + \xi_i \\ \\ \xi_i \sim N(0, \tau^2) \end{array} \right.$$

All the ε_i and ξ_i are considered independent. From the Bayesian perspective, a number of unknown parameters, σ_i^2, μ, and τ^2 have to be estimated, and therefore probability distributions, in this context called prior distributions, must be specified for them. In effect, these latter probability specifications constitute a third stage of the model.

A key implication of the two-stage model is that there is information about a particular θ_i from all of the other study-specific effects, $\{\theta_1, \ldots, \theta_k\}$. Specifically, an estimate of θ_i is

$$\hat{\theta}_i(\tau) = (1 - B_i(\tau)) \, T_i + B_i(\tau) \, \hat{\mu}(\tau), \qquad (22.1)$$

a weighted average of the observed ith study-specific effect, T_i, and the estimate of the population mean effect, $\hat{\mu}(\tau)$, where the weights are $B_i(\tau) = \sigma_i^2/(\sigma_i^2 + \tau^2)$. Although most meta-analyses are interested in learning about μ and τ^2, it is informative to examine the behavior of equation 22.1 as a function of τ. The weighting factor $B(\tau)$ controls how much the estimate of $\hat{\theta}_i$ shrinks toward the population mean effect. When τ is taken to be 0, $B(\tau)$ is 1 and the two-stage model is equivalent to a fixed-effects model. Small values of τ describe situations in which studies can borrow strength from each other and information is meaningfully shared across studies. Very large values of τ on the other hand, imply that $B(\tau)$ is close to 0, so that not much is gained in combining the studies. The Bayesian approach to meta-analysis, particularly through the specification of a probability distribution for τ, provides a formalism for investigating how similar the studies are and for sharing information among the studies when they are not identical. To investigate the sensitivity of the posterior estimates of the study effects to τ, William Du-Mouchel and Sharon-Lise Normand (2000) recommended constructing a plot of the conditional posterior mean study effect, $\hat{\theta}_i(\tau)$, conditioned on a range of likely values of τ to see how $\hat{\theta}_i(\tau)$ varies as a function of τ. They call this a trace plot and illustrate its use as well as discuss many other practical issues in the implementation of Bayesian hierarchical models in meta-analysis.

22.4.3.1 Specification of Prior Distributions The models considered here for combining information from different studies include the assumption that the various studies are samples from a large population of possible studies and that the studies to be combined are exchangeable (Draper et al. 1993). Knowledge and assumptions about such study distributions, whether derived from experts or from ancillary sources, are incorporated into Bayesian models as prior distributions. Various choices for prior distributions and the consequences of such choices are discussed here.

In the use of Bayesian methods in practice, there is a concern that inferences based on a posterior distribution will be highly sensitive to the specification (or misspecification) of the prior distributions. A formal attempt to address this concern is the use of robust Bayesian methods. Instead of worrying about the choice of a single correct prior distribution, the robust Bayesian approach considers a range of specifications and assesses how sensitive inferences are to the different priors (see, for example, Kass and Greenhouse 1988; Spiegelhalter, Abrams, and Myles 2004). In other words, if the inferences do not change as the prior distributions vary over a wide range of possible distributions, then it is safe to conclude that the inferences are robust to the prior specifications. E. E. Kaizar and colleagues illustrated the use of Bayesian sensitivity analysis in a meta-analysis investigating the risk of antidepressant use and suicidality in children (2006).

The specification of a prior distribution for the between-study component of variance, τ^2, is often challenging. In general, an improper prior distribution, that is, one that does not have a finite integral, is a poor choice for the hierarchical models used here, because the posterior distribution may also be improper, precluding appropriate inference and model selection. One choice of prior distribution for the between study variance is to use a member of the family of distributions that is conjugate to the distribution that models the effects of the individual studies. For example, the inverse gamma distribution is the conjugate distribution for the variance of a normal distribution (see, for example, Gilks, Richardson, and Spiegelhalter 1996; for an alternative view, see Gelman 2006). With conjugate distributions, it is often possible to sample directly from the full conditional distribution simplifying software implementation (Robert 1994, 97–106). Conjugate prior distributions are appropriate in two situations. The first is when the expert opinion concerning the possible distribution of study effects can be approximately represented by some distribution within the conjugate family. Second is when there is little background information about possible study to study variability, a weakly informative prior distribution can often be chosen from the conjugate family.

A variety of nonconjugate prior distributions may also be chosen to help improved representation of prior beliefs, despite the greater computational complexity they require. The DuMouchel prior distribution (DuMouchel and Normand 2000), also called the log-logistic distribution, is diffuse but proper. It has monotonically decreasing density, and takes the form $\pi(\tau) = \sqrt{s_0}/(\sqrt{s_0} + \tau)^2$, where τ^2 is the study-to-study variance, and s_0 is a parameter relating the variances of the individual studies. DuMouchel recommends $s_0 = \sqrt{\dfrac{k}{\Sigma\, s_i^{-1}}}$, where k is the number of studies and s_i is the variance of the ith study. This choice results in the median of $\pi(\tau)$ being equal to s_0, and the first and third quartiles being equal to $\sqrt{s_0}/3$ and $3\sqrt{s_0}$ respectively. With a normal distribution for the study effects, the resulting shrinkage of the posterior estimates from the individual data values toward the mean is such that there is roughly a 25 percent probability of essentially no shrinkage (each study stands on its own), a 25 percent probability of essentially complete shrinkage (all studies have an equivalent outcome effect), and a 50 percent chance that the studies borrow strength from each other, that is, their posterior distributions represent a compromise between the individual study effect and the mean of all studies. Michael Daniels noted that the DuMouchel prior has similar properties to the uniform shrinkage prior (1999).

DuMouchel and Jeffrey Harris provided an early example of sensitivity analysis for a meta-analysis using Bayesian methods (1983). Their aim was to combine information from studies of the effects of carcinogens on humans and various animal species. Their data set consisted of a matrix (rows for species, columns for carcinogens) of the estimated slope of the dose-response lines from each experiment. They postulated a linear model for the logarithm of these slopes. In addition, they adopted a hierarchical model for the parameters of the linear model. That is, the parameters of the linear model were themselves random variables with certain specified prior distributions. In this context, an example of a sensitivity analysis that they undertook is to vary the prior distribution to see how the inferences are affected. They also used a variant of the leave-one-out method by leaving out an entire row or an entire column from their data matrix to investigate the fit of their model.

22.4.4 Other Methods

There are, of course, many other analytic methods used in research synthesis that we could consider from the viewpoint of sensitivity analysis, for example, regression analysis. Our aim in this section is to alert the reader to the need for sensitivity analysis at the data combining stage of a research synthesis. The details depend on the methods used to combine data. For meta-regression analysis,

for example, there are a number of diagnostic tools, such as residual analysis and case analysis, that can be applied (meta-regression, Van Houwelingen, Arends, and Stijnen 2002; diagnostic tools, Weisberg 1985; Hedges and Olkin 1985, chapter 12). In addition, as we have indicated, there are open research opportunities for the development of new methods for sensitivity analysis.

22.5 PUBLICATION BIAS

One criticism of meta-analysis is that the available studies may not be representative of all studies addressing the research question. Publication bias refers to the claim that studies with statistically significant results are more likely to be published than studies that do not. Robert Rosenthal imagined that these unpublished studies ended up in investigators' files, and dubbed this phenomenon the file drawer problem (1979). Many sources of this problem have been identified. One is reporting bias, which arises in published studies where authors do not report results that are not statistically significant or provide insufficient information about them (Hedges 1988). Another is retrieval bias, which is due to the inability to retrieve all of the research, whether published or not (Rosenthal 1988; for a discussion of other sources see Rothstein, Sutton, and Borenstein 2005).

22.5.1 Assessing the Presence of Publication Bias

Figure 22.4 is an example of a funnel plot which Richard Light and David Pillemer introduced for the graphical detection of publication bias (1984). A funnel plot is a scatterplot of sample size versus estimated effect size for a group of studies. Because small studies will typically show more variability among the effect sizes than larger studies, and there will be fewer of the latter, the plot should look like a funnel, hence its name. When there is publication bias against, say, studies showing small effect sizes, a bite will be taken out of a part of this plot. In figure 22.4, for example, the plot is skewed to the right, and there appears to be a sharp cutoff of effect sizes at 0.30, suggesting possible publication bias against small or negative effect sizes. Figure 22.5, on the other hand, shows a funnel plot for a meta-analysis where there is no publication bias: it is a plot of 100 simulated studies comparing treatment to control. For the ith study, there are n_i observations per group, where n^i span the range 10 to 100. The population effect size was set at 0.50. Notice the funnel

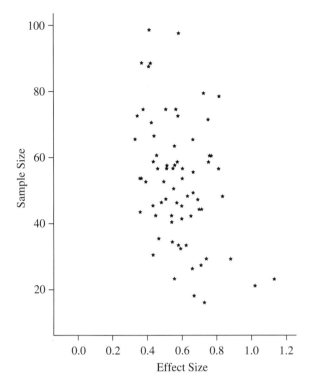

Figure 22.4 Funnel Plot, Only Studies Statistically Significant at the 0.05 Level Reported

SOURCE: Authors' compilation.

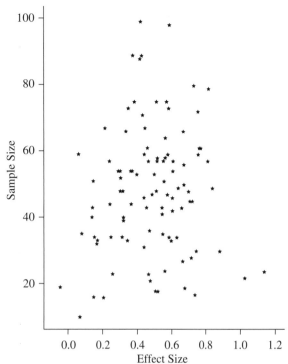

Figure 22.5 Funnel Plot for All Studies

SOURCE: Authors' compilations.

shape. The funnel plot is thus a useful visual diagnostic for informally assessing the file drawer problem.

Two simple biasing mechanisms are depicted in figures 22.4 and 22.6. The data in both are derived from that of figure 22.5. In figure 22.4, all studies that report statistically significant results at the 0.05 level are shown. In figure 22.6, significant results have an 80 percent chance of being reported, and the rest only a 10 percent chance (for illustrations of other mechanisms, see chapter 23, this volume).

The funnel plot, along with its variations, is certainly a useful visual aid for detecting publication bias (Galbraith 1988; Vandenbroucke 1988). Susan Duval and Richard Tweedie (2000a, 2000b) noted that these variants can give very different impressions of the extent of publication bias and proposed a quantitative complement to the funnel plot known as the trim-and-fill method. They used the fact that when there is no bias, the estimates should be symmetric about the true effect size. They use this sym-

metry to estimate the number of missing studies, and then impute those studies to get less biased estimates of the effect size (see, however, Terrin et al. 2003; Copas and Shi 2000). The trim-and-fill method has become rather popular, in part because of its simplicity (for further discussion, see chapter 23, this volume).

Having used this plot as a diagnostic tool to indicate the presence of publication bias, the next step is to decide just how to adjust the observed combined effect size estimate for the bias. Suppose, for the moment, that an adjustment method is agreed upon. A sensitivity analysis would find the adjusted values for a variety of possible selection mechanisms. If these values are not far from the unadjusted average, the latter is deemed robust, or insensitive to publication bias. Otherwise, the unadjusted average should not be trusted unless a subsequent search for more studies still supports it. In practice, using a single adjustment method is usually not warranted. It is better to apply a range of plausible adjustment methods to see how robust the observed value is.

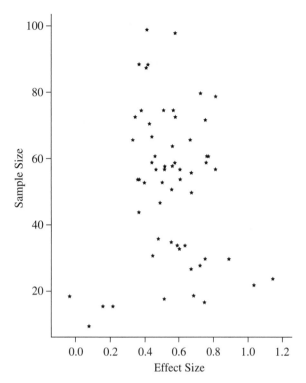

Figure 22.6 Funnel Plot for a Biased Reporting Mechanism

SOURCE: Authors' compilation.
NOTE: See text for details.

22.5.2 Assessing the Impact of Publication Bias

Publication bias can be dealt with in several ways once it has been detected, or suspected. One approach explicitly incorporates the biasing or selection mechanism into the model to get an adjusted overall estimate of the effect size. For instance, suppose that the biasing mechanism favors publication of statistically significant results, and that the primary studies report one-sided z-tests using the 0.05 level, which has a cutoff of $z = 1.645$. That is, the probability, $w(z)$, of publication of a study's results increases with its z-value. An extreme case is when only the statistically significant results are published, so that

$$w(z) = \begin{cases} 0 & \text{if } z < 1.645 \\ 1 & \text{if } z \geq 1.645. \end{cases}$$

The probability $w(z)$ is known as a weight function (Rao 1985). In this approach, the weight function is used to construct the likelihood function based on the pub-lished studies. Adjusted estimates of the effect size are then derived using standard maximum likelihood or Bayesian methods (for technical details, see the appendix and chapter 23, this volume).

Several comments about this approach are in order. First, it is flexible in the sense that it can also be used to model other types of biases, such as retrieval bias. For example, if p_1 and p_2 are the probabilities of retrieving a published and an unpublished study, respectively, the probability of a study entering a meta-analysis at all may be modeled as $p_1 w(z)$ for a published study and as $p_2 w(z)$ for an unpublished one. Several empirical studies of the retrieval process have been done, and may be useful in providing rough guidelines for determining ranges for p_1 and p_2 (Chalmers, Frank, and Reitman 1990; Dickersin et al. 1985). Next, because the weight function is typically not known, it is useful to do the analysis of the weighted models with various plausible families of weight functions. If the resulting effect size estimates do not change much, such a sensitivity analysis will give more credence to the inferences based on the retrieved studies (Iyengar and Greenhouse 1988).

There are several theoretical and empirical studies of weight functions. Satish Iyengar and Ping Zhao (1994) studied how the MLE for the normal and student t distri-butions changes as the weight function varies. In particular, they showed that for commonly used weight function families, the MLE varies monotonically with a parameter that controls the amount of selection bias within that family, so computing the range of the estimates is easy. Other studies of this approach include Iyengar, Paul Kvam, and Singh, who considered the effect of weight functions on the Fisher information of the effect size (1998). This in turn dictates the accuracy of the MLE; and Larry Hedges and Jack Vevea, who provided extensive empirical stud-ies of the use of weighted distributions when selection is related both to the estimated effect size and its estimated standard deviation (2005).

We end this discussion with a numerical example il-lustrating the use of weighted distributions for the data in table 22.1. Suppose that all studies use a two-sided z-test at the 0.05 level, which has cutoffs -1.96 and 1.96. Suppose further that the bias is such that a study with a significant result will definitely be published, and that other studies will be published with some probability less than 1. A weight function that models this situation is

$$w(z;b) = \begin{cases} e^{-\beta} & \text{if } -1.96 < z < 1.96 \\ 1 & \text{else.} \end{cases}$$

This choice of *w* is actually a member of a family of weight functions parameterized by a nonnegative quantity ß, indicating uncertainty about the probability of publication when an effect size is not significant; the larger the value of ß, the greater the degree of publication bias. Maximum likelihood estimation of the two parameters δ, the true effect size, and ß, the bias parameter yields -0.006 (SE $= 0.033$) and 0.17 (SE $= 0.192$), respectively. Because the estimate of the true effect size based on the unweighted model is -0.008, we conclude that the result is not sensitive to this choice of publication bias mechanism. The estimate of ß indicates that the probability of publishing a statistically nonsignificant result is about 0.85. Note, however, that the standard error for the estimate of ß is relatively large, indicating that the data provide little information about that parameter.

22.5.3 Other Approaches

Other approaches to sensitivity analysis for dealing with publication bias include the fail-safe sample size approach of Robert Rosenthal (1979). This uses combined significance levels rather than effect size estimates. The fail-safe N is then the smallest number of studies showing nonsignificant studies to overturn the result based on the retrieved studies (for a critique of and for variations on this approach, see Iyengar and Greenhouse 1988; for a Bayesian approach, see Bayarri 1988). Finally, John Copas and J. Q. Shi proposed a latent variable that models the publication process: the variable depends linearly on the reciprocal of the study's standard error, it is correlated with the effect size estimate, and allows the study to be seen when it is positive (2000). They then varied the slope and intercept of the model which tune the degree of publication bias to assess the effects on the overall effect size.

22.6 SUMMARY

Sensitivity analysis is a systematic approach to address the question, "What happens if some aspect of the data or the analysis is changed?" Because at every step of a research synthesis decisions and judgments are made that can affect the conclusions and generalizability of the results, we believe that sensitivity analysis has an important role to play in research synthesis. For the results of a meta-analysis to be convincing to a wide range of readers, it is important to assess the effects of these decisions and judgments. One way to do this is to demonstrate that a broad range of viewpoints and statistical analyses yield

substantially the same conclusions. Sensitivity analysis and diagnostics as described in this chapter provide a conceptual framework and the technical tools for doing just that. On the other hand, a meta-analysis is useful even if the conclusions are sensitive to such alternatives, for then the sensitivity analysis points to the features of the problem that deserve further attention.

22.7 APPENDIX

Suppose that $w(z)$ is interpreted as the probability of publishing a study when the effect size estimate from that study is z. If an effect size estimate is normally distributed, the published effect size estimate from a study is a random variable from the weighted normal probability density

$$f(z;\delta,w) = \frac{\phi(z-\delta;\sigma^2/n)w(x)}{A(\delta,\sigma^2/n;w)}$$

where $f(z;s^2)$ is the density of the normal distribution with mean zero and variance s^2,

$$A(\delta,\sigma^2;w) = \int_{-\infty}^{\infty} \phi(t-\delta;\sigma^2)w(t)\,dt$$

is the normalizing constant, and n and δ are the sample size and the true effect size for that study, respectively. Inference about the effect size can then be based on this weighted model, using maximum likelihood or Bayes estimates. Because the weighted model is also a member of an exponential family, an iterative procedure for computing the estimate is relatively easy. Often, primary studies report t statistics with their degrees of freedom; in that case, a noncentral t density replaces the normal density above, and the computations become considerably more difficult (for an example involving the combination of t statistics, see Iyengar and Greenhouse 1988). When the degrees of freedom within each primary study are fairly large, however, the normal approximation is quite accurate.

22.8 REFERENCES

Bayarri, Mary J. 1988. "Discussion." *Statistical Science* 3(1): 109–35.

Borenstein, Michael, and Hannah Rothstein. 1999. *Comprehensive MetaAnalysis: A Computer Program for Research Synthesis.* Englewood, N.J.: Biostat, Inc. http://www. meta-analysis.com.

Carlin, John B. 1992. "Meta-Analysis for 2×2 Tables: A Bayesian Approach." *Statistics in Medicine* 11(2): 141–58.

Chalmers, Thomas, C. Frank, and D. Reitman. 1990. "Minimizing the Three Stages of Publication Bias." *Journal of the American Medical Association.* 263(10): 1392–95.

Copas, John B. and Shi, Jian Qing. 2000. "Meta-Analysis, Funnel Plots and Sensitivity Analysis." *Biostatistics* 1(3): 247–62.

Daniels Michael J. 1999. "A Prior for the Variance in Hierarchical Models." *The Canadian Journal of Statistics* 27(3): 567–78.

Dempster, Arthur, and Louise M. Ryan. 1985. "Weighted Normal Plots." *Journal of the American Statistical Association* 80(392): 845–50.

DerSimonian, Rebecca, and Nan Laird. 1986. "Meta-Analysis in Clinical Trials." *Controlled Clinical Trials* 7(3): 177–88.

Dickersin, Kay, P. Hewitt, L. Mutch, Iain Chalmers, and Thomas Chalmers. 1985. "Perusing the Literature: Comparison of MEDLINE Searching with a Perinatal Trials Database." *Controlled Clinical Trials* 6(4): 306–17.

Draper, David, James S. Hodges, Colin L. MallowsL, and Daryl L. Pregibon. 1993. "Exchangeability and Data Analysis (with discussion)." *Journal of the Royal Statistical Society, A* 156: 9–37.

DuMouchel, William, and Jeffrey Harris. 1983. "Bayes Methods for Combining the Results of Cancer Studies in Humans and other Species." *Journal of the American Statistical Society* 78(382): 293–315.

DuMouchel, William, and Sharon-Lise Normand. 2000. "Computer-Modeling and Graphical Strategies for Meta-Analysis." In *Meta-Analysis in Medicine and Health Policy*, edited by Dalene Stangl and Don Berry. New York: Marcel Dekker.

Duval, Sue and Richard Tweedie. 2000a. "A Non-Parametric 'Trim and Fill' Method of Assessing Publication Bias in Meta-Analysis." *Journal of the American Statistical Association* 95(449): 89–98.

Duval, Susan, and Richard Tweedie. 2000b. "Trim and Fill: A Simple Funnel Plot Based Method of Testing and Adjusting for Publication Bias in Meta-Analysis." *Biometrics* 56(2): 455–63.

Fienberg, Stephen, Matthew Johnson, and Brian Junker. 1999. "Classical Multilevel and Bayesian Approaches to Population Size Estimation using Multiple Lists." *Journal of the Royal Statistical Society*, Ser. A, 162(3): 383–405.

Galbraith, R. 1988. "A Note on Graphical Presentation of Estimated Odds Ratios from Several Clinical Trials." *Statistics in Medicine* 7(8): 889–94.

Gaver, Donald P., David Draper, Prem K. Goel, Joel B. Greenhouse, Larry V. Hedges, Carl N. Morris, and Christine Waternaux. 1992. *Combining Information: Statistical Issues and Opportunities for Research.* Washington, D.C.: National Academy Press.

Gelman, Andrew. 2006. "Prior Distributions for Variance Parameters in Hierarchical Models." *Bayesian Analysis* 1(3): 515–33.

Gilks, Walter R., Sylvia Richardson, and David J. Spiegelhalter, eds. 1996. *Markov Chain Monte Carlo in Practice.* London: Chapman & Hall.

Gilks, Walter R., A. Thomas, and David J. Spiegelhalter. 1994. "A Language and Program for Complex Bayesian Modeling." *The Statistician: Journal of the Institute of Statisticians* 43(1): 169–77.

Greenhouse, Joel B., Davida Fromm, Satish Iyengar, Mary A. Dew, Audrey Holland, and Robert Kass. 1990. "The Making of a Meta-Analysis: A Quantitative Review of the Aphasia Treatment Literature." In *The Future of Meta-Analysis*, edited by Kenneth W. Wachter and Miron L. Straf. New York: Russell Sage Foundation.

Hedges, Larry V. 1988. "Directions for Future Methodology." In *The Future of Meta-Analysis.* edited by Kenneth W. Wachter and Miron L. Straf. New York: Russell Sage Foundation.

———. 1998. "Bayesian Approaches to Meta-Analysis." In *Recent Advances in the Statistical Analysis of Medical Data*, edited by Brian Everitt and Graham Dunn. London: Edward Arnold.

Hedges, Larry V., and Ingram Olkin. 1985. *Statistical Methods for Meta-Analysis.* Academic Press. New York.

Hedges, Larry V., and Jack Vevea. 2005. "Selection Model Approaches to Publication Bias." In *Publication Bias in Meta-Analysis: Prevention, Assessment and Adjustments*, edited by Hannah Rothstein, Alexander J. Sutton, and Michael Borenstein. New York: John Wiley & Sons.

Hedges, Larry V., Rose Giaconia, and Nathaniel Gage. 1981. *The Empirical Evidence on the Effectiveness of Open Education.* Stanford, Calif.: Stanford University.

Higgins, Julian, and Simon Thompson. 2002. "Quantifying Heterogeneity in a Meta-Analysis." *Statistics in Medicine* 21(11): 1539–58.

Hill, Bruce M. 1965. "Inference about Variance Components in the One-Way Model." *Journal of the American Statistical Association* 60(311): 806–25.

Iyengar, Satish, and Joel B. Greenhouse. 1988. "Selection Models and the File-Drawer Problem (with discussion)." *Statistical Science* 3(1): 109–35.

Iyengar, Satish, Paul Kvam, and Harshinder Singh. 1999. "Fisher Information in Weighted Distributions." *Canadian Journal of Statistics* 27(4): 833–41.

Iyengar, Satish, and Ping L. Zhao. 1994. "Maximum Likelihood Estimation for Weighted Distributions." *Probability and Statistics Letters* 21: 37–47.

Kaizar, Eloise E., Joel B. Greenhouse, Howard Seltman, Kelly Kelleher. 2006. "Do Antidepressants Cause Suicidality in Children? A Bayesian Meta-Analysis (with discussion)." *Clinical Trials: Journal of the Society for Clinical Trials* 3(2): 73–98.

Kass Rob, and Joel B. Greenhouse. 1989. "Comment on 'Investigating therapies of potentially great benefit: ECMO' by J. Ware." *Statistical Science* 4(4): 310–17.

Koopmans, Lambert. 1987. *Introduction to Contemporary Statistical Methods.* Boston, Mass.: Duxbury Press.

Light, Richard, and David Pillemer. 1984. *Summing Up.* Cambridge, Mass.: Harvard University Press.

Normand, Sharon-Lise. 1999. "Meta-Analysis: Formulating, Evaluating, Combining, and Reporting." *Statistics in Medicine* 18(3): 321–59.

Rao, C. Radhakrishna. 1985. "Weighted Distributions Arising out of Methods of Ascertainment: What Population Does a Sample Represent?" In *A Celebration of Statistics,* edited by Anthony C. Atkinson and Stephen E. Fienberg. New York: Springer.

Raudenbush, Stephen W., and Anthony S. Bryk. 2002. *Hierarchical Linear Models: Applications and Data Analysis Methods,* 2nd ed. Newbury Park, Calif.: Sage Publications.

Robert, Christian P. 1994. *The Bayesian Choice.* New York: Springer-Verlag.

Rosenthal, Robert. 1979. "The 'File-Drawer' Problem and Tolerance for Null Results." *Psychological Bulletin* 86: 638–41.

———. 1988. "An Evaluation of Procedures and Results." In *The Future of Meta-Analysis,* edited by Kenneth W. Wachter and Miron L. Straf. New York: Russell Sage Foundation.

Rothstein, Hannah, Alexander J. Sutton, and Michael Borenstein, eds. 2005. *Publication Bias in Meta-Analysis: Prevention, Assessment and Adjustments.* New York: John Wiley & Sons.

Spiegelhalter, David J., Keith R. Abrams, and Jonathan P. Myles. 2004. *Bayesian Approaches to Clinical Trials and Health-Care Evaluation.* Chichester: John Wiley & Sons.

Stangl, Dalene K., and David A. Berry, eds. 2000. *Meta-Analysis in Medicine and Health Policy.* New York: Marcel Dekker.

Sutton, Alexander J., Keith R. Abrams, David R. Jones, Trevor A. Sheldon, and Fujian Song. 2000. *Methods for Meta-analysis in Medical Research.* Chichester: John Wiley & Sons.

Terrin, Norma, Chris Schmid, Joe Lau, and Ingram Olkin. 2003. "Adjusting for Publication Bias in the Presence of Heterogeneity." *Statistics in Medicine* 22(13): 2113–26.

Van Houwelingen, Hans C., Lidia R. Arends, and Theo Stijnen. 2002. "Advanced Methods in Meta-Analysis: Multivariate Approach and Meta-Regression." *Statistics in Medicine* 21(4): 589–624.

Vandenbroucke, Jan P. 1988. "Passive Smoking and Lung Cancer: A Publication Bias?" *British Medical Journal* 296: 391–92.

Wachter, Kenneth, and Miron Straf. 1990. *The Future of Meta-Analysis.* New York: Russell Sage Foundation.

Weisberg, Sanford. 1980. *Applied Linear Regression.* New York: John Wiley & Sons.

———. 1985. *Applied Linear Regression,* 2nd. ed. Chichester: John Wiley & Sons.

23

PUBLICATION BIAS

ALEX J. SUTTON

CONTENTS

23.1 INTRODUCTION

That the published scientific literature documents only a proportion of the results of all research carried out is a perpetual concern. Further, there is good direct and indirect evidence to suggest that the unpublished proportion may be systematically different from the published, since selectivity may exist in deciding what to publish. (Song et al. 2000; Dickersin 2005) For example, researchers may choose not to write up and submit studies with uninteresting findings (such as nonstatistically significant effect sizes), or such studies may not be accepted for publication. Even if a study is published, there may be selectivity in which aspects are written up. For example, significant outcomes (Chan et al. 2004) or subgroups may be given precedent over nonsignificant ones. There is also evidence to suggest that studies with significant outcomes are published more quickly than those with nonsignificant outcomes (Stern and Simes 1997). Additionally, there are concerns that research with positive and statistically significant results are published in more prestigious places and cited more times, making it more visible, and hence easier to find. Indeed, the publication process should be thought of as a continuum and not a dichotomy (Smith 1999). In keeping with the previous literature, these (whole) publication and related biases will be referred to simply as publication bias throughout the chapter, unless a more specific type of bias is being discussed, though dissemination bias is perhaps a more accurate name for the collection (Song al. 2000).

In areas where any such selectivity exists, a synthesis based on only the published results will be biased. Publication bias is therefore a major threat to the validity of meta-analysis and other synthesis methodologies. However, it should be pointed out that this is not an argument against systematic synthesis of evidence, because publication bias affects the literature and hence any methodology used to draw conclusions from the literature (using one or more studies) will be similarly threatened. Thus, publication bias is fundamentally a problem with the way individual studies are conducted, disseminated, and interpreted. By conducting a meta-analysis, we can at least attempt to identify and estimate the likely effect of such bias by considering the information contained in the distribution of effect sizes from the available studies. This is the basis of the majority of statistical methods described later.

There is agreement that prevention is the best solution to the problem of selectively reported research. Indeed, with advances in electronic publishing making the presentation of large amounts of information more economically viable than traditional paper-based publishing methods, there is some hope that the problem will diminish, if not disappear. No evidence yet exists to indicate this is the case, however, nor is this a solution to the suppression of information due to vested economic interests (Halpern and Berlin 2005).

The use of prospective registries of prospectively studies for selecting studies to be included in systematic reviews has been suggested (Berlin and Ghersi 2005), because it provides an unbiased sampling frame guaranteeing the elimination of publication bias (relating to the suppression of whole studies at least). However, trial registration, per se, does not guarantee availability of data and an obligation to disclose results in a form that is accessible to the public is also required. Further, registries exist for randomized controlled trials in numerous medical areas. Such a solution will not be feasible, however, even in the long term, for some forms of research, including that relating to analysis of (routine) observational data, where the notion of a study that can be registered before analysis may be nonexistent. The notion of prospectively designing multiple studies with the intention of carrying out a meta-analysis in the future has also been put forward as a solution to the problem (Berlin and Ghersi 2005), but again, this may be difficult to orchestrate in some situations.

Carrying out as comprehensive a search as possible when obtaining literature for a synthesis will help minimize the influence of publication bias. In particular, this may involve searching the grey literature for studies not formally published (see chapter 6 in this volume; Hopewell, Clarke, and Mallett 2005). Since the beginning of the Internet, the feasibility and accessibility of publication by means other than commercial publishing houses has greatly increased. This implies that the importance of searching the grey literature may increase in the future.

Despite researchers' best efforts, at least in the current climate, alleviation of the problem of publication bias may not be possible in many areas of science. In such instances, graphical and statistical tools have been developed to address publication bias within a meta-analysis framework. The remainder of this chapter provides an overview of these methods, which aim to detect and adjust for such biases. It should be pointed out that if a research synthesis does not contain a quantitative synthesis (for example, if results are considered too heterogeneous, or the data being synthesized are not quantitative) then, though publication bias may still be a problem, methods to deal with it are still very limited.

23.2 MECHANISMS

Discussion in the literature about the precise nature of the mechanisms that lead to suppression of whole studies and other forms of publication bias is considerable. If these mechanisms could be accurately specified and quantified, then, just like any bias, the appropriate adjustments to a meta-analytic dataset to counteract might be fairly straightforward. Unfortunately, measuring such effects directly, globally, or in specific datasets is difficult and, of course, the mechanisms may vary with dataset and subject area.

What is generally agreed is that the statistical significance, effect size, and study size could all influence the likelihood of a study being published, although this list is probably not exhaustive (Begg and Berlin 1988). The sections that follow outline methods to identify and adjust for publication bias. It is important to note that these assume different underlying mechanisms for publication bias—some that suppression is only a function of effect size, others that it is only a function of statistical significance (whether this is assumed one or two sided is open to debate), and still others that it is a function of these or other factors, such as origin of funding for a study and the like.

Although the primary goal here is to provide a pragmatic and up-to-date overview of the statistical methods to address publication bias in meta-analysis, an emphasis has been placed on highlighting the assumptions made regarding the underlying suppression mechanisms assumed by each method. It is hoped that insight into how the different methods relate to each other can be facilitated through this approach. That the actual mechanisms are usually unknown makes being absolutist about recommendations on use of the different methods difficult. This issue is considered further in the discussion.

23.3 METHODS FOR IDENTIFYING PUBLICATION BIAS

23.3.1 The Funnel Plot

The funnel plot is generally considered a good exploratory tool for investigating publication bias and, like the more commonly used forest plot, for a visual summary of a meta-analytic dataset (Sterne, Becker, and Egger 2005;

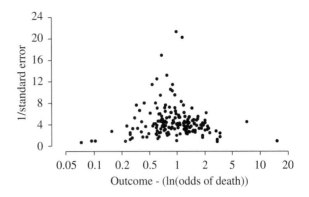

Figure 23.1 Symmetric Funnel Plot, Overall Mortality
SOURCE: Bown et al. 2002.

see chapter 26, this volume). It is essentially a scatter plot of a measure of study size against a measure of effect sizes. The expectation is that is should appear symmetric with respect to the distribution of effect sizes and funnel shaped if no bias is present. The premise for this appearance is that effect sizes should be evenly distributed (that is, symmetric) around the underlying true effect size, with more variability in the smaller studies than the larger ones because of the greater influence of sampling error (that is, producing a narrowing funnel shape as study precision increases). If publication bias is present, we might expect some suppression of smaller, unfavorable, and nonsignificant studies that could be identified by a gap in one corner of the funnel and hence could induce asymmetry in the plot.

Example plots are provided in figures 23.1 and 23.2. Note that in these plots the outcome measure is plotted on the x-axis, the most common way such plots are presented, though it goes against the convention that the unknown quantity is plotted on the y-axis. Funnels in both orientations exist in the literature. These are real data relating to total mortality and mortality during surgery following ruptured abdominal aortic aneurysm repair from published (observational) patient case-series (Bown et al. 2002; Jackson, Copas, and Sutton 2005). They are included here because they exemplify a reasonably symmetrical funnel plot (figure 23.1) and a more asymmetric one (figure 23.2) with a lack of small studies with small effects in the bottom right-hand corner. Also interesting is that this would appear to be a good example of outcome reporting bias given that only seventy-seven studies report mortality during surgery (figure 23.2) out of the

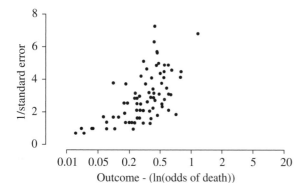

Figure 23.2 Asymmetric Funnel Plot, Mortality During Surgery

SOURCE: Author's compilation.

171 reports included in figure 23.1 for total mortality. The gap at the bottom right of the surgery mortality plot (figure 23.2) suggests that studies in which mortality during surgery had high rates were less likely to report this outcome than studies with lower rates.

For this example, because the data are noncomparative, statistical significance of outcome is not defined and hence it is assumed that suppression is related to magnitude of outcome (example funnel plots for comparative data where statistical significance potentially influences publication are considered below). It is important to emphasize that funnel plots are often difficult to interpret. For example, close scrutiny of figures 23.1 and 23.2 suggests that, though producing a reasonably symmetric funnel in outline, the density of points is not consistent within the funnel, a higher density being apparent on the right side in figure 23.1, particularly for the lower third of the plot, suggesting some bias.

It is interesting to note that study suppression caused by study size, effect size, or statistical significance (one sided), either individually or in combination, could produce an asymmetric funnel plot. A further possibility is that two-sided statistical significance suppression mechanisms (that is, significant studies in either direction are more likely to be published than those that are not) could create a tunnel or hole in the middle of the funnel, particularly when the underlying effect size is close to 0.

Certain issues need to be considered when constructing and interpreting funnel plots. Generally, the more studies in a meta-analysis, the easier it is to detect gaps and hence potential publication bias. If numerous studies

are included, density problems may arise (see figures 23.1 and 23.2). The spread of the size of the studies is another important factor. A funnel would be difficult to interpret, for example, if all the studies were a similar size. Ultimately, any assessment of whether publication bias is present using a funnel plot is subjective (Lau et al. 2006). A recent empirical evaluation also suggests that interpretation of such a plot can be limited (Terrin, Schmid, and Lau 2005).

There has been debate over the most appropriate axes for the funnel plot (Sterne and Egger 2001), particularly with respect to the y-axis (the measure of study precision), with sample size, variance (and its inverse) and standard error (and its inverse) all being options. This choice can affect the appearance of the plot considerably. For instance, if the variance or standard error is used, then the top of the plot is compressed while the bottom of the plot is expanded compared to when using inverse variance, inverse standard error, or sample size. This has the effect of giving more space to the smaller studies in which publication bias is more likely. Example comparative plots have been published elsewhere (Sterne and Egger 2001). If variance or standard error or their inverses are used, it is possible to construct expected 95 percent confidence intervals around the pooled estimate that form guidelines for the expected shape of the funnel (under a fixed effect assumption) that can aid the interpretation of the plot. Such guidelines should be thought of as approximate because they are constructed around the meta-analytic pooled estimate, that may itself be biased because of publication bias. The assistance in interpreting the plot is in indicating regions where studies are to be expected if no suppression exists. These guidelines can also help to indicate how heterogeneous the studies are (see chapter 24, this volume) by observing how dispersed the studies are with respect to the guidelines.

It is important to appreciate that funnel plot asymmetry may not be related to publication bias at all, but to other factors instead. Any factor associated both with study size and effect size could potentially confound the observed relationship. For example, smaller studies may be carried out under less rigorous conditions than larger studies. These poorer quality smaller studies may be more prone to bias and thus result in exaggerated effect size estimates. Conversely, smaller studies could be conducted under more carefully controlled experimental conditions than the larger studies, resulting in differences in effect sizes. Further, there may be situations where a higher intensity of the intervention were possible for the smaller studies

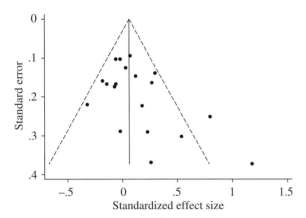

Figure 23.3 Funnel Plot, Teacher Expectancy on Pupil IQ, 95% Guidelines

SOURCE: Author's compilation.

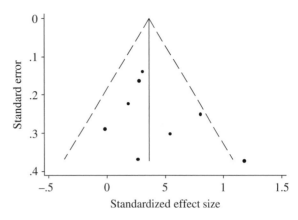

Figure 23.4 Funnel Plot, Teacher Expectancy on Pupil IQ, prior contact One Week or Less

SOURCE: Author's compilation.

that lead to the true effect sizes in the smaller studies being more efficacious than those for the larger studies. If situations like these exist, then there is a relationship between-study size and effect size that would cause a funnel plot of the studies to be asymmetric, which could be wrongly diagnosed as publication bias. A graphical way of exploring potential causes of heterogeneity in a funnel plot would be to plot studies with different values for a covariate of interest (for example, study quality) using a different symbol or using multiple funnel plots of subgrouped data and exploring whether this could explain differences between study results. For some outcome measures, such as odds ratios, the estimated effect size and its standard error are correlated and the more extreme the effect size is, the stronger the correlation. This implies that the funnel plot on such a scale will not be symmetrical even if no publication bias is present. The induced asymmetry, however, is quite small for small and moderate effect sizes (Peters et al. 2006).

Example: Effects of Teacher Expectancy on Pupil IQ. Figure 23.3 presents a funnel plot with the approximate expected 95 percent confidence intervals around the pooled estimate for the expected shape of the funnel under a fixed effect assumption plotting standard error on the y-axis. The guidelines are straight lines and thus the theoretical underlying shape of the funnel is also straight sided. Because the majority of the points (seventeen of nineteen) lie within the guidelines, there is little evidence of heterogeneity between studies. There is, though, some

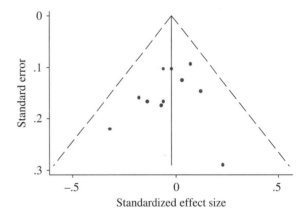

Figure 23.5 Funnel Plot, Teacher Expectancy on Pupil IQ, prior contact More Than One Week

SOURCE: Author's compilation.

suggestion of a gap in the bottom left-hand corner of the funnel, making it asymmetric, that is, possibly a symptom of publication bias. However, the weeks of teacher-student contact before expectancy induction is a known modifier for this dataset, where effect sizes are systematically higher in the studies with lower previous contact (see chapter 16, this volume). Figures 23.4 and 23.5) present funnel plots of the subgroups for prior contact of one week or less and more than one week respectively. These funnels are centered around quite different effect sizes and there is little evidence of asymmetry in either plot, though splitting the data reduces numbers of studies in each plot, which does hamper interpretation. Hence it would appear that asymmetry in figure 23.3 was probably caused by confounding by the moderator, prior contact time, rather than publication bias (Sterne et al. 2005).

In the next section, we present formal statistical tests of asymmetry in a funnel plot. These get around some of the issues relating to subjectivity of assessment of funnel plots and induced asymmetry on some outcome scales. They are all, however, tests for asymmetry and not publication bias per se given that they do not (automatically) adjust for confounding variables.

23.3.2 Statistical Tests

All the tests described provide a formal statistical test for asymmetry of a funnel plot, and should be considered complementary to it, and are based on the premise of testing for an association between observed effect size and (a function of) its standard error or study sample size.

23.3.2.1 Nonparametric Correlation Test   The rank correlation test examines the association between effect estimates and their variances as outlined (Begg and Mazumdar 1994). Define the standardized effect sizes of the k studies to be combined to be

$$\overline{T}_i^* = (T_i - \overline{T}_.)/(\tilde{v}_i^*)^{1/2}, \qquad (23.1)$$

where T_i and v_i are the estimated effect size and sampling variance from the ith study, and the usual fixed-effect estimate of the pooled effect is

$$T_. = \left(\sum_{j=1}^{k} v_j^{-1} T_j\right)\bigg/\sum_{j=1}^{k} v_j^{-1}, \qquad (23.2)$$

and

$$\tilde{v}_i^* = v_i - \left(\sum_{j=1}^{k} v_j^{-1}\right)^{-1}, \qquad (23.3)$$

which is the variance of $(T_i - \overline{T}_.)$.

The test is based on deriving a rank correlation between T_i^* and $\tilde{v}_i^*$, achieved by comparing the ranks of the two quantities. Having assigned each T_i^* and $\tilde{v}_i^*$ a rank (largest value gets rank 1 and so on), it is then necessary to evaluate all $k(k-1)/2$ possible pairs of the k studies. Defining the number of all possible pairings in which one factor is ranked in the same order as the other as P, (that

is the ranks of both T_i^* and $\tilde{v}_i^*$ are higher or lower for one study compared to the other), and the number in which the ordering is reversed as Q, (that is the rank of T_i^* is higher than that of the paired study while the rank of $\tilde{v}_i^*$ is lower, or vice versa). A normalized test statistic is obtained by calculating

$$Z = (P - Q)/[k(k-1)(2k+5)/18]^{1/2}, \qquad (23.4)$$

which is the normalized Kendall rank correlation test statistic and can be compared to the standardized normal distribution. Any effect-size scale can be used as long as it is assumed distributed asymptotic normal.

The test is known to have low power for meta-analyses that include few studies (Sterne, Gavaghan, and Egger 2000). From comparisons between this and the linear regression test described later (Sterne, Gavaghan, and Egger 2000), it would appear that the linear regression test is more powerful than the rank correlation test, though results of the two tests can sometimes be discrepant.

Example. Effects of Teacher Expectancy on Pupil IQ. Applying the rank correlation test to this dataset produces a test statistic of $Z = 1.75$, leading to $p = 0.08$. Although not formally significant at the 5 percent level, such a result should—on the basis of the known low power of the test—be treated with concern. However, because such a test does not take into account the modifier (previous contact with student), such an analysis is of limited use in this context.

23.3.2.2 Linear Regression Tests   An alternative parametric test based on linear regression has been described by Matthias Egger and his colleagues (1997). This regresses the standard normal deviate, z_i (i.e $z_i = T_i/\sqrt{v_i}$), against $1/\sqrt{v_i}$:

$$z_i = \beta_0 + \beta_1(1/\sqrt{v_i}) + \varepsilon_i, \qquad (23.5)$$

Where $\varepsilon_i \sim N(0, \sigma^2)$. This regression can be thought of as fitting a line to Galbraith's radial plot, in which the fitted line is not constrained to go through the origin (1994; see also chapter 22, this volume). The intercept provides a measure of asymmetry, and the standard two-sided test of whether $\beta_0 = 0$ (that is, $\beta_0/se(\beta_0)$ compared to a t-distribution with $n-2$ degrees of freedom) is Egger's test of asymmetry. A negative intercept indicates that smaller studies are associated with bigger effects. This un-weighted regression is identical to the weighted regression:

$$T_i = \beta_1 + \beta_0\sqrt{v_i} + \varepsilon_i \cdot \sqrt{v_i}, \qquad (23.6)$$

with weights $1/v_i$. Thus, it is possible to plot this regression line on a funnel plot. Note that this model is neither the standard fixed nor random effect meta-regression model commonly used (Thompson and Sharp 1999; see also chapter 16, this volume). Here the error term includes a multiplicative overdispersion parameter rather than assuming the usual additive between-study variance component to account for heterogeneity. There would appear to be no theoretical justification for this model over the more commonly used meta-regression formulations.

Egger's test has highly inflated type 1 errors in some circumstances when binary outcomes are considered, at least in part because of the correlation between estimated (log) odds ratios and their standard errors discussed earlier (Macaskill, Walter, and Irwig 2001; Schwarzer, Antes, and Schumacher 2002; Peters, Sutton 2006). This has led to a number of modified tests being developed for binary outcomes (Macaskill, Walter, and Erwig 2001; Harbord, Egger, and Sterne 2006; Peters et al. 2006; Schwarzer, Antes, and Schumacher 2007) and more generally (Stanley 2005). A comparative evaluation of these modified tests is required before guidance can be given on which test is optimal in a given situation.

If a meta-analysis dataset displays heterogeneity that can be explained by including study-level covariates, these should be accounted for before an assessment of publication bias is made because their influence can distort a funnel plot and thus any statistical methods based on it, as discussed earlier. One way of doing this is to extend the regression models to include study-level covariates. Evaluating the performance of such an approach is ongoing work. Alternatively, if covariates are discrete, separate assessments can be made for each grouping. Splitting the data in this way will reduce the power of the tests considerably, however.

Use of these regression tests is becoming commonplace in the medical literature but is much less common in other areas, including the social sciences. Although these tests are generally more powerful than the nonparametric alternative, they still have low power for meta-analyses of small to moderate numbers of studies. Because these are typical in the medical literature, caution should be exerted in interpreting nonsignificant test results. The tests will have much improved performance characteristics with larger numbers of studies, as typically found in the social sciences. Regardless of the number of studies in the meta-analysis, caution should also be exerted when interpreting significant test results because,

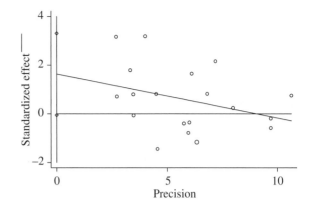

Figure 23.6 Egger's Regression Line on Galbraith Plot, Teacher Expectancy on Pupil IQ

SOURCE: Author's compilation.

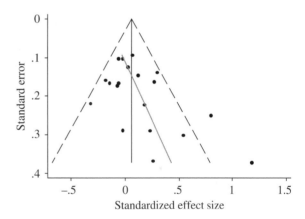

Figure 23.7 Egger's Regression Line on Funnel Plot, Teacher Expectancy on Pupil IQ

SOURCE: Author's compilation.

as discussed, detection of funnel asymmetry does not necessarily imply publication bias.

Example: Effects of Teacher Expectancy on Pupil IQ. The Egger's regression test is applied to the teacher expectancy example (and because the outcome is measured on a continuous scale the modified alternatives are not applicable). Figures 23.6 and 23.7 are two graphical representations of the fitted regression line on a Galbraith plot and funnel plot, respectively. On the Galbraith plot, inferences are made on whether the intercept is different from 0, though it is whether the gradient of the line is different from 0 that is of interest on the funnel plot (though, of course, estimation is equivalent). The p-value for this test is 0.057. As for the rank-based test discussed earlier, this does not take into account the known modifier, prior contact time, and hence is of limited value. However further covariates can be included in the regression model allowing known confounders such as this to be adjusted for. In this example, it would be possible to include prior contact time as either a continuous covariate, as it was recorded, or, dichotomized, as done to plot the subgrouped funnels in figures 23.4 and 23.5. A scatter plot of effect size against prior contact time suggests a nonlinear relationship, and hence prior contact time is dichotomized, as before, for simplicity, and included in the regression model. When this is done the p-value for the test statistic increases to 0.70, removing a lot of the suspicion of publication bias, though caution should be exerted that reduction in significance is not due to collinearity of covariates (Sterne and Egger 2005).

It is important to point out that testing for publication bias in no way deals with the problem if bias exists. Indeed, what should be concluded if a significant asymmetry test result is obtained for a meta-analysis? Clearly caution in the conclusions of the meta-analysis can be expressed because the pooled estimate may be biased. Further, whether or not the asymmetry is caused by publication bias, the usual meta-analysis fixed and random effect modeling assumptions are probably violated (see chapter 14, this volume). If funnel symmetry exists for studies on a certain topic today, chances are that asymmetry will not go away in the future. If it does, it is likely that it will be completely due to beneficial or statistically significant studies being published more quickly than those with unfavorable or nonsignificant results—a phenomena sometimes called pipeline bias effects. If the aim of a meta-analysis is ultimately to inform policy related decisions, then simply flagging a meta-analysis as being potentially biased is not ultimately helpful. The next section outlines methods for assessing the impact of possible publication bias on a meta-analysis and examines methods for adjusting for such bias.

23.4 METHODS FOR ASSESSING THE IMPACT OF PUBLICATION BIAS

Here we examine three approaches to assessing the impact of publication bias on a meta-analysis. The first provides an estimate of the number of missing studies that would need to exist to overturn the current conclusions.

The second and third outline nonparametric and parametric methods of adjusting the pooled meta-analytic estimate.

23.4.1 Fail Safe N

This simple method was one of the first described to address the problem of publication bias. It considers the question of how many new studies averaging a null result are required to bring the overall treatment effect to nonsignificance (Rosenthal 1979).

The method is based on combining the normal z-scores (the Z_i) corresponding to the p-values observed for each study. The overall z-score can be calculated by:

$$Z = \sum_{i=1}^{k} Z_i \bigg/ \sqrt{k}, \qquad (23.7)$$

where k is the number of studies. This sum of z-scores is a z-score itself and the combined z-scores are considered significant if $Z > Z_{\alpha/2}$. The $\alpha/2$ percentage point of a standardized normal distribution, which is 1.96 for $\alpha = 5$ percent.

The approach determines the number of unpublished studies, with an average observed effect of zero, there would need to be to reduce the overall z-score to nonsignificance. Define k_0 to be the additional number of studies required such that

$$\sum_{i=1}^{k} Z_i \bigg/ \sqrt{k + k_0} < Z_{\alpha/2}. \qquad (23.8)$$

Rearranging this gives us

$$k_o > -k + \left(\sum_{i=1}^{k} Z_i\right)^2 \bigg/ (Z_{\alpha/2})^2. \qquad (23.9)$$

After k_0 is calculated, one can judge whether it is realistic to assume that this many studies exist unpublished in the research domain under investigation. The estimated fail-safe N should be considered in proportion to the number of published studies (k). Rosenthal suggested that the fail-safe N may be considered as being unlikely, in reality, to exist if it is greater than a tolerance level of $5k + 10$ (1979).

This method should be considered nothing more than a crude guide due to the following shortcomings (Begg and Berlin 1988). One of these is that combining Z-scores does not directly account for the sample sizes of the studies. Another is that the choice of zero for the average effect of the unpublished studies is arbitrary. It could easily be less than zero. Hence, in practice, many fewer studies might be required to overturn the meta-analysis result. As a result, it has lead to unjustified complacency about publication bias. Still another shortcoming is that the method does not adjust or deal with treatment effects—just p-values (and thus no indication of the 'true' effect size is obtained). Yet another is that heterogeneity among the studies is ignored (Iyengar and Greenhouse 1988). Last is that the method is not influenced by the shape of the funnel plot (Orwin 1983).

Example: Effects of Teacher Expectancy on Pupil IQ. Individual z scores are computed by dividing the effect sizes by the corresponding standard errors. This leads to $\Sigma Z_i = 11.03$. Using equation 23.9 gives $k_0 > -19 + (11.03/1.96)^2$, leading to $k_0 \geq 13$. This implies that if thirteen or more studies are unretrieved with zero effect (compared to the nineteen published ones in the dataset), on average, then the meta-analysis would be nonsignificant at the 5 percent level. However, as before, the effect of previous contact with student has not been adjusted for in the analysis and hence these results should be treated with caution. Furthermore, extensions to allow for including covariates have not been developed. Failsafe N could be calculated for separate subgroups of data (defined by covariates) but this is not pursued here.

Several extensions and variations of Rosenthal's file-drawer method have been developed that include alternative file-drawer calculations (Orwin 1983; Iyengar and Greenhouse 1988; Becker 2005; Gleser and Olkin 1996). One is based on effect-size rather than p-value reduction (Orwin 1983) and methods for estimating the number of unpublished studies (Gleser and Olkin 1996). This latter method can be seen as a halfway house between simple file-drawer approaches and the selection modeling approaches outlined below.

23.4.2 Trim and Fill

The nonparametric trim and fill method for adjusting for publication bias was developed as a simpler alternative to parametric selection models (section 23.4.3) for adjusting the problem of publication bias (Duval and Tweedie 2000a, 2000b). It is indeed now the most popular method for adjusting for publication bias (Borenstein 2005). Also, because the method is a way of formalizing the use of a funnel plot, the ideas underlying the statistical calculations can be communicated and understood visually, increasing its accessibility.

The method is based on rectifying funnel plot asymmetry, and assumes that suppression is working on the

effect size scale. It also assumes that the studies to the far left-hand-side of the funnel are missing (the number is determined by the method) and hence is a one-sided procedure. If missing studies are expected on the other side of the funnel, the data can be flipped—by multiplying the effect sizes by -1—before applying the method. In short, the method calculates how many studies would need to be trimmed off the right side of the funnel, to leave a symmetric center. The adjusted pooled effect size is then calculated. The trimmed studies are reinstated together with the same number of filled studies that are imputed as mirror images. This augmented symmetrical dataset is used to calculate the variance of the pooled estimate.

An iterative procedure is used to obtain an estimate of the number of missing studies. Three estimators were originally investigated, two shown through simulation to be superior (Duval and Tweedie 2000a, 2000b). Initially, the pooled effect size is subtracted from each study's effect size estimate. Ranks are then assigned according to their absolute values. It is then helpful to construct the vector of rankings, ordered by effect size, indicating those estimates that are less than the pooled effect size by adding a negative indicator. For example, if a vector $(-6, -5, -2, 1, 3, 4, 7, 8, 9, 10, 11)$ were obtained, it would indicate that three of the eleven estimates are less than the pooled effect size, and have absolute effect magnitudes ranked two, five, and six in the dataset (higher ranks indicating studies with estimates that deviate the most from the pooled effect estimate). The first estimator of the number of missing studies is defined as

$$R_0 = \gamma^* - 1, \tag{23.10}$$

where γ^* is the rightmost run of consecutive ranks of the vector defined—that is, for the example given, $\gamma^* = 5$ because the rightmost run of ranks is of length five, determined by the run seven, eight, nine, ten, and eleven, hence $R_0 = 4$.T. The estimated R_0 studies with the largest effect sizes are removed (trimmed) and the process continues iteratively until the estimate of the number of missing studies converges. The second estimate is

$$L_0 = [4T_n - n(n + 1)]/[2n - 1], \tag{23.11}$$

where T_n is the sum of the ranks of the studies whose effect sizes are greater than the pooled estimate (the Wilcoxon statistic for the dataset) and n indicates the number of studies. In the example vector given, $T_n = 1 + 3 + 4 + 7 + 8 + 9 + 10 + 11 = 53$. It follows that $L_0 = [4(53) - 11(12)]/[2(11) - 1] = 3.81$, which is rounded to four missing studies estimated. Like, R_0, L_0 is calculated iteratively until convergence is achieved. Full worked examples of these iterative calculations are available elsewhere (Duval and Tweedie 1998; Duval 2005). Since nether estimate is superior in all situations, use of both as a sensitivity analysis has been recommended (Duval 2005).

In practice, either a fixed or random effect meta-analysis estimate can be used for the iterative trimming and filling parts of this method. In the initial exposition, random effect estimates were used (Duval and Tweedie 2000a, 2000b). Random effect estimators can be more influenced by publication bias, however, given that the less precise studies—those more prone to publication bias—are given more weight in relative terms (Poole and Greenland 1999). Using fixed-effect methods to trim studies should therefore be considered and has recently been evaluated, though it does not always outperform the use of a random-effect model (Peters et al. 2007).

The performance of this method was initially evaluated using simulation studies under conditions of study homogeneity, and under the suppression mechanism the method assumes, that is, the n-most extreme effect sizes are suppressed (Duval and Tweedie 2000a, 2000b). Under these conditions, it performed well. However, further simulations suggest that it performs poorly in the presence of among-study heterogeneity when no publication bias is present (Terrin et al. 2003). Further evaluation of the method into scenarios where suppression is assumed to be based on statistical significance, with and without heterogeneity present, suggests that it outperforms non-adjusted meta-analytic estimates but can provide corrections that are far from perfect (Peters et al. 2007).

As for interpretation of the funnel plot, as discussed, the asymmetry this model rectifies may be caused by factors other than publication bias, which should be borne in mind when interpreting the results. A limitation of the method is that it does not allow for the inclusion of study level covariates as moderator variables.

Example: Effects of Teacher Expectancy on Pupil IQ. For illustration, Trim and fill, using a fixed effect model and the L_0 estimator is applied to the teacher expectancy dataset. Three studies are estimated as missing and imputed as shown in figure 23.8 (the triangles). The fixed effect pooled estimate for the original dataset was 0.060 (95 percent CI −0.011 to 0.132) and this reduced to 0.025 (95 percent CI −0.045 to 0.095) when the three studies were filled. Because this analysis ignores the covariate—previous contact with student—it should be interpreted cautiously. Although trim and fill does not allow modera-

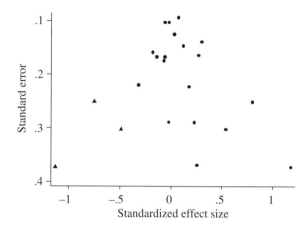

Figure 23.8 Trim and Fill Analysis, Teacher Expectancy on Pupil IQ

SOURCE: Author's compilation.
KEY: Circles are original studies, triangles are "filled" imputed studies

tor variables, separate analyses of subgroups of the data are possible. If this is done here, no studies are filled for prior contact longer than a week subgroup and one study is imputed for contact of a week or less subgroup, which changes the pooled estimate by only a small amount (0.357 (0.203 to 0.511) to 0.319 (0.168 to 0.469).

23.4.3 Selection Modeling

Selection models can be used to adjust a meta-analytic dataset by specifying a model that describes the mechanism by which effect estimates are assumed to be either suppressed or observed. This is used with an effect size model that describes what the distribution of effect sizes would be if there were no publication bias to correct for it.

If the selection model is known, use of the method is relatively straightforward but the precise nature of the suppression will usually be unknown. If this is the case, alternative candidate selection models can be applied as a form of sensitivity analysis (Vevea and Woods 2005), or the selection model can actually be estimated from the observed meta-analytic dataset. Because the methods are fully parametric, they can include study-level moderator variables. Although complex to implement, they have been recommended over other methods, which potentially produce misleading results, in scenarios where effect sizes are heterogeneous (Terrin et al. 2003). A potential

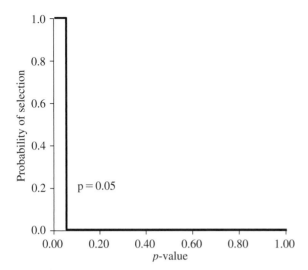

Figure 23.9 Simple Weight Function, Published if Significant

SOURCE: Author's compilation.

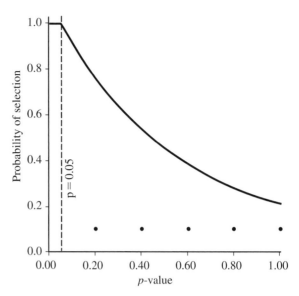

Figure 23.10 Example Weight Function, Probability of Publication Diminishes

SOURCE: Author's compilation.

drawback is that they perform poorly if the number of observed studies is small.

A comprehensive review of selection models has recently been published elsewhere and is recommended reading for an expanded and more statistically rigorous account particularly regarding the implementation of these methods that require complex numerical methods (Hedges and Vevea 2005).

Two classes of explicit selection models have been developed: those that model suppression as a function of an effect size's p-value and and those that model suppression as a function of a study's effect size and standard error simultaneously. Both are implemented through weighted distributions that give the likelihood a given effect estimate will be observed if it occurs. Each are discussed in turn.

23.4.3.1 Suppression as a Function of p-Value Only
The simplest selection model suggested is one that assumes all studies with statistically significant effect sizes are observed (for example $p < 0.05$ two-tailed, or $p > 0.975$ or $p < 0.025$ one-tailed) and all others are suppressed (Lane and Dunlap 1978; Hedges 1984). The associated weight function is represented graphically in figure 23.9, where the likelihood a study is observed if it occurs is given on the y-axis ($w(p)$) and the p-value scale is along the x-axis. Using this model, any effect size with a

p-value < 0.05 has a probability of one (certainty) of being observed and zero otherwise. Perhaps it is more realistic to assume that studies that are not statistically significant have a fixed probability of less than one that they are published—that is, not all nonsignificant studies are suppressed. Alternatively, it could be assumed that the probability a nonsignificant study is published is some continuous decreasing function of the p-value—that is, the larger the p-value, the less chance a study has of being published (Iyengar and Greenhouse 1988). Figure 23.10 illustrates a possible weight function in this class with the probability of observing an effect size diminishing exponentially as the p-value increases. More realistic still may be to assume a step function with steps specified as occurring at perceived milestones in statistical significance—based on the psychological perception that a p-value of 0.049 is considerably different from one of 0.051 (Hedges 1992; Vevea, Clements, and Hedges 1993; Hedges and Vevea 1996). Figure 23.11 provides a graphical representation of a weight function of this type. Finally, it is possible that the location and the height of the steps in such a weight function can be determined from the data (Dear and Begg 1992). It should be noted, however, that when the selection model is left to be determined by the data, it

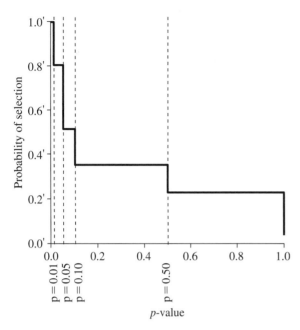

Figure 23.11 Example Psychological steps Weight Function

SOURCE: Author's compilation.

is often estimated poorly because the observed effects include little information on the selection process. It is possible to include study level covariates in such models (Vevea and Hedges 1995). This can be done to remove confounding effects in the distribution of effect sizes as discussed when considering alternative reasons for funnel plot asymmetry.

23.4.3.2 Suppression as a Function of Effect Size and Its Standard Error These models assume the probability that a result is suppressed is a function of both effect size and standard error, with the probability of suppression increasing as effect size decreases and standard error increases (Copas and Li 1997; Copas and Shi 2000). Although such a model is perhaps more realistic than one modeling suppression as related to p-value, it does mean there are identifiability problems, implying parameters cannot be estimated from the effect size data alone. This in turn means that if you believe this model to be plausible, correcting for publication bias is not possible without making assumptions that cannot be tested (Copas and Shi 2001). It has thus been recommended to evaluate such models over the plausible regions of values for the model

parameters that determine the suppression process (Copas and Shi 2000, 2001). To this end, a bound for the bias as a function of the fraction of missing studies has been proposed (Copas and Jackson 2004).

23.4.3.3 Simulation Approaches One recent development is a simulated pseudo-data approach to correct for publication bias (Bowden, Thompson, and Burton 2006). This is not really a different type of selection model, but rather an alternative way of evaluating them. It requires the selection model to be fully specified, but does allow arbitrarily complex selection processes to be modeled.

23.4.3.4 Bayesian Approaches Several authors have considered implementing selection models from a Bayesian perspective (Givens, Smith, and Tweedie 1997; Silliman 1997). One feature of this approach is that it can provide an estimate of the number of suppressed studies and their effect size distribution.

Example: Effects of Teacher Expectancy on Pupil IQ. A detailed analysis of applying different selection models to this dataset—including those that depend on the p-value with and without the weight function estimated from the data, and those depending on effect size and standard error simultaneously—is available elsewhere (Hedges and Vevea 2005). Here, I summarize one of these analyses as an example of how such models can be applied.

This analysis uses a fixed effect model including prior contact time as a dichotomous moderator variable, as defined earlier. The weight function has two cut points at $p = 0.05$ and 0.30 with weights estimated from the data. Although using more cut points may be desirable, the relatively small number of studies mean parameters would not be estimable. The estimated weight function is presented in figure 23.12. Here the probability of selection is set to one for p-values between 0 and ≤ 0.05 and it is estimated as 0.89 (SE 0.80) and 0.56 (SE 0.64) for the intervals 0.05 to < 0.30 and 0.30 to 1.00 respectively. Note the large standard errors on these weights, which is indicative of models in which the weights are defined by the data and particularly when the number of studies is not large. The estimated pooled effect for less than one week of prior contact changed from 0.349 (SE 0.079) to 0.321 (SE 0.111) when the selection model was applied and the covariate effect similarly changed from -0.371 (SE 0.089) to -0.363. 0.109), indicating that the conclusions appear to be relatively robust to the effects of publication bias.

A summary of the results from applying the different methods to this dataset is presented in table 23.1 and shows broad consistency. When prior contact is not ac-

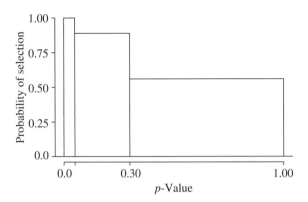

Figure 23.12 Estimate Weight Function, Teacher Expectancy Example

SOURCE: Hedges and Veva selection method approaches.

counted for in the model, a suggestion of asymmetry or bias is evident, but when this possible confounder is included, the threat of bias appears much lessened.

23.4.4 Excluding Studies Based on Their Size

A simple alternative to adjusting for publication bias is to assume studies of a certain size, regardless of their results, will be published and hence only include these in the meta-analysis. This effectively gives smaller studies zero weight in the meta-analysis. If publication bias only exists, or is most severe in the small studies (an assumption that underlies several of the statistical models presented) this will generally result in less biased pooled estimates, however the choice of cut-off for study inclusion/exclusion is arbitrary.

23.5 METHODS TO ADDRESS SPECIFIC DISSEMINATION BIASES

As mentioned earlier, incomplete data reporting may occur at various levels below the suppression of whole studies. Analysis of specific outcomes or subgroups, for example, may have been conducted but not been written up and hence available for meta-analysis. Similarly, all the details of an analysis required for meta-analysis may not be reported. For example, the standard error of an effect size or a study level covariate required for meta-regression may not have been published.

This latter category of missing data may be entirely innocent and be attributable simply to lack of journal space or a lack of awareness about the importance of reporting such information. Such missingness may thus not be related to outcome and data can be assumed missing (completely) at random. If this is the case, standard methods for dealing with missing data can be applied to the meta-analytic dataset, though this is rarely done in practice

Table 23.1 Summary, Methods to Address Publication Bias

Method	Findings
Funnel plot of all the data	Funnel visibly asymmetric
Funnel plot subgrouped by prior contact	Little evidence of asymmetry for either funnel
Egger's regression test ignoring prior contact data	$p = 0.057$ (i.e. some evidence of bias)
Egger's regression test including prior contact covariate	$p = 0.70$ (i.e. little evidence of bias)
File drawer method ignoring prior contact data	13 studies would have to be missing (compared to 19) to overturn conclusions
Trim and Fill ignoring prior contact data	3 studies estimated missing. Effect size changes from 0.060 (s.e. 0.036) to 0.025 (s.e. 0.036)
Trim and Fill subgrouped by prior contact	For > 1 week prior contact, 0 studies estimated missing For ≤ 1 week prior contact 1, study estimated missing. Effect size changes from 0.357 (s.e. 0.079) to 0.319 (s.e. 0.077)
Selection model including prior contact covariate (fixed effect, 2 cut points ($p = 0.05$ and 0.30) estimated from data)	For > 1 week prior contact, effect size changes from -0.371 (s.e. 0.089) to -0.363 (0.109) For ≤ 1 week prior contact, effect size changes from 0.349 (s.e. 0.079) to 0.321 (s.e. 0.111)

SOURCE: Author's compilation.

(Little and Rubin 1987; Pigott 2001; Sutton and Pigott 2004). More ad hoc methods can also be applied as necessary (Song et al. 1993; Abrams, Gillies, and Lambert 2005).

If the data missing is due to less innocent reasons then it is probably safest to assume data are not missing at random. This is the typical assumption made when addressing publication bias, though it is often not framed this way. Because outcomes and subgroups and the like may be suppressed under similar mechanisms to those acting on whole studies, such missingness may manifest itself in a similar way and hence the methods covered thus far will be appropriate to address it. There may be advantages, however, of developing and applying methods that address specific forms of missing information. This area of research is in its infancy.

23.5.1 Outcome Reporting Biases

The issue of outcome reporting bias has received considerable attention in recent years, and empirical research indicates that it is a serious problem, at least for randomized controlled trials in medicine (Hahn, Williamson, and Hutton 2002; Chan, Hróbjartsson et al. 2004; Chan, Krleza-Jeric et al. 2004; Chan and Altman 2005). This evidence lends weight to the premise that if a study measures multiple outcomes, those that are statistically significant are more likely to be published than those that are not. Such bias could be expected to be detected as general publication bias and dealt with in that way. Such an analysis, however, would ignore the patterns of missing data across the multiple outcomes measured across all studies and information, such as levels of statistical significance, across all reported outcomes. A method has been developed that does consider such patterns and adjusts estimates under the assumption that the most statistically significant of a fixed number of identically distributed independent outcomes are reported (Hutton and Williamson 2000; Williamson and Gamble 2005). Although its assumptions are unrealistically strict, the model does provide an upper bound on the likely impact of outcome reporting bias and could help determine whether contacting study investigators for the potentially missing data would be cost effective.

A selection model has also been developed for a context-specific application—success rates of surgery for emergency aneurysm repair—to address outcome reporting bias based on the dataset plotted in figures 23.1 and 23.2 (Jackson, Copas, and Sutton 2005). Here an outcome, assumed reported with no bias, provided information on a further outcome where selectivity in reporting was evident. Assuming all bias was due to outcome reporting, that is, there were no completely missing studies, the extra information gained from the outcome for which publication bias was assumed not present meant that the selection model was identifiable and a unique solution could be gained without further assumptions.

23.5.2 Subgroup Reporting Biases

Subgroup analysis suppression could exist under a similar mechanism to that for outcomes; potentially many subgroups could be compared, and only those interesting or statistically significant reported. A sensitivity analysis approach similar to that for outcome reporting bias has been suggested that involves data imputation for missing subgroup data under the assumption that the results that are not published tend to be nonsignificant (Hahn et al. 2000).

23.5.3 Time-Lag Bias

Time-lag bias may exist if research with large effect sizes or significant results is published more quickly than that with small effect sizes or insignificant results. When such a bias operates, the first published studies show that systematically greater effect sizes compare to subsequently published investigations (Trikalinos and Ioannidis 2005). The summary effect would thus diminish over time as more data accumulates. Evidence such a bias exists in the field of genetic epidemiology is strong (Ioannidis et al. 2001). Methods to detect such biases have been described elsewhere (Trikalinos and Ioannidis 2005).

23.6 DISCUSSION

The aim of this chapter is to provide an overview of the problem of publication bias with a focus on the methods developed to address it in an research synthesis context. Whatever approach is taken, the fact remains that publication bias is a difficult problem to deal with because the mechanisms causing the bias are usually unknown, and the merit of any method to address it depends on how close the assumptions the method makes are to the truth. Methods developed for adjusting pooled results have therefore been recommended only as sensitivity analyses. Although this would seem sensible advice, the context of the meta-analysis would also seem an important issue given that sensitivity analysis is of limited use in a decision making context. That is, you have to make one decision based on the synthesis regardless of how many sensitivity analyses are performed.

Software has been developed that implements many of the methods described here. An overview and comparison of software functionality in addressing publication bias is available elsewhere (Borenstein 2005). Exceptions are the selection models and methods to deal specifically with outcome, subgroup, and time-lag biases. If these methods are to be used more routinely, more user-friendly software clearly needs to be developed.

Novel and emerging areas of research in publication bias not covered here include an investigation of how publication bias affects the estimation of between-study heterogeneity parameter in a random effects meta-analysis model (Jackson 2006), allowing for differential levels of publication bias across studies of different study designs (Sutton, Abrams, and Jones 2002), and capture-recapture methods across electronic databases to estimate the number of missing studies (Bennett et al. 2004).

Future investigation may prove fruitful in at least three areas. One involves developing a unified approach for dealing with data missing due to different selection mechanisms, such as simultaneously adjust for missing studies and missing outcomes for known studies separately. Another entails using meta-epidemiological research to develop global estimates of bias as a function of study precision that might in turn be used to adjust future meta-analyses by incorporating them into distributions for a Bayesian analysis (Sterne et al. 2002). A third is developing methods to address publication bias for statistical models that are more complicated than a standard meta-analysis, where construction of a funnel plot is often not possible (Ades and Sutton 2006).

23.7. REFERENCES

Abrams, Keith R., Clare L. Gillies, and Paul C. Lambert. 2005. "Meta-Analysis of Heterogeneously Reported Trials Assessing Change from Baseline." *Statistics in Medicine* 24(24): 3823–44.

Ades, Anthony, and Alexander J. Sutton. 2006. "Multiple Parameter Evidence Synthesis in Epidemiology and Medical Decision-Making: Current Approaches." *Journal of the Royal Statistical Society Series A* 169(1): 5–35.

Becker, Betsy J. 2005. "Failsafe *N* or File-Drawer Number." In *Publication Bias in Meta-Analysis: Prevention, Assessment and Adjustments*, edited by Hannah R. Rothstein, Alexander J. Sutton, and Michael Borenstein. Chichester: John Wiley & Sons.

Begg, Colin B., and Jesse A. Berlin. 1988. "Publication Bias: A Problem in Interpreting Medical Data (with discussion)." *Journal of the Royal Statistics Society Series A.* 151(3): 419–63.

Begg, Colin B., and Madhuchhanda Mazumdar. 1994. "Operating Characteristics of a Rank Correlation Test for Publication Bias." *Biometrics* 50(4): 1088–101.

Bennett, Derrick A., Nancy K. Latham, Caroline Stretton, and Craig S. Anderson. 2004. "Capture-Recapture Is a Potentially Useful Method for Assessing Publication Bias." *Journal of Clinical Epidemiology* 57(4): 349–57.

Berlin, Jesse A., and Davina Ghersi. 2005. "Preventing Publication Bias: Registries and Prospective Meta-Analysis." In *Publication Bias in Meta-Analysis: Prevention, Assessment and Adjustments*, edited by Hannah R. Rothstein, Alexander J. Sutton, and Michael Borenstein. Chichester: John Wiley & Sons.

Borenstein, Michael. 2005. "Software for Publication Bias." In *Publication Bias in Meta-Analysis: Prevention, Assessment and Adjustments*, edited by Hannah R. Rothstein, Alexander J. Sutton and Michael Borenstein. Chichester: John Wiley & Sons.

Bowden, Jack, John R. Thompson, and Paul Burton. 2006. "Using Pseudo-Data to Correct for Publication Bias in Meta-Analysis." *Statistics in Medicine* 25(22): 3798–3813.

Bown, Matthew. J., Alexander J. Sutton, Peter R. Bell, and Robert D. Sayers. 2002. "A Meta-Analysis of 50 Years of Ruptured Abdominal Aortic Aneurysm Repair." *British Journal of Surgery* 89(6): 1–18.

Chan, An-Wen, and Douglas G. Altman. 2005. "Identifying Outcome Reporting Bias in Randomised Trials on PubMed: Review of Publications and Survey of Authors." *British Medical Journal* 330(7494): 753.

Chan, An-Wen, Asbjorn Hrobjartsson, Mette T. Haahr, Peter C. Gotzsche, and Douglas G. Altman. 2004. "Empirical Evidence for Selective Reporting of Outcomes in Randomized Trials. Comparison of Protocols to Published Articles." *Journal of the American Medical Association* 291(20): 2457–65.

Chan, An-Wen, Karmela Krleza-Jeric, Isabelle Schmid, and Douglas G. Altman. 2004. "Outcome Reporting Bias in Randomized Trials Funded by the Canadian Institutes of Health Research." *Canadian Medical Association Journal* 171(7): 735–40.

Copas, John B., and Daniel Jackson. 2004. "A Bound for Publication Bias Based on the fraction of Unpublished Studies." *Biometrics* 60(1): 146–53.

Copas, John B., and Hu G. Li. 1997. "Inference for Non-Random Samples." *Journal of the Royal Statistical Society, Series B* 59(1): 55–95.

Copas, John B., and Jian Q. Shi. 2000. "Reanalysis of Epidemiological Evidence on Lung Cancer and Passive Smoking." *British Medical Journal* 320(7232): 417–18.

———. 2001. "A Sensitivity Analysis for Publication Bias in Systematic Reviews." *Statistical Methods in Medical Research* 10(4): 251–65.

Dear, Keith B.G., and Colin B. Begg. 1992. "An Approach for Assessing Publication Bias Prior to Performing a Meta-Analysis." *Statistical Science* 7(2): 237–45.

Dickersin, Kay. 2005. "Publication Bias: Recognizing the Problem, Understanding its Origins and Scope, and Preventing Harm." In *Publication Bias in Meta-Analysis: Prevention, Assessment and Adjustments*, edited by Hannah R. Rothstein, Alexander J. Sutton, and Michael Borenstein. Chichester: John Wiley & Sons.

Duval, Sue. 2005. "The Trim and Fill Method." In *Publication Bias in Meta-Analysis: Prevention, Assessment and Adjustments*, edited by Hannah R. Rothstein, Alexander J. Sutton, and Michael Borenstein. Chichester: John Wiley & Sons.

Duval, Sue, and Richard Tweedie. 1998. "Practical Estimates of the Effect of Publication Bias in Meta-Analysis." *Australasian Epidemiologist* 5(449): 14–17.

———. 2000. "A NonParametric 'Trim and Fill' Method of Assessing Publication Bias in Meta-Analysis." *Journal of the American Statistical Association* 95(449): 89–98.

———. 2000. "Trim and Fill: A Simple Funnel Plot Based Method of Testing and Adjusting for Publication Bias in Meta-Analysis." *Biometrics* 56(2): 455–63.

Egger, Matthias, George Davey Smith, Martin Schneider, and Christoph Minder. 1997. "Bias in Meta-Analysis Detected by a Simple, Graphical Test." *British Medical Journal* 315(7109): 629–34.

Galbraith, Rex. F. 1994. "Some Applications of Radial Plots." *Journal of the American Statistical Association* 89(428): 1232–42.

Givens, Geof H., David D. Smith, and Richard L Tweedie. 1997. "Publication Bias in Meta-Analysis: A Bayesian Data-Augmentation Approach to Account for Issues Exemplified in the Passive Smoking Debate." *Statistical Science* 12(4): 221–50.

Gleser, Leon J, and Ingrim Olkin. 1996. "Models for Estimating the Number of Unpublished Studies." *Statistics in Medicine* 15(23): 2493–2507.

Hahn, Seokyung, Paula R. Williamson, and Jane L. Hutton. 2002. "Investigation of Within-Study Selective Reporting in Clinical Research: Follow-Up of Applications Submitted to a Local Research Ethics Committee." *Journal of Evaluation in Clinical Practice* 8(3): 353–59.

Hahn, S., Paula R. Williamson, Jane L. Hutton, Paul Garner, and E. Victor Flynn. 2000. "Assessing the Potential for Bias in Meta-Analysis Due to Selective Reporting of Subgroup Analyses Within Studies." *Statistics in Medicine* 19(24): 3325–36.

Halpern, Scott D., and Jesse A. Berlin. 2005. "Beyond Conventional Publication Bias: Other Determinants of Data Suppression." In *Publication Bias in Meta-Analysis: Prevention, Assessments and Adjustments*, edited by Hannah R. Rothstein, Alexander J. Sutton, and Michael Borenstein. Chichester: John Wiley & Sons.

Harbord, Roger M., Matthias Egger, and Jonathan A.C. Sterne. 2006. "A Modified Test for Small-Study Effects in Meta-Anlayses of Controlled Trials with Binary Endpoints." *Statistics in Medicine* 25(20): 3443–57.

Hedges, Larry V. 1984. "Estimation of Effect Size under Nonrandom Sampling: The Effects of Censoring Studies Yielding Statistically Insignificant Mean Differences." *Journal of Educational Statistics* 9(1): 61–85.

———. 1992. "Modeling Publication Selection Effects in Meta-Analysis." *Statistical Science* 7(2): 246–55.

Hedges, Larry V., and Jack L. Vevea. 1996. "Estimating Effects Size Under Publication Bias: Small Sample Properties and Robustness of a Random Effects Selection Model." *Journal of Educational and Behavioral Statistics* 21(4): 299–333.

———. 2005. "Selection Method Approaches." In *Publication Bias in Meta-Analysis: Prevention, Assessments and Adjustments*, edited by Hannah R. Rothstein, Alexander J. Sutton, and Michael Borenstein. Chichester: John Wiley & Sons.

Hopewell, Sally, Michael Clarke, and Sue Mallett. 2005. "Grey Literature and Systematic Reviews." In *Publication Bias in Meta-Analysis: Prevention, Assessment and Adjustments*, edited by Hannah R. Rothstein, Alexander J. Sutton, and Michael Borenstein. Chichester: John Wiley & Sons.

Hutton, Jane L., and Paula R. Williamson. 2000. "Bias in Meta-Analysis due to Outcome Variable Selection Within Studies." *Applied Statistics* 49(3): 359–70.

Ioannidis, John P., Evangelina E. Ntzani, Thomas A. Trikalinos, and Despina G. Contopoulos-Ioannidis. 2001. "Replication Validity of Genetic Association Studies." *Nature Genetics* 29(3): 306–9.

Iyengar, Satish, and Joel B. Greenhouse. 1988. "Selection Models and the File Drawer Problem." *Statistical Science* 3(1): 109–35.

Jackson, Daniel. 2006. "The Implication of Publication Bias for Meta-Analysis' Other Parameter." *Statistics in Medicine* 25(17): 2911–21.

Jackson, Daniel, John Copas, and Alexander Sutton. 2005. "Modelling Reporting Bias: The Operative Reporting Rate for Ruptured Abdominal Aortic Aneurysm Repair." *Journal of the Royal Statistical Society Series A* 168(4): 737–52.

Lane, David M., and William P. Dunlap. 1978. "Estimating Effect-Size Bias Resulting from Significance Criterion in Editorial Decisions." *British Journal of Mathematical and Statistical Psychology* 31: 107–12.

Lau, Joseph, John P. A. Ioannidis, Norma Terrin, Christopher H Schmid, and Ingram Olkin. 2006. "The Case of the Misleading Funnel Plot." *British Medical Journal* 333(7568): 597–600.

Little, Roderick J. A., and Don. B. Rubin. 1987. *Statistical Analysis with Missing Data*. New York: John Wiley & Sons.

Macaskill, Petra, Stephen D. Walter, and Lesley Irwig. 2001. "A Comparison of Methods to Detect Publication Bias in Meta-Analysis." *Statistics in Medicine* 20(4): 641–54.

Orwin, Robert G. 1983. "A Fail-Safe N for Effect Size in Meta-Analysis." *Journal of Educational Statistics* 8(2): 157–59.

Peters, Jaime L., Alexander J. Sutton, David R. Jones, Keith R. Abrams, and Lesley Rushton. 2006. "Comparison of Two Methods to Detect Publication Bias in Meta-Analysis." *Journal of the American Medical Association* 295(6): 676–80.

———. 2007. "Performance of the Trim and Fill Method in the Presence of Publication Bias and Between-Study Heterogeneity." *Statistics in Medicine* 26(25): 4544–62.

Pigott, Therese D. 2001. "Missing Predictors in Models of Effect Size." *Evaluation and the Health Professions* 24(3): 277–307.

Poole, Charles, and Sander Greenland. 1999. "Random-Effects Meta-Analysis Are not Always Conservative." *American Journal of Epidemiology* 150(5): 469–75.

Rosenthal, Robert. 1979. "The File Drawer Problem and Tolerance for Null Results." *Psychological Bulletin* 86: 638–41.

Schwarzer, Guido, Gerd Antes, and Martin Schumacher. 2002. "Inflation of Type 1 Error Rate in Two Statistical Tests for the Detection of Publication Bias in Meta-Analyses with Binary Outcomes." *Statistics in Medicine* 21(17): 2465–77.

———. 2007. "A Test for Publication Bias in Meta-Analysis with Sparse Binary Data." *Statistics in Medicine* 26(4): 721–33.

Silliman, Nancy P. 1997. "Hierarchical Selection Models with Applications in Meta-Analysis." *Journal of the American Statistical Association* 92(429): 926–36.

Smith, Richard. 1999. "What Is Publication? A Continuum." *British Medical Journal* 318(7177): 142.

Song, Fujan, Alison Easterwood, Simon Guilbody, Lelia Duley, and Alexander J. Sutton. 2000. "Publication and Other Selection Biases in Systematic Reviews." *Health Technology Assessment* 4(10): 1–115.

Song, Fujan, Nick Freemantle, Trevor A. Sheldon, Allan House, Paul Watson, and Andrew Long. 1993. "Selective Serotonin Reuptake Inhibitors: Meta-Analysis of Efficacy and Acceptability." *British Medical Journal* 306(6879): 683–87.

Stanley, Tom D. 2005. "Beyond Publication Bias." *Journal of Economic Surveys* 19(3): 309–45.

Stern, Jerome M., and R. John Simes. 1997. "Publication Bias: Evidence of Delayed Publication in a Cohort Study of Clinical Research Projects." *British Medical Journal* 315(7109): 640–45.

Sterne, Jonathan A.C., and Matthias Egger. 2001. "Funnel Plots for Detecting Bias in Meta-Analysis: Guidelines on Choice of Axis." *Journal of Clinical Epidemiology* 54(10): 1046–55.

———. 2005. "Regression Methods to Detect Publication and Other Bias in Meta-Analysis." In *Publication Bias in Meta-Analysis: Prevention, Assessments and Adjustments*, edited by Hannah R. Rothstein, Alexander J. Sutton, and Michael Borenstein. Chichester: John Wiley & Sons.

Sterne, Jonathan A.C., Betsy J. Becker, and Matthias Egger. 2005. "The Funnel Plot." In *Publication Bias in Meta-Analysis*, edited by Hannah R. Rothstein, Alexander J. Sutton and Michael Borenstein. Chichester: John Wiley & Sons.

Sterne, Jonathan A.C., David Gavaghan, and Matthias Egger. 2000. "Publication and Related Bias in Meta-Analysis: Power of Statistical Tests and Prevalence in the Literature." *Journal of Clinical Epidemiology* 53(11): 1119–29.

Sterne, Jonathan A.C., Peter Juni, Kenneth F. Schulz, Douglas G. Altman, Christopher Bartlett, and Matthias Egger. 2002. "Statistical Methods for Assessing the Influence of Study Characteristics on Treatment Effects in 'Meta-Epidemiological' Research." *Statistics in Medicine* 21(11): 1513–24.

Sutton, Alexander J., and Therese D. Pigott. 2004. "Bias in Meta-Analysis Induced by Incompletely Reported Studies." In *Publication Bias in Meta-Analysis: Prevention, Assessments and Adjustments*, edited by Hannah Rothstein, Alexander J. Sutton, and Michael Borenstein. Chichester: John Wiley & Sons.

Sutton, Alexander J., Keith R. Abrams, and David R. Jones. 2002. "Generalized Synthesis of Evidence and the Threat of Dissemination Bias: The Example of Electronic Fetal Heart Rate Monitoring (EFM)." *Journal of Clinical Epidemiology* 55(10): 1013–24.

Terrin, Norma, Christopher H. Schmid, and Joseph Lau. 2005. "In an Empirical Evaluation of the Funnel Plot, Researchers Could Not Visually Identify Publication Bias." *Journal of Clinical Epidemiology* 58(9): 894–901.

Terrin, Norma, Christopher H. Schmid, Joseph Lau, and Ingram Olkin. 2003. "Adjusting for Publication Bias in the Presence of Heterogeneity." *Statistics in Medicine* 22(13): 2113–26.

Thompson, Simon G., and Stephen J. Sharp. 1999. "Explaining heterogeneity in meta-analysis: a comparison of methods." *Statistics in Medicine* 18(20): 2693–2708.

Trikalinos, Thomas A., and John P.A. Ioannidis. 2005. "Assessing the Evolution of Effect Sizes Over Time." In *Publication Bias in Meta-Analysis: Prevention, Assessment and Adjustments*, edited by Hannah R. Rothstein, Alexander J. Sutton, and Michael Borenstein. Chichester: John Wiley & Sons.

Vevea, Jack L., and Larry V. Hedges. 1995. "A General Linear Model for Estimating Effect size in the Presence of Publication Bias." *Psychometrika* 60(3): 419–35.

Vevea, Jack L. and Carol M. Woods. 2005. "Publication Bias in Research Synthesis: Sensitivity Analysis Using a Priori Weight Functions." *Psychological Methods* 10(4): 428–43.

Vevea, Jack L., Nancy C. Clements, and Larry V Hedges. 1993. "Assessing the Effects of Selection Bias on Validity Data for the General Aptitude Test Battery." *Journal of Applied Psychology* 78(6): 981–87.

Williamson, Paula R., and Carol Gamble. 2005. "Identification and Impact of Outcome Selection Bias in Meta-Analysis." *Statistics in Medicine* 24(10): 1547–61.

TYING RESEARCH SYNTHESIS TO SUBSTANTIVE ISSUES

24

ADVANTAGES OF CERTAINTY AND UNCERTAINTY

WENDY WOOD
Duke University

ALICE H. EAGLY
Northwestern University

CONTENTS

24.1 INTRODUCTION

A research synthesis typically is not an endpoint in the investigation of a topic. Rarely does a synthesis offer a definitive answer to the theoretical or empirical question that inspired the investigation. Instead, most research syntheses serve as way stations along a sometimes winding research path. The goal is to describe the status of a research literature by highlighting what is unknown as well as what is known. It is the unknowns that are especially likely to suggest useful directions for new research.

The purpose of this chapter is to delineate the possible relations between research syntheses, theory development, and future empirical research. By indicating the weaknesses as well as the strengths of existing research, a synthesis can channel thinking about research and theory in directions that would improve the next generation of empirical evidence and theoretical development. In general, scientists' belief about what the next steps are for empirical research and theorizing following a synthesis depends on the level of certainty that they accord its findings (see Cooper and Rosenthal 1980). In this context, certainty refers to the confidence with which the scientific community accepts empirical associations between variables and the theoretical explanations for these associations.

As we explain in this chapter, the certainty accorded to findings is influenced by the perceived validity of the research. In a meta-analysis, invalidity can arise at the level of the underlying studies as well as at the meta-analytic, aggregate level. Threats to validity, as William Shadish, Thomas Cook, and Donald Campbell defined it, reduce scientists' certainty about the empirical relations and theoretical explanations considered in a meta-analysis, and this uncertainty in turn stimulates additional research (2002). Empirical relations and theoretical explanations in research syntheses range from those that the scientific community can accept with high certainty to those it can hold with little certainty only. Highly certain conclusions suggest that additional research of the kind evaluated is not required to substantiate the focal effect or validate the theoretical explanation, whereas less certain conclusions suggest a need for additional research.

Our proposal to evaluate the uncertainty that remains in empirical relations and in interpretations of those relations challenges more standard ways of valuing research findings. Generally, evaluation of research, be it by journal editors or other interested scientists, favors valid empirical findings and well-substantiated theories. In this conventional approach, science progresses through the cumulation of well-supported findings and theories. Although it is standard practice for researchers to call for additional investigation at the end of a written report, such requests are often treated as rhetorical devices that have limited meaning in themselves. In our experience, editors' publication decisions and manuscript reviewers' evaluations are rarely influenced by such indicators of sources of uncertainty that could be addressed in future investigation.

Following scientific convention, research syntheses would be evaluated favorably to the extent that they produce findings and theoretical statements that appear to contribute definitively to the canon of scientific knowledge. In striving to meet such criteria, the authors of syntheses might focus on what is known in preference to what is unknown. In contrast, we believe that, by highlighting points of invalidity in a research literature, research syntheses offer an alternative, generative route for scientific progress. This generative route does not produce valid findings and theoretical statements but instead identifies points of uncertainty and thus promising avenues for subsequent research and theorizing that can reduce the uncertainty. When giving weight to the generative contribution of a synthesis, journal editors and manuscript reviewers would consider how well a synthesis frames questions. By this metric, research syntheses can contribute significantly to scientific progress even when the empirical findings or theoretical understanding of them can be accorded only limited certainty.

Certainty is not attached to a synthesis as a whole but rather to specific claims, such as generalizations, which vary in their truth value. This chapter will help readers identify where uncertainty is produced in synthesis findings and how it can be addressed in future empirical research and theorizing. We evaluate two goals for research syntheses and consider the guidance that they provide to subsequent research and theory. One is to integrate studies that examined a relation between variables in order to establish the size and direction of the relation. Another is to identify conditions that modify the size of the relation between two variables or processes that mediate the relation of interest.

Meta-analyses that address the first goal are designed to aggregate results pertaining to relations between variables so as to assess the presence and magnitude of those relations in a group of studies. Hence these investigations include a mean effect size and confidence interval and a test of the null hypotheses, and consider whether the relation might be artifactual. In these analyses, the central focus is on establishing the direction and magnitude of a relationship. This approach is deductive to the extent that the examined relation is predicted by a theoretical account. It is inductive to the extent that syntheses are designed to establish an empirical fact as, for example, in some meta-analytic investigations of sex differences and of social interventions and medical and psychological treatments. As we explain, uncertainty in such syntheses could arise in the empirical relation itself, as occurs when the outcomes of available studies are inconclusive about the existence of a relationship, as well as in ambiguous explanations of the relation of interest.

Meta-analyses that address the second goal use characteristics of studies to identify moderators or mediators of an empirical relation. Moderating variables account for variability in the size and direction of a relation between two variables, whereas mediators intervene between a predictor and outcome variable and represent a causal mechanism through which the relation occurs (see Kenny, Kashy, and Bolger 1998). Like investigations of main effects, the investigation of moderators and mediators is deductive when the evaluated relations are derived from theories, and inductive when the moderators and mediators are identified by examining the findings of the included studies. Uncertainty in these meta-analyses can emerge in the examined relations and in the explanations of findings yielded by tests of moderators and mediators.

We conclude with a discussion of how research syntheses use meta-analytic estimates of mean effect sizes and evaluations of mediators and moderators in order to test competing theories. Just as testing theories about psychological and social processes lends primary research its complexity and richness, meta-analyses also can be designed to examine the plausibility of one theoretical account over others. Meta-analyses with this orientation proceed by examining the overall pattern of empirical findings in a literature, including the relation of interest and potential mediators and moderators of it, with the purpose of testing competing theoretical propositions. Uncertainty in this type of meta-analysis emerges to the extent that the data evaluated cannot firmly discriminate between the theories based on their empirical support.

24.2 ESTIMATING RELATIONS BETWEEN VARIABLES

For many meta-analyses, a critical focus is to demonstrate that two variables are related. This relation is known as a main effect when the impact of one variable on another is evaluated independently of other possible predictors. Of course, some single syntheses report multiple sets of relationships between variables. For example, in a synthesis of the predictors of career success, Thomas Ng and his colleagues related eleven human capital variables, four organizational sponsorship variables, four sociodemographic variables, and eight stable individual differences variables to three forms of career success: salary, promotion, and career satisfaction (2005). Each of these relationships—for example, between organizational tenure and salary—could be considered a meta-analysis in and of itself. Including all of these meta-analyses in a single article allows readers to compare the success with which many different variables predict a set of outcomes.

Syntheses designed to detect empirical relations in these ways yield estimates that can be accepted with high degrees of certainty when the relation between the variables is consistent across the included studies. Such syntheses provide a relatively precise estimate of the magnitude of the relation in the specified population, regardless of whether that magnitude is large, small, or even null. The synthesist might first calculate a confidence interval around the mean effect to determine whether it is statistically significant, as indicated by an interval that does not include zero. He or she also would assess the consistency of the findings by computing one of the homogeneity tests of meta-analytic statistics (see chapter 14, this volume). Failure to reject the hypothesis of homogeneity suggests that study findings estimate a common population effect size. Certainty about the conclusions from the estimate of the size, of the confidence interval, and the test of homogeneity of effects requires that these calculations be based on enough studies to provide statistically powerful tests. Moreover, the studies should represent the various conditions under which the empirical relation is expected to obtain, including different laboratories, subject populations, and measuring instruments. Findings that are homogeneous given adequate power and an appropriate range of conditions suggest an empirical result that can be accepted with some certainty.

The specific inferences that can be made from homogeneous effect sizes depend on whether the analytic strategy involves fixed or random effects models for estimating effect size parameters (see chapters 15 and 16, this volume). One error model involves fixed-effect calculations in which studies in a synthesis differ from each other only in error attributable to the sampling of units, such as participants, within each study. Fixed-effect model results are generalizable only to studies similar to those included in the synthesis. Random-effects models offer the alternative assumption that studies in the meta-analysis are drawn randomly from a larger population of relevant research. In this case, variability arises from the sampling of studies from a broader population of studies as well as from the sampling of participants within studies. Random-effects model results can be generalized to the broader population of studies from which the included studies were drawn. Thus, certainty concerning a homogeneous effect can be held with respect to the specific studies in a meta-analysis (fixed-effects models) or a broader set of possible participants, settings, and measures that represent the full population of potential research studies (random-effects models).

Some meta-analyses not only address the size of a relation but also present the relation in specific metrics that can be interpreted in real-world terms. Pharmacological investigations, for example, sometimes report dose-response relationships that describe the specific outcomes obtained at different dosage levels of a drug or treatment. Even if no single study explored multiple treatment levels, studies with similar magnitudes of a treatment can be grouped and comparisons made across magnitude groupings to identify specific dose-response relations. Illustrating this approach, Luc Dauchet, Philippe Amouyel, and Jean Dallongeville meta-analyzed seven studies that estimated the effects of fruit and vegetable consumption on risk of stroke across an average 10.7 years (2005). For studies that assessed fruit consumption, each additional portion of fruit per day was associated with an 11 percent reduction in the risk of stroke. For studies that assessed fruit and vegetable consumption, each additional portion was associated with a 5 percent reduction, and for studies that assessed vegetable consumption, each portion was associated with only a 3 percent reduction. Furthermore, fruit consumption, though not vegetable or fruit and vegetable combined, had a linear relation to stroke risk, consistent with a causal effect of dose on response. The reported homogeneity in effects can be interpreted with some certainty despite the small number of studies because each study used large numbers of respondents, thereby providing relatively stable estimates, and because the studies included samples from Europe, Asia, and America and therefore the different diets typical of each region. Adding to the apparent validity of study findings, the effects for fruit consumption remained even after controlling for a number of lifestyle stroke risk factors such as exercise and smoking. Because the results can be accepted with some certainty, additional research to estimate the relation in comparable populations and settings is not likely to provide further insight. Nonetheless, the specific protective mechanism of fruit consumption was not identified, and the synthesists noted the need for additional research to identify the protective mechanism involved. As we explain in section 24.3, this issue could be addressed in syntheses or in primary research identifying mediators and moderators of the effects of food consumption on stroke risk.

Certainty also is accorded to the slightly more complex outcome in which homogeneity is obtained after the removal of a small number of outlying effect sizes. The empirical relation could still warrant high certainty if homogeneous findings emerged in all except a few studies. Given the vagaries of human efforts in science, a few discrepant outcomes should not necessarily shake one's confidence in a coherent underlying effect. For example, Alex Harris synthesized the literature on expressive writing about stressful experiences and identified one anomalously large effect in the fourteen available studies (2006). After it was removed, a homogeneous result was obtained such that writing experiences significantly reduced healthcare use in healthy samples. Sometimes synthesists can detect the reasons for aberrant findings, such as atypical participant samples or research procedures. Other times, as with Harris's synthesis, the reasons for the discrepancy are not clear. In section 24.2.2, we explain that future research would be useful to identify why some outcomes do not fit the general pattern. In general, findings that are homogeneous except for a few outlying effect sizes suggest that further research in the same paradigm would be redundant, especially if the causes of the discrepant findings are discernable and irrelevant to the theories or practice issues under consideration.

Clear outcomes of the types we have discussed were obtained in Marinus van Ijzendoorn, Femmie Juffer, and Caroline Poelhuis's investigation of the effects of adoption on cognitive development and intellectual abilities (2005). This policy-relevant question motivated a focus on effect magnitude. When these synthesists compared the adopted children with their unadopted peers left behind in an institution or birth family, the adopted children showed marked gains in IQ and school achievement. These effects were significantly different from zero and proved homogeneous across the included studies. Thus, adopted children were cognitively advanced compared with their unadopted peers, and this conclusion can be accepted with some certainty given the conditions represented in the included studies. Whether the cognitive benefits of adoption were enough to advance achievement levels to equal those of environmental peers, such as siblings in the adopting family, was less clear, however. As we explain in section 24.3, the uncertainty associated with this aspect of the findings indicates a need for additional investigation.

24.2.1 Implications of Relations Held with Certainty

Meta-analytic investigations produce empirical relations associated with high certainty when they meet the noted criteria, that is, that they have ample power to detect heterogeneity and that they aggregate across a representative sampling of settings, participants, and methods for the particular research question. In such cases, meta-analyses provide guidance for estimating the expected magnitude of effects in subsequent primary studies and meta-analyses. Researchers armed with knowledge of the size of effect typical in a research paradigm, for example, can use appropriate numbers of participants in subsequent studies to ensure adequate power to detect effects of that size.

Especially useful for allowing researchers to gauge the magnitude of effects are syntheses that aggregate the findings of multiple existing meta-analyses. For example, F. D. Richard, Charles Bond, and Juli Stokes-Zoota compiled 322 meta-analyses that had addressed social psychological phenomena (2003). Their estimates suggest the effect size that can be expected in particular domains of research. Another aggregation of meta-analytic effects, reported by David Wilson and Mark Lipsey, investigated how various features of study methodology influenced the magnitude of effects in research on psychological, behavioral, and educational interventions (2001; Lipsey and Wilson 1993). Their estimates across more than 300 meta-analyses revealed the outcomes that could be expected from methodological features of different research designs.

When meta-analyses provide estimates held with a high degree certainty, they can inform future research by providing information to calculate power analyses relevant to follow-up in primary research as well as meta-analytic investigations (Cohen 1988; Hedges and Pigott 2001). Once the expected effect size is known, power calculations can identify with some precision the sample size needed to detect effects of a given size in future research using similar operations and subject samples. With respect to subsequent meta-analyses, power calculations

for tests of effect magnitude indicate the required number of studies to detect an effect, given the average number of participants per study in that literature. Meta-analyses producing highly certain conclusions also provide heterogeneity estimates useful to calculate the number of studies appropriate to detect heterogeneity in subsequent syntheses. In this way, effects held with high certainty provide standards of magnitude and variability that can inform subsequent research designed to estimate empirical relations.

24.2.2 Uncertainty Attributable to Unexplained Heterogeneity

Meta-analyses designed to identify relations between variables often find that the included effect sizes are heterogeneous, and homogeneity is not established by removal of relatively few outlying effect sizes. In other words, studies have produced heterogeneous estimates of a given relation. Significant heterogeneity that arises from unexplained differences between findings threatens statistical conclusion validity, defined as the accuracy with which investigators can estimate the presence and strength of the relation between two variables (Shadish, Cook, and Campbell 2002). Heterogeneity also threatens the construct validity of a meta-analyzed relation, defined as how accurately generalizations can be made about theoretical constructs on the basis of research operations. Threats to construct validity lower confidence in the theoretical interpretation of the results, because it is unclear whether the various research findings represent a common underlying phenomenon.

Illustrating lowered confidence that comes from threats to construct validity is Daphne Oyserman, Heather Coon, and Marcus Kemmelmeier's meta-analysis of the cultural orientations of individualism and collectivism (2002). The synthesists compared three methods for operationalizing these constructs: direct assessments by psychological scales administered to individuals, use of the culture or country values of these constructs as identified by Geert Hofstede (1980), and priming by experimental manipulations to enhance the accessibility of one or the other construct. The three operations did not always produce the same effects, and, when common effects did arise, they could not with any certainty be attributed to the same psychological mechanism. Given this pattern of findings, Oyserman and her colleagues wisely called for research and theorizing that differentiated among the various meanings of individualism and collectivism and that developed appropriate operations to assess each of these.

It is unknown whether many early syntheses produced unexplained variability in effect sizes because these often did not report homogeneity statistics. However, computer software now provides synthesists with ready access to such techniques for assessing uniformity of synthesized findings. Correspondingly, standards have changed so that most published meta-analyses now contain reports of the variability in study outcomes and use this information to interpret results.

When faced with significant unexplained variability in effects, meta-analysts have several options. These begin with whether fixed- or random-effects models will be used to estimate effect size parameters (see chapter 3, this volume). When analyses were initially conducted using a fixed-effects model, one option was to rethink the assumptions of the analysis and, if appropriate, to use a random effects approach that assumed more sources of variance but still a single underlying effect size. With random effects models, the mean effect sizes, confidence intervals, significance tests, and homogeneity analyses are all computed with an error estimate that reflects the random variability that arises from both within-studies and between-studies sources. Thus, even if the hypothesis of homogeneity is rejected in the fixed effects model, the data may prove consistent with the assumption of a single underlying effect size in a random effects model, and the conceptual interpretation of the effect can be held with some certainty.

A second, separate option to achieve an interpretable empirical relation when study outcomes are heterogeneous is to evaluate features of the research that moderate, or covary with, the effect of interest. Moderator analyses are common within both a fixed-effects and a random-effects analyses. As we explain in section 24.3, when the systematic variability in the study findings can be satisfactorily explained merely by including such third variables, relatively high certainty can be accorded the obtained empirical relation as it is moderated by another variable or set of variables.

24.2.3 Uncertainty Attributable to Lack of Generalizability

Even when a meta-analysis produces an empirical relation for which the homogeneity of effects cannot be rejected, high certainty should be attached only within the limits of the included studies. Typically, these limits are not extremely narrow because most meta-analyses aggregate across multiple operationalizations of constructs, multiple periods, and multiple samples. They thus possess greater external validity than any single study. External validity refers to the generalizability of an effect to other groups of people or other settings (Shadish, Cook, and Campbell 2002).

The studies within a synthesis may, however, share one or more limitations of their design, execution, or reporting. For example, they may all use the same limited participant population (such as whites or males), or they may all be conducted in a single setting (such as experimental laboratories). Given such sample limitations, the external validity of the empirical relations demonstrated by the synthesis may be compromised.

Limitations in the sample of studies that threaten external validity can be revealed by presenting study attributes in tables or other displays to show their distribution (see chapter 26, this volume). Systematic presentation of study attributes also can reveal limits in the methods of operationalizing variables within a literature. Because it is rare that any one method of operationalizing a variable completely captures the meaning of a given concept, syntheses generally possess poor construct validity when studies uniformly use a particular measurement technique or experimental manipulation (see chapter 2, this volume). Given that the resulting meta-analytic findings may be limited to that particular operationalization of the construct and may not be obtained otherwise, they should be accepted with little certainty.

Meta-analyses can usefully guide subsequent research by revealing uncertain conclusions due to limited variations in study design and implementation or to limited samples intended to represent broader target populations. Subsequent primary research might examine whether the effect extends to underinvestigated populations of participants and research settings. Emphasizing this concern, John Weisz, Amanda Doss, and Kristin Hawley noted that a number of meta-analyses of youth psychotherapy have documented substantial evidence of success in treating diverse problems and disorders (2005). The empirical base for such conclusions remains somewhat limited, however, with, for example, 60 percent of the studies Weisz and his colleagues identified failing to report sample ethnicity (2005), and those doing so including approximately 57 percent white participants, 27 percent black, 5 percent Latino, 1 percent Asian, and 9 percent other ethnicities. Also, more than 70 percent of the studies provided no information about sample income or socioeconomic status. Thus it remains uncertain whether the treatment gains documented in recent meta-analyses of youth psychotherapy generalize across ethnic groups and the socioeconomic spectrum.

Another reason that meta-analyses may have limited certainty is that synthesists have confined their tests to one or a small subset of the possible operationalizations of conceptual variables. These omissions can help to guide subsequent research. Even though high certainty might be achieved with the specific operations available, follow-up research could be designed to represent conceptual variables even more optimally. This strategy is illustrated by Marinus van Ijzendoorn and his colleagues' meta-analysis of the cognitive developmental effects of adoption on children (2005). Some of their estimates of the effects compared an adopted sample's IQ with the population value of 100 for IQ. However, this population value may be an underestimate because intelligence tests tend to become outdated as people from Western countries increasingly profit from education and information and thus increase the population value beyond the presumed average score of 100. As a result, population IQ values may underestimate actual peer IQ. The synthesists speculated that this phenomenon explains the apparent presence of developmental delays when comparing adoptees with contemporary peers and its absence when comparing them with population values. Given this validity threat, subsequent research would need to assess peer intelligence in relation to IQ norms obtained from contemporaneous peers, such as siblings in the adopting family.

Sample limitations that threaten the external validity of meta-analytic findings also can arise from meta-analysts' inadequate searching of the available research literature or from limits in the available literature stemming from biases in publication. Search procedures are likely to be inadequate when synthesists search only accessible studies. Typically, accessibility is greatest for studies in the published literature, and, at least for Western scientists, studies in the English-language press. A wider selection of databases can yield a more representative sample of studies. Also, the most common databases may not access the grey literature (or fugitive literature) of studies that have not been published in journals (see chapter 6, this volume). Such literature includes doctoral dissertations, many of which remain unpublished because of statistically nonsignificant findings. Unpublished studies also can be accessed, at least in part, through locating reports of studies presented at conferences. Ordinarily,

conference presentations include studies that will never be submitted for publication, perhaps because they do not meet the high standards of journal publication. By accessing this grey or fugitive literature, synthesists would obtain a more representative sample of the studies that have been conducted.

Even if multiple databases are searched and grey or fugitive literature is included, some studies remain inaccessible because their findings were never written up, generally because their findings were not promising. Researchers' anticipation of publication bias in favor of statistically significant findings lowers their motivation to write up studies (Greenwald 1975), and, even without the operation of a real publication bias in the decisions of manuscript reviewers or editors, produces a bias in the sample of available studies. To address this issue of file-drawer studies, meta-analysts wisely turn to quantitative methods of detecting publication bias (for example, Kromrey and Rendina-Gobioff 2006; see also chapter 23, this volume). In these ways, limitations in search strategies or in the sample of studies evaluated can produce uncertainty in the conclusions drawn about synthesis effects.

A different problem that creates uncertainty about an empirical relation can arise when a meta-analysis uncovers evidence of serious methodological weaknesses in a research literature. Under such circumstances, synthesists may decide that a portion of the studies in the research provided invalid tests of the hypothesis of interest and thus contaminate the population of studies that provided valid tests. This situation arose in Alice Eagly and her colleagues' (1999) examination of the classic hypothesis that memory is better for information that is congenial with people's attitudes (or proattitudinal) than for uncongenial (or counterattitudinal). Despite several early confirmations of this hypothesis, many later researchers failed to replicate the finding. The meta-analysis produced an overall congeniality effect but also evidence of methodological weaknesses in the assessment of memory in many early studies (for example, failure to use masked coding of free recall memory measures). Because of the incomplete reporting of measurement details in many studies, it was unclear whether removal of these methodological sources of artifact would leave even a small congeniality effect. In follow-up experiments that corrected the methodological weaknesses, Eagly and other colleagues in a subsequent study obtained consistently null congeniality effects under a wide range of conditions, thus clarifying the findings of the earlier meta-analysis (2000).

Process data revealed that thoughtful counterarguing of uncongenial messages enhanced memory for them, despite their unpleasantness and inconsistency with prior attitudes.

In summary, the results of meta-analyses designed to evaluate relations between two variables can guide subsequent research in various ways, depending on the certainty accorded to the synthesis findings. Effects interpreted with high certainty provide a benchmark for effective design of subsequent research, especially for identifying appropriate sample sizes in subsequent designs to detect effects. Effects interpreted with limited certainty highlight the need for additional research to address the specific aspect of invalidity that threatens certainty. When certainty is lowered by poor external validity, subsequent research might establish that the effect occurs in new samples and settings. When certainty is lowered by poor construct validity, subsequent research might demonstrate that the effect emerges with multiple operations, especially those that are the most appropriate representations of the underlying constructs.

24.3 EVALUATING MODERATORS AND MEDIATORS OF AN EFFECT

Many meta-analyses are designed to examine not just a simple relation between two variables but also the role of additional variables that may moderate or mediate that relation. Such investigations have an empirical focus to the extent that they deductively evaluate patterns of relations among variables and a theoretical focus to the extent that they inductively examine theoretically relevant variables that might moderate or mediate the relation of interest. As with a meta-analysis focused on the relation between two variables, the certainty of the conclusions about moderators or mediators depends on the validity of the demonstrated empirical relations as well as on the theoretical interpretation of those relations.

Interpretations of empirical relations should be accorded little certainty when moderator or mediator variables are identified from post hoc or unanticipated discoveries in the data analysis. Uncertainty also is attached to moderating and mediating variables that are not sufficient to account for heterogeneity in study outcomes. When unexplained heterogeneity remains in predictive models, the certainty of scientists' theoretical interpretations declines because of threats to construct validity of the moderating and mediating variables.

24.3.1 Moderators of Main-Effect Findings

In meta-analysis, as in primary research, findings can reflect post hoc decisions that are organized by patterns that emerge in the data or can reflect theoretically grounded, a priori decisions. In one post hoc method, potential moderators of effect sizes are identified empirically from the study outcomes themselves. For example, Larry Hedges and Ingram Olkin suggested clustering effect sizes of similar magnitude and then evaluating the clusters of outcomes for study characteristics that might covary with the groupings (1985).

Illustrating this clustering approach is Donald Sharpe and Lucille Rossiter's analysis of the psychological functioning of the siblings of children with a chronic illness (2002). The synthesists noted an overall negative effect on sibling functioning along with significant heterogeneity in study outcomes that was not accounted for by a priori predictors. They empirically clustered studies into illness types, differentiating between those illnesses associated with statistically significant debilitating effects and those that obtained the contrary finding of a positive but nonsignificant effect. The illnesses yielding debilitating effects (cancer, diabetes, anemia, bowel disease) tended to be treated with intrusive ongoing methods that would likely be more restrictive of sibling school and play activities than the illnesses yielding null effects (cardiac abnormalities, craniofacial anomalies, infantile hydrocephalus). The post hoc classification of studies into illness types, however, renders these moderator findings somewhat uncertain and suggests the need for follow-up research to validate the moderating relation.

The distinction between a priori predictions and post hoc accounts of serendipitous relations is often not sharp in practice, either because investigators portray post hoc findings as a priori or because they have theoretical frameworks that are flexible enough to encompass a wide range of obtained relations. Nonetheless, moderator analyses not derived from theory can be unduly influenced by chance data patterns and thereby threaten statistical conclusion validity. Even though meta-analytic statistical techniques can adjust for the fact that post hoc tests capitalize on chance variability in the data, less certainty should be accorded to the empirical relations established with moderators derived from post hoc tests than to those developed within an explicit theoretical framework (for post hoc comparisons, see Hedges and Olkin 1985). This low level of certainty points to the need for follow-up research to validate the moderating relation in question.

Moderators also have uncertain empirical status when they are unable to account for all of the variability in a given empirical relation. Even Sharpe and Rossiter's empirical approach by which they identified the type of illness moderator from study outcomes failed to account adequately for the variability across included studies (2002). The remaining unexplained variability within the moderator groupings provides an additional reason for follow-up research.

As we argued with respect to estimation of relations between variables, uncertainty in the empirical evidence for an effect inherently conveys uncertainty in interpretation. However, uncertainty in interpretation can emerge from a variety of sources in addition to empirical uncertainty. Uncertain interpretations of moderating variables could emerge even when moderators are identified a priori and successfully account for the empirical variability in study outcomes. This kind of conceptual ambiguity is especially likely when moderators are identified on a between-studies basis—that is, when some studies represent one level of the moderator, and others represent other levels. With this strategy, the moderator variable typically is an attribute of a whole research report, and participants in the primary research have not been randomly assigned to levels of the moderator. Analyses on such synthesis-generated variables (see chapter 2, this volume) do not have strong statistical conclusion validity because the moderator variable may be confounded with other attributes of studies that covary with it.

Moderators identified on a between-study basis can be contrasted to within-study moderator analyses in which two or more levels of the moderator are present in individual studies. Analyses of within-study moderators can be interpreted with greater certainty, especially when researchers manipulated the moderators in the underlying primary studies and participants were assigned randomly to levels of the moderator variable. Under these circumstances, moderator effects can be attributed to the manipulated variables.

Especially low levels of certainty in theoretical explanation are associated with moderators that serve as proxy variables for other, causal factors. For example, synthesists have sometimes examined how the effects reported in included studies vary with those reports' publication year and identity of authors. Although the empirical relation between such variables and the effect of interest may be well established, its interpretation usually is not clear because the moderating variable cannot itself be regarded as causal but instead is a stand-in for the actual causal

factors. For example, an effect of publication year might be a proxy for temporal changes in the cultural or other factors that affect the phenomenon under investigation. This mechanism is especially plausible as an explanation for relations between publication year of the research reports and the size and magnitude of sex differences in social behavior. In recent years, women's increased participation in organized sports, employment outside the home, and education has co-occurred with reduced tendencies for research to report that men take greater risks than women (Byrnes, Miller, and Schafer 1999) and that men are more assertive than women (Twenge 2001). Other sex differences, however, have varied with year in a nonlinear fashion that does not correspond in any direct way with social change or with known variation in research methods—for example, the sex difference in smiling was smaller in data collected during the 1960s and 1980s than in other periods (LaFrance, Hecht, and Paluck 2003).

Other effects of publication year could be attributable to publication practices in science. For example, Michael Jennions and Anders Møller evaluated the effects of publication year in forty-four meta-analyses covering a wide range of topics in ecological and evolutionary biology (2002). They reported that across meta-analyses, effect size was related to year of publication such that more recently published studies reported weaker findings. These year effects could reflect that larger, statistically significant study outcomes within a research literature are published earlier, given that they have shorter time-lags in submission, reviewing, and editorial decisions. As Jennions and Møller argued, these shorter time lags could reflect publication biases in science that favor statistically significant findings. They noted that such a publication bias could also explain why the effect sizes were smaller in the studies with larger samples. Presumably, studies with small samples and small effects were missing because their statistically nonsignificant findings render them unlikely to meet the standards for publication. In this account, year and sample size serve as proxies for the true causal factor, which is the publication bias shared by authors, peer reviewers, and journal editors. Despite the plausibility of this account, proxy variables such as year cannot establish the effects with a high level of certainty.

24.3.2 Strategies Given Moderators of Low Certainty

We have argued that uncertainty can arise in moderator analyses when moderators are identified post hoc, are proxies for other causal factors, or are tested by between-study comparisons. These approaches yield instability in the obtained empirical relations as well as ambiguity in the correct conceptual interpretation of the moderator. Subsequent research can be designed to address these sources of uncertainty.

When uncertainty arises because moderators are assessed on a between-studies basis, sometimes such analyses can be validated within the same synthesis by within-study analyses. These opportunities arise when a few studies in a literature manipulate a potential moderator and the remaining studies can be classified as possessing a particular level of that moderator. For example, Wendy Wood and Jeffrey Quinn's synthesis of research on the effects of forewarning of impending influence appeals revealed that research participants tended to shift their attitudes toward an appeal before it is delivered, apparently in a pre-emptive attempt to reduce the appeal's influence (2003). Between-study moderator analyses suggested that this shift emerged primarily with message topics that were not highly involving. However, the effects of involvement could be established with limited certainty in the between-study comparisons; only a single study used a highly involving topic. Fortunately, Wood and Quinn were able to validate the involvement effect with supplemental within-study analyses that compared effect sizes in the high and low involvement conditions of studies that experimentally varied participants' involvement with the topic of the appeal. When corroborated in this way, greater certainty can be accorded to the empirical relations established by between-study moderator analyses and to their conceptual interpretation.

Interpretations given to proxy variables have particularly low certainty because they are both post hoc moderators and evaluated typically on a between-studies basis. To address these points of uncertainty, subsequent research is needed to demonstrate the replicability of the variable's covariation with study outcomes as well as to clarify its interpretation. Explaining the impact of a proxy variable can be quite difficult because it might reflect a wide range of other variables. There may be few guides for selecting any particular candidates to pursue in subsequent research. As we explain in the next section, one possible strategy is to examine study attributes or other variables that might plausibly mediate the effects of the proxy variable on study outcome.

When uncertainty arises in the conceptual interpretation of moderators, the optimal strategy for follow-up research is experimentation to manipulate the constructs

that have been identified. For example, Alice Eagly and Steven Karau found that the tendency for men to emerge as leaders more than women in initially leaderless small groups was stronger with masculine than feminine group tasks (1991). This finding, however, presented some ambiguity because the gender-typing of tasks was confounded with other features of studies' methods, and the primary researchers had determined gender typing in differing ways. Barbara Ritter and Janice Yoder therefore conducted an experiment in which mixed-sex pairs expected to perform a planning task that was identical except for its masculine, feminine, or gender-neutral topic (2004). The usual gender gap favoring male leadership was present with masculine and gender-neutral tasks but reversed to favor female leadership with feminine tasks. This finding confirmed the moderation of leader emergence by task sex-typing and secured the conceptual interpretation that access to leadership depends on the congruity of the particular leadership role with gender stereotypical expectations.

Sometimes the correct interpretation of any single moderator in a given synthesis is uncertain, but greater certainty emerges when the results are viewed across a set of moderating variables. This situation occurred in Wood and Quinn's synthesis (2003). They evaluated how warning effects on attitudes are moderated by a number of factors associated with participants' active defense of their existing views. The included research varied a diverse set of factors to encourage defensive thought about the appeal, including instructing participants to list their thoughts before receiving the message and using message topics high in personal involvement. Consistent with the interpretation that these experimental variations did in fact increase recipients' counterarguing to the impending message, they all increased attitudinal resistance to the appeal. This convergence of effects across a number of theoretically relevant variables promotes greater certainty in interpreting moderating relations than would be possible with examination of any single variable alone.

In summary, moderator variables that synthesists discover through empirical clustering on a post hoc basis or test through between-studies analyses can be informative, particularly because such analyses can demonstrate relations not investigated in primary research. Nonetheless, substantial uncertainty about the stability of the empirical relations can arise from the threats to statistical conclusion validity inherent in such analyses. Also, uncertainty in theoretical accounts can arise from the low construct validity inherent in such analyses, given the possible confounding influences of unacknowledged variables. Especially low levels of certainty should be accorded to the interpretation of variables such as author sex and year of publication that serve as proxies for other, more proximal predictors. Given moderators that yield low empirical certainty, one strategy is to replicate the moderator's impact in within-study moderator tests, and another is to manipulate the moderating variables in subsequent experimentation. A third strategy is to examine convergence of effects across sets of possible moderators. Yet a fourth strategy, as we explain in the next section, is to conduct mediational analyses that examine study characteristics that covary with the moderator.

24.3.3 Mediators of Main-Effect Findings

Meta-analyses can meet the dual goals of clarifying empirical relations and identifying theoretical interpretations by evaluating the mediators, or causal mechanisms, which underlie an effect. A strong demonstration of mediation requires a closely specified set of relations between the mediator and the main-effect relation to be explained (see Kenny, Kashy, and Bolger 1998). In brief, a mediating effect can be said to obtain when the following pattern occurs: an independent variable such as a health intervention program significantly influences a dependent variable such as exercise activity, a third, potentially mediating variable, such as self-efficacy to exercise also influences the exercise dependent variable, and the effect of the intervention program in increasing exercise is at least partially accounted for by increases in self-efficacy. Whenever several primary studies within a literature have evaluated a common mediating process, meta-analytic models can be constructed to address each of the relations involved in a mediation model (see chapter 20, this volume).

Illustrating mediation tests with meta-analytic data, Marta Durantini and her colleagues examined the mechanisms through which highly expert sources exert influence on health topics, in particular condom use (2006). In general, expert source messages increased condom use more than less expert ones. To identify the underlying psychological processes, the synthesists predicted the magnitude of the expertise effect from the possible mediating beliefs recipients held. The influence of expert sources on recipients' behavior change proved mediated by changes in recipients' perceptions of norms, perceived control, behavioral skills, and HIV knowledge. Experts' ability to bring about behavior change can thus be attrib-

uted to their success at changing these particular beliefs with respect to condom use. However, despite the strong empirical support for mediation, the models were tested with correlational data. The potential threat to statistical conclusion validity in correlational designs suggests the need for additional research to manipulate the mechanism in question.

Mediational analyses also can be applied to account for the effects of proxy variables such as date of publication and researcher identity. In a demonstration of how the effects of year of publication or data collection can reflect changes over time in the phenomenon being investigated, Brooke Wells and Jean Twenge reported significant shifts over time in sexual activity, especially among women (2005). From 1950 to 1999, the percentage of sexually active young people in the United States and Canada increased, and the age at first intercourse decreased. Wells and Twenge explored whether generational challenges, such as increasing divorce rates, are responsible for these changes. They found a positive relation between incidence of divorce and percentage of young people who were sexually active and approved of premarital sex. The generational changes did not have uniform effects across all indicators of sexuality, however, and divorce had little relation to age at first intercourse. The inconsistent evidence that cohort changes are linked to sexuality indicators renders the cause of the year effects uncertain, and future research is required to assess additional causal factors.

When the proxy variable that predicts study outcomes is research group, a promising approach to explaining these findings is to demonstrate that research group covaries with different research practices such as the preferred methods, data analytic procedures, or strategy for reporting findings. In the absence of such a demonstration, an innovative procedure for disentangling discrepancies between research groups' findings entails joint design of crucial experiments by the antagonists, a procedure Gary Latham, Miriam Erez, and Edwin Locke used to successfully resolve conflicting findings on the effect of workers' participation on goal commitment and performance (1988). Drawing on the spirit of scientific cooperation, the researchers who produced the discrepant findings in this area cooperated in the design and execution of experiments in order to identify critical points of divergence that accounted for the effects of research group. Especially subtle differences in methods not discernible from research reports can be revealed by such joint primary research.

Unfortunately, synthesists rarely have many opportunities to assess mediation. A set of primary studies assessing a common mediating process is most likely to be found when research has been conducted within a dominant theoretical framework. More commonly, because only a few studies in a literature yield data concerning common mediators of an effect, only a few relations among mediators and the effect can be evaluated. In this circumstance, mediator findings provide only suggestive evidence for underlying mechanisms. This situation arose in a meta-analysis of the relation between self-esteem and how easily people can be influenced (Rhodes and Wood 1992). A number of studies in the synthesis assessed message recipients' recall of the persuasive arguments in the influence appeal as well as opinion change to the appeal. Lower self-esteem was associated with lesser recall as well as with lesser opinion change compared with results for moderate self-esteem. People with low esteem were apparently too withdrawn and distracted to attend to and comprehend the messages and were not influenced by them. The parallel effects for recall and opinions provide suggestive support for models that emphasize message reception as a mediator of opinion change (McGuire 1968). However, because enough data were not available to examine the link between reception and influence, some uncertainty clouds this interpretation. In general, the internal validity of conclusions about a mediating process is threatened to the extent that a research synthesis is unable to examine the full range of relations involved in the hypothesized mediational process.

Sometimes the likely mechanism of mediation in a sythnesis is suggested by the pattern of study outcomes. For example, in a meta-analysis of epidemiological studies examining the effects of tea consumption on colorectal cancer risk, Can-Lan Sun and her colleagues reported sex differences in the protective effects of black tea (2006). That is, a significant risk reduction emerged for women but a nonsignificant risk enhancement effect emerged for men. The synthesists speculated that the effects for gender were attributable to hormonal mediators, given that in direct tests, sex hormones have been implicated in protection against colorectal cancer. Such a mediational model would need to be tested directly, however, to be accepted with any certainty.

In general, when studies do not provide measures of mediation, several indirect meta-analytic methods can be used to test hypotheses concerning mediation. One such method involves constructing general process models at the meta-analytic level from studies that have examined components of the process relations. For example, Beth Lewis and her colleagues examined mediators of the effects of health interventions to increase exercise on people's actual exercise activity (2002). Although they did not have enough data from individual studies to build complete mediational models, they evaluated the two components of mediation. First, they examined the effects of the interventions on potential mediators, including behavioral processes (such as substituting alternative responses, stimulus control), self-efficacy beliefs, cognitive processes (such as increasing knowledge, understanding risks), social support, and perceived pros versus cons of exercise. Second, they examined whether changes in the mediators coincided with changes in exercise activity. They concluded that the most consistent evidence supported behavioral processes as a mediator of successful interventions. However, in the absence of mediator data that examined the full sequence of events involved in the causal model, a meta-analytic combination of studies assessing component parts does not conclusively demonstrate the full proposed model. Because building process models at a meta-analytic level from separate tests of a model's components threatens internal validity, such analyses cannot be interpreted with high certainty.

Another indirect method for evaluating mediators involves judges' coding or rating of process-relevant aspects of the synthesized studies. Process ratings are especially useful when the included studies are drawn from diverse theoretical paradigms, and no single mediational process was assessed by more than a few studies. In this case, process ratings may make it possible to test a general causal model across the full range of included experimental paradigms. For example, David Wallace and his colleagues used process ratings to test the idea that the directive effects of attitudes on behavior are weakened by situational constraints, such as perceived social pressure and perceived difficulty of performance (2005). Constraints were estimated through contemporary students' ratings of the likelihood that others close to them would want them to engage in each of the behaviors in the included research and that each behavior would be difficult to perform. As expected, attitudes were not as closely related to behavior when the student judges reported stronger social pressure or difficulty. The synthesists concluded that behavior is a product of both situations and attitudes, and the relative impact of any one factor depends on the other. Of course, process ratings do not directly substantiate mediating processes. Although such judgments may have high reliability and stability, and their empirical relations to effect sizes may be strong, they are not necessarily unbiased. Such judgments may not accurately estimate the processes that took place for participants of studies conducted at earlier dates. For these and other reasons, judges' ratings of mediators do not yield high certainty in theoretical interpretation.

Uncertainty concerning the empirical status of mediating processes underlying an effect is a frequent occurrence in research syntheses for the reasons we have stated. We discussed three problems that may create uncertainty about mediation in meta-analytic syntheses. Mediational models may rely exclusively on correlational data, be constructed from primary research that tested only subsets of the overall model, and use judges' ratings to represent mediating variables.

24.3.4 Strategies Given Low Certainty About Mediators

When mediational models cannot be established with high certainty, the causal status of the mediator with respect to the relation of interest remains uncertain. Primary research is then needed to substantiate the suggested process models.

In research on eyewitness facial identification, investigators have pursued this strategy of following up with experimentation on the mediational models suggested in research syntheses. A meta-analytic synthesis by Peter Shapiro and Steven Penrod suggested that reinstatement of the context in which a face was initially seen facilitated subsequent recognition of the face (1986). The mechanisms through which context had this impact could not be evaluated in the synthesis, and subsequent primary research addressed this issue. By experimentally varying features of the target stimulus and features of the context, follow-up research indicated that reinstatement of context enhances recognition accuracy because it functions as a memory retrieval cue (Cutler and Penrod 1988; Cutler, Penrod, and Martens 1987). Follow-up research thus demonstrated with some certainty a mediational model that could account for the meta-analytic findings, specifically that context reinstatement provided memory cues to assist recognition.

In general, we advocate a two-pronged strategy to document mediational models with sufficient certainty. First, meta-analytic methods of evaluating the mediational process are devised and correlational analyses conducted to demonstrate that the mediator covaries as expected with the effect of interest. Because mediational models claim

to identify causal relations, we also advocate a second step of directly manipulating the hypothesized mediating process in primary research to document its causal impact on the dependent variable. Inviting such follow-up research is, for example, Sun and her colleague's finding about colorectal cancer risk, gender, and black tea (2006). If the protective effects are attributable to female sex hormones, then follow-up research with animals that directly manipulates tea consumption and hormone levels and examines the effects on cancer incidence should provide evidence of the mediating role of hormones on risk reduction. These two research strategies establish mediation with a high level of certainty by providing complementary evidence: correlational assessment of the mediator demonstrates that it varies in the predicted manner with the relation of interest, and experimental manipulation of the mediator demonstrates its causal role. Accomplishing the correlational assessment in a meta-analysis takes advantage of this technique's ability to document main-effect relations and to suggest possible process models. Manipulating the mediator in subsequent primary experiments capitalizes on the strong internal validity of this type of research.

24.4 USING MAIN EFFECTS, MODERATORS, AND MEDIATORS TO TEST THEORIES COMPETITIVELY

In recent years, research syntheses have begun to be used to test alternative theoretical accounts for particular phenomena. In competitive theory testing, the analysis strategy is not inductive but is driven largely by competing theoretical propositions. This approach benefits from the ready availability of meta-analytic statistical techniques and computing programs that facilitate theory testing, such as programs designed to estimate path models. Such meta-analyses sometimes proceed by examining main effect relations but more often by examining potential mediators and moderators of effects that are deduced directly from theoretical frameworks.

In meta-analyses, certainty in competitive theory testing is threatened when theoretically derived variables fail to account adequately for the patterns of observed effects. We have discussed some of these sources of uncertainty with respect to main effects, moderators, and mediators, especially the poor construct validity that emerges when theoretically defined variables are insufficient to account for the variability in effect size outcomes. Additional limits to construct validity arise when the included literature does not sample the full range of the theoretical variables or when the pattern of empirical effects is consistent with multiple theoretical models. In addition, uncertainty is suggested by threats to internal validity, such as insufficient evidence of causality among theoretical constructs.

24.4.1 Mechanisms of Competitive Theory Testing

Competitive theory testing about main effect relations might occur when one theory favors certain predictors of an outcome, whereas another theory favors other predictors. Illustrating this approach, Thomas Ng and his colleague's synthesis tested whether career success is better accounted for by human capital theory emphasizing employees' experience and knowledge as opposed to sociopolitical theory emphasizing employees' networking ties and visibility in the company or field (2005). Success in terms of salary was best predicted by human capital variables, presumably because they indicate individuals' worth to an organization. In contrast, success in terms of promotion to higher job levels was best predicted by social factors, presumably because of the political nature of promotion decisions. In this analysis, then, different theories appeared to account for different job outcomes.

Illustrating competitive theory testing using moderator analyses, Jean Twenge, Keith Campbell, and Craig Foster evaluated theoretical explanations for the reduction in marital satisfaction that occurs with the advent of children (2003). They considered how well each theory accounted for the pattern of moderating variables in the included studies. In the synthesis, marital satisfaction decreased more for mothers than fathers, those of higher than lower SES, and parents of infants than older children. This moderator pattern was consistent with two of the theories but not with others. Specifically, the moderation could be explained by a role conflict account, given that high SES women with infants are especially likely to experience conflict as parenthood causes family roles to be reorganized along traditional lines. Alternatively, the moderation could reflect a restriction of freedom account, if, as the synthesists assumed, high SES women with infants experience the greatest changes with the onset of motherhood, because their time becomes more restricted. With this failure of the meta-analytic data to arbitrate between the two competing theories, it is unclear how to interpret mothers' SES as it influences marital satisfaction. Given this threat to the conceptual validity of the SES measure, any one account needs to be accorded a relatively low

level of certainty as an explanation of the obtained pattern of findings.

Sometimes it is possible to arbitrate between competing theories by using statistical procedures that identify how well each set of theoretical relations fit the obtained meta-analytic findings. This approach was taken in Judith Ouellette and Wendy Wood's synthesis comparing decision making theories of action, in which behavior is a product of people's explicit intentions to respond, with theories of habitual repetition, in which repeated behavior is cued by stable features of contexts in which it is routinely performed (1998). These synthesists constructed path models to compare intentions and past behavior as predictors of future behavior. Consistent with decision-making models, behaviors such as getting flu shots, which are unlikely to be repeated with enough regularity to form habits, proved primarily a product of people's intentions. However, consistent with a habit model, behaviors such as wearing seat belts, which can be repeated with enough regularity in stable contexts to form habits, were primarily repeated without influence from intentions. This analysis strategy of computing path models in different groupings of studies and then comparing the models is essentially a correlational approach. Given the threat to internal validity in such a design, additional experimental research is needed to test the underlying causal model.

An additional illustration of statistical procedures to test theories competitively comes from Daniel Heller, David Watson, and Remus Ilies's investigation of whether individual differences in life satisfaction reflect the top-down effects of temperament and personality traits that yield chronic affective tendencies or the bottom-up mechanisms that reflect the impact of particular life events such as work and marriage (2004). Competitive tests of these accounts estimated meta-analytic relations between personality (neuroticism, conscientiousness), general life satisfaction, and specific satisfaction with work and marriage. A completely top-down model with personality predicting each of the types of satisfaction failed to fit the data. Instead, the data suggested some bottom-up input in the form of direct relations among the types of satisfaction themselves. However, the synthesists noted that their use of job and marital satisfaction as proxy variables for actual job and marital outcomes threatens the conceptual validity of their bottom-level measures. As a result, only low certainty can be accorded the conclusions based on the effects of these substitutes for real life events, and a more definitive test of the theory awaits more optimal operationalizations. Additional uncertainty in the conclusions arises from reliance on cross-sectional correlational studies. Such designs provide little basis for inferring causality and do not rule out the possibility that life satisfaction and life events influence personality, despite the synthesists' claim that personality is to a large extent innate and unlikely to be influenced by bottom-up events in this manner. Thus, this threat to internal validity further lowers certainty in the synthesis conclusions.

24.4.2 Strategies Given Low Certainty About Competing Theories

The appropriate strategy to follow up a competitive theory testing meta-analysis depends on where uncertainty emerges. As illustrated in Jean Twenge and her colleagues' investigation of the decrease in marital satisfaction with parenthood, the moderating variables predicting reduced marital satisfaction in the synthesis could plausibly be interpreted in terms of multiple theories (2003). Thus, follow-up meta-analytic or primary research is needed to assess directly the moderating and mediating variables uniquely indicative of each theoretical account.

Uncertainty in theory testing also arises from correlational designs that do not provide enough evidence of the direction of the causal relation among theoretical constructs. For example, Ouellette and Wood's meta-analysis used path models to differentiate between habits and intentions as predictors of future behavior (1998). To address the potential threat to validity, Jeffrey Quinn, Anthony Pascoe, David Neal, and Wendy Wood established habits through simple response repetition in an experimental paradigm and were able to demonstrate the causal impact of habitual tendencies as opposed to intentions in guiding future action (2009). Considerable theoretical certainty is established by this strategy of conducting experiments to evaluate the causal mechanisms thought to underlie correlational findings from a meta-analysis.

Uncertain meta-analytic conclusions also were produced by the cross-sectional correlational design of Heller, Watson, and Ilies's analysis of bottom-up versus top-down models of life satisfaction (2004). It is interesting to compare this approach with Sonja Lyubomirsky, Laura King, and Ed Diener's related meta-analysis addressing the top-down effects of dispositional happiness on success in specific life domains of marriage, friendship, and work (2005). Lyubomirsky and her colleagues conducted several analyses, each synthesizing a specific research design. The analysis of cross-sectional designs demonstrated the existence of relations between the happy dis-

positions and success in specific domains. The analysis of longitudinal designs provided clearer evidence of causal direction by demonstrating that happiness precedes success at work life, social relationships, and physical well-being. The analysis of experimental designs potentially could yield even greater certainty about causality, though the findings from this paradigm were mixed, with experimental inductions of happy mood associated with greater sociality and health but not always more effective problem solving. Lyubormirsky and her colleague's use of these study designs increased the certainty of their conclusions concerning the top-down influences of dispositions on life outcomes of sociality and health.

In summary, subsequent primary research or meta-analyses can be devised that use research designs with high internal validity to address the uncertainty in the causal direction underlying predicted relations.

24.5 CERTAIN AND UNCERTAIN FINDINGS DIRECT SUBSEQUENT RESEARCH

The scientific contributions of research syntheses with highly certain findings are widely acknowledged. Syntheses that produce highly certain research findings and theoretical accounts contribute to the canon of known scientific facts and theoretical models. These syntheses can be considered foundational in that they provide a base of accepted knowledge from which to launch new empirical investigations and develop new explanatory frameworks.

A less well recognized role for research synthesis is to identify points of uncertainty in empirical findings and theoretical explanations. Synthesis findings contribute to the progression of science when they take stock of what is not yet known across a body of literature. This generative type of synthesis identifies uncertainties that can catalyze subsequent empirical data collection and theory development. Generative syntheses provide guideposts indicating what issues need subsequent investigation. To assume this role, meta-analyses need to clearly identify the sources of uncertainty in the covered literature. Through the many examples in this chapter, we identified particular threats to validity that generate uncertainty in empirical and theoretical accounts. Subsequent research and theoretical development can then target these sources of uncertainty.

In contrast to these positive functions of the identification of uncertainties, syntheses also could produce uncertain findings with little understanding of where the uncertainty emerges. When uncertainty arises from ambiguous or unknown sources, syntheses fail as generative guides to subsequent research and theorizing. For example, syntheses that are conceptualized extremely broadly and that include studies in a wide variety of paradigms may not have much success in predicting study outcomes because of the many specific considerations that determine the size of effects within particular paradigms. In addition, the wide range of possible theoretical explanations for such diverse effects can make it difficult to test any particular theoretical model for the findings. To produce results that highlight specific sources of uncertainty, smaller, subsidiary meta-analyses examining predictors and mediators specific to each experimental paradigm would be necessary.

The two roles that syntheses can play with respect to subsequent research, providing both a foundation of accepted knowledge and a generative impetus to further investigation, have important implications for the way that research findings are portrayed. If scientific payoff through publication and citation is believed to flow primarily from highly certain empirical findings and theoretical explanations, synthesists may fail to capitalize on the generative role that syntheses can play in highlighting promising avenues for subsequent research. This role is not met through the conventional call for additional research that can be found at the end of many published studies. Instead, generative syntheses systematically assess the points of invalidity in an existing literature and identify the associated uncertainty with which scientists should accept empirical findings and theoretical accounts. We have argued for the importance of generative syntheses throughout this chapter, and we note that such analyses already appear in top journal outlets in psychology. Illustrating such a synthesis is the Oyserman, Coon, and Kemmelmeier analysis of the limited construct validity of individualism and collectivism in cultural psychology (2002).

The belief that syntheses are evaluated primarily for producing highly certain findings might also encourage synthesists inappropriately to portray their work as producing clear outcomes that definitively answer research questions. Conclusions that should be drawn with considerable uncertainty, because of either insufficient empirical documentation or unclear theoretical accounts, could be portrayed instead as highly certain. In the earlier edition of this handbook, we suggested that false certainty is conveyed when synthesists fail to test for homogeneity of effects or when they fail to consider adequately the implications of rejection of homogeneity for the effects they are summarizing (Eagly and Wood 1994). These problems are especially likely to occur in meta-analyses ori-

ented toward identifying empirical relations between variables. False certainty in meta-analyses that competitively test theories is likely to emerge when synthesists fail to acknowledge the limited conceptual validity of their variables or the poor internal validity that might, for example, impede identification of causal mechanisms in the theoretical analysis.

Syntheses that falsely emphasize certainty can discourage subsequent investigation and distort the scientific knowledge base. To the extent that these problems occur, quantitative synthesis fails to fulfill its promise of furthering progress toward the twin goals of ascertaining empirical relations and determining accurate theoretical explanations for these relations.

24.6 CONCLUSION

By emphasizing the varying levels of certainty that the scientific community should accord to the conclusions of syntheses, we encourage the authors and the consumers of meta-analyses to adopt a critical stance in relation to research syntheses. Syntheses should be critically evaluated with respect to specific goals. We differentiated between the goal of establishing particular relations and the goal of evaluating mediators and moderator of those relations. We noted that fulfilling either of these goals would allow meta-analyses to compare the adequacy of theoretical accounts for phenomena, including testing them competitively.

By contributing to foundational knowledge in a discipline and by generating new avenues for subsequent research, syntheses do not represent the final phase of a field's consideration of a hypothesis. Instead, the two goals of syntheses we considered represent potentially helpful aids to sifting through and evaluating evidence of empirical relations and theoretical explanations.

We intend for this chapter to encourage the authors of research syntheses to provide careful analyses of the certainty with which they accept empirical and theoretical conclusions. This chapter also promotes syntheses' generative role by suggesting areas of subsequent empirical research and theory development. Perhaps the perception that journal publication requires clear-cut findings has sometimes encouraged the authors of syntheses to announce their conclusions more confidently than meta-analytic evidence warrants. To counter this perception, we encourage journal editors to foster appropriate exploration of uncertainty and to favor publication of research syntheses that do not obscure or ignore the uncertainty that should be applied to their conclusions. Careful analysis of the certainty that should be accorded to the set of conclusions produced by a meta-analysis fosters a healthy interaction between research syntheses and future primary research and theory development.

24.7 REFERENCES

Byrnes, James P., David C. Miller, and William D. Schafer. 1999. "Gender Differences in Risk Taking: A Meta-Analysis." *Psychological Bulletin,* 125(3): 367–83.

Cohen, Jacob. 1988. *Statistical Power Analysis for the Behavioral Sciences.* Hillsdale, N.J.: Lawrence Erlbaum.

Cooper, Harris M., and Robert Rosenthal. 1980. "Statistical Versus Traditional Procedures for Summarizing Research Findings." *Psychological Bulletin,* 87(3): 442–49.

Cutler, Brian L., and Steven D. Penrod. 1988. "Improving the Reliability of Eyewitness Identification: Lineup Construction and Presentation." *Journal of Applied Psychology* 73(2): 281–90.

Cutler, Brian L., Steven D. Penrod, and Todd K. Martens. 1987. "Improving the Reliability of Eyewitness Identification: Putting Context into Context." *Journal of Applied Psychology* 72(4): 629–37.

Dauchet, Luc, Philippe Amouyel, and Jean Dallongeville. 2005. "Fruit and Vegetable Consumption and Risk of Stroke: A Meta-Analysis of Cohort Studies." *Neurology* 65(8): 1193–97.

Durantini, Marta R., Dolores Albarracín, Amy L. Mitchell, Allison N. Earl, and Jeffrey C. Gillette. 2006. "Conceptualizing the Influence of Social Agents of Behavior Change: A Meta-Analysis of the Effectiveness of HIV-Prevention Interventionists for Different Groups." *Psychological Bulletin* 132(2): 212–48.

Eagly, Alice H., Serena Chen, Shelly Chaiken, and Kelly Shaw-Barnes. 1999. "The Impact of Attitudes on Memory: An Affair to Remember." *Psychological Bulletin* 125(1): 64–89.

Eagly, Alice H., Patrick Kulesa, Laura A. Brannon, Kelly Shaw, and Sarah Hutson-Comeaux. 2000. "Why Counterattitudinal Messages are as Memorable as Proattitudinal Messages: The Importance of Active Defense Against Attack." *Personality and Social Psychology Bulletin* 26(11): 1392–1408.

Eagly, Alice H., and Steven J. Karau. 1991. "Gender and the Emergence of Leaders: A Meta-Analysis." *Journal of Personality and Social Psychology* 60(5): 685–710.

Eagly, Alice H., and Wendy Wood. 1994. "Using Research Syntheses to Plan Future Research." In *The Handbook of Research Synthesis,* edited by Harris Cooper and Larry V. Hedges. New York: Russell Sage Foundation.

Greenwald, Anthony G. 1975. "Consequences of Prejudice Against the Null Hypothesis." *Psychological Bulletin* 82(1): 1–20.

Harris, Alex H.S. 2006. "Does Expressive Writing Reduce Health Care Utilization? A Meta-Analysis of Randomized Trials." *Journal of Consulting and Clinical Psychology* 74(2): 243–52.

Hedges, Larry V., and Ingram Olkin. 1985. *Statistical Methods for Meta-Analysis.* Orlando, Fl.: Academic Press.

Hedges, Larry V., and Therese D. Pigott. 2001. "The Power of Statistical Tests in Meta-Analysis." *Psychological Methods* 6(3): 203–17.

Heller, Daniel, David Watson, and Remus Ilies. 2004. "The Role of Person Versus Situation in life Satisfaction: A Critical Examination." *Psychological Bulletin* 130(4): 574–600.

Hofstede, Geert. 1980. *Culture's Consequences.* Beverly Hills, Calif.: Sage Publications.

Jennions, Michael D., and Anders P. Møller. 2002. "Relationships Fade with Time: A Meta-Analysis of Temporal Trends in Publication in Ecology and Evolution." *Proceedings of the Royal Society of London* 269: 43–48.

Kenny, David. A., Deborah Kashy, and Nils Bolger. 1998. "Data Analysis in Social psychology. In *Handbook of Social Psychology,* 4th ed., edited by Daniel Gilbert, Susan Fiske, and Gordon Lindzey. New York: McGraw-Hill.

Kromrey, Jeffrey D., and Gianna Rendina-Gobioff. 2006. "On Knowing What We Do Not Know: An Empirical Comparison of Methods to Detect Publication Bias in Meta-Analysis." *Educational and Psychological Measurement* 66(3): 357–73.

LaFrance, Marianne, Marvin A. Hecht, and Elizabeth L. Paluck. 2003. "The Contingent Smile: A Meta-Analysis of Sex Differences in Smiling." *Psychological Bulletin* 129(2): 305–34.

Latham, Gary P., Miriam Erez, and Edwin A. Locke. 1988. "Resolving Scientific Disputes by the Joint Design of Crucial Experiments by the Antagonists: Application to the Erez-Latham Dispute Regarding Participation in Goal Setting." *Journal of Applied Psychology* 73(4): 753–72.

Lewis, Beth A., Bess H. Marcus, Russell R. Pate, and Andrea L. Dunn. 2002. "Psychosocial Mediators of Physical Activity Behavior Among Adults and Children." *American Journal of Preventive Medicine* 23(2S): 26–35.

Lipsey, Mark W., and David B. Wilson. 1993. "The Efficacy of Psychological, Educational, and Behavioral Treatment: Confirmation from Meta-Analysis." *American Psychologist,* 48(12): 1181–1209.

Lyubomirsky, Sonja, Laura King, and Ed Diener. 2005. "The Benefits of Frequent Positive Affect: Does Happiness Lead to Success?" *Psychological Bulletin* 131(6): 803–55.

McGuire, William J. 1968. "Personality and Susceptibility to Social Influence." In *Handbook of Personality Theory and Research*, edited by E. F. Borgatta and W. W. Lambert. Chicago: Rand McNally.

Ng, Thomas W. H., Lillian T. Eby, Kelly L. Sorensen, and Daniel C. Feldman. 2005. "Predictors of Objective and Subjective Career Success: A Meta-Analysis." *Personnel Psychology* 58(2): 367–408.

Ouellette, Judith A., and Wendy Wood. 1998. "Habit and Intention in Everyday Life: The Multiple Processes by Which Past Behavior Predicts Future Behavior." *Psychological Bulletin* 124(1): 54–74.

Oyserman, Daphne, Heather M. Coon, and Markus Kemmelmeier. 2002. "Rethinking Individualism and Collectivism: Evaluation of Theoretical Assumptions and Meta-Analyses." *Psychological Bulletin* 128(1): 3–72.

Quinn, Jeffrey M., Anthony Pascoe, David Neal, and Wendy Wood. 2009. *Strategies of Self-Control: Habits Are Not Tempting.* Manuscript under editorial review.

Rhodes, Nancy D., and Wendy Wood. 1992. "Self-Esteem and Intelligence Affect Influenceability: The Mediating Role of Message Reception." *Psychological Bulletin* 111(1): 156–71.

Richard, F. D., Charles F. Bond, Jr., and Juli J. Stokes-Zoota. 2003. "One Hundred Years of Social Psychology Quantitatively Described." *Review of General Psychology* 7(4): 331–63.

Ritter, Barbara A., and Janice D. Yoder. 2004. "Gender Differences in Leader Emergence Persist Even for Dominant women: An Updated Confirmation of Role Congruity Theory." *Psychology of Women Quarterly* 28(3): 187–93.

Shadish, William, R., Thomas D. Cook, and Donald T. Campbell. 2002. *Experimental and Quasi-Experimental Designs for Generalized Causal Inference.* Boston, Mass.: Houghton Mifflin.

Shapiro, Peter N., and Steven D. Penrod. 1986. "Meta-Analysis of Facial Identification Studies." *Psychological Bulletin* 100(2): 139–56.

Sharpe, Donald, and Lucille Rossiter. 2002. "Siblings of Children with a Chronic Illness: A Meta-Analysis." *Journal of Pediatric Psychology* 27(8): 699–710.

Sun, Can-Lan, Jian-Min Yuan, Woon-Puay Koh, and Mimi C. Yu. 2006. "Green Tea, Black Tea and Colorectal Cancer Risk: A Meta-Analysis of Epidemiologic Studies." *Carcinogenesis* 27(7): 1301–09.

Twenge, Jean M. 2001. "Changes in Women's Assertiveness in Response to status and Roles: A Cross-temporal Meta-Analysis, 1931–1993." *Journal of Personality and Social Psychology* 81(1): 133–45.

Twenge, Jean M., W. Keith Campbell, and Craig A. Foster. 2003. "Parenthood and Marital Satisfaction: A Meta-Analytic Review." *Journal of Marriage and the Family* 65(3): 574–83.

Van Ijzendoorn, Marinus H., Femmie Juffer, and Caroline W.K. Poelhuis. 2005. "Adoption and Cognitive Development: A Meta-Analytic Comparison of Adopted and Nonadopted Children's IQ and School Performance." *Psychological Bulletin* 131(2): 301–16.

Wallace, David S., René M. Paulson, Charles G. Lord, and Charles F. Bond Jr. 2005. "Which Behaviors do Attitudes Predict? Meta-Analyzing the Effects of Social Pressure and Perceived Difficulty." *Review of General Psychology* 9(3): 214–27.

Weisz, John R., Amanda J. Doss, and Kristin M. Hawley. 2005. "Youth Psychotherapy Outcome Research: A Review and Critique of the Evidence Base." *Annual Review of Psychology* 56: 337–63.

Wells, Brooke E., and Jean M. Twenge. 2005. "Changes in Young People's Sexual Behavior and Attitudes, 1943–1999: A Cross-Temporal Meta-Analysis." *Review of General Psychology* 9(3): 249–61.

Wilson, David B., and Mark W. Lipsey. 2001. "The Role of Method in Treatment Effectiveness Research: Evidence from Meta-Analysis." *Psychological Methods* 6(4): 413–29.

Wood, Wendy, and Jeffrey M. Quinn. 2003. "Forewarned and Forearmed? Two Meta-Analytic Syntheses of Forewarnings of Influence Appeals." *Psychological Bulletin* 129(1): 119–38.

25

RESEARCH SYNTHESIS AND PUBLIC POLICY

DAVID S. CORDRAY
Vanderbilt University

PAUL MORPHY
Vanderbilt University

CONTENTS

Raising the drinking age has a direct effect on reducing alcohol-related traffic accidents among youth affected by the laws, on average, across the states.

U.S. General Accounting Office (1987)[1]

More time spent doing homework is positively related to achievement, but only for middle and high school students.

Harris Cooper, Jorgianne Robinson, and Erika Patall (2006),
Harris Cooper and Jeffrey Valentine (2001)

Radiofrequency denervation can relieve pain from neck joints, but may not relieve pain originating from lumbar discs, and its impact on low-back pain is uncertain.

Leena Niemisto, Eija Kalso, Antti Malmivaara,
Seppo Seitsalo, and Heikki Hurri (2003)

25.1 INTRODUCTION

These three statements are based on some form of research syntheses. Each was motivated by different interests but all share a purpose—they were conducted to inform policy makers and practitioners about what works. Rather than relying on personal opinion, a consensus of experts, or available research, each statement is based on the results of a systematic, quantitatively based synthesis. The purpose of this chapter is to describe the past, present and future roles of research synthesis in public policy arenas like the ones represented in our three examples. Because meta-analysis is the chief research synthesis method for quantitatively integrating prior evidence and is an essential focus of this handbook we do not dwell on the technical aspects of meta-analysis. Instead, we focus on contemporary features of the policy context and the potential roles of research synthesis, especially meta-analysis, as a tool for novice, journeyman, and expert policy analysts. In doing so, we briefly describe what is generally meant by the phrase *public policy*, highlighting its breadth and who might be involved. The discussion concludes with a characterization of how research synthesis practices and results can be influenced by the political processes associated with public policy making. Stated bluntly, politics can affect the way that analysts plan, execute, and report the findings of their studies. The end products represent a mixture of evidence and political ideology, thereby minimizing the value of what is supposed to be an orderly and relatively neutral view of evidence on a particular policy topic.

25.2 PUBLIC POLICY, POLICY MAKERS, AND THE POLICY CONTEXT

It is relatively easy to recognize when statements, opinions, and actions could be construed to represent instances of public policy action. However, the phrase *public policy* is amorphous. Strangely enough, it is rarely defined, even in books devoted to the topic. As such, we take a common sense approach to its definition by turning to the Merriam-

Webster dictionary for guidance. Looking at the dictionary definitions of each term separately, we find that the term *public* is defined as "of, relating to, or affecting the people as a whole" and the term policy is defined as "a definite course or method of action selected to guide and determine present and future decisions." A secondary definition indicates that policy entails "wisdom in the management of affairs." Putting these terms together, along with explicit and implicit meanings behind their definitions, *public policy* refers to a deliberate course of action that affects the people as a whole, now and in the future.

Public policies come in many shapes, sizes, and forms. They can range from broad policies such national medical insurance to constrained or small-scale policies such as a city ordinance to curb littering. Further, many broad policies unleash a cascade of "smaller *p*" policies. For example, provisions in the 2001 reauthorization of Title I of Elementary and Secondary Education Act (known popularly as No Child Left Behind) call for states to use scientifically based educational programs and practices. In turn, this law required additional smaller policy decisions such as defining what qualifies as scientific evidence.

The origins of policies can result from partisan sea changes or bipartisan collaboration. However, more often than not, they are the result of incremental changes and adjustments in existing public policies. These adjustments are likely to be the result of interactions among an elaborate network of policy actors. In a common scenario, policy makers (for example, members of the legislature) are likely to be approached by other members of the policy shaping communities (for example, lobbyists and special interest groups) about the merits or demerits of impending courses of action. Those seeking to influence a primary policy maker may use empirical evidence in supporting their positions. This is often an ongoing process, with input from multiple sources. At some point in this process, differences of opinion become evident, and the policy-relevant question becomes "who is right" about "what works." For example, before reauthorization hearings, the credibility of the evidence underlying the positions of proponents and opponents became key to the fate of the Special Supplemental Food Program for Women, Infant and Children (or WIC) program. These were the policy questions that framed one of GAO's first policy-related syntheses (GAO 1984).

25.2.1 Who Are the Policy Makers?

The foregoing discussion provides a classic view of policy making and implies an aristocratic view of the policy maker, with a single or few actors making decisions about the fate of programs (for example, the president, members of congress). Other views are more egalitarian about the composition of the policy community. For example, Lee Cronbach notes: "In a politically lively situation . . . a program becomes the province of a *policy-shaping community*, not of a lone decision maker or tight-knit group" (1982, 6, emphasis added).

In addition, the checks and balances built into democratic governing structures in the United States and elsewhere introduce a dizzying interplay of policy makers and actors within these policy shaping communities. Officials within the executive, legislative, and judicial sectors of federal, state, and local governments enact, implement, monitor, and revise policies. These actions stem from laws, regulations, executive orders, and judicial opinions, to name a few. However, as we will see, the array of policy actors is not limited to officials in these three branches of government. The private sector, including businesses, foundations, associations, and the public at large, often play key roles in policy decisions. In turn, these unofficial policy actors can be essential to the specification of research questions within meta-analytic studies, affecting what is and is not examined, how, and why.

The definition of a policy-shaping community for Cronbach is left to the particular policy circumstance. However, officials at the Agency for Healthcare Research Quality (AHRQ, formerly the Agency for Health Care Policy and Research, or AHCPR), opened the doors widely in defining the policy-shaping community associated with its twelve evidence-based practice centers. According to AHRQ, these centers were charged with developing

> evidence-based reports and technology assessments on topics relevant to clinical, social/behavioral, economic, and other health care organization and delivery issues—specifically those that are common, expensive, and/or significant for the Medicare and Medicaid populations. With this program, AHRQ became a 'science partner' with private and public organizations in their efforts to improve the quality, effectiveness, and appropriateness of health care by synthesizing and facilitating the transition of evidence-based research findings. Topics are nominated by non-federal partners such as professional societies, health plans, insurers, employers, and patient groups. (2006, 1)

It is significant that the AHRQ statement appears to redefine the conventional notion of who are policy makers. The quote strongly emphasizes partnerships among public and private organizations and citizens. Affirming this broader representation of policy makers, David Sackett, William Rosenberg, Muir Gray, Brian Haynes, and Scott Richardson refer to evidence-based medicine as a "hot topic for clinicians, public health practitioners, purchasers, planners, and the public" (1996, 71). The current interest in evidence-based practice shatters the old notion of policy making as a top-down enterprise by broadly defining potential users of research and syntheses. Stated differently, policy-oriented research syntheses can involve public policy issues that are intended for consumption by an array of public- and private-sector interests.

25.2.2 Competing Agendas and Perspectives

Public policy entails such fundamental issues as how to allocate resources, how to protect the public from natural and man-made harm, how to promote health, education, and welfare, and how to equalize opportunities. Whereas Cronbach's image of the policy shaping community appears polite and orderly, the actual process is likely to be less rational. In arriving at answers to these questions, the collection of actors within policy shaping communities (legislators, judges, lobbyists, and policy analysts) can, and often do, bring to the table a host of conflicting views, ideologies, and agendas. As a result of these differences, public policy actions often represent negotiated settlements across these multiple perspectives, but unilateral, mandated policies can also be an outcome of the process. Admittedly, neither is the most rational means of optimizing public policy, but public policy is at the heart of democracy.

As much as we may favor democracy, inherent in the idea of collective action is the realization that conflicting views, ideologies, and agendas bring with them the opportunities for the introduction of biases. Meta-analysts are not immune to the influence of biases introduced by their own predispositions or those of the organization within which they work (Wolf 1990). We return to this issue after an overview of research synthesis within the policy context, in the next section.

25.3 SYNTHESIS IN POLICY ARENAS

The statistical roots of meta-analysis date back to the early 1900s, when Karl Pearson attempted to aggregate the results of statistical studies on the effectiveness of a vaccine against typhoid (1904). Given that the vaccine was to be given to British soldiers at risk of typhoid fever, this could be regarded as the first use of meta-analytic

techniques in a public policy context. Routine use of meta-analytic techniques, in general, did not appear until after classic papers by Gene Glass and Richard Light and Paul Smith were published (Glass 1976; Light and Smith 1974). Some policy-oriented analysts did not waste much time adopting and adapting Gene Glass's methods of meta-analysis to policy-oriented topics. As such, various forms of quantitative research synthesis have been used within the policy context for nearly three decades (see Cordray and Fischer 1995; Hunt 1997; Wachter and Straf 1990).

A number of books are particularly enlightening about the emergence of meta-analysis and research synthesis in policy areas. The first was a joint venture between the National Research Council's Committee on National Statistics (CNSTAT) and the Russell Sage Foundation (RSF). The volume, titled *The Future of Meta-Analysis* and edited by Kenneth Wachter and Miron Straf, is the result of a workshop to examine the strengths and weaknesses of meta-analysis, with particular attention to its application to public policy areas (1990). The workshop was convened by CNSTAT when it became apparent that the translation of meta-analytic methods to policy areas may require additional vigilance over the integrity of all phases of the research process.

The second book, *How Science Takes Stock: The Story of Meta-Analysis*, also sponsored by the Russell Sage Foundation, was written by Morton Hunt. In telling the early story of meta-analysis, Hunt chronicled numerous examples of how meta-analysis was used in examining the effects of policy topics in education, medicine, delinquency, and public health (1997). The book provides some rare and interesting behind-the-scenes accounts of these early applications. Many of these accounts are derived from candid interviews with authors of meta-analytic studies sponsored by the foundation and summarized in *Meta-Analysis for Explanation* (Cook et al. 1992).

Richard Light and David Pillemer provided a general overview of the science of synthesizing research (1984). However, their introduction set the stage for the use of these techniques in policy areas. The approach to the topic also is readily seen in GAO's early edition of their report on the Evaluation Synthesis methodology (see Cordray 1990; GAO 1992). Other books that highlight the use of research synthesis in policy-like contexts include Harris Cooper (1989), John Hunter and Frank Schmidt (1990), and John Hunter, Frank Schmidt, and Gregory Jackson (1982). Harris Cooper and Larry Hedges, whose work the Russell Sage Foundation also supported as part of its efforts to infuse meta-analysis into the policy community and beyond, provided a technical and practical approach to research synthesis (1994). In addition, as described later in some detail, the Cochrane Collaboration represents a major driving force behind contemporary interest in research synthesis in the public sector (www.cochrane .org). Not only have many systematic reviews been conducted under the auspices of the Cochrane Collaboration, its organizational structure, procedures, and policies have served as an important prototype for other organizations such as the Campbell Collaboration and IES's What Works Clearinghouse (www.campbellcollaboration.org; www.ed.gov/ies/whatworks). Without meaning to overstate the case, it could be argued that the cumulative effect of these varied resources have significantly facilitated the use of quantitative methods in policy making endeavors. The next section takes a broad look at the role of syntheses in such efforts.

25.3.1 Some Historical Trends

How prevalent are policy-oriented quantitative syntheses such as meta-analyses?[2] Although there are no official registries of policy-oriented quantitative syntheses, we attempted to examine their popularity by examining data bases on published and unpublished literature, searching the World Wide Web, and looking at specialized archives such as the *Congressional Record*. As seen in this section, policy-oriented research syntheses (more specifically, meta-analyses) are not easily identified by conventional search methods. For example, a search of thirty-four electronic data bases available through the Vanderbilt University libraries identified fewer than fifty entries when the search terms were *meta-analysis and public policy*. Other searches produced similar results.

Figure 25.1 shows three trend lines for references to policy, meta-analysis, and the joint occurrence of policy and meta-analysis in the Web of Science database for 1980 to 2006.[3] Given the relative youth of research synthesis and, in particular, meta-analysis, it is not surprising to see that before 1990, fewer than seventy-five meta-analytic studies were captured in this bibliometric archive.[4] After 1990, the growth in number of references was dramatic; by the year 2006, more than 3,000 references to meta-analysis were identified. A parallel growth is seen in the number of references to studies, syntheses, and other publications pertaining to policy; as with meta-analysis, the peak year for references to policy is 2006. In that year alone, a total of more than 12,000 references were found.

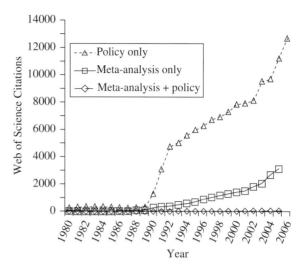

Figure 25.1 Comparison of Publications

SOURCE: Authors' compilation.

Despite the growth in published studies of meta-analysis, in general, and studies directed at policy issues, the explicit presence of meta-analysis within policy studies is strikingly low. We found no references to policy and meta-analysis in the Web of Science searchers before 1990; by 2006, the joint use of these keywords increased to only seventy-one references. We are aware that restricting the search to the term *meta-analysis* per se probably underrepresents the true presence of meta-analytic-type syntheses in policy areas. On the other hand, the omission of either term suggests that the potential policy analytic role of meta-analysis has not yet taken a firm hold in the hearts and minds of analysts.

Other evidence of the infancy of meta-analysis in the macro-policy context is evident from the infrequent mention of meta-analysis per se in congressional hearings. A LexisNexis search of the *Congressional Record* of hearings from 1976 to 2006 showed a modest presence of the term in hearings. At its peak in 1998, it was mentioned in about sixty hearings. Given that there are, on average, about 1,500 hearings per year, meta-analysis was mentioned at most in about 4 percent of the hearings.

Although there has been interest in the use of meta-analysis and other forms of quantitative synthesis (like GAO's evaluation synthesis) in policy areas, its presence is not dominant, as indexed in readily available resources such as the Web of Science and the *Congressional Record*.

As previously noted, this absence could be attributed to the relative youth of meta-analysis, but it also reflects a lack of policy-identity on the part of meta-analyst. Contemporary meta-analysts may see their work contributing to the knowledge-base as the primary goal; as a secondary goal, it is available to others who may want to use the results in the policy process. In later sections, we will see that there is a great deal of work being done with research synthesis methods in policy areas. We get the impression that meta-analysis is on the brink of becoming a major part of the evidence base underlying important questions about which programs and policies work (or not).

25.4 POTENTIAL BIASES

Although minimizing the influence of bias in all phases of the research process is a common goal for all investigators, Wachter and Straf and Frederic Wolf reminded us of the need to minimize biases in: framing the questions, selecting studies for inclusion, decisions regarding analysis, and reporting the results (Wachter and Straf (1990; Wolf 1990). David Cordray identified specific biases that can enter into these phases of the research process (1994). Deliberate efforts to minimizing bias are especially important when social science methods are used to answer public policy questions. In fact, as described later, the importance of minimizing several forms of bias has resulted in the development of detailed procedures and policies for the conduct of policy-oriented syntheses by major organizations that promote the use of syntheses to inform policy makers and practitioners about what works, such as the Cochrane Collaboration. In this section, we present an extended discussion of an extraordinary example of how competing agendas and their associated biases can play-out when meta-analysis is conducted on a highly volatile public policy topic. The specific case involves the Environmental Protection Agency's (EPA) analysis of the effects of secondhand smoke on health outcomes. Other examples reinforce the idea that politics and ideologies can affect synthesis practices.

25.4.1 The Environmental Protection Agency and Secondhand Smoke

In 1992, the Environmental Protection Agency issued a report that causally linked exposure to environmental tobacco smoke (ETS, or secondhand smoke) with the increased risk of lung cancer. The EPA concluded pointedly

that "ETS is a Group A human carcinogen" that causes the cancer deaths of 3,000 Americans annually (1992). Using the Radon Gas and Indoor Air Quality Research Act of 1986 as justification for its study, EPA analysts and expert consultants summarized available studies through September of 1991. To estimate the risk of ETS on health outcomes, the EPA synthesized reports from twenty-seven case-control studies and four prospective cohort studies conducted across eight countries. These studies examined the effects of spousal smoking on lung cancer rates for nonsmoking females married to smoking spouses. Combining the results of epidemiological studies resulted in a relative risk (RR) ratio of 1.19. The 90 percent confidence interval for this estimate did not include 1.0, so EPA declared the RR as statistically significant. In combination with their analysis of bioplausibility, EPA argued that the significant RR ratio allowed them to assert that secondhand smoke was a cause of adverse health outcomes. On release, this report had an immediate effect on policy. Actions included introduction of the Ban on Smoking in Federal Buildings Act and the introduction of the Smoke-Free Environment Act (HR 3434), eliminating smoking in public, nonresidential buildings.

The EPA ETS study was the subject of at least five congressional hearings, received substantial criticism, and resulted in an extended legal battle. Among the extended list of objections, critics first insisted that EPA altered its criteria for statistical significance, using a 90 percent confidence interval, when an earlier statistical analysis using the conventional $\alpha = 0.05$ (two-tailed) failed to reject the null hypothesis. Here, opponents insisted that the EPA bent the rules of analysis to fit its own ends (Gori 1994b) or stated that they performed an unorthodox reanalysis (Coggins 1994). Second, critics questioned the adequacy of EPS's assessment and management of confounding variables (Coggins 1994; Gravelle and Zimmerman 1994b; Redhead and Rowberg 1998). One analyst even suggested that control for confounds was impossible, given the nonexperimental nature of the primary studies (Gori 1994b). It is significant that critics accused the EPA of cherry-picking data (studies) to support a ban on smoking in the work place, that it, the EPA, ignored or only partially reported disaffirming findings (Coggins 1994; Gori 1994a; Gravelle and Zimmerman 1994a). The EPA report failed to mention the absence of findings for work place risk or cancer risk, and the study results pertain only to exposure in the home. In an effort to sort out the merits of these criticisms, Congress asked the Congressional Research Service to examine the EPA report; they agreed with prior criticisms and raised numerous additional problems with EPA's ETS study (1995).

After several rounds of criticisms and rebuttals EPA found itself in an extended legal battle with an important member of the policy-shaping community—the tobacco industry (see Farland, Bayard, and Jinot 1994; Jinot and Bayard 1994). Ultimately, a federal district judge—Judge William Osteem—invalidated the study because of procedural and technical flaws (1998). On the procedural side, Osteem concluded in his ruling that EPA came to its conclusion about the danger of secondhand smoke long before, in 1989, they collected and synthesized the available empirical evidence. Further, in his ruling, Judge Osteem concluded that EPA did not follow the law [the Radon Research Act] which requires them "to seek the advice of a representative committee during the research" (89). He explicitly noted that because the tobacco industry was not represented, the "assessment would very possibly not have been conducted in the same manner nor reached the same conclusions" (89). On the technical side, the judge relied on published criticisms, using classic social science notions, that "EPA disregarded information and made findings on selective information. EPA deviated from its own Risk Assessment Guidelines by failing to disclose important findings and reasoning; and left significant questions without answers" (88). Finally, changing the significance testing criteria from .05 to .10 was deemed inappropriate because it gave the impression that EPA was tailoring their analysis (adopting a more liberal standard) to support their 1989 position on the cancerous effects of secondhand smoke.

The Osteem ruling includes several points that highlight the EPA's flaunting of scientific conventions and general wisdom. Large among them were adjusting significance a posteriori from .05 to .10 and excluding the tobacco industry from the development of their review. As a policy directed meta-analysis, these two decisions were damaging, first, as an affront to the division of exploratory and confirmatory analysis, and, second, by restricting the protective peer review process to those with whom one already agrees. Proper peer-review alone would have corrected both problems. It is not surprising, therefore, that the OMB would institute policy guidelines addressing these issues in government funded research (Office of Management and Budget 2004). Variously, the new policy demanded consideration beyond what the data show to how the data might be made use of, balancing of oversight committees, and not least that "transparency and openness, avoidance of real or perceived conflicts of interest, a workable process for public comment and involvement," along with adherence to predefined procedures (12).

In this example, EPA had a stake in the outcome of the synthesis. As documented in the Judge Osteem's opinion, important procedural and technical decisions were influenced by EPA's policy agenda. For our purposes, the EPA-ETS example shows that meta-analytic studies can serve as an integral part of the policy-making process. However, it also shows that it can be subverted in the policy arena. The lesson to be drawn from this example is that checks and balances are needed to create a firewall between the science of research synthesis and the policy process.

25.4.2 National Institute of Education, Desegregation, and Achievement

In less dramatic examples, the effects of an analysts' perspective on research synthesis (and meta-analytic) results also is evident in the case of syntheses of the effectiveness of desegregation on student achievement. For example, the (then) National Institute of Education became worried about the radically different conclusion that experts in desegregation came to after synthesizing the literature. Because of these disparities, in 1982 the National Institute of Education (NIE) established a panel of six desegregation experts to examine if meta-analytic methods could be used to resolve prior differences among experts. Although the conclusions showed less variability, the role of prior perspective was still evident. Jeffrey Schneider provided an overview of the NIE project (1990). Harris Cooper's research on the NIE panel provided direct evidence of the influence of prior perspectives on meta-analytic practices and results (1986). That is, the meta-analytic results and conclusions presented by each expert closely paralleled the individual positions that each had offered in prior research and writing.

25.4.3 National Reading Panel and Phonics

In another example, the National Reading Panel (NRP) was charged with determining what practices work in classroom reading instruction. One of the six areas examined was phonics instruction. Using meta-analysis methods, the researchers found that systematic phonics instruction was more effective than other instructional methods (National Reading Panel 2000). Gregory Camilli, Sadako Vargas, and Michele Yurecko, however, took issue with this finding (2003). They replicated and extended the NPR meta-analysis, finding that the NRP study overlooked other methods of instruction that were as or more effective than systematic phonics instruction. Here, the NRP's narrow focus on a specific mode of instruction appears to have resulted in national policy about phonics that downplayed the importance of other instructional practices. Harris Cooper rightfully noted that the work of Camilli and others was guided by their own set of biases in defining the scope of their research (2005). However, there is an extensive literature on the so-called reading wars (see Chall 1983, 1992, 2000) supporting the appropriateness of using the broader definitional framework used by Camilli and others (2003). In other words, even a quick synthesis of the literature and the debates it contains would have alerted members of the National Reading Panel that there are different perspectives (and stakeholders) on what counts as reading and instructional practices. A broader set of inclusion rules might have avoided some of the criticism that the NRP received on its report.

These examples, to varying degrees, suggest that meta-analysis methods, when pressed into public service, may not always serve the public well. The next two sections focus on how the effects of politics, personal perspectives, and methodological biases can be minimized. We focus on three major organizations that have been instrumental in devising procedures to maximize the quality of research syntheses, but before doing so, we describe the role of the evidence-based practice movement as a dominant force behind the use of research synthesis in policy areas.

25.5 EVIDENCE-BASED PRACTICES: FUEL FOR POLICY-ORIENTED META-ANALYSES

Interest in policy-oriented meta-analyses has been fueled by two recent and interrelated developments. The first represents a broad-based desire on the part of professionals (initially in medicine and now in other policy areas such as education and justice), federal and state legislators, and agency executives (in both the public and private sectors) to make the decisions about policies, programs, products, and practices based on evidence of their effectiveness. Although this perspective has been referred under varying labels (scientifically based practices, evidence-based medicine, evidence-based policy), the core idea is that programs supported by federal funds should be based on credible, empirical evidence. The second development is related to the first in hat it defines what we mean by credible, empirical evidence. Following the lead from medicine, there has been a profound increase in the

preference for the use of randomized control trials (RCTs) to assess the effects of what works (see Whitehurst 2003). This is a natural outgrowth of the interest in evidence-based practices (EBP) because RCTs provide the most trustworthy basis for estimating the causal effects of program, policies, and practices (Boruch 1997; Boruch and Mosteller 2001). For the purposes of understanding the role of meta-analysis in policy arenas, we focus on the role of the EBP movement.

25.5.1 Organizational Foundations for Evidence-Based Practices

Evidence-based practice has become a popular refrain in the policy world (Cottingham, Maynard, and Stagner 2005). David Sackett and his colleagues traced the idea in medicine back to the mid-nineteenth century (1996). In support of this interest, a network of public and private organizations has appeared on the policy scene to assist in identifying relevant research, screening it for quality, synthesizing results across studies, disseminating results, and translating research evidence into practice guidelines. The organizations we highlight here are dedicated to promoting the production of unbiased syntheses of the effects of programs (interventions), practices (medical and otherwise), policies, and products. A premium is placed on the integrity of the synthesis process. They have established procedures that make the processes of selecting studies, evaluating their quality, and summarizing the results as transparent to the reader as is possible.

25.5.1.1 Cochrane Collaboration At the forefront of identifying practices that work was the Cochrane Collaboration (www.cochrane.org). Established in 1993, the Cochrane Collaboration, a nonprofit, worldwide organization named after epidemiologist Archie Cochrane, produces and disseminates systematic reviews of evidence from randomized trials on the effects of healthcare interventions (www.cochrane.org/docs/newcomersguide.htm). The productivity of Cochrane investigators (a largely volunteer workforce) has been nothing short of spectacular; as of this writing, 2,893 Cochrane reviews have been completed on an array of health-care topics. Because the Cochrane Collaboration is well established, previous reviews are regularly updated, providing consumers with the most recent evidence on what works (or does not work) in hundreds of health-care subspecialties. At the time of writing, the Cochrane Central Registry of Controlled Trials contains an astonishing 479,462 entries. Beyond its stunning level of production, the Cochrane Collaboration serves as a model for the structure and operation of many organizations that followed in its footsteps.

25.5.1.2 Campbell Collaboration Named after one of the most influential figures in social science research, Donald T. Campbell, the Campbell Collaboration (C2) was founded in 2000 (www.campbellcollaboration.org).[5] As with the Cochrane Collaboration, it is composed of a largely voluntary, worldwide network of researchers who direct their efforts at identifying effective social and educational interventions. In light of concerns about the influence of personal or organizational agendas, a well-structured synthesis process has been adopted. Synthesis protocols are critiqued by experts to minimize oversights in question formation, searches and quality control. The Campbell Collaboration's raison d'être is to produce research syntheses where the methodology is transparent, that minimize the influence of conflicts of interest, and that are based on the best evidence (that is, its preference for and reliance on evidence from RCTs and high quality quasi-experiments). It shares these goals with its sibling organization—the Cochrane Collaboration. Both organizations have been the inspiration for the structure and functioning of numerous public agencies, such as the Institute of Education Sciences' What Works Clearinghouse, established in 2002 (www.ed.gov/ies/whatworks) and private organizations dedicated to promoting evidence-based practices, such as the Coalition for Evidence-based Policies (www.evidencebasedprograms.org).

25.5.1.3 IES's What Works Clearinghouse Established in 2002, the What Works Clearinghouse (WWC) is a major organizational initiative supported by the Institute of Education Sciences (IES) within the U.S. Department of Education (www.whatworks.ed.gov). Its major goal is to synthesize what is known about topic areas that are nationally important to education.[6] Studies associated with these areas are systematically gathered and subjected to a comprehensive three-stage critiquing process. Again, the focus of the WWC is on the effects of interventions—programs, policies, practices, and products. Studies are screened for their relevance to the focal topic, quality of outcome measures, and adequacy of the data reported. Studies that meet these initial screens are assessed for the strength of the evidence underlying the study's findings. There are three classifications for strength of the evidence—meet evidence standards, meet evidence standards with reservations, and does not meet standards. The standards are clearly spelled out for each category and, similar to the Campbell Collaboration guidelines, evidence based on well-executed RCTs is given the highest

ratings. In addition, results based on high quality quasi-experiments (for example, regression-discontinuity designs) and well-executed single-subject (case) designs also are viewed as meeting evidential standards. At this stage, studies that meet standards and meet standards with reservations are summarized, effect sizes are calculated, and an index of intervention improvement is calculated. Intervention reports are prepared that document, in great detail, all inclusion and exclusion decisions, ratings for standards of evidence, the data that are used in the calculation of effect sizes, and references to all documents considered in the synthesis are provided (sorted by whether they meet or did not meet the standards of evidence). The third rating examines the strength of the evidence across studies, for a given intervention. Considerations for these ratings entail a combination of design quality and the degree of consistency of results across studies. To date, the WWC has produced thirty-one intervention reports on several different topic areas. The depth and clarity of these reports are remarkable.

25.5.1.4 Organizations Around the World The Cochrane Collaboration has a network of centers like the Canadian Cochrane Network and Centre (http://cochrane.mcmaster.ca) located across the globe. Not surprising, in addition to the Cochrane Collaboration, there are numerous public and private organizations with long standing commitments to evidence-based practices and the use of systematic reviews to identify what works around the world. Edward Mullen has compiled a listing of web resources that identify organizations that foster evidence-based policies and practices in, for example, Australia, Canada, Latin America, Sweden, United Kingdom, United States, and New Zealand (2004). Some of these organizations were founded at about the same time as the Cochrane Collaboration. The EPPI-Centre at the University of London, to take one example, was founded in 1990.

25.5.2 Additional Government Policies and Actions

During the past decade, the U.S. Congress and executive branch departments and agencies have set the stage for greatly enhanced use of meta-analytic methods in policy areas. In particular, there has been a promulgation of laws—for example, the No Child Left Behind legislation calling for educational programs based on "scientifically based research"—and policies requiring the use of practices that have been scientifically shown to be effective. Agencies such as the Agency for Healthcare Research

Quality, SAMHSA's National Registry of Evidence-based Programs and Practices or NREPP, the joint venture of Agency for Healthcare Research Quality (AHRQ), the American Medical Association (AMA) and the American Public Health Association's (APHA) National Guideline Clearinghouse, and the Center for Disease Control's Community Guides have devoted significant resources to identifying evidence-based practices.

Interest in improving the efficiency, effectiveness and cost-effectiveness of programs supported by subnational entities also is evident. For example, the Washington State legislature has appropriated funds to support analyses of the effectiveness and cost-effectiveness of programs, policies, and practices. To this end, the Washington State Institute for Public Policy has issued several reports summarizing the results of their synthesis of a range of programs that operate within the state of Washington (for example, Aos et al. 2004). The analysis is exemplary for the technical care with which Steve Aos and his colleagues conducted their analysis of both the cost and the effects of programs and also for the clarity of the presentation. They provide a model for how complex and technically sophisticated policy-oriented syntheses can be reported to meet the potential needs of a variety of potential consumers.

The foregoing shows that a broadly represented *policy shaping community,* including members of Congress, governmental agencies, courts, universities, and professional associations, has been at work promoting and facilitating the use of RCTs and methods for identifying evidence-based practices that work. We would be remiss if we did not acknowledge the important role that private foundations have played in the promotion of meta-analysis as a tool for aiding public policy analysis. The Russell Sage Foundation has made a considerable commitment to understanding how to optimize meta-analytic methods in the policy context. The Smith-Richardson and the Hewlett Foundations have provided funds to investigators involved with the Campbell Collaboration (see Cottingham, Maynard, and Stagner 2005). The Coalition for Evidence-Based Policy has been supported by the Hull Family Foundation, the Rockefeller Foundation, and the William T. Grant Foundation.

25.6 GENERAL TYPES OF POLICY SYNTHESES

The predominant questions within policy-oriented research syntheses revolve around questions of the effects of programs, policies, practices, and products. In the policy context, at least two types of research syntheses are directed at these types of questions. One of these entails questions from identifiable members of the policy-making community. Another addresses policy-relevant questions without explicit input from an identifiable member of the policy-making community. Each type implies a different approach to the formulation of the questions to be addressed. We refer to the former as *policy-driven* because it entails translating a request by policy makers into researchable questions that can be answered, at least in principle, with available research synthesis techniques. We refer to the latter as *investigator-initiated.* Such syntheses address topics that are potentially policy-relevant, ranging from simple main effect questions like "What works?" to more comprehensive questions of "What works, for whom, under what circumstances, how and why"?[7] Here, the analyst has considerable degrees of freedom in specifying the nature and scope of the investigation. Of course, investigator-initiated syntheses must make a series of assumptions about the informational needs (now and in the future) of members of the policy community. We describe the question formulation process of each type of policy synthesis in turn.

25.6.1 Policy-Driven Research Syntheses

Policy-driven research syntheses, those entailing a direct request from policy makers, are relatively rare. These requests can be broad or narrow. For example, the mandate issued to the Washington State Institute for Public Policy by the Washington legislature directed them to "identify evidence-based programs that can lower crime and give Washington taxpayers a good return on their money" (2005, 1). Here, analysts were required to estimate the effects of programs, their cost, and their cost-effectiveness. Other policy-driven requests are more narrowly focused on questions about specific programs, policies, or practices that are under active policy consideration. GAO's evaluation synthesis method was designed explicitly for this purpose (see Cordray 1990; GAO 1992).[8] Despite their rarity, the available examples of policy-driven studies undertaken in direct response to questions raised by specific policy makers are instructive in illustrating the kinds of issues that typify policy debates.

As noted earlier, questions of who is right about what works arise and can be profitably addressed by research synthesis methods. Here, GAO has undertaken several policy-driven syntheses. In addition to its synthesis of the WIC program, the credibility of evidence was focal to their analysis of debates on federal and state initiatives to increase the legal drinking age (GAO 1984, 1987). Not all of GAO's syntheses were as highly focused on resolving disputes about the methodological adequacy of earlier studies. In its synthesis on the effects of housing allowance programs, debates centered on the effects on individuals—such as participation rates, equity of access, and rent burden—versus the effects on the housing markets, such as housing availability and quality (GAO 1986). Further, in an unusual application of synthesis methods, Senator John Chafee (RI) asked GAO to determine if there was enough empirical evidence to support the enactment of the teen pregnancy programs specified in two proposed pieces of legislation (GAO 1986). This application of synthesis methods was formalized by GAO as a Prospective Evaluation Synthesis methodology (Cordray and Fischer 1995; GAO 1990).

By design, GAO structures its syntheses around the issues defining all sides of the policy debate. GAO's approach to its synthesis of the effects of WIC was directed at the policy debate about its effectiveness versus counter claims of methodological inadequacies. In the end, both sides of the debate were correct. Some credible evidence supported the contention that WIC was effective, and some that methodological inadequacies were associated with many previous evaluations of WIC (GAO 1984). GAO's contribution to the debate was that it addressed the policy concerns directly and was able to provide the policy makers with a balanced analysis of what is known about the effects of WIC and what is not, either because studies had not been conducted or because they were too flawed to be relied on.

Unlike the question formulation process in most types of research, where the investigator has latitude in determining what will be examined and how, in policy-driven syntheses the questions are predetermined by, and often negotiated with, actual policy makers. As seen in previous examples, these types of syntheses often involve a desire on the part of the policy maker, to know more than answers to questions of what works. Additional questions seek differentiated answers about program effects for different demographic subgroups, effects on individuals and higher levels of aggregation (for example, the market place), and effects across different intervention models or program components (for example, what works best).

25.6.2 Investigator-Initiated Policy Syntheses

Investigator-initiated syntheses can be regarded as potentially relevant to policy questions because they exist largely as the result of the efforts of organizations described in the last section. That is, many of these potentially relevant syntheses are readily available in electronic archives, such as those by Campbell and Cochrane collaborators, and ready to be used as policy issues arise for a variety of policy makers. Whereas the policy-driven syntheses are reactive to the needs of specific policy makers, those labeled potentially relevant can be viewed as prospective or proactive, attempting to answer policy questions before they emerge in the policy context. They are considered potentially useful because they are often completed without an explicit connection to a specific policy issue or debate. As such, their utility depends on the intuitions of the investigator about what is important and their skills formulating the questions to be addressed (see Cottingham, Maynard, and Stagner 2005).

25.7 MINIMIZING BIAS AND MAXIMIZING UTILITY

Because the policy-driven form of synthesis is rare and generally depends idiosyncratically on the needs of those requesting the synthesis, the issues addressed in this section are most applicable to investigator-initiated syntheses. Although these types of syntheses are often obliged to follow the standards and practices developed by one of the major synthesis organizations, such as the Campbell Collaboration, the investigator still has some degrees of freedom in specifying the scope of their synthesis (Cottingham, Maynard, and Stagner 2005).

Earlier we provided examples of some of the ways policy-oriented syntheses can be adversely influenced by political pressures, ideologies, and wishful thinking of analysts. We suggested the need for a firewall to insulate as much as possible the science of synthesizing evidence from these political forces. It is not possible to provide an exact set of blueprints for constructing such a firewall. On the other hand, it is possible to highlight aspects of the synthesis process that are vulnerable to biases. The most vulnerable aspects discussed here include formulating questions, searching literature, performing analyses, and reporting results.

25.7.1 Principles for Achieving Transparency

The policies and practices promulgated and followed by the Cochrane Collaboration and the Campbell Collaboration serve as the building blocks for the firewall needed to minimize the influence of politics on policy syntheses. Both organizations place premiums on conducting reviews and syntheses that are as transparent as possible throughout the synthesis process. To achieve this transparency, both organizations have established processes that allow for other experts in the field to evaluate the scope and quality of the study. These peer reviews are conducted at the time the study is proposed and before it is accepted as a Cochrane or Campbell Collaboration product.

To ensure that relevant issues are considered before previous studies are synthesized, Cochrane and Campbell investigators are asked to submit the proposed protocol they will use to execute their synthesis. The items to be included in the protocol include statements about the need for the synthesis, a synopsis of any substantive debates in the literature (or policy communities), and a detailed plan for study selection, retrieval, data extraction, and analysis. The proposal also requests synthesists to state in writing any real or perceived conflicts of interest pertaining to the substance of the work. Finalized protocol proposals are published within the archives. The final report for each synthesis is also examined to ensure that the study plan was followed and that analyses are fair and balanced.

25.7.2 Question Specification

Questions about public policies can take on a number of forms and require the use of a variety of social science methods. Although meta-analytic methods have been used to answer questions like what the prevalence of a problem is (see Lehman and Cordray 1993), the recent interest in evidence-based practices is squarely focused on identifying programs (interventions), policies, practices and products that work. To answer these questions, several aspects of the synthesis need to be specified and several key decisions affect the nature and scope of the synthesis. Chief among these decisions is the specification of what we mean by the what of the what works question. Several alternative formulations of what will affect the synthesis and its utility to the intended audience.

A range of advice has been offered in defining the object (the what) of policy-oriented syntheses. Phoebe Cottingham, Rebecca Maynard, and Matt Stagner held that "policy-makers and practitioners often view programs and policies differently" (2005, 285). For policy makers, a program may represent a funding stream, with multiple components being implemented in an effort to achieve a specific goal, such as improving achievement in U.S. public schools. To be responsive and useful, a synthesis directed at this type of policy maker would require that all programs and other policy actions be included in the funding stream. Aos and his colleagues (2004) structured their synthesis across a number of policy goals, identifying individual programs directed at the broad goals.

On the other hand, practitioners such as teachers and principals may define a program by the precise shape of its content (Cottingham, Maynard, and Stagner 2005). In an effort to generate and use evidence to guide public policy, Phoebe Cottingham and her colleagues recommended that syntheses be structured around well-defined synthesis topics more specifically focused on particular interventions (2005). In light of the emphasis on evidence-based practices in legislation like No Child Left Behind, the differences between policy makers and practitioners in views of programs is not as stark as Cottingham and her colleagues initially portrayed (2005). Rather than focusing on broad policy goals, policy makers in Congress sided with the needs of practitioners.

25.7.2.1 Optimizing Utility The different emphases of policy makers and practitioners suggest that the utility of a given synthesis could be jeopardized for one of these two distinct audiences, depending on whether the definition of the policy object was broadly or narrowly defined. Several practices have been used to optimize the utility of a given synthesis across potential audiences. For example, in their study of the effects of Comprehensive School Reform (CSR) on student achievement, Geoffrey Borman and his colleagues constructed the meta-analysis so that they could estimate the overall effects of CSR as a policy and the separate effects of the twenty-nine CSR program models that had been implemented in schools (2003). The Institute of Education Sciences' (IES) What Works Clearinghouse develops two types of reports to maximize the utility of the work for multiple audiences. *Intervention Reports* are targeted to specific programs (for example, Too Good 4 Drugs) and products (publisher's reading programs like Read Well). *Topic Reports* are designed to examine effects of multiple programs, practices, and products on a focal area in education. This combination of products meets the needs of multiple audiences. Borman and his colleagues' analysis of the effects of different comprehensive school reform models and the overall effects across models is another example of how the effects can be presented to meet multiple information needs of users (2003).

Of course, examining effects across interventions is not simply a matter of aggregating the results from separate intervention reports. Between-study (for example, quality of the research design, types of outcomes, and implementation fidelity) and between-intervention (for example, types of clients served, costs, and intervention duration) factors need to be accounted for in the coding protocol and analyses. Sandra Wilson and Mark Lipsey provided an example of this form of synthesis (2006). Mark Lipsey's analysis of juvenile delinquency intervention models shows the differential effects of conceptually distinct types of intervention models, such as behavioral and affective (1992). These types of syntheses need to be conceptualized as such early in the synthesis process. The coding protocol for syntheses are much more complex than those for studies that focus on estimating simple average effects based on the unit of assignment to conditions, for example, WWC Intervention Reports (see Wilson and Lipsey 2006; Lipsey (1992)). This approach is often referred to as an intent-to-treat analysis, retaining all participants in the analysis regardless of whether they did not comply with their treatment assignment (for example, drop-outs, cross-overs, incomplete receipt of the intended treatment dose).

25.7.3 Searches

EPA's meta-analysis on the effects of second-hand smoke on health status did not include every relevant study (1992). Representatives of the tobacco industry were quick to point out this omission. Had the results of those studies been included, the aggregate results would have shown even weaker effects of secondhand smoke. The exclusion of other interventions (beyond systematic phonics instruction) in the meta-analysis conducted for the National Reading Panel (2000) was spotted by Camilli and others (2003). Further, their reanalysis of the meta-analytic data base found additional relevant studies that had not been included in the original NRP syntheses. It also showed different results and implied different policy conclusions than those offered by the NRP. Both of these examples highlight the critical importance of a comprehensive search, retrieval, coding, and analysis process. The comprehensiveness of the search process is as important to policy-oriented syntheses as traditional ones. Is there anything unique about the search process in policy-oriented syntheses? An examination of the search strategies in such studies suggests a few issues that deserve attention.

Disputes about the credibility of evidence underlying a program's effects need to be carefully examined. These can arise from differences in perspectives about the definition of evidence, they can arise from selective use of the literature by opponents and proponents of the policy or program, and from differences in which and how outcomes are measured (GAO 1984, 1986, 1987). Thus the first consideration in specifying the parameters of a search is a thorough understanding of the policy-debates and the evidence used by parties in those debates.

A broader basis for the search is likely to be needed, relative to a more traditional synthesis of the academic literature base. Lipsey and Wilson provided a broad-based, policy-relevant list of data bases and electronic sources (2001, appendix A). Their list reminds us that evaluation reports do not always become part of the published literature. Borman and his colleagues used two databases developed by a research firm (American Institutes for Research, or AIR) and a regional education lab supported by the U.S. Department of Education (Northwest Regional Education Laboratory, or NWREL) to identify districts that implemented and evaluated Comprehensive School Reform models (2003). GAO relied on agency field offices in its search for local evaluations of the WIC program (1984). Early syntheses are another albeit obvious resource. The Campbell Collaboration now houses SPECTR, a repository of experiments in education, justice, and other social areas.

Contacting experts and program developers, looking at program websites and agency officials to ensure the comprehensiveness of lists of prior studies is becoming common practice. The foregoing is not to imply that all studies must be included in the synthesis but instead that all relevant studies must be considered within the synthesis. Those that fail to meet quality standards can be identified and legitimately excluded from further consideration. Agreement on the exclusion of studies, however, is not widespread.

25.7.4 Screening Studies

As previously described, the WWC uses a well-structured set of procedures to screen studies on their relevance to synthesis goals. Screening studies for adequate population coverage, the intervention (for example, Wilson and Lipsey 2006), outcomes, geography, and time frame are particularly important when the policy questions entail changes to existing laws or practices connected to the governance structure of a particular country. Cottingham and her colleagues noted that the welfare system in the United Kingdom is different enough from its counterpart in the United States and that search strategies need to be delimited to studies from a given country (2005). These considerations depend on the policy topic. For example, studies on the effects of specific medical treatments, such as secondhand smoke and radiofrequency denervation, may not require such delimitations.

Screening for policy relevance is the first part of the study selection task. The final set of studies to be included in a synthesis will be those that pass screens on their relevance to policy area and the quality of evidence standards.

25.7.5 Standards of Evidence

Within the policy context, the credibility of evidence underlying the studies included in a synthesis is fundamental. We saw that concerns about the inadequate quality of the epidemiological studies, mainly case-control studies, were, in part, the basis for discrediting the meta-analysis by EPA on the effects of secondhand smoke (EPA 1992). Concerns about the quality of evidence by opponents of the WIC program were pivotal in instigating Senator Helms' request for the synthesis by GAO (1984). In both these cases, the questions turned around whether the research designs were adequate to support causal claims of the effects of the program (WIC) or purported causal agent (secondhand smoke). Concern about the quality of evidence underlying claims of scientifically based practices was important enough to the framers of the reauthorization of ESEA Title I (No Child Left Behind) to compel them to repeat their request more than 100 times throughout the legislation. Interest in evidence-based practices has prompted a rash of meta-analyses that redo prior syntheses in an effort to isolate effects based on randomized experiments rather than all available literature.

The major organizations conducting policy-oriented research syntheses—the Cochrane Collaboration, the Campbell Collaboration, and the What Works Clearinghouse within the Institute of Education Sciences—follow the premise that evidence from randomized controlled trials should be the gold standard for judging whether an intervention works. An intent-to-treat estimate from well-executed RCTs is regarded as the most compelling evidence of the direction and the magnitude of intervention's effect. These estimates provide a statistically sound basis for meta-analysis and other forms of quantitative synthesis (for example, cost-effective).

On the other hand, requiring that interventions be evaluated with RCTs is a high standard that can affect the utility of policy-oriented syntheses. To the extent that earlier studies fail to meet this standard, the policy community has to revert to decision making based on nonempiri-

cal considerations. This prospect has resulted in a vigorous discussion about how to accommodate the trade-off between wanting to rely on high standards of evidence and wanting to contribute to the policy discussion in a useful and timely fashion. Three general approaches have been used. The first entails classifying studies based on the degree to which they meet technical standards. We refer to this as the level of evidence approach. The second approach can be thought of as a variation of the first and was coined best-evidence synthesis by Robert Slavin (1986, 2002). The third uses statistical models (for example, meta-regression) to control for the influence of study quality. This approach is well entrenched in the short methodological history of meta-analysis. We refer to this as the *meta-regression adjustment* approach.

25.7.5.1 Level of Evidence Approach The main idea behind the level of evidence approach is that some designs produce estimates that are more or less trustworthy. As such, the synthesis should take these different levels of credibility into account. We identify two forms of this approach. First is the purest synthesis, which is based only on the results from randomized control trials (RCTs). A variation segments or disaggregates the results into more than the two categories implied in the purist case (that is, meets evidence standards or not).

Several syntheses undertaken within the Campbell Collaboration guidelines, exemplify the purist approach (Petronsino, Turpin-Petrosino, and Buehler 2003; Nye, Turner, and Schwartz 2003; Visher, Winterfield, and Coggeshall 2006). Not all syntheses conducted under the auspices of the Campbell Collaboration hold as firmly to this purest notion (for example, Wilson and Lipsey 2006). The synthesis by Anthony Petrosino and his colleagues examined the effects of Scared Straight and other juvenile awareness programs for preventing juvenile delinquency (2003). They uncovered 487 studies, of which thirty were evaluations and eleven potentially randomized trials. After additional screenings, they included seven studies in their meta-analysis. Here, the consequence of holding high standards is a reduced number of available studies. This can become a severe problem.

Visher and her colleagues found only eight RCTs, but only after expanding the scope of the synthesis (2006). Nye and his colleagues on parent involvement on achievement had somewhat more success in finding RCTs, in that nineteen met criteria (2003). This synthesis, though focused on intent-to-treat estimates in its synthesis, departed from the purest point of view by incorporating a series of interesting post hoc and moderator analyses.

25.7.5.2 Segmentation by Credibility or Strength of Evidence Rather than creating a dichotomy like acceptable or unacceptable, the technical adequacy of a study can be characterized in a more differentiated fashion, allowing for degrees of acceptability. As described earlier, IES and the What Works Clearinghouse have developed three categories of evidence: meets standards, meets standards with reservations, and does not meet standards. The intervention reports classify each study according to this scheme and report on the effects of studies in the first two categories. To date, the WWC synthesists have completed reports for thirty-one specific interventions in six policy areas. Of these, eighteen were based on at least one study that met standards evidence. The evidence was based on a well-executed randomized control trial or, a quality quasi-experiment, or a quality single subject research design. In thirteen reports, the highest level of design quality was meets standards with reservations. Often, an intervention involved a mixture of studies that met standards or met standards with reservations. Only ten reports were based on at least one study that met the highest standards and did not rest on any other studies that met standards but with reservations. It should be apparent that this form of segmentation preserves the goal of identifying interventions that are based on the most trustworthy evidence (RCTs, high quality quasi-experiments, and well-executed single subject or case studies), but also provides policy makers and practitioners with an overview of what is known from studies with less credibility. Although not operating under the guidelines of any organization, Borman and his colleagues used a segmentation scheme that could be plainly understood by a variety of user groups (2003). They classified program models into four groups: strongest evidence, highly promising, promising, and in greatest need for additional research. With proper caveats, the segmentation of interventions by variation in the strength of evidence on its effectiveness gives the users something to base their decisions on, rather than having to rely on expert or personal opinion in picking an intervention or to stay the course.

25.7.5.3 Best-Evidence Syntheses Robert Slavin describes best-evidence synthesis as an alternative to meta-analysis and traditional reviews (1986, 2002). It uses two of the major features of meta-analysis—quantification of results as effect sizes and systematic study selection—but argues that the synthesis should be based on the best available evidence. If RCTs are not available, the next best evidence would be used. Over time, as results from RCTs become available, the effects of an intervention would be reexamined in light of this new and more compelling evidence base. Although Slavin's ideas had not yet appeared in the literature at the time GAO conducted its research synthesis of the WIC program, analysts used a form of best-evidence synthesis in selecting the subsets of studies that were used to derive their conclusions for each question posed by Senator Helms. GAO analysts and their consultants (David Cordray and Richard Light) rated studies on ten dimensions. Studies deemed adequate on four of these dimensions were used in the synthesis. None of the available studies were rated a 10. There were too few studies (for example, six for the main outcome on birth weight) to segment them into good, better, and best categories.

25.7.5.4 Meta-Regression Adjustment Approach One of the most common forms of synthesis entails the use of regression models, hence the term meta-regression. This represents a general analytic approach to account for between study differences in methodology, measurement, participants, and attributes of the intervention under consideration. Rather than segmenting the studies into levels of evidence, design differences (and more) are examined empirically. In their initial analysis, Borman and his colleagues included all studies, regardless of the method, from pre-post only through randomized and varying types of quasi-experiments (2003). Their approach also examined the effects of other methods variables, reform attributes (parental governance), cost, and other nonmethods factors that could affect the results, such as the degree of developer support available in each study.

As noted, not all syntheses conducted under the auspices of the Campbell Collaboration are restricted to RCTs. Wilson and Lipsey included studies with control groups, both random or nonrandom (2006). In their study, the nonrandom studies needed to show evidence of initial equivalence on demographic variables. They found eighty-nine eligible reports and included seventy-three in their synthesis. Their analyses controlled for four methods factors. In a unique approach to identifying potential mediator variables, they examined the influence of program and participant variables on the outcome (effect sizes), controlling for the influence of between-study methods artifacts. Moderator variables that survived this screen were included in their final models. They reported no differences in effects between studies based on randomized designs and those based on well-controlled, nonrandom designs. Had they found differences between these two types of designs, additional controls would have been needed (for example, additional stratifications or sensitivity analyses).

25.7.5.5 Combined Strategies Using two or more of the three approaches outlined above to control or adjust for differential design quality does not appear to be as pervasive in the recent synthesis literature as is warranted. However, as noted, one good example of how to meet the information needs of multiple perspectives on a given policy or program is provided by Borman and his colleagues (2003). They first examined the overall effects of the comprehensive school reform models as a broad-based program. Using a meta-regression approach, they included all studies, regardless of the research design, and segmented the models according to the overall strength of the evidence upon which they rested. They offered extended analyses of those models based on the best-evidence approach. Fortunately, unlike the 1984 GAO analysis of WIC, several models had been evaluated with the best types of research designs. The usefulness of the results was enhanced by other variables being included in the meta-regression. These other policy-oriented variables provided the reader with an idea of whether the features of the program made a difference. We use the term *idea* to highlight the fact that these results are inherently correlational unless key features of the treatment are randomly assigned to persons within conditions.

25.7.6 Moderator Variables

Within the evidence-based practices movement, the emphasis is on answering the causal question, "what works?" Reliance on RCTs seems to have diverted attention from answering an important broader set of policy questions about for whom, under what circumstances, how and why. These are given less priority because they are hard to answer within the paradigm focused on the use of RCTs. Potential moderator variables like implementation fidelity, attributes of the participants, and context factors are not randomly assigned to studies, and their influence is not easy to estimate because of specification errors.

Including moderator variables ought to be encouraged because they are potentially valuable to policy makers, researchers, program developers, and other practitioners. However included, it is important that readers are provided a thorough discussion of the strengths and weaknesses of results presented. With that caveat in mind, we present a few examples that show the potential value of analyses of moderator variables to a range of policy makers. Nye and others showed that length of parent involve-

ment interventions was not related to effects but that the type of parent involvement was (2003). Here, providing incentives to parents made a difference, as did parent education and training. Their analyses also revealed effects for reading and math achievement, but not for science. In another example, Harris Cooper, Jorgianne Robinson, and Erika Patall showed that part of the confusion that pervaded the literature about the effects of homework on achievement could be resolved by differentiating effects by grade and development stage (2006). Cooper and his colleagues' analysis is a very good example of how meta-analyses not exclusively focused on main effects, such as intent-to-treat estimates, can be used to generate important hypotheses for future research.

Given the breadth of policy issues, it is difficult to say much about the types of moderator variables that should or could be included in policy-oriented syntheses. However, based on the kinds of issues raised by policy makers and practitioners, a few seem as if they would be generically policy relevant. Our short list of potentially relevant moderator variables include implementation fidelity; duration, frequency, and intensity of interventions; setting within which the intervention was tested (demonstration school versus regular school); who tested the intervention (developer or third-party); intervention (program) cost; and risk categories for participants.

25.7.7 Reporting and Communicating Results

Effect sizes have been the workhorse of syntheses. They have definable statistical qualities. They have a certain degree of intuitive appeal. It is relatively easy to understand odds ratios. The Glass-Hedges effect size calculation is not complex and can be easily interpreted within normative frameworks that depend on sensible conventions (Cohen 1988). On the other hand, their intuitive appeal is not universal. The translation of an effect size into more meaningful indicators has received some well-deserved attention. For example, the improvement index used by the What Works Clearinghouse tells the reader how much difference an intervention is expected to make relative to the average of the counterfactual group. Rather than being expressed as an effect size, the index is expressed as a percentage of improvement that ranges between -50 and $+50$. Whether this index is meaningful to policy makers and other users is not yet clear.

In another effort to translate effect sizes into meaning units, Wilson and Lipsey calibrated the meaning of an effect sizes relative to typical levels of aggressive behavior

(2006). Using this survey data from the Centers for Disease Control and Prevention as a benchmark, they were able to derive a prevalence rate for population members as if they were given the intervention. They found that 15 percent of adolescents would exhibit unwanted behavior (if untreated), given an effect size of 0.21, the expected population rate would be 8 percent if the intervention was implemented.

Several other approaches offered by specialists in meta-analysis are available. Cottingham and her colleagues (2005) made a strong case for using natural units wherever possible. That is, they advised against automatically converting results into effect sizes. William Shadish and Keith Haddock provided guidance on aggregating evidence in natural units (1994). A second approach to improving communication of policy-relevant results is through the use of visual presentations (see Light, Singer, and Willett 1994). Funnel, stem-and-leaf plots, figures displaying case-by-case representations of effect sizes and 95 percent confidence intervals, and ordinary plots of the linear, interactive, and other nonlinear effects of moderators can go a long way toward making the results of meta-analysis comprehensible to a wide range of policy makers and practitioners.

25.8 CONCLUSION

Extensive financial and intellectual resources have been directed toward identifying policies, programs, and practice based on quality evidence. Given the relative newness of interest in evidence-based practices (EBP), it is not surprising that policy-oriented syntheses are only now beginning to be made public. It would appear that the EBP movement has afforded meta-analysis a permanent seat at the policy table.

The current low profile of meta-analysis in traditional resources such as bibliometric databases could be simply because of the relatively newness of these forms of inquiry. After all, it was only a little more than fifteen years ago that major efforts to synthesize RCTs in medicine were initiated, the Cochrane Collaboration and EPPI-Centre at the University of London, to take two examples, were established in the early 1990s.

More important than the volume of policy-oriented syntheses currently available is the development of guidelines and procedures to make them as fair and honorable as possible. Given that politics and personal agendas can influence the quality of policy-oriented syntheses, efforts

to minimize their influence have been mounted by major organizations (Campbell Collaboration, Cochrane Collaboration, and WWC) in the form of well-articulated policies and practices. The synthesis methods developed by these organizations are an important step toward creating the much-needed firewall between the science of the syntheses and the political pressures inherent in the policy context.

This chapter offers three essential messages. First is that the manner in which research syntheses are conceptualized, conducted, and reported depend on what we mean by the phrase *policy makers and practitioners*. In discussions of the potential role of meta-analysis or, more broadly, research synthesis, there is a tendency to not differentiate these two groups. It is as if we are referring to a homogeneous collection of individuals (that is, "policy-makersandpractitioners"). Second is that the goals for all research syntheses should be to minimize the influence of personal and methodological biases. Although we should minimize the influence of bias, analysts also need to maximize the utility of their product by sharply framing their questions in light of contemporary policy issues or debates (that is, optimizing relevance) and crafting their synthesis methodology so that answers can be provided in meaningful and accurate ways. Finally, though it is unlikely that any single policy-oriented research synthesis can satisfy the information needs of all members of a policy shaping community, there is a set of methods that can enhance the utility of a research synthesis. Approaches to enhancing the utility of syntheses across diverse members of the policy shaping community include the segmentation (or differentiation) of results by differences in the quality of evidence available, the use of a set of analytic approaches that include intent-to-treat estimates and follow-up analyses of policy-relevant moderator variables, and the use of policy-meaningful indices of effects (such as natural units of measurement) and common language descriptions of results.

25.9 ACKNOWLEDGMENTS

Paul Morphy's contribution to this chapter was supported by grant number R305B0110 (Experimental Education Research Training or ExpERT, David S. Cordray, Training Director) from the U.S. Department of Education, Institute of Education Sciences (IES). The opinions expressed in this chapter are our own and do not reflect those of the funding agency.

25.10 NOTES

1. The U.S. Government Accountability Office's (GAO, formerly the U.S. General Accounting Office).

2. We recognize that variety of narrative syntheses of evidence related to policy-oriented topics have been undertaken throughout contemporary history. In this section we focus on the prevalence of quantitative syntheses in policy areas.

3. The Web of Science searches used the following parameters: Meta-analysis = (metaanalysis OR meta-analysis); policy = ((policy* OR policies) OR (legislat* OR legal); meta-analysis and policy = (metaanalysis OR meta-analysis) AND (policy* OR policies). *Broader specification for meta-analysis-like topics that could appear in the literature (for example, quantitative synthesis and research synthesis) did not increase the number of instances that we found.* Including the term "synthesis" brought into the listing such topics "chemical synthesis" were too ill-targeted.

4. The Web of Science may have changed its methods of data capture between early and later years. The figures (for example, journal coverage) from 1990 forward are likely to be more accurate.

5. The Campbell Collaboration was founded under the able leadership of Robert F. Boruch, University of Pennsylvania.

6. An anonymous reviewer noted that some of the policy organizations that we discuss later in the chapter use a limited set of the array of procedures that customarily define a meta-analysis. For example, the What Works Clearinghouse, supported by the Institute of Education Sciences (IES) simply reports the separate effect sizes from studies of educational programs and products that they identify, screen, and rate methodological quality. These practices are due, in part, to the simple fact that they have found a relatively small number of studies (within program and product categories) that are judged to be adequate to support a causal inference of effectiveness. This does not mean when more evidence is available that the full repertoire of meta-analytic methods will not be employed.

7. In primary research studies the analyses often address questions beyond whether an intervention "works". In policy syntheses the goal is to identify quality studies (for example, RCTs) and, if possible, aggregate the effect estimates that are extracted. For example, at this point in its development, the What Works Clearing-house simply reports effect estimates without any further analysis of between study differences or subgroup analyses.

8. By way of some background, the majority leader of committees of the U.S. Congress frequently ask the U.S. Government Accountability Office to examine what is known (and not known) about the effects or cost-effectiveness of government programs, policies and practices. As a research-arm of Congress, GAO is organizationally independent of the administrative branch programs it evaluates.

25.11 REFERENCES

Aos, Steve, Roxanne Lieb, Jim Mayfield, Marna Miller, and Annie Pennucci. 2004. *Benefits and Costs of Prevention and Early Intervention Programs for Youth.* Olympia: Washington State Institute for Public Policy.

Borman, Geoffrey D., Gina M. Hewes, Laura T. Overman, and Shelly Brown. 2003. "Comprehensive School Reform and Achievement: A Meta-Analysis." *Review of Educational Research* 73(2): 125–230.

Boruch, Robert F. 1997. *Randomized Experiments for Planning and Evaluation: A Practical Guide.* Thousand Oaks, Calif.: Sage Publications.

Boruch, Robert F., and Frederick Mosteller, eds. 2001. *Evidence Matters.* Washington, D.C.: Brookings Institution Press.

Camilli, Gregory, Sadako Vargas, and Michele Yurecko. 2003. "Teaching Children to Read: The Fragile Link Between Science and Federal Education Policy." *Education Policy Analysis Archives* 11(15). http://epaa.asu.edu/epaa/v11n15/

Chall, Jeanne S. 1983. *Learning to Read: The Great Debate*, 2nd ed. New York: McGraw-Hill.

———. 1992. "The New Reading Debates: Evidence from Science, Art, and Ideology." *The Teachers College Record* 94(2): 315–28.

———. 2000. *The Academic Achievement Challenge: What Really Works in the Classroom.* New York: Guilford Press.

Cohen, Jacob. 1988. *Statistical Power Analysis for the Behavioral Science*, 2nd ed. Hillsdale, N.J.: Lawrence Erlbaum.

Coggins, Christopher. 1994. *Statement of the Tobacco Institute.* Testimony before the U.S. Congress. Senate Committee on Environment and Public Works. Subcommittee on Clean Air and Nuclear Regulation, 103rd Cong., 2nd sess. May 11, 1994.

Cook, Thomas D., Harris Cooper, David S. Cordray, Heidi Hartmann, Larry V. Hedges, Richard J. Light, Thomas A. Louis, and Frederick Mosteller. 1992. *Meta-Analysis for Explanation: A Casebook*. New York: Russell Sage Foundation.

Cooper, Harris M. 1986. "On the Social Psychology of Using Research Reviews: The Case of Desegregation and Black Achievement." In *The Social Psychology of Education*, edited by Richard Feldman. Cambridge: Cambridge University Press.

———. 1989. *Integrative Research: A Guide for Literature Reviews*, 2nd ed. Newbury Park, Calif.: Sage Publications.

———. 2005. "Reading Between the Lines: Observations on the Report of the National Reading Panel and Its Critics." *Phi Delta Kappan* 86(6): 456–61.

Cooper, Harris M., and Larry V. Hedges, eds. 1994. *The Handbook of Research Synthesis*. New York: Russell Sage Foundation.

Cooper, Harris M., and Jeffrey C. Valentine. 2001. "Using Research to Answer Practical Questions about Homework." *Educational Psychologist* 36(3): 143–54.

Cooper, Harris M., Jorgianne C. Robinson, and Erika A. Patall. 2006. "Does Homework Improve Academic Achievement? A Synthesis of Research 1987–2003." *Review of Educational Research* 76(1): 1–62.

Cordray, David S. 1990. "The Future of Meta-Analysis: An Assessment from the Policy Perspective." In *The Future of Meta-Analysis*, edited by Kenneth W. Wachter and Miron L. Straf. New York: Russell Sage Foundation.

———. 1994. "Strengthening Causal Interpretation of Non-Experimental Data: The role of Meta-Analysis." *New Directions in Program Evaluation* 60: 59–96.

Cordray, David S., and Robert L. Fischer. 1995. "Evaluation Synthesis." In *Handbook of Practical Program Evaluation*, edited by Joseph Wholey, Harry Hatry, and Katherine Newcomer. San Francisco, Calif.: Jossey-Bass.

Cottingham, Phoebe, Rebecca Maynard, and Matt Stagner. 2005. "Generating and Using Evidence to Guide Public Policy and Practice: Lessons from the Campbell Test-Bed Project. *Journal of Experimental Criminology* 1(3): 279–94.

Cronbach, Lee J. 1982. *Designing Evaluations of Educational and Social Programs*. San Francisco, Calif.: Jossey-Bass.

Environment Protection Agency 1992. *Respiratory Health Effects of Passive Smoking: Lung Cancer and Other Disorders*. Washington, D.C.: United States Environmental Protection Agency.

Environmental Protection Agency 1986. *Risk Assessment Guidelines*. 51 Fed. Reg. 33992.

Farland, William, Steven Bayard, and Jennifer Jinot. 1994. "Environmental Tobacco Smoke: A Public Health Conspiracy? A Dissenting View." *Journal of Clinical Epidemiology* 47(4): 335–37.

Glass, Gene V. 1976. "Primary, Secondary, and Meta-Analysis Research." *Educational Researcher* 5(10): 3–8.

Gori, Gio Batta. 1994a. "Science, Policy, and Ethics." *The Journal of Clinical Epidemiology* 47(4): 325–34.

———. 1994b. "Reply to the Preceding Dissents." *The Journal of Clinical Epidemiology*, 47: 351–53.

Gravelle, Jane G., and Dennis Zimmerman. 1994a. *Evaluation of ETS Exposure Studies*. Testimony before the U.S. Congress. Senate Committee on Environment and Public Works. Subcommittee on Clean Air and Nuclear Regulation, 103rd Cong., 2nd sess. May 11, 1994.

———. 1994b. *Cigarette Taxes to Fund Health Care Reform: An Economic Analysis*. Report prepared for the U.S. Congress. Senate Committee on Environment and Public Works. Subcommittee on Clean Air and Nuclear Regulation, 103rd Cong., 2nd sess. March 8, 1994.

Hunt, Morton. 1997. *How Science Takes Stock*. New York: Russell Sage Foundation.

Hunter, John, and Frank Schmidt. 1990. *Methods of Meta-Analysis: Correction Error and Bias in Research Findings*. Newbury Park, Calif.: Sage Publications.

Hunter, John, Frank Schmidt, and Gregory Jackson. 1982. *Meta-Analysis: Cumulating Research Findings Across Studies*. Beverly Hills, Calif.: Sage Publications.

Jinot, Jennifer, and Steven Bayard. 1994. "Respiratory Health Effects of Passive Smoking: EPA's Weight of Evidence Analysis." *The Journal of Clinical Epidemiology* 47(4): 339–49.

Landenberger, Nana A., and Mark W. Lipsey. 2005. "The Positive Effects of Cognitive-Behavioral Programs for Offenders: A Meta-Analysis of Factors Associated with Effective Treatment." *Journal of Experimental Criminology* 1(1): 451–76.

Lehman, Anthony, and David S. Cordray. 1993. "Prevalence of Alcohol, Drug, and Mental Health Disorders Among the Homeless: A Quantitative Synthesis." *Contemporary Drug Problems* 1993(Fall): 355–83.

Lipsey, Mark W. 1992. "Juvenile Delinquency Treatment: A Meta-Analytic Inquiry into the Variability of Effects." In *Meta-Analysis For Explanation: A Casebook*, edited Thomas D. Cook, Harris M. Cooper, David S. Cordray, Heidi Hartmann,, Larry V. Hedges, Richard J. Light, Thomas A. Louis, and Frederick Mosteller. New York: Russell Sage Foundation.

Lipsey, Mark W., and David Wilson. 2001. *Practical Meta-Analysis*. Newbury Park, Calif.: Sage Publications.

Light, Richard J., and David Pillemer. 1984. *Summing Up: The Science of Reviewing Research*. Cambridge, Mass.: Harvard University Press.

Light, Richard J., and Paul Smith. 1971. "Accumulating Evidence: Procedures for Resolving Contradictions Among Different Studies." *Harvard Educational Review* 41(4): 429–71.

Light, Richard J., Judith Singer, and John Willett. 1994. "The Visual Presentation and Interpretation of Meta-Analyses." In *The Handbook of Research Synthesis*, edited by Harris M. Cooper and Larry V. Hedges. New York: Russell Sage Foundation.

Mullen, Edward J. 2004. "Evidence-based Policy and Practice." and Columbia University in the City of New York. http://www.columbia.edu/cu/musher/Internet percent 20Links percent20for percent20Bibliography.htm

National Reading Panel. 2000. *Report of the National Reading Panel: Teaching Children to Read—An Evidence-Based Assessment of the Scientific Research Literature on Reading and its Implications for Reading Instruction*. Washington, D.C.: National Institute of Child Health and Human Development.

Niemisto, Leena, Eija Kalso, Antti Malmivaara, Seppo Seitsalo, and Heikki Hurri. 2003. "Radio Frequency Denervation for Neck and Back Pain." *Cochrane Database of Systematic Reviews* 2003(1): CD00400048.

Nye, Chad, Herb Turner, and Jamie Schwartz. 2003. "Approaches to Parent Involvement for Improving the Academic Performance of Elementary School-Age Children." In *Campbell Collaboration Reviews of Intervention and Policy Evaluations* (C2-RIPE). Philadelphia, Pa.: Campbell Collaboration.

Office of Management and Budget. 2004. "Final Information Quality Bulletin for Peer Review." http://www.whitehouse.gov/omb/inforeg/peer2004/peer_bulletin.pdf

Osteem, William L. 1998. Opinion. In Flue-cured Tobacco Cooperative Stabilization Corporation et ux. v. United States Environmental Protection Agency. 4 F. Supp. 2d 435: 1998 US Dist. LEXUS 10986; 1998 OSHD (CCH) P31, 3614:28 ELR 21445. United States District Court for the Middle District of North Carolina.

Pearson, Karl. 1904. "Antityphoid Inoculation." *British Medical Journal* 2(2294): 1243–246.

Petrosino, Anthony, Carolyn Turpin-Petrosino, and John Buehler. 2003. "'Scared Straight' and Other Juvenile Awareness Programs for Preventing Juvenile Delinquency: Campbell Review Update." In *Campbell Collaboration Reviews of*

Intervention and Policy Evaluations (C2-RIPE), November 2003. Philadelphia, Pa.: Campbell Collaboration.

Redhead, C. Stephen, and Richard E. Rowberg. 1998. "CRS Report for Congress: Environmental Tobacco Smoke and Lung Cancer Risk." Washington, D.C.: Congressional Research Service.

Sackett, David L., William M. C. Rosenberg, J. A. Muir Gray, R. Brian Haynes, and W. Scott Richardson. 1996. "Evidence-Based Medicine: What It Is and What It Isn't." *British Journal of Medicine* 312(7023): 71–72.

Schneider, Jeffrey M. 1990. "Research, Meta-Analysis, and Desegregation Policy." In *The Future of Meta-Analysis*, edited by Kenneth W. Wachter and Miron L. Straf. New York: Russell Sage Foundation.

Shadish, William R., and C. Keith Haddock. 1994. "Combining Estimates of Effect Size." In *The Handbook of Research Synthesis*, edited by Harris M. Cooper and Larry V. Hedges. New York: Russell Sage Foundation.

Slavin, Robert E. 1986. "Best-Evidence Synthesis: An Alternative to Meta-Analysis and Traditional Reviews." *Educational Researcher* 15(9): 5–11.

———. 2002. "Evidence-Based Education Policies: Transforming Educational Practice and Research." *Educational Researcher* 31(7): 15–21.

U.S. General Accounting Office. 1984. *WIC Evaluations Provide Some Favorable but No Conclusive Evidence on Effects Expected for the Special Supplemental Program for Women, Infants, and Children*. GAO/PEMD-84–4. Washington, D.C.: U.S. Government Printing Office.

———. 1986. *Housing Allowances: An Assessment of Program Participation and Effects*. GAO/PEMD-86–5. Washington, D.C.: U.S. Government Printing Office.

———. 1987. *Drinking Age Laws: An Evaluation Synthesis of Their Impact on Highway Safety*. GAO/PEMD-87–10. Washington, D.C.: U.S. Government Printing Office.

———. 1990. *Prospective Evaluation Synthesis*. Washington, D.C.: U.S. Government Printing Office.

———. 1992. *The Evaluation Synthesis*. GAO/PEMD-10.1.2. Washington, D.C.: U.S. Government Printing Office.

Visher, Christy A., Laura Winterfield, and Mark B. Coggeshall. 2006. "Systematic Review of Non-Custodial Employment Programs: Impact on Recidivism Rates of Ex-Offenders." In *Campbell Collaboration Reviews of Intervention and Policy Evaluations (C2-RIPE)*, February 2006. Philadelphia, Pa.: Campbell Collaboration.

Wachter, Kenneth W., and Miron L. Straf, eds. *The Future of Meta-Analysis*. New York: Russell Sage Foundation.

Whitehurst, Grover. 2003. "The Institute of Education Sciences: New Wine, New Bottles." Paper presented at the Annual Meeting of the American Education Research Association. Chicago (April 21–25, 2003).

Wilson, Sandra J., and Lipsey, Mark W. 2006. "The Effects of School-Based Social Information Processing Interventions on Aggressive Behavior (Part 1): Universal Programs." In *Campbell Collaboration Reviews of Intervention and Policy Evaluations (C2-RIPE)*, March 2006. Philadelphia, Pa.: Campbell Collaboration.

Wolf, Fredric M. 1990. "Methodological Observations on Bias." In *The Future of Meta-Analysis*, edited by Kenneth W. Wachter and Miron L. Straf. New York: Russell Sage Foundation.

PART

REPORTING THE RESULTS

26

VISUAL AND NARRATIVE INTERPRETATION

GEOFFREY D. BORMAN
University of Wisconsin, Madison

JEFFREY A. GRIGG
University of Wisconsin, Madison

CONTENTS

26.1 INTRODUCTION

Meta-analyses in the social sciences and education often provide policy makers, practitioners, and researchers with critical new information that summarizes the central quantitative findings from a particular research literature. However, meta-analytic articles and reports are also rather technical pieces of work that can leave many readers without a clear sense of the results and their implications. In this respect, both visual and narrative interpretations of research syntheses are important considerations for presenting and describing complex results in more informative and accessible ways.

Typically, readers of meta-analyses in the social sciences and education are offered two types of summary tables to help them make sense of the results. First, a table presents the overall summary effect size for the entire meta-analysis. This table may also include mean effect sizes for subgroups of studies defined by important substantive or methodological characteristics of the included studies (for instance, whether the study employed an experimental or nonexperimental design). Reports and articles often present a second table that includes a lengthy catalog of all studies included in the meta-analysis, along with such information as their publication dates, the authors' names, sample sizes, effect sizes, and, perhaps, differences among the studies along key substantive or methodological dimensions such as the ages of participants or the nature of the control group employed. Usually the material in these sorts of tables is not arrayed in a particularly informative order but instead simply sorted chronologically by publication year or alphabetically by the authors' last names. As we show in this chapter, however, the situation is improving.

Such tables can be useful to describe the individual data elements—the outcomes gleaned from each of the studies included in the meta-analysis—and can help present information about the overall effects observed and how those effects might vary by interesting methodological and substantive dimensions. However, we assert that data from meta-analyses can be presented in ways that facilitate understanding of the outcomes and that communicate more information than have been conventionally employed in the literature. Rather than limiting presentation and interpretation of meta-analytic data to simple tables and descriptive text, we argue for greater consideration of figures and displays.

This chapter also examines how the careful integration of quantitative and narrative approaches to research syn-

thesis can enhance the interpretation of meta-analyses. We outline three key circumstances under which one should consider applying narrative techniques to quantitative reviews. First, some of the earliest critics of meta-analysis suggested that the methodology applied a mechanistic approach to research synthesis that sacrificed most of the information contributed by the included studies and glossed over salient methodological and substantive differences (Eysenck 1978; Slavin 1984). We believe that many contemporary applications of meta-analysis have overcome these past criticisms through more rigorous analyses of effect-size variability and modeling of the substantive and methodological moderators that might explain variability in the observed outcomes. However, we argue that integration of narrative forms of synthesis can help further clarify and describe a variegated research literature. Second, there may be evidence that some literature, though of general interest and importance, is not amenable to meta-analysis. The typical meta-analysis would simply exclude this literature and focus only on the results that can be summarized using quantitative techniques. We believe, though, that such a situation may call for blending meta-analytic and narrative synthesis so that a more comprehensive body of research can be considered. Finally, beyond descriptive data and significance tests of the effect sizes, we contend that researchers should also use narrative techniques for interpreting effect sizes, primarily in terms of understanding the practical or policy importance of their magnitudes. In these ways and others, research syntheses that offer the best of both meta-analysis and narrative review offer great promise.

26.2 DISPLAYING EFFECT SIZES

26.2.1 Current Practices

To address the question of how the results of meta-analyses are commonly reported in the social sciences and education, we examined all meta-analyses published in the 2000 through 2005 issues of *Psychological Bulletin* and *Review of Educational Research*. We hoped to determine both how the presentation of findings may have changed since the last edition of this handbook as well as how the presentation of findings may differ between these two journals. Table 26.1 displays the results of this examination and compares them to the previous findings reported by Richard Light, Judith Singer, and John Willett (1994). The first and second columns of the table report the number of meta-analyses published in a given journal and

Table 26.1 Types of Tabular and Graphic Summaries Used in All Meta-Analyses

Volume	Meta-Analyses (n)	Table		Figure					
		Raw Effect Size	Summary Effect Size	Any Graph	Effect Size Dist.	Funnel Plot	Other Graphs	Represents Uncertainty	Represents Summary Effect
Psychological Bulletin									
1985–1991	74	41	57	14	8	1	8	NA	NA
Percentage of Total		*55%*	*77%*	*19%*	*11%*	*1%*	*11%*		
Psychological Bulletin									
2000	11	7	10	6	2	1	5	2	1
2001	5	3	4						
2002	11	9	11	7	6		3	1	
2003	14	10	11	9	6	2	1	1	
2004	9	9	8	5	3	4	2	1	1
2005	10	5	9	4	2	1	4	2	1
2000–2005	60	43	53	31	19	8	15	7	3
Percentage of Total		*72%*	*88%*	*52%*	*32%*	*13%*	*25%*	*12%*	*5%*
Review of Educational Research									
2000	2	2	2	1					
2001	4	2	4	2	2		1		
2002	2	2	1	2	2		1		
2003	3	3	3	2	2				
2004	4	3	4	3	3	1	1		1
2005	5	4	4	3	1	1	1		
2000–2005	20	16	18	13	10	2	4	0	1
Percentage of Total		*80%*	*90%*	*65%*	*50%*	*10%*	*20%*	*0%*	*5%*

SOURCE: Authors' compilation (Light, Singer, and Willet 1994).

year, columns 3 and 4 summarize the use of tables in the articles, columns 5 through 8 report the types of graphs used in the articles, and columns 9 and 10 indicate how the graph was used.

Our examination revealed a number of relevant observations. As one might expect from two respected journals focusing on publishing research syntheses in the social sciences and education, meta-analyses remain prominent in *Psychological Bulletin* and appear consistently in *Review of Educational Research*. Compared to the results from *Psychological Bulletin* articles published from 1985 through 1991, the more recently published articles from 2000 through 2005 have presented data through the use of a somewhat different assortment of figures. Further, in comparing the 2000 through 2005 articles published in

Psychological Bulletin and *Review of Educational Research*, we see modest differences between the two journals.

Using tables to present summary findings remains the most common method for presenting results, and the 2000 through 2005 totals show that this is becoming a nearly universal practice. Offering readers a comprehensive table that tallies the effect sizes gleaned from each of the studies included in the meta-analysis has also become a common practice. These tables are presented alphabetically by author name more often than not. Some, though, use other organization schemes, including sorting by the year in which the included studies were published, by the magnitude of the effect size reported, by methodological features of the studies, or by a key substantive variable.

As with tables, the use of graphs and figures has increased since the late 1980s and early 1990s. Only 19 percent (fourteen of seventy-four) of the *Psychological Bulletin* articles from 1985 through 1991 used graphs of any kind to present findings, versus 52 percent of *Psychological Bulletin* articles from 2000 through 2005. At 65 percent (thirteen of twenty), the proportion of articles including graphs or figures from the *Review of Educational Research*, though from a smaller sample, was even higher. As the results in table 26.1 suggest, graphs have been used for a number of purposes: representing the distribution of effect sizes across the examined studies by using a stem-and-leaf plot or a histogram; showing the relationship between effect size and sample size—often with a funnel plot; or displaying the relationship between effect size and a substantive or methodological characteristic of the examined studies—customarily with a line graph.[1] These types of displays are used more frequently in recent than in past issues of *Psychological Bulletin*, as reviewed by Richard Light and his colleagues (1994), and are reported with considerable prevalence in both of the journals.

Graphs and figures can also be used to represent the degree of uncertainty of the reported effect sizes, generally with 95 percent confidence intervals, or to represent an overall summary of the effects reported. As the results in the ninth column of table 26.1 suggest, some articles used graphics to represent uncertainty in estimates, though this practice was noted only in the articles published in *Psychological Bulletin*. We were somewhat surprised to observe that few articles in either journal took advantage of the opportunity to visually represent the summary effect-size estimate.

From the data reported in table 26.1, we suggest that graphic displays of some meta-analytic data have become more common and consistent in the social sciences, but that authors do not routinely take full advantage of the potential offered by graphic representation. Tables and graphics can be used to represent the findings in powerful ways and to test the assumptions of a study, as we hope to illustrate in the following sections.

26.2.2 Presenting Data in Tables

We concur with Light and his colleagues on the value of presenting raw data to readers and of ordering the data in a substantively informative fashion (1994). Our survey of the published meta-analyses in *Psychological Bulletin* and *Review of Educational Research* suggested that the statement that "very few [tables] are ordered with the intention of conveying a substantive message" may not be as accurate as it once was, at least in the field of psychology (Light, Singer, and Willett 1994, 443). Of the sixty published meta-analyses we examined in *Psychological Bulletin*, twenty-nine included tables with effect sizes listed alphabetically by the first author's last name (three articles listed the effect sizes chronologically, by year of publication). An alphabetic sort can make it easier for readers to find a specific study included in the synthesis—or to identify ones that were omitted. Twenty of the articles went beyond this typical strategy to include tables sorted by a substantive or methodological characteristic. In fact, three articles in *Psychological Bulletin* included tables ordered by two measures. For example, Elizabeth Gershoff, in a meta-analysis of corporal punishment and child behavior, first divided the included studies into those that measured the outcome in childhood and those that measured the outcome in adulthood (2002). Within each of these categories, the studies were grouped by child construct, and only then listed alphabetically by author surname. Providing this structure to the accounting of the studies included in the review gives the reader a sense of the research terrain and reveals the analyst's orientation to the data. To easily seek out a particular study, one need only consult the list of references. To communicate multiple classifications to readers, analysts may also consider coding the included studies for their meaningful characteristics and presenting the codes as columnar data in the table.

Similar to our discovery that published articles in *Psychological Bulletin* were more likely than those in *Review of Educational Research* to include graphs that represented uncertainty, our examination of the two journals suggested that the tables in *Review of Educational Research* were ordered more conventionally—that is, alphabetically by author last name or chronologically by publication year. Thirteen of the articles sorted the studies by the author's name and one article sorted the studies by year. Only three, however, used any other kind of sorting mechanism, such as the effect size or the substantive characteristics of the studies. As meta-analyses become increasingly ambitious and include longer tables or appendices of included studies, it would behoove authors to organize these data in meaningful and more accessible ways.

Effective tables provide two core functions in meta-analytic studies. First, they account for the raw data from the original studies examined by the authors, and offer the potential to communicate substantive differences between the studies in addition to the effect size gleaned from each one. Such presentations help readers recognize the origins of each data point presented in the meta-analysis, provide an account of the coding assumptions made by the analyst, and can facilitate replications of the synthesis in the future. Well-crafted tables can also describe overall effects and can document uncertainty through presentation of the standard errors and the 95 percent confidence intervals of the effect sizes. Tables that provide such information are fairly well represented in published research syntheses, and are far more prevalent than the graphic displays, which we turn to now.

26.2.3 Using Graphs

In *The Commercial and Political Atlas*, William Playfair asserted that from a chart "as much information may be *obtained in five minutes as would require whole days to imprint on the memory, in a lasting manner, by a table of figures*" (1801/2005, xii). We believe that Playfair's claims that graphics can serve as powerful tools for the transmission and retention of information are as true today as they were two centuries ago. Graphics are used for a number of purposes within the context of meta-analysis, including testing assumptions and serving as diagnostic tools, representing the distribution of effect sizes, and illustrating sources of variation in effect size data. Graphs can also represent summary effect sizes and their uncertainty, although this purpose appears underutilized. Different types of graphs appear to be better suited for particular purposes, but some display types emergent in the social sciences and education literature simultaneously attempt to achieve a combination of goals.

26.2.3.1 Assessing Publication Bias Graphs can be used to diagnose whether an analysis complies with the assumptions on which it is based, including, most notably, that the effect sizes are normally distributed and are not subject to publication or eligibility bias.

If published reports do not provide enough information to derive an effect size and are omitted from a synthesis, or if unpublished reports are not adequately represented, then the meta-analytic results may be biased. Funnel plots of the effect sizes of the included studies are frequently used to evaluate subjectively whether the effect sizes from the included studies are, as Light and Pillemer put it, well behaved (1984).

By plotting the effect sizes of the studies against the sample size, standard error, or precision of the included

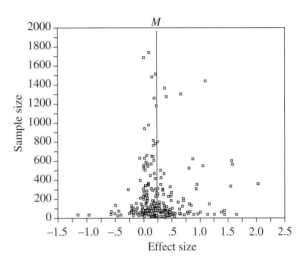

Figure 26.1 Funnel Plot of Effect Size Against Sample Size
SOURCE: Albarracín et al. 2005.
NOTE: Funnel plot. Two effects with extremely large sample sizes were excluded to make the shape of the plot more apparent. These large sample groups had average effect sizes.

studies, one can determine the relationship between the effect-size estimates and the presumed underlying parameter. If bias is not present, the plot takes the form of a funnel centered on the mean or summary effect size. If bias is present, then the plot is distorted, either by assuming an asymmetrical pattern or by having a void. Some statistical tests for detecting funnel plot asymmetry are available, namely Matthias Egger's linear regression test and Colin Begg's rank correlation test, but these have low statistical power and are rarely used in syntheses (Egger et al. 1997; Begg and Mazumdar 1994). Another similar statistical method, the trim-and-fill procedure, can be used as a sensitivity analysis to correct for model-identified bias. As illustrated in Christopher Moyer, James Rounds, and James Hannum's study, this method can generate a new summary effect size with a confidence interval to reflect possibly missing studies (2004; for a discussion of methods for detecting publication bias, see chapter 23, this volume).

Dolores Albarracín and her colleagues conducted a meta-analysis of 194 studies of HIV-prevention interventions and included a funnel plot that we reproduce as figure 26.1 (2005). First we consider the orientation of the graph's axes. Funnel plots can be arranged in a number of ways, but in this case the sample size is on the Y-axis with the largest value at the top, though some authors choose to rank the sample sizes in reverse order, allowing for larger studies—those with the greatest weight and presumably more reliable estimates—to be closest to the effect size values represented on the x-axis. Other authors choose to put the sample size on the x-axis, resulting in a horizontal funnel, though a vertical funnel is conventionally preferred. In any case, the scale—particularly when sample size is reported instead of the standard error of the estimate—sometimes does not gracefully include the largest studies, as is the case here. This problem can be addressed by omitting the observations that do not fit, as these authors did, or by representing a break in the scale, but one should consider Jonathan Sterne and Matthias Egger's suggestion that the standard error has optimal properties as the scale representing the size of the study (2001). Albarracín and her colleagues also provided an indication of the overall summary effect size, in the form of a vertical line marked M (2005). This simple addition to the figure clearly presents the summary effect size, yet was not common in the articles we reviewed. The plot shown in figure 26.1 suggests that, despite some evidence of a positive skew, these data are indeed well behaved.

If the subjective evaluation of the shape of a funnel plot does not seem rigorous enough, a normal quantile plot may also be used to represent potential bias (Wang and Bushman 1998). Tarcan Kumkale and Dolores Albarracín conducted a meta-analysis of the long-term change of persuasion (a sleeper effect) of individuals who receive information accompanied by a cue that the source may not be credible (a discounting cue) (2004). Their analysis included a number of exemplary figures, including the normal quantile plot we reproduce here as figure 26.2. The observed standard effect-size estimate is plotted along one axis against the expected values of the effect sizes had they been drawn from a normal distribution. The deviation from the normal distribution is represented as the distance of the point from the straight diagonal line, which is surrounded by two lines that represent the 95 percent confidence interval. If publication bias is present, the points representing the effect-size estimates will deviate from the normal distribution, will not follow a straight line, and will fall outside of the confidence interval (Wang and Bushman 1998). Although the figure might be improved by presenting both axes on the same range and scale, figure 26.2 suggests that the meta-analysis successfully captured the existing research on their targeted construct in a relatively unbiased fashion.

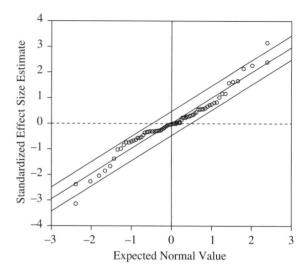

Figure 26.2 Normal Quantile Plot Used to Evaluate Publication Bias

SOURCE: Kumkale and Albarracín 2004.
NOTE: Normal quantile plot of the effect sizes representing the magnitude of change in persuasion in discounting-cue conditions from the immediate to the delayed posttest.

26.2.3.2 Representing Effect-Size Distributions

In addition to testifying to the presence or absence of publication bias, graphs can efficiently represent the distribution of effect sizes, particularly the central tendency, spread, and shape of the effects. The principal tools to achieve this are stem-and-leaf plots and box or schematic plots. In a stem-and-leaf plot, the first decimal place of the effect size serves as the stem and the second decimal place appears as the leaf. In a box plot or schematic plot, the median effect size is the middle of the box, the upper and lower quartiles are the ends of the box, the maximum and minimum effect sizes are the whiskers, and outliers are represented by single data points placed outside the whiskers (for more on these two modes of representation, see Tukey 1977, 1988).

Our examination of recent meta-analyses revealed that stem-and-leaf plots have become a popular method for presenting data. Twenty-nine percent of the articles in *Psychological Bulletin* and 30 percent of the articles in *Review of Educational Research* used stem-and-leaf displays. A few of these plots used more than one set of leaves on a given stem to show how the distributions vary by notable characteristics of the studies. Linnea

Ehri and her colleagues, for instance, conducted a meta-analysis of thirty-eight phonics reading instruction experiments and presented the pool of estimates comprising the analysis as three side-by-side stem-and-leaf plots (2001), which we present as figure 26.3. The three separate leaves off the central stem in this figure correspond to differences among the studies in terms of the timing of the posttest after the phonics intervention. By virtue of the method of presentation, it is indeed apparent that most of the instruction experiments lasted less than one year and were associated with heterogeneous effects, that longer exposure to instruction was associated with more consistently positive effects, and that the follow-up estimates provide evidence of lasting effects (Ehri et al. 2001, 414). The authors then examined potential moderating variables to account for the heterogeneity in short-term effects.

Richard Light and his colleagues had urged meta-analysts to consider using stem-and-leaf and schematic plots to present the distribution of effect sizes (1994). To some extent, this advice has been followed, as stem-and-leaf plots have become more prevalent in recent published syntheses. Schematic plots, on the other hand, are still quite uncommon—we observed only one in the eighty articles we examined.

26.2.3.3 Representing Systematic Variation in Effect Size An emerging and effective practice is to use graphs to illustrate how effect sizes vary according to substantive issues or methodological features of the studies constituting the meta-analysis. These findings are characteristically presented as summary tables, but can also be effectively presented graphically. Parallel schematic plots, bar graphs, and scatterplots can be used. The choice of format depends to a great extent on the type of variable that is the source of variation in the outcome.

To present the influence of a categorical variable, a parallel schematic plot (also known as a box and whisker plot) or a histogram can be used. A meta-analysis by Eric Turkheimer and Mary Waldron of the research literature on nonshared environment, which offers one possible explanation for how siblings can have remarkably different outcomes from one another, presented the effect sizes from the primary studies in a number of parallel schematic plots, including one organized by the source of the data: observation, the parent, the child, or an aggregate of measures (2000). The graph, which we present here as figure 26.4, reveals both the variation in effect sizes by reporter and—by virtue of a single line connecting the median points—the difference in medians across types.

Stem	Leaf End of Instruction (1 yr)[a]	Leaf End of Instruction (>1 year)[b]	Leaf Followup[c]
3.7	1		
2.2	7		
2.1			
2.0			
1.9	9		
1.8			
1.7			
1.6			
1.5			
1.4	12		
1.3			
1.2			
1.1	9		
0.9	1		
0.8	4		
0.7	0236	5	6
0.6	001233	47	
0.5	00133	24	6
0.4	345789		
0.3	23367889	6	238
0.2	014457	48	8
0.1	23469	7	
+0.0	01344479	0	
−0.0	7		
−0.1	1		
−0.2	05		
−0.3	3		
−0.4	7		7

Figure 26.3 Stem-and-Leaf Plot of Effect Sizes of Systematic Phonics instruction on Reading by Timing of Posttest

SOURCE: Ehri et al. 2001. Reprinted by permission of SAGE Publications.

NOTE: Stem-and-leaf plot showing the distribution of mean effect sizes of systematic phonics instruction on reading measured at the end of instruction and following a delay.

[a]End of instruction or end of Year 1 when instruction lasted longer.
[b]End of instruction which lasted between 2 and 4 years.
[c]Followup tests were administered 4 to 12 months after instruction ended.

This figure shows that correlations using observer reports are high and variable, that parental estimates are lower, that estimates based on child reports were higher, and that studies using reports of a child's environment from more

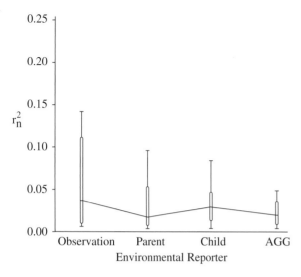

Figure 26.4 Parallel Schematic Plot of Median Effect Size by Environmental Reporter

SOURCE: Turkheimer and Waldron 2000.
NOTE: Median effect size by environmental reporter. AGG indicates aggregate of more than one reporter. (See figure 26.5 caption for explanation of box and whisker plots.)

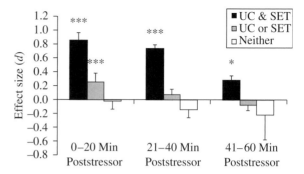

Figure 26.5 Mean Cortisol Effect Size for Performance Tasks with Both Social-Evaluative Threat (SET) and Uncontrollability (UC), Either SET or UC, and Neither by Timing of Poststressor Measurement

SOURCE: Dickerson and Kemeny 2004
Note the effective use of shading as well as the error bars and p-value indicators.
NOTE: Mean ($\pm$ *SEM*) cortisol effect size (d) for performance tasks with both social-evaluative threat (SET) and uncontrollability (UC), performance tasks with either SET or UC, and performance tasks without either component or passive tasks (Neither) during intervals 0–20, 21–40, and 41–60 min poststressor. *$p < .05$. ***$p < .001$.

than one source revealed correlations that were low but the least variable of all.

Bar charts or histograms can also be effectively used to illustrate systematic sources of variation, as in figure 26.5. Sally Dickerson and Margaret Kemeny, in an examination of cortisol responses to stressors, presented their findings in a diagram, which is organized simultaneously by two measures: the type of outcome that was measured and the time lag between the stressor and the outcome measurement (2004). The erosion of the effect over time is immediately evident, as are the varying effect sizes graphed within each of the three categories of poststressor lag. Across all three categories of poststressor lag, when both social-evaluative threat and uncontrollability tasks are set, the cortisol level is considerably higher than when either or none of the conditions are present. The error bars and the asterisks indicating, respectively, uncertainty and the statistical significance of the findings offer an added degree of sophistication and clarity to the figure with the addition of a few simple visual elements.

The preceding examples consider categorical substantive variables, but the influence of continuous moderator variables can also be effectively represented using graphs,

as in figure 26.6. In a meta-analysis of racial comparisons of self-esteem, Bernadette Gray-Little and Adam Hafdahl presented this scatterplot of the effect size against the average age in years of the participants in the studies (2000). The varying sizes of the points show the relative weights of the studies, and the linear regression line reveals the trend between effect size and age. This graph efficiently communicates to the reader that the majority of studies—and especially those granted the most weight in the analysis—included children between the ages of ten and eighteen, that both positive and negative effects were observed, and that more positive effects sizes were reported in studies with older participants.

Some more recently developed graphical techniques endeavor to accomplish a number of purposes at once. Two notable formats are the radial plot and the forest plot, which has sometimes been referred to as a confidence interval plot (Galbraith 1988a, 1988b, 1994).[2] The radial plot is a highly technical figure that can sometimes pose a challenge to the reader, but its virtue and application to meta-analysis is that it explicitly compares estimates that have different precisions. Two essential pieces of information are needed to create a radial plot: each estimate's precision, which is calculated as the inverse or reciprocal

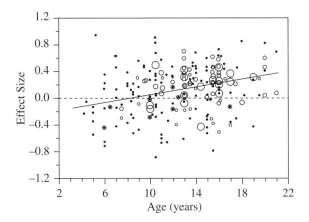

Figure 26.6 Scatterplot of Effect Size by Average Participant Age in Study

SOURCE: Gray-Little and Hafdahl 2000
NOTE: Scatterplot of Hedges's d against average participant age in the study. Effect size (ES) weights are restricted to 700. Symbol area is proportional to ES weight. Dotted horizontal line represents no effect ($d = 0.0$). Solid line represents weighted linear regression equation for all independent ESs with available data ($k = 257$).

of its standard error ($x_i = 1/\sigma_i$), and the standardized effect estimate ($y_i = z_i/\sigma_i$). These data are plotted against one another to show how the standardized deviated estimates vary according to their precision, with the more precise estimates placed further from the origin. The radial element is introduced to recover the original z_i estimates by extrapolating a line from the origin through the (x_i, y_i) point to the circular scale (Galbraith 1988a). Most earlier applications use a radial plot to summarize odds-ratio results from clinical trials (see Galbraith 1988b, 1994; Whitehead and Whitehead 1991; Schwarzer, Antes, and Schumacher 2002).[3]

Rex Galbraith applied a radial plot to meta-analytical data, which we show here as figure 26.7, and which is in many ways a remarkable figure (1994). First note that the studies included in the meta-analysis are visually coded in two ways: by the numbers identifying the individual studies at the top and by the substantive characteristics—drug versus diet and primary versus secondary—represented by the shape and coloring of the icons. This coding scheme, though shown in the context of a radial plot, would be beneficial in any form of visual representation. The data are presented in the middle of the figure with the inverse of the standard error as the x-axis and the standardized estimate as the y-axis; a two-unit change in the

standardized estimate is equivalent to the 95 percent confidence interval. The data are odds ratios for fatal coronary disease; negative estimates or ratios less than one are associated with reduced mortality. Two studies—identified as eleven and twenty-one—report significant reductions in mortality, but their precision is on the lower end of the distribution. One can see that though most of the diet studies report reduced mortality, none are especially precise, and that the drug studies farther to the right are more equivocal. One can recover the original odds ratio estimates from the circular scale to the right, but the strength of this method is to represent the relative weights and values of the estimates and to distinguish the amount of variation in the sample.

The radial plot is a powerful way to represent data and is favored among some statisticians, but for the uninitiated it is not as immediately evident in the same way that other graphic representations can be (see, for example, Whitehead and Whitehead 1991). Furthermore, though the radial plot effectively represents variation in estimates and their relation to their standard errors, this emphasis seems to come at the expense of clearly representing a summary measure—for example, the absolute efficacy of the cholesterol-lowering drugs is not evident in figure 26.7. Another form of representation, known generally as the forest plot, combines images with text to present meta-analytic information in a more accessible form, though we grant that the variation in standard errors is not presented as precisely as in a radial plot.[4]

26.2.3.4 The Forest Plot If sheer frequency of use is a suitable measure of acceptance, then the display known as the forest plot is the new standard for the presentation of meta-analytical findings. The large volume of existing forest plots is related in large part to its widespread adoption in medical research, particularly as represented in the Cochrane Collaboration's Cochrane Library, which has published thousands of syntheses of medical trials that employ forest plots.[5] Although the origin of the term *forest plot* is uncertain (Lewis and Clarke 2001), J. A. Freiman and colleagues were the first to display the point estimates of studies with horizontal lines to represent the confidence interval (1978). Lewis and Ellis introduced the summary effect measure at the bottom of the plot for use in a meta-analysis (1982; Lewis and Clarke 2001).

The forest plot has not yet been widely adopted in the social sciences, however. The Campbell Collaboration, the emerging social sciences counterpart to the Cochrane Collaboration, frequently includes syntheses with forest plots, but its collection does not universally use the practice.[6]

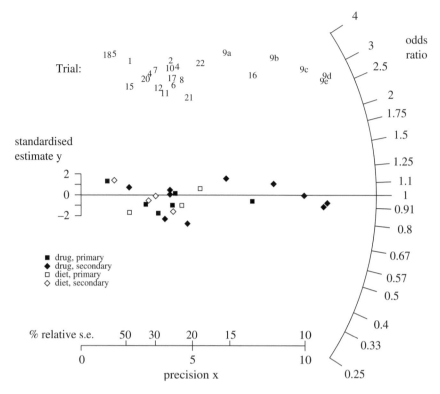

Figure 26.7 Radial Plot of Fatal Coronary Heart Disease Odds Ratios

SOURCE: Galbraith 1994. Reprinted with permission from the *Journal of the American Statistical Association.* © 1994. All rights reserved.

NOTE: A Radial Plot of Fatal Coronary Heart Disease Odds Ratios (Treated/Control) for 22 Cholesterol Lowering Trials Using z_i and σ_i Given by (14). Data from Ravnskov (1992, table 1).

Notably, not one of the meta-analyses published from 2000 to 2005 in either *Psychological Bulletin* or *Review of Educational Research* included a forest plot. Although it was not called a forest plot at the time, Light and his colleagues (1994) offered such a display by Salim Yusuf and his colleagues (1985) as an exemplary method to present the findings from a meta-analysis, and encouraged its application in the social sciences. We hope to reinforce that suggestion.

A forest plot combines text with graphics to display both the point estimates with confidence intervals as well as an estimate of the overall summary effect size. As a result, it simultaneously represents uncertainty and the summary effect and indicates the extent to which each study contributes to the overall result. The forest plot follows a number of conventions. The point estimate of each study is represented as a square or a circle, with its size

often varying in proportion to the weight the individual study is accorded in the analysis. At the bottom of the figure, a diamond represents the overall summary effect size. The center of the diamond shows the point estimate of the summary effect size, and its width represents the limits of its 95 percent confidence interval. A vertical line extending through the figure can represent the point estimate of the summary effect size or the point of a null effect. In either case, one can compare how the effect sizes across the studies in the meta-analysis are distributed about the point of reference. A plotted effect size is statistically different from the overall or null effect if the diamond representing it does not intersect the vertical line in the figure.

Figures 26.8 and 26.9 show how the data from the teacher expectancy meta-analysis used in this volume (see appendix) can be displayed using a forest plot. Figure 26.8 shows a forest plot we created using the user-

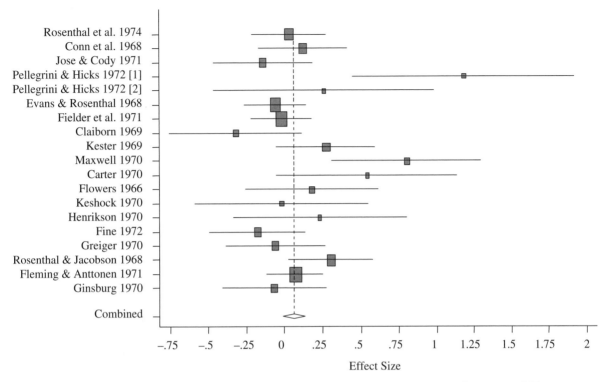

Figure 26.8 Forest Plot of Effect Sizes Derived from Eighteen Studies of Teacher Expectancy Effects on Pupil IQ

SOURCE: Author's compilation.

written metagraph command in Stata.[7] Note that the figure includes the name of each study, a representation of the point estimate, its 95 percent confidence interval, and a dotted vertical line at the overall combined mean effect estimate. In a single figure, one can present the summary effect size, its uncertainty, and its statistical significance, the distribution of effects across the included studies, the relative contribution of each study to the overall effect estimate, and the degree of uncertainty for each of the individual effect estimates from the studies. The size of the squares represents the weight of each effect estimate, which tends to counteract the visual attention commanded by the estimates with large confidence intervals. In this case, the inverse variance weights are used.

The data in figure 26.8 are sorted as originally presented (Raudenbush and Bryk 1985). The order, in and of itself, does not offer much interpretative value and certainly does not suggest a systematic relationship with an underlying construct. However, these data offer an impor-

tant moderator variable: the estimated weeks of teacher-student contact prior to expectancy induction. Figure 26.9, which takes advantage of this additional substantive information, was produced using Comprehensive Meta Analysis Version 2.[8]

Figure 26.9 gains from the forest plot's potential to serve as both a figure and a table. One can elect to present a number of statistics in the middle columns. In this case, we have presented the value of the moderator variable, the standardized effect size, the standard error, and the p-value of the estimate. In both figures, the left-hand tip of the diamond representing the summary effect crosses the zero point, indicating that the overall effect is not statistically different from zero. More important, the ordering of these data illustrates the relationship between the teachers' previous experience with their students and the standardized mean differences between the treatment and control groups. That is, when teachers had more previous experience with students, the researchers' efforts to

Study name	Prior Weeks	Statistics for each study			Standardized difference in means and 95% Cl
		Standardized difference in means	Standard error	p	
Pellegrini & Hicks 1972 [1]	0	1.180	0.373	0.002	
Pellegrini & Hicks 1972 [2]	0	0.260	0.369	0.481	
Kester 1969	0	0.270	0.164	0.100	
Carter 1970	0	0.540	0.302	0.074	
Flowers 1966	0	0.180	0.223	0.420	
Maxwell 1970	1	0.800	0.251	0.001	
Keshock 1970	1	−0.020	0.280	0.945	
Rosenthal & Jacobson 1968	1	0.300	0.139	0.031	
Rosenthal et al. 1974	2	0.030	0.125	0.810	
Henrikson 1970	2	0.230	0.290	0.428	
Fleming & Anttonen 1971	2	0.070	0.094	0.456	
Evans & Rosenthal 1968	3	−0.060	0.103	0.560	
Greiger 1970	5	−0.060	0.167	0.719	
Ginsburg 1970	7	−0.070	0.174	0.687	
Fielder et al. 1971	17	−0.020	0.103	0.846	
Fine 1972	17	−0.180	0.159	0.258	
Jose & Cody 1971	19	−0.140	0.167	0.402	
Conn et al. 1968	21	0.120	0.147	0.414	
Claiborn 1969	24	−0.320	0.220	0.146	
Overall		0.060	0.036	0.698	

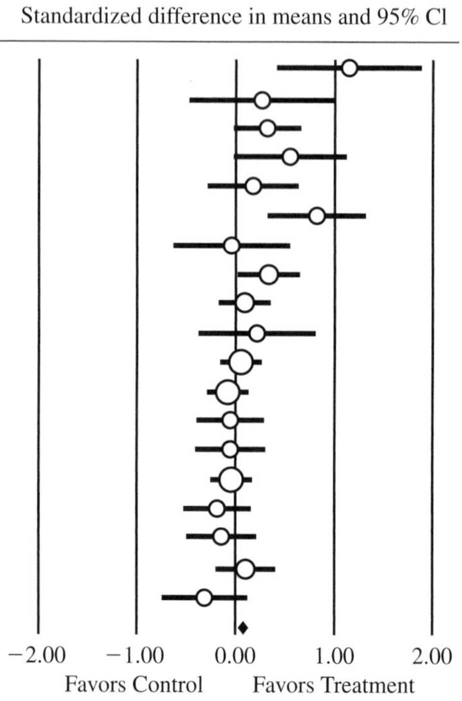

−2.00   −1.00   0.00   1.00   2.00
Favors Control   Favors Treatment

Figure 26.9 Forest Plot of Effect Sizes Derived from Eighteen Studies of Teacher Expectancy Effects on Pupil IQ Sorted by Weeks of Teacher-Student Contact Prior to Expectancy Induction

SOURCE: Authors' compilation using data from Raudenbush and Bryk 1985.
NOTE: "Prior Weeks" are estimated weeks of teacher-student contact prior to expectancy induction.

influence the teachers' expectations tended to be less effective. This single plot illustrates the degree of uncertainty for each study, the contribution of each study to the overall estimate, the overall estimate itself, and the relationship between an important moderator variable and the effect size.

As another example of the use of the forest plot, we use the meta-analytic data from Peter Cohen on the correlations between student ratings of the instructor and student achievement (1981, 1983), which we show as figure 26.10. This figure was also produced using Comprehensive Meta-Analysis Version 2. In this case, the moderator variable is categorical, so we group the data according to the method used to manage pre-intervention differences on student ability, either covariate adjustment or random assignment. The method of ability control was a source of systematic variation across the primary studies and can be clearly illustrated in this graphic representation.

The fourteen covariate-adjustment studies are grouped at the top of the display and followed by a row that presents the mean effect estimate and confidence interval for this subgroup of studies. The random assignment studies, in turn, are grouped with their own subgroup estimates. At the bottom of the figure, the overall effect size and its confidence interval are shown. In addition to portraying the particular outcomes of each study included in the meta-analysis, the figure demonstrates that the effect sizes for both subgroups of studies and for the overall summary

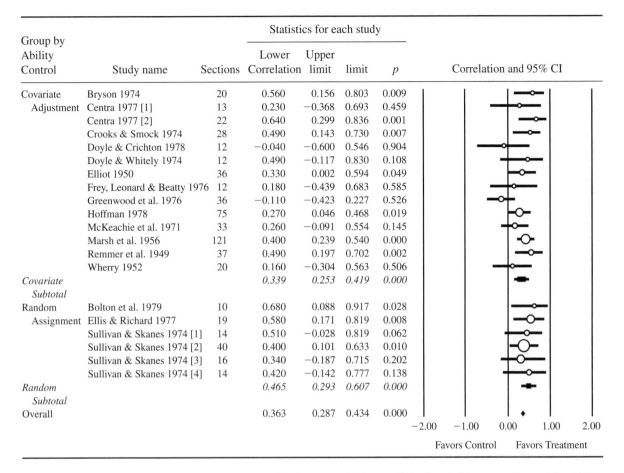

| Group by Ability Control | Study name | Sections | Statistics for each study | | | | Correlation and 95% CI |
			Correlation	Lower limit	Upper limit	p	
Covariate Adjustment	Bryson 1974	20	0.560	0.156	0.803	0.009	
	Centra 1977 [1]	13	0.230	−0.368	0.693	0.459	
	Centra 1977 [2]	22	0.640	0.299	0.836	0.001	
	Crooks & Smock 1974	28	0.490	0.143	0.730	0.007	
	Doyle & Crichton 1978	12	−0.040	−0.600	0.546	0.904	
	Doyle & Whitely 1974	12	0.490	−0.117	0.830	0.108	
	Elliot 1950	36	0.330	0.002	0.594	0.049	
	Frey, Leonard & Beatty 1976	12	0.180	−0.439	0.683	0.585	
	Greenwood et al. 1976	36	−0.110	−0.423	0.227	0.526	
	Hoffman 1978	75	0.270	0.046	0.468	0.019	
	McKeachie et al. 1971	33	0.260	−0.091	0.554	0.145	
	Marsh et al. 1956	121	0.400	0.239	0.540	0.000	
	Remmer et al. 1949	37	0.490	0.197	0.702	0.002	
	Wherry 1952	20	0.160	−0.304	0.563	0.506	
Covariate Subtotal			*0.339*	*0.253*	*0.419*	*0.000*	
Random Assignment	Bolton et al. 1979	10	0.680	0.088	0.917	0.028	
	Ellis & Richard 1977	19	0.580	0.171	0.819	0.008	
	Sullivan & Skanes 1974 [1]	14	0.510	−0.028	0.819	0.062	
	Sullivan & Skanes 1974 [2]	40	0.400	0.101	0.633	0.010	
	Sullivan & Skanes 1974 [3]	16	0.340	−0.187	0.715	0.202	
	Sullivan & Skanes 1974 [4]	14	0.420	−0.142	0.777	0.138	
Random Subtotal			*0.465*	*0.293*	*0.607*	*0.000*	
Overall			0.363	0.287	0.434	0.000	

−2.00   −1.00   0.00   1.00   2.00

Favors Control   Favors Treatment

Figure 26.10 Forest Plot of Validity Studies Correlating Instructor Ratings and Student Achievement Grouped by Ability Control

SOURCE: Authors' compilation using data from Cohen 1983.
NOTE: Cohen (1983); studies with sample sizes less than ten were excluded.

effect estimate are positive and statistically different from zero. Also, one can see that the effect size for the random-assignment studies is somewhat larger than the estimate for the studies using covariate adjustment. The 95 percent confidence intervals for the effect sizes for random-assignment and covariate-adjustment studies overlap, however, suggesting that the difference in ability control is substantively, but not statistically, significant.

The same principles that apply to ordering studies in tables can also be fruitfully applied to ordering them in figures, namely forest plots. Depending on the data, one can enhance the potential meaning conveyed by the figure by ordering the studies according to a single underlying scale or by grouping them by a pertinent substantive or methodological characteristic. In addition to the moderating variables we discussed earlier, one could similarly order studies by number of participants, dose of intervention, or other pertinent variables.

These illustrations highlight the utility of the forest plot format. The format, though, does have its limitations. Other forms of representation, such as the funnel plot or the radial plot, address publication bias assumptions or effect size variations especially well. Also, a different form of representation may be more appropriate for an especially large meta-analysis, particularly to portray the overall variation in effect sizes across the studies. For example, Geoffrey Borman and his colleagues used a stem-and-leaf plot to show the distribution of 1,111 effect sizes gleaned from the comprehensive school reform literature (2003). A forest plot with that many observations would clearly not be feasible. One could, however, divide the numerous observations along key substantive or methodological lines and present them in a single summary figure that displays the outcomes for the constituting components. Subsequent figures could elaborate on each of the parts that comprise the whole. Clarity and accuracy are paramount, and an especially large dataset may call for other measures—such as multiple plots, a stem-and-leaf plot, or a box plot—to be most instructive to the reader.

26.2.4 Visual Display Summary

In the context of a meta-analysis, figures can serve diagnostic purposes, such as assessing publication bias. They can convey information about the individual attributes of studies that provide the data points of the analysis such as their standard error. They can provide information regarding the central tendency of the effect-size estimate, its statistical significance, and its variability. They can show how differences in effect size are related to systematic differences across the studies. They can display powerful combinations of meta-analytic information. Used thoughtfully, displays can translate complex data from research syntheses to help researchers, policy makers, practitioners, and lay audiences alike see and understand the core outcomes from the work. Indeed, with the growing demand for quantitative evidence to inform social, educational, and health policy, efficient and effective presentation of meta-analytic data is likely to become increasingly important. As Playfair suggested:

> Men of high rank, or active business, can only pay attention to general outlines; nor is attention to particulars of use, any further than as they give a general information; it is hoped that, with the assistance of these Charts, such information will be got, without the fatigue and trouble of studying the particulars of which it is composed. (1801/2005, xiv–xv)

More succinctly, Richard Light and his colleagues noted that "simple graphic displays of effect sizes in a meta-analysis can impose order onto what at first may seem like a morass of data" (1994, 451). They also made some sound recommendations for producing high-quality and informative figures. We echo many of their points and add some suggestions of our own.

The Spirit of Good Graphics Can Even Be Captured in a Table of Raw Effect Sizes, Providing They Are Presented in a Purposeful Way. The majority of published meta-analyses we examined presented catalogues of raw effect-size data or summary effect-size information. A comprehensive table that presents the meta-analytic data gleaned from each of the studies included in the meta-analysis is a useful feature that can facilitate potential replication and reveal the attributes of the individual studies that lie behind the overall summary statistics. Although the prominence of these tables in recently published studies is well deserved, we contend that they are not always used to their full potential. Rather than simply presenting the results from the included studies in alphabetical order by author name, tables can often be ordered more meaningfully and instructively, such as by the effect sizes or by the value of a key substantive or methodological moderator variable. Such reordering often can convey much more useful information than conventional alphabetical or chronological listings. These general principles of useful ordering of effect size data can also be profitably applied to forest plots, given their similarity to tables.

Graphic Displays Should be Used to Emphasize the Big Picture. Authors making effective use of graphics should start by making clear the main take-home message of the meta-analysis. This can be accomplished by displaying the summary effect size and its variability, or can demonstrate the importance a moderator plays in shaping the results across the studies. Generally, stem-and-leaf plots and schematic plots are especially effective for presenting the big picture. Both figures provide an intuitive display of the distribution of a large number of effect sizes and they facilitate identification of outliers. Once the larger picture is established, the reader can pursue the more detailed questions. Forest plots with summary effect-size indicators can depict the big picture and convey many of the smaller details.

Graphics Are Especially Valuable for Communicating Findings from a Meta-Analysis When the Precision or Confidence of Each Study's Outcome is Depicted. A number of the formats we have reviewed address the issue of the error inherent in any given study. Whether using a standard error scale in a funnel plot, a precision scale in a radial plot, or a confidence interval bar in a forest plot, recent developments in the representation of meta-analytic findings expressly present both the estimate offered by a single study as well as its error.

Graphic Displays Should Encourage the Eye to Compare Different Pieces of Information. When important moderators influence the outcomes across the studies in the meta-analysis, when distributions of effect sizes found in distinct subgroups of studies appear quite different, and when different analyses of the data reveal variable outcomes, graphic displays can effectively portray these distinctions. The side-by-side graphical representations of two or more subgroups of studies or from several analyses of the same sample of studies can make highly technical issues more intuitive and clear to the reader.

The Selection of Graphical Representation Should be Informed by One's Purpose and Data. Just as the type of variable in question may inform the decision to use a line graph or a bar graph, the graphs we have reviewed in this chapter perform differently under different conditions. For example, whether the effect is expressed in terms of odds or as a correlation may influence the form of representation one chooses. Funnel plots aptly represent the distribution of effects and are well served when called on to address questions of publication bias. Stem-and-leaf and schematic plots—either alone or in multiple parallel plots—portray the overall results of a meta-analysis and compare key subgroups of studies efficiently. They are particularly useful when considering an especially large group of studies. A forest plot can also illustrate the uncertainty of each effect estimate, the contribution of each study to the overall estimate, the overall estimate itself, the relationship between a relevant variable and the effect size, and some of the information presented in the often-used table of primary studies.

We hope that by presenting this review of the strengths and limitations of a comprehensive set of graphic approaches to portraying meta-analytic data we will have helped readers make informed decisions about how best to communicate their results (for further reading on the principles and history of statistical graphics, see Bertin 1983; Cleveland 1994; Harris 1999; Tufte 1990, 1997, 2001, 2006; Wainer 1984, 1996; Wainer and Velleman 2001).

26.3 INTEGRATING META-ANALYTIC AND NARRATIVE APPROACHES

26.3.1 Strengths and Weaknesses

Among the key strengths of meta-analysis are parsimony, precision, objectivity, and replicability. Using a well-de-

fined process for selecting the primary studies and an accepted methodology for generating summary statistics, a large number of empirical studies in a given substantive area can effectively be reduced to a single estimate of the overall magnitude of an effect or relationship. Two reviewers using the same meta-analytic procedures should develop the same statistical summary. If they use different techniques, the descriptions of the methods should make the differences explicit. When two reviews differ in their results and conclusions, the reader will understand why. Narrative reviews, on the other hand, can describe the results from both quantitative and qualitative studies in a given area, easily discuss the various caveats of the individual studies, and trace the overall evolution of the empirical and theoretical work in the area. What narrative reviews lack in terms of parsimony, precision, and replicability, they can provide in terms of rich and careful description of the various methodological and substantive differences found across a research literature.

26.3.2 Reasons for Combining Meta-Analytic and Narrative Methods

As Harris Cooper and his colleagues suggested, an effective research synthesis should include both analytic and synthetic components (2000). It should make sense of the literature analytically by identifying theoretical, methodological, and practical inconsistencies found across the primary studies and should describe the evolution of the empirical literature and help the reader understand how the current research synthesis advances that evolution. These analytical contributions may take the form of narrative descriptions, but they might also suggest empirical assessments of how such inconsistencies in the literature contribute to differences in the observed effect sizes. The synthetic component of the research synthesis draws together all evidence in the domain under investigation to provide parsimonious and precise answers to questions about the overall effects found across the included studies and about how the various methodological, theoretical, and practical differences from the studies moderate the observed outcomes.

Several scholars have called for narrative and meta-analytic methods to be integrated and some have produced syntheses that have embodied characteristics of both traditions. Indeed, such discussions have taken place since the late 1970s and early 1980s when meta-analysis was a nascent field. Most notably, Richard Light and David Pillemer argued that reviews organized to ally both qualitative and quantitative forms of information would ultimately maximize knowledge (1982). Thomas Cook and Laura Leviton took a similar stance:

> Science needs to know its stubborn, dependable, general 'facts,' and it also needs data-based, contingent puzzles that push ahead theory. Our impression is that meta-analysts stress the former over the latter, and that many qualitative reviewers stress the latter more than the former (or at least more than meta-analysts do). Of course, neither the meta-analyst nor the qualitative reviewer *needs* to make either prioritization. Each can do both and any one reviewer can consciously use both qualitative and quantitative techniques in the same review. Indeed, s/he should. (1980, 468)

Beyond this rhetoric, there are clear examples in the literature of researchers who have since developed the mixed-method approaches these earlier authors advocated. Robert Slavin, for instance, has suggested an alternative form of research synthesis that he calls best-evidence synthesis (1986). This method uses the systematic inclusion criteria and effect-size computations typical of meta-analyses, but discusses the findings of critical studies in a form more typical of narrative reviews. Slavin has applied this method to the synthesis of a number of educational practices and interventions, including, most recently, the language of instruction for English language learners (Slavin and Cheung 2005). He proposed the method largely in response to the problem of mixing apples and oranges some critics of meta-analysis had voiced. The best evidence approach avoids aggregating findings of varying phenomena and methodology and focuses on only the best evidence to address the synthetic questions. Further, narrative exposition details the results of the included studies so that the reader gains a better appreciation for the context and meaning of the synthetic findings. This method also seems especially relevant when discussing the relative effect sizes of various interventions that might be of interest to policy makers and practitioners. Clear description of the programmatic methods behind the summary measures can enable the practices that produced positive results in previous trials to be replicated and the proper context to supporting implementation of the intervention to be configured.

A second example comes in a monograph describing the meta-analytic results from a synthesis of various types of summer school programs to promote improved student outcomes, especially achievement (Cooper et al. 2000). In this monograph, Cooper and his colleagues illustrated the use of a combination of novel methods for conducting and describing meta-analytic findings. Rather than restricting their synthesis to a somewhat homogeneous collection of the best evidence, as Slavin advised (1986), the researchers applied narrative description and meta-analytic integration to the larger literature. In this case, the authors illustrated how the analyst can consider all of the evidence and, using both statistical and narrative exposition, identify those studies that provide the best answer to the broader synthetic question of the overall effect of summer school, as well as respond to the analytical challenge of identifying the features of summer programs most consistently associated with improved student outcomes.

Though both methods are good examples of existing frameworks for integrating meta-analytic and narrative reviews, we contend that there has been little direct guidance to help researchers understand the rationales and methods for achieving this integration. In the sections that follow, we describe three central reasons why the analyst may choose such an approach to research synthesis and provide examples of how one may accomplish it.

26.3.2.1 Describing Results The first notable problem one might encounter when attempting to synthesize a body of evidence is that not all studies are necessarily amenable to traditional forms of meta-analysis. If one encounters an important study, or set of studies, that do not provide the details necessary to compute an effect size, what is the suggested course of action? What about studies that provide descriptive statistics or qualitative description and no effect-size data? Typically, such studies are omitted and may remain unreported by the researcher, regardless of their importance and potential contributions to understanding the phenomena under review. We maintain, however, that these studies might offer important information that can and should be included in research syntheses, of which a meta-analysis is a chief element.

Narrative description of some research that is not amenable to meta-analysis can help the reader understand the broader literature in an area in a way that supplements or surpasses the overall mean effect size. First, in an area with limited research evidence, discussion of these studies and presentation of some synthetic information regarding, for instance, the observed direction of effects, can offer the reader some preliminary and general ideas about the past findings. It does not offer the same precision of a weighted effect size, but in combination with other evidence may be valuable and informative. For instance, the Cooper and colleagues synthesis of the summer school literature included information gleaned from thirty-nine reports that did not include enough statistical data to calculate an effect size (2000). The authors, however, productively included this research by generating information concerning the general direction of the achievement effects observed in the primary studies (in this case, all positive, mostly positive, even, mostly negative, and all negative). This directional information reinforced the meta-analytic finding that summer school positively influences student achievement. Although such vote-counting procedures can be misleading on their own (Hedges and Olkin 1980), by combining careful narrative description with vote-count estimates of effect size, the authors were able to assemble enough analytical and synthetic details to provide evidence about overall effects and the potential moderators of those effects. The narrative description of these results can also identify the limitations of the previous research base and suggest new directions researchers might take to improve the rigor of future studies, so that traditional meta-analytic procedures can be applied to new work in the area.

Detailed narrative descriptions of the specific interventions or phenomena under consideration can also facilitate future replications of the synthesized effects by both practitioners and researchers. Describing exemplary practices used in particular studies can help explain differences in the effects observed and can open the black box to provide practitioners and policy makers with clear suggestions of how they might improve related policies and programs.

26.3.2.2 Providing Context In some circumstances, authors are faced with the challenge of making sense of the results gleaned from studies included in a given substantive area that differ in terms of many attributes, including the: publication type (journal article, dissertation, external evaluation, or internal evaluation); methods (pretest-posttest comparison, experimental comparison, or nonequivalent control-group design); samples (at-risk, average, or gifted students); actual characteristics of the phenomena or interventions (academic remediation or enrichment program); and indicators of the outcomes of the phenomena or interventions (teacher-constructed assessments or standardized tests). Differences such as these are commonly found in the social sciences and in education, and pose methodological challenges for the analyst. Some choose to pool findings from numerous available primary studies with little regard to the varying methodologies and relevance to the issue at hand. Others, using highly restrictive criteria for the inclusion of primary

studies, simply choose to omit large portions of the literature. How can the analyst contend with these differences but also take advantage of this natural variation across studies to describe how methodological and substantive differences may shape the outcomes?

On one hand, Gene Glass maintained that "it is an empirical question whether relatively poorly designed studies give results significantly at variance with those of the best designed studies" (1976, 4). On the other, Slavin argued that "far more information is extracted from a large literature by clearly describing the best evidence on a topic than by using limited journal space to describe statistical analyses of the entire methodologically and substantively diverse literature" (1986, 7). Should the researcher combine studies that used varying methods and are characterized by varying substantive characteristics or should one focus only on what is considered the best evidence?

In truth, the answer probably depends most on the meta-analyst's research questions. If one wants to know the effect of an intervention for a specific period and based on data from only the most rigorous experimental studies, the analyst should choose the small number of studies that fit that description. However, if one were interested in studying how much the effect varied depending on treatment duration, and in assessing the difference in effect estimates from experimental and quasi-experimental studies, then the analyst can learn considerably more by systematically investigating the variance in effect magnitudes across a group of studies that differ along these dimensions.

By integrating narrative and quantitative techniques for research synthesis, we believe that authors can clearly outline and discuss the complexities of a literature, and describe how the known differences across the studies relate to variation in the effect estimates. In this way, authors can productively use the variability across studies to qualify the conditions under which effects are larger or smaller, and, through synthetic analyses of moderators of effect size, can quantify precisely how substantial the differences in the outcome are.

Two recent applications of this approach are noteworthy. First, in their meta-analysis of the achievement effects of summer school programs for K–12 students, Harris Cooper and his colleagues assumed that the summer programs to be included in their synthesis would vary, and that the research methods for studying these summer programs would as well (2000). Indeed, they hypothesized about the potential differences in effect estimates associated with the numerous identifiable (and unidentifiable) variations noted in the summer school literature: "We could continue to posit predictions. Most of the multitude of permutations and interacting influences, however, have no or few empirical testings described in the research literature" (Cooper et al 2000, 17). How they dealt with this randomness, the numerous moderators of effect sizes, and the general complexities of this literature are instructive.

First, they contrasted underused (at least in the social sciences) random-effects analyses of their data with more conventional fixed-effects analyses. Second, they used an infrequently used technique for adjusting effect sizes to remove the covariance between methodological and substantive characteristics. Third, they showed how an analyst may take advantage of subgroups of studies that contain substantively interesting variations within them. For instance, in looking at a subgroup of studies that measured both math and reading outcomes, the authors were able to draw stronger inferences about differing impacts of summer school programs on math and reading achievement because most, if not all, other variations were held constant (for a discussion of the advantages of within-study moderators, see chapter 24, this volume). Finally, the authors clearly outlined and discussed the complexities of this literature, and described how the known differences across the studies affected summer school effect estimates, thereby enriching the quantitative results with narrative description. In this way, the authors productively used the variability across studies to qualify the conditions under which summer programs are more successful and to quantify precisely how much more successful they are. This combination of conventional meta-analytic procedures, relatively novel integrative procedures, and narrative description provides a strong practical example of how a complex literature should be integrated and reported.

As Glass suggested, by empirically examining a diverse range of studies, we may assess how and to what extent methodological differences across the primary studies are associated with differences in the observed outcomes (1976). When outcomes are robust across studies of varying methodologies, one can be more confident in the conclusions. On the other hand, if studies differ in terms of both rigor and results, then one can focus on the subset of more rigorous studies when formulating conclusions. This analysis of the consequences of methodological variations for the estimation of the effects, which is unique to the enterprise of meta-analysis, allows meth-

odologists and consumers of the research literature to recognize the biases in the literature and to understand empirically both their frequency and magnitude.

The Borman et al. meta-analysis of research on the achievement effects of twenty-nine of the most widely disseminated comprehensive school reform (CSR) models is a noteworthy example of this approach (2003). The authors began by assessing how and to what extent various methodological differences (for example, experimental or nonexperimental designs, and internal or third-party evaluations of the reform programs) shaped understandings of the overall achievement effects of CSR. Specifically, the preliminary analysis empirically identified and quantified the methodological biases in the literature and, in general, characterized the overall quality of the research evidence.

After characterizing the overall CSR research base through a combination of narrative description and synthetic analysis, Borman and his colleagues focused on only the subgroup of studies that provided the best evidence for evaluating the effectiveness of each of the twenty-nine CSR models: experimental and matched comparison group studies that were conducted by third-party evaluators who were independent from the CSR program developers (2003). The authors determined which studies provided the best evidence, not by a priori judgments or by other potentially subjective criteria, but by empirical analyses and narrative description of the CSR literature's methodological biases. The final analyses relied on only the best evidence, but the additional analyses and descriptions were also instructive because the methodological biases were clearly identified, quantified, and discussed.

These two examples suggest how analysts can not only contend with a highly variegated research literature, but also effectively model and describe its differences to help advance the field. Cooper et al. used a combination of narrative description and a series of synthetic analyses to help describe how practitioners can make the most of summer school programs for students (2000). Borman et al. productively modeled and described the contextual and methodological variability within the school reform literature to better understand the conditions under which the policy was most productive and to evaluate empirically and discuss the methodological biases within the research base (2003). Both syntheses helped move their respective fields forward in ways that would not have been possible had the researchers simply engaged in a synthesis of a relatively homogeneous set of studies meeting specific a priori criteria.

26.3.2.3 Interpreting the Magnitude of Effect Sizes

Beyond testing the statistical significance of empirical findings, it has become increasingly common practice to report the magnitude of an effect or the strength of a relationship using effect sizes. Indeed, the most recent American Psychological Association publication manual lists the failure to report effect sizes as a defect in reporting research (American Psychological Association 2001). Relative to reporting only the statistical significance of research findings, reporting the magnitude of effects and relationships is certainly an advance. However, how should one interpret and describe the practical significance of effect sizes? Typically, effect-size estimates are interpreted in two ways. One is to rely on commonly accepted benchmarks that differentiate small, medium, and large effects. A second method is to explicitly compare the reported effect size to those reported in earlier but similar studies.

The most established benchmarks are those presented by Jacob Cohen, where an effect size, d, of 0.20 is interpreted as a small effect 0.50 equates to a medium effect, and effects larger than 0.80 are regarded as large effects (1988). Cohen provided these effect size values "to serve as operational definitions of the quantitative adjectives 'small,' 'medium,' and large'" (12). However, even Cohen has expressed reservations about the general application of these benchmarks:

> These proposed conventions were set forth throughout with much diffidence, qualifications, and invitations not to employ them if possible. The values chosen had no more reliable a basis than my own intuition. (1988, 532)

Despite these provisos, Lipsey's ambitious summary of the distribution of mean effect sizes from 102 meta-analyses of psychological, educational, and behavioral treatments provided consistent empirical support for Cohen's self-confessed "intuitions" (1990). Lipsey's findings suggested that the bottom third of this distribution ranged from $d = 0.00$ to $d = 0.32$, the middle third ranged from $d = 0.33$ to $d = 0.55$, and the top third ranged from $d = 0.56$ to $d = 1.20$. The median effect sizes for the low, middle, and top third of the distribution were $d = 0.15$, $d = 0.45$, and $d = 0.90$, respectively, which might also be used as general standards.

Although the Jacob Cohen and Mark Lipsey benchmarks provide relatively consistent advice to researchers to help them describe in more practical terms the magnitude of an effect size, we believe that more nuanced de-

scriptions are possible (Cohen 1988; Lipsey 1990)). One alternative would be to compare an effect size to earlier reported effect sizes from studies involving similar samples, comparable outcomes, and related phenomena. For instance, a researcher might study the impact of an innovative classroom-based intervention for struggling readers that, when contrasted to a no-treatment control condition, has an effect size of $d = 0.20$—a small effect size based on Cohen's benchmarks. The savvy reader, however, may be particularly interested in how this intervention's effect compares to those of other interventions, including competing classroom-based reading programs, or an alternate policy such as one-on-one tutoring in literacy. If these alternate interventions produce roughly similar effects, the reading program in question could be interpreted as having rather typical, or medium, effects. Another consideration, particularly in the policy arena, is cost. If the reading program and the one-on-one tutoring program appear to produce similar outcomes, but the per-pupil cost of tutoring is substantially higher, then the reading program's effects may no longer appear small nor typical, but optimal. Thus, the judgment of magnitude can be enriched or even altered by comparing the meta-analytic results with a wide swath of prior research.

Indeed, Gene Glass, Barry McGaw, and Mary Lee Smith were particularly supportive of this approach, arguing that the effectiveness of a particular intervention can be interpreted only in relation to other interventions that seek to produce the same effect (1981). In addition, Cooper suggested that the researcher should use multiple yardsticks to help frame and interpret the practical importance of the noted effects (1981). More recently, Carolyn Hill and her colleagues argued that instead of universal standards, such as the Cohen benchmarks, one must interpret magnitudes of effects in context of the interventions being studied, of the outcomes being measured, and of the samples or subsamples being examined (2007). They also suggested different frames of reference, including information about effect size distributions, other external performance criteria, expected or normative change on a given outcome, and existing performance gaps among subgroups (for example, the magnitude of the black-white achievement gap, or the female-male wage gap).

We concur that these types of comparisons to other benchmarks or yardsticks are very informative for helping readers understand the practical significance of reported effect sizes. Whenever possible, such information should be discussed and be reported as part of the narrative description of the effect of a phenomenon to help contextualize and describe its impacts.

26.3.3 Integrating Approaches Summary

Bodies of empirical research in the social sciences and education are often characterized by caveats and complexities that make solely synthetic approaches to research synthesis overly reductive and simplistic. In an effort to summarize key findings from the literature that are not amenable to meta-analytic integration, to further clarify and describe a variegated research literature, and to offer greater clarity and interpretation on the magnitude and practical significance of effect-size data, we believe that research syntheses can often benefit from the integration of narrative and quantitative forms of synthesis.

We offer several key recommendations for augmenting synthetic procedures with narrative methods of research synthesis:

Be Mindful When Considering Whether to Drop a Study from the Review Sample, Because a Study Can Be Useful Even if It Does Not Offer an Effect Size for Inclusion in a Meta-Analysis. Bad evidence is worse than no evidence, but even when a literature provides few results that are amenable to traditional forms of meta-analysis, the analyst can produce some gains in the accumulation of knowledge by carefully synthesizing the results that do exist and by describing its limitations so that it will be properly interpreted and so that future research in the area may be better designed.

Heed the 1981 Advice of Glass and His Colleagues to Analyze the Full Complement of Potential Studies, Consider How Their Differences May Inform the Results, and Then Describe to the Reader How Those Results Came to Be. Meta-analysis can help researchers understand variability in effects noted across a given substantive area, but this understanding is incomplete without careful blending of narrative description of the methodological and substantive factors that are associated with the differences. Although some analysts cull potential studies on the basis of substantive moderators and potential sources of bias, we generally favor including studies that might influence the results of a synthesis and then empirically examining their impact. Including them not only identifies how sensitive the results are to the composition of the analytical sample, but also can lead to a fruitful discussion of the influence of moderating variables (which can, of course, then be illustrated visually). However, this objective examination of all the available evidence in lieu of applying

subjective criteria for eliminating studies necessitates narrative description of how and why such study differences contribute to variability in effect size data. When this is appropriate and when it is done well, analysts can contribute far more to the field than they might have by censoring the included research.

Help Readers Understand Effect Size Results by Placing Them in Context. A meta-analysis that uses narrative description to help interpret effect-size data in context offers far more to the field than simple summaries and comparisons to general standards of large, medium, and small effects. We suggest various yardsticks for understanding the magnitudes of effect sizes (Cooper 1981), including: Cohen's U_3 (1981); information gleaned from compendia of effect size data from particular fields of study, such as those Lipsey and Wilson summarize (1993); effects of other interventions designed for groups or individuals who are similar to those included in the current meta-analysis; standardized mean differences that exist among pertinent subgroups in similar untreated samples or populations (for example, the black-white test score gap); existing standards for normative growth or change in the outcome that may be derived from data for similar untreated samples or populations (for example, the expected achievement gain associated with one grade level or school year); and the use of cost data in conjunction with effect size information.

26.4 CONCLUSION

This chapter has outlined how recent meta-analytic work has used graphical representations and narrative descriptions to enhance the interpretation of effect size data. Although optimal use of figures and integration of narrative synthesis has not been commonplace in the meta-analytic literature, we have observed several recent developments that suggest that future work may more regularly feature these elements. First, with respect to the graphical representation of effect size data, we are encouraged to note that published meta-analyses in *Psychological Bulletin* have more regularly included figures from 2000 through 2005 than they did from 1985 through 1991 reported previously by Light, Singer, and Willett (1994).

Second, use of high-quality figures that effectively present the big picture is also on the rise in certain contexts. Most notably, the Cochrane Collaboration and its social sciences counterpart, the Campbell Collaboration, have helped advance the use of the forest plot in their systematic reviews. We found no published examples of forest plots in *Psychological Bulletin* and *Review of Educational Research* but are hopeful that the work of the Campbell Collaboration and its increasing influence on the research communities from the social sciences and education will affect future published work. The forest plot, which is effective for presenting a comprehensive picture along with the finer details, at least for a manageable number of studies included in the analysis, is the current state-of-the-art graphic representation of effect size data.

We believe that meta-analysis has largely overcome early criticisms suggesting that the methodology failed to distinguish and account for important methodological and substantive differences that can affect the synthetic outcomes. Instead, high-quality syntheses, when appropriate, explicitly model the effects of these methodological and substantive moderators and interpret them. These types of analyses are important contributions of meta-analysis that extend beyond simple reports of mean effect sizes and contribute stronger understandings of how and why variegated research literatures yield differential effect estimates. Further, not all results are amenable to conventional meta-analytic procedures. In some cases, though, one may learn a great deal by summarizing these outcomes and by describing the contexts under which they were observed. If nothing else, these cases can be illustrative to the field in suggesting what is needed in the future to help produce better evidence that is cumulative.

Finally, along with graphical representation of effect size data, narrative interpretation of the magnitude and practical significance of effect size data is a key consideration to help make meta-analytic results more informative, intuitive, and policy-relevant. We urge those conducting research syntheses in the social sciences to not only develop the big picture and its finer details through the use of figures, but also to take great care to put the effects in context. Although Jacob Cohen's original benchmarks for judging the magnitudes of effect sizes have provided some useful guidance, researchers must go beyond these very general yardsticks and consider others, such as the effects of comparable phenomena on similar outcomes involving similar populations along with other estimates of expected or normative change on the outcomes of interest (1988).

26.5 NOTES

1. Other graphs included scatter plots, line graphs, box or schematic plots, quantile plots, and path analysis diagrams.

2. The radial plot is sometimes referred to as a Galbraith Plot, after the statistician who created it.

3. The L'Abbé plot (1987) is another method of representing dichotomous outcome data that has gained favor in the clinical and epidemiological literature (see, for example, Jiménez et al. 1997; Song 1999).

4. The word occasionally appears as Forrest plot, for reasons that are unknown and perhaps apocryphal (Lewis and Clarke 2001). Here we follow the widely used forest plot spelling.

5. The Cochrane Library is a collection of evidence-based medicine databases that can be accessed at http://www.thecochranelibrary.com or http://www.cochrane.org. As of May 2008, the Cochrane Library's database included 5,320 records, consisting of 3,464 complete reviews and protocols for a further 1,856 reviews (http://www3.interscience.wiley.com/cgi-bin/mrwhome/106568753/ProductDescriptions.html). The Cochrane Review Manager (RevMan), which performs the analysis and generates graphic representations, is available free of charge to those who prepare reviews for the Cochrane Library and to others who wish to use it for noncommercial purposes. Active editorial oversight, strict design parameters, and a standardized tool account for the uniformity of the reviews in the Cochrane Library.

6. The Campbell Collaboration can be accessed at http://www.campbellcollaboration.org.

7. Metagraph was written by Dr. Adrian Mander, senior biostatistician at MRC Human Nutrition Research, Cambridge, U.K. Stata includes a suite of commands for conducting meta-analysis and representing findings, including: meta, metan, and metareg; see Sterne et al. (2007) for more infomation.

8. The Comprehensive Meta Analysis (CMA) software is produced by Biostat, Inc. A stand-alone plotting engine for graphics is anticipated in 2007 that will allow for the user to modify rows, cells, and columns, create .pdf files, and allows users to work with values computed in other programs; an updated version of the CMA program is expected to be released in 2008. A free trial of the software is available for download at http://www.Meta-Analysis.com.

26.6 REFERENCES

Albarracín, Dolores, Jeffrey C. Gillette, Allison N. Earl, Laura R. Glasman, Marta R. Durantini, and Moon-Ho Ho. 2005. "A Test of Major Assumptions About Behavior Change: A

Comprehensive Look at the Effects of Passive and Active HIV-Prevention Interventions Since the Beginning of the Epidemic." *Psychological Bulletin* 131(6): 856–97.

American Psychological Association. 2001. *Publication Manual of the American Psychological Association*, 5th ed. Washington, D.C.: American Psychological Association.

Begg, Colin B., and Madhuchhanda Mazumdar. 1994. Operating characteristics of a rank correlation test for publication bias. *Biometrics* 50(4): 1088–101.

Bertin, Jacques. 1983. *Semiology of Graphics*, translated by William J. Berg. Madison: University of Wisconsin Press.

Borman, Geoffrey D., Gina M. Hewes, Laura T. Overman, and Shelly Brown. 2003. "Comprehensive School Reform and Achievement: A Meta-Analysis." *Review of Educational Research* 73(2): 125–230.

Cohen, Jacob. 1988. *Statistical Power Analysis for the Behavioral Sciences* (2nd ed.). Hillsdale, NJ: Lawrence Erlbaum.

Cohen, Peter A. 1981. "Student Ratings of Instruction and Student Achievement: A Meta-Analysis of Multisection Validity Studies." *Review of Educational Research* 51(3): 281–309.

———. 1983. Comment on "A selective review of the validity of student ratings of teaching." *Journal of Higher Education* 54(4): 449–58.

Cook, Thomas D., and Laura Leviton. 1980. "Reviewing the Literature: A Comparison of Traditional Methods with Meta-analysis." *Journal of Personality* 48(4): 449–72.

Cooper, Harris M. 1981. "On the Significance of Effects and the Effects of Significance" *Journal of Personality and Social Psychology* 41(5): 1013–18.

Cooper, Harris M., Kelly Charlton, Jeff C. Valentine, and Laura Muhlenbruck. 2000. "Making the Most of Summer School: A Meta-Analytic and Narrative Review." *Monographs of the Society for Research in Child Development* 65(1): 1–118.

Cleveland, William S. 1994. *The Elements of Graphing Data*, rev. ed. Summit, N.J.: Hobart Press.

Dickerson, Sally S., and Margaret E. Kemeny. 2004. "Acute Stressors and Cortisol Responses: A Theoretical Integration and Synthesis of Laboratory Research." *Psychological Bulletin* 130(3): 355–91.

Egger, Matthias, George Davey Smith, Martin Schneider, and Christoph Minder. 1997. "Bias in Meta-Analysis Detected by a Simple, Graphical Test." *British Medical Journal* 315(7109): 629–34.

Ehri, Linnea C., Simone R. Nunes, Steven A. Stahl, and Dale M. Willows. 2001. "Systematic Phonics Instruction Helps Students Learn to Read: Evidence from the National Rea-ding Panel's Meta-Analysis." *Review of Educational Research* 71(3): 393–447.

Eysenck, Hans J. 1978. "An Exercise in Mega-Silliness. *American Psychologist* 33(5): 517.

Freiman, J. A., T. C. Chalmers, H. Smith, and R. R. Kuebler. 1978. "The Importance of Beta, The Type II Error and Sample Size in the Design and Interpretation of the Randomized Control Trial: Survey of 71 'Negative Trials.'" *New England Journal of Medicine* 299 (13): 690–94.

Galbraith, Rex F. 1988a. "Graphical Display of Estimates Having Different Standard Errors." *Technometrics* 30(3): 271–81.

———. 1988b. "A Note on Graphical Presentation of Estimated Odds Ratios from Several Clinical Trials." *Statistics in Medicine* 7(8): 889–94.

———. 1994. "Some Applications of Radial Plots." *Journal of the American Statistical Association* 89(428): 1232–42.

Gershoff, Elizabeth Thompson. 2002. "Corporal Punishment by Parents and Associated Child Behaviors and Experiences: A Meta-Analytic and Theoretical Review." *Psychological Bulletin* 128(4): 539–79.

Glass, Gene V. 1976. "Primary, Secondary, and Meta-Analysis Research." *Educational Researcher* 5(10): 3–8.

Glass, Gene V., Barry McGaw, and Mary Lee Smith. 1981. *Meta-Analysis in Social Research*. Beverly Hills, Calif.: Sage Publications.

Gray-Little, Bernadette, and Adam R. Hafdahl. 2000. "Factors Influencing Racial Comparisons of Self-Esteem: A Quantitative Review." *Psychological Bulletin* 126(1): 26–54.

Harris, Robert L. 1999. *Information Graphics: A Comprehensive Illustrated Reference*. New York: Oxford University Press.

Hedges, Larry V., and Ingram Olkin. 1985. *Statitical Methods for Meta-analysis*. Orlando, Fl.: Academic Press.

Hill, Carolyn J., Howard S. Bloom, Alison Rebeck Black, and Mark W. Lipsey. 2007. "Empirical Benchmarks for Interpreting Effect Sizes in Research." *MDRC* Working Papers on Research Methodology. New York: MRDC. http://www.mdrc.org/publications/459/full.pdf

Jiménez, F. Javier, Eliseo Guallar, and José M. Martín-Moreno. 1997. "A Graphical Display Useful for Meta-Analysis." *European Journal of Public Health* 7(1): 101–05.

Kumkale, G. Tarcan, and Albarracín, Dolores. 2004. "The Sleeper Effect in Persuasion: A Meta-Analytic Review." *Psychological Bulletin* 130(1): 143–72.

L'Abbé, Kristan A., Allan S. Detsky, and Keith O'Rourke. 1987. "Meta-Analysis in Clinical Research." *Annals of Internal Medicine* 107(2): 224–33.

Lewis, Steff, and Mike Clarke. 2001. "Forest Plots: Trying to See the Wood and the Trees." *British Medical Journal* 322(7300): 1479–80.

Lewis, J. A., and S. H. Ellis. 1982. "A Statistical Appraisal of Post-Infarction Beta-Blocker Trials." *Primary Cardiology* 1(Suppl.): 31–37.

Light, Richard J., and David B. Pillemer. 1984. *Summing Up: The Science of Reviewing Research*. Cambridge, Mass.: Harvard University Press.

Light, Richard J., Judith D. Singer, and John B. Willett. 1994. "The Visual Presentation and Interpretation of Meta-Analyses." In *The Handbook of Research Synthesis*, edited by Harris Cooper and Larry V. Hedges. New York: Russell Sage Foundation.

Lipsey, Mark W. 1990. *Design Sensitivity: Statistical Power for Experimental Research*. Newbury Park, Calif.: Sage Publications.

Moyer, Christopher A., James Rounds, and James W. Hannum. 2004. "A Meta-Analysis of Massage Therapy Research. *Psychological Bulletin* 130(1): 3–18.

Playfair, William. 1801/2005. *The Commercial and Political Atlas and Statistical Breviary*, edited by Howard Wainer and Ian Spence. New York: Cambridge University Press.

Raudenbush, Stephen W. 1984. "Magnitude of Teacher Expectancy Effects on Pupil IQ as a Function of the Credibility of Expectancy Induction: A Synthesis of Findings From 18 Experiments." *Journal of Educational Psychology* 76(1): 85–97.

Raudenbush, Stephen W., and Anthony S. Bryk. 1985. "Empirical Bayes Meta-Analysis." *Journal of Educational Statistics* 10(2): 75–98.

Schwarzer, Guido, Gerd Antes, and Martin Schumacher. 2002. "Inflation of Type I Error Rate in Two Statistical Tests for the Detection of Publication Bias in Meta-Analyses with Binary Outcomes." *Statistics in Medicine* 21: 2465–77.

Slavin, Robert E. 1984. "Meta-Analysis in Education: How Has It Been Used?" *Educational Researcher* 18(8): 6–27.

———. 1986. "Best-Evidence Synthesis: An Alternative to Meta-Analytical and Traditional Reviews." *Educational Researcher* 9(15): 5–11.

Slavin, Robert E., and Alan Cheung. 2005. "A Synthesis of Research on Language of Reading Instruction for English Language Learners." *Review of Educational Research* 75(2): 247–84.

Song, Fujian. 1999. "Exploring Heterogeneity in Meta-Analysis: Is the L'Abbé Plot Useful?" *Journal of Clinical Epidemiology* 52(8): 725–30.

Sterne, Jonathan A. C., and Matthias Egger. 2001. "Funnel Plots for Detecting Bias in Meta-Analysis: Guidelines on Choice of Axis." *Journal of Clinical Epidemiology* 54(10): 1046–55.

Tufte, Edward R. 1990. *Envisioning Information*. Cheshire, Conn.: Graphics Press.

———. 1997. *Visual Explanations: Images and Quantities, Evidence and Narrative*. Cheshire, Conn.: Graphics Press.

———. 2001. *The Visual Display of Quantitative Information*, 2nd ed. Cheshire, Conn.: Graphics Press.

———. 2006. *Beautiful Evidence*. Cheshire, Conn.: Graphics Press.

Tukey, John W. 1977. *Exploratory Data Analysis*. Reading, Mass.: Addison-Wesley.

———. 1988. *The Collected Works of John W. Tukey, Volume V: Graphics 1965–1985*, edited by William S. Cleveland. Pacific Grove, Calif.: Wadsworth and Brooks.

Turkheimer, Eric, and Mary Waldron. 2000. "Nonshared Environment: A Theoretical, Methodological, and Quantitative review." *Psychological Bulletin* 126(1): 78–108.

Wang, Morgan C., and Brad J. Bushman. 1998. "Using the Normal Quantile Plot to Explore Meta-Analytic Data Sets." *Psychological Methods* 3(1): 46–54.

Wainer, Howard. 1984. "How to Display Data Badly." *The American Statistician* 38(2): 137–47.

———. 1996. "Depicting Error." *The American Statistician* 50(2): 101–11.

Wainer, Howard, and Paul F. Velleman. 2001. "Statistical Graphics: Mapping the Pathways of Science." *Annual Review of Psychology* 52: 305–35.

Whitehead, Anne, and John Whitehead. 1991. "A General Parametric Approach to the Meta-Analysis of Randomized Clinical Trials." *Statistics in Medicine* 10(11): 1665–77.

Yusuf, Salim, Richard Peto, John Lewis, Rory Collins, and Peter Sleight. 1985. "Beta Blockage During and After Myocardial Infarction: An Overview of the Randomized Trials." *Progress in Cardiovascular Diseases* 27(5): 335–71.

27

REPORTING FORMAT

MIKE CLARKE

CONTENTS

27.1 INTRODUCTION

The other chapters in this handbook discuss the complexities and intricacies of conceptualizing and conducting research syntheses or systematic reviews. If these pieces of research are to be of most use to readers, they need to be reported in a clear way. The authors should provide a level of detail that allows readers to assess the methods used for the review, as well as the robustness and reliability of its findings, any statistical analyses, and the conclusions. If readers have come to the review for information to help make a decision about future action, they need to be presented with the necessary knowledge clearly and unambiguously and to be able to judge the quality of the conduct of the review.

This chapter summarizes and focuses on the need for the authors to set out what they did, what they found, and what it means. This is as true of a research synthesis as it is of any article reporting a piece of research. Much of what is written here seeks to reinforce the ideas of good writing practice in research, making it specific to reporting research synthesis. It builds on the chapter in the first edition by Katherine Halvorsen (1994). Readers of this chapter who want more detailed guidance might wish to consult additional sources such as the QUOROM statement for journal reports of systematic reviews of randomized trials (Moher et al. 1999) and the *Cochrane Handbook for Systematic Reviews of Interventions* (Higgins and Green 2008). Furthermore, the output of the Working Group on American Psychological Association Journal Article Reporting Standards (the JARS Group) outlined in table 27.1, provides guidance on reporting research synthesis in the social sciences (see section 27.9).

The structure adopted for Cochrane Collaboration systematic reviews (www.cochrane.org, www.thecochrane library.com) has also been influential in this chapter. These reviews, of which there are more than 3,500 by late 2008, all follow the same basic structure using a standard set of headings and following strong guidance on the information that should be presented within each section (Allen, Clarke, and Tharyan 2007). All Cochrane reviews are preceded by a published protocol setting out the methods for the research synthesis in advance. In several parts of this chapter, I refer to the possibility of referring readers back to the protocol as a way to provide detail on the methods of the research synthesis without adding to the length of its report. Even if a formal protocol is not published for a research synthesis, the synthesists might wish to prepare one as part of their own planning of their proj-

ect and then make this available to provide full details of their methods.

Cochrane reviews assess the effects of health-care interventions, but the basic concepts can be applied to any research synthesis, regardless of whether it is of the effects of interventions or centered on health care. In considering the content of the report of a research synthesis, one of the main constraints might be the limits on article length and content (such as tables and figures) by journals. However, with the availability of web repositories for longer versions of articles, these constraints should be less of a problem for research synthesists.

This chapter takes the reader through the various issues to consider when preparing a report of a research synthesis. It also serves as a guide to what a reader of a review might expect to find within such a structured report. When considering the perspective of the reader of their review, synthesists also need to consider that there may be multiple audiences for their findings. Findings may have relevance to people making decisions at a regional or national level, as well as to individual practitioners and members of the public deciding about actions of direct relevance to themselves or individuals they are caring for. The review might also be a source of knowledge for other researchers planning new studies or other reviews. And, by providing a synthesis of existing research in a topic area, these reports can be of interest to people wishing to find out more about how the area of research has evolved over time. Those who have done the research synthesis should therefore decide whether to try to speak to all of these audiences in a single report, target one audience only, or prepare a series of separate, adequately cross-referenced, publications.

If you are reading this chapter as a guide to reporting your own research synthesis, you should consider who you are expecting to read and be influenced by your findings—using the cited examples as a starting point—before deciding on the key issues to describe in most detail in the report. For example, a report meant for academics and researchers should be very detailed about the methods of the review and of the relevant studies. A report meant for people making decisions about interventions investigated in the review will need to present the findings of the review clearly and in enough detail for these readers to identify the elements of most importance to their decisions. A report aimed at policy makers might be shorter—providing less detail about the methods of the review, summarizing the overall results of the review, giving more consideration to the implications of the findings

Table 27.1 Meta-Analysis Reporting Standards (MARS)

Paper Section and Topic	Description
Title	• Make it clear that the report describes a research synthesis and include meta-analysis, if applicable Footnote funding source(s)
Abstract	• The problem or relation(s) under investigation • Study eligibility criteria • Type(s) of participants included in primary studies • Meta-analysis methods (indicating whether a fixed or random model was used) • Main results (including the more important effect sizes and any important moderators of these effect sizes) • Conclusions (including limitations) • Implications for theory, policy and/or practice

Introduction	• Clear statement of the question or relation(s) under investigation
	  ○ Historical background
	  ○ Theoretical, policy and/or practical issues related to the question or relation(s) of interest
	  ○ Rationale for the selection and coding of potential moderators and mediators of results
	  ○ Types of study designs used in the primary research, their strengths and weaknesses
	  ○ Types of predictor and outcome measures used, their psychometric characteristics
	  ○ Populations to which the question or relation is relevant
	  ○ Hypotheses, if any

Method	• Operational characteristics of independent (predictor) and dependent (outcome) variable(s)
Inclusion and exclusion criteria	• Eligible participant populations • Eligible research design features (random assignment only, minimal sample size) • Time period in which studies needed to be conducted • Geographical and/or cultural restrictions

Moderator and mediator analyses	• Definition of all coding categories used to test moderators or mediators of the relation(s) of interest
	• Reference and citation databases searched
	• Registries (including prospective registries) searched
	  ○ Keywords used to enter databases and registries
	  ○ Search software used and version (Ovid)
	• Time period in which studies needed to be conducted, if applicable

Search strategies	• Other efforts to retrieve all available studies,
	  ○ Listservs queried
	  ○ Contacts made with authors (and how authors were chosen)
	  ○ Reference lists of reports examined
	• Method of addressing reports in languages other than English
	• Process for determining study eligibility
	  ○ Aspects of reports were examined (i.e., title, abstract, and/or full text)
	  ○ Number and qualifications of relevance judges
	  ○ Indication of agreement
	    • How disagreements were resolved
	• Treatment of unpublished studies
	• Number and qualifications of coders (e.g., level of expertise in the area, training)
	• Intercoder reliability or agreement
	• Whether each report was coded by more than one coder and if so, how disagreements were resolved

(Continued)

Table 27.1 (*Continued*)

Paper Section and Topic	Description

Coding procedures	• Assessment of study quality
	  ○ If a quality scale was employed, a description of criteria and the procedures for application
	  ○ If study design features were coded, what these were
	• How missing data were handled
	• Effect size metric(s)
	  ○ Effect sizes calculating formulas (means and SDs, use of univariate F to r transform, and so on)
	  ○ Corrections made to effect sizes (small sample bias, correction for unequal ns, and so on)
	• Effect size averaging or weighting method(s)
	• How effect-size confidence intervals (or standard errors) were calculated
	• How effect-size credibility intervals were calculated, if used
	• How studies with more than one effect size were handled

Statistical methods	• Whether fixed or random effects models were used and the model choice justification • How heterogeneity in effect sizes was assessed or estimated • Means and SDs for measurement artifacts, if construct-level relationships were the focus • Tests and any adjustments for data censoring (publication bias, selective reporting) • Tests for statistical outliers • Statistical power of the meta-analysis • Statistical programs or software packages used to conduct statistical analyses

Results	• Number of citations examined for relevance
	• List of citations included in the synthesis
	• Number of citations relevant on many but not all inclusion criteria excluded from the meta-analysis
	  ○ Number of exclusions for each exclusion criteria (effect size could not be calculated), with examples
	• Table giving descriptive information for each included study, including effect size and sample size
	• Assessment of study quality, if any
	• Tables and/or graphic summaries
	  ○ Overall characteristics of the database (number of studies with different research designs)
	  ○ Overall effect size estimates, including measures of uncertainty (confidence and/or credibility intervals)
	• Results of moderator and mediator analyses (analyses of subsets of studies)
	  ○ Number of studies and total sample sizes for each moderator analysis
	  ○ Assessment of interrelations among variables used for moderator and mediator analyses
	  ○ Assessment of bias including possible data censoring

Discussion	• Statement of major findings
	• Consideration of alternative explanations for observed results
	  ○ Impact of data censoring
	• Generalizability of conclusions,
	  ○ Relevant populations
	  ○ Treatment variations
	  ○ Dependent (outcome) variables
	  ○ Research designs, and so on
	• General limitations (including assessment of the quality of studies included)
	• Implications and interpretation for theory, policy, or practice
	• Guidelines for future research

SOURCE: Reprinted with permission from JARSWG 2008

of the review for practice, and placing these in the appropriate context for decision makers who might simply be seeking an answer to the question of what to do based on the research that has already been done. This wide variety of audiences and the range of ways in which syntheses can be published means that there is no single recipe for the structure of a report. This chapter is intended to cover all of the key areas and recognizes that not all reports of research syntheses will need to give the same emphasis to each section.

27.2 TITLE FOR RESEARCH SYNTHESIS

The starting point for any research synthesis should be a clearly formulated question. This sets out the scope for the review. Although it may be similar to the question for a new study, the synthesist needs to recognize that she is not planning a new study, in which she would be responsible for the design, recruitment and follow-up procedures. She is simply setting the boundaries for the studies that have already been done by others. The question provides a guide to what will, and will not, be eligible once these studies are identified. Having formulated a clear question and conducted the review, the title of its report needs to capture the essence of this, so that people who would benefit from reading the review are more likely to find it and, having found it, to read it.

As with any research article, a good title needs to contain the key pieces of information about, at least, the design of the research being described. As a starting point, it needs to identify the article as a research synthesis, systematic review, or meta-analyses. These words and phrases will have different meanings to different authors and readers. If the synthesis sets out to systematically search for, appraise, and present eligible studies (without necessarily doing any statistical analyses on these studies) then perhaps research synthesis or systematic review would be most appropriate. If it contains a meta-analysis, then this term might be used on its own. However, it is worth remembering that meta-analysis can be done by combining the results of studies without the formal process of a systematic review. For example, a research group might present a meta-analysis combining the results of a series of trials they have done, without drawing in the results of other, similar trials by others. Therefore, combining the phrases to produce "A research synthesis, including a meta-analysis of" might capture better the content of a report of a research synthesis that includes the mathematical combination of the results of eligible studies.

Having set down the study design for the research article (that is, that it is a research synthesis), the remainder of the title can usually be constructed much as those for studies that are eligible to the review. It might show all or some of the population being studied, what happened to them, the outcome of most interest, and the types of study design eligible for the review. For example, "A research synthesis of [study design] of [intervention] for [population] to assess the effect on [outcome]." Adopting this structure, embedding key descriptors of the review in the title, should also help people find the review when they search bibliographic databases or scan contents lists of journals. This is increasingly important because the explosion in research articles, syntheses, and systematic reviews means that finding these in the literature—amidst the reports of the individual studies, discussion pieces, and reviews that are not systematic—is a challenge for users of reviews, which the author of the research synthesis can help overcome when deciding on a title for the synthesis (Moher et al. 2007).

The authors of the research synthesis need to minimize the possibility that someone who should read the report is misled and skips over it or, conversely, that a reader will be misled into going further into a report not relevant to their interests. For example, a title such as "A research synthesis of randomized trials of antibiotics for preventing postoperative infections" might, at first sight, seem fairly complete. However, such a title could cover a review of the role of antibiotics compared to placebo to prevent infection (to identify whether it is better to use antibiotics), direct comparisons of different antibiotics (to identify the best one), or comparisons of antibiotics against all other forms of infection control (to find which is the best intervention to adopt). The underlying question and scope for each of these three reviews are fundamentally different, but the general title fails to capture this aspect for the reader. In such circumstances, titles such as "A research synthesis of randomized placebo controlled trials of antibiotics versus placebo for preventing postoperative infections," "A research synthesis of randomized trials comparing different antibiotics for preventing postoperative infections," and "A research synthesis of randomized trials of antibiotics versus other strategies for preventing postoperative infections" might be more appropriate.

Some authors might wish to use a declarative title, summarizing the main finding of their review, such as "Tamoxifen improves the chances of surviving of breast cancer: a research synthesis of randomized trials." Other authors might choose a more neutral title, such as "A research synthesis and meta-analysis of tamoxifen for the treatment of women with breast cancer." To some extent, the type of title might be determined by the policy of the journal or other place of publication for the review. For example, the titles of all Cochrane reviews follow a particular structure, are nondeclarative and because they are Cochrane reviews, do not include words or phrases such as systematic review, research synthesis, or meta-analysis.

27.3 QUESTION FOR RESEARCH SYNTHESIS

The authors of a research synthesis need a well-formulated question before embarking on their work. This is one of the hallmarks of the systematic approach, compared to something more haphazard in which biases have not been minimized. This helps guide the various steps of the research synthesis such as where and how to search, what studies are eligible, and what the main emphasis of the findings should be. Similarly, the reader of the eventual review will benefit from a clear summary of the question to be addressed in the review. Formulating and clarifying the question both at the outset of the work on the review and at the start of the report of its findings is important.

One of the key things to do for readers within the question is distinguish those systematic reviews that set out to obtain clear answers to guide future actions from those where the intention was to describe the existing research to help design a new, definitive study. An example of the former might be "What are the effects of one-to-one teaching on the ability of teenagers to learn math?" and of the latter, "What research has been done into ways of teaching math to teenagers?" The reader should be told which of these two approaches is to be taken. This will allow the reader to decide if the review is likely to provide an answer to the questions the reader has. Someone looking for a single, numeric estimate of the effect of one-to-one teaching based on existing research might find it in the former whereas, in the latter, they are more likely to find only descriptive accounts of the different pieces of research that have been done.

A research synthesis might contain more than one question and comparison. This should also be made clear early in the report. For example, the earlier description of three reviews of antibiotics for the prevention of postoperative infection might be done within a single review. Making it clear to the readers that this is the case early on in the review will help them to decide whether to continue reading it and, if they are particularly interested in one of the questions, to identify that part.

It is also important to be careful that the question for the review is a fair reflection of its content. For example, posing the question for a review as "Does healthy eating advice reduce the risk of a heart attack?" might seem reasonable but is probably not a fair guide to the answers that will come from the review. At its simplest, the available answers would be yes, no, and don't know, where no covers both no effect on the risk of a heart attack and an increased risk. It is unlikely that the review would be this simple. It is more likely that the review will seek to quantify the risk and to provide this estimate regardless of its direction. Thus, a better question might be "What is the effect of healthy eating advice on the risk of a heart attack?"

However, even with that expansion, the question currently does make the intentions for the review clear. Remembering that research syntheses are retrospective and that readers of reviews might have their own preconceptions about what studies it will contain, additional information is needed. This additional information should clarify what comparisons will be examined in the review, because it will not necessarily be a comparison of interventions. Using the healthy eating advice example, the simple question could be expanded to "What is the effect of healthy eating advice compared to smoking cessation advice on the risk of a heart attack?" or "What is the effect of healthy eating advice compared to providing a healthy diet on the risk of a heart attack?" or "What is the effect of healthy eating advice on the risk of a heart attack compared to a stroke?" The eligible studies, review methods and findings for these questions would be quite different and the report should minimize ambiguity by being clear in its question.

27.3.1 Background

Using the background section to set the scene for the review helps orient the reader to the topic. It is also an opportunity to convey the importance and relevance of the review. However, in the face of limitations on the length of a report, it is usually one of the sections that can easily be cut back. It might provide details on how common the circumstances or condition under investigation are, how widely the intervention is used (if the review is about the effects of an intervention), and how this intervention might work. If some of the research to be included in the review is already likely to be well known, that it is should be mentioned in the background. Some information on the consequences and severity of the outcomes being assessed might also be helpful.

As noted, reviews are by nature retrospective and the reviewers should therefore use the background section to describe any influences on their reasons for doing the review. If they chose a particular focus, perhaps concentrating on one population subgroup, one type of intervention from a class of interventions, or one possible relationship, they should explain that they have done so. This is especially important if readers might be surprised by the focus or concerned that it arose from bias in the formulation of the question, such that the synthesists had narrowed their project in a particular way to reach a preformed conclusion.

If a protocol describing the methods for the review had been prepared in advance, that it was should be mentioned in either the background or the methods section. Likewise, if the review had been registered prospectively, this also should be noted. Both steps help in planning and drawing attention to the review, and also serve to minimize bias and the perception of bias in its conduct.

27.3.2 Objectives

By setting out the question for their review, the authors will have summarized the scope of their review. In describing their objectives, they can elaborate on what they are hoping to achieve. For example, if the question for the review is to estimate the relative effects on repeat offences of two policies for dealing with car thieves, the objectives may be to identify which of the policies would be the most appropriate in a particular city.

27.3.3 Eligibility Criteria

As with any scientific research in which information is to be collected from past events, a research synthesis needs to describe how it did this and provide enough information to reassure readers that biases that might have arisen from prior knowledge of the existing studies were minimized. In the context of a review, the past events or observations are the previous studies and the eligibility criteria for the review should describe those features that determined whether studies would be included or excluded.

In the report of the research synthesis, the authors should describe what they sought. This is not necessarily the same as what they found, which should be described

in the results section. For example, the original intention might have been to include studies done anywhere in the world, but all the studies identified were conducted in the United States. The synthesists might have wished to study all antibiotics and searched for all drugs, but find that only one or two had been studied. Or the eligibility criteria for the review might have encompassed children of all ages but researchers might have restricted their studies to those aged between eight and twelve years.

For most syntheses, the eligibility criteria can be set out under four domains:

Types of participants: who will have been in the eligible studies? This might include information about sex and age, past experiences, diagnosis (and perhaps the reliability required for this diagnosis) of an underlying condition, and location.

Types of interventions: what will have been done to, or by, the participants? This should be described in operational terms. If the synthesis is assessing the effects of an intervention, the authors should ensure that they also provide adequate information on the comparison condition (that is, what kinds of interventions are eligible for the synthesis). If, for example, the comparison is with usual care, it should be made clear whether this is what the original researchers regarded as usual care when they did their study or whether it is what the synthesists would regard as usual care at the present time.

Types of outcome measures: what happened to the participants? The synthesists might wish to divide these into primary and secondary outcome measures, to highlight those outcomes (that is, the primary outcomes), which they judge to be especially important in assessing whether or not an intervention is effective. They should also distinguish between outcomes or constructs (for example, anxiety) and measures of these (for example, a particular scale to measure anxiety). It may also be important to state the timing of the measurement of an outcome. For example, if postoperative pain is to be an outcome, what time points are relevant to the synthesis? In a research synthesis, the outcomes that are available for the report will depend on whether they were measured in the original studies. This means that the synthesists might not be able to present data on their primary outcomes but they should still make these clear in their report—they should not redefine their primary outcomes on the basis of the availability of data.

Types of studies: what research designs will be eligible? This might also include a discussion of the strategies the synthesists will use to assess study quality, including any characteristics of the studies, beyond their design, which might influence their decision on whether to include it. They should describe how they will obtain the information needed to make these judgments, because it will not necessarily be available in the study's published report. For example, if the eligibility criteria require that only randomized trials in which allocation concealment was adequate are included, the synthesists should outline what they will do if this information is not readily available, bearing in mind that research has shown that the lack of this information in a report does not necessarily indicate its absence from the study (Soares et al. 2004).

Under these four domains, the authors might use a mixture of inclusion and exclusion criteria to describe what would definitely be eligible for their synthesis, or what would definitely be ineligible. For example, they might include only randomized trials and exclude studies in which the outcomes were not measured for a minimum period after the intervention. Authors should avoid listing mirror-image inclusion and exclusion criteria. Little is gained, for example, from writing both that only prospective cohort studies are eligible and that retrospective case control studies are ineligible.

27.4 METHODS

The methods section of the synthesis needs to describe what was done, but it should also provide information on what the synthesists had intended to do but were not able to do because of the studies they found. If a protocol for their synthesis is available, the authors might refer to that, rather than repeating material—especially if the protocol includes plans that could not be implemented in the synthesis (for example, because studies of a certain type were not identified or there were inadequate data to perform subgroup analyses). The methods section should cover four topics: search strategy, application of eligibility criteria, data extraction or collection, and statistical methods. These are discussed in the following section. Furthermore, any changes to the methods between the original plans for the synthesis and the final synthesis should be noted and explained.

27.4.1 Search Strategy

The studies in a research synthesis can be considered its raw material or building blocks, and the first step in bringing these together is the search strategy used to find them. Including the search strategy in the report of the synthesis is important for transparency but also to allow others to reproduce the synthesis. It is unlikely that readers will be able to replicate fully the search done by the synthesists because of additions and other changes to any bibliographic databases in the intervening period. However, the search strategy should be described in enough detail to allow it to be run again. This will make it easier for someone who wishes to update the synthesis to do so by using the same search strategy to find studies that will have become available since the original search. It will also be helpful to people who are planning a research synthesis of a similar topic, because sections of the search strategy might be applicable to such a synthesis.

In describing their search strategy, authors should list the terms used, how these were combined, whether any limitations were used, where they searched, when, and who did the searching. If journals, conference proceedings, or other sources apart from electronic databases were searched, these should be listed, along with information on any searches for unpublished material. Details should be given of how many people did the searches, whether they worked independently, and the procedures for resolving any disagreements. This information will help the reader assess the thoroughness of the searching for the synthesis but, given its technical nature and the potential for the details to be long and unwieldy, authors might chose to include simply a short list of which databases were searched and when the searches were done.

Including the actual search strategies used for bibliographic databases (which might run to hundreds of lines covering both free text searches and the use of index terms) in full in the synthesis will depend in large part on space limitations and the target audience for the synthesis. A reader who is most interested in the implications for practice of a research synthesis will be much less interested in the details of the search strategy than someone who wishes to use parts of the search strategy for a synthesis of a related topic. Consideration should also be given to whether the typical reader of the synthesis will wish to be confronted by many pages of search terms. The authors might chose to provide the set of terms used for only one database within their report, and note how modifications were made for other databases to reflect different indexing strategies. The full search strategies might be given in appendices, on a website, or made available to readers on request. If the number of records retrieved with each term is included in these lists, future synthesists can consider the usefulness of the individual terms when designing a search for a synthesis covering a similar topic.

27.4.2 Application of Eligibility Criteria

Information should be given on whether one or more people applied the eligibility criteria, whether they worked independently to begin with, and the procedures for resolving any disagreements between them. Some indication of whether people working independently tended to agree or disagree on their decisions will help readers judge the ease or difficulty of applying the eligibility criteria, and the possibility that another team of authors might have reached different decisions on what, belongs within the synthesis. If bias might have been introduced because of prior knowledge of potentially eligible studies or because of knowledge gathered while searching or applying eligibility criteria, any procedures to minimize this bias should be described. For example, if the reports of the studies were modified to prevent the assessors from identifying the original researchers, their institution or the source of the article, that they were should be described. This might be achieved by selective printing of pages of the original reports, by blacking out key information, or by cutting text out and making photocopies. Such efforts might be particularly time consuming and their value in minimizing bias has not been demonstrated (Berlin 1997). If the quality of the studies were assessed as part of the application of the eligibility criteria, such procedures should also be described.

27.4.3 Data Extraction or Collection

As with the application of the eligibility criteria, the report of the research synthesis should include information on the number of people who did the data extraction, whether they worked independently, in duplicate, or as a team, and procedures for resolving any disagreements. Including data on the initial agreement rates between the extractors may help readers assess the difficulty of the data extraction process. This process relates to extracting information from the reports of the potentially eligible studies for use in the research synthesis. These data are not simply the numeric results of the study but also in-

clude descriptive information about the design and interpretations of the information in the report (for example, to determine if the study is eligible or of adequate quality).

If some of the eligible studies were reported more than once, the decision rules used to determine whether a single report or a combination of information from more than one report was used should be stated. Furthermore, where a study had not reported enough of the information needed for the synthesis, this section should describe any steps taken to obtain unpublished data. This will be a challenge for most research syntheses and not all will be able to obtain the missing data (Clarke and Stewart 1994). The unpublished data might be for a study as a whole or for parts of a study (such as unreported subgroups or outcome measures). If attempts to obtain unpublished information failed, the synthesists' opinions on why they failed might be helpful. For example, do they believe it was because of too few contact details, too little time, not enough necessary data, or unwillingness to provide the data?

27.4.4 Statistical Methods

The methods actually used should be described, along with any methods that were planned but could not be implemented because of limitations in the data available, unless the reader can be pointed to the description of these in an earlier protocol. Any nonstandard statistical methods should be described in detail, as should the synthesists' rationale for any choices made between alternative standard methods. Because of the risk of bias arising from the conduct of post hoc subgroup or sensitivity analyses, the authors should clearly indicate whether their analyses were predefined, perhaps by reference to the protocol for the synthesis if one is available. However, it is worth remembering that subgroup analyses that were a priori for the research synthesis might have been influenced by existing knowledge of the findings of the included studies. This reinforces the importance of specifying the rationale for these analyses in the report and, ideally, showing how they are based on evidence that is independent of the studies being included.

Sometimes, synthesists will wish to do subgroup or sensitivity analyses that were not planned but arose from the data encountered during the synthesis. These analyses need to be interpreted with added caution and the authors should clearly identify any such analyses to the reader.

If the synthesists decide that a quantitative summary of the results of the included studies is not warranted, they should still justify any statistical methods used to analyze the studies at the individual level. If their decision not to do a quantitative synthesis is based on too much heterogeneity among the studies, they should define clearly what they mean by too much and how this was determined, for example, whether it was done without any consideration of the results of the studies, on the basis of an inspection of the confidence intervals for each study or a statistical test such as I^2 (Higgins et al 2003). This explanation should distinguish between design heterogeneity (for example, because the interventions tested in each study are very different, the populations vary too greatly or the outcome measures are not combinable) and statistical heterogeneity (that is, variation in the results of the studies).

27.4.5 Changes Between Design or Protocol and Synthesis

Whether or not the proposed methods for the synthesis were written in a formal protocol, the authors should report on any changes they made to their plans after they started work on the synthesis. These changes might relate to the eligibility criteria for the studies or aspects of the methods used. The reasons for the changes should be given and, if necessary, discussed in the context of whether the changes might have introduced bias to the synthesis. If sensitivity analyses are possible to examine directly the impact of the changes, these analyses should be described either in this section or under statistical methods (see chapter 22, this volume).

27.5 RESULTS

The raw material for the synthesis is the studies that have been identified. The results section should describe what was found for the synthesis, separating studies judged eligible, ineligible, and eligible but ongoing (with results not available). Studies identified but not classified as definitely eligible or ineligible should be described.

In describing the studies, the minimum information to provide in the report would, of course, be a source of additional information for each study. This might be a reference to a publication or contact details for the original researchers. However, this will make it difficult for readers to understand the similarities and differences between studies. A description of each included study is preferable. This is especially true if the synthesis draws conclusions that depend on characteristics of the included studies, so that the reader can determine whether they agree or disagree with the decisions the synthesists made. If the

results for each individual study are included in the report, readers will be able to perform quantitative syntheses for themselves, perhaps reconfiguring the groupings of studies the synthesists used. The description of each study could be provided in tabular form to make it easier for readers to compare and contrast designs. It should include details within the four domains noted for setting the eligibility criteria for the review, as well as details of the number of people in each study. It is not usually necessary to include the conclusions of each study.

Any studies judged eligible but excluded, or studies readers might have expected to have been included that were excluded because the synthesists found them ineligible, should be reported and the reason for the exclusion given. If well-known studies excluded from the synthesis are not mentioned, some readers might assume that the apparent failure of the synthesists to find them casts doubt on the credibility of the synthesis. In some reports of research syntheses, the excluded studies might be mentioned at the same time as the eligibility criteria for the synthesis, particularly if there are specific reasons relating to these criteria that led to the exclusion. Otherwise, they should be discussed within the results section.

It is most important that the authors devote adequate space to reporting the primary and secondary analyses they set out in their methods section, including those with nonsignificant results. This also relates to any a priori subgroup or sensitivity analyses. One challenge synthesists face is overcoming the publication biases of the original researchers. Synthesists should not perpetuate such bias by selectively reporting outcomes measures on the basis of the results they obtain during the synthesis. Synthesists should report all prespecified outcome measures or make this information available in a report supplement. If doing so is not possible, the synthesists should state that it is not possible explicitly.

Where quantitative syntheses are reported or shown, the meta-analysts should be explicit about the amount of data in each of these combined analyses. This relates to the number of studies, the number of participants, and the numbers of events in each analysis, and the relative contribution of each study to the combined analysis. This might be done by presenting the separate results of each included study, listing the studies and noting how many participants and events each study contributes to each analysis, and giving the confidence intervals for the effect estimates. This is particularly important for meta-analyses in which it is not possible to include all eligible studies in all analyses. For example, if the research synthesis includes ten studies with 2,000 participants, the reader should not be misled into believing that all of these contribute to every analysis if they do not, when some analyses might only be able to draw on a fraction of the 2,000. Being specific about which studies are in each meta-analysis presented in the report also helps readers make their own judgments about the applicability of the results by consulting the details provided for these studies.

Synthesists should give careful consideration to the order in which they present the included studies within their own analyses (see chapter 26, this volume). There is no single correct approach, but some of the options are presenting studies in chronological, alphabetic, or size order. Similarly, the synthesists should give careful consideration to the order in which they present their analysis. Options here include the order of the importance of the outcomes, statistical power, size of the effect estimate, or likely impact of the result on practice.

27.6 DISCUSSION

As with any research article, synthesists should be cautious about introducing evidence into the discussion section not presented earlier in the report. They should avoid using the discussion to provide answers to the question that they cannot support with evidence from the studies they found, but may need to contextualize the findings of the included studies with additional information from elsewhere. Two examples are discussing the difficulties of implementing an intervention outside a research setting, and contrasting the strength of a relationship identified within the populations in the included studies with its likely strength in other populations eligible for the research synthesis but which did not feature in any of the included studies.

Any shortcomings of the synthesis, such as the impact of limitations within the search strategy, should be mentioned. These might include restrictions to specific languages or a failure to search the grey literature, which might make the synthesis particularly susceptible to publication bias (Hopewell et al. 2007). Other limitations that should be discussed include difficulties in obtaining all of the relevant data from the eligible studies and the potential impact of not obtaining unpublished data. The findings of sensitivity analyses should be discussed, in particular if these indicate that the results of the synthesis would have been different had different methods been used.

If eligible studies were identified but not included, for example, because they were ongoing, the potential for these to change the results in any important ways should be discussed.

27.6.1 Conclusions and Implications

As in the discussion, synthesists should try to avoid introducing new evidence in the conclusion to answer the question for their synthesis. In some circumstances, it will be necessary to draw on evidence that does not come from the studies and analyses presented in the results section. In these circumstances, such evidence should ideally be included in the background or discussion first. Any such evidence should be widely accepted and robust. For example, if a synthesis finds that a particular analgesic is beneficial for treating a mild headache but no evidence was found in the included studies that directly compared different doses of the drug, it would be legitimate to give information on the standard doses available. Similarly, if a drug is known to be hazardous for people with kidney problems, the synthesists might include the caution in their conclusions.

The authors should be explicit about any values or other judgments they use in making their recommendations. If they conclude that one intervention is better than another, they should be explicit about whether this arises because of their personal balancing of opposing effects on particular outcomes. They might decide, for example, that a slight increase in an adverse outcome is more than offset by the benefits on the primary outcome. Readers might not share this judgment, however. In the context of a synthesis in which different studies might contribute to different outcomes, the authors should be clear about how a lack of data for some of the outcomes might have influenced their judgments. For example, the review might contain reliable and extensive data in favor of an intervention but the adverse outcomes might not have been investigated or reported in the same depth in the included studies.

The conclusions for the synthesis can be organized as two separate sections, setting out the authors' opinions on the implications for practice and the implications for research. The former should include any limitations to the conclusions if, for example, the general implications would not apply for some people. The implications for research should include some comment on whether, in the presence of the called for research, the research synthesis will be updated. The implications should be presented in a structured way, to help the reader understand the recommendations and to minimize ambiguity. This is particularly true of the implications for research. Synthesists should avoid simplistic statements such as "better research is needed." They need to specify what research is needed, perhaps by following the same structure of interventions, participants, outcome measures, and study designs used in the eligibility criteria (Brown et al. 2006; Clarke, Clarke, and Clarke 2007).

Simple statements such as "We are unable to recommend" should be avoided. They are unlikely to be helpful, because of their inherent ambiguity, and might arise for a variety of reasons that are less likely in reports of individual studies. For example, the synthesis might reveal a lack of adequate good quality evidence to make any recommendation. A statement such as "There is not enough evidence to make any recommendation about the use of" would be preferable. Another possibility is that the authors are not allowed to make any recommendations, regardless of the results, perhaps because of rules at the organization they work for. In such circumstances, the statement might best be phrased, "We are not permitted to make recommendations." Alternatively, and akin to what might happen in an individual study, the authors might feel unable to recommend a particular course of action because their synthesis has shown it to be ineffective or harmful. If this is the case, it would be preferable for the authors to be explicit and write something such as "We recommend that the intervention not be used."

In a complex or large synthesis of the effects of interventions, one way to summarize the conclusions is to group the implications on the basis of which interventions have been shown to be beneficial, harmful, lacking in evidence, needing more research, and probably not warranting more research. If the synthesis is not about interventions but instead covers, for example, the relationship between characteristics of people and their behavior, a similar categorization of the conclusions but on a different basis—on the basis of confidence about a positive or negative relationship, one that is unproven and worthy of more research, and one that is unproven but does not justify more research—could be used.

Synthesists should avoid broad statements that there is no evidence for an effect or a relationship if they actually mean that there is no evidence from the types of study they have synthesized. This is important if other types of evidence would have shown an effect or relationship. As an example, there might be no randomized trials for a particular treatment despite its effectiveness having been shown by other types of study. Similarly, synthesists should avoid confusing no evidence of an effect or rela-

tionship with evidence of no effect or relationship. The former can arise from a paucity of data in the included studies, and the latter usually requires significant data to rule out a small effect.

27.6.2 Updating the Research Synthesis

The first time that the synthesis is done, this section might include details about the synthesists' plans to update it (if an update is being considered). This could include any decision rules to initiate the update, such as the passage of a certain amount of time, the availability of a new study, or the availability of enough new evidence to possibly change the synthesis' findings. After a synthesis has been updated, this section could also be used to highlight how the update had changed from earlier versions, and any plans for future updates. This section is also an opportunity for the synthesists to state whether they wish to be informed of eligible studies they have not identified.

27.7 AUTHOR NOTES

27.7.1 Author Contributions

In the interests of transparency, the work done by each of the authors should be described. This should cover both the preparation of any preceding protocol and the report of the synthesis. It also needs to cover the various tasks necessary to undertake the synthesis. These include issues relating to the design of the research synthesis, such as the initial formulation of the question and the selection of the eligibility criteria, as well as the conduct of the synthesis. Within the conduct of the synthesis, tasks such as the design of the search strategy, the identification of studies and the assessment of their eligibility, the extraction of data, its analysis, and conclusions drawn might be undertaken by different people within the author team. All authors should agree on the content of the final report, but one or more should be identified as its guarantors, because some authors might not feel able to take full responsibility for technical aspects that are outside of their skill base. As an example, an information specialist might feel competent to take responsibility for the searching and a statistician for the meta-analyses, but not the reverse.

27.7.2 Acknowledgments

Some people may have worked on the synthesis, or may have provided information for it, but have not made a substantial enough contribution to be listed as an author. Such individuals should be acknowledged and, if named, should give their permission to be named. In research synthesis, this might include the original researchers who responded to questions about their studies or provided additional data, or people within the research synthesis team who worked on one specific aspect of the project, such as the design and conduct of electronic searches, but did not otherwise contribute.

27.7.3 Funding

Sources of funding for the synthesis should be listed. If any conditions were associated with this funding that might have influenced the conduct of the synthesis, its findings, or its conclusions, these should be stated so as to inform readers of the potential impact of the funding. The sources of funding might be separated into those provided explicitly for the preparation of the synthesis; those supporting a synthesist's work, which includes the time needed to do research; and indirect funding that supported editorial or other input.

27.7.4 Potential Conflict of Interest

The authors should describe any conflicts of interest that might have influenced the conduct, findings, conclusions or publication of the synthesis. It is important to discuss any potential conflicts of interest that readers might perceive to have influenced the synthesis, to provide reassurance that this is not the case. In a research synthesis, this might include the involvement of the authors in any of the studies considered for the synthesis, because their special knowledge of these studies and access to the data is likely to be different to what was available for the other studies.

27.8 ABSTRACT AND OTHER SUMMARIES

When a synthesis is published, its abstract is likely to appear at the beginning. But, as with most articles, the abstract is likely to be one of the last things to be written. Hence the position of this guidance near the end of this chapter. In preparing any summary of a piece of research, the authors need to give careful consideration to the audience and whether they are likely to read (or even have access to the full synthesis). As noted earlier, many different audiences for a research synthesis are possible—people making decisions relating directly to themselves or other individuals, people making policy decisions at a regional or national level, and people planning new research. The abstract can be used to highlight whether the different potential audiences will find things of relevance within the full report. Alternatively, the authors might need to prepare more than one abstract or summary to cater to the different audiences. Ideally, the abstract should be structured to draw attention to the inclusion criteria for the review, the scope of the searching, the quantity of research evidence included, the results of any meta-analyses of the primary outcomes, and the main implications.

27.9 SUMMARY CHECKLIST

The Working Group on American Psychological Association Journal Article Reporting Standards (the JARS Group) arose out of a request from the APA Publications and Communications Board to examine reporting standards for both individual research studies and meta-analysis published in APA journals. For meta-analysis, the JARS Group began by considering several efforts to establishing reporting standards for meta-analysis in health care. These included QUOROM or Quality of Reporting of Meta-analysis (Moher et al. 1999) and the Potsdam consultation on meta-analysis (Cook, Sackett, and Spitzer 1995). The JARS Group developed a combined list of elements from these, rewrote some items with social scientists in mind, and added a few suggestions of its own. After vetting their effort, the JARS Group arrived at the list of recommendations contained in table 1 (Journal Article Reporting Standards Working Group 2007).

27.10 REFERENCES

Allen, Claire, Mike Clarke, and Prathap Tharyan. 2007. "International Activity in the Cochrane Collaboration with Particular Reference to India." *National Medical Journal of India* 20(5): 250–55.

Berlin, Jesse A. 1997. "Does Blinding of Readers Affect the Results of Meta-Analyses? University of Pennsylvania Meta-Analysis Blinding Study Group." *Lancet* 350(9072): 185–86.

Brown, Paula, Karla Brunnhuber, Kalipso Chalkidou, Iain Chalmers, Mike Clarke, Mark Fenton, Carol Forbes, Julie Glanville, Nicholas J. Hicks, Janet Moody, Sarah Twaddle, Hazim Timimi, and Paula Young. 2006. "How to Formulate Research Recommendations." *British Medical Journal* 333(7572): 804–6.

Clarke, Lorcan, Mike Clarke, and Thomas Clarke. 2007. "How Useful Are Cochrane Reviews in Identifying Research Needs?" *Journal of Health Services Research and Policy* 12(2): 101–3.

Clarke, Mike J., and Lesley A. Stewart. 1994. "Obtaining Data from Randomised Controlled Trials: How Much Do We Need to Perform Reliable and Informative Meta-Analyses?" *British Medical Journal* 309(6960): 1007–10.

Cook, Deborah J., David L. Sackett, and Walter O. Spitzer. 1995. "Methodologic Guidelines for Systematic Reviews of Randomized Control Trials in Health Care from the Potsdam Consultation on Meta-Analysis." *Journal of Clinical Epidemiology* 48(1): 167–71.

Halvsorsen, Katherine T. 1994. "The Reporting Format." In *The Handbook of Research Synthesis*, edited by Harris Cooper and Larry V. Hedges. New York: Russell Sage Foundation.

Higgins, Julian P.T., and Sally Green, eds. *Cochrane Handbook for Systematic Reviews of Interventions*, vers. 5. Chichester: John Wiley & Sons. http://www.cochrane.org/resources/handbook/hbook.htm

Higgins, Julian P.T., Simon G. Thompson, Jonathan J. Deeks, and Douglas G. Altman. 2003. "Measuring Inconsistency in Meta-Analyses." *British Medical Journal* 327(7414): 557–60.

Hopewell, Sally, Steve McDonald, Mike Clarke, and Matthias Egger. 2007. "Grey Literature in Meta-Analyses of Randomized Trials of Health Care Interventions." *Cochrane Database of Systematic Reviews* 2:MR000010.

Moher, David, Deborah J. Cook, Simon Eastwood, Ingrim Olkin, Drummond Rennie, and David F. Stroup. 1999. "Improving the Quality of Reports of Meta-Analyses of Randomised Controlled Trials: The QUOROM Statement." *Lancet* 354(9193): 1896–1900.

Moher, David, Jennifer Tetzlaff, Andrea C. Tricco, Margaret Sampson, and Douglas G. Altman. 2007. "Epidemiology and Reporting Characteristics of Systematic Reviews." *PLoS Medicine* 4(3): e78.

Journal Article Reporting Standards Working Group [JARSWG]. 2008. "Reporting Standards for Research in Psychology: Why Do We Need Them? What Might They Be?" Washington, D.C.: American Psychological Association.

Soares, Helios P., Stephen Daniels, Ambuj Kumar, Mike Clarke, Charles Scott, Sandra Swann, and Benjamin Djulbegovic. 2004. "Bad Reporting Does Not Mean Bad Methods for Randomized Trials: Observational Study of Randomised Controlled Trials Performed by the Radiation Therapy Oncology Group." *British Medical Journal* 328(7430): 22–25.

PART
XI

SUMMARY

28

THREATS TO THE VALIDITY OF GENERALIZED INFERENCES

GEORG E. MATT
San Diego University

THOMAS D. COOK
Northwestern University

CONTENTS

28.1 INTRODUCTION

This chapter provides a nonstatistical way of summarizing many of the main points in the preceding chapters. In particular, it takes the major assumptions outlined and translates them from formal statistical notation into ordinary English. The emphasis is on expressing specific violations of formal meta-analytic assumptions as concretely labeled threats to valid inference. This explicitly integrates statistical approaches to meta-analysis with a falsificationist framework that stresses how secure knowledge depends on ruling out alternative interpretations. Thus, we aim to refocus readers' attention on the major rationales for research synthesis and the kinds of knowledge meta-analysts seek to achieve.

The special promise of meta-analysis is to foster empirical knowledge about general associations, especially causal ones, that is more secure than what other methods typically warrant. In our view, no rationale for meta-analysis is more important than its ability to identify the realm of application of a knowledge claim—that is, identifying whether the association holds with specific populations of persons, settings, times and ways of varying the cause or measuring the effect; holds across different populations of people, settings, times, and ways of operationalizing a cause and effect; and can even be extrapolated to other populations of people, settings, times, causes, and effects than those studied to date. These are all generalization tasks that researchers face, perhaps no one more explicitly than the meta-analyst.

It is easy to justify why we translate violated statistical assumptions into threats to validity, particularly threats to the validity of conclusions regarding the generality of an association. The past twenty-five years of meta-analytic practice have amply demonstrated that primary studies rarely present a census or even a random sample of the populations, universes, categories, classes, or entities (terms we use interchangeably) about which generalizations are sought. The salient exception is when random sampling occurs from some clearly designated universe, a procedure that does warrant valid generalization to the population from which the sample was drawn, usually a human population in the social sciences. But most surveys take place in decidedly restricted settings (a living room, for instance) and at a single time, and the relevant cause and effect constructs are measured without randomly selecting items. Moreover, many people are not interested in the population a particular random sample represents, but ask instead whether that same association hold with a different kind of person, in a different setting, at a different time, or with a different cause or effect. These questions concern generalization as extrapolation rather than representation (Cook 1990). How can we extrapolate from studied populations to populations with many, few, or even no overlapping attributes?

The sad reality is that the general inferences meta-analysis seeks to provide cannot depend on formal sampling theory alone. Other warrants are also needed. This chapter assumes that ruling out threats to validity can serve as one such warrant. Doing so is not as simple or as elegant as sampling with known probability from a well-designated universe, but it is more flexible and has been used with success to justify how manipulations or measures are chosen to represent cause and effect constructs (that is, construct validity). If meta-analysis is to deal with generalization understood as both representation and extrapolation, we need ways of using a particular database to justify reasonable conclusions about what the available samples represent and how they can be used to extrapolate to other kinds of persons, settings, times, causes, and effects.

This chapter is not the first to propose that a framework of validity threats allows us to probe the validity of research inferences when a fundamental statistical assumption has been violated. Donald Campbell introduced his internal validity threats for instances when primary studies lack random assignment, creating quasi-experimental design as a legitimate extension of the thinking R. A. Fisher had begun (Campbell 1957; Campbell and Stanley 1963). Similarly, this chapter seeks to identify threats to valid inferences about generalization that arise in meta-analyses, particularly those that follow from infrequent random sampling. Of course, Donald Campbell and Julian Stanley also had a list of threats to external validity, and these also have to do with generalization (1963). But their list was far from complete and was developed more with primary studies in mind than with research syntheses. The question this chapter asks is how one can proceed to justify claims about the generality of an association when the within-study selection of persons, settings, times, and measures is almost never random and when it is also not even reasonable to assume that the available

sample of studies is itself unbiased. This chapter proposes a threats-to-validity approach rooted in a theory of construct validity as one way to throw provisional light on how to justify general inferences.

28.1.1 Why We Conduct Research Syntheses

At the core of every research synthesis is an association about which we want to learn something that cannot be gleaned from a single existing study. These associations can be of many kinds, such as the relationship between a risk factor (for example, secondhand smoke exposure) and a disease outcome (for example, lung cancer), between a treatment (for example, antidepressant medication) and an effect (for example, suicide), between a predictor (for example, SAT) and criterion (for example, college GPA), or between a dose (for example, hours of psychotherapy) and either a single response (for example, psychological distress) or a response change (for example, obesity levels).

Many primary studies have relatively small sample sizes and thus low statistical power to detect a meaningful population effect. It is thus commonly hoped that meta-analysis will increase the precision with which an association is estimated. The relevant intuition here is that combining estimates from multiple parallel studies (or exact replications) will increase the total number of primary units for analysis (for example, persons), thus reducing the sampling error of an association estimate. However, parallel studies are rare. More typical are studies of an association that differ in many ways, including how samples of persons were selected, such that it is unreasonable to assume they come from the same population. So we need to investigate how inferences about associations are threatened when a meta-analysis is conducted. This is especially important when the studies are few in number and heterogeneous in the samples of persons, settings, interventions, outcomes, and times. In our experience, most associations with which meta-analysts deal are causal, of the form that manipulating one entity (that is, treatment) will induce variability in the other (that is, outcome). So a second reason for conducting research syntheses is to examine how causal inference is threatened or enhanced in meta-analyses. A third rationale is to strengthen the generalization of an association (causal or otherwise) that has been examined across the available primary studies, each with its own time frame, means of selecting persons and settings, and of implementing the treatment and outcome measures.

In meta-analytic practice, conclusions are rarely warranted by statistical sampling theory or even strong explanatory theories. Instead, heavy reliance is placed on less familiar principles for exploring generalizations about specific target universes, for identifying factors that might moderate and mediate an association, and for extrapolating beyond whatever universes are in the available database (Cook 1990, 1993). We need to examine these other, less well-known principles and the threats to validity that can be extracted from them.

28.1.2 Validity Threats

In the meta-analytic context, validity threats describe factors that can induce spurious inferences about the existence or size of an association, about the likelihood it is causal, or about its generalizability. Meta-analytic inferences are more compelling, the better the identified threats have been ruled out, either because they are empirically implausible or because other evidence indicates they did not operate in the research under consideration. Ruling out alternatives requires a framework of validity threats, some taken from formal statistical models, others from theories about generalization, and others emerging as empirical products born of critical reflection on the conduct of research syntheses. Any list of validity threats so constructed is bound to be subject to change. Indeed, such change is desirable, reflecting improvements in theories of method and in critical discourse about the practice of research synthesis. New threats should emerge as others die out, because they are rare in actual research practice. Most of the threats listed here are in the previous lists of Georg Matt and Thomas Cook (1994) and William Shadish, Thomas Cook, and Donald Campbell (2002), but some are new and a few have dropped out.

Some threats relevant to research syntheses also apply to individual primary studies, such as unreliability in measuring outcomes. Others are extensions to threats that apply to primary studies, but are now reconfigured to reflect how a threat is distributed across studies, such as the failure to find any studies that assign treatment at random. Other threats are unique to meta-analyses because they depend on how estimates of association are synthesized across single studies, such as publication bias. These last have higher-order analogs in general principles of research design so that publication bias can be considered a particular instance of the more general threat of sampling bias. However, publication bias is application focused, one of the highly specific ways in which meta-analysts

encounter sampling bias. So when two labels are possible we prefer to formulate a threat in the form closest to actual meta-analytic practice.

A list of validity threats can be used in two major ways. Retrospectively, it can help discipline claims emerging from a research synthesis by identifying which threats are plausible in light of current knowledge and by indicating the additional evidence needed to assess the plausibility of a threat and thereby evaluate how well the causal claim is justified. Second, validity threats can also be used prospectively for planning a research synthesis. The purpose then is to anticipate as many potential threats as possible and to design the research so as to rule them out. Identifying threats prospectively is better than retrospectively, though each is valuable.

28.1.3 Generalized Inferences

The generalized inferences possible in a research synthesis are different from those in a primary study because, in the latter, inferences are inextricably limited to the relatively narrow ways in which times, research participants, settings, and cause and effects are usually sampled or measured. In contrast, research syntheses involve multiple samples of diverse persons and settings as well as numerous treatment implementations and outcome measures, collected at unique times and over varying periods. This potential for heterogeneous sampling provides the framework for generalizing to broader classes or universes than a single study makes possible. But why is this, given that the most secure path to generalization—formal sampling theory—does not generate the heterogeneity that meta-analysts typically find and use for generalization purposes? Interestingly, even advocates of random selection use purposive selection for choosing the items they put into their surveys. In so doing they rely on theories of construct validity to justify generalizing from the items selected to the more general constructs these measures are intended to represent. Our theory of generalization in the meta-analytic context depends on factors ultimately derived from theories of construct validity rather than formal sampling. Each of these theories uses sampling particulars to justify inferences about abstract entities. Construct validity uses systematic sampling to warrant inferences about the constructs a measure or manipulation represents. Sampling theory uses random sampling to warrant generalizing from a sample to the populations it represents. How then does construct validity justify general inferences? Thomas Cook proposed five useful principles (1990, 1993).

28.1.3.1 Demonstrating Proximal Similarity

Donald Campbell first introduced the principle of proximal similarity in the context of construct validity (1986). For research syntheses, proximal similarity refers to the correspondence in attributes between an abstract entity, such as an outcome construct or a class of settings, about which inferences are sought and the particular instances of the entity captured by the studies included in the meta-analysis. To make inferences about target entities, meta-analysts first seek to identify which of the potentially relevant studies are indeed members of the target class. That is, their attributes include those that theoretical analysis indicates are prototypical of the construct being targeted (Rosch 1973, 1978; Smith and Medin 1981). For example, if the target population is of persons—let us say Asian Americans—then we need to know whether the participants have ancestors who come from certain nations. If we want to generalize to factories, we want to know whether the research actually took place in factory settings. Do the studies all occur in the 1980s—the historical period to which generalization is sought? Are the interventions instances of the category labeled decentralized decision making, in that all affected parties are represented in deliberations about policy? Are the outcome operations plausible measures of the higher order construct productivity, in that a physical output is measured and related to standards about quantity or quality? These kinds of questions about category membership are obvious and the sine qua non of any theory of representation. Alas, though, category membership justified by proximal similarity is not sufficient for inferring that the sampled instances represent the target class. More is needed.

28.1.3.2 Exploring Heterogeneous and Substantively Irrelevant Third Variables

In his theory of construct validity, Donald Campbell also required ruling out theoretical and methodological irrelevancies (1986). These are features of the sampled particulars that substantive experts deem irrelevant to a category but that can be part of the sampling particulars used to represent that category. Analysts must therefore have a theory of which aspects of a target entity are prototypical, or central to its understanding, as well as aspects that are peripheral and do not make much of a conceptual difference. If it can be shown that these substantive irrelevancies make no difference, then an association is deemed robust with respect to these features. As a result, the generalization is strengthened because the association being tested does not depend on largely or totally irrelevant features that happen to be correlated with the prototypical features. Research syntheses benefit from heterogeneous implementations across primary studies because this makes the irrelevancies associated with each construct heterogeneous. In the relatively rare case when a synthesis depends on primary studies that share the same substantive irrelevancy, then that attribute does not vary and is confounded with the target class. For instance, if all the Asian Americans in a particular set of studies were Vietnamese or male, then the general Asian American category would be confounded with a single nation of origin or with a single gender. Similarly, if an association is limited to studies of public schools that are all in affluent suburbs with volunteer participants, then public schools are confounded with both the suburban location and the volunteer respondents. If all the measures of productivity tap into the quantity but not quality of what has been produced, then the productivity concept would be confounded with its quantity component. The obvious requirement for unconfounding the targets of interest and the way they are assessed is ensuring that the studies sampled for meta-analysis vary in the country of origin among Asian Americans, in location among public schools, and in how productivity is measured.

28.1.3.3 Probing Discriminant Validity

Even ruling out the role of theoretical irrelevancies is not enough. Any relationship that is not robust across substantively relevant differences invites the analyst to identify the conditions that moderate or mediate the association. Research syntheses include primary studies with their own populations of persons, their own categories of setting, their own period, and their own theoretical descriptions of what a treatment or outcome class is and how it should be manipulated or measured. The issue is not then to generalize to a single common population; it is to probe how general the relationship is across different populations and to determine the specific conditions under which it varies. Does the effect of antidepressant medication hold in children, adolescents, adults, and the elderly? Is the effect of secondhand smoke different in pre- and postmenopausal women? Is the effect of psychotherapy conducted in clinical practice settings different from that in university research settings? Questions such as these address the generalizability of meta-analytic findings by probing their discriminant validity.

28.1.3.4 Studying Empirical Interpolation and Extrapolation

The preceding discussion deals only with situations where the synthesized studies contain instances of the classes about which we want to generalize.

However, in some cases, we seek to extrapolate beyond the available data and draw conclusions about domains not yet studied (Cronbach 1982). This is an intrinsically difficult enterprise, but even here it is plausible to assert that research syntheses have advantages over primary studies. Consider the case of a relationship that is demonstrably robust across a wide range of conditions. The analyst may then want to induce that the same relationship is likely to hold in other, as-yet unstudied, circumstances given that it has held so robustly over the wide variety of circumstances already examined. Consider, next, the case in which some specific causal contingencies have been identified for which the association does and does not hold. This knowledge can then be used to help identify specific situations in which extrapolation is not warranted, the more so if a general theory can be adduced to explain the variation in effectiveness. Now, imagine a sample of studies that is relatively homogeneous on some or all attributes. There is then no evidence of the association's empirical robustness or of the specific circumstances that moderate when it is observed, making inductive extrapolation to other universes insecure. Of course, conceptual examination is always required to detect moderating variables, even with a heterogeneous collection of studies. For instance, many pharmaceutical drugs are submitted to regulatory agencies for approval with adults and may not have been tested with adolescents, young children, or the elderly. Yet individual physicians make off-label prescriptions where they extrapolate from approved groups, such as adults, to those that have not been studied, such as adolescents, children, and the elderly. Indeed, some experts estimate that 80 to 90 percent of pediatric patients are prescribed drugs off-label, requiring thoughtful physicians to assume that the relevant physiological processes do not vary by age—an assumption that may or may not be true for a given pharmaceutical agent (Tabarrok 2000). Extrapolation depends here, not just on the robustness of sampling plans and results, but also on substantive theory about age differences in physiological processes.

28.1.3.5 Building on Causal Explanations In the experimental sciences, generalization is justified, not from sampling theory, but from attempts to achieve complete understanding of the processes causally mediating an association. The crucial assumption is that, once these processes have been identified, they can be set in motion by multiple causal agents, including some not yet studied but plausible on theoretical or pragmatic grounds. Indeed, the public is often asked to pay for basic science on the

grounds that ameliorative processes will be discovered that can be instantiated in many ways, making them viable across a wide range of persons and settings today and tomorrow. Research syntheses can enhance our understanding of conditions mediating the magnitude and direction of an association because the variety of persons, settings, times and measures available in the database will be greater than that any single study can capture. The crucial issue then becomes what the obstacles (that is, threats) are that arise when using meta-analysis to extrapolate findings by identifying the processes by which a class of interventions produces an effect. This is a difficult issue to pursue because explicit examples of causal explanation remain rare in meta-analysis because the concern has been more with identifying causal relationships and some of their moderating rather than mediating factors (for example, Becker 2001; Cook et al. 1992; Harris and Rosenthal 1985; Premack and Hunter 1988; Shadish and Sweeney 1991). But even so, moderation provides clues to mediation through the way results are patterned; and many meta-analyses do include individual studies making claims about causal mediation. It would be stretching the point to claim that detecting causal mediation is a strength of meta-analysis today. It is not overly ambitious, however, to contend that the potential for explanation is greater in meta-analyses than in individual studies, even explicitly explanatory ones.

The five principles we have just enumerated for strengthening generalized inference in meta-analyses are not independent. None is sufficient for generalized causal inference as understood here, and probably only proximal similarity is necessary. However, taken together, they describe current efforts at generalization within a synthesis framework, and they are warranted through their link to the generalization afforded by theories of construct validity rather than formal sampling or well-corroborated and well-demarcated substantive theory. Taken together, the principles suggest feasible strategies for exploring and justifying conclusions about the generalization of an association. They are more relevant for conclusions about what the sampling particulars represent and what the limits of generalization might be, however, than they are for the intrinsically more problematic task of extrapolating to unstudied entities.

Our discussion follows from the benefits research synthesis promise. We first discuss threats relevant to meta-analyses that primarily seek to describe the degree of association between two variables, and those to the inference that a relationship is causal in the manipulability or activity theory sense (Collingwood 1940; Gasking 1955; Whitbeck 1977). We then turn our attention to generalization, beginning with those validity threats that apply when generalizing to particular target populations, constructs, categories, and the like. We examine generalization threats pertinent to moderator variable analyses that concern the empirical robustness of an association and thus affect its strength and even direction (Mackie 1974). Finally, we examine those generalization threats that apply when meta-analysis is used to extrapolate to novel, as-yet unstudied universes. Threats discussed in the earlier sections almost always apply to later sections that are devoted to the increasingly more complex inferential tasks of causation and generalization. To avoid repetitiveness, we discuss only the unique threats relevant to the later sections.

28.2 THREATS TO INFERENCES ABOUT THE EXISTENCE OF AN ASSOCIATION BETWEEN TREATMENT AND OUTCOME CLASSES

The following threats deal with issues that may lead to erroneous conclusions about the existence of a relationship between two classes of variables, including an independent (that is, treatment) and dependent variable (that is, outcome). See table 28.2 for a list of these threats. In hypothesis testing language, these threats lead to type I or type II errors emanating from deficiencies in either the primary studies or the research synthesis process. A solitary deficiency in a single study is unlikely to jeopardize meta-analytic conclusions in any meaningful way. More problematic is whether a specific deficiency operates across all or most of the studies being reviewed or whether different deficiencies fail to cancel each other out across studies, such that one direction of bias predominates, either to over- or underestimate an association (see chapter 8, this volume). The need to rule out these possibilities leads to the list of validity threats in table 28.1.

28.2.1 Unreliability in Primary Studies

Unreliability in implementing or measuring treatments and in measuring outcomes attenuates the effect-size estimates from primary studies and the average effect-size estimates computed across such studies. Attenuation corrections have been proposed and, if their assumptions are accepted, these corrections may be used to provide estimates that approximate what would have happened had treatments and outcomes been observed without error, or

Table 28.1 Threats to Inferences About Existence of Association Between Treatment and Outcome Classes

(1)	Unreliability in primary studies
(2)	Restriction of range in primary studies
(3)	Missing effect sizes in primary studies
(4)	Unreliability of codings in meta-analyses
(5)	Capitalizing on chance in meta-analyses
(6)	Biased effect size sampling
(7)	Publication bias
(8)	Bias in computing effect sizes
(9)	Lack of statistical independence among effect sizes
(10)	Failure to weight effect sizes proportional to their precision
(11)	Underjustified use of fixed or random effects models
(12)	Lack of statistical power for detecting an association

SOURCE: Authors' compilation.

had the error been constant across studies (Hunter and Schmidt 1990; Rosenthal 1991; Wolf 1990); see also chapter 17, this volume). There are limits, however, to what can be achieved because the implementation of interventions is less understood than the measurement of outcomes, and is rarely documented well in primary studies. Some primary researchers do not even report the reliability of the scores on the outcomes they measured. These practical limitations impede comprehensive attempts to correct individual effect size estimates, and force some meta-analysts either to ignore the impact of unreliability or to generate (untested) assumptions about what the missing reliability might have been in a given study.

28.2.2 Restriction of Range in Primary Studies

When the range of a treatment or predictor variable is restricted, effect-size estimates are reduced. In contrast, if the outcome variable range is restricted, the within-group variability is reduced. Restricting range on the outcome will otherwise decrease the denominator of the effect size estimate, d, and thus increase effect-size estimates. For example, if a weight-loss researcher limits participation to subjects who are at least 25 percent but not more than 50 percent overweight, within-group variability in weight loss will likely drop, increasing the effect size estimate by decreasing its denominator. Thus, range restrictions in primary studies can attenuate or inflate effect-size estimates, depending on which variable is restricted. As Wolfgang Viechtbauer pointed out, two standardized effect sizes based on the same outcome measures from two studies could be incommensurable if the samples were drawn from populations with different variances (2007). This problem can be avoided if the raw units are consistent across the studies of a meta-analysis, making it unnecessary to standardize the effect size—the case when Jean Twenge and Keith Campbell used the Rosenberg Self-Esteem Scale and the Coopersmith Self-Esteem Inventory to conduct a meta-analysis of birth cohort differences in adults and children, respectively (2001).

Given the possibility of counterbalancing range restriction effects within individual studies and across all studies in a research synthesis, it is not easy to estimate the overall impact of range restriction. Nonetheless, meta-analysts have suggested adjustments to mean estimates of correlations (and their standard error) to control for the attenuation due to range restriction (see chapter 17, this volume). When valid estimates of the population range or variance are available, these adjustments provide reasonable estimates of the correlation that would be observed in a population with such variance. Adjusting for range restriction in a manipulated variable (that is, a carefully planned intervention) for which population values may not be known, however, is more problematic.

28.2.3 Missing Effect Sizes in Primary Studies

Missing effect sizes occur when study reports fail to include findings for all the sampled groups, outcome measures, or times. This sometimes happens because of space limitations in a journal or because a research report is restricted to only part of the overall study. Some effect sizes also go unreported because the findings were not statistically significant, thus inflating the average effect size from a meta-analysis. Otherwise, the impact of unreported effect sizes will vary, depending on why an author decided to include some findings and exclude others.

To prevent this kind of bias, it is important to code the most complete version of a report (such as dissertations and technical reports rather than published articles) and to contact study authors to obtain additional information, as Steven Premack and John Hunter (1988) and George Kelley, Kristi Kelley, and Zung Vu Tran (2004) did with some success. However, this last strategy is not always feasible if authors cannot be found or no longer have the needed information. Meta-analysts must then code research reports for evidence of missing effect sizes and use sensitivity analyses to explore how the data known to be missing might have affected overall findings. This can

involve examining whether studies with little or no missing data yield comparable findings to studies with more missing data or missing data patterned a particular way.

Another approach is to pursue the imputation strategies discussed in chapter 21 of this handbook, though it is clear that they hold only under certain model assumptions. When these assumptions cannot be convincingly justified, the impact of these procedures must be questioned. An example comes from early meta-analyses of the effects of psychotherapy, where it was assumed that the most likely estimate for a missing effect size was zero, based on the assertion that researchers failed to report only those outcomes that did not differ from zero (that is, were not statistically significant). Though this assumption has never been comprehensively tested, David Shapiro and Diana Shapiro coded 540 unreported effect sizes as zero and then added them to 1,828 reported effects, reducing the average effect from .93 to .72 (1982). But effect sizes may be unreported for many other reasons—for example, when they are negative, or positive but associated with unreliable measures—raising concern about the assumption that they average zero. For a general introduction to multiple imputation methods, see Roderick Little and Donald Rubin (2002), Donald Rubin (1987), and Sandip Sinharay et al. (2001). For applications in meta-analysis, see Alexander Sutton (2000), George Kelley (2004), William Shadish et al. (1998), and Carolyn Furlow and Natasha Beretvas (2005).

28.2.4 Unreliability of Codings in Meta-Analyses

Meta-analytic data are the product of a coding process susceptible to human error. Unreliability at the level of research synthesis (unreliable determination of means, standard deviations, and sample sizes, for example) is not expected to bias average effect size estimates. This is because in classical measurement theory, measurement error is independently distributed and uncorrelated with true scores. It will, however, inflate the variance of the observed effect sizes, increasing estimates of standard error and reducing statistical power for hypothesis tests. David Wilson discusses several strategies for controlling and reducing error in coding (chapter 9, this volume). In our experience pilot testing the coding protocol, comprehensive coder training, engaging coders with expertise in the substantive area being reviewed, consulting external literature, contacting primary authors, using reliability estimates as controls, generating confidence ratings for individual codings, and conducting sensitivity analyses are all helpful strategies for reducing and controlling for error in data coding.

28.2.5 Capitalizing on Chance in Meta-Analyses

Although research syntheses may combine findings from hundreds of studies and thousands of respondents, they are not immune to inflated type I error when many statistical tests are conducted without adequate control for error rate—a problem exacerbated in research syntheses with few studies. Typically, meta-analysts conduct many analyses as they probe the robustness of an effect size across various methodological and substantive characteristics that might moderate effect sizes, and as they otherwise explore the data. To reduce capitalizing on chance, researchers must adjust error rates, examine families of hypotheses in multivariate analyses, or stick to a small number of a priori hypotheses.

28.2.6 Biased Effect-Size Sampling

Research reports frequently present more than one estimate, especially when there are multiple outcome measures, multiple treatment and control groups, and multiple delayed assessment time points. Some of these effect estimates may be irrelevant for a particular topic, and some of the relevant ones will be substantively more important than others. Meta-analysts must then decide which estimates will enter the meta-analysis. Bias occurs when estimates are selected that are as substantively relevant as those not selected but that have different average effect sizes. Georg Matt discovered this when three independent coders recoded a subsample of the studies in Mary Lee Smith, Gene Glass, and Thomas Miller's meta-analysis of the benefits of psychotherapy (Matt 1989; Smith, Glass, and Miller 1980). Following what seemed to be the same rules as Smith and her colleagues for selecting and calculating effect estimates, the recoders extracted almost three times as many effect estimates whose mean effect size was approximately .50, compared to the .90 from the Smith, Glass, and Miller original codings. Rules are required in each meta-analysis that clearly indicate which effect estimates to include; and best practice is to specify these rules before data collection begins. If several plausible rules are identified, it is then important to examine whether they lead to the same results. Although specifying such rules is relatively easy, implementing them validly (and reliably) depends on training coders to identify relevant effect sizes within studies. Data analyses should also explore for possible coder differences in results.

28.2.7 Publication Bias

Some studies are conducted but never written up; of those written up, some are not submitted for publication; and of those submitted, some are never published. Publication bias exists when the average effect estimate from published studies differs from that of the population of studies ever conducted on the topic. Anthony Greenwald and Robert Rosenthal have both argued that published studies in the behavioral and social sciences are likely to be a biased sample of all the studies actually carried out, because studies with statistically significant findings supporting a study's hypotheses are more likely to be submitted for publication and ultimately published, given reviewer and editor biases against the null hypothesis (Greenwald 1975; Rosenthal 1979). An extension of this bias may even operate among published studies if those that are easier to identify and retrieve have different effect sizes than other studies. This could arise because of the greater visibility of studies in major publications that are well abstracted by the major referencing services, because authors in major outlets are better connected in professional networks, or because the paper's title or abstract is more likely to contain keywords relevant for a particular meta-analysis. Unsuccessful replications by lesser-known researchers are not likely to appear in major journals, however appropriate they might be for a conference or a more obscure publication.

Strenuous attempts should be made to find unpublished or difficult-to-retrieve studies, and separate effect-size estimates should be calculated for published and unpublished studies as well as for studies that varied in how difficult they were to locate. The chapters in this volume on scientific communication (4), reference databases (5), the fugitive literature (6), and publication bias (23) all provide additional suggestions for dealing with publication bias (see also Rothstein, Sutton, and Borenstein 2005).

Several important developments should be stressed. First, research registries are particularly promising for avoiding publication biases, especially in research domains under the influence of regulatory agencies, such as the Food and Drug Administration. Second, the Cochrane Collaboration has taken the idea of research registries one step further, developing a registry for prospective research syntheses, in which eligible studies for a meta-analysis are identified and evaluated before their findings are even known. Third, meta-analysts can now rely on a set of powerful exploratory data analysis methods to detect biases in published studies (Egger et al. 1997; Sterne and Egger 2001; Sutton, Duval et al. 2000; Sutton, Song et al. 2000). And, finally, if the evidence indicates that effect estimates depend on publication history, the likely consequences of this bias should be assessed with a variety of methods, including Larry Hedges and Jack Vevea's selection method approach (Vevea and Hedges 1995; Hedges and Vevea 1996; Vevea, Clements, and Hedges 1993; Vevea and Woods 2005), and Susan Duval's trim and fill method (Duval and Tweedie 2000; Sutton, Duval et al. 2000).

28.2.8 Bias in Computing Effect Sizes

Meta-analyses often require transforming findings from primary studies into a common metric such as the correlation coefficient, a standardized mean difference, or an odds ratio (see chapter 12, this volume). This is required because studies differ in the type of quantitative information they originally provide, and bias results if some types of transformation lead to systematically different estimates of average effect size or standard error when compared to others.

The best-understood case of transformation bias concerns the situation in which probability levels are aggregated across studies and some are truncated (Rosenthal 1990; Wolf 1990); for instance, when a study reports a group difference statistically significant at $p < .05$ without specifying the exact probability level. A conservative estimate of the observed effect size can be obtained by assuming that $p = .05$ then finding the relevant critical value of the test statistic that would have been observed (given the appropriate degrees of freedom) at $p = .05$. Because it is known that the actual probability value was smaller than the one assumed, the transformed effect size is known to be conservatively biased.

Another case of potential bias involves studies that used dichotomous outcomes measures or continuous variables that have been dichotomized (for example, improved versus not improved, convicted versus not convicted). Julio Sánchez-Meca, Fulgencio Marín-Martínez, and Salvador Chaon-Moscoso compared seven approaches to converting effect sizes derived from continuous and dichotomous outcome variables into a common metric. They concluded that the Cox and Probit based effect-size indices showed the least bias across simulated population effect sizes, δ, ranging from 0.2 to 0.8 (2003). Whenever

possible, meta-analysts are advised to calculate effect sizes directly from means, standard deviations, or the like rather than from approximations such as truncated p-levels or proportions improved. This will be possible for many studies, though not all. Simulation studies of the performance of different indicators like that of Sánchez-Meca and his colleagues can help make an informed choice, but when they are not available, meta-analysts should empirically examine whether estimates differ by the type of effect size transformation used.

28.2.9 Lack of Statistical Independence

Stochastic dependencies among effect sizes may influence average effect estimates and their precision (see chapter 19, this volume). The effect size estimates in a meta-analysis may lack statistical independence for at least four reasons: collecting data on multiple outcomes for the same respondents; comparing different interventions to a single control group, or different controls to a single intervention; calculating an effect estimate for each of several sub samples of person within the same study (for example, women and men); and the same research team conducts multiple studies on the same topic (Hedges 1990). These situations can be conceptualized hierarchically—for instance, as multiple outcomes or repeated assessments of the same measure nested within each study. Ignoring or misspecifying the resulting covariance structure of effect sizes can lead to invalid estimates of mean effect sizes and their standard errors (see also chapter 18, this volume).

Leon Gleser and Ingram Olkin discuss several approaches to deal with such dependencies (chapter 19, this volume). The most simple involves analyzing for each study only one of the set of possible correlated effects, for example, the mean or median of all effects, a randomly selected effect estimate, or the most theoretically relevant estimate (Lipsey and Wilson 2001). Another involves Bonferroni and "ensemble adjusted" p-values. Although these strategies are relatively simple to apply, each is conservative and fails to take into account all of the available data. Larry Hedges and Ingram Olkin therefore developed a multivariate statistical framework in which dependencies can be directly modeled (Hedges and Olkin 1985; see also Raudenbush, Becker, and Kalaian 1988; Rosenthal and Rubin 1986). Bayesian and hierarchical linear modeling approaches have also been successfully applied to deal with multiple effect size within studies (Raudenbush and Bryk 1985; van Houwelingen, Arends, and Stijnen

2002; Saleh et al. 2006; Eddy, Hasselblad, and Shachter 1990; Sutton and Abrams 2001; Scott et al. 2007; Nam, Mengersen, and Garthwaite 2003; Hox and de Leeuw 2003; Prevost, Abrams, and Jones 2000; see also chapter 26, this volume). These multivariate models are more complex than the previously discussed univariate techniques. Some require estimates of the covariance structure among the correlated effect sizes that may be difficult to obtain, in part because of missing information in primary studies. Moreover, the gains in estimation due to using these multivariate techniques will often be small.

28.2.10 Failure to Weight Effect Sizes Proportionally to Their Precision

Everything else being equal, studies with larger sample sizes yield effect-size estimates that are more precise (that is, have smaller standard errors) than studies with smaller ones. Simply averaging effect sizes of different precision may yield biased average effects and sampling errors even if each study's estimates are themselves unbiased (see chapter 14, this volume). Therefore, when effect sizes are combined, it has become common practice to use weighted averages, allowing more precise estimates to have a stronger influence on the overall findings. Larry Hedges and Ingram Olkin have shown that the optimal weight is the inverse of the sampling variance of an effect size (1985). This validity threat was common in early meta-analyses of psychotherapy outcomes but has become increasingly rare in recent years in meta-analyses using common effect-size metrics, for example, d, r, odds ratios (Shapiro and Shapiro 1982; Smith, Glass, and Miller 1980).

28.2.11 Underjustified Use of Fixed or Random Effects Models

When analyzing effect sizes, Larry Hedges and Ingram Olkin stressed the importance of deciding on a model with fixed or random effects (1985). Perhaps the most important difference between the two models concerns the inferences they allow. Fixed-effect models justify inferences about treatment effects only for the studies at hand. In their simplest form, they assume that all the studies in a meta-analysis involve independent samples of participants from the same population, and so it is legitimate to postulate a single but unknown effect size, estimates of which differ only as a result of sampling variability. In contrast, random effects models justify inferences about a

universe of studies from which the studies at hand have been sampled. Each study is assumed to have its own unique "true" effect and to be a sample realization from a universe of related, yet distinct studies. Thus, observed differences in treatment effects have two sources, which are different samples of participants (as in the fixed-effect model) and true differences in treatment effects between studies (compared to a single true effect size in the fixed model). This means that in the random-effects model, treatment effects are best represented as a distribution of true effects represented by their expected value and variance.

These two models have important implications for the analysis and interpretation of effect size estimates (see chapters 15 and 16, this volume). The fixed-effect model is analytically simpler, requiring estimates of the specific fixed effects of interest (for example, mean effect sizes for treatment A, B, and C) and their precision. The random-effects model requires estimating the population variance associated with the universe of treatment effects given the estimates observed in a sample of treatments and their precision. As a consequence of the different underlying statistical models, fixed-effects analyses limit inferences to the specific fixed levels of a factor that were included in a meta-analysis (for example, treatments A, B, and C). In contrast, random effects models strive for inferences about the population of levels of a factor, given the samples of levels that were included in a meta-analysis (for example, population of treatments consisting of A, B, C, F, . . . , Z).

The decision to assume a fixed- or random-effects model is primarily influenced by the substantive assumptions meta-analysts make about the processes generating an effect and about the desired inferences. That is, are the treatments, settings, outcomes, and participants examined in different studies sufficiently standardized and similar that they should be considered equivalent (that is, fixed) for purposes of interpreting treatment effects? Are we interested only in drawing inferences about the specific instances of treatments, settings, outcomes, and participants that were studied (that is, fixed-effect model)? Or are we interested in drawing more general inferences about the classes of treatments, settings, and outcomes to which the specific instances that were studied belong (that is, random-effect model)? Only secondarily is the decision influenced by the heterogeneity observed in the data and the results of a homogeneity test. This is particularly true because tests of homogeneity tend to have low statistical power in situations commonly found in research syntheses, for example, modest heterogeneity, unequal sample sizes, and small within-study sample sizes (Hardy and Thompson 1998; Harwell 1997; Hedges and Pigott 2001; Jackson 2006).

Assuming excellent statistical power, a fixed model seems more reasonable if the hypothesis of homogeneity is not rejected and if no third variables can be adduced from theory or experience that may have caused (unreliable) variation in effect size estimates. However, if theory suggests a fixed population effect but a homogeneity test indicates otherwise, then two possible interpretations follow. One is that the assumption of a fixed population is wrong and that a random-effects model is more appropriate; the second is that the fixed-effects model is true, but variables responsible for the increased variability are not known or adequately modeled and as such, in this research case, cannot account for the observed heterogeneity. Thus, there is no simple indicator of which model is correct. However, it is useful to examine as critically as possible all the substantive theoretical reasons for inferring a fixed- over a random-effects model, whether the homogeneity assumption is rejected or not.

28.2.12 Lack of Statistical Power for Detecting an Association

Although the focus of meta-analyses is on estimating the magnitude of effects and their precision rather than null hypothesis testing, meta-analysts often report findings from hypothesis tests about the existence of an association. Under most circumstances, the statistical power for detecting an association in a meta-analysis is influenced by the number of studies, the sample sizes within studies, the type of assignment of units to experimental conditions in the primary studies (for example, cluster randomization), and (for analyses employing random effects assumptions) the between-studies variance component (Hedges and Pigott 2001; Donner, Piaggio, and Villar 2003; Cohn and Becker 2003). When compared to statistical analyses in primary studies, tests of mean effect sizes will typically be more powerful in meta-analyses, particularly in fixed-effect models estimating the average effect of a class of interventions based on many similar studies. There are some paradoxical exceptions to this rule in random-effect models, for instance, when small studies add more between-studies variance than they compensate for by adding information about the mean effect (Hedges and Pigott 2001). Similarly, research syntheses of cluster

Table 28.2 Threats to Inferences About Causal Nature of Relationship Between Treatment and Outcome Classes

(1) Absence of studies with successful random assignment
(2) Primary study attrition

SOURCE: Author's compilation.

randomization trials must take design effects into account to obtain correct estimates of statistical power (Donner, Piaggio, and Villar 2001, 2003).

Some meta-analysts are limited to a few studies, each having small sample sizes. Others are interested in examining effect sizes for subclasses of treatments and outcomes, different types of settings, and different subpopulations. Careful power analyses can help clarify the type II error rate of a test failing to reject a null hypothesis. The meta-analyst then has to decide which trade-off to make between the number and type of statistical tests and the statistical power of these tests.

28.3 THREATS TO INFERENCES ABOUT THE CAUSAL NATURE OF AN ASSOCIATION BETWEEN TREATMENT AND OUTCOME CLASSES

Threats to inferences about whether an association between treatment and outcome classes is causal or spurious arise mostly out of the designs of primary studies. It is at the level of the individual study that the temporal sequence of cause and effect, randomization of units to conditions, and other design features aimed at strengthening causal inferences are implemented.

The following threats are in addition to those already presented that refer to the likelihood of an association between treatment and outcome classes. See table 28.3 for a list of these threats. The logic of causal inference is fairly straightforward at the primary study level, but is complicated at the research synthesis level because findings from partially flawed primary studies often have to be combined. Inferences about causation are not necessarily jeopardized by deficiencies in primary studies because— at least in theory—individual sources of bias in the primary studies may cancel each other out exactly when aggregated in a research synthesis. So at the research synthesis level, a threat arises only if the deficiencies within each primary study combine across studies to create a predominant direction of bias.

28.3.1 Absence of Studies with Successful Random Assigment

In primary studies, unbiased causal inference requires establishing the direction of causality and ruling out third-variable alternative explanations. Inferring the direction of causality is easy in experimental and quasi-experimental studies where knowledge is usually available about the temporal sequence from manipulating the treatment to measuring its effects. In theory, third-variable alternative explanations are ruled out when participants in primary studies are randomly assigned to treatment conditions or if a regression-discontinuity study is done (Shadish, Cook, and Campbell 2002; West, Biesanz, and Pitts 2000).

In research syntheses of well-implemented randomized trials, the strong causal inferences at the primary study level are transferred to the research synthesis level. This is why, for example, the Cochrane Collaboration generally restricts its syntheses to randomized experiments. If randomization was poorly implemented or not done at all, then causal inferences are ambiguous. Additionally, the pervasive possibility of third-variable explanations arise, as in quasi-experimental studies of early childhood interventions and juvenile delinquency, where those participating in programs are often more disadvantaged and likely to mature at different rates than those in control groups (Campbell and Boruch 1975). If lack of random assignment in primary studies yields a predominant bias across these studies, causal inference at the level of the research synthesis is jeopardized.

One approach to this threat compares the average effect estimates from studies with random assignment to those studies on the same question with more systematic assignment. If effect estimates differ, causal primacy must then be given to findings from randomized designs, assuming that the randomized and nonrandomized studies are equivalent on other characteristics. Although randomized experiments and quasi-experiments sometimes result in similar estimates, those that do not are notable (Shadish, Luellen, and Clark 2006; Jacob and Ludwig 2005; Boruch 2005; Lipsey 1992; Smith, Glass, and Miller 1980; Wittmann and Matt 1986; Chalmers et al. 1983). A second strategy involves the careful explication of possible biases caused by flawed randomization or associated with a particular quasi-experimental design in each primary study. Based on an adequate understanding of possible biases, adjustments for pretreatment differences between groups can sometimes be made to project causal effects, controlling for the identified biases. The problem here, of course, is justifying that all important biases have been identified and their operation has been correctly modeled. This latter approach is particularly useful when the first approach fails if too few primary studies with randomized designs exist.

28.3.2 Primary Study Attrition

Attrition of participants from treatment or measurement is common in even the most carefully designed randomized experiments and quasi-experiments. If attrition in these primary studies is differential across treatment groups, effect estimates may then be biased, inflating or deflating causal estimates. If the biases operating in each direction cancel each other out, then causal inference at the research synthesis level is not affected. However, if predominant bias persists across the primary studies, then causal inferences from the synthesis are called into question. For instance, Mark Lipsey's meta-analysis of juvenile delinquency interventions found that the more amenable juveniles might have dropped out of treatment groups and the more delinquent juveniles out of control groups (1992). The net effect at the level of the research synthesis is a potential bias.

To address this threat, it is critical to code, for each primary study, information about the level and differential nature of attrition from the treatment groups. The latter information cannot be inferred from the level of attrition, but requires the primary study to have reported the possible differential attrition of participants. If this information is available, the analyst can then examine the effect of attrition on effect estimates by disaggregating studies according to their level of total and differential attrition. In addition, sensitivity analyses can be conducted to explore whether the observed levels of attrition present plausible threats.

28.4 THREATS TO GENERALIZED INFERENCES

Although some of the threats discussed in the previous sections are germane to generalization (such as underjustified use of fixed and random effects models), we now turn explicitly to generalization issues.

28.4.1 Sampling Biases Associated with Persons, Treatments, Outcomes, Settings, and Times

Statistical sampling theory is sometimes invoked as the justification for generalizing from the obtained instances

Table 28.3 Threats to Generalized Inferences in Research Syntheses

Inferences to Target Constructs and Universes
  (1) Sampling biases associated with the persons, treatments, outcomes, settings, and times
  (2) Underrepresentation of prototypical attributes
  (3) Restricted heterogeneity of substantively irrelevant third variables
  (4) Mono-operation bias
  (5) Mono-method bias
  (6) Rater drift
  (7) Reactivity effects

Inferences about Robustness and Moderating Conditions
  (8) Restricted heterogeneity in universes of persons, treatments, outcomes, settings, and times
  (9) Moderator variable confounding
  (10) Failure to test for homogeneity of effect sizes
  (11) Lack of statistical power for homogeneity tests
  (12) Lack of statistical power for studying disaggregated groups
  (13) Misspecification of causal mediating relationships

Extrapolations to Novel Constructs and Universes
  (14) Misspecification of models for extrapolation

SOURCE: Authors' compilation.

(or samples) of persons, treatments, outcomes, and settings to the universes or domains they are thought to represent, and about which researchers want to draw inferences. However, it is rare in individual primary studies for instances to be selected at random from their target universes, just as it is rare in research syntheses to randomly sample primary studies, treatments, outcomes, or settings from the universe of ever conducted research on a given topic. Some exceptions are noteworthy, as when Robert Orwin and David Cordray (1985) and Georg Matt (1989) randomly sampled studies from those included in the Smith, Glass, and Miller (1980) meta-analysis of the psychotherapy outcome—assuming that Smith and her colleagues had a census of all studies or a random sample from the census. Perhaps the most feasible random selection in meta-analysis is when, as some methodologists have suggested (Lipsey and Wilson 2001), only one effect size is sampled per study so as to provide an unbiased estimate of the mean effect size per study.

More common study selection strategies are for meta-analysts to seek collecting the population of published and unpublished studies on a topic, or restricting the se-

lection of studies to all those with specific person, treatment, outcome, setting, or time characteristics of substantive importance (Mick et al. 2003; Moyer et al. 2002). In either of these circumstances, inferences from the samples to their target universes can be biased because the studies that were excluded may have different average effect sizes than those that were included. As a result, generalizations can easily be overstated, even if they are supported by data from hundreds of studies and thousands of research participants.

28.4.2 Underrepresentation of Prototypical Attributes

Research syntheses should start with the careful explication of the target constructs about which inferences are to be drawn, at a minimum identifying their prototypical attributes and any less central features at the boundaries with other constructs. Thus, it can be that the collection of primary studies in the research synthesis does not contain representations of all the prototypical elements. This was the case when William Shadish and his colleagues attempted to investigate the effects of psychotherapy in clinical practice (2000). They observed that many studies included a subset of prototypical features of clinical practice, but no single study included all of its features as they defined them. As a result, generalization to real-world psychotherapy practice is problematic, even though there are literally thousands of studies of the effectiveness of psychotherapy practice. In such a situation, the operations implemented in the primary studies force us to revise the construct about which inferences are possible, reminding us that we cannot generalize to the practice of psychotherapy as it is commonly conducted in the United States. An important task of meta-analysis is to inform the research community about the underrepresentation of prototypical elements of core constructs in the literature on hand, turning attention to the need to incorporate them into the sampling designs of future studies.

28.4.3 Restricted Heterogeneity of Substantively Irrelevant Third Variables

Even if sampling from the universes about which generalized inferences are sought were random, and if important prototypical elements were represented in the reviewed studies, a threat arises if a research synthesis cannot demonstrate that the causal association is robust and holds across substantively irrelevant characteristics. For instance, if the reviewed studies on the effectiveness of homework in middle school were conducted by just one research team, relied on voluntary participation by students, or depended on teachers being highly motivated, the threat would then arise that all conclusions about the general effectiveness of homework are confounded with substantively irrelevant aspects of the research context. To give an even more concrete example, if private schools were disclosed to be those where school revenues come from student fees, donations, and the proceeds on endowments (rather than from taxes), it is irrelevant whether the schools are parochial or nonparochial, Buddhist, Catholic, Jewish, Muslim, or Protestant, military, or elite academic. To generalize to private schools in the abstract requires being able to show that relationships are not limited to one or a few of these contexts—say, parochial or military schools.

Limited heterogeneity of universes will also impede the transfer of findings to new universes (that is, extrapolation), because it hinders the ability to demonstrate the robustness of a causal relationship across substantive irrelevancies of design, implementation, or measurement method. The wider the range and the larger the number of substantively irrelevant aspects across which a finding is robust, and the better moderating influences are understood, the stronger the belief that the finding will also hold under the influence of not yet examined contextual irrelevancies.

28.4.4 Mono-Operation Bias

The coding and rating systems of research syntheses often rely exclusively on single items to measure such complex target constructs as setting, diagnosis, or treatment type. It is well known from the psychometric literature that single-item measures have poor measurement properties. They are notoriously unreliable, tend to underrepresent a construct, and are often confounded with irrelevant constructs. To address and improve on common measurement limitations of meta-analytic coding manuals, rigorous interrater reliability studies and standard scale development procedures have to become standard practice to allow valid inferences about target constructs of a meta-analysis.

28.4.5 Mono-Method Bias

If the measurement method in a meta-analysis relies on a single coder who reads and codes a research publication,

following the operations delineated in a coding manual, bias is possible. To avoid it, coding procedures for meta-analyses have to incorporate multimethod coding approaches. This could include having multiple coders (perhaps with different substantive backgrounds related to the research topic) code all items of a coding manual, contacting the original authors to provide clarification and additional information, obtaining additional write-ups of a study, and relying on external, supplementary sources, such as describing the psychometric properties of an outcome measure or allegiance and experience of a researcher (Robinson, Berman, and Neimeyer 1990). In the absence of such improvements, conclusions about important target constructs relying on single coder ratings remain suspect.

28.4.6 Rater Drift

Reading, understanding, and coding publications of multifaceted primary studies involve many cognitively challenging tasks. Over time, coders learn through practice, develop heuristics to simplify complex tasks, fatigue and become distracted, and may change their cognitive schemas as a result of exposure to study reports. As a consequence, the same coder may unknowingly change over time such that earlier codings of the same evidence differ from later ones. To address this validity threat, rater drift needs to be monitored as part of a continuing coder training program. Moreover, changes in coding manuals should be made publicly to reflect changes in the understanding of a code, which may necessitate recoding studies that were examined under the earlier coding rules.

28.4.7 Reactivity Effects

A measure is said to be reactive if the measurement process itself influences the outcome (Webb et al. 1981). In the case of a research synthesis, the coding process itself may inadvertently influence the coding outcome. For instance, knowing that the author of a study is a well-known expert in the field, rather than an unknown novice, may predispose a coder to rate research design characteristics more favorably for the expert than for the novice. Similarly, knowing that the treatment yielded no benefits over the control condition may bias a coder to rate the implementation of the treatment condition more critically. To minimize reactivity biases, it is desirable to mask raters to any influence that could bias their codings, including authorship and study results. Further, as much as possible, attempts should be made to avoid reactivity biases, including codings that require as little rater inference as possible. For example, rather than asking raters to arrive at an overall code describing study quality, it would be better to ask them to code specific aspects of studies that pertain to quality (such as the method of allocating participants to groups and attrition), which are less likely to be reactive (see chapters 7 and 9, this volume).

28.4.8 Restricted Heterogeneity in Universes of Persons, Treatments, Outcomes, Settings, and Times

Inferences about the conditions moderating the magnitude and direction of an association are facilitated if the relationship can be studied for a large number of diverse persons, treatments, outcomes, settings, and times. This is the single most important potential strength of research syntheses over individual studies. Although meta-analyses of standardized treatments and specifically designated outcomes, populations, and settings may increase the precision of some effect size estimates, such restrictions hamper our ability to better understand the conditions under which such relationships can and cannot be observed. Rather than limiting a meta-analysis to a review of a single facet of the universe of interest or lumping together a heterogeneous set of facets, we encourage meta-analysts—sample sizes permitting—to explicitly represent and take advantage of such heterogeneities.

Such advice has implications for those observers who have suggested that causal inferences in general—and causal moderator inferences in particular—could be enhanced if meta- analysts relied only on studies with superior methodology, particularly randomized experiments with standardized treatments, manifestly valid outcomes, and clearly designated settings and populations (Chalmers et al. 1989; Chalmers and Lau 1992; Sacks et al. 1987; Slavin 1986). Such a strategy appears to be useful in areas where research is fairly standardized. However, other limitations arise because this standardization limits the heterogeneity in research designs, treatment implementations, outcome measures, recruitment strategies, subject characteristics, and the like. Hence it is not possible to examine empirically how robust a particular effect is that has been obtained in a restricted, standardized context.

Meta-analysis has the potential to increase one's confidence in generalizations to new universes (that is, extrapolation) if findings are robust across a wide range and large number of different universes. The more robust the findings and the more heterogeneous the populations, settings, treatments, outcomes, and times in which they were observed, the greater the belief that similar findings will be observed beyond the populations studied. If the evidence for stubborn empirical robustness can be augmented by evidence for causal moderating conditions, the novel universes in which a causal relationship is expected to hold can be even better identified.

The logical weakness of this argument lies in its inductive basis. That psycho-educational interventions with adult surgical patients have consistently shown earlier release from the hospital across a broad range of major and minor surgeries, diverse respondents, and treatment providers throughout the 1960s, 1970s, and 1980s (Devine 1992) cannot logically guarantee the same effect will hold for as yet unstudied surgeries, and in the future. However, the robustness of the relationship does strengthen the belief that psycho-educational interventions will have beneficial effects with new groups of patients and novel surgeries in the near future (that is, extrapolation). Homogeneous populations, treatments, outcomes, settings, and times limit the examination of causal contingencies and robustness, and consequently, impede inferences about the transfer of findings to novel contexts.

28.4.9 Moderator Variable Confounding

At the synthesis level, moderator variables describe characteristics of the classes of treatment, outcomes, settings, populations, or times across which the magnitude or direction of a causal effect differs—a generalizability question. Claims about moderator variables are involved when a research synthesis concludes that treatment class A is less effective in population C than D, stronger with outcome of type E than F, weaker in setting G than H or positive in past times but not recently. Moderator variables are even involved when the claim is made in a synthesis that treatment type A is superior to treatment type B, because this assumes that the studies of A are equivalent to those of B on everything correlated with the outcome other than A versus B. Any uncertainty about this is pertains to the role that average study differences might have played in achieving the obtained difference between studies of A and B; thus it is an issue of moderator variables.

Threats to valid inference about the causal moderating role in research syntheses are pervasive (Lipsey 2003; Shadish and Sweeney 1991). This is because participants in primary studies are rarely assigned to moderator conditions at random (for example, to outcome type E versus F or to settings G versus H). Claims about the causal role of moderators are often questionable in syntheses, however, because any one moderator is often confounded with other study characteristics. For instance, studies of smoking cessation in hospital primary care facilities are likely to attract older participants than similar studies on college campuses. Likewise, behavioral treatments tend to involve behavioral outcome measures of a narrow target behavior, whereas client-centered interventions are likely to rely on broader measures of well-being and personal growth. If the moderator variable (for example, primary care setting versus college campus) is confounded with characteristics of the design, setting, or population (for example, age), differences in the size or direction of a treatment effect that might be attributable to the moderator are instead confounded with other attributes.

To deal with this possibility, meta-analysts should examine within-study comparisons of the moderator effect because these are obviously not prone to between-study confounds. For instance, if the moderating role of treatment types A, B, and C is at stake, a meta-analysis can be conducted of all the studies with internal comparisons of the treatment types (see Shapiro and Shapiro 1982). If the moderating role of subject gender is of interest, inferences about gender differences might depend on within-study contrasts of males and females rather than on comparisons of studies with only males or females or that assess the percentage of males. Statistical modeling provides another popular approach (Shadish and Sweeney 1991). The validity of the causal inferences based on such models depends on the ability of the meta-analyst to identify and reliably measure all confounding variables. Multivariate statistical adjustments can be informative here, but not definitive because of the difficult task of conceptualizing all such confounds and measuring them well across the studies under review.

28.4.10 Failure to Test for Homogeneity of Effect Sizes

Under a fixed-effect statistical model, a statistical test for homogeneity assesses whether the variability in effect estimates exceeds that expected from sampling error alone (Hedges 1982; Rosenthal and Rubin 1982). If the null hypothesis of homogeneous effect sizes is retained, a single population effect size provides a parsimonious model of the data, and the weighted mean effect size d provides an adequate estimate. If the null hypothesis is rejected, the

implication is that subclasses of studies may exist that differ in population effect size, triggering the search to identify the nature of such subclasses. Hence, heterogeneity tests play an important role in examining the robustness of a relationship and in initiating the search for potential moderating influences.

The failure to test for heterogeneity may result in lumping manifestly different subclasses of persons, treatments, outcomes, settings, or times into one class. This problem is frequently referred to as the apples and oranges problem of meta-analysis. However, Gene Glass, Robert Rosenthal, and Georg Matt have all argued that apples and oranges should indeed be mixed if the interest is in generalizing to such higher-order characteristics as fruit or whatever else inheres in an array of different treatments, outcomes, settings, persons, and times (Glass 1977; Smith, Glass, and Miller 1980; Rosenthal 1990; Matt 2003, 2005). We should indeed be willing to combine studies of manifestly different subclasses of persons, treatments, outcomes, settings, or times if they yield equivalent results in reviews. In this context, the homogeneity test indicates when studies yield such different results that a single, common average effect size needs to be disaggregated through blocking on study characteristics that might explain the observed variance in effect sizes.

Coping with this threat is relatively simple, and homogeneity tests have become standard both as a way of identifying the most parsimonious statistical model to describe the data and of testing model specification (see chapter 15, this volume). However, the likelihood of design flaws in primary studies, of publication biases, and the like makes the interpretation of homogeneity tests more complex (Hedges and Becker 1986; Schulze 2004). If all the studies being meta-analyzed share the same flaw, or if studies with zero and negative effects are less likely to be published, then a consistent bias results across studies and can make the effect sizes appear more homogeneous than they really are. Further, even if the collection of studies is not biased, a failure to reject the null-hypothesis that observed effect sizes are a sample realization of the same population effect does not prove it. Conversely, if all the studies have different design flaws, effect sizes could be heterogeneous even though they share the population effect. Obviously, the causes of heterogeneity that are of greatest interest are substantive rather than methodological. Consequently, it is useful to differentiate between homogeneity tests conducted before and after the assumption, that all study-level differences in methodological irrelevancies have been accounted for, has been defended.

28.4.11 Lack of Statistical Power for Homogeneity Tests

A homogeneity test examines the null hypothesis that two or more population effect sizes are equal, thus probing whether the observed variability in effect sizes is more than would be expected from sampling error alone. It is thus the gatekeeper test for deciding to continue the search for variables that moderate the average effect size obtained in a review—a generalization task. But when these homogeneity tests have little statistical power, as is usually the case, their type II error rate will be large and lead to the erroneous conclusions that the search for moderator variables should be abandoned (Jackson 2006; Hedges and Pigott 2001; Harwell 1997). In sum, statistically nonsignificant homogeneity tests provide inclusive evidence and should not be relied on exclusively to justify conclusions about the robustness of an association.

28.4.12 Lack of Statistical Power for Studying Disaggregated Groups

If there is reason to believe that treatment effects may differ across types of treatments, outcomes, persons, settings, or over time (that is, moderator effects), highly aggregated classes (for example, therapy or well-being) have to be disaggregated to examine the conditions under which an effect changes direction or magnitude. Such subgroup analyses rely on a smaller number of studies than main effect analyses and involve additional statistical tests that may necessitate procedures for type I error control, lowering the statistical power for the subanalyses in question. Consequently, the likelihood of finding statistically significant differences are reduced, even if such differences exist in the population. The flaws of this erroneous inference are compounded if a meta-analyst then concludes that an effect generalizes across subgroups of the universe because no statistically significant differences were found. Large samples of studies mitigate this problem to some extent, though the power in research syntheses is more complex than in primary studies. This is because power depends not only on the effect size, type I error rate, and sample size of primary study participants, but also on the number of studies and the underlying statistical model (Cohn and Becker 2003; Hedges and Pigott 2004).

28.4.13 Misspecification of Causal Mediating Relationships

Mediating relationships are examined to shed light on the processes by which a causal agent, such as secondhand smoke, transmits its effects on an outcome, such as asthma exacerbation. They provide highly treasured explanations of otherwise merely descriptive causal associations. Instead of simply noting that a treatment affects an outcome, mediating relationships inform us why and how a treatment affects the outcome. Explanatory models rely on causal mediating relationships to justify extrapolations and to specify causal contingencies. Both are important generalization tasks.

Few meta-analyses of mediating processes exist, and those that do use quite different strategies. Monica Harris and Robert Rosenthal's meta-analysis of the mediation of interpersonal expectancy effects used the available experimental studies that had information on at least one of the four mediational links abstracted from relevant substantive theory, which were never tested together in any individual study (1985). Steven Premack and John Hunter's meta-analysis of mediation processes underlying individual decisions to form a trade union relied on combining correlation matrices from studies on subsets of variables believed to mediate a cause-effect relationship (1988). The entire mediation model had not been probed in any individual study. Betsy Becker's meta-analysis relied on the collection of individual correlations or correlation matrices to generate a combined correlation matrix (1992). That is, individual studies may contribute evidence for one or more of the different causal links postulated in the mediational model. All these approaches examine mediational processes where the data about links between variables come from within-study comparisons. However, a fourth mediational approach infers causal links from between-study comparisons. William Shadish's research on the mediation of the effects of therapeutic orientation on treatment effectiveness is a salient example (1992; Shadish and Sweeney 1991).

Although the four approaches differ considerably, they all struggle with the same practical problem—how to test mediational models involving multiple causal connections when only few (or no) studies are available in which all the connections have been examined within the same study. A major source of threats to the meta-analytic study of mediation processes arises when the different mediational links are supported by different quantities and qualities of evidence, perhaps because of excessive missing data on some links, the operation of a predominant direction of bias for other links, or publication biases involving yet other parts of the mediational model.

On the basis of the few available meta-analyses of causal mediational processes, it seems reasonable to hypothesize that missing data are likely to be the most pervasive reason for misspecifying causal models (Cook et al. 1992). Missing data may lead the meta-analyst to omit important mediational links from the model that is actually tested and may also prevent consideration of alternative plausible models. Although there are many substantive areas with numerous studies probing a particular descriptive causal relationship, there are far fewer with as many that provide a broad range of information on mediational processes, and even fewer where two or more causal mediating models are directly compared. There are many reasons for these deficiencies. Process models are derived from substantive theory, and these theories change over time as they are improved or as new theories emerge. Obviously, the novel constructs in today's theories are not likely to have been measured well in past work. Moreover, today's researchers are reluctant to measure constructs from past theories that are now out of fashion or obsolete. The dynamism of theory development thus does not fit well the meta-analytic requirement for a stable set of mediating constructs or the simultaneous comparison of several competing substantive theories. Another relevant factor is that, even if a large number of mediational variables have been measured, they are often measured less well and documented less thoroughly than measures of cause and effect constructs. Measures of mediating variables are given less attention and are sometimes not analyzed unless a molar causal relationship has been established. Given the pressure to publish findings in a timely fashion and to write succinct rather than comprehensive reports, the examination of mediating processes is often postponed, limited to selected findings, or based on inferior measures.

28.4.14 Misspecification of Models for Extrapolation

The most challenging generalization involves inferences to new populations, treatments, outcomes, and settings. Such extrapolations to novel conditions have to rely on the available empirical evidence and the strength of the generalized inferences this evidence allows to the target universes represented in the studies. Therefore, all threats discussed up to this point for generalized inferences to target universes (28.4.1 to 28.4.7) and robustness and moderators (28.4.8 to 28.4.13) necessarily apply and affect extrapolations to novel constructs and universes.

Based on the available evidence, extrapolations rely on explicit and implicit models or informal heuristics to project treatment effects under yet-unstudied conditions, such as the effects of a new cancer drug in pediatric patients based on evidence in adult populations or the effects of psychotherapy in clinical practice based on studies conducted in research settings. At issue, then, is the validity of the model used to project the effects of an off-label use of a drug (for example, based on weight, metabolism, development stage of organs) or the effects of psychotherapy in practice (for example, based on caseload of therapists, patient mix, manualized or eclectic therapy). Under the best conditions, these models are informed by a comprehensive understanding of the causal mediating processes and empirical evidence about the robustness of effects and their moderating conditions across a broad range of substantively relevant and irrelevant variables. To the extent that this is not the case, the threat arises that the explicit or implicit extrapolation models may be misspecified, creating the risk for incorrect projections of effects under novel conditions.

Our discussion has made it clear that extrapolations to novel conditions carry a larger degree of uncertainty than generalized inferences to universes from which particulars have been studied. In this situation, the delicate goal of the meta-analyst is to communicate the existing evidence and the models from which an extrapolation is made, the assumptions built into such models, the uncertainty of the inference, and the conditions necessary to reduce the uncertainty. For the user of a meta-analysis, it may then be possible to weigh the risks and harm of an incorrect extrapolation against the likelihood and benefits of a correct extrapolation.

28.5 CONCLUSIONS

We have argued that the unique and major promise of research syntheses lies in strengthening empirical generalizations of associations between classes of treatments and outcomes. The history of meta-analytic practice, however, has demonstrated that this promise is threatened, because meta-analysts cannot rely on the two major scientific warrants for generalizability claims, random sampling from designated populations and strong causal explanations. This chapter offers an alternative approach, a threats-to validity framework, to explore and make a case for generalized inferences in meta-analysis when the established models cannot be applied.

Following Donald Campbell and Lee Cronbach, we distinguish between three types of generalized inferences central to meta-analyses (Campbell and Stanley 1963; Cook and Campbell 1979; Cronbach 1980). The first deals with inferences about an association in target universes of persons, treatments, outcomes, and settings based on particulars sampled from these universes. The second concerns the robustness and conditional nature of an association across heterogeneous, substantively relevant and irrelevant conditions. The third concerns extrapolating or projecting an association to novel, as-yet unstudied universes. The proposed threats-to-validity framework seeks to assist meta-analysts in exploring alternative explanations to the generalized inferences of interest. Because meta-analysts cannot realistically rely on the elegance and strength of sampling theory to warrant generalizability claims, the proposed framework offers a falsificationist approach to identify and rule out plausible alternative explanations to justify generalized inferences in meta-analyses.

Although the threats-to-validity framework makes no assumptions about random sampling or comprehensive causal explanations, it does require that we critically investigate all plausible concerns that can lead to spurious inferences and rule out each concern on grounds that it is implausible or based on evidence that it did not operate in a specific situation. Similar to the causal inferences based on Campbell's threats to validity for quasi-experimental designs, the generalized inferences based on the proposed threats to validity for meta-analyses are tentative and only as good as the critical evaluation of the plausible threats. The threats we have presented are, in large measure, a summary of what other scholars have identified since Gene Glass's pioneering work (Glass 1976; Smith, Glass, and Miller 1980). We expect the list to continue to change at the margin as new threats are identified, current ones are seen as less relevant than we now think, and as meta-analysis is put to new purposes.

To the fledgling meta-analyst, our list of validity threats may appear daunting and the proposed remedies overwhelming. More experienced practitioners will be less intimidated by the numerous threats because they operate from an implicit theory about the differential seriousness and prevalence of these threats and can recognize when the needed substantive and technical expertise and resources are on hand. Even so, all research syntheses have to make important trade-offs between partially conflicting goals. Thus, should methodologically less rigorous studies be excluded to strengthen causal inferences if this also limits the ability to explore potential moderating conditions and the empirical robustness of effects? Should resources be allocated to code fewer study features with greater validity or to code more study features that might capture even more potentially important reasons why effect sizes differ? When should one stop searching for more relevant studies if they are increasingly fugitive, or stop trying to obtain missing data from primary study authors, or stop checking coder reliability and drift? To the experienced practitioner, the definitive research synthesis is perhaps even more of a fiction than the definitive primary study. The goal is research syntheses that answer important questions about the generalizability of an association while making explicit the limitations of the findings claimed, raising new and important questions about the boundaries of our understanding, while also setting the agenda for the next generation of research.

In the first edition of this handbook, we called for the development of a viable new theory of generalization that could guide the design of research syntheses. We believe that the principles proposed by Thomas Cook and further elaborated by William Shadish and his colleagues can serve as a starting point for such a theory (Cook 1990, 1993; Shadish et al. 2002). These principles provide practical guidelines for exploring and building arguments around a generalizability claim, for which meta-analyses can then provide empirical warrants. They are not, though, principles firmly ensconced in statistical theory, as is the case with probability sampling. But probability sampling has a limited reach across the persons, settings, times, and instances of the cause and effect that are necessarily involved in generalizing from sampling details to causal claims. For all its undisputed theoretical elegance, probability sampling is most relevant to generalizing to populations of units and, sometimes, to settings. Its practical relevance to the other entities involved in causal generalization is much less clear. Although significant progress has been made since modern research synthesis methods were introduced more than thirty years ago, significant additional progress is needed to achieve a more complete understanding of how research syntheses can achieve even better causal generalization. The necessary condition for this is the existence of one or more comprehensive and internally cogent theories of causal generalization. But this we do not yet have in any form that leads to novel practical actions when conducting meta-analyses.

28.6 REFERENCES

Becker, Betsy J. 1992. "Models of Science Achievement: Forces Affecting Male and Female Performance in School Science." In *Meta-Analysis for Explanation: A Casebook*, edited by Thomas D. Cook, Harris Cooper, David S. Cordray, Heidi Hartmann, Larry V. Hedges, Richard J. Light, Thomas A. Louis, and Frederick Mosteller. New York: Russell Sage Foundation.

———. 2001. "Examining Theoretical Models Through Research Synthesis: The Benefits of Model-Driven Meta-Analysis." *Evaluation and the Health Professions.* 24(2): 190–217.

Boruch, Robert F. 2005. "Comments on 'Can the Federal Government Improve Education Research?' by Brian Jacob and Jacob Ludwig." *Brookings Papers on Education Policy* 2005(1): 67–280.

Campbell, Donald T. 1957. "Factors Relevant to the Validity of Experiments in Social Settings." *Psychological Bulletin* 54(4): 297–312.

———. 1986. "Relabeling Internal and External Validity for Applied Social Scientist." In *Advances in Quasi-Experimental Design and Analysis*, edited by William M.K. Trochim. San Francisco: Jossey-Bass.

Campbell, Donald T., and Robert F. Boruch. 1975. "Making the Case for Randomized Assignment to Treatments by Considering the Alternatives: Six Ways in which Quasi-Experimental Evaluations Tend to Underestimate Effects." In *Evaluation and Experiment: Some Critical Issues in Assessing Social Programs*, edited by C. A. Bennett and A. A. Lumsdaine. New York: Academic Press.

Campbell, Donald T., and Julian C. Stanley. 1963. *Experimental and Quasi-Experimental Designs for Research.* Chicago: Rand McNally.

Chalmers, Thomas C., P. Celano, Henry S. Sacks, and H. Smith Jr. 1983. "Bias in Treatment Assignment in Controlled Clinical Trials." *New England Journal of Medicine* 309(22): 1358–61.

Chalmers, Thomas C., Peg Hewett, Dinah Reitman, and Henry S. Sacks. 1989. "Selection and Evaluation of Empirical Research in Technology Assessment." *International Journal of Technology Assessment in Health Care* 5(4): 521–36.

Chalmers, Thomas C., and Joseph Lau. 1992. "Meta-Analysis of Randomized Control Trials Applied to Cancer Therapy." In *Important Advances in Oncology, 1992*, edited by Vincent T. Devita, Samuel Hellman and Steven A. Rosenberg. Philadephia, Pa.: Lippincott.

Cohn, Lawrence D., and Betsy J. Becker. 2003. "How Meta-Analysis Increases Statistical Power." *Psychological Methods* 8(3): 243–53.

Collingwood, Robin G. 1940. *An Essay on Metaphysics*. Oxford: Clarendon Press.

Cook, Thomas D. 1990. "The Generalization of Causal Connections." In *Research Methodology: Strengthening Causal Interpretations of Nonexperimental Data*, edited by Lee Sechrest, E. Perrin and J. Bunker. Washington, D.C.: U.S. Department of Health and Human Services.

———. 1993. "Understanding Causes and Generalizing About Them." In *New Directions in Program Evaluation*, edited by Lee B. Sechrest and A. G. Scott. San Francisco: Jossey Bass.

Cook, Thomas D., and Donald T. Campbell. 1979. *Quasi-Experimentation: Design and Analysis Issues for Field Settings.* Chicago: Rand McNally.

Cook, Thomas D., Harris Cooper, David S. Cordray, Heidi Hartmann, Larry V. Hedges, Richard J. Light, Thomas A. Louis, and Frederick Mosteller, eds. 1992. *Meta-Analysis for Explanation: A Casebook.* New York: Russell Sage Foundation.

Cronbach, Lee J. 1980. *Toward Reform of Program Evaluation*. San Francisco: Jossey-Bass.

———. 1982. *Designing Evaluations of Educational and Social Programs*. San Francisco: Jossey-Bass.

Devine, Elizabeth C. 1992. "Effects of Psychoeducational Care with Adult Surgical Patients: A Theory Probing Meta-Analysis of Intervention Studies." In *Meta-Analysis for Explanation: A Casebook*, edited by Thomas D. Cook, Harris Cooper, David S. Cordray, Heidi Hartmann, Larry V. Hedges, Richard J. Light, Thomas A. Louis, and Frederick Mosteller. New York: Russell Sage Foundation.

Donner, Allan, Gilda Piaggio, and José Villar. 2001. "Statistical Methods for the Meta-Analysis of Cluster Randomization Trials." *Statistical Methods in Medical Research* 10(5): 325–38.

———. 2003. "Meta-Analyses of Cluster Randomization Trials: Power Considerations." *Evaluation and the Health Professions* 26(3): 340–51.

Duval, Sue, and Richard Tweedie. 2000. "Trim and Fill: A Simple Funnel-Plot-Based Method of Testing and Adjusting for Publication Bias in Meta-Analysis." *Biometrics* 56(2): 455–63.

Eddy, David M., Vic Hasselblad, and Ross Shachter. 1990. "An Introduction to a Bayesian Method for Meta-Analysis: The Confidence Profile Method." *Medical Decision Making* 10(1): 15–23.

Egger, Matthias, George Davey Smith, Martin Schneider, and Christoph Minder. 1997. "Bias in Meta-Analysis Detected by a Simple, Graphical Test." *British Medical Journal* 315(7109): 629–34.

Furlow, Carolyn F., and S. Natasha Beretvas. 2005. "Meta-Analytic Methods of Pooling Correlation Matrices for Structural Equation Modeling Under Different Patterns of Missing Data." *Psychological Methods* 10(2): 227–54.

Gasking, Douglas. 1955. "Causation and Recipes." *Mind* 64: 479–87.

Glass, Gene V. 1976. "Primary, Secondary, and Meta-Analysis of Research." *Educational Researcher* 5(10): 3–8.

———. 1977. "Integrating Findings: The Meta-Analysis of Research." *Review of Research in Education* 5(1): 351–79.

Greenwald, Anthony G. 1975. "Consequences of Prejudice Against the Null Hypothesis." *Psychological Bulletin* 82(1): 1–20.

Hardy, Rebecca J., and Simon G. Thompson. 1998. "Detecting and Describing Heterogeneity in Meta-Analysis." *Statistics in Medicine* 17(8): 841–56.

Harris, Monica J., and Robert Rosenthal. 1985. "Mediation of Interpersonal Expectancy Effects: 31 Meta-Analyses." *Psychological Bulletin* 27(3): 363–86.

Harwell, Michael. 1997. "An Empirical Study of Hedge's Homogeneity Test." *Psychological Methods* 2(2): 219–31.

Hedges, Larry V. 1982. "Estimation of Effect Sizes from a Series of Independent Experiments." *Psychological Bulletin* 92(2): 490–99.

———. 1990. "Directions for Future Methodology." In *The Future of Meta-Analysis*, edited by Kenneth W. Wachter and Miron L. Straf. New York: Russell Sage Foundation.

Hedges, Larry V., and Betsy Jane Becker. 1986. "Statistical Methods in the Meta-Analysis of Research on Gender Differences." In *The Psychology of Gender: Advances Through Meta-Analysis*, edited by Janet Shibley Hyde and Marcia C. Linn. Baltimore, Md.: Johns Hopkins University Press.

Hedges, Larry V., and Ingram Olkin. 1985. *Statistical Methods for Meta-Analysis*. Orlando, Fl.: Academic Press.

Hedges, Larry V., and Therese D. Pigott. 2001. "The Power of Statistical Tests in Meta-Analysis." *Psychological Methods* 6(3): 203–17.

———. 2004. "The Power of Statistical Tests for Moderators in Meta-Analysis." *Psychol Methods* 9(4): 426–45.

Hedges, Larry V., and Jack L. Vevea. 1996. "Estimating Effect Size Under Publication Bias: Small Sample Properties and Robustness of a Random Effects Selection Model." *Journal of Educational and Behavioral Statistics* 21(4): 299–332.

Hox, Joop J., and Edith D. de Leeuw. 2003. "Multilevel Models for Meta-Analysis." In *Multilevel Modeling: Methodological Advances, Issues, and Applications*, edited by Steven P. Reise and Naihua Duan. Mahwah, N.J.: Lawrence Erlbaum.

Hunter, John E., and Frank L. Schmidt. 1990. *Methods of Meta-Analysis: Correcting Error and Bias in Research Findings*. Newbury Park, Calif.: Sage Publications.

Jackson, Dan. 2006. "The Power of the Standard Test for the Presence of Heterogeneity in Meta-Analysis." *Statistics in Medicine* 25(15): 2688–99.

Jacob, Brian, and Jens Ludwig. 2005. "Can the Federal Government Improve Education Research?" *Brookings Papers on Education Policy* 2005(1): 47–87.

Kelley, George A., Kristi S. Kelley, and Zung Vu Tran. 2004. "Retrieval of Missing Data for Meta-Analysis: A Practical Example." *International Journal of Technology Assessment in Health Care* 20(3): 296–99.

Lipsey, Mark W. 1992. "Juvenile Delinquency Treatment: A Meta-Analytic Inquiry into the Variability of Effects." In *Meta-Analysis for Explanation: A Casebook*, edited by Thomas D. Cook, Harris Cooper, David S. Cordray, Heidi Hartmann, Larry V. Hedges, Richard J. Light, Thomas A. Louis, and Frederick Mosteller. New York: Russell Sage Foundation.

———. 2003. "Those Confounded Moderators in Meta-Analysis: Good, Bad, and Ugly." *Annals of the American Academy of Political and Social Science* 587(May, Special Issue): 69–81.

Lipsey, Mark W., and David B. Wilson. 2001. *Practical Meta-Analysis*. Applied Social Research Methods Series, vol. 49. Thousand Oaks, Calif.: Sage Publications.

Little, Roderick J. A., and Donald B. Rubin. 2002. *Statistical Analysis with Missing Data*, 2nd ed. New York: John Wiley & Sons.

Mackie, John L. 1974. *The Cement of the Universe: A Study of Causation*. Oxford: Oxford University Press.

Matt, Georg E. 1989. "Decision Rules for Selecting Effect Sizes in Meta-Analysis: A Review and Reanalysis of Psychotherapy Outcome Studies." *Psychological Bulletin* 105(1): 106–15.

———. 2003. "Will it Work in Münster? Meta-Analysis and the Empirical Generalization of Causal Relationships." In *Meta-analysis*, edited by H. Holling, D. Böhning and R. Schulze. Göttingen: Hogrefe and Huber.

———. 2005. "Uncertainty in the Multivariate World: A Fuzzy Look through Brunswik's Lens." In *Multivariate*

Research Strategies. A Festschrift in Honor of Werner W. Wittmann, edited by Andre Beauducel, Bernhard Biehl, Michael Bosniak, Wolfgang Conrad, Gisela Schönberger and Deitrich Wagener. Maastricht: Shaker Publishing.

Matt, Georg E., and Thomas D. Cook. 1994. "Threats to the Validity of Research Syntheses." In *The Handbook of Research Synthesis.*, edited by Harris Cooper and Larry V. Hedges. New York: Russell Sage Foundation.

Mick, Eric, Joseph Biederman, Gahan Pandina, and Stephen V. Faraone. 2003. "A Preliminary Meta-Analysis of the Child Behavior Checklist in Pediatric Bipolar Disorder." *Biological Psychiatry* 53(11): 1021–27.

Moyer, Anne, John W. Finney, Carolyn E. Swearingen, and Pamela Vergun. 2002. "Brief Interventions for Alcohol Problems: A Meta-Analytic Review of Controlled Investigations in Treatment-Seeking and NonTreatment-Seeking Populations." *Addiction* 97(3): 279–92.

Nam, In-Sum, Kerrie Mengersen, and Paul Garthwaite. 2003. "Multivariate Meta-Analysis." *Statistics in Medicine* 22(14): 2309–33.

Orwin, Robert G., and David S. Cordray. 1985. "The Effects of Deficient Reporting on Meta-Analysis: A Conceptual Framework and Reanalysis." *Psychological Bulletin* 97(1): 134–47.

Premack, Steven L., and John E. Hunter. 1988. "Individual Unionization Decisions." *Psychological Bulletin* 103(2): 223–34.

Prevost, Teresa C., Keith R. Abrams, and David R. Jones. 2000. "Hierarchical Models in Generalized Synthesis of Evidence: An Example Based on Studies of Breast Cancer Screening." *Statistics in Medicine.* 19(24): 3359–376.

Raudenbush, Stephen W., Betsy J. Becker, and Hripsime Kalaian. 1988. "Modeling Multivariate Effect Sizes." *Psychological Bulletin* 103(1): 111–20.

Raudenbush, Stephen W., and Anthony S. Bryk. 1985. "Empirical Bayes Meta-Analysis." *Journal of Educational Statistics* 10(2): 75–98.

Robinson, Leslie A., Jeffrey S. Berman, and Robert A. Neimeyer. 1990. "Psychotherapy for the Treatment of Depression: A Comprehensive Review of Controlled Outcome Research." *Psychological Bulletin* 108(1): 30–49.

Rosch, Eleanor. 1973. "Natural Categories." *Cognitive Psychology* 4: 328–50.

———. 1978. "Principles in Categorization." In *Cognition and Categorization*, edited by Eleanor Rosch and Barbara B. Lloyd. Hillsdale, N.J.: Lawrence Erlbaum.

Rosenthal, Robert. 1979. "The File Drawer Problem and Tolerance for Null Results." *Psychological Bulletin* 86(3): 638–41.

———. 1990. "An Evaluation of Procedure and Results." In *The Future of Meta-Analysis*, edited by Kenneth W. Wachter and Miron L. Straf. New York: Russell Sage Foundation.

———. 1991. *Meta-Analytic Procedures for Social Research*, 2nd ed. Newbury Park, Calif.: Sage Publications.

Rosenthal, Robert, and Donald B. Rubin. 1982. "Comparing Effect Sizes of Independent Studies." *Psychological Bulletin* 92(2): 500–4.

———. 1986. "Meta-Analytic Procedures for Combining Studies with Multiple Effect Sizes." *Psychological Bulletin* 99(3): 400–6.

Rothstein, Hannah R., Alexander J. Sutton, and Michael Borenstein, eds. 2005. *Publication Bias in Meta-Analysis: Prevention, Assessment and Adjustments.* New York: John Wiley & Sons.

Rubin, Donald B. 1987. *Multiple Imputation for Nonresponse in Surveys.* New York: John Wiley & Sons.

Sacks, H. S., J. Berrier, D. Reitman, V. A. Ancona-Berk, and T. C. Chalmers. 1987. "Meta-Analyses of Randomized Controlled Trials." *The New England Journal of Medicine* 316(8): 450–55.

Saleh, A. K. Ehsanes, Khatab M. Hassanein, Ruth S. Hassanein, and H. M. Kim. 2006. "Quasi-Empirical Bayes Methodology for Improving Meta-Analysis." *Journal of Biopharmaceutical Statistics* 16(1): 77–90.

Sánchez-Meca, Julio, Fulgencio Marín-Martínez, and Salvador Chacon-Moscoso. 2003. "Effect-Size Indices for Dichotomized Outcomes in Meta-Analysis." *Psychological Methods* 8(4): 448–67.

Schulze, Ralf. 2004. *Meta-Analysis: A Comparison of Approaches.* New York: Hogrefe and Huber.

Scott, J. Cobb, Steven Paul Woods, Georg E. Matt, Rachel A. Meyer, Robert K. Heaton, J. Hampton Atkinson, and Igor Grant. 2007. "Neurocognitive Effects of Methamphetamine: A Critical Review and Meta-Analysis." *Neuropsychology Review* 17(3): 275–97.

Shadish, William R. 1992. "Do Family and Marital Therapies Change What People Do? A Meta-Analysis of Behavioral Outcomes." In *Meta-Analysis for Explanation: A Casebook*, edited by Thomas D. Cook, Harris Cooper, David S. Cordray, Heidi Hartmann, Larry V. Hedges, Richard J. Light, Thomas A. Louis, and Frederick Mosteller. New York: Russell Sage Foundation.

Shadish, William R., Thomas D. Cook, and Donald T. Campbell. 2002. *Experimental and Quasi-Experimental Design for Generalized Causal Inference.* Boston, Mass.: Houghton Mifflin.

Shadish, William R., Xiangen Hu, Renita R. Glaser, Richard Kownacki, and Seok Wong. 1998. "A Method for Exploring the Effects of Attrition in Randomized Experiments with Dichotomous Outcomes." *Psychological Methods* 3(1): 3–22.

Shadish, William R., Jason K. Luellen, and M. H. Clark. 2006. "Propensity Scores and Quasi-Experiments: A Testimony to the Practical Side of Lee Sechrest." In *Strengthening Research Methodology: Psychological Measurement and Evaluation*, edited by Richard Bootzin and Patrick E. McKnight. Washington, D.C.: American Psychological Association.

Shadish, William R., Georg E. Matt, Ana M. Navarro, and Glenn Phillips. 2000. "The Effects of Psychological Therapies Under Clinically Representative Conditions: A Meta-Analysis." *Psychological Bulletin* 126(4): 512–29.

Shadish, William R., and Rebecca B. Sweeney. 1991. "Mediators and Moderators in Meta-Analysis: There's a Reason We Don't Let Dodo Birds Tell Us Which Psychotherapies Should Have Prizes." *Journal of Consulting Clinical Psychology* 59(6): 883–93.

Shapiro, David A., and Diana Shapiro. 1982. "Meta-Analysis of Comparative Therapy Outcome Studies: A Replication and Refinement." *Psychological Bulletin* 92(3): 581–604.

Sinharay, Sandip, Hal S. Stern, and Daniel Russell. 2001. "The Use of Multiple Imputation for the Analysis of Missing Data." *Psychological Methods* 6(4): 317–29.

Slavin, Robert E. 1986. "Best-Evidence Synthesis: An Alternative to Meta-Analytic and Traditional Reviews." *Educational Researcher* 15(9): 5–11.

Smith, Edward E., and Douglas L. Medin. 1981. *Categories and Concepts*. Cambridge, Mass.: Harvard University Press.

Smith, Mary Lee, Gene V. Glass, and Thomas I. Miller. 1980. *The Benefits of Psychotherapy*. Baltimore, Md.: Johns Hopkins University Press.

Sterne, Jonathan A. C., and Matthias Egger. 2001. "Funnel Plots for Detecting Bias in Meta-Analysis: Guidelines on Choice of Axis." *Journal of Clinical Epidemiology* 54(10): 1046–55.

Sutton, Alexander J. 2000. *Methods for Meta-Analysis in Medical Research*. Wiley Series in Probability and Mathematical Statistics. New York: John Wiley & Sons.

Sutton, Alexander J., and Keith R. Abrams. 2001. "Bayesian Methods in Meta-Analysis and Evidence Synthesis." *Statistical Methods in Medical Research* 10(4): 277–303.

Sutton, Alexander J., Susan J. Duval, Richard L. Tweedie, Keith R. Abrams, and David R. Jones. 2000. "Empirical Assessment of Effect of Publication Bias on Meta-Analyses." *British Medical Journal* 320(7249): 1574–77.

Sutton, Alexander J., Fujian Song, Simon M. Gilbody, and Keith R. Abrams. 2000. "Modelling Publication Bias in Meta-Analysis: A Review." *Statistical Methods in Medical Research* 9(5): 421–45.

Tabarrok, Alexander T. 2000. "Assessing the FDA via the Anomaly of Off-Label Drug Prescribing. *The Independent Review* V(1): 25–53.

Twenge, Jean M., and W. Keith Campbell. 2001. "Age and Birth Cohort Differences in Self-Esteem: A Cross-Temporal Meta-Analysis." *Personality and Social Psychology Review* 5(4): 321–44.

van Houwelingen, Hans C., Lidia R. Arends, and Theo Stijnen. 2002. "Advanced Methods in Meta-Analysis: Multivariate Approach and Meta-Regression." *Statistics in Medicine* 21(4): 589–624.

Vevea, Jack L., and Larry V. Hedges. 1995. "A General Linear Model for Estimating Effect Size in the Presence of Publication Bias." *Psychometrika* 60(3): 419–35.

Vevea, Jack L., and Carol M. Woods. 2005. "Publication Bias in Research Synthesis: Sensitivity Analysis Using a Priori Weight Functions." *Psychological Methods* 10(4): 428–43.

Vevea, Jack L., Nancy C. Clements, and Larry V. Hedges. 1993. "Assessing the Effects of Selection Bias on Validity Data for the General Aptitude Test Battery." *Journal of Applied Psychology* 78(6): 981–87.

Viechtbauer, Wolfgang. 2007. "Confidence Intervals for the Amount of Heterogeneity in Meta-Analysis." *Statistics in Medicine* 26(1): 37–52.

Webb, Eugene J., Donald T. Campbell, Richard D. Schwartz, Lee Sechrest, and Janet B. Grove. 1981. *Nonreactive Measures in the Social Sciences*, 2nd ed. Chicago: Rand McNally.

West, Stephen G., Jeremy C. Biesanz, and Steven C. Pitts. 2000. "Causal Inference and Generalization in Field Settings : Experimental and Quasi-Experimental Designs." In *Handbook of Research Methods in Social and Personality Psychology*, edited by H.T. Reid and C.M. Judd. New York: Cambridge University Press.

Whitbeck, Caroline. 1977. "Causation in Medicine: The Disease Entity Model." *Philosophy of Science* 44(4): 619–37.

Wittmann, Werner W., and Georg E. Matt. 1986. "Meta-Analyse als Integration von Forschungsergebnissen am Beispiel deutschsprachiger Arbeiten zur Effektivität von Psychotherapie." *Psychologische Rundschau* 37(1): 20–40.

Wolf, Frederic M. 1990. "Methodological Observations on Bias." In *The Future of Meta-Analysis*, edited by Kenneth W. Wachter and Miron L. Straf. New York: Russell Sage Foundation.

29

POTENTIALS AND LIMITATIONS

HARRIS COOPER
Duke University

LARRY V. HEDGES
Northwestern University

CONTENTS

29.1 INTRODUCTION

We suspect that readers of this volume have reacted to it in one of two ways: they have been overwhelmed by the number and complexity of the issues that face a research synthesist or they have been delighted to have a manual to help them through the synthesis process. We further suspect that which reaction it was depended on whether the book was read while thinking about or while actually performing a synthesis. In the abstract, the concerns raised in this handbook must seem daunting. Concretely, however, research syntheses have been carried out for decades and will be for decades to come, and the problems encountered in their conduct do not really go away when they are ignored. Knowledge builders need a construction manual to accompany their blueprints and bricks.

Further, there is no reason to believe that carrying out a sound research synthesis is any more complex than carrying out sound primary research. The analogies between synthesis and survey research, content analysis, or even quasi-experimentation fit too well for this not to be the case. Certainly, each type of research has its unique characteristics. Still, when research synthesis procedures have become as familiar as primary research procedures much of what now seems overwhelming will come to be viewed as difficult, but manageable.

Likewise, expectations will change. As we mentioned in chapter 1, we do not expect a primary study to provide unequivocal answers to research questions, nor do we expect it to be performed without flaw. Research syntheses are the same. The perfect synthesis does not and never will exist. Synthesists who set standards that require them to implement all of the most rigorous techniques described in this book are bound to be disappointed. Time and other resources (not to mention the logic of discovery) will prevent such an accomplishment. The synthesist, like the primary researcher, must balance methodological rigor against practical feasibility.

29.2 UNIQUE CONTRIBUTIONS OF RESEARCH SYNTHESIS

29.2.1 Increased Precision and Reliability

Some comparisons were made between primary research and research synthesis. It was suggested that the two forms of study have much in common. In contrast, when we compare research synthesis as it was practiced before meta-analytic techniques were popularized the dissimi-

larities are what capture attention. Perhaps most striking are the improvements in precision and reliability that new techniques have made possible.

Precision in literature searches has been improved dramatically by the procedures described here. Yesterday's synthesist attempted to uncover a literature by examining a small journal network and making a few telephone calls. This strategy is no longer viable. The proliferation of journals and e-journals and the increase in the sheer quantity of research renders the old strategy inadequate. Today, the computer revolution, the development and maintenance of comprehensive research registries, and the emergence of the Internet mean that a synthesist can reach into thousands of journals and individual researcher's data files. The present process is far from perfect, of course, but it is an enormous improvement over past practice.

The coding of studies and their evaluation for quality have also come a long way. The old process was rife with potential bias. In 1976, Gene Glass wrote this:

> A common method for integrating several studies with inconsistent findings is to carp on the design or analysis deficiencies of all but a few studies—those remaining frequently being one's own work or that of one's students and friends—and then advancing the one or two "acceptable" studies as the truth of the matter. (4)

Today, research synthesists code their studies with considerably more care to avoid bias. Explicit, well-defined coding frames are used by multiple coders. Judgments about quality of research are sometimes eschewed entirely, allowing the data to determine if research design influences study outcomes.

Finally, the process of integrating and contrasting primary study results has taken a quantum leap forward. Gone are the days when a vote count of significant results decided an issue. It is no longer acceptable to string together paragraph descriptions of studies, with details of significance tests, and then conclude that "the data are inconsistent, but seem to indicate." Today's synthesist can provide confidence intervals around effect size estimates for separate parts of a literature distinguished by both methodological and theoretical criteria calculated several different ways using different assumptions about the adequacy of the literature search and coding scheme, and alternative statistical models.

These are only a few of the ways that the precision and reliability of current syntheses outstrip those of their predecessors. The increases in precision and reliability that a synthesist can achieve when a body of literature grows from two to twenty to two hundred studies were squandered with yesterday's methods.

29.2.2 Testing Generalization and Moderating Techniques

Current methods not only capitalize on accumulating evidence by making estimates more precise, but also permit the testing of hypotheses that may have never been tested in primary studies. This advantage is most evident when the objective of the synthesist is to establish the generality or specificity of results. The ability of a synthesist to test whether a finding applies across settings, times, people, and researchers surpasses by far the ability of most primary studies. However, for yesterday's synthesist, the variation among studies was a nuisance. It clouded interpretation. The old methods of synthesis were handicapped severely when it came to judgments of whether a set of studies revealed consistent results and, if not, what might account for the inconsistency. Yesterday's synthesist often chose the wrong datum (significance tests rather than effect sizes) and the wrong evaluation criteria (implicit cognitive algebra rather than formal statistical test) on which to base these judgments. For today's synthesist, variety is the spice of life. The methods described in this handbook make such analyses routine. When the outcomes of studies prove too discrepant to support a claim that one estimate of a relationship underlies them all, current techniques provide the synthesist with a way of systematically searching for moderating influences, using consistent and explicit rules of evidence.

29.2.3 Asking Questions About the Scientific Enterprise

Finally, the practice of research synthesis permits scientists to study their own enterprise. Questions can be asked about trends in research over time, both trends in results and trends in how research is carried out. Kenneth Ottenbacher and Harris Cooper, for example, examined the effects of special versus regular class placement on the social adjustment of children with mental handicaps (1984). They found that studies of social adjustment conducted after 1968, when mainstreaming became the law, favored regular class placement, whereas studies before 1968 favored special classes.

Synthesists often uncover results that raise interesting questions about the relationship between research and the general societal conditions in which it takes place. By facilitating this form of self-analysis, current synthesis methods can be used to advance not only particular problem areas but also entire disciplines, by addressing more general questions in the psychology and sociology of science.

29.3 LIMITATIONS OF RESEARCH SYNTHESIS

29.3.1 Correlational Nature of Synthesis Evidence

Research syntheses have limitations as well. Among the most critical is the correlational nature of synthesis-generated evidence. As detailed in chapter 2, study-generated evidence appears when individual studies contain results that directly test the relation under consideration. Synthesis-generated evidence appears when the results of studies using different procedures to test the same hypothesis are compared. As Harris Cooper notes in chapter 2, only study-generated evidence can be used to support claims about causal relations and only when experimental designs are used in them. Specific confounds can be controlled statistically at the level of synthesis-generated evidence, but the result can never lead to the same confidence in inferences produced by study-generated evidence from investigations employing random assignment. This is an inherent limitation of research syntheses. It also provides the synthesist with a fruitful source of suggestions concerning directions for future primary research.

29.3.2 Post Hoc Nature of Synthesis Tests

Researchers interested in conducting syntheses almost always set upon the process with some, if not considerable, knowledge of the empirical data base they are about to summarize. They begin with a good notion of what the literature says about which interventions work, which moderators operate, and which explanatory variables are helpful. This being so, the synthesist cannot state a hypothesis so derived and then also conclude that the same evidence supports the hypothesis. As Kenneth Wachter and Miron Straf pointed out, once data have been used to develop a theory they cannot be used to test it (1990). Synthesists who wish to test specific, a priori hypotheses in meta-analysis must go to extra lengths to convince their audience that the generation of the hypothesis and the data used to test it are truly independent.

29.3.3 Need for New Primary Research

Each of these limitations serves to underscore an obvious point: a research synthesis should never be considered a replacement for new primary research. Primary research and research synthesis are complementary parts of a common practice; they are not competing alternatives. The additional precision and reliability brought to syntheses by the procedures described in this text do not make research integrations substitutes for new data-gathering efforts. Instead, they help ensure that the next wave of primary research is sent off in the most illuminating direction. A synthesis that concludes a problem is solved, that no future research is needed, should be viewed with extreme skepticism. Even the best architects can see how the structures they have built can be improved.

29.4 DEVELOPMENTS IN RESEARCH SYNTHESIS METHODOLOGY

29.4.1 Improving Research Synthesis Databases

Two problems that confront research synthesists recurred throughout this text. They concern the comprehensiveness of literature searches, especially as searches are influenced by publication bias, and missing data in research reports. Each of these issues was treated in its own chapter, but many authors felt the need to consider their impact in the context of other tasks the synthesist carried out. The correspondence between the studies that can be accessed by the synthesist and the target population of studies is an issue confronted during data collection, analysis, and interpretation. Missing data is an issue in data coding, analysis, and interpretation.

There is little need to reiterate the frustration synthesists feel when they contemplate the possible biases in their conclusions caused by studies and results they cannot retrieve. Instead, we note that the first edition of this book made a call for improvements in research synthesis procedures related to ways to restructure the scientific information delivery systems so that more of the needed information becomes available to the research synthesist. This has occurred, as has improvements in techniques for estimating and addressing problems in missing data.

Still, the problem is far from solved. In the first edition of this handbook, we suggested that research synthesists should call for an upgrading of standards for reporting of primary research results in professional journals. We also suggested that journals need to make space for authors to report the means and standard deviations for all conditions, for the reliability of measures (and for full texts of measures when they are not previously published), and for range restrictions in samples. Further, as primary statistical analyses become more complex, the requirements of adequate reporting will become more burdensome. Although means and standard deviations are adequate for simple analyses, full correlation matrices (or even separate correlation matrices for each of several groups studied) may also be required for adequate reporting of multivariate analyses. Reporting this detailed information in journals, however, is less feasible because it would require even more journal space. Yet primary analyses are not fully interpretable (and synthesis is certainly not possible) without them. Therefore, meta-analysts need some system whereby authors submit extended results sections along with their manuscripts and these sections are made available on request.

Two recent developments have greatly improved the availability of comprehensive descriptions of primary research, both of which go considerably beyond our proposals. The first involves the establishment of standards for reporting of studies in scientific journal articles. For example, spurred on by a concern regarding the adequacy of reporting of randomized clinical trials in medicine and health care, the Consolidated Standards for Reporting Trials, or CONSORT, statement was developed by a group of international clinical trialists, statisticians, epidemiologists, and biomedical editors. It has since been adopted as a reporting standard for numerous journals in medicine and public health and by a growing number of journals in the social sciences. CONSORT consists of a checklist and flow diagram for reporting studies that employ random assignment of participants to two treatment groups. Data collected by the CONSORT Group does indicate that adoption of the standards improves the reporting of study design and implementation (Moher, Schulz, and Altman 2001).

The CONSORT statement was shortly followed by an effort to improve the reporting of quasi-experiments. Brought together by the Centers for Disease Control, a group of editors of journals that published HIV behavioral intervention studies developed and proposed the use of a checklist they named TREND (Transparent Reporting of Evaluations with Nonrandomized Designs) (Des Jarlais, Lyons, and Crepaz 2004). The TREND checklist consists of twenty-two items. Its impact on reporting practices is yet to be ascertained.

The American Educational Research Association (AERA) has recently provided its own set of standards for reporting on empirical social science research (AERA 2006). These standards are meant to encompass the broadest range of research approaches, from experiments using random assignment to qualitative investigations. The finally, the American Psychological Association offered a set of reporting recommendations for psychology (APA Publications and Communications Board Working Group on Journal Article Reporting Standards 2008).

The second development with enormous potential for resolving the problem of missing data is the establishment of auxiliary web sites by journals on which authors archive information about their research that could not be included in journal reports because of space limitations. For example, the American Psychological Association provides its journals with websites for this purpose. The sites are referred to in the published articles and access is free to anyone who reads the journal.

Journal reporting standards and auxiliary data storage, however, are only measures for addressing the problem of missing data. They do not address the issue of publication bias. A partial solution to publication bias is the continued development of research registries. Because research registers attempt to catalog investigations both when they are initiated and when they are completed, they are a unique opportunity for overcoming publication bias. In addition, they may be able to provide meta-analyses with more thorough reports of study results.

Of course, registers are not a panacea. Registers are helpful only if the researchers who register their studies keep and make available the results of their studies to synthesists who request them. Thus we renew our call in the earlier edition for professional standards that promote data sharing of even unpublished research findings.

Still, forward-looking researchers, publishers, and funding agencies would be well advised to look toward adopting reporting standards and creating auxiliary websites and research registers as some of the most effective ways to foster the accumulation of knowledge. Given the pervasiveness of the publication bias and missing data problems, these efforts seem to hold considerable promise for helping research synthesists find the missing science.

29.4.2 Improving Analytic Procedures

Many aspects of analytic procedures in research synthesis have undergone rapid development in the last two decades. However, there are still many areas in which the development of new statistical methods or the adaptation of methods used elsewhere are needed. Perhaps as important as the addition of new methods is the wise use of the considerable repertoire of analytic methods already available to the research synthesist. We mention here a few analytic issues that arise repeatedly throughout this handbook. The first two relate to the database problems just mentioned.

29.4.2.1 Missing Data Perhaps the most pervasive practical problem in research synthesis is that of missing data, which obviously influences any statistical analyses. Even though missing data as a general problem was addressed directly in chapter 21, and the special case of missing data attributable to publication bias in chapter 23, the problems created by missing data arose implicitly throughout this handbook. For example, the prevalence of missing data on moderator and mediating variables influences the degree to which the problems investigated by a synthesis can be formulated. Missing data influences the data collection stage of a research synthesis by affecting the potential representativeness of studies in which complete data are available. It also affects coding certainty and reliability in studies with partially complete information. Missing data provides an especially pernicious problem as research syntheses turn to increasingly sophisticated analytic methods to handle dependencies and multivariate causal models (see chapters 19 and 20, this volume).

We believe that one of the most important emerging areas in research synthesis is the development of analytic methods to better handle missing data. Much of this work will likely be in the form of adapting methods developed in other areas of statistics to the special requirements of research synthesis. These methods will produce more accurate analyses when data are not missing completely at random (that is, are not a random sample of all data), but are well enough related to observed study characteristics that they can be predicted reasonably well with a model based on data that are observed (that is, are what Little and Rubin would call missing at random). The methods for model-based inferences with missing data using the EM algorithm are discussed in section 21.5.1 and those based on multiple imputation in section 21.5.2 of this volume (see also Little and Rubin 2002; Rubin 1987).

29.4.2.2 Publication Bias Publication bias is a form of missing data less easy to address with analytic approaches. In the last decade progress in the development of analytic models for studying and correcting publication bias by analytic means has been substantial. As might be expected, publication bias is easier to detect than to correct (at least to correct in a way that inspires great confidence).

Also as expected, most methods for either detecting or correcting for publication bias are rather weak unless the number of studies is large. Moreover, because publication bias results in what Rod Little and Donald Rubin would call nonignorable missing data, methods to correct for publication bias necessarily depend on the form of the model for the publication selection process that produces the missing data (2002). Even the most sophisticated and least restrictive of these methods (the nonparametric weight function methods, for example) make the assumption that selection depends only on the p-value (for a discussion of these methodologies, see chapter 23, this volume).

These emerging methods are promising, but even their most enthusiastic advocates suggest that it is better not to have to correct for publication bias in the first place. Thus the best remedy for publication bias is the development of invariant sampling frames that would arise through research registers. This nonanalytic development in methodology could be a better substitute for improved statistical corrections in areas in which it is feasible.

29.4.2.3 Fixed- Versus Random-Effects Models

The contrast between fixed- and random-effects models has been drawn repeatedly throughout this handbook—directly in chapters 14, 15, and 16; indirectly in chapters 19, 20, 21, and 23; and implicitly in chapters 17 and 27. The issues in choosing between fixed- and random-effects models are subtle, and sound choices often involve considerable subjective judgment. In all but the most extreme cases, reasonable and intelligent people might disagree about their choices of modeling strategy. However, there is much agreement about criteria that might be used in these judgments.

The criterion offered in chapter 14 and reiterated in chapter 16 for the choice between fixed- and random-effects models is conceptual. Meta-analysts should choose random-effects models if they think of studies as different from one another in ways too complex to capture by a few single study characteristics. If the object of the analysis is to make inferences about a universe of such diverse studies, the random-effects approach is conceptually sensible. Synthesists should choose fixed-effects models if they believe that the important variation in effects is a consequence of a small number of study characteristics. In the fixed-effects analysis the object is to draw inferences about studies with similar characteristics.

A practical criterion for choosing between fixed- and random-effects models is offered in several chapters in

this handbook (14, 15, and 16)—namely, the criterion of empirical adequacy of the fixed effects model. In each of these chapters, the authors refer to test of homogeneity, model specification, of goodness of fit of the fixed-effects model. These tests essentially compare the observed (residual) variation of effect-size estimates with the amount that would be expected if the fixed-effect model fit the parameter values exactly. All of the authors argued that the use of fixed-effect models was more difficult to justify when tests of model specification showed substantial variation in effect sizes that was not explained by the fixed effects model as specified. Conversely, fixed effects models were easier to justify when tests of model specification showed no excess (residual) variation. As noted in section 14.3, this empirical criterion, though important, is secondary to the theoretical consideration of how best to conceive the inference.

Some authors prefer the use of fixed-effects models under almost any circumstance, regarding unexplained variation in effect-size parameters as seriously compromising the certainty of inferences drawn from a synthesis (see Peto 1987; section 15.1, this volume). Other authors advocate essentially a random-effects approach under all circumstances. The random-effects model underlies all the methods discussed in chapter 16 of this volume, and is consistent with the Bayesian methods also discussed. In both cases, the implicit argument was conceptual, namely, that studies constituted exchangeable units from some universe of similar studies. In both cases, the object was to discover the mean and variance of effect-size parameters associated with this universe.

One difference is that the Bayesian interpretation of random-effects models draws on the idea that between-studies variation can be conceived as uncertainty about the effect-size parameter (see section 16.2.2). This interpretation of random-effects models was also used in the treatment of synthetic regression models in chapter 20 (section 20.3.4.4), where the use of a random-effects model was explicitly referred to as adding a between-studies component of uncertainty to the analysis.

Yet another empirical criterion, offered in chapter 16, was based on the number of studies. Random-effects models are advocated when the number of studies is large enough to support such inferences. The reason for choosing random-effects models (when empirically feasible) was conceptual—the belief that study effect sizes would be different across studies and that those differences would be controlled by a large number of unpredictable influences.

Thus, though there are differences among the details of the criteria given, the positions of these authors reflect a consensus on criteria. Conceptual criteria would be applied first, with the model chosen according to the goals of the inference. Only after considering the objective of the inference would empirical criteria influence modeling strategy, perhaps by forcing a reformulation of the question to be addressed in the synthesis. The empirical finding of persistent heterogeneity among the effect sizes might force reconsideration of a fixed-effects analysis. Similarly, the ambition to use a random-effects approach to draw inferences about a broad universe of studies might need to be reconsidered if the collection of empirical research results was too small for meaningful inference, to be replaced by fixed-effects comparisons among particular studies.

A particularly interesting area of development is applying random-effects models to the problem estimating the results of populations of studies that may not be identical in composition to the study sample (see U.S. General Accounting Office 1992; Boruch and Terhanian 1998). For example, in medical research, the patients (or doctors or clinics) selected to participate in clinical trials are typically not representative of the general patient population. Although the studies as conducted may be perfectly adequate for assessing whether a new treatment can produce treatment effects that are different from zero, they may not give a representative sample for assessing the likely effect in the nation if the new treatment became the standard. Simply extrapolating from the results of the studies conducted would not be warranted if the treatment was less effective with older or sicker patients than were typically included in the trials. Efforts to use information about systemic variation in effect sizes to predict the impact of a new medical treatment (or any other innovation) if it were adopted are an exciting application of research synthesis.

29.4.2.4 Bayesian Methods There has been an increasing interest in Bayesian methods in research synthesis. One reason is the recognition that random-effects analyses depend on the (unknown) effect-size variance component, and that classical analyses are based on a single (estimated) value of that variance component. Yet predicating an analysis on any single estimated value inadequately conveys the uncertainty of the estimated variance component, which is often quite large. Bayesian analyses can incorporate that uncertainty by computing a weighted average over the likely values of the variance component with weight proportional to the likelihood of

that value, that is, the posterior probability (see DuMouchel 1990; Hedges 1998).

A second reason for the appeal of Bayesian methods is that research synthesis, like any scientific endeavor, has many subjective aspects. Bayesian statistical analyses, designed to formally incorporate subjective knowledge and its uncertainty, have become increasingly important in many areas of applied statistics. We expect research synthesis to be no exception. As quantitative research syntheses are more widely used, we expect to see a concomitant development of Bayesian methods suited to particular applications. One advantage to Bayesian methods is that they can be used to combine hard data on some aspects of a problem with subjective assessments concerning other aspects of the problem about which there are no hard data. One such situation arises when the inference that is desired depends on several links in a causal chain, but not all of the links have been subject to quantitative empirical research (see Eddy, Hasselblad, and Schacter 1992).

29.4.2.5 Causal Modeling in Research Synthesis Structural equation modeling is destined to become more important in research synthesis for two reasons. First, the use of multivariate causal models in primary research will necessitate their greater use in research synthesis. Second, as research synthesists strive for greater explanation, it seems likely that they will turn to structural equation models estimated using underlying matrices constructed through the meta-analytic integration of its elements whether or not they have been used in the underlying primary research literature (see Cook et al. 1992). The basic methodology for the meta-analysis of models is given in chapter 20. We expect the work given there to be the basis of many meta-analyses designed to probe the fit between proposed causal explanations and the observed data (for a discussion of the relationship between synthesis-generated evidence and inferences about causal relationships, see chapter 2).

A different and entirely new aspect of this methodology is that it has been used to examine the generalizability of research findings over facets that constitute differences in study characteristics (Becker 1996). For example, the method has been used to estimate the contribution of between-measures variation to the overall uncertainty (standard error) of the relationship or partial relationship between variables. This application of models in synthesis is exciting because it provides a mechanism for evaluating quantitatively the effect of construct breadth on generalizability of research results. Because broader constructs (those admitting a more diverse collection of operations) will typically lead to operations variation, syntheses based on broader constructs will typically be associated with greater overall uncertainty. However, broader constructs will have a larger range of application. Quantitative analyses of generalizability provide a way not only of understanding the body of research that is synthesized, but also of planning the measurement strategy of future research studies to ensure adequate generalizability across the constructs of interest.

29.5 CRITERIA FOR JUDGING THE QUALITY OF RESEARCH SYNTHESES

In chapter 28, Georg Matt and Thomas Cook began the process of extracting wisdom from technique. They attempted to help synthesists take the broad view of their work, to hold it up to standards that begin answering the question, "What have I learned?"

The criteria for quality syntheses can vary with the needs of the reader. Susan Cozzens found that readers using literature reviews to follow developments within their area of expertise valued comprehensiveness and did not consider important the reputation of the author (1981). However, if the synthesis was outside the reader's area, the expertise of the author was important, as was brevity. Kenneth Strike and George Posner also addressed the quality of knowledge synthesis with a wide-angle lens (1983). They suggested that a valuable knowledge synthesis must have both intellectual quality and practical utility. A synthesis should clarify and resolve issues in a literature rather than obscure them. It should result in a progressive paradigm shift, meaning that it brings to a theory greater explanatory power, to a practical program expanded scope of application, and to future primary research an increased capacity to pursue unsolved problems. Always, a knowledge synthesis should answer the question asked; readers should be provided a sense of closure.

In fact, readers have their own criteria for what makes a good synthesis. Harris Cooper asked post–masters degree graduate students to read meta-analyses and to evaluate five aspects of the papers (1986). The results revealed that readers' judgments were highly intercorrelated: syntheses perceived as superior on one dimension were also perceived as superior on others. Further, the best predictor of the readers' quality judgments was their confidence in their ability to interpret the results. Syntheses seen as more interpretable were given higher quality ratings. Readers were also asked to provide open-ended comments

on papers, which were subjected to an informal consent analysis. The dimensions that readers mentioned most frequently were organization, writing style, clarity of focus, use of citations, attention to variable definitions, attention to study methodology, and manuscript preparation.

Interestingly, researchers in medicine and public health have focused more attention on the development of consumer-oriented evaluation devices for meta-analysis than social scientists have. Still, the resulting scales and checklists are not numerous and efforts to construct them have been only partly successful. For example, after a thorough search of the literature a report prepared by the Agency for Healthcare Research and Quality (see West et al. 2002) identified one research synthesis quality scale (Oxman and Guyatt 1991) and ten sources labeled checklists, though two of these sources are essentially textbooks on conducting research syntheses for medical and health researchers (Clarke and Oxman 1999; Khan et al. 2000; see table 29.1).

Cooper also developed an evaluative checklist for consumers of syntheses of social science research (2007). Following the plan of this handbook, the process of research synthesis was divided into six stages:

1. defining the problem,

2. collecting the research evidence,

3. evaluating the correspondence between the methods and implementation of individual studies and the desired inferences of synthesis,

4. summarizing and integrating the evidence from individual studies,

5. interpreting the cumulative evidence, and

6. presenting the research synthesis methods and results.

Table 29.2 presents the twenty questions Cooper offered as most essential to evaluating social science research syntheses (2007). They are written from the point of view of a synthesis consumer and each is phrased so that an affirmative response means confidence could be placed in that aspect of the synthesis' methodology. Although the list is not exhaustive, most of the critical issues discussed in this volume are expressed in the questions, as are the dimensions used in medical and health checklists that seem most essential to work in the social sciences.

It is important to make three additional points about the checklist. First, it does not use a scaling procedure

Table 29.1 Scales and Checklists for Evaluating Research Syntheses Located by the Agency for Healthcare Research and Quality

Auperin, A., J. P. Pignon, and T. Poynard. 1997. "Review Article: Critical Review of Meta-Analyses of Randomized Clinical Trials in Hepatogastroenterology." *Alimentary Pharmacology and Therapeutics* 11(2): 215–25.

Beck, C. T. 1997. "Use of Meta-Analysis as a Teaching Strategy in Nursing Research Courses." *Journal of Nursing Education* 36(2): 87–90.

Clarke, Mike, and Andy D. Oxman, eds. 1999. *Cochrane Reviewer's Handbook 4.0.* Oxford: The Cochrane Collaboration.

Harbour, R., and J. Miller. 2001. "A New System for Grading Recommendations in Evidence-Based Guidelines." *British Medical Journal* 323(7308): 334–36.

Irwig, L., A.N.A. Tosteson, C. Gatsonis, J. Lau, G. Colditz, T. C. Chalmers, and F. Mosteller. 1994. "Guidelines for Meta-Analyses Evaluating Diagnostic Tests." *Annals of Internal Medicine* 120(8): 667–76.

Khan, K. S., G. Ter Riet, J. Glanville, A. J. Sowden, and J. Kleijnen. 2000. *Undertaking Systematic Reviews of Research on Effectiveness: CRD's Guidance for Carrying Out or Commissioning Reviews.* York: University of York, NHS Centre for Reviews and Dissemination.

Oxman, Andy D., and Gordon H. Guyatt. 1991. "Validation of an Index of the Quality of Review Articles." *Journal of Clinical Epidemiology* 44(11): 1271–78.

Sacks, H. S., J. Berrier, D. Reitnam, V. A. Ancona-Berk, and T. C. Chalmers. 1987. "Meta-Analyses of Randomized Controlled Trials." *New England Journal of Medicine* 316(8): 450–55.

Scottish Intercollegiate Guidelines Network. 2001. *Sign 50: A Guideline Developers' Handbook. Methodology Checklist 1: Systematic Reviews and Meta-Analyses.* Edinburgh: Scottish Intercollegiate Guidelines Network. http://www.sign.ac.uk/guidelines/fulltext/50/checklist1.html

Smith, Andrew. F. 1997. "An Analysis of Review Articles Published in Four Anaesthesia Journals." *Canadian Journal of Anaesthesia* 44(4): 405–9.

West Suzanne, Valerie King, Timothy S. Carey, Kathleen N. Lohr, Nikki McKoy, Sonya F. Sutton, and Linda Lux. 2002. *Systems to Rate the Strength of Scientific Evidence.* Evidence Report/Technology Assessment 47. *AHRQ* Publication 02-E016. Rockville, Md.: Agency for Healthcare Research and Quality.

that could in turn be used to calculate a numerical score for a synthesis, such that higher scores might indicate more trustworthy synthesis. This approach was rejected because of concerns raised about similar scales used to evaluate the quality of primary research (see Valentine and Cooper 2008; chapter 7, this volume), in particular, that single quality scores are known to generate wildly different conclusions depending on what dimensions of quality are included and on how the different items on the scale are weighted. Also, when summary numbers are generated, studies with very different profiles of strengths and limitations can receive similar scores, making syntheses with very different validity characteristics appear more similar than they actually are.

Second, some questions on the list lead to clearer prescriptions than others do of what constitutes good synthesist practices. This occurs for two reasons. First, the definition will depend at least somewhat on the topic under consideration. For example, the identification of information channels to search to locate studies relevant to a synthesis topic and what terms to use when searching reference databases clearly depend on topic decisions (see chapter 5). The checklist can therefore only suggest that syntheses based on complementary sources and proper and exhaustive database search terms should be considered by consumers as more trustworthy; it cannot specify what these sources and terms might be. Second, as many of the chapters in this book suggest, a consensus has not yet emerged on some judgments about the adequacy of synthesis methods, even among expert synthesists.

Finally, a distinction is made in the checklist between questions that relate to the conduct of research synthesis in general and to meta-analysis in particular. This is because not all research areas will have the needed evidence base to conduct a meta-analysis that produces interpretable results. This does not mean, however, that other

Table 29.2 Checklist of Questions for Evaluating Research Syntheses

Defining the problem
    1. Are the variables of interest given clear conceptual definitions?
    2. Do the operations that empirically define the variables of interest correspond to the variables' conceptual definitions?
    3. Is the problem stated so that the research designs and evidence needed to address it can be specified clearly?
    4. Is the problem placed in a meaningful theoretical, historical, or practical context?

Collecting the research evidence
    5. Were complementary searching strategies used to find relevant studies?
    6. Were proper and exhaustive terms used in searches and queries of reference databases and research registries?
    7. Were procedures employed to assure the unbiased and reliable application of criteria to determine the substantive relevance of studies, and retrieval of information from study reports?

Evaluating the correspondence between the methods and implementation of individual studies and the desired inferences of the synthesis
    8. Were studies categorized so that important distinctions could be made among them regarding their research design and implementation?
    9. If studies were excluded from the synthesis because of design and implementation considerations, where these considerations explicitly and operationally defined, and consistently applied to all studies?

Summarizing and integrating the evidence from individual studies
    10. Was an appropriate method used to combine and compare results across studies?
    11. If effect sizes were calculated, was an appropriate effect size metric used?
    12. If a meta-analysis was performed were average effect sizes and confidence intervals reported, and was an appropriate model used to estimate the independent effects and the error in effect sizes?
    13. If a meta-analysis was performed, was the homogeneity of effect sizes tested?
    14. Were study design and implementation features (as suggested by question 8) along with other critical features of studies, including historical, theoretical and practical variables (as suggested by question 4) tested as potential moderators of study outcomes?

Interpreting the cumulative evidence
    15. Were analyses carried out that tested whether results were sensitive to statistical assumptions and, if so, where these analyses used to help interpret the evidence?
    16. Did the research synthesists discuss the extent of missing data in the evidence base and examine its potential impact on the synthesis' findings?
    17. Did the research synthesists discuss the generality and limitations of the synthesis findings?
    18. Did the synthesists make the appropriate distinction between study-generated and synthesis-generated evidence when interpreting the synthesis' results?
    19. If a meta-analysis was performed, did the synthesists contrast the magnitude of effects with other related effect sizes and/or present a practical interpretation of the significance of the effects?

Presenting the research synthesis methods and results
    20. Were the procedures and results of the research synthesis clearly and completely documented?

SOURCE: Cooper 2007. Reprinted with permission from World Education.

aspects of sound synthesis methodology can be ignored (for example, clear definition of terms, appropriate literature searches).

In discussing several meta-analyses of the desegregation literature, Kenneth Wachter and Miron Straf pointed out that sophisticated literature searching procedures, data quality controls, and statistical techniques can bring a research synthesist only so far (1990). Eventually, Wachter

wrote, "there doesn't seem to be a big role in this kind of work for much intelligent statistics, as opposed to much wise thought" (182). These authors were correct in emphasizing the importance of wisdom in research integration. Wisdom is essential to any scientific enterprise, but wisdom starts with sound procedure. Synthesists must consider a wide range of technical issues as they piece together a research domain. This handbook is meant to help research synthesists build well-supported knowledge structures. But structural integrity is a minimum criterion for synthesis to lead to scientific progress. A more general kind of design wisdom is also needed to build knowledge structures that people want to view, visit, and, ultimately, live within.

29.6 REFERENCES

American Educational Research Association. 2006. "Standard for Reporting on Empirical Social Science Research in AERA Publications." Washington, D.C.: American Educational Research Association. http://www.aera.net/uploaded Files/Opportunities/StandardsforReportingEmpiricalSocial Science_PDF.pdf

American Psychological Association (APA) Publications and Communications Board Working Group on Journal Article Reporting Standards, 2008. "Reporting Standards for Research in Psychology: Why Do We Need Them? What Might They Be?" *American Psychologist* 63(8): 1–13.

Becker, Betsy J. 1996. "The Generalizability of Empirical Research Results." In *From Psychometrics to Giftedness: Papers in Honor of Julian Stanley*, edited by Camilla P. Benbow and David Lubinsky. Baltimore, Md.: Johns Hopkins University Press.

Boruch, Robert, and George Terhanian. 1998. "Cross Design Synthesis." In *Advances in Educational Productivity*, edited by Douglas M. Windham and David W. Chapman. London: JAI Press.

Clarke, Mike, and Andy D. Oxman, eds. 1999. *Cochrane Reviewer's Handbook 4.0*. Oxford: The Cochrane Collaboration.

Cook, Thomas D., Harris Cooper, David S. Cordray, Heidi Hartmann, Larry V. Hedges, Richard R. Light, Thomas A. Louis, and Frederick Mosteller. 1992. *Meta-Analysis for Explanation: A Casebook*. New York: Russell Sage Foundation.

Cooper, Harris 1982. "Scientific Guidelines for Conducting Integrative Research Reviews." *Review of Educational Research* 52(2, Summer): 291–302.

———. 1986. "On the Social Psychology of Using Integrative Research Reviews: The Case of Desegregation and Black Achievement." In *The Social Psychology of Education*, edited by Robert Feldman. Cambridge: Cambridge University Press.

———. 2007. *Evaluating and Interpreting Research Synthesis in Adult Learning and Literacy*. Boston, Mass. : National College Transition Network, New England Literacy Resource Center/World Education, Inc.

Cozzens, Susan E. 1981. *Users Requirements for Scientific Reviews*. Grant report No. IST-7821947. Washington, D.C.: National Science Foundation.

Des Jarlais, Don C., Cynthia Lyons, and Nicole Crepaz. 2004. "Improving the Reporting Quality of Nonrandomized Evaluations of Behavioral and Public health Interventions: The TREND statement." *American Journal of Public Health* 94(3): 361–66.

DuMouchel, Walter H. 1990. "Bayesian Meta-Analysis." In *Statistical Methodology in Pharmaceutical Sciences*, edited by Donald A. Berry. New York: Marcel Dekker.

Eddy, David M., Vic Hasselblad, and Ross Schacter. 1992. *Meta-Analysis by the Confidence Profile Method*. San Diego: Academic Press.

Glass, Gene V. 1976. "Primary, Secondary, and Meta-Analysis of Research." *Educational Researcher* 5(10): 3–8.

Hedges, Larry V. 1998. "Bayesian Meta-Analysis." In *Statistical Analysis of Medical Data: New Developments*, edited by Brian S. Everitt and Graham Dunn. London: Arnold.

Kahn, Khalid S., Gerben Ter Riet, Julie Glanville, Amanda J. Sowden, and Jos Kleijnen. 2000. *Undertaking Systematic Reviews of Research Effectiveness: CRD's Guidance for Carrying Out or Commissioning Reviews*. York: University of York, NHS Centre for Review and Dissemination.

Little, Roderick J.A., and Donald Rubin, B. 2002. *Statistical Analysis with Missing Data*, 2nd ed. New York: John Wiley & Sons.

Moher, David, Kenneth F. Schulz, and Douglas G. Altman. 2001. "The CONSORT Statement: Revised Recommendations for Improving the Quality of Reports of Parallel-Group Randomised Trials." *Lancet* 357(9263): 1191–94.

Ottenbacher, Kenneth, and Harris Cooper. 1984. "The Effects of Class Placement on the Social Adjustment of Mentally Retarded Children." *Journal of Research and Development in Education* 17: 1–14.

Oxman, Andy D., and Gordon H. Guyatt. 1991. "Validation of an Index of the Quality of Review Articles." *Journal of Clinical Epidemiology* 44(11): 1271–78.

Peto, Richard. 1987. "Why Do We Need Systematic Overviews of Randomized trials (with Discussion)." *Statistics in Medicine* 6(3): 233–44.

Rubin, Donald B. 1987. *Multiple Imputation for Nonresponse in Surveys*. New York: John Wiley & Sons.

Strike, Kenneth, and George Posner. 1983. "Types of Syntheses and Their Criteria." In *Knowledge Structure and Use: Implications for Synthesis and Interpretation*, edited by Spencer Ward and Linda J. Reed. Philadelphia, Pa.: Temple University Press.

U.S. General Accounting Office. 1992. *Cross Design Synthesis: A New Strategy for Medical Effectiveness Research.* GAO/PEMD No. 92-18. Washington, D.C.: U.S. Government Printing Office.

Valentine, Jeffrey C., and Harris Cooper. 2008. "A Systematic and Transparent Approach for Assessing the Methodological Quality of Intervention Effectiveness Research: The Study Design and Implementation Assessment Device (Study DIAD)." *Psychological Methods* 13(2): 130–49.

Wachter, Kenneth W., and Miron L. Straf, eds. 1990. *The Future of Meta-Analysis.* New York: Russell Sage Foundation.

West, Suzanne, Valerie King, Timothy Carey, Kathleen N. Lohr, Nikki McKoy, Sonya F. Sutton, and Linda Lux. 2002. *Systems to Rate the Strength of Scientific Evidence.* Evidence Report/Technology Assessment 47. *AHRQ* Publication No. 02-E016. Rockville, Md.: Agency for Healthcare Research and Quality.

GLOSSARY

A Posteriori Tests: Tests that are conducted after examination of study results. Used to explore patterns that seem to be emerging in the data. The same as *Post Hoc tests.*

A Priori Tests: Tests that are planned before the examination of the results of the studies under analysis. The same as *Planned Tests.*

Aggregate Analysis: The integration of evidence across studies when a description of the quantitative frequency or level of an event is the focus of a research synthesis.

Agreement Rate: The most widely used index of interrater reliability in research synthesis. Defined as the number of observations agreed on divided by the total number of observations. Also called percentage agreement.

Apples and Oranges Problem: A metaphor for studies that appear related, but are actually measuring different things. A label sometimes given as a criticism of meta-analysis because meta-analysis combines studies that may have differing methods and operational definitions of the variables involved in the calculation of effect sizes.

Artifact Correction: The modification of an estimate of effect size (usually by an artifact multiplier) to correct for the effects of an artifact. See also *Artifact Multiplier.*

Artifact Distribution: A distribution of values for a particular artifact multiplier derived from a particular research literature.

Artifact Multiplier: The factor by which a statistical or measurement artifact changes the expected observed value of a statistic.

Artifacts: Statistical and measurement imperfections that cause observed statistics to depart from the population (parameter) values the researcher intends to estimate.

Artificial Dichotomization: The arbitrary division of scores on a measure of a continuous variable into two categories.

Attenuation: The reduction or downward bias in the observed magnitude of an effect size produced by methodological limitations in a study such as measurement error or range restriction.

Available Case Analysis: Also called pairwise deletion, a method for missing data analysis that uses all available data to estimate parameters in a distribution so that, for example, all available pairs of values for two variables are used to estimate a correlation.

Bayes Posterior Coverage: A Bayesian analogue to the confidence interval.

Bayes's Theorem: A method of incorporating data into a prior distribution to obtain a posterior distribution.

Bayesian Analysis: An approach that incorporates a prior distribution to express uncertainty about population parameter values.

Best-Evidence Synthesis: Research syntheses that rely on study results that are the best available, not an a priori or idealized standard of evidence.

Between-Study Moderators: Third variables or sets of variables that affect the direction or magnitude of relations between other variables. Identified on a between-studies basis when some reviewed studies represent one level of the moderator and other studies represent other levels. See *Within-Study Moderators.*

Between-Study Predictors: Measured characteristics of studies hypothesized to affect true effect sizes.

Between-Study Sample Size: The number of studies in a meta-analysis.

Bibliographic Database: A machine-readable file consisting at minimum of entries sufficient to identify documents. Additional information may include subject indexing, abstracts, sponsoring institutions, and so on. Typically can be searched on a variety of fields (for example, author's name, keywords in title, year of publication, journal name). See also *Reference Database*.

Bibliographic Search: An exploration of literature to find reports relevant to the research topic. A search is typically conducted by consulting sources such as paper indexes, reference lists of relevant documents, contents of relevant journals/books, and electronic bibliographic databases.

Birge Ratio: The ratio of certain chi-square statistics (used in tests for heterogeneity of effects, model fit, or that variance components are zero) to their degrees of freedom. Used as an index of heterogeneity or lack of fit. Has an expected value of one under correct model specification or when the between-studies variance component is zero.

Buck's Method: A method for missing data imputation that replaces missing observations with the predicted value from a regression of the missing variable on the completely observed variable; first suggested by Buck (1960).

Categorical (Grouping) Variable: A variable that can take on a finite number of values used to define groups of studies.

Certainty: The confidence with which the scientific community accepts research conclusions, reflecting the truth value accorded to conclusions.

Citation Search: A literature search in which documents are identified based on their being cited by other documents.

Classical Analysis: A analysis that assumes that parameter values are fixed, unknown constants and that the data contain all the information about these parameters.

Coder: A person who reads and extracts information from research reports.

Coder Reliability: The equivalence with which coders extract information from research reports.

Coder Training: A process for describing and practicing the coding of studies. Modification of coding forms, coding conventions, and code book may occur during the training process.

Coding Conventions: Specific methods used to transform information in research reports into numerical form. A primary criterion for choosing such conventions is to retain as much of the original information as possible.

Coding Forms: The physical records of coded items extracted from a research report. The forms, and items on them, are organized to assist coders and data entry personnel.

Combined Procedure: A procedure that combines effect size estimates based on vote-counting procedures and effect size estimates based on effect size procedures to obtain an overall estimate of the population effect size.

Combining Results: Putting effect sizes on a common metric (for example, r or d) and calculating measures of location (for example, mean, median) or combining tests of significance.

Comparative Study: A study in which two or more groups of individuals are compared. The groups may involve different participants, such as when the two groups of patients receive two different treatments, or the same group being tested or administered treatment at two separate times, such as when two diagnostic tests that measure the same parameter are administered at separate times to the same group.

Comparisons: Linear combinations of means (of effect sizes), often used in the investigation of patterns of differences among three or more means (of effect sizes). The same as Contrasts.

Complete Case Analysis: A method for analysis of data with missing observations, which uses only cases that observe all variables in the model under investigation.

Conceptual (or Theoretical) Definition: A description of the qualities of the variable that are independent of time and space but which can be used to distinguish events that are and are not relevant to the concept.

Conditional Distribution: The sampling distribution of a statistic (such as an effect size estimate) when a parameter of interest (such as the effect size parameter) is held constant.

Conditional Exchangeability of Study Effect Size: A property of a set of studies that applies when the investigator has no a priori reason to expect the true effect size of any study to exceed the true effect size of any other study in the set, given the two studies share certain identifiable characteristics.

Conditional Mean Imputation: Another term for Buck's method or regression imputation, a method for missing data imputation that replaces missing observations with the predicted value from a regression of the missing variable on the completely observed variable; first suggested by Buck (1960). See also *Buck's Method*.

Conditional Variance: Variability due to sampling error that associated with estimation of an effect size. Same as *Estimation Variance*.

Confidence Interval (CI): The interval within which a population parameter is expected to lie.

Confidence Ratings: A method for coders to directly rate their confidence level in the accuracy, completeness, and so on, of the data being extracted from primary studies. Can establish a mechanism for discerning high-quality from lesser-quality information in the conduct of subsequent analyses.

Construct Validity: The extent to which generalizations can be made about higher-order constructs on the basis of research operations.

Controlled Vocabulary: The standard terminology defined by an indexing service for a reference database and made available through a *Thesaurus*.

Correlation Coefficient: An index that gives the magnitude of linear relationship between two quantitative variables. It can range from −1 (perfect negative relationship) to +1 (perfect positive relationship), with 0 indicating no linear relationship.

Correlation Matrix: A matrix (rectangular listing arrangement) of the zero-order correlations among a set of p variables. The diagonal elements of the correlation matrix equal one because they represent the correlation of each variable with itself. Within a single study the off-diagonal elements are denoted r_{ij}, for $i, j = 1$ ti p, $i \neq j$. The correlation r_{ij} is the correlation between variables i and j.

Covariates: Covariates are variables that are predictive of the outcome measure and are incorporated into the analysis of data. In randomized intervention studies, they are sometimes used to improve the precision with which key parameters are estimated and to increase the statistical power of significance tests. In nonrandomized studies, they are used to potentially reduce the bias caused by confounding, that is, the presence of characteristics associated with both the exposure variable and the outcome variable.

Data Coding: The process of obtaining data. In meta-analysis it consists of reading studies and recording relevant information on data collection forms.

Data Coding Protocol: A manual of instruction for data coders, which explains how to code data items, how to handle exceptions, and when to consult the project manager to resolve unexpected situations.

Data Entry: The process of transferring information from data collection forms into a database.

Data Reduction: The process of combining data items in a database to produce a cases-by-variables file for statistical analysis.

Density Function: A mathematical function that defines the probability of occurrence of values of a random variable.

Descriptive Analysis: Descriptive statistics that characterize the results and attributes of a collection of research studies in an synthesis.

Descriptors: Subject headings used to identify important concepts included in a document.

Differences Between Proportions: The sample statistic D, population parameter Δ. Computed as $D = p_1 - p_2$, where p is a proportion.

Disattenuation: The process of correcting for the reduction or downward bias in an effect size produced by attenuation.

Effect Size: A generic term that refers to the magnitude of an effect or more generally to the size of the relation between two variables. Special cases include standardized mean difference, correlation coefficient, odds ratio, or the raw mean difference.

Effect-Size Items (coding): Items related to an effect size. Can include the nature and reliability of outcome an predictor measures, sample size(s), means and standard deviations, and indices of association.

Eligibility Criteria: Conditions that must be met by a primary study in order for it to be included in the research synthesis. Also called *Inclusion Criteria*.

EM Algorithm: An iterative estimation procedure that cycles between estimation of the posterior expectation (E) of a random variable (or collection of variables) and the maximization (M) of its posterior likelihood. Useful for maximum likelihood estimation with missing data.

Error of Measurement: Random errors in the scores produced by a measuring instrument.

Estimation Variance, v_i: The variance of the effect size estimator given a fixed true effect size. The same as *Conditional Variance*.

Estimator of Effect Size, T_i: The estimator of the true effect size based on collecting a sample data.

Evaluation Synthesis: A method of synthesizing evidence about programs and policies developed by the U.S. General Accountability Office.

Evidence-Based Practices: Programs, products and policies that have been shown to be effective using quality research designs and methods.

Exclusion Criteria: See *Ineligibility Criteria*.

Experiment: An investigation of the effects of manipulating a variable.

Experimental Units: The smallest division of the experimental (or observational) material such that any two units may receive different treatments.

Explicit Truncation: A process by which all scores in a distribution in a certain range (for example, all scores below $z = -0.10$) are eliminated from the data used in the data analysis. See *Range Restriction*.

Exploratory Data Analysis: A descriptive (as opposed to inferential) analysis of data that often supplements numerical summaries with visual displays.

External Validity: The value of a study or set of studies for generalizing to individuals, settings, or procedures. Studies with high external validity can be generalized to a larger number of individuals, settings, or procedures than can studies with lower external validity. Largely determined by the sampling of individuals, settings, or procedures employed in the investigations from which generalizations are to be drawn.

Extrinsic Factors: The characteristics of research other than the phenomenon under investigation or the methods used to study that phenomenon, for example, the date of publication and the gender of the author.

Falsificationist Approach: A framework that stresses how secure knowledge depends on identifying and ruling out plausible alternative interpretations.

File-Drawer Problem: The situation in which study results go unreported (and thus are left in file drawers) when tests of hypotheses do not show statistically significant results. See also *Publication Bias*.

Fixed Effects: Effects (effect sizes or components of a model for effect size parameters) that are unknown constants. Compare with *Random Effects*.

Fixed-Effects Model: A model for combining effect sizes that assumes all effect sizes estimate a common population parameter, and that observed effect sizes differ from that parameter only by virtue of sampling error. Compare with *Random-Effects Model*.

Flat-File: In data management, a file that constitutes a collection of records of the same type that do not contain repeating items. Can be represented by two-dimensional array of data items, that is, a case-by variables data matrix.

Footnote Chasing: A technique for discovering relevant documents by tracing authors' footnotes (more broadly, their citations) to earlier documents. Also known as the *Ancestry Approach*. See also *Forward Citation Searching*.

Forward Citation Searching: A way of discovering relevant documents by looking up a known document in a citation index and finding the later documents that have cited it. See also *Footnote Chasing*.

Fourfold Table: A table that reports data concerning the relationship between two dichotomous factors. See *Odds Ratio*, *Log Odds Ratio*, and *Rate Difference*.

Free-Text Search: A search in which the computer matches the user's search terms against all words in a document (title, abstract, subject indexing, and full text). The search terms may be either controlled vocabulary or natural language, but the latter is usually implied.

Fugitive Literature: Papers or articles produced in small quantities, not widely distributed, and usually not listed in commonly used abstracts and indexes. See also *Grey Literature*.

Full-Text Database: A machine-readable file that consists of the complete texts of documents as opposed to merely citations and abstracts.

Funnel Plot: A graphic display of sample size plotted against effect size. When many studies come from the same population, each estimating the same underlying parameter, the graph should resemble a funnel with a single spout.

Gaussian Distribution: Data distributed according to the normal bell-shaped curve.

Generalization: Important purpose and promise of a meta-analysis. Refers to empirical knowledge about a general association. It can involve identifying whether an association holds with specific populations of persons, settings, times and ways of varying the cause or measuring the effect; holds across different populations of people, settings, times and ways of operationalizing a cause and effect; or can be extrapolated to other populations of people, settings, times, causes and effects than those that have been studied to date.

Generalized Least Squares (GLS) Estimation: An estimation procedure similar to the more familiar ordinary least squares method. A sum of squared deviations between paramaters and data is minimized, but GLS allows the data points for which paramaters are being estimated to have unequal population variances and nonzero covariances (that is, to be dependent).

Grey Literature: Literature that is produced on all levels of government, academia, business and industry in print and electronic formats, but which is not controlled by commercial publishers. See also *Fugitive Literature*.

Heterogeneity: The extent to which observed effect sizes differ from one another. In meta-analysis, statistical tests allow for the assessment of whether the variability in observed effect sizes is greater than would be expected given chance (that is, sampling error alone). If so, then the observed effects are said to be heterogeneous.

Heterogeneity-Homogeneity of Classes: The range-similarity of classes of persons, treatments, outcomes, settings, and times included in the studies that have been reviewed.

High Inference Codes: Codes of study characteristics that require the synthesis to infer the correct value. See also *Low Inference Codes*.

Homogeneity: A condition under which the variability in observed effect sizes is not greater than would be expected given sampling error.

Homogeneity Test: A test that a collection of effect size estimates exhibit greater variability than would be expected if their corresponding effect size parameters were identical.

Hypothesis: A research problem containing a prediction about a particular link between the variables—based on theory or previous observation.

Identification Items (coding): The category of coded items that document the research reports that are synthesized. Author, coder, country, and year and source of publication are typical items.

Ignorable Response Mechanism: When data are either missing completely at random (MCAR) or are missing at random (MAR), the mechanism that causes missing data is called ignorable because the response mechanism does not have to be modeled when using maximum likelihood or multiple imputation for missing data.

Inclusion Criteria: See *Eligibility Criteria*.

Indirect Relation: A relationship between two variables where a third (or more) variable intervenes. A variable can have both direct and indirect relations to an outcome; its indirect relations are achieved by way of one more intermediate (mediator) variables. See *Mediator Variable*.

Ineligibility Criteria: Conditions or characteristics that render a primary study ineligible for inclusion in the research synthesis. The same as *Exclusion Criteria*.

Instrumentalism: A philosophical view that scientific theories or their concepts and measures have no necessary relation to external validity.

Intercoder Correlation: An index of interrater reliability for continuous variables, based on the Pearson correlation coefficient. Used in research synthesis to estimate interrater reliability, analogous to its use in education to estimate test reliabilities when parallel forms are available.

Interim Report: A report written on completion of a portion of a study. A series of interim reports may be generated during the course of a single study.

Internal Validity: The value of a study or set of studies for concluding that a causal relationship exists between variables, that is, that one variable affects another. Studies with high internal validity provide a strong basis for making a causal inference. Largely determined by the control of alternative variables that could explain the relationship found within a study.

Interrater Reliability: The extent to which different raters rating the same studies assign the same rating to the coded variables.

Intraclass Correlation: Computed as the ratio of the variance of interest over the sum of the variance of interest plus error, the intraclass correlation is a family of indicies of agreement (or consistency) for continuous variables. The intraclass correlation can be used to isolate different sources of variation in measurement and to estimate their magnitude using the analysis of variance. The intraclass correlation is also used to describe the amount of clustering in a population in clustered or multilevel samples where it is the ratio of between-cluster variance to total variance.

Invisible College: A geographically dispersed network of scientists or scholars who share information in a particular research area through personal communication.

Kappa: A versatile family of indices of interrater reliability for categorical data, defined as the proportion o the best possible improvement over chance that is actually obtained by the raters. Generally superior to other indices designed to remove the effects of chance agreement, kappa is a true reliability statistic that in large samples is equivalent to the intraclass correlation coefficient.

Large-Sample Approximation: The statistical theory that is valid when each study has a large (within-study) sample size. The exact number of cases required to quality as large depends on the effect size index used. In meta-analysis, most statistical theory is based on large-sample approximations.

Linked Meta-Analysis: A meta-analysis in which series of (often bivariate) meta-analyses are used to represent a sequence of relationships.

Listwise Deletion: A method for analyzing data with missing observations by analyzing only those cases with complete data; also termed complete case analysis.

Log Odds: See *Logit*.

Log Odds Ratio: The sample statistic, l, population parameter λ. Computed as the natural logarithm of the odds ratio $\ln(o)$. See *Fourfold Table*.

Logistic Regression: A statistical model in which the logit of a probability is postulated to be a linear function of a set of independent variables.

Logit: The logarithm of the odds value associated with a probability. If Π is a probability, the logit is $\ln \Pi/(1 - \Pi)$, where ln denotes natural logarithm.

Logit Transformation: A transformation commonly used with proportions or other numbers that have values between zero and one. The logit is the logarithmic transformation of the ratio $[p/(1 - p)]$, that is, $\ln [p/(1 - p)]$.

Low Inference Codes: Codes of study characteristics that require the synthesists only to locate the needed information in the research report and transfer it to the synthesis database. See also *High Inference Codes*.

Mantel-Haenszel Statistics: A set of statistics for combining odds ratios and testing the statistical significance of the resulting average odds ratio.

Marginal Distribution: The distribution of a statistic (such as an effect size estimate) when relevant parameters (such as the effect size parameter) are allowed to vary according to the prior distribution.

Maximum Likelihood: A commonly used method for obtaining an estimate for an unknown parameter from a population distribution.

Maximum Likelihood Methods for Missing Data: Methods for analysis of data with missing observations that provide maximum likelihood estimates of model parameters.

Mean of Random Effects: The mean of the true effect sizes in the random-effects model.

Measurement Error: Random departure of an observed score of an individual from his/her actual (true) score. Sources of measurement error include random response error, transient error, and specific factor error.

Mediating Variable: Third variables or sets of variables that provide a causal account of the mechanisms underlying the relationship between other variables. Transmits a cause-effect relationship in the sense that it is a consequence of a more distal cause and a cause of more distal measured effects. A fundamental property of a mediator is that it is related to both a precursor and an outcome, while the precursor may show only an indirect relation to the outcome via the mediator.

Meta-Analysis: The statistical analysis of a collection of analysis results from individual studies for the purpose of integrating the findings.

Meta-Regression: The use of regression models to assess the influence of variation in contextual, methodological, participant, and program attributes on effect size estimates.

Method of Maximum Likelihood: A method of statistical estimation in which one chooses as the point estimates of a set of parameters values that maximize the likelihood of observing the data in the sample.

Method of Moments: A method of statistical estimation in which the sample moments are equated to their expectations and the resulting set of equations solved for the parameters of interest.

Metrics of Measurement: The indices used to quantify variables and the relationships between variables.

Missing at Random: Observations that are missing for reasons related to completely observed variables that are included in the model and not to the value of the missing observation.

Missing Completely at Random: Observations that are missing for reasons unrelated to any variables in the data. Missing observations occur as if they were deleted at random.

Missing Data: Data representing either study outcomes or study characteristics that are unavailable to the synthesist.

Model-Based Meta-Analysis: A meta-analysis that analyzes complex chains of events such as the prediction of behaviors based on a set of variables, and models the intercorrelations among variables.

Moderator Variable: A variable or set of variables that affect the direction or magnitude of relations between other variables. Variables such as gender, type of outcome, and other study features are often identified as moderator variables.

Multinomial Distribution: A probability distribution associated with the independent classification of a sample of subjects into a series of mutually exclusive and exhaustive categories. When the number of categories is two, the distribution is a binomial distribution.

Multiple Imputation: A method for data analysis with missing observations that generates several possible substitutes for each missing data point based on models for the reasons for the missing data.

Multiple Operationalism: The use of several measures that share a conceptual definition but that have different patterns of irrelevant components.

Multiple-Endpoint Studies: Studies that measure the outcome on the same subject at different points in time.

Nonignorable Response Mechanism: A term that describes observations that are missing for reasons related to unobserved variables. Occurs when observations are missing because of their unknown values or because of other unknown variables.

Nonparametric Statistical Procedures: Statistical procedures (tests or estimators) whose distributions do not require knowledge of the functional form of the probability distribution of the data. Also called distribution free statistics.

Normal Deviate: The location on a standard normal curve given in Z-score form.

Normal Distribution: A bell-shaped curve that is completely described by its mean and standard deviation.

Not Missing at Random (NMAR): A type of missing data where the probability of observing a value depends on the value itself; occurs in censoring mechanisms when, for example, high values on a variable are more likely to be missing than moderate or low values.

Odds Ratio: A comparison of odds values for two groups. When the odds ratio is equal to one, the event is equally likely to occur for both groups.

Odds Value: A probability divided by its complement. If Π is a probability, its associated odds value is $\Pi(1 - \Pi)$.

Omnibus Tests of Significance: Significance tests that address unfocused questions, as in F tests with more than 1 degree of freedom (df) for the numerator or in chi-square tests with more than 1 df.

Operational Definition: A definition of a concept that relates the concept to observable events.

Operational Effect Size: The effect size actually estimated in the study as conducted; for example, the (attenuated) effect size computed using an unreliable outcome measure. Compare with *Theoretical Effect Size*.

Overdispersion: A larger than expected variance.

p-**Value:** The probability associated with a statistical hypothesis test. The *p*-value is the probability of obtaining a sample result at least as large as the one obtained given a true null hypothesis (that is, given that sampling error is the only reason that the observed result deviates from the hypothesized parameter value).

Pairwise Deletion: See available case analysis.

Parallel Randomized Experiments: A set of randomized experiments testing the same hypothesis about the relationship between treatments (independent variables) and outcomes (dependent variables).

Parameter: The population value of a statistic.

Partial Relations: Relationships between two variables where a third (or more) variable has been statistically controlled, such as in a multiple regression equation.

Partial Truncation: A process by which the frequencies of scores in certain parts of the range are reduced without completely eliminating scores in these ranges. For example, if one limits the data to high school graduates and above, the frequency of people with IQs below 85 will be reduced relative to the entire population. See *Range Restriction*.

Pearson Correlation: See *Correlation Coefficient*.

Peer Review: The review of research by an investigator's peers, usually conducted at the request of a journal editor prior to publication of the research and for purposes of evaluation to ensure that the research meets generally accepted standards of the research community.

Phi Coefficient (φ): A measure of association between two binary variables that cross-classified against each other. The phi coefficient is calculated as the product moment correlation coefficient between a pair of binary random variables.

Placebo: A pseudo-treatment administered to subjects in a control group.

Planned Tests: Tests that are planned before the examination of the results of the data under analysis. The same as *A Priori Tests*.

Policy-Shaping Community: The full array of individuals and organizations with a vested interest in a policy issue.

Policy-Driven Synthesis: A synthesis that is requested by members of a specific policy-shaping community.

Policy-Oriented Quantitative Synthesis: Research syntheses that use aspects of meta-analysis to answer questions about programs, products and policies.

Pooling Results: Using data or results from several primary studies to compute a combined significance test or estimate.

Post Hoc Tests: Tests that are conducted after examination of the data to explore patterns that seem to be emerging in the data. The same as *A Posteriori Tests*.

Posterior Distribution: An updated prior distribution that incorporates observed data (also posterior mean, posterior variance, etc.)

Power: See *Statistical Power*.

Power Analysis: See *Statistical Power Analysis*.

Precision: A measure used in evaluating literature searches. The ratio of documents retrieved and deem relevant to total documents retrieved. See *Recall*.

Precision of Estimation: Computationally, the inverse of the variance of an estimator. Conceptually, larger samples estimate population parameters more precisely than smaller samples. Results from larger samples will have smaller confidence intervals compared to results from smaller samples, all else being equal.

Pretest-Posttest Design: A study in which participants are tested on the outcome variable, exposed to an intervention, and tested again on the outcome variable. Often this design leaves so many validity threats plausible that it is difficult to interpret the results.

Primary Study: A report of original research, usually published in a technical journal or appearing as a thesis or dissertation; refers to the original research reports collected for a research synthesis.

Prior Distribution: An expression of the uncertainty of parameter values as a statistical distribution before observing the data (also prior mean, prior variance, etc.).

Prior Odds: The relative prior likelihood of pairs of parameter values.

Prospective Registration: The compilation of collation of initiated research projects while they are still ongoing. See *Retrospective Registration*.

Prospective Research Synthesis: Eligible studies for a meta-analysis are identified and evaluated before their findings are even known.

Proximal Similarity: The similarity in manifest characteristics (for example, prototypical attributes) between samples of persons, treatments, outcomes, settings, and times and their corresponding universes.

Proxy Variables: Variables, such as year of study publication, that serve as surrogates for the true, causal variables, such as developments in research methods or changes in data-reporting practices.

Public Policy: A deliberate course of action that affects the people as a whole, now and in the future.

Publication Bias: The tendency for studies with statistically significant results to have a greater chance of being published than studies with non-statistically significant results. Because of this, research syntheses that fail to include unpublished studies may exaggerate the magnitude of a relationship.

Quasi-Experiment: An experiment in which the method of allocation of observations to study conditions is not done by chance. See *Randomized experiment.*

Random Effects: Effects that are assumed to be sampled from a distribution of effects. Compare with *Fixed Effects.*

Random Effects in a Regression Model for Predicting Effect Size: The deviation of study *i*'s true effect size from the value predicted on the basis of the regression model.

Random-Effects Model: A model for combining effect sizes under which observed effect sizes may differ from each other because of both sampling error and true variability in population parameters.

Random-Effects Variance Component: A variance of the true effect size viewed as either the variance of true effect sizes in a population of studies from which the synthesized studies constitute a random sample or the degree of the investigator's uncertainly about the proposition that the true effect size of a particular study is near a predicted value.

Random Assignment: The allocation of individuals, communities, groups, or other units of study to an experimental intervention, using a method that assures that assignment to any particular group is by chance.

Random Sampling: The allocation of individuals, communities, groups, or other units in a population to a sample, using a method that assures that assignment to the sample is by chance.

Randomized Experiment: An experiment that employs random assignment of observations to study conditions. See *Quasi-Experiment.*

Range Restriction: A situation in a data set in which a pre-selection process causes certain scores or ranges of scores to be missing from the data. For example, all people with scores below the mean may have been excluded from the data. Range restriction is produced by both *Explicit truncation* and *Partial truncation.*

Rank Correlation Test: A nonparametric statistical test in which the correlation between two factors is assessed by comparing the ranking of the two samples.

Rate Difference: See *Difference Between Proportions* and *Fourfold Table.*

Rate Ratio–Risk Ratio: The ratio of two probabilities, *RR*, is a popular measure of the effect of an intervention. The names given to this measure in the health sciences, the *risk* ratio or relative *risk*, reflect the fact that the measure is not symmetric, but is the ratio of the first group's probability of an undesirable outcome to the second's probability, recognizing that the ratio could also be expressed in terms of the favorable outcome, at least in theory. In fact, in the health sciences, the term *rate* is often reserved for quantities involving the use of *person-time*, that is, the product of the number of individuals and their length of follow-up.

Raw Mean Difference: The difference between the means of the intervention and comparison groups.

Recall: A measure used in evaluating literature searches. The ratio of relevant documents retrieved to total documents deemed relevant in a collection. See *Precision.*

Reduction of Random Effects Variance: The difference between the baseline and the residual random effects variance.

Reference Database (aka Bibliographic Database):   A repository of citations to documents, for example, Psyc-INFO.

Regression Discontinuity Design: The allocation of individuals, communities, groups, or other units of study to intervention conditions, using a specific cutoff score on an assignment variable.

Regression Imputation: Also called Buck's (1960) method or regression imputation; A method for missing data imputation that replaces missing observations with the predicted value from a regression of the missing variable on the completely observed variable; first suggested by Buck (1960).

Relative Risk: See *Rate Ratio.*

Relevance: The use of construct and external validity as entry criteria for selecting studies for a synthesis.

Reliability: Generally, the repeatability of measurement. More specifically, the proportion of observed variance of scores on a measuring scale that is not due to the variance of random measurement errors; or the proportion of observed variance of scores that is due to variance of true scores. Higher values indicate less measurement error and vice versa.

Replication: The repetition of a previously conducted study.

Research Problem: A statement of what variables are to be related to one another and an implication of how the relationship can be tested empirically.

Research Quality: Factors such as study design or implementation features that affect the validity of a study or set of studies.

Research Register: A database of research studies (planned, active, and/or completed), usually oriented around a common feature of the studies such as subject matter, funding source, or design.

Research Synthesis: A review of a clearly formulated question that uses systematic and explicit methods to identify, select and critically appraise relevant research, and to collect and analyze data from the studies that are included in the synthesis. Statistical methods (meta-analysis) may or may not be used to analyze and summarize the results of the included studies. Systematic review is a synonym of research synthesis.

Residual Random-Effects Variance: The residual variance of the true random effects after taking into account a set of between-studies predictors.

Response Mechanism: Term used to describe the hypothetical reasons for missing data.

Retrieved Studies: Primary studies identified in the literature search phase of a research synthesis and whose research report is obtained in full by the research synthesis investigators.

Review-Generated Evidence: In a meta-analysis, evidence that arises from studies that do not directly test the relation being considered. Compare with *Study-Generated Evidence*.

Robust Methods: Techniques that are not overly sensitive to the choice of a prior or sampling distribution or to outliers in the data.

Robustness: Consistency in the direction and/or magnitude of effect size for studies involving different classes of persons, treatments, settings, outcomes, or times.

Samples: Constituents or members of a universe, class, population, category, type, or entity.

Sampling Error: The difference between an effect size statistic and the population parameter that it estimates that occurs due to employing a sample instead of the entire population.

Search Strategy: The formal query that retrieves documents through term-matching. Consists of controlled vocabulary or natural language or both, and may make use of logical operators such as OR, AND, and NOT. Natural-language searches may further include word proximity operators, phrase-binding, word-stemming devices, wild-card characters, and so on.

Search Terms: Words used to identify the key aspects of a subject for use in searching for relevant documents.

Segmentation: Subclassifying study results to reflect important methodological or policy-relevant issues.

Selection Bias: An error produced by systematic differences between individuals, entities, or studies selected for analysis and those not selected.

Sensitivity Analysis: An analysis used to determine whether and how sensitive the conclusions of the analysis are violations of assumptions or decision rules used by the synthesist. For example, a sensitivity analysis may involve eliminating dependent effect sizes to see if the average effect size changes.

Sign Test: A test of whether the probability of getting a positive result is greater than chance (that is, 50/50).

Significance Level: See *p-Value*.

Significance Test: See *Statistical Significance Test*.

Single-Case Research: Research that studies how individual units change over time, and what causes this change.

Single-Value Imputation: A method for missing data analysis that replaces missing observations with one value such as the complete case mean or the predicted value from a regression of variables with missing data on variables completely observed.

Square Mean Root Sample Size: An average sample size for a group of studies with unequal sample sizes. It is not as influenced by extreme values as the arithmetic mean is.

Standardized Mean Difference: An effect size that expresses the difference (in standard deviation units) between he means of two groups. The d-index is one standardized mean difference effect size.

Statistical Power: The probability that a statistical test will correctly reject a false null hypothesis.

Statistical Power Analysis: A statistical analysis that estimates the likelihood of obtaining a statistically significant relation between two variables, given assumptions about the sample size employed, the desired type I error rate, and the true magnitude of the relationship.

Statistical Significance Test: A statistical test that gives evidence about the plausibility of the hypothesis of no relationship between variables (that is, the null hypothesis).

Stem-and-Leaf Display: A histogram in which the data values themselves are used to characterize the distribution of a variable. Useful for describing the center, spread, and shape of a distribution.

Study Descriptors: Coded variables that describe the characteristics of research studies other than their results or outcomes. For example, the nature of the research design and procedures used, attributes of the subject sample, and features of the setting are all study descriptors.

Study Protocol: A written document that defines the concepts used in the research and that describes study procedures in as much detail as possible before the research begins.

Study-Generated Evidence: In a meta-analysis, evidence that arises from studies that directly test the relation being considered. For example, several studies might compare the relative effects of attending summer school (compared to control students) on boys versus girls. A meta-analysis of these comparisons is considered study generated evidence. Compare with *Synthesis-Generated Evidence*.

Subject Subsamples: Groupings of a subject sample by some characteristic – for example, gender or research site.

Substantive Factors: Characteristics of research studies that represent aspects of the phenomenon under investigation, for example, the nature of the treatment conditions applied and the personal characteristics of the subjects treated.

Substantively Irrelevant Study Characteristic: Characteristics of studies that are irrelevant to the major hypotheses under study, but are inevitably incorporated into research operations to make them feasible. The analyst's goal is to show that tests of the hypothesis are not confounded with such substantive irrelevancies.

Sufficient Statistics: Term used for the statistics needed to estimate a parameter; the sum of the values of a variable is the sufficient statistic for computing the mean of a distribution.

Synthesis-Generated Evidence: In a meta-analysis, evidence that arises from studies that do not directly test the relation being considered. For example, some studies might examine the effect of attending summer school for low-SES students, while others might examine this relation for middle-SES students. A meta-analysis of the differential effectiveness of summer school for low vs. middle-SES students is considered synthesis generated evidence because the studies do not directly test this relation. Compare with *Study Generated Evidence*.

Synthetic Correlation Matrix: A matrix of correlations, each synthesized from one or more studies.

Systematic Artifacts: Statistical and measurement imperfections that bias observed statistics (statistical estimates) in a systematic direction (either upward or downward). See *Artifacts*.

Test of Significance: See *Statistical Significance Test*.

Theoretical Effect Size: The effect size that would be obtained in a study that differs in some theoretically defined way from one that is actually conducted. Typically, theoretical effect sizes are estimated from operational effect sizes with the aid of auxiliary information about study designs or measurement methods. For example, the (disattenuated) correlation coefficient corrected for the effects of measurement unreliability is a theoretical effect size. Compare to *Operational Effect Size*.

Theory Testing: The empirical validation of hypothesized relationships, which may involve the validation of a single relationship between two variables or a network of relations between several variables. Often validation depends on observing the consequences implied by, but far removed from, the basic elements of the theory.

Thesaurus: A list of subject terms in the controlled vocabulary authorized by subject experts to index the subject content of citations in a database.

Threat to Validity: A likely source of bias, alternative explanation, or plausible rival hypothesis for the research findings.

Time Series Design: Research design in which units are tested at different times, typically equal intervals, during the course of a study.

Total Variance of an Effect Size Estimate: The sum of the random effects variance and the estimation (or fixed effects) variance. The same as *Unconditional Variance*.

Transformation: The application of some arithmetic principle to a set of observations to convert the scale of measurement to one with more desirable characteristics, for example, normality, homogeneity, or linearity.

Treatment-by-Studies Interaction: The varying effect of a treatment across different studies; equivalently, the effect of moderators on the true effect size in a set of studies of treatment efficacy.

Trial Register: A database of research studies, planned, active and or completed, usually oriented around a common feature of the studies such as the subject matter, funding source or design.

Trim and Fill Method: A funnel-plot-based method of testing and adjusting for publication bias in meta-analysis.

True Effect Size: The effect magnitude that an investigator seeks to estimate when planning a study.

True Score: Conceptually, the score an individual would obtain in the complete absence of measurement error. A true score is defined as the average score that a particular individual would obtain if he or she could be independently measured with an infinite number of equivalent measures, with random measurement error being eliminated by the averaging process.

Unconditional Mean Imputation: Also called mean imputation, a method for missing data analysis that replaces missing observations with the complete case mean for that variable.

Unconditional Variance: An estimate of the total variability of a set of observed effect sizes, presumed to be due to both within-study and between-study variance.

Unit of Analysis: The study unit that contributes to the calculation of an effect size and its standard error within a study. This unit may be a single individual or a collection of individuals, like a classroom or dyad.

Universe: The population (or hyperpopulation) of studies to which one wished to generalize.

Unpublished Literature: Papers or articles not made available to the public through the action of a publisher.

Unsystematic Artifacts: Statistical imperfections, such as sampling error, which cause statistical estimates to depart randomly from the population parameters that a researcher intends to estimate. See *Artifacts*.

Utility: The extent to which a synthesis provides a full array of policy questions.

Validity Threat: See *Threat to Validity*.

Variance Components: The sources into which the variance of effect sizes is partitioned. Also, the random variables whose variances are the summands for the total variance.

Variance-Covariance Matrix: A matrix (rectangular listing arrangement) that contains the variances and covariacnes for a set of p variables. Within a single study the diagonal elements of the covariance matrix are the variances S_{ii}^2 for variables 1 to p. The covariances S_{ij}^2 for $i, j = 1$ to p, $i \neq j$, are the off-diagonal elements of the matrix.

Vote-Counting Procedure: is a procedure in which one simply tabulates the number if studies with significant positive results, the number with significant negative results, and the number with nonsignificant results. The category with the most votes presumably provides the best guess about the direction of the population effect size. More modern vote-counting procedures provide an effect size estimate and confidence interval.

Weighted Least Squares Regression in Research Synthesis: A technique for estimating the parameters of a multiple regression model wherein each study's contribution to the sum of products of the measured variables is weighted by the precision of that study's effect size estimator.

Weighting: The process of allowing observations with certain desirable characteristics to contribute more to the estimation of a parameter. In meta-analysis, each study's contribution is often weighted by the precision of that study's effect size estimator. Occasionally study results are weighted by different factors, such as study quality.

Within-Study Moderators: Moderators are identified on a within-study basis when two or more levels of the moderator are present in individual reviewed studies. See *Between-Studies Moderators* and *Moderator*.

Within-Study Sample Size: The number of units within a primary study.

Z-Transformed Correlation Coefficient: The sample statistic z, population parameter ζ. The Fisher Z transformation of a correlation r is $.5 \ln[(1 + r)/(1 - r)]$.

APPENDIX A

DATA SETS

Three data sets have been used repeatedly throughout this handbook. This appendix provides brief descriptions of the data sets, references for the research studies that reported the primary data analyses, and integrative research reviews that examined the studies.

1. DATA SET I: STUDIES OF GENDER DIFFERENCES IN CONFORMITY USING THE FICTITIOUS NORM GROUP PARADIGM

Alice Eagly and Linda Carli (1981) conducted a comprehensive review of experimental studies of gender differences in conformity. The studies in this data set are a portion of the studies that Eagly and Carli examined, the studies that they called "other conformity studies." Because all of these studies use an experimental paradigm involving a nonexistent norm group, we call them "fictitious norm group" studies.

These studies measure conformity by examining the effect of knowledge of other people's responses on an individual's response. Typically, an experimental subject is presented with an opportunity to respond to a question of opinion. Before responding, the individual is shown some "data" on the responses of other individuals. The "data" are manipulated by the experimenters, and the "other individuals" are the fictitious norm group. For example, the subject might be asked for an opinion on a work of art and told that 75 percent of Harvard undergraduates liked the work "a great deal."

Several methodological characteristics of studies that might mediate the gender difference in conformity were examined by Eagly and Carli (1981) and by Betsy J. Becker (1986). Two of them are reported here: the percentage of male authors and the number of items on the instrument used to measure conformity. Eagly and Carli reasoned that male experimenters might engage in sex typed communication that transmitted a subtle message to female subjects to conform. Consequently, studies with a large percentage of male experimenters would obtain a higher level of conformity from female subjects. Becker reasoned that the number of items on the measure of conformity might be related to effect size for two reasons: Because each item is a stimulus to conform, instruments with a larger number of items produce a greater press to conform; instruments with a larger number of items would also tend to be more reliable, and hence they would be subject to less attenuation due to measurement unreliability.

A summary of the ten studies and the effect size estimates calculated by Eagly and Carli, and in some cases recalculated by Becker, is given in table A.1. The effect size is the standardized difference between the mean for males and that for females, and a positive value implies that females are more conforming than males. The total sample size (the sum of the male and female sample sizes), the percentage of male authors, and the number of items on the outcome measure are also given.

Table A.1 Data Set I: Studies of Gender Differences in Conformity

Study	Total Sample Size	Effect Size Estimate	Male Authors (%)	Items (n)
King 1959	254	.35	100	38
Wyer 1966	80	.37	100	5
Wyer 1967	125	−.06	100	5
Sampson and Hancock 1967	191	−.30	50	2
Sistrunk 1971	64	.69	100	30
Sistrunk and McDavid 1967 [1]	90	.40	100	45
Sistrunk and McDavid 1967 [4]	60	.47	100	45
Sistrunk 1972	20	.81	100	45
Feldman-Summers et al. 1977 [1]	141	−.33	25	2
Feldman-Summers et al. 1977 [2]	119	.07	25	2

NOTE: These data are from Eagly and Carli 1981 and Becker 1986.

Table A.2 Data Set II: Studies of the Effects of Teacher Expectancy on Pupil IQ

Study	Estimated Weeks of Teacher-Student Contact Prior to Expectancy Induction	d	Standard Error
Rosenthal et al. 1974	2	.03	.125
Conn et al. 1968	21	.12	.147
Jose and Cody 1971	19	−.14	.167
Pellegrini and Hicks 1972 [1]	0	1.18	.373
Pellegrini and Hicks 1972 [2]	0	.26	.369
Evans and Rosenthal 1968	3	−.06	.103
Fielder et al. 1971	17	−.02	.103
Claiborn 1969	24	−.32	.220
Kester 1969	0	.27	.164
Maxwell 1970	1	.80	.251
Carter 1970	0	.54	.302
Flowers 1966	0	.18	.223
Keshock 1970	1	−.02	.289
Henrikson 1970	2	.23	.290
Fine 1972	17	−.18	.159
Greiger 1970	5	−.06	.167
Rosenthal and Jacobson 1968	1	.30	.139
Fleming and Anttonen 1971	2	.07	.094
Ginsburg 1970	7	−.07	.174

NOTE: These data are from Raudenbush and Bryk 1985; see also Raudenbush 1984.

2. DATA SET II: STUDIES OF THE EFFECTS OF TEACHER EXPECTANCY ON PUPIL IQ

Stephen Raudenbush (1984) reviewed randomized experiments on the effects of teacher expectancy on pupil IQ (see also Raudenbush and Bryk 1985). The experiments usually involve the researcher administering a test to a sample of students. A randomly selected portion of the students (the treatment group) are identified to their teachers as "likely to experience substantial intellectual

Table A.3 Data Set III: Validity Studies Correlating Student Ratings of the Instructor with Student Achievement

Study	Sample	Ability Control	n^a	r
Bolton et al. 1979	General psychology	Random assignment	10	.68
Bryson 1974	College algebra	CIAT pretest	20	.56
Centra 1977 [1]	General biology	Final exam for prior	13	.23
Centra 1977 [2]	General psychology	semester of same course	22	.64
Crooks and Smock 1974	General physics	Math pretest	28	.49
Doyle and Crichton 1978	Introductory communications	PSAT verbal score	12	−.04
Doyle and Whitely 1974	Beginning French aptitude test	Minnesota Scholastic	12	.49
Elliot 1950	General chemistry	College entrance exam scores	36	.33
Ellis and Richard 1977	General psychology	Random assignment	19	.58
Frey, Leonard and Beatty 1975	Introductory calculus	SAT math scores	12	.18
Greenwood et al. 1976	Analytic geometry and calculus	Achievement pretest	36	−.11
Hoffman 1978	Introductory math	SAT math scores	75	.27
McKeachie et al. 1971	General psychology	Intelligence test scores	33	.26
Marsh et al. 1956	Aircraft mechanics	Hydraulics pretest	121	.40
Remmer et al. 1949	General chemistry	Orientation test scores	37	.49
Sullivan and Skanes 1974 [1]	First-year science	Random assignment	14	.51
Sullivan and Skanes 1974 [2]	Introductory psychology	Random assignment	40	.40
Sullivan and Skanes 1974 [3]	First-year math	Random assignment	16	.34
Sullivan and Skanes 1974 [4]	First-year biology	Random assignment	14	.42
Wherry 1952	Introductory psychology	Ability scores	20	.16

NOTE: These data are from Cohen 1983; studies with sample sizes less than ten were excluded.
[a] Number of sections.

growth." All students are tested again, and the standardized difference between the mean treatment group and that of the other students is the effect size. The theory underlying the treatment is that the identification of students as showing promise of exceptional growth changes teacher expectations about those students, which may in turn effect outcomes such as intelligence (possibly through intermediate variables). Raudenbush proposed that an important characteristic of studies that was likely to be related to effect size was the amount of contact that teachers had with the students prior to the attempt to change expectations (1984). He reasoned that teachers who had more contact would have already formed more stable expectations of students-expectations that would prove more difficult to change via the experimental treatment. A summary of the studies and their effect sizes is given in Table A.2. The effect size is the standardized difference between the means of the expectancy group and the control group on an intelligence test. Positive values of the effect size imply a higher mean score for the experimental (high-expectancy) group.

3. DATA SET III: VALIDITY STUDIES CORRELATING STUDENT RATINGS OF THE INSTRUCTOR WITH STUDENT ACHIEVEMENT

Peter Cohen reviewed studies of the validity of student ratings of instructors (1981, 1983). The studies were usually conducted in multisection courses in which the sections had different instructors but all sections used a common final examination. The index of validity was a correlation coefficient (a partial correlation coefficient, controlling for a measure of student ability) between the section mean instructor ratings and the section mean examination score. Thus, the effective sample size was the number of sections. Table A.3 gives the effect size (product moment correlation coefficient) and a summary of each study with a sample size of ten or greater.

4. REFERENCES

Becker, Betsy J. 1986. "Influence Again." In *The Psychology of Gender: Progress Through Meta-Analysis,* edited by

Janet S. Hyde and Marcia L. Linn. Baltimore: Johns Hopkins University Press.

Cohen, Peter A. 1981. "Student Ratings of Instruction and Student Achievement: A Meta-Analysis of Multisection Validity Studies." *Review of Educational Research* 51(3): 281–309.

———. 1983. "Comment on 'A Selective Review of the Validity of Student Ratings of Teaching.'" *Journal of Higher Education* 54: 449–58.

Eagly, Alice H., and Linda L. Carli. 1981. "Sex of Researchers and Sex Typed Communication as Determinants of Sex Differences in Influenceability: A Meta-Analysis of Social Influence Studies." *Psychological Bulletin* 90: 1–20.

Raudenbush, Stephen W. 1984. "Magnitude of Teacher Expectancy Effects on Pupil IQ as a Function of the Credibility of Expectancy Induction: A Synthesis of Findings from 18 Experiments." *Journal of Educational Psychology* 76(1): 85–97.

Raudenbush, Stephen W., and Anthony S. Bryk. 1985. "Empirical Bayes Meta-Analysis." *Journal of Educational Statistics* 10(2): 75–98.

4.1 Data Set I: Studies Reporting Primary Results

Feldman-Summers, Shirley, Daniel E. Montano, Danuta Kasprzyk, and Beverly Wagner. 1977. *Influence When Competing Views Are Gender Related: Sex as Credibility.* Unpublished manuscript, University of Washington.

King, B. T. 1959. "Relationships Between Susceptibility to Opinion Change and Child Rearing Practices." In *Personality and Persuasibility,* edited by Carl I. Hovland and Irving L. Janis. New Haven: Yale University Press.

Sampson, Edward E., and Francena T. Hancock. 1967. "An Examination of the Relationship Between Ordinal Position, Personality, and Conformity." *Journal of Personality and Social Psychology* 5(4): 398–407.

Sistrunk, Frank. 1971. "Negro-White Comparisons in Social Conformity." *Journal of Social Psychology* 85: 77–85.

———. 1972. "Masculinity-Femininity and Conformity." *Journal of Social Psychology* 87(1): 161–2.

Sistrunk, Frank, and John W. McDavid. 1971. "Sex Variable in Conformity Behavior." *Journal of Personality and Social Psychology* 17(2): 200–7.

Wyer, Robert S., Jr. 1966. "Effects of Incentive to Perform Well, Group Attraction, and Group Acceptance on Conformity in a Judgmental Task." *Journal of Personality and Social Psychology* 4(1): 21–26.

———. 1967. "Behavioral Correlates of Academic Achievement: Conformity under Achievement-AffiliationIncen-

tive Conditions." *Journal of Personality and Social Psychology* 6(3): 255–63.

4.2 Data Set II: Studies Reporting Primary Results

Carter, D. L. 1971. "The Effect of Teacher Expectations on the Self-Esteem and Academic Performance of Seventh Grade Students *Dissertation Abstracts International* 31 (University Microfilms No. 7107612): 4539–A.

Claiborn, William. 1969. "Expectancy Effects in the Classroom: A Failure to Replicate." *Journal of Educational Psychology* 60(5): 377–83.

Conn, L. K., C. N. Edwards, Robert Rosenthal, and D. Crowne. 1968. "Perception of Emotion and Response to Teachers' Expectancy by Elementary School Children. *Psychological Reports* 22(1): 27–34.

Evans, Judith, and Robert Rosenthal. 1969. "Interpersonal Self-Fulfilling Prophecies: Further Extrapolations from the Laboratory to the Classroom." *Proceedings of the 77th Annual Convention of the American Psychological Association* 4: 371–2.

Fielder, William R., R. D. Cohen, and S. Feeney. 1971. "An Attempt to Replicate the Teacher Expectancy Effect." *Psychological Reports* 29(3, pt. 2): 1223–28.

Fine, L. 1972. "The Effects of Positive Teacher Expectancy on the Reading Achievement of Pupils in Grade Two." *Dissertation Abstracts International* 33(University Microfilms No. 7227180): 1510–A.

Fleming, Elyse, and Ralph Anttonen. 1971. Teacher Expectancy or My Fair Lady." *American Educational Research Journal* 8(2): 241–52.

Flowers, C. E. 1966. "Effects of an Arbitrary Accelerated Group Placement on the Tested Academic Achievement of Educationally Disadvantaged Students." *Dissertation Abstracts International* 27(University Microfilms No. 6610288): 991–A.

Ginsburg, R. E. 1971. "An Examination of the Relationship Between Teacher Expectations and Student Performance on a Test of Intellectual Functioning." *Dissertation Abstracts International* 31(University Microfilms No. 710922): 3337–A.

Grieger, R. M., II. 1971. "The Effects of Teacher Expectancies on the Intelligence of Students and the Behaviors of Teachers." *Dissertation Abstracts International* 31(University Microfilms No. 710922): 3338–A.

Henrikson, H. A. 1971. "An Investigation of the Influence of Teacher Expectation upon the Intellectual and Academic Performance of Disadvantaged Children." *Dissertation

Abstracts International 31(University Microfilms No. 7114791): 6278–A.

Jose, Jean, and John Cody. 1971. "Teacher-Pupil Interaction as It Relates to Attempted Changes in Teacher Expectancy of Academic Ability Achievement." *American Educational Research Journal* 8: 39–49.

Keshock, J. D. 1971. "An Investigation of the Effects of the Expectancy Phenomenon upon the Intelligence, Achievement, and Motivation of Inner-City Elementary School Children." *Dissertation Abstracts International* 32(University Microfilms No. 7119010): 243–A.

Kester, S. W. 1969. "The Communication of Teacher Expectations and Their Effects on the Achievement and Attitudes of Secondary School Pupils." *Dissertation Abstracts International* 30(University Microfilms No. 6917653): 1434–A.

Maxwell, M. L. 1971. "A Study of the Effects of Teachers' Expectations on the IQ and Academic Performance of Children." *Dissertation Abstracts International* 31(University Microfilms No. 7101725): 3345–A.

Pellegrini, Robert, and Robert Hicks. 1972. "Prophecy Effects and Tutorial Instruction for the Disadvantaged Child." *American Educational Research Journal* 9(3): 413–19.

Rosenthal, Robert, S. Baratz, and C. M. Hall, 1974. "Teacher Behavior, Teacher Expectations, and Gains in Pupils' Rated Ccreativity." *Journal of Genetic Psychology* 124(1): 115–21. Rosenthal, Robert, and Lenore Jacobson. 1968. *Pygmalion in the classroom.* New York: Holt, Rinehart and Winston.

4.3 Data Set III: Studies Reporting Primary Results

Bolton, Brian, Dennis Bonge, and John Man. 1979. "Ratings of Instruction, Examination Performance, and Subsequent Enrollment in Psychology Courses." *Teaching of Psychology* 6(2): 82–85.

Bryson, Rebecca. 1974. "Teacher Evaluations and Student Learning: A Reexamination." *Journal of Educational Research* 68(1): 12–14.

Centra, John A. 1977. "Student Ratings of Instruction and Their Relationship to Student Learning." *American Educational Research Journal* 14(1): 17–24.

Crooks, T. J., and H. R. Smock. 1974. Student Ratings of Instructors Related to Student Achievement. Urbana: University of Illinois, Office of Instructional Resources.

Doyle, Kenneth O., and Leslie I. Crichton. 1978. "Student, Peer, and Self-Evaluations of College Instructors." *Journal of Educational Psychology* 70(5): 815–26.

Doyle, Kenneth O., and Susan Whitely. 1974. "Student Ratings as Criteria for Effective Teaching." *American Educational Research Journal* 11(3): 259–74.

Elliott, D. N. 1950. "Characteristics and Relationships of Various Criteria of College and University Teaching." *Purdue University Studies in Higher Education* 70: 5–61.

Ellis, Norman R., and Henry C. Rickard. 1977. "Evaluating the Teaching of Introductory Psychology." *Teaching of Psychology* 4(3): 128–32.

Frey, Peter W., Dale W. Leonard, and William Beatty. 1975. "Student Ratings of Instruction: Validation Research." *American Educational Research Journal* 12(4): 435–47.

Greenwood, G. E., A. Hazelton, A. B. Smith, and W. B. Ware. 1976. "A Study of the Validity of Four Types of Student Ratings of College Teaching Assessed on a Criterion of Student Achievement Gains." *Research in Higher Education* 5(2): 171–78.

Hoffman, R. Gene. 1978. "Variables Affecting University Student Ratings of Instructor Behavior." *American Educational Research Journal* 15(2): 287–99.

McKeachie, W. J., Y. G. Lin, and W. Mann. 1971. "Student Ratings of Teacher Effectiveness: Validity Studies." *American Educational Research Journal* 8(3): 435–45.

Morsh, Joseph E., George G. Burgess, and Paul N. Smith. 1956. "Student Achievement as a Measure of Instructor Effectiveness." *Journal of Educational Psychology* 47: 79–88.

Remmers, H. H., F. D. Martin, and D. N. Elliott. 1949. Are Students Ratings of Instructors Related to Their Grades? *Purdue University Studies in Higher Education* 66: 1726.

Sullivan, Arthur M., and Graham R. Skanes. 1974. "Validity of Student Evaluation of Teaching and the Characteristics of Successful Instructors." *Journal of Educational Psychology* 66(4): 584–90.

Wherry, R. J. 1952. *Control of Bias in Ratings* (PRS Reports Nos. 914, 915 919, 920, and 921). Washington, D.C.: Department of the Army, Adjutant General's Office.

APPENDIX B

TABLES

Table B.1 Probability that a Standard Normal Deviate Exceeds Z

Z	Second Digit of Z									
	.00	.01	.02	.03	.04	.05	.06	.07	.08	.09
.0	.5000	.4960	.4920	.4880	.4840	.4801	.4761	.4721	.4681	.4641
.1	.4602	.4562	.4522	.4483	.4443	.4404	.4364	.4325	.4286	.4247
.2	.4207	.4168	.4129	.4090	.4052	.4013	.3974	.3936	.3897	.3859
.3	.3821	.3783	.3745	.3707	.3669	.3632	.3594	.3557	.3520	.3483
.4	.3446	.3409	.3372	.3336	.3300	.3264	.3228	.3192	.3156	.3121
.5	.3085	.3050	.3015	.2981	.2946	.2912	.2877	.2843	.2810	.2776
.6	.2743	.2709	.2676	.2643	.2611	.2578	.2546	.2514	.2483	.2451
.7	.2420	.2389	.2358	.2327	.2296	.2266	.2236	.2206	.2177	.2148
.8	.2119	.2090	.2061	.2033	.2005	.1977	.1949	.1922	.1894	.1867
.9	.1841	.1814	.1788	.1762	.1736	.1711	.1685	.1660	.1635	.1611
1.0	.1587	.1562	.1539	.1515	.1492	.1469	.1446	.1423	.1401	.1379
1.1	.1357	.1335	.1314	.1292	.1271	.1251	.1230	.1210	.1190	.1170
1.2	.1151	.1131	.1112	.1093	.1075	.1056	.1038	.1020	.1003	.0985
1.3	.0968	.0951	.0934	.0918	.0901	.0885	.0869	.0853	.0838	.0823
1.4	.0808	.0793	.0778	.0764	.0749	.0735	.0721	.0708	.0694	.0681
1.5	.0668	.0655	.0643	.0630	.0618	.0606	.0594	.0582	.0571	.0559
1.6	.0548	.0537	.0526	.0516	.0505	.0495	.0485	.0475	.0465	.0455
1.7	.0446	.0436	.0427	.0418	.0409	.0401	.0392	.0384	.0375	.0367
1.8	.0359	.0351	.0344	.0336	.0329	.0322	.0314	.0307	.0301	.0294
1.9	.0287	.0281	.0274	.0268	.0262	.0256	.0250	.0244	.0239	.0233
2.0	.0228	.0222	.0217	.0212	.0207	.0202	.0197	.0192	.0188	.0183
2.1	.0179	.0174	.0170	.0166	.0162	.0158	.0154	.0150	.0146	.0143
2.2	.0139	.0136	.0132	.0129	.0125	.0122	.0119	.0116	.0113	.0110
2.3	.0107	.0104	.0102	.0099	.0096	.0094	.0091	.0089	.0087	.0084
2.4	.0082	.0080	.0078	.0075	.0073	.0071	.0069	.0068	.0066	.0064
2.5	.0062	.0060	.0059	.0057	.0055	.0054	.0052	.0051	.0049	.0048
2.6	.0047	.0045	.0044	.0043	.0041	.0040	.0039	.0038	.0037	.0036
2.7	.0035	.0034	.0033	.0032	.0031	.0030	.0029	.0028	.0027	.0026
2.8	.0026	.0025	.0024	.0023	.0023	.0022	.0021	.0021	.0020	.0019
2.9	.0019	.0018	.0018	.0017	.0016	.0016	.0015	.0015	.0014	.0014
3.0	.0013	.0013	.0013	.0012	.0012	.0011	.0011	.0011	.0010	.0010
3.1	.0010	.0009	.0009	.0009	.0008	.0008	.0008	.0008	.0007	.0007
3.2	.0007									
3.3	.0005									
3.4	.0003									
3.5	.00023									
3.6	.00016									
3.7	.00011									
3.8	.00007									
3.9	.00005									
4.0	.00003									

SOURCE: Reproduced from S. Siegel, *Nonparametric Statistics* (New York: McGraw-Hill, 1956), p. *247,* with the permission of the publisher.

Table B.2 Critical Values of the Chi-Square Distribution

Degrees of Freedom	Left Tail Area						
	0.50	0.75	0.90	0.95	0.975	0.99	0.995
1	0.4549	1.323	2.706	3.841	5.025	6.635	7.879
2	1.386	2.773	4.605	5.991	7.378	9.210	10.60
3	2.366	4.108	6.251	7.815	9.348	11.34	12.84
4	3.357	5.385	7.779	9.488	11.14	13.28	14.86
5	4.351	6.626	9.236	11.07	12.83	15.09	16.75
6	5.348	7.841	10.64	12.59	14.45	16.81	18.55
7	6.346	9.037	12.02	14.07	16.01	18.48	20.28
8	7.344	10.22	13.36	15.51	17.53	20.09	21.95
9	8.343	11.39	14.68	16.92	19.02	21.67	23.59
10	9.342	12.55	15.99	18.31	20.48	23.21	25.19
11	10.34	13.70	17.28	19.68	21.92	24.72	26.76
12	11.34	14.85	18.55	21.03	23.34	26.22	28.30
13	12.34	15.98	19.81	22.36	24.74	27.69	29.82
14	13.34	17.12	21.06	23.68	26.12	29.14	31.32
15	14.34	18.25	22.31	25.00	27.49	30.58	32.80
16	15.34	19.37	23.54	26.30	28.85	32.00	34.27
17	16.34	20.49	24.77	27.59	30.19	33.41	35.72
18	17.34	21.60	25.99	28.87	31.53	34.81	37.16
19	18.34	22.72	27.20	30.14	32.85	36.19	38.58
20	19.34	23.83	28.41	31.41	34.17	37.57	40.00
21	20.34	24.93	29.62	32.67	35.48	38.93	41.40
22	21.34	26.04	30.81	33.92	36.78	40.29	42.80
23	22.34	27.14	32.01	35.17	38.08	41.64	44.18
24	23.34	28.24	33.20	36.42	39.36	42.98	45.56
25	24.34	29.34	34.38	37.65	40.65	44.31	46.93
26	25.34	30.43	35.56	38.89	41.92	45.64	48.29
27	26.34	31.53	36.74	40.11	43.19	46.96	49.64
28	27.34	32.62	37.92	41.34	44.46	48.28	50.99
29	28.34	33.71	39.09	42.56	45.72	49.59	52.34
30	29.34	34.80	40.26	43.77	46.98	50.89	53.67
31	30.34	35.89	41.42	44.99	48.23	52.19	55.00
32	31.34	36.97	42.58	46.19	49.48	53.49	56.33
33	32.34	38.06	43.75	47.40	50.73	54.78	57.65
34	33.34	39.14	44.90	48.60	51.97	56.06	58.96
35	34.34	40.22	46.06	49.80	53.20	57.34	60.27
36	35.34	41.30	47.21	51.00	54.44	58.62	61.58
37	36.34	42.38	48.36	52.19	55.67	59.89	62.88
38	37.34	43.46	49.51	53.38	56.90	61.16	64.18
39	38.34	44.54	50.66	54.57	58.12	62.43	65.48
40	39.34	45.62	51.81	55.76	59.34	63.69	66.77
50	49.33	56.33	63.17	67.50	71.42	76.15	79.49
60	59.33	66.98	74.40	79.08	83.30	88.38	91.95
70	69.33	77.58	85.53	90.53	95.02	100.4	104.2
80	79.33	88.13	96.58	101.9	106.6	112.3	116.3
90	89.33	98.65	107.6	113.1	118.1	124.1	128.3
100	99.33	109.1	118.5	124.3	129.6	135.8	140.2

(Continued)

Table B.2 (*Continued*)

Degrees of Freedom	Left Tail Area						
	0.50	0.75	0.90	0.95	0.975	0.99	0.995
110	109.3	119.6	129.4	135.5	140.9	147.4	151.9
120	119.3	130.1	140.2	146.6	152.2	159.0	163.6
130	129.3	140.5	151.0	157.6	163.5	170.4	175.3
140	139.3	150.9	161.8	168.6	174.6	181.8	186.8
150	149.3	161.3	172.6	179.6	185.8	193.2	198.4
160	159.3	171.7	183.3	190.5	196.9	204.5	209.8
170	169.3	182.0	194.0	201.4	208.0	215.8	221.2
180	179.3	192.4	204.7	212.3	219.0	227.1	232.6
190	189.3	202.8	215.4	223.2	230.1	238.3	244.0
200	199.3	213.1	226.0	234.0	241.1	249.4	255.3

SOURCE: Reprinted from Hedges and Olkin with permission of the publisher.

Table B.3 Fisher's *z* Transformation of *r*

r	Second Digit of *r*									
	.00	.01	.02	.03	.04	.05	.06	.07	.08	.09
.0	.000	.010	.020	.030	.040	.050	.060	.070	.080	.090
.1	.100	.110	.121	.131	.141	.151	.161	.172	.182	.192
.2	.203	.213	.224	.234	.245	.255	.266	.277	.288	.299
.3	.310	.321	.332	.343	.354	.365	.377	.388	.400	.412
.4	.424	.436	.448	.460	.472	.485	.497	.510	.523	.536
.5	.549	.563	.576	.590	.604	.618	.633	.648	.662	.678
.6	.693	.709	.725	.741	.758	.775	.793	.811	.829	.848
.7	.867	.887	.908	.929	.950	.973	.996	1.020	1.045	1.071
.8	1.099	1.127	1.157	1.188	1.221	1.256	1.293	1.333	1.376	1.422

r	Third Digit of *r*									
	.000	.001	.002	.003	.004	.005	.006	.007	.008	.009
.90	1.472	1.478	1.483	1.488	1.494	1.499	1.505	1.510	1.516	1.522
.91	1.528	1.533	1.539	1.545	1.551	1.557	1.564	1.570	1.576	1.583
.92	1.589	1.596	1.602	1.609	1.616	1.623	1.630	1.637	1.644	1.651
.93	1.658	1.666	1.673	1.681	1.689	1.697	1.705	1.713	1.721	1.730
.94	1.738	1.747	1.756	1.764	1.774	1.783	1.792	1.802	1.812	1.822
.95	1.832	1.842	1.853	1.863	1.874	1.886	1.897	1.909	1.921	1.933
.96	1.946	1.959	1.972	1.986	2.000	2.014	2.029	2.044	2.060	2.076
.97	2.092	2.109	2.127	2.146	2.165	2.185	2.205	2.227	2.249	2.273
.98	2.298	2.323	2.351	2.380	2.410	2.443	2.477	2.515	2.555	2.599
.99	2.646	2.700	2.759	2.826	2.903	2.994	3.106	3.250	3.453	3.800

SOURCE: Reprinted from Snedecor and Cochran with permission of the publisher.

NOTE: *z* is obtained as $ \ln \frac{1+r}{1-r}$

INDEX

Numbers in **boldface** refer to figures and tables.